Metals

Metalloids

Nonmetals

10 8B	11 1B	12 2B	13 3A	14 4A	15 5A	16 6A	17 7A	18 8A
								2 **He** 4.00 helium
			5 **B** 10.81 boron	6 **C** 12.01 carbon	7 **N** 14.01 nitrogen	8 **O** 16.00 oxygen	9 **F** 19.00 fluorine	10 **Ne** 20.18 neon
			13 **Al** 26.98 aluminum	14 **Si** 28.09 silicon	15 **P** 30.97 phosphorus	16 **S** 32.06 sulfur	17 **Cl** 35.45 chlorine	18 **Ar** 39.95 argon
28 **Ni** 58.69 nickel	29 **Cu** 63.55 copper	30 **Zn** 65.39 zinc	31 **Ga** 69.72 gallium	32 **Ge** 72.63 germanium	33 **As** 74.92 arsenic	34 **Se** 78.97 selenium	35 **Br** 79.90 bromine	36 **Kr** 83.80 krypton
46 **Pd** 106.42 palladium	47 **Ag** 107.87 silver	48 **Cd** 112.41 cadmium	49 **In** 114.82 indium	50 **Sn** 118.71 tin	51 **Sb** 121.75 antimony	52 **Te** 127.60 tellurium	53 **I** 126.90 iodine	54 **Xe** 131.29 xenon
78 **Pt** 195.08 platinum	79 **Au** 196.97 gold	80 **Hg** 200.59 mercury	81 **Tl** 204.38 thallium	82 **Pb** 207.2 lead	83 **Bi** 208.98 bismuth	84 **Po** (209) polonium	85 **At** (210) astatine	86 **Rn** (222) radon
110 **Ds** (281) darmstadtium	111 **Rg** (280) roentgenium	112 **Cn** (285) copernicium	113 **Nh** (284) nihonium	114 **Fl** (289) flerovium	115 **Mc** (288) moscovium	116 **Lv** (292) livermorium	117 **Ts** (292) tennessine	118 **Og** (294) oganesson

64 **Gd** 157.25 gadolinium	65 **Tb** 158.93 terbium	66 **Dy** 162.50 dysprosium	67 **Ho** 164.93 holmium	68 **Er** 167.26 erbium	69 **Tm** 168.93 thulium	70 **Yb** 173.04 ytterbium	71 **Lu** 174.97 lutetium
96 **Cm** (247) curium	97 **Bk** (247) berkelium	98 **Cf** (251) californium	99 **Es** (252) einsteinium	100 **Fm** (257) fermium	101 **Md** (258) mendelevium	102 **No** (259) nobelium	103 **Lr** (260) lawrencium

A PEARSON AUSTRALIA CUSTOM BOOK

Fundamentals of Chemistry

Second Edition

Compiled by Dr Kate Rowen
Reviewed by Dr Leonie Hughes, Dr LanChi Koenigsberger and Dr Kate Rowen

This custom book is compiled for Murdoch University from:

INTRODUCTORY CHEMISTRY
5TH EDITION
N. J. TRO

MODULE 12 ORGANIC COMPOUNDS
WRITTEN BY KATE ROWEN FOR THIS CUSTOM EDITION

Pearson Australia
707 Collins Street
Melbourne VIC 3008
Ph: 03 9811 2400

www.pearson.com.au

Project Management Team Leader:	Jill Gillies
Custom Portfolio Manager:	Lucie Bartonek
Production Manager:	Lisa D'Cruz
Project Editor:	Catherine du Peloux Menagé
Indexer:	Mary Coe
Production Controller:	Dominic Harman

Print ISBN: 978 1 4886 2308 0
eBook ISBN: 978 1 4886 2321 9

Typeset by S4Carlisle Publishing Services

BRIEF TABLE OF CONTENTS

TABLE OF CONTENTS

PREFACE

This custom textbook has been specially prepared for students studying Fundamentals of Chemistry at Murdoch University. All the content you need for the unit is in this book, every Module is relevant. In developing this textbook we have applied our extensive involvement in teaching this level of chemistry. Collectively we have over four decades of experience and a thorough understanding of your learning needs.

Modules 1 – 11 were sourced from an American book, Introductory Chemistry by Tro, but we have adapted them extensively to the way this level of chemistry is taught in Australia. Textbooks for this kind of chemistry unit typically use a particular style of problem solving, but in our lectures we show what we believe to be better methods using equations and other types of reasoning. The demonstration of different problem solving approaches between lectures and the textbook naturally causes confusion for students who may already be anxious about studying chemistry for the first time. We have adapted the sourcebook material for Modules 1 – 11 so the textbook and lectures show the same problem solving techniques.

Fundamentals of Chemistry also provides an introduction to the diverse and amazing world of organic compounds, which is covered in Module 12. Despite our best efforts, we have never been able to source textbook material that succinctly covers the curriculum topics for this module to the depth required. Students would have to rely on lecture material and other disparate resources that we brought together to support learning about organic compounds. We decided the only thing to do was to write Module 12 specifically for this book. Now all the content you need for Fundamentals of Chemistry is in one tailored source.

The source textbook is American, so there are many references to everyday life in America and less commonly used units of measurement. Therefore we have endeavoured to ensure that most of the units of measurement referred to in the book are the metric units that are used in Australia and the international scientific community.

We want your first experience with chemistry to be a positive one, where you will learn much about the world around you in addition to the basic chemistry you need to support your further study. This is what has motivated us to work towards providing you with the best textbook possible for Fundamentals of Chemistry.

Dr Kate Rowen, Dr Leonie Hughes, Dr LanChi Koenigsberger.

▲ Everything that you can see in this room is made of matter. As students of chemistry, we are interested in how the differences between different kinds of matter are related to the differences between the molecules and atoms that compose the matter. The molecular structures shown here are water molecules on the left and carbon atoms in graphite on the right.

Matter 1

"Thus, the task is, not so much to see what no one has yet seen; but to think what nobody has yet thought, about that which everybody sees." —Erwin Schrödinger (1887–1961)

MODULE OUTLINE

1.1 What Is Matter?

LO: Define matter, atoms, and molecules.

We define **matter** as anything that occupies space and has mass. Some types of matter—such as steel, water, wood, and plastic—are readily visible to our eyes. Other types of matter—such as air or microscopic dust—are not visible to our naked eyes. Matter may sometimes appear smooth and continuous, but actually it is not. Matter is ultimately composed of **atoms**, submicroscopic particles that are the fundamental building blocks of matter ▶ Figure 1.1(a). In many cases, these atoms are bonded together to form **molecules**, two or more atoms joined to one another in specific geometric arrangements ▶ Figure 1.1(b). Recent advances in microscopy have allowed us to image the atoms (▶ Figure 1.2) and molecules (▶ Figure 1.3) that compose matter, sometimes with stunning clarity.

1.2 Classifying Matter According to Its State: Solid, Liquid, and Gas

LO: Classify matter as solid, liquid, or gas.

The common **states of matter** are **solid**, **liquid**, and **gas** (▶ Figure 1.4). In solid matter, atoms or molecules pack close to each other in fixed locations. Although neighboring atoms or molecules in a solid may vibrate or oscillate, they do not move around each other, giving solids their familiar fixed volume and rigid shape. Ice, diamond, quartz, and iron are examples of solid matter. Solid matter may be **crystalline**, in which case its atoms or molecules arrange in geometric patterns with long-range, repeating order ▶ Figure 1.5(a), or it may be **amorphous**, in which case its atoms or molecules do not have long-range order ▶ Figure 1.5(b). Examples of *crystalline* solids include salt (▶ Figure 1.6) and diamond; the well-ordered, geometric shapes of salt and diamond crystals reflect the well-ordered geometric arrangement of their atoms. Examples of *amorphous* solids include glass, rubber, and plastic.

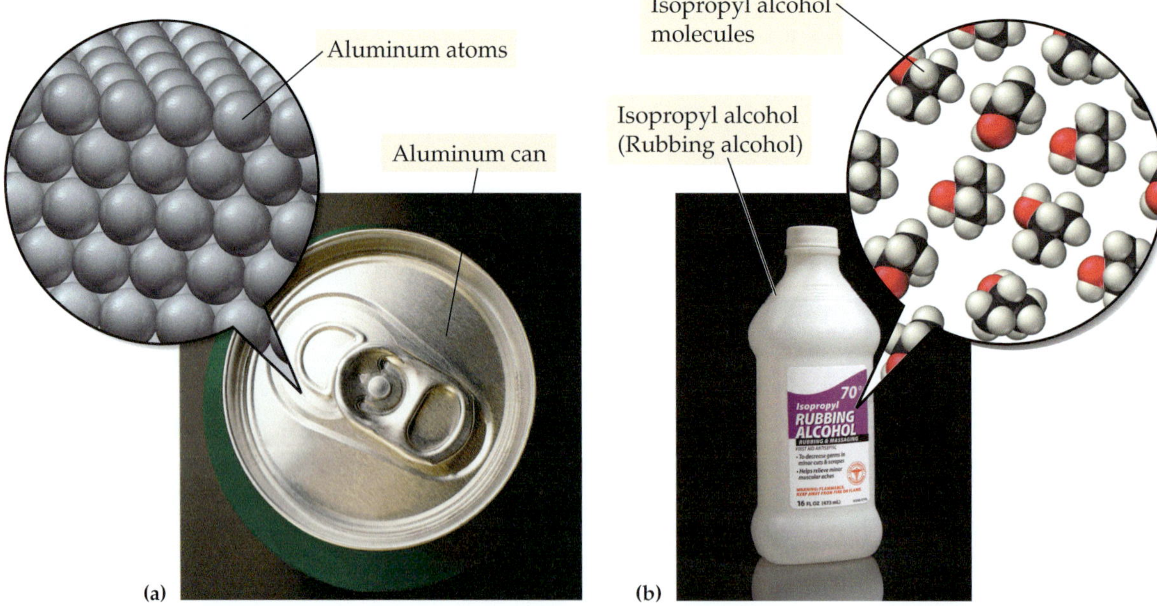

▲ **FIGURE 1.1 Atoms and molecules** All matter is ultimately composed of atoms. **(a)** In some substances, such as aluminum, the atoms exist as independent particles. **(b)** In other substances, such as rubbing alcohol, several atoms bond together in well-defined structures called molecules. © 1.1a Photodisc/Getty Image; 1.1b Editorial Image, LLC/Alamy.

▲ **FIGURE 1.2 Scanning tunneling microscope image of nickel atoms** A scanning tunneling microscope (STM) creates an image by scanning a surface with a tip of atomic dimensions. It can distinguish individual atoms, seen as blue bumps, in this image. (*Source:* Reprint Courtesy of International Business Machines Corporation, copyright © International Business Machines Corporation.) © IBM Research, Almaden Research Center.

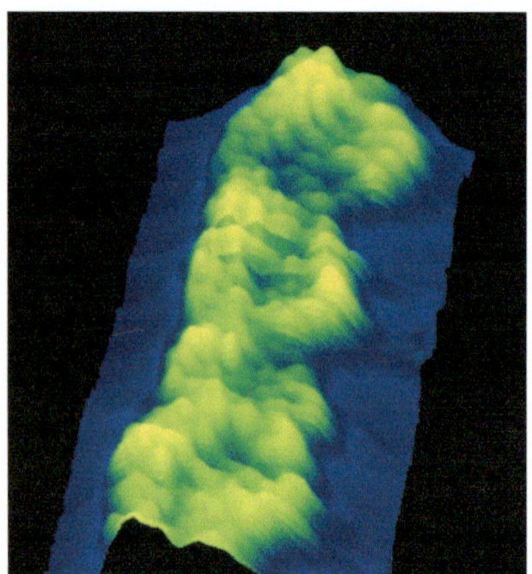

▲ **FIGURE 1.3 Scanning tunneling microscope image of a DNA molecule** DNA is the hereditary material that encodes the operating instructions for most cells in living organisms. In this image, the DNA molecule is yellow, and the double-stranded structure of DNA is discernible. © Driscoll,Youngquist, & Baldeschwieler/Caltech/Science Source.

In liquid matter, atoms or molecules are close to each other (about as close as molecules in a solid), but they are free to move around and past each other. Like solids, liquids have a fixed volume because their atoms or molecules are in close contact. Unlike solids, however, liquids assume the shape of their container because the atoms or molecules are free to move relative to one another. Water, gasoline, alcohol, and mercury are examples of liquid matter.

In gaseous matter, atoms or molecules are separated by large distances and are free to move relative to one another. Because the atoms or molecules that compose

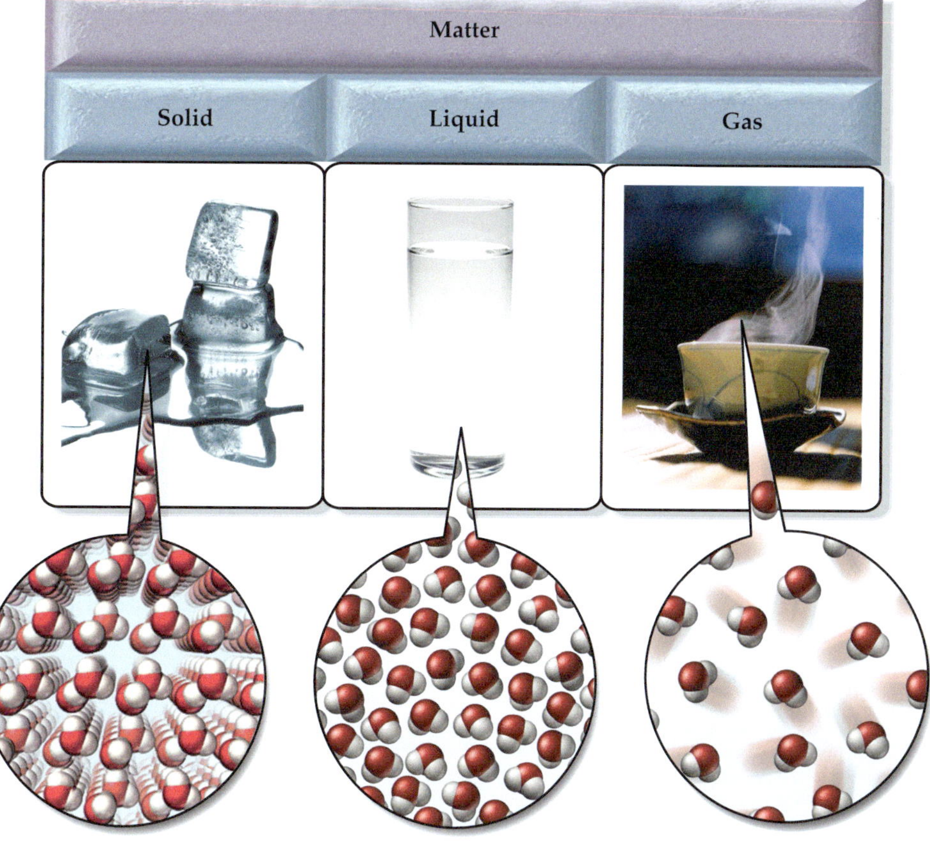

▶ **FIGURE 1.4 Three states of matter** Water exists as ice (solid), water (liquid), and steam (gas). In ice, the water molecules are closely spaced and, although they vibrate about a fixed point, they do not generally move relative to one another. In liquid water, the water molecules are also closely spaced but are free to move around and past each other. In steam, water molecules are separated by large distances and do not interact significantly with one another. © Mark Downey/Photodisc/Getty Images.

(a) Crystalline solid

(b) Amorphous solid

▲ **FIGURE 1.5 Types of solid matter** **(a)** In a crystalline solid, atoms or molecules occupy specific positions to create a well-ordered, three-dimensional structure. **(b)** In an amorphous solid, atoms do not have any long-range order. © Marek CECH/Shutterstock.

▲ **FIGURE 1.6 Salt: a crystalline solid** Sodium chloride is an example of a crystalline solid. The well-ordered, cubic shape of salt crystals is due to the well-ordered, cubic arrangement of its atoms. © Natural History Museum/Alamy.

gases are not in contact with one another, gases are **compressible** (▶ Figure 1.7). When we inflate a bicycle tire, for example, we push more atoms and molecules into the same space, compressing them and making the tire harder. Gases always assume the shape and volume of their containers. Oxygen, helium, and carbon dioxide are all examples of gases. Table 1.1 summarizes the properties of solids, liquids, and gases.

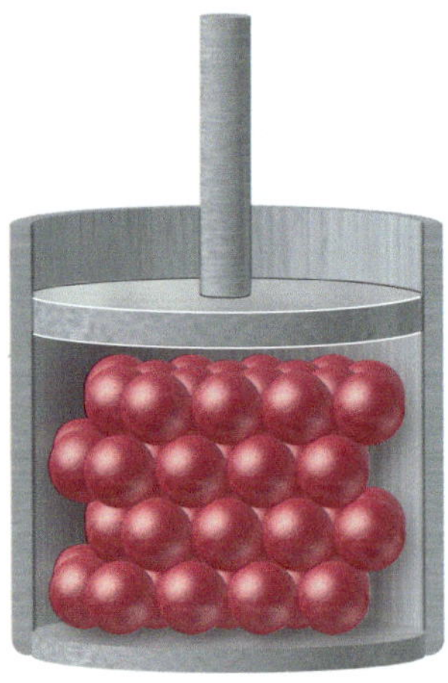

Solid—not compressible

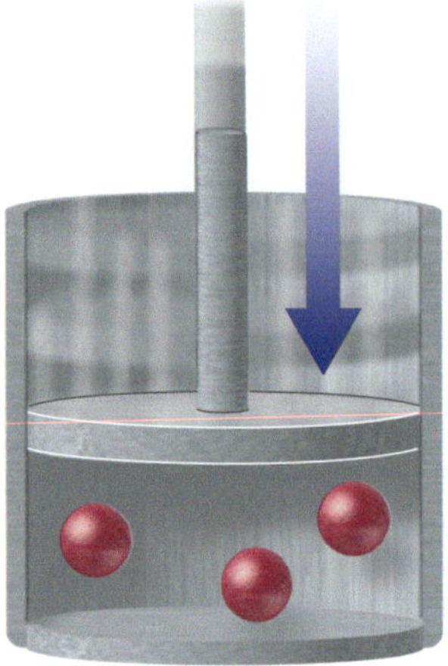

Gas—compressible

▲ **FIGURE 1.7 Gases are compressible** Because the atoms or molecules that compose gases are not in contact with one another, we can compress gases.

TABLE 1.1 Properties of Liquids, Solids, and Gases

State	Atomic/ Molecular Motion	Atomic/ Molecular Spacing	Shape	Volume	Compressibility
Solid	Oscillation/ vibration about fixed point	Close together	Definite	Definite	Incompressible
Liquid	Free to move relative to one another	Close together	Indefinite	Definite	Incompressible
Gas	Free to move relative to one another	Far apart	Indefinite	Indefinite	Compressible

CONCEPTUAL CHECKPOINT 1.1

Which image best represents matter in the gas state?

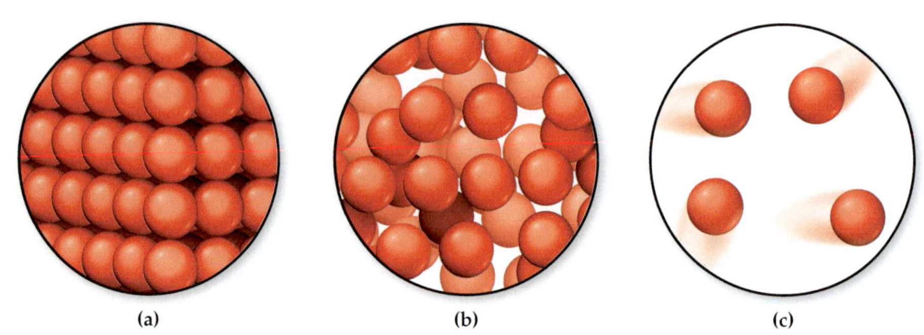

Note: You can find the answers to all Conceptual Checkpoints at the end of the module.

1.3 Classifying Matter According to Its Composition: Elements, Compounds, and Mixtures

LO: Classify matter as element, compound, or mixture.

In addition to classifying matter according to its state, we can classify it according to its composition (▶ Figure 1.8). Matter may be either a **pure substance**, composed of only one type of atom or molecule, or a **mixture**, composed of two or more different types of atoms or molecules combined in variable proportions.

Pure substances are composed of only one type of atom or molecule. Helium and water are both pure substances. The atoms that compose helium are all helium atoms, and the molecules that compose water are all water molecules—no other atoms or molecules are mixed in.

Pure substances can themselves be divided into two types: elements and compounds. Copper is an example of an **element**, a substance that cannot be broken down into simpler substances. The graphite in pencils is also an element—carbon. No chemical transformation can decompose graphite into simpler substances; it is pure carbon. All known elements are listed in the periodic table on the inside front cover of this book and in alphabetical order on the inside back cover of this book.

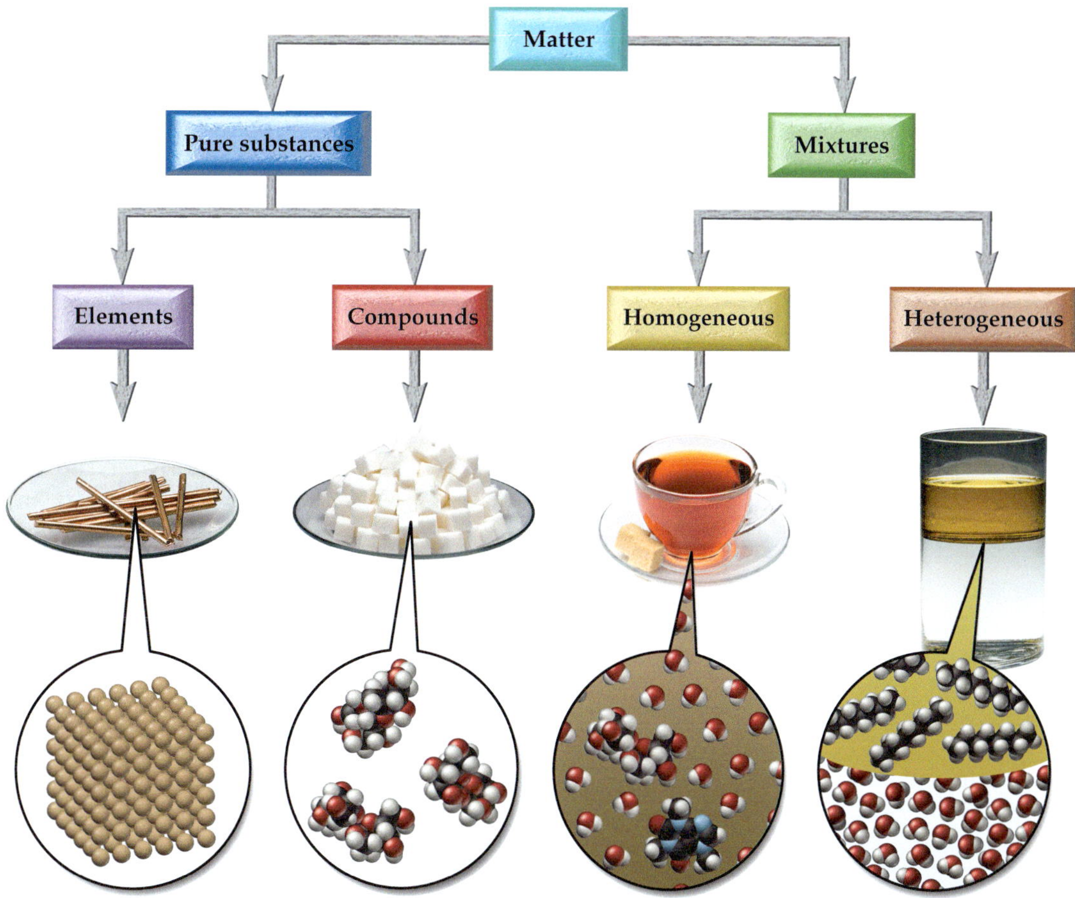

▲ **FIGURE 1.8 Classification of matter** Matter may be a pure substance or a mixture. A pure substance may be either an element (such as copper) or a compound (such as sugar), and a mixture may be either homogeneous (such as sweetened tea) or heterogeneous (such as gasoline and water). © (centre right): /iStockphoto/Getty Images. (right): Kip Peticolas/Fundamental Photographs.

A compound is composed of atoms from different elements that are chemically united (bonded). A mixture is composed of different substances that are not chemically united, but simply mixed together.

A pure substance can also be a **compound**, a substance composed of two or more elements in fixed definite proportions. Compounds are more common than pure elements because most elements are chemically reactive and combine with other elements to form compounds. Water, table salt, and sugar are examples of compounds; they can all be decomposed into simpler substances. If you heat sugar on a pan over a flame, you decompose it into several substances including carbon (an element) and gaseous water (a different compound). The black substance left on your pan after burning is the carbon; the water escapes into the air as steam.

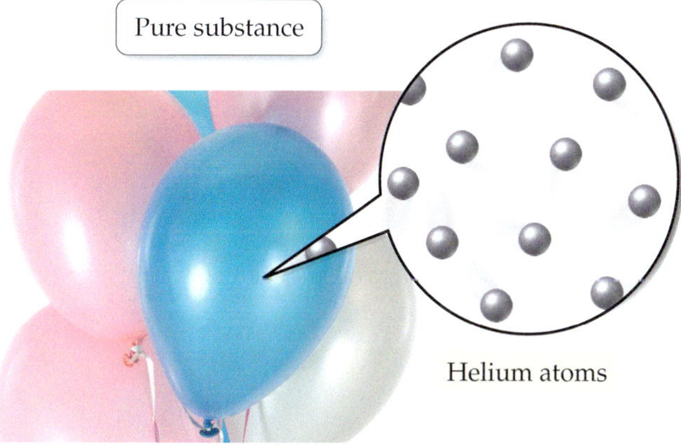

▲ Helium is a pure substance composed only of helium atoms.

▲ Water is a pure substance composed only of water molecules. © Martial Colomb/Photodisc/Getty Images.

The majority of matter that we encounter is in the form of mixtures. Apple juice, a flame, salad dressing, and soil are all examples of mixtures; they each contain several substances with proportions that vary from one sample to another. Other common mixtures include air, seawater, and brass. Air is a mixture composed primarily of nitrogen and oxygen gas, seawater is a mixture composed primarily of salt and water, and brass is a mixture composed of copper and zinc. Each of these mixtures can have different proportions of its constituent components. For example, metallurgists vary the relative amounts of copper and zinc in brass to tailor the metal's properties to its intended use—the higher the zinc content relative to the copper content, the more brittle the brass.

We can also classify mixtures according to how uniformly the substances within them mix. In a **heterogeneous mixture**, such as oil and water, the composition varies from one region to another. In a **homogeneous mixture**, such as salt water or sweetened tea, the composition is the same throughout. Homogeneous mixtures have uniform compositions because the atoms or molecules that compose them mix uniformly. Remember that the properties of matter are determined by the atoms or molecules that compose it.

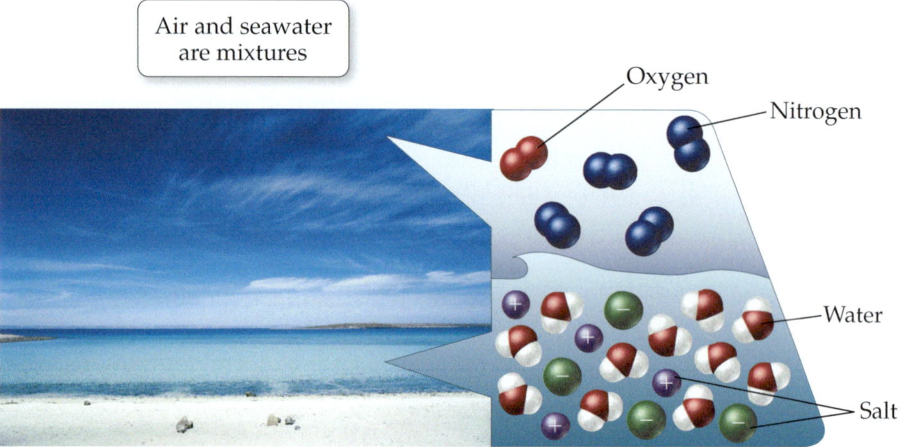

▲ Air and seawater are examples of mixtures. Air contains primarily nitrogen and oxygen. Seawater contains primarily salt and water. © Getty Images.

To summarize, as shown in Figure 1.8 (on p.5):

- Matter may be a pure substance, or it may be a mixture.
- A pure substance may be either an element or a compound.
- A mixture may be either homogeneous or heterogeneous.
- Mixtures may be composed of two or more elements, two or more compounds, or a combination of both.

EXAMPLE 1.1 CLASSIFYING MATTER

Classify each type of matter as a pure substance or a mixture. If it is a pure substance, classify it as an element or a compound; if it is a mixture, classify it as homogeneous or heterogeneous.

(a) a lead weight
(b) seawater
(c) distilled water
(d) Italian salad dressing

SOLUTION

Begin by examining the alphabetical listing of pure elements on the inside back cover of this book. If the substance appears in that table, it is a pure substance and an element. If it is not in the table but is a pure substance, then it is a compound. If the substance is not a pure substance, then it is a mixture. Think about your everyday experience with each mixture to determine if it is homogeneous or heterogeneous.

(a) Lead is listed in the table of elements. It is a pure substance and an element.
(b) Seawater is composed of several substances, including salt and water; it is a mixture. It has a uniform composition, so it is a homogeneous mixture.
(c) Distilled water is not listed in the table of elements, but it is a pure substance (water); therefore, it is a compound.
(d) Italian salad dressing contains a number of substances and is therefore a mixture. It usually separates into at least two distinct regions with different composition and is therefore a heterogeneous mixture.

▶ SKILLBUILDER 1.1 | Classifying Matter

Classify each type of matter as a pure substance or a mixture. If it is a pure substance, classify it as an element or a compound. If it is a mixture, classify it as homogeneous or heterogeneous.

(a) mercury in a thermometer
(b) exhaled air
(c) chicken noodle soup
(d) sugar

▶ FOR MORE PRACTICE Example 1.4; Problems 16, 17, 18, 19, 20, 21.

Note: The answers to all Skillbuilders appear at the end of the module.

1.4 Differences in Matter: Physical and Chemical Properties

LO: Distinguish between physical and chemical properties.

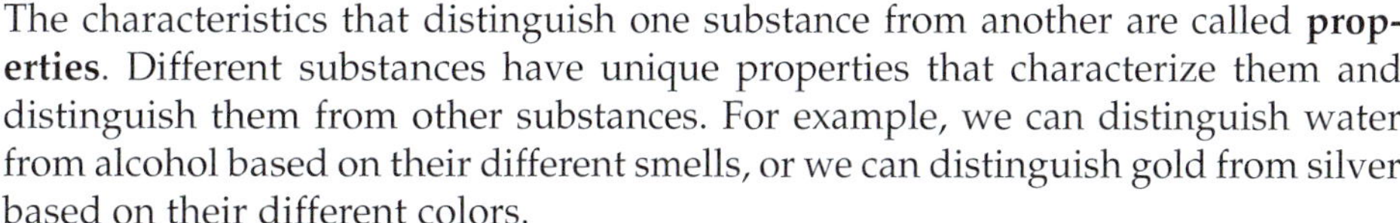

The characteristics that distinguish one substance from another are called **properties**. Different substances have unique properties that characterize them and distinguish them from other substances. For example, we can distinguish water from alcohol based on their different smells, or we can distinguish gold from silver based on their different colors.

In chemistry, we categorize properties into two different types: physical and chemical. A **physical property** is one that a substance displays without changing its composition. A **chemical property** is one that a substance displays only through changing its composition. For example, the characteristic odor of gasoline is a physical property—gasoline does not change its composition when it exhibits its odor. On the other hand, the flammability of gasoline is a chemical property—gasoline does change its composition when it burns.

The atomic or molecular composition of a substance does not change when the substance displays its physical properties. For example, the boiling point of water—a physical property—is 100 °C. When water boils, it changes from a liquid to a gas, but the gas is still water (◀ Figure 1.9). On the other hand, the

◀ **FIGURE 1.9 A physical property** The boiling point of water is a physical property, and boiling is a physical change. When water boils, it turns into a gas, but the water molecules are the same in both the liquid water and the gaseous steam.
© Michael Dalton/Fundamental Photographs.

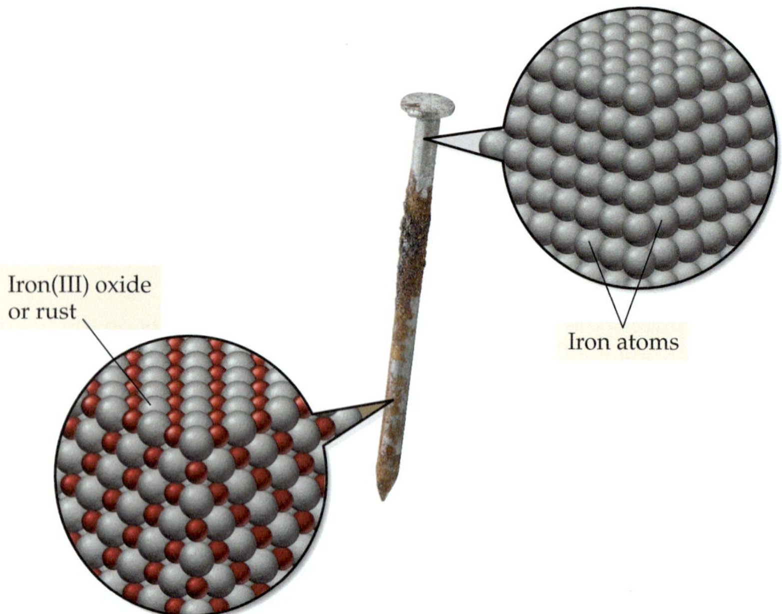

▲ **FIGURE 1.10 A chemical property** The susceptibility of iron to rusting is a chemical property, and rusting is a chemical change. When iron rusts, it turns from iron to iron(III) oxide.

susceptibility of iron to rust is a chemical property—iron must change into iron(III) oxide to display this property (▲ Figure 1.10). Physical properties include odor, taste, color, appearance, melting point, boiling point, and density. Chemical properties include corrosiveness, flammability, acidity, and toxicity.

EXAMPLE 1.2 PHYSICAL AND CHEMICAL PROPERTIES

Classify each property as physical or chemical.

(a) the tendency of copper to turn green when exposed to air
(b) the tendency of automobile paint to dull over time
(c) the tendency of gasoline to evaporate quickly when spilled
(d) the low mass (for a given volume) of aluminum relative to other metals

SOLUTION

(a) Copper turns green because it reacts with gases in air to form compounds; this is a chemical property.
(b) Automobile paint dulls over time because it can fade (decompose) due to sunlight or it can react with oxygen in air. In either case, this is a chemical property.
(c) Gasoline evaporates quickly because it has a low boiling point; this is a physical property.
(d) Aluminum's low mass (for a given volume) relative to other metals is due to its low density; this is a physical property.

▶SKILLBUILDER 1.2 | Physical and Chemical Properties

Classify each property as physical or chemical.

(a) the explosiveness of hydrogen gas
(b) the bronze color of copper
(c) the shiny appearance of silver
(d) the ability of dry ice to sublime (change from solid directly to vapor)

▶FOR MORE PRACTICE Example 1.5; Problems 22, 23, 24, 25.

1.5 Changes in Matter: Physical and Chemical Changes

LO: Distinguish between physical and chemical changes.

Every day, we witness changes in matter: Ice melts, iron rusts, and fruit ripens, for example. What happens to the atoms and molecules that make up these substances during these changes? The answer depends on the kind of change. In a **physical change**, matter changes its appearance but not its composition. For example, when ice melts, it looks different—water looks different from ice—but its composition is the same. Solid ice and liquid water are both composed of water molecules, so melting is a physical change. Similarly, when glass shatters, it looks different, but its composition remains the same—it is still glass. Again, this is a physical change. On the other hand, in a **chemical change**, matter *does* change its composition. For example, copper turns green upon continued exposure to air because it reacts with gases in air to form new compounds. This is a chemical change. Matter undergoes a chemical change when it undergoes a **chemical reaction**. In a chemical reaction, the substances present before the chemical change are called **reactants**, and the substances present after the change are called **products**:

$$\text{Reactants} \xrightarrow[\text{Change}]{\text{Chemical}} \text{Products}$$

We will cover chemical reactions in much more detail in Module 8.

State changes—transformations from one state of matter (such as solid or liquid) to another—are always physical changes.

The differences between physical and chemical changes are not always apparent. Only chemical examination of the substances before and after the change can verify whether the change is physical or chemical. For many cases, however, we can identify chemical and physical changes based on what we know about the changes. Changes in state, such as melting or boiling, or changes that involve merely appearance, such as those produced by cutting or crushing, are always physical changes. Changes involving chemical reactions—often evidenced by heat exchange or color changes—are always chemical changes.

The main difference between chemical and physical changes is related to the changes at the molecular and atomic level. In physical changes, the atoms that compose the matter *do not* change their fundamental associations, even though the matter may change its appearance. In chemical changes, atoms do change their fundamental associations, resulting in matter with a new identity. *A physical change results in a different form of the same substance, while a chemical change results in a completely new substance.*

Consider physical and chemical changes in liquid butane, the substance used to fuel butane lighters. In many lighters, you can see the liquid butane through the plastic case of the lighter. If you push the fuel button on the lighter without turning the flint, some of the liquid butane *vaporizes* (changes from liquid to gas). You cannot see the gaseous butane, but if you listen carefully you can usually hear hissing as it leaks out (◄ Figure 1.11). Since the liquid butane and the gaseous butane are both composed of butane molecules, the change is physical. On the other hand, if you push the button *and* turn the flint to create a spark, a chemical change occurs. The butane molecules react with oxygen molecules in air to form new molecules, carbon dioxide and water (◄ Figure 1.12). The change is chemical because the molecular composition changes upon burning.

▼ **FIGURE 1.11 Vaporization: a physical change** If you push the button on a lighter without turning the flint, some of the liquid butane vaporizes to gaseous butane. Since the liquid butane and the gaseous butane are both composed of butane molecules, this is a physical change.

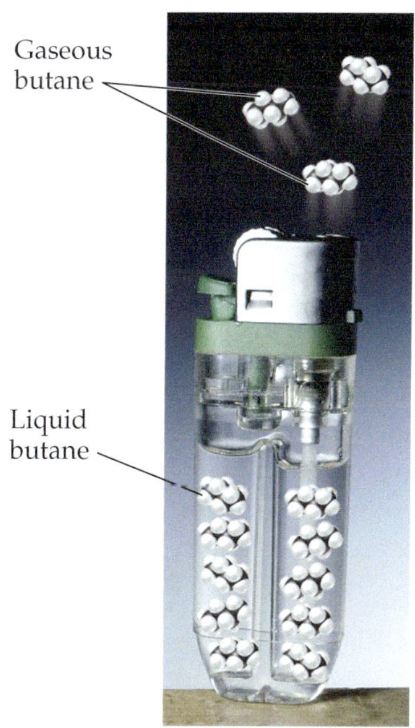

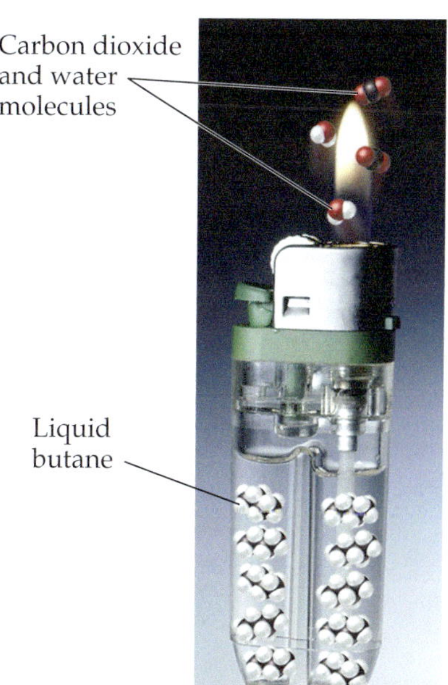

◄ **FIGURE 1.12 Burning: a chemical change** If you push the button *and* turn the flint to create a spark, you produce a flame. The butane molecules react with oxygen molecules in air to form new molecules, carbon dioxide and water. This is a chemical change. © (bottom left): Richard Megna/Fundamental Photographs. (bottom right): Richard Megna/Fundamental Photographs.

EXAMPLE 1.3 PHYSICAL AND CHEMICAL CHANGES

Classify each change as physical or chemical.

(a) the rusting of iron
(b) the evaporation of fingernail-polish remover (acetone) from the skin
(c) the burning of coal
(d) the fading of a carpet upon repeated exposure to sunlight

SOLUTION

(a) Iron rusts because it reacts with oxygen in air to form iron(III) oxide; therefore, this is a chemical change.
(b) When fingernail-polish remover (acetone) evaporates, it changes from liquid to gas, but it remains acetone; therefore, this is a physical change.
(c) Coal burns because it reacts with oxygen in air to form carbon dioxide; this is a chemical change.
(d) A carpet fades on repeated exposure to sunlight because the molecules that give the carpet its color are decomposed by sunlight; this is a chemical change.

▶SKILLBUILDER 1.3 | Physical and Chemical Changes

Classify each change as physical or chemical.

(a) copper metal forming a blue solution when it is dropped into colorless nitric acid
(b) a train flattening a penny placed on a railroad track
(c) ice melting into liquid water
(d) a match igniting a firework

▶FOR MORE PRACTICE Example 1.6; Problems 26, 27, 28, 29.

CONCEPTUAL CHECKPOINT 1.2

In this figure liquid water is vaporizing into steam. © Etienne du Preez/Shutterstock.

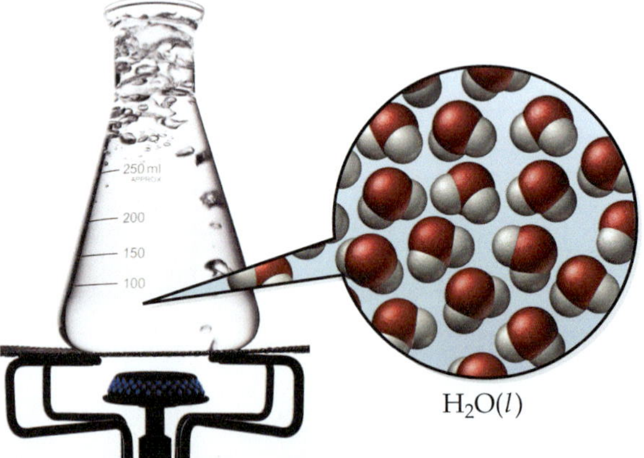

Which diagram best represents the molecules in the steam?

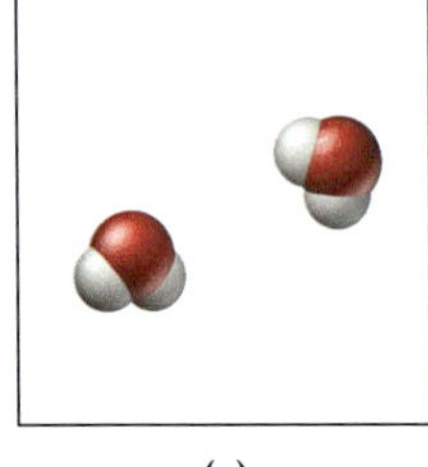

(a)

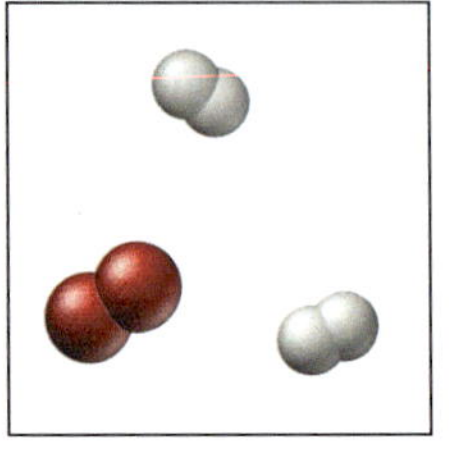

(b)

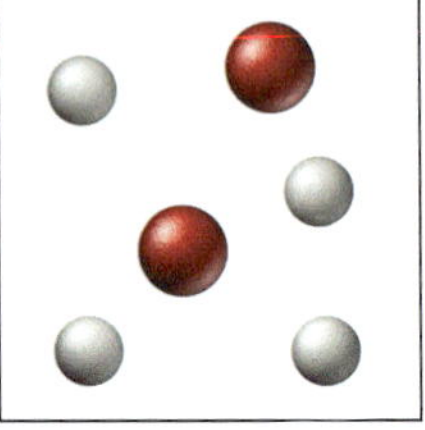

(c)

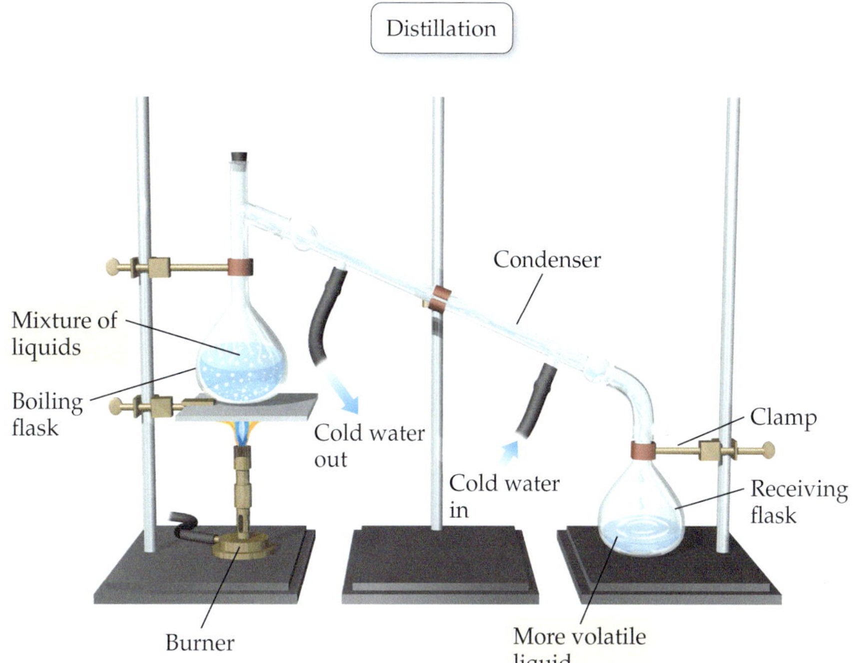

▶ **FIGURE 1.13 Separating a mixture of two liquids by distillation** The liquid with the lower boiling point vaporizes first. The vapors are collected and cooled (with cold water) until they condense back into liquid form.

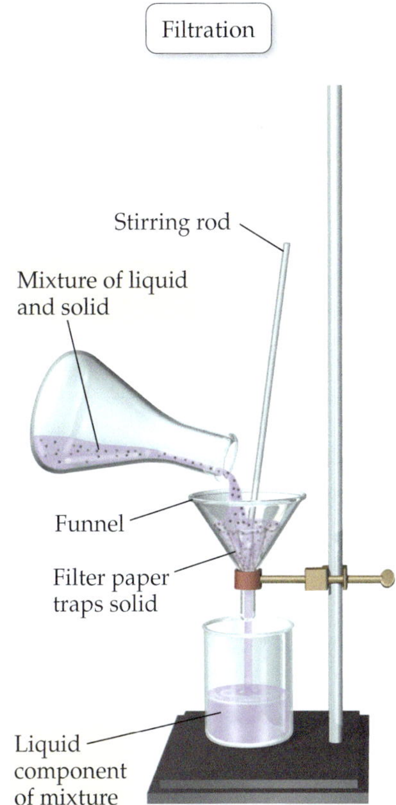

▲ **FIGURE 1.14 Separating a solid from a liquid by filtration**

Separating Mixtures Through Physical Changes

Chemists often want to separate mixtures into their components. Such separations can be easy or difficult, depending on the components in the mixture. In general, mixtures are separable because the different components have different properties. We can use various techniques that exploit these differences to achieve separation. For example, oil and water are immiscible (do not mix) and have different densities. For this reason, oil floats on top of water and we can separate it from water by **decanting**—carefully pouring off—the oil into another container. We can separate mixtures of miscible liquids by **distillation**, in which we heat the mixture to boil off the more **volatile**—the more easily vaporizable—liquid. We then recondense the volatile liquid in a condenser and collect it in a separate flask (▲ Figure 1.13). If a mixture is composed of a solid and a liquid, we can separate the two by **filtration**, in which we pour the mixture through filter paper usually held in a funnel (◀ Figure 1.14).

MODULE IN REVIEW

Self-Assessment Quiz

Q1. Which substance is a pure compound?

(a) Gold
(b) Water
(c) Milk
(d) Fruit cake

Q2. Which property of trinitrotoluene (TNT) is most likely a chemical property?

(a) Yellow color
(b) Melting point is 80.1 °C
(c) Explosive
(d) None of the above

Q3. Which change is a chemical change?

(a) The condensation of dew on a cold night
(b) A forest fire
(c) The smoothening of rocks by ocean waves
(d) None of the above

Answers: 1:b, 2:c, 3:b

Chemical Principles

Matter: Matter is anything that occupies space and has mass. It is composed of atoms, which are often bonded together as molecules. Matter can exist as a solid, a liquid, or a gas. Solid matter can be either amorphous or crystalline.

Classification of Matter: We can classify matter according to its composition. Pure matter is composed of only one type of substance; that substance may be an element (a substance that cannot be decomposed into simpler substances), or it may be a compound (a substance composed of two or more elements in fixed definite proportions). Mixtures are composed of two or more different substances, the proportions of which may vary from one sample to the next. Mixtures can be either homogeneous, having the same composition throughout, or heterogeneous, having a composition that varies from region to region.

Properties and Changes of Matter: We can divide the properties of matter into two types: physical and chemical. The physical properties of matter do not involve a change in composition. The chemical properties of matter involve a change in composition. We can divide changes in matter into physical and chemical. In a physical change, the appearance of matter may change, but its composition does not. In a chemical change, the composition of matter changes.

Relevance

Matter: Everything is made of matter—you, me, the chair you sit on, and the air we breathe. The physical universe basically contains only two things: matter and energy. We begin our study of chemistry by defining and classifying these two building blocks of the universe.

Classification of Matter: Since ancient times, humans have tried to understand matter and harness it for their purposes. The earliest humans shaped matter into tools and used the transformation of matter—especially fire—to keep warm and to cook food. To manipulate matter, we must understand it. Fundamental to this understanding is the connection between the properties of matter and the molecules and atoms that compose it.

Properties and Changes of Matter: The physical and chemical properties of matter make the world around us the way it is. For example, a physical property of water is its boiling point at sea level—100 °C. The physical properties of water—and all matter—are determined by the atoms and molecules that compose it. If water molecules were different—even slightly different—water would boil at a different temperature. Imagine a world where water boiled at room temperature.

Chemical Skills

LO: Classify matter as element, compound, or mixture (Section 1.3).

Begin by examining the alphabetical listing of elements on the inside back cover of this book. If the substance is listed in that table, it is a pure substance and an element.

If the substance is not listed in that table, refer to your everyday experience with the substance to determine whether it is a pure substance. If it is a pure substance not listed in the table, then it is a compound.

If it is not a pure substance, then it is a mixture. Refer to your everyday experience with the mixture to determine whether it has uniform composition (homogeneous) or nonuniform composition (heterogeneous).

LO: Distinguish between physical and chemical properties (Section 1.4).

To distinguish between physical and chemical properties, consider whether the substance changes composition while displaying the property. If it *does not* change composition, the property is physical; if it *does*, the property is chemical.

Examples

EXAMPLE 1.4 CLASSIFYING MATTER

Classify each type of matter as a pure substance or a mixture. If it is a pure substance, classify it as an element or a compound. If it is a mixture, classify it as homogeneous or heterogeneous.

(a) pure silver
(b) swimming-pool water
(c) dry ice (solid carbon dioxide)
(d) blueberry muffin

SOLUTION

(a) Pure element; silver appears in the element table.
(b) Homogeneous mixture; pool water contains at least water and chlorine, and it is uniform throughout.
(c) Compound; dry ice is a pure substance (carbon dioxide), but it is not listed in the table.
(d) Heterogeneous mixture; a blueberry muffin is a mixture of several things and has nonuniform composition.

EXAMPLE 1.5 DISTINGUISHING BETWEEN PHYSICAL AND CHEMICAL PROPERTIES

Classify each property as physical or chemical.

(a) the tendency for platinum jewelry to scratch easily
(b) the ability of sulfuric acid to burn the skin
(c) the ability of hydrogen peroxide to bleach hair
(d) the density of lead relative to other metals

SOLUTION

(a) Physical; scratched platinum is still platinum.
(b) Chemical; the acid chemically reacts with the skin to produce the burn.
(c) Chemical; the hydrogen peroxide chemically reacts with hair to change the hair.
(d) Physical; you can determine the density of lead by measuring the volume and mass of a lead sample.

LO: Distinguish between physical and chemical changes (Section 1.5).

To distinguish between physical and chemical changes, consider whether the substance changes composition during the change. If it *does not* change composition, the change is physical; if it *does*, the change is chemical.

EXAMPLE 1.6 DISTINGUISHING BETWEEN PHYSICAL AND CHEMICAL CHANGES

Classify each change as physical or chemical.

(a) the explosion of gunpowder in the barrel of a gun
(b) the melting of gold in a furnace
(c) the bubbling that occurs when you mix baking soda and vinegar
(d) the bubbling that occurs when water boils

SOLUTION

(a) Chemical; the gunpowder reacts with oxygen during the explosion.
(b) Physical; the liquid gold is still gold.
(c) Chemical; the bubbling is a result of a chemical reaction between the two substances to form new substances, one of which is carbon dioxide released as bubbles.
(d) Physical; the bubbling is due to liquid water turning into gaseous water, but it is still water.

KEY TERMS

amorphous **[1.2]**
atoms **[1.1]**
chemical change **[1.5]**
chemical property **[1.4]**
compound **[1.3]**
compressible **[1.2]**
crystalline **[1.2]**
decanting **[1.5]**
distillation **[1.5]**
element **[1.3]**
filtration **[1.5]**
gas **[1.2]**
heterogeneous mixture **[1.3]**
homogeneous mixture **[1.3]**
liquid **[1.2]**
matter **[1.1]**
mixture **[1.3]**
molecule **[1.1]**
physical change **[1.5]**
physical property **[1.4]**
property **[1.4]**
pure substance **[1.3]**
solid **[1.2]**
state of matter **[1.2]**
volatile **[1.5]**

EXERCISES

QUESTIONS

1. Define matter and list some examples.
2. What is matter composed of?
3. What are the three states of matter?
4. What are the properties of a solid?
5. What is the difference between a crystalline solid and an amorphous solid?
6. What are the properties of a liquid?
7. What are the properties of a gas?
8. Why are gases compressible?
9. What is a mixture?
10. What is the difference between a homogeneous mixture and a heterogeneous mixture?
11. What is a pure substance?
12. What is an element? A compound?
13. What is the difference between a mixture and a compound?
14. What is the definition of a physical property? What is the definition of a chemical property?
15. What is the difference between a physical change and a chemical change?

PROBLEMS

CLASSIFYING MATTER

16. Classify each pure substance as an element or a compound.
- **(a)** aluminum
- **(b)** sulfur
- **(c)** methane
- **(d)** acetone

17. Classify each pure substance as an element or a compound.
- **(a)** carbon
- **(b)** baking soda (sodium bicarbonate)
- **(c)** nickel
- **(d)** gold

18. Classify each mixture as homogeneous or heterogeneous.
- **(a)** coffee
- **(b)** chocolate sundae
- **(c)** apple juice
- **(d)** gasoline

19. Classify each mixture as homogeneous or heterogeneous.
- **(a)** baby oil
- **(b)** chocolate chip cookie
- **(c)** water and gasoline
- **(d)** wine

20. Classify each substance as a pure substance or a mixture. If it is a pure substance, classify it as an element or a compound. If it is a mixture, classify it as homogeneous or heterogeneous.
- **(a)** helium gas
- **(b)** clean air
- **(c)** rocky road ice cream
- **(d)** concrete

21. Classify each substance as a pure substance or a mixture. If it is a pure substance, classify it as an element or a compound. If it is a mixture, classify it as homogeneous or heterogeneous.
- **(a)** urine
- **(b)** pure water
- **(c)** Snickers™ bar
- **(d)** soil

PHYSICAL AND CHEMICAL PROPERTIES AND PHYSICAL AND CHEMICAL CHANGES

22. Classify each property as physical or chemical.
- **(a)** the tendency of silver to tarnish
- **(b)** the shine of chrome
- **(c)** the color of gold
- **(d)** the flammability of propane gas

23. Classify each property as physical or chemical.
- **(a)** the boiling point of ethyl alcohol
- **(b)** the temperature at which dry ice sublimes (turns from a solid into a gas).
- **(c)** the flammability of ethyl alcohol
- **(d)** the smell of perfume

24. Which of the following properties of ethylene (a ripening agent for bananas) are physical properties, and which are chemical?
- colorless
- odorless
- flammable
- gas at room temperature
- 1 L has a mass of 1.260 g under standard conditions
- mixes with acetone
- polymerizes to form polyethylene

25. Which of the following properties of ozone (a pollutant in the lower atmosphere but part of a protective shield against UV light in the upper atmosphere) are physical, and which are chemical?
- bluish color
- pungent odor
- very reactive
- decomposes on exposure to ultraviolet light
- gas at room temperature

26. Classify each change as physical or chemical.
- **(a)** A balloon filled with hydrogen gas explodes upon contact with a spark.
- **(b)** The liquid propane in a barbecue evaporates away because someone left the valve open.
- **(c)** The liquid propane in a barbecue ignites upon contact with a spark.
- **(d)** Copper metal turns green on exposure to air and water.

27. Classify each change as physical or chemical.
- **(a)** Sugar dissolves in hot water.
- **(b)** Sugar burns in a pot.
- **(c)** A metal surface becomes dull because of continued abrasion.
- **(d)** A metal surface becomes dull on exposure to air.

28. A block of aluminum is **(a)** ground into aluminum powder and then **(b)** ignited. It then emits flames and smoke. Classify **(a)** and **(b)** as chemical or physical changes.

29. Several pieces of graphite from a mechanical pencil are **(a)** broken into tiny pieces. Then the pile of graphite is **(b)** ignited with a hot flame. Classify **(a)** and **(b)** as chemical or physical changes.

HIGHLIGHT PROBLEMS

30. Classify each as a pure substance or a mixture.

(a) (b)

(c) (d)

31. Classify each as a pure substance or a mixture. If it is a pure substance, classify it as an element or a compound. If it is a mixture, classify it as homogeneous or heterogeneous.

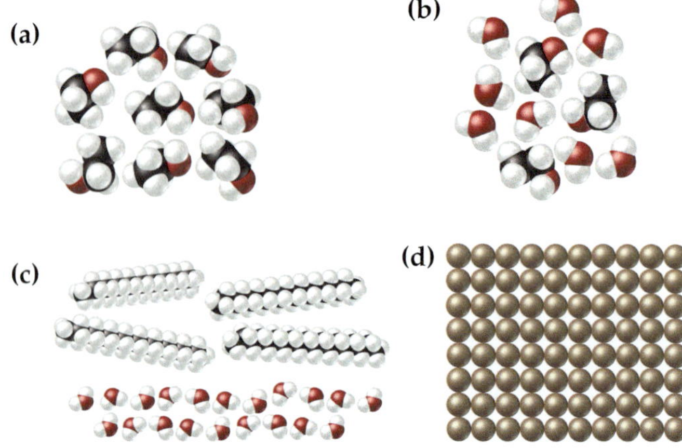

32. This molecular drawing shows images of acetone molecules before and after a change. Was the change chemical or physical?

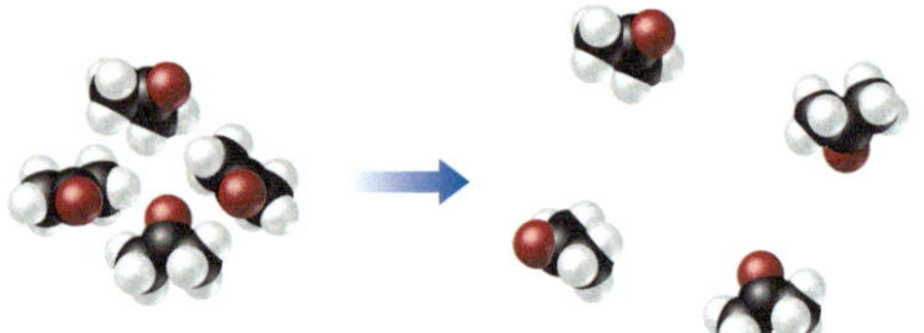

33. This molecular drawing shows images of methane molecules and oxygen molecules before and after a change. Was the change chemical or physical?

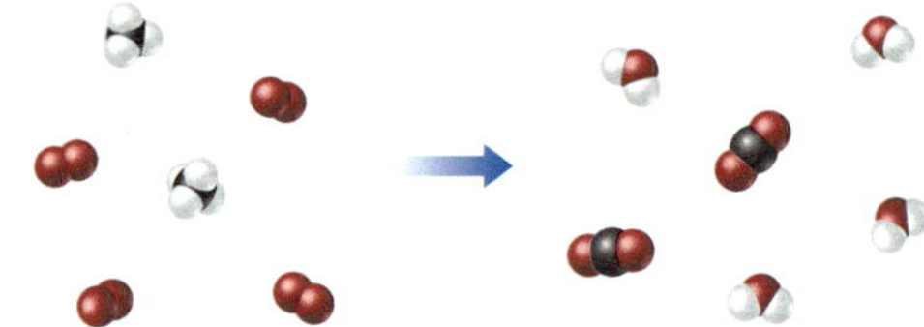

Answers to Skillbuilder Exercises

Skillbuilder 1.1

(a) pure substance, element
(b) mixture, homogeneous
(c) mixture, heterogeneous
(d) pure substance, compound

Skillbuilder 1.2

(a) chemical
(b) physical
(c) physical
(d) physical

Skillbuilder 1.3

(a) chemical
(b) physical
(c) physical
(d) chemical

Answers to Conceptual Checkpoints

1.1 (c) The particles are far apart and moving relative to one another.

1.2 (a) Vaporization is a physical change, so the water molecules are the same before and after the boiling.

▲ Seaside rocks are typically composed of silicates, compounds of silicon and oxygen atoms. Seaside air, like all air, contains nitrogen and oxygen molecules, and it often also contains substances called amines. The amine shown here is triethylamine, which is emitted by decaying fish. Triethylamine is one of the compounds responsible for the fishy smell of the seaside.

Elements 2

"Nothing exists except atoms and empty space; everything else is opinion." —
Democritus (460–370 B.C.)

MODULE OUTLINE

2.1 The Nuclear Theory of the Atom

LO: Describe the respective properties and charges of electrons, neutrons, and protons.

The **nuclear theory of the atom** has three basic parts:

1. Most of the atom's mass and all of its positive charge are contained in a small core called the *nucleus*.
2. Most of the volume of the atom is empty space through which the tiny, negatively charged *electrons* are dispersed.
3. There are as many negatively charged electrons outside the nucleus as there are positively charged particles *(protons)* inside the nucleus, so that the atom is electrically neutral.

In addition to the points above that are the key components of the nuclear theory of the atom, note the following:

The nucleus may also contain uncharged particles called *neutrons*.

Atoms combine in simple, whole-number ratios to form compounds.

If the nucleus of the atom were the size of this dot ·, the average electron would be about 10 m away. Yet the dot would contain almost the entire mass of the atom. What if matter were composed of atomic nuclei piled on top of each other like marbles? Such matter would be incredibly dense; a single grain of sand composed of solid atomic nuclei would have a mass of 5 million kg. Astronomers believe that black holes and neutron stars are composed of this kind of incredibly dense matter.

Protons and neutrons have very similar masses. In SI units, the mass of the proton is 1.67262×10^{-27} kg, and the mass of the neutron is a close 1.67493×10^{-27} kg. A more common unit to express these masses, however, is the **atomic mass unit (amu)**, defined as one-twelfth of the mass of a carbon atom containing 6 protons and 6 neutrons. A proton has a mass of 1.0073 amu, and a neutron has a mass of 1.0087 amu. Electrons, by contrast, have an almost negligible mass of 0.00091×10^{-27} kg, or approximately 0.00055 amu.

EVERYDAY CHEMISTRY

▶ Solid Matter?

If matter really is mostly empty space, then why does it appear so solid? Why can you tap your knuckles on the table and feel a solid thump? Matter appears solid because the variation in the density is on such a small scale that our eyes can't see it. Imagine a jungle gym 100 stories high and the size of a football field. It is mostly empty space. Yet if you were to view it from an airplane, it would appear as a solid mass. Matter is similar. When you tap your knuckles on the table, it is much like one giant jungle gym (your finger) crashing into another (the table). Even though they are both primarily empty space, one does not fall into the other.

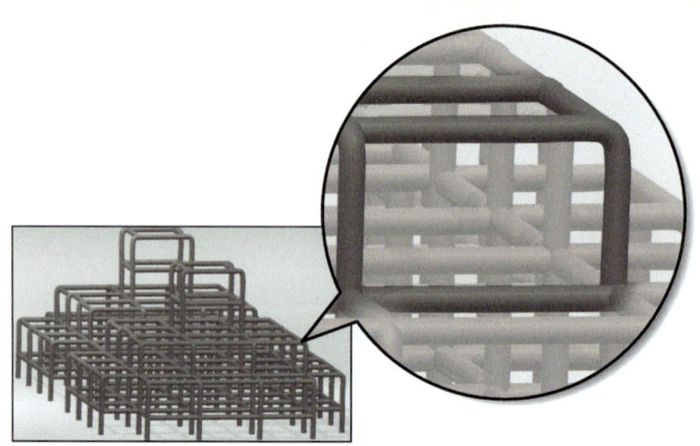

▲ Matter appears solid and uniform because the variation in density is on a scale too small for our eyes to see. Just as this scaffolding appears solid at a distance, so matter appears solid to us.

◀ If a proton had the mass of a baseball, an electron would have the mass of a rice grain. The proton is nearly 2000 times as massive as an electron. © Maxwell Art And Photo/Pearson.

The proton and the electron both have electrical **charge**. The proton's charge is 1+ and the electron's charge is 1–. The charges of the proton and the electron are equal in magnitude but opposite in sign, so that when the two particles are paired, the charges exactly cancel. The neutron has no charge.

What is electrical charge? Electrical charge is a fundamental property of protons and electrons, just as mass is a fundamental property of matter. Most matter is charge-neutral because protons and electrons occur together and their charges cancel. However, you may have experienced excess electrical charge when brushing your hair on a dry day. The brushing action results in the accumulation of electrical charge on the hair strands, which then repel each other, causing your hair to stand on end.

We can summarize the nature of electrical charge as follows (◀ Figure 2.1).

- Electrical charge is a fundamental property of protons and electrons.
- Positive and negative electrical charges attract each other.
- Positive–positive and negative–negative charges repel each other.
- Positive and negative charges cancel each other so that a proton and an electron, when paired, are charge-neutral.

Note that matter is usually charge-neutral due to the canceling effect of protons and electrons. When matter does acquire charge imbalances, these imbalances usually equalize quickly, often in dramatic ways. For example, the shock you receive when touching a doorknob during dry weather is the equalization of a charge imbalance that developed as you walked across the carpet. Lightning is an equalization of charge imbalances that develop during electrical storms.

If you had a sample of matter—even a tiny sample, such as a sand grain—that was composed of only protons or only electrons, the forces around that matter would be extraordinary, and the matter would be unstable. Fortunately, matter is not that way—protons and electrons exist together, canceling

◀ **FIGURE 2.1 The properties of electrical charge**

each other's charge and making matter charge-neutral. Table 2.1 summarizes the properties of protons, neutrons, and electrons (subatomic particles).

TABLE 2.1 Subatomic Particles

	Mass (kg)	Mass (amu)	Charge
proton	1.67262×10^{-27}	1.0073	1+
neutron	1.67493×10^{-27}	1.0087	0
electron	0.00091×10^{-27}	0.00055	1–

▶ Matter is normally charge-neutral, having equal numbers of positive and negative charges that exactly cancel. When the charge balance of matter is disturbed, as in an electrical storm, it quickly rebalances, often in dramatic ways such as lightning. © Jeremy Woodhouse/Photodisc/Getty Images.

CONCEPTUAL CHECKPOINT 2.1

An atom composed of which of these particles would have a mass of approximately 12 amu and be charge-neutral?

(a) 6 protons and 6 electrons

(b) 3 protons, 3 neutrons, and 6 electrons

(c) 6 protons, 6 neutrons, and 6 electrons

(d) 12 neutrons and 12 electrons

2.2 Elements: Defined by Their Numbers of Protons

LO: Determine the atomic symbol and atomic number for an element using the periodic table.

We have seen that atoms are composed of protons, neutrons, and electrons. However, it is the number of protons in the nucleus of an atom that identifies it as a particular element. For example, atoms with 2 protons in their nucleus are helium atoms, atoms with 13 protons in their nucleus are aluminum atoms, and atoms with 92 protons in their nucleus are uranium atoms. The number of protons in an atom's nucleus defines the element (▶ Figure 2.2). Every aluminum atom has 13 protons in its nucleus; if it had a different number of protons, it would be a different element. The number of protons in the nucleus of an atom is its **atomic number** and is represented with the symbol Z.

The periodic table of the elements (▶ Figure 2.3) lists all known elements according to their atomic numbers. Each element is represented by a unique **chemical symbol**, a one- or two-letter abbreviation for the element that appears directly below its atomic number on the periodic table. The chemical symbol for helium is He; for aluminum, Al; and for uranium, U. The chemical symbol and the atomic number always go together. If the atomic number is 13, the chemical symbol must

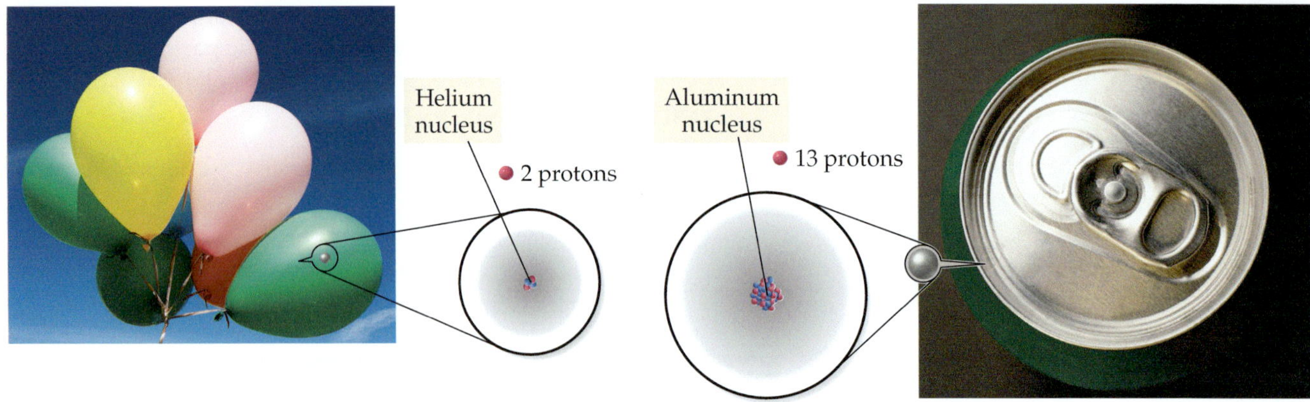

▲ **FIGURE 2.2 The number of protons in the nucleus defines the element** © (left) www.photos.com/Jupiter Images/Getty Images; (right) John A. Rizzo/Digital Vision/Getty Images.

be Al. If the atomic number is 92, the chemical symbol must be U. This is just another way of saying that the number of protons defines the element.

Most chemical symbols are based on the English name of the element. For example, the symbol for carbon is C; for silicon, Si; and for bromine, Br. Some elements, however, have symbols based on their Latin names. For example, the symbol for potassium is K, from the Latin *kalium*, and the symbol for sodium is Na, from the Latin *natrium*. Additional elements with symbols based on their Greek or Latin names include:

lead	Pb	*plumbum*
mercury	Hg	*hydrargyrum*
iron	Fe	*ferrum*
silver	Ag	*argentum*
tin	Sn	*stannum*
copper	Cu	*cuprum*

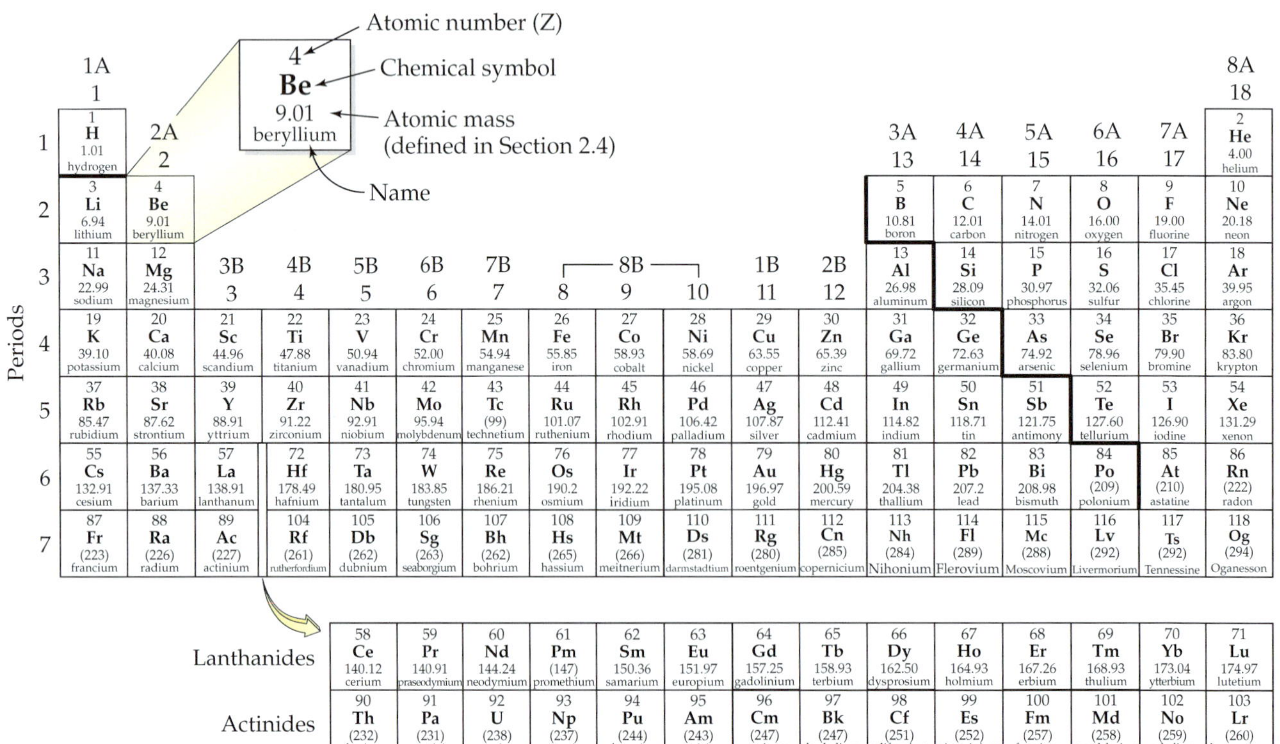

Periods	1A 1	2A 2	3B 3	4B 4	5B 5	6B 6	7B 7	8B 8	8B 9	8B 10	1B 11	2B 12	3A 13	4A 14	5A 15	6A 16	7A 17	8A 18
1	1 H 1.01 hydrogen																	2 He 4.00 helium
2	3 Li 6.94 lithium	4 Be 9.01 beryllium											5 B 10.81 boron	6 C 12.01 carbon	7 N 14.01 nitrogen	8 O 16.00 oxygen	9 F 19.00 fluorine	10 Ne 20.18 neon
3	11 Na 22.99 sodium	12 Mg 24.31 magnesium											13 Al 26.98 aluminum	14 Si 28.09 silicon	15 P 30.97 phosphorus	16 S 32.06 sulfur	17 Cl 35.45 chlorine	18 Ar 39.95 argon
4	19 K 39.10 potassium	20 Ca 40.08 calcium	21 Sc 44.96 scandium	22 Ti 47.88 titanium	23 V 50.94 vanadium	24 Cr 52.00 chromium	25 Mn 54.94 manganese	26 Fe 55.85 iron	27 Co 58.93 cobalt	28 Ni 58.69 nickel	29 Cu 63.55 copper	30 Zn 65.39 zinc	31 Ga 69.72 gallium	32 Ge 72.63 germanium	33 As 74.92 arsenic	34 Se 78.96 selenium	35 Br 79.90 bromine	36 Kr 83.80 krypton
5	37 Rb 85.47 rubidium	38 Sr 87.62 strontium	39 Y 88.91 yttrium	40 Zr 91.22 zirconium	41 Nb 92.91 niobium	42 Mo 95.94 molybdenum	43 Tc (99) technetium	44 Ru 101.07 ruthenium	45 Rh 102.91 rhodium	46 Pd 106.42 palladium	47 Ag 107.87 silver	48 Cd 112.41 cadmium	49 In 114.82 indium	50 Sn 118.71 tin	51 Sb 121.75 antimony	52 Te 127.60 tellurium	53 I 126.90 iodine	54 Xe 131.29 xenon
6	55 Cs 132.91 cesium	56 Ba 137.33 barium	57 La 138.91 lanthanum	72 Hf 178.49 hafnium	73 Ta 180.95 tantalum	74 W 183.85 tungsten	75 Re 186.21 rhenium	76 Os 190.2 osmium	77 Ir 192.22 iridium	78 Pt 195.08 platinum	79 Au 196.97 gold	80 Hg 200.59 mercury	81 Tl 204.38 thallium	82 Pb 207.2 lead	83 Bi 208.98 bismuth	84 Po (209) polonium	85 At (210) astatine	86 Rn (222) radon
7	87 Fr (223) francium	88 Ra (226) radium	89 Ac (227) actinium	104 Rf (261) rutherfordium	105 Db (262) dubnium	106 Sg (263) seaborgium	107 Bh (262) bohrium	108 Hs (265) hassium	109 Mt (266) meitnerium	110 Ds (281) darmstadtium	111 Rg (280) roentgenium	112 Cn (285) copernicium	113 Nh (284) Nihonium	114 Fl (289) Flerovium	115 Mc (288) Moscovium	116 Lv (292) Livermorium	117 Ts (292) Tennessine	118 Og (294) Oganesson

Lanthanides	58 Ce 140.12 cerium	59 Pr 140.91 praseodymium	60 Nd 144.24 neodymium	61 Pm (147) promethium	62 Sm 150.36 samarium	63 Eu 151.97 europium	64 Gd 157.25 gadolinium	65 Tb 158.93 terbium	66 Dy 162.50 dysprosium	67 Ho 164.93 holmium	68 Er 167.26 erbium	69 Tm 168.93 thulium	70 Yb 173.04 ytterbium	71 Lu 174.97 lutetium
Actinides	90 Th (232) thorium	91 Pa (231) protactinium	92 U (238) uranium	93 Np (237) neptunium	94 Pu (244) plutonium	95 Am (243) americium	96 Cm (247) curium	97 Bk (247) berkelium	98 Cf (251) californium	99 Es (252) einsteinium	100 Fm (257) fermium	101 Md (258) mendelevium	102 No (259) nobelium	103 Lr (260) lawrencium

▲ **FIGURE 2.3 The periodic table of the elements**

▲ The name *bromine* originates from the Greek word *bromos*, meaning "stench." Bromine vapor, seen as the red-brown gas in this photograph, has a strong odor. © Charles D. Winters/Science Source.

Early scientists often gave newly discovered elements names that reflected their properties. For example, *argon* originates from the Greek word *argos*, meaning "inactive," referring to argon's chemical inertness (it does not react with other elements). *Bromine* originates from the Greek word *bromos*, meaning "stench," referring to bromine's strong odor. Scientists named other elements after countries. For example, polonium was named after Poland, francium after France, and americium after the United States of America. Still other elements were named after scientists. Curium was named after Marie Curie, and mendelevium after Dmitri Mendeleev. You can find every element's name, symbol, and atomic number in the periodic table (inside front cover) and in an alphabetical listing (inside back cover) in this book.

Curium
96
Cm
(247)

▲ Curium is named after Marie Curie (1867–1934), a chemist who helped discover radioactivity and also discovered two new elements. Curie won two Nobel Prizes for her work. © Library of Congress Prints and Photographs Division.

EXAMPLE 2.1 ATOMIC NUMBER, ATOMIC SYMBOL, AND ELEMENT NAME

List the atomic symbol and atomic number for each element.

(a) silicon
(b) potassium
(c) gold
(d) antimony

SOLUTION

As you become familiar with the periodic table, you will be able to quickly locate elements on it. At first you may find it easier to locate them in the alphabetical listing on the inside back cover of this book, but you should become familiar with their positions in the periodic table.

Element	Symbol	Atomic Number
silicon	Si	14
potassium	K	19
gold	Au	79
antimony	Sb	51

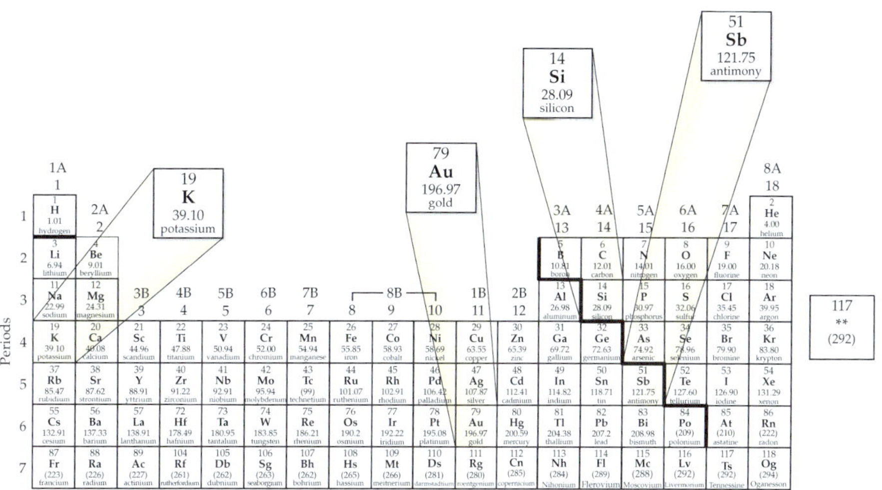

▶SKILLBUILDER 2.1 | Atomic Number, Atomic Symbol, and Element Name

Find the name and atomic number for each element.

(a) Na
(b) Ni
(c) P
(d) Ta

▶FOR MORE PRACTICE Problems 42, 43, 44, 45, 46, 47, 48, 49.

2.3 Isotopes: When the Number of Neutrons Varies

LO: Determine atomic numbers, mass numbers, and isotope symbols for an isotope.

LO: Determine number of protons and neutrons from isotope symbols.

All atoms of a given element have the same number of protons; however, they do not necessarily have the same number of neutrons. Because neutrons and protons have nearly the same mass (approximately 1 amu), and the number of neutrons in the atoms of a given element can vary, all atoms of a given element *do not* have the same mass. For example, all neon atoms in nature contain 10 protons, but they may have 10, 11, or 12 neutrons (▼ Figure 2.4). All three types of neon atoms exist, and each has a slightly different mass. Atoms with the same number of protons but different numbers of neutrons are **isotopes**. Some elements, such as beryllium (Be) and aluminum (Al), have only one stable naturally occurring isotope, while other elements, such as neon (Ne) and chlorine (Cl), have two or more.

For a given element, the relative amounts of each different isotope in a naturally occurring sample of that element are always the same. For example, in any natural sample of neon atoms, 90.48 % of the atoms are the isotope with 10 neutrons, 0.27 % are the isotope with 11 neutrons, and 9.25 % are the isotope with 12 neutrons as summarized in Table 2.2. This means that in a sample of 10,000 neon atoms, 9048 have 10 neutrons, 27 have 11 neutrons, and 925 have 12 neutrons. These percentages are the **percent natural abundance** of the isotopes. The preceding numbers are for neon only; each element has its own unique percent natural abundance of isotopes.

Recent studies have shown that for some elements, the relative amounts of each different isotope vary depending on the history of the sample. However, these variations are usually small and beyond the scope of this book.

Percent means "per hundred." 90.48 % means that 90.48 atoms out of 100 are the isotope with 10 neutrons.

TABLE 2.2 Neon Isotopes

Symbol	Number of Protons	Number of Neutrons	A (Mass Number)	Percent Natural Abundance
Ne-20 or $^{20}_{10}\text{Ne}$	10	10	20	90.48 %
Ne-21 or $^{21}_{10}\text{Ne}$	10	11	21	0.27 %
Ne-22 or $^{22}_{10}\text{Ne}$	10	12	22	9.25 %

The sum of the number of neutrons and protons in an atom is its **mass number** and is given the symbol **A**.

$$\text{A} = \text{Number of protons} + \text{Number of neutrons}$$

For neon, which has 10 protons, the mass numbers of the three different naturally occurring isotopes are 20, 21, and 22, corresponding to 10, 11, and 12 neutrons, respectively.

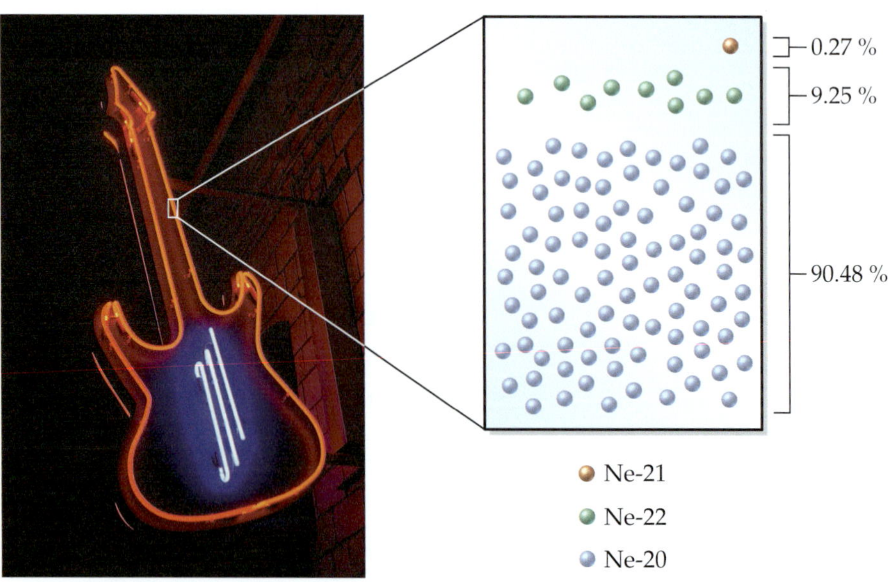

▶ **FIGURE 2.4 Isotopes of neon** Naturally occurring neon contains three different isotopes: Ne-20 (with 10 neutrons), Ne-21 (with 11 neutrons), and Ne-22 (with 12 neutrons). © U.S. Department of Energy

We often symbolize isotopes in the following way:

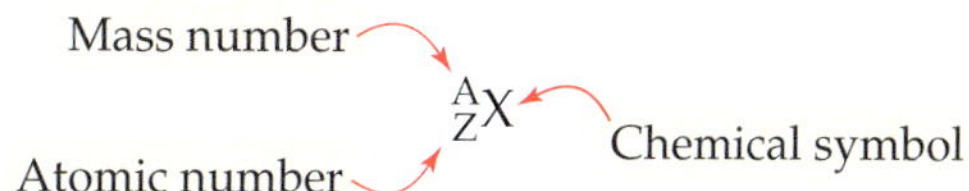

where X is the chemical symbol, A is the mass number, and Z is the atomic number.

For example, the symbols for the neon isotopes are:

$$^{20}_{10}\text{Ne} \quad ^{21}_{10}\text{Ne} \quad ^{22}_{10}\text{Ne}$$

Notice that the chemical symbol, Ne, and the atomic number, 10, are redundant. If the atomic number is 10, the symbol must be Ne, and vice versa. The mass numbers, however, are different, reflecting the different number of neutrons in each isotope.

CONCEPTUAL CHECKPOINT 2.2

Carbon has two naturally occurring isotopes: $^{12}_{6}\text{C}$ and $^{13}_{6}\text{C}$. Using circles to represent protons and squares to represent neutrons, draw the nucleus of each isotope.

A second common notation for isotopes is the chemical symbol (or chemical name) followed by a hyphen and the mass number of the isotope:

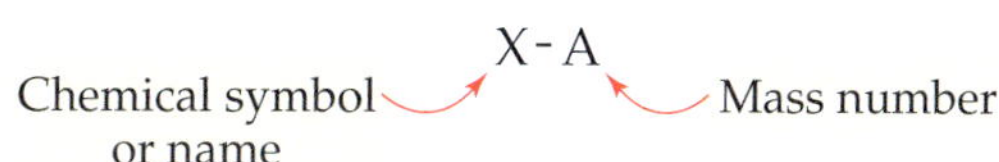

In this notation, the neon isotopes are:

Ne-20	neon-20
Ne-21	neon-21
Ne-22	neon-22

Notice that all isotopes of a given element have the same number of protons (otherwise they would be a different element). Notice also that the mass number is the *sum* of the number of protons and the number of neutrons. The number of neutrons in an isotope is the difference between the mass number and the atomic number.

In general, mass number increases with increasing atomic number.

EXAMPLE 2.2 ATOMIC NUMBERS, MASS NUMBERS, AND ISOTOPE SYMBOLS

What are the atomic number (Z), mass number (A), and symbols of the carbon isotope that has 7 neutrons?

SOLUTION

You can determine that the atomic number (Z) of carbon is 6 (from the periodic table). This means that carbon atoms have 6 protons. The mass number (A) for the isotope with 7 neutrons is the sum of the number of protons and the number of neutrons.

$$A = 6 + 7 = 13$$

So, Z = 6, A = 13, and the symbols for the isotope are C-13 and $^{13}_{6}\text{C}$.

▶SKILLBUILDER 2.2 | Atomic Numbers, Mass Numbers, and Isotope Symbols

What are the atomic number, mass number, and symbols for the chlorine isotope with 18 neutrons?

▶FOR MORE PRACTICE Example 2.17; Problems 52, 54, 56, 57.

EXAMPLE 2.3 NUMBERS OF PROTONS AND NEUTRONS FROM ISOTOPE SYMBOLS

How many protons and neutrons are in the chromium isotope $^{52}_{24}Cr$?

	SOLUTION
The number of protons is equal to Z (lower left number).	$\#p^+ = Z = 24$
The number of neutrons is equal to A (upper left number) –Z (lower left number).	$\#n = A - Z$ $= 52 - 24$ $= 28$

▶**SKILLBUILDER 2.3 | Numbers of Protons and Neutrons from Isotope Symbols**

How many protons and neutrons are in the potassium isotope $^{39}_{19}K$?

▶**FOR MORE PRACTICE** Example 2.18; Problems 58, 59.

CONCEPTUAL CHECKPOINT 2.3

If an atom with a mass number of 27 has 14 neutrons, it is an isotope of which element?

(a) silicon

(b) aluminum

(c) cobalt

(d) niobium

CONCEPTUAL CHECKPOINT 2.4

Throughout this book, we represent atoms as spheres. For example, we represent a carbon atom by a black sphere as shown here. In light of the nuclear theory of the atom, when represented this way, would C-12 and C-13 look different? Why or why not?

Carbon

2.4 Atomic Mass: The Average Mass of an Element's Atoms

LO: Calculate atomic mass from percent natural abundances and isotopic masses.

As we just learned, the atoms of a given element may have different masses (because of isotopes). Therefore, mass number is not always a useful quantity describing a significant quantity of an element. We can, however, calculate an average mass—called the **atomic mass**—for each element. You can find the atomic mass of each element in the periodic table directly beneath the element's symbol; it represents the average mass of the atoms that compose that element. For example, the periodic table lists the atomic mass of chlorine as 35.45 amu. Naturally occurring chlorine consists of 75.77 % chlorine-35 (mass 34.97 amu) and 24.23 % chlorine-37 (mass 36.97 amu). Its atomic mass is:

Some books use the terms *average atomic mass* or atomic weight instead of simply *atomic mass*.

$$\text{Atomic mass} = (0.7577 \times 34.97 \text{ amu}) + (0.2423 \times 36.97 \text{ amu}) \\ = 35.45 \text{ amu}$$

CHEMISTRY **IN THE ENVIRONMENT**

▶ Radioactive Isotopes at Hanford, Washington

The nuclei of the isotopes of a given element are not all equally stable. For example, naturally occurring lead is composed primarily of Pb-206, Pb-207, and Pb-208. Other isotopes of lead also exist, but their nuclei are unstable. Scientists can make some of these other isotopes, such as Pb-185, in the laboratory. However, within seconds Pb-185 atoms emit a few energetic subatomic particles from their nuclei and change into different isotopes of different elements (which are themselves unstable). These emitted subatomic particles are called **nuclear radiation**, and we call the isotopes that emit them **radioactive**. Nuclear radiation, always associated with unstable nuclei, can be harmful to humans and other living organisms because the emitted energetic particles interact with and damage biological molecules. Some isotopes, such as Pb-185, emit significant amounts of radiation only for a very short time. Others remain radioactive for a long time—in some cases millions or even billions of years.

The nuclear power and nuclear weapons industries produce by-products containing unstable isotopes of several different elements. Many of these isotopes emit nuclear radiation for a long time, and their disposal is an environmental problem. For example, in Hanford, Washington, which for 50 years produced fuel for nuclear weapons, 177 underground storage tanks contain 200 million litres of highly radioactive nuclear waste. Certain radioactive isotopes within that waste will produce nuclear radiation for the foreseeable future. Unfortunately, some of the underground storage tanks in Hanford are aging, and leaks have allowed some of the waste to seep into the environment. While the danger from short-term external exposure to this waste is minimal, ingestion of the waste through contamination of drinking water or food supplies would pose significant health risks. Consequently, Hanford is now the site of the largest environmental cleanup project in U.S. history, involving 11,000 workers. The U.S. government expects the project to last for decades, and current costs are about \$2 billion per year.

Radioactive isotopes are not always harmful, however, and many have beneficial uses. For example, physicians give technetium-99 (Tc-99) to patients to diagnose disease. The radiation emitted by Tc-99 helps doctors image internal organs or detect infection.

B2.1 CAN YOU ANSWER THIS? *Give the number of neutrons in each of the following isotopes: Pb-206, Pb-207, Pb-208, Pb-185, Tc-99.*

◀ Storage tanks at Hanford, Washington, contain 200 million litres of high-level nuclear waste. Each tank pictured here holds 4 million litres.

Notice that the atomic mass of chlorine is closer to 35 than 37 because naturally occurring chlorine contains more chlorine-35 atoms than chlorine-37 atoms. Notice also that when we use percentages in these calculations, we must always convert them to their decimal value. To convert a percentage to its decimal value, we divide by 100. For example:

$$75.77\ \% = 75.77/100 = 0.7577$$

$$24.23\ \% = 24.23/100 = 0.2423$$

In general, we calculate atomic mass according to the following equation:

$$\begin{aligned}\text{Atomic mass} = &(\text{Fraction of isotope 1} \times \text{Mass of isotope 1}) + \\ &(\text{Fraction of isotope 2} \times \text{Mass of isotope 2}) + \\ &(\text{Fraction of isotope 3} \times \text{Mass of isotope 3}) + \ldots\end{aligned}$$

where the fractions of each isotope are the percent natural abundances converted to their decimal values. Atomic mass is useful because it allows us to assign a characteristic mass to each element and, as we will see in Module 5, it allows us to quantify the number of atoms in a sample of that element.

EXAMPLE 2.4 CALCULATING ATOMIC MASS

Gallium has two naturally occurring isotopes: Ga-69 with mass 68.9256 amu and a natural abundance of 60.11 %, and Ga-71 with mass 70.9247 amu and a natural abundance of 39.89 %. Calculate the atomic mass of gallium.

	SOLUTION
Remember to convert the percent natural abundances into decimal form by dividing by 100.	$\text{Fraction Ga-69} = \frac{60.11}{100} = 0.6011$ $\text{Fraction Ga-71} = \frac{39.89}{100} = 0.3989$
Use the fractional abundances and the atomic masses of the isotopes to calculate the atomic mass according to the atomic mass definition.	$\text{Atomic mass} = (0.6011 \times 68.9256 \text{ amu}) + (0.3989 \times 70.9247 \text{ amu})$ $= 41.4312 \text{ amu} + 28.2919 \text{ amu}$ $= 69.7231 = 69.72 \text{ amu}$

▶SKILLBUILDER 2.4 | Calculating Atomic Mass

Magnesium has three naturally occurring isotopes with masses of 23.99, 24.99, and 25.98 amu and natural abundances of 78.99 %, 10.00 %, and 11.01 %. Calculate the atomic mass of magnesium.

▶FOR MORE PRACTICE Example 2.19; Problems 62, 63.

CONCEPTUAL CHECKPOINT 2.5

A fictitious element is composed of isotopes A and B with masses of 61.9887 and 64.9846 amu, respectively. The atomic mass of the element is 64.52. What can you conclude about the natural abundances of the two isotopes?

(a) The natural abundance of isotope A must be greater than the natural abundance of isotope B.

(b) The natural abundance of isotope B must be greater than the natural abundance of isotope A.

(c) The natural abundances of both isotopes must be about equal.

(d) Nothing can be concluded about the natural abundances of the two isotopes from the given information.

2.5 Ions: Losing and Gaining Electrons

LO: Determine ion charge from numbers of protons and electrons.

LO: Determine the number of protons and electrons in an ion.

The charge of an ion is indicated in the upper right corner of the symbol.

In chemical reactions, atoms often lose or gain electrons to form charged particles called **ions**. For example, a neutral lithium (Li) atom contains 3 protons and 3 electrons; however, in reactions, a lithium atom loses one electron (e^-) to form a Li^+ ion.

$$Li \rightarrow Li^+ + e^-$$

The Li^+ *ion* contains 3 protons but only 2 electrons, resulting in a net charge of 1+. We usually write ion charges with the magnitude of the charge first followed by the sign of the charge. For example, we write a positive two charge as 2+ and a negative two charge as 2–. The charge of an ion depends on how many electrons were gained or lost and is given by the formula:

$$\begin{aligned} \text{Ion charge} &= \text{number of protons} - \text{number of electrons} \\ &= \#p^+ - \#e^- \end{aligned}$$

where p^+ stands for *proton* and e^- stands for *electron.*

For the Li^+ ion with 3 protons and 2 electrons the charge is:

$$\text{Ion charge} = 3 - 2 = 1+$$

A neutral fluorine (F) atom contains 9 protons and 9 electrons; however, in chemical reactions a fluorine atom gains 1 electron to form F^- ions:

$$F + e^- \rightarrow F^-$$

The F^- *ion* contains 9 protons and 10 electrons, resulting in a 1– charge.

$$\begin{aligned} \text{Ion charge} &= 9 - 10 \\ &= 1- \end{aligned}$$

Positively charged ions, such as Li^+, are **cations**, and negatively charged ions, such as F^-, are **anions**. Ions behave very differently than the atoms from which they are formed. Neutral sodium atoms, for example, are extremely reactive, interacting violently with most things they contact. Sodium cations (Na^+), on the other hand, are relatively inert—we eat them all the time in sodium chloride (table salt). In nature, cations and anions always occur together so that, again, matter is charge-neutral. For example, in table salt, the sodium cation occurs together with the chloride anion (Cl^-).

EXAMPLE 2.5 DETERMINING ION CHARGE FROM NUMBERS OF PROTONS AND ELECTRONS

Determine the charge of each ion.

(a) a magnesium ion with 10 electrons
(b) a sulfur ion with 18 electrons
(c) an iron ion with 23 electrons

SOLUTION

To determine the charge of each ion, use the ion charge equation.

$$\text{Ion charge} = \#p^+ - \#e^-$$

You are given the number of electrons in the problem. You can obtain the number of protons from the element's atomic number in the periodic table.

(a) Magnesium's atomic number is 12.

$$\text{Ion charge} = 12 - 10 = 2+ \ (Mg^{2+})$$

(b) Sulfur's atomic number is 16.

$$\text{Ion charge} = 16 - 18 = 2- \ (S^{2-})$$

(c) Iron's atomic number is 26.

$$\text{Ion charge} = 26 - 23 = 3+ \ (Fe^{3+})$$

▶SKILLBUILDER 2.5 | Determining Ion Charge from Numbers of Protons and Electrons

Determine the charge of each ion.

(a) a nickel ion with 26 electrons
(b) a bromine ion with 36 electrons
(c) a phosphorus ion with 18 electrons

▶FOR MORE PRACTICE Example 2.20; Problems 66, 67.

EXAMPLE 2.6 DETERMINING THE NUMBER OF PROTONS AND ELECTRONS IN AN ION

Determine the number of protons and electrons in the Ca^{2+} ion.

The periodic table indicates that the atomic number for calcium is 20, so calcium has 20 protons. You can find the number of electrons using the ion charge equation.

SOLUTION

$$\text{Ion charge} = \#p^+ - \#e^-$$

$$2+ = 20 - \#e^-$$

$$\#e^- = 20 - 2 = 18$$

The number of electrons is 18.
The Ca^{2+} ion has 20 protons and 18 electrons.

▶SKILLBUILDER 2.6 | Determining the Number of Protons and Electrons in an Ion

Determine the number of protons and electrons in the S^{2-} ion.

▶FOR MORE PRACTICE Example 2.21; Problems 68, 69.

Ions and the Periodic Table

For many main-group elements (Section 2.7), we can use the periodic table to predict how many electrons tend to be lost or gained when an atom of that particular element ionizes. The number associated with the letter A above each *main-group* column in the periodic table—1 through 8—gives the number of *valence electrons* for the elements in that column. We will discuss the concept of valence electrons more fully in the next section in this module; for now, you can think of valence electrons as the outermost electrons in an atom. Because oxygen is in column 6A, we can deduce that it has 6 valence electrons; because magnesium is in column 2A, it has 2 valence electrons, and so on. An important exception to this rule is helium—it is in column 8A, but has only 2 valence electrons. Valence electrons are particularly important because, as we shall see in Module 3, these electrons are the ones that are most important in chemical bonding.

We can predict the charge acquired by a particular element when it ionizes from its position in the periodic table relative to the noble gases.

> Main-group elements tend to form ions that have the same number of valence electrons as the nearest noble gas.

1A	2A		3A	4A	5A	6A	7A	8A
Li^+	Be^{2+}				N^{3-}	O^{2-}	F^-	
Na^+	Mg^{2+}		Al^{3+}			S^{2-}	Cl^-	
K^+	Ca^{2+}		Ga^{3+}			Se^{2-}	Br^-	
Rb^+	Sr^{2+}	Transition metals form cations with various charges	In^{3+}			Te^{2-}	I^-	
Cs^+	Ba^{2+}							

▲ **FIGURE 2.5 Elements that form predictable ions**

For example, the closest noble gas to oxygen is neon. When oxygen ionizes, it *acquires* two additional electrons for a total of 8 valence electrons—the same number as neon. When determining the closest noble gas, we can move either forward or backward on the periodic table. For example, the closest noble gas to magnesium is also neon, even though neon (atomic number 10) falls before magnesium (atomic number 12) in the periodic table. Magnesium *loses* its 2 valence electrons to attain the same number of valence electrons as neon.

In accordance with this principle, the alkali metals (Group 1A) tend to lose 1 electron and form 1+ ions, while the alkaline earth metals (Group 2A) tend to lose 2 electrons and form 2+ ions. The halogens (Group 7A) tend to gain 1 electron and form 1– ions. The groups in the periodic table that form predictable ions are shown in ▲ Figure 2.5. Become familiar with these groups and the ions they form. Later in this module (Section 2.9), we will examine a theory that more fully explains why these groups form ions as they do.

EXAMPLE 2.7 CHARGE OF IONS FROM POSITION IN PERIODIC TABLE

Based on their position in the periodic table, what ions do barium and iodine tend to form?

SOLUTION

Because barium is in Group 2A, it tends to form a cation with a 2+ charge (Ba^{2+}). Because iodine is in Group 7A, it tends to form an anion with a 1– charge (I^-).

▶SKILLBUILDER 2.7 | Charge of Ions from Position in Periodic Table

Based on their position in the periodic table, what ions do potassium and selenium tend to form?

▶FOR MORE PRACTICE Problems 126, 127.

CONCEPTUAL CHECKPOINT 2.6

Which pair of ions has the same total number of electrons?

(a) Na^+ and Mg^{2+}

(b) F^- and Cl^-

(c) O^- and O^{2-}

(d) Ga^{3+} and Fe^{3+}

2.6 Arrangements of Electrons Around the Nucleus

LO: Write electron configurations and orbital diagrams for atoms.

The main points made about electrons so far is that they are negatively charged, have negligible mass, and they are dispersed throughout the volume of an atom outside the nucleus. In this section the way that electrons are arranged around the nucleus of an atom is introduced. The current understanding of these electron arrangements is called the quantum-mechanical model.

The number and arrangement of electrons in atoms dictates the chemical properties of that element. Experimental data has led to the discovery that electrons occupy discreet energy levels around the nucleus. These energy levels are called **orbitals**. When energy is put into an atom, electrons are excited to higher-energy orbitals. When an electron in an atom relaxes from a higher-energy orbital to a lower-energy orbital, the atom emits light. The energy (and therefore the colour) of the emitted light corresponds to the energy difference between the two orbitals in the transition.

In this section, we examine quantum-mechanical orbitals and electron configurations. An electron configuration is a compact way to specify the occupation of quantum-mechanical orbitals by electrons.

Quantum-Mechanical Orbitals

$n = 4$
$n = 3$
$n = 2$
$n = 1$
Energy

▲ **FIGURE 2.6 Principal quantum numbers** The principal quantum numbers ($n = 1, 2, 3\ldots$) determine the energy of the hydrogen quantum-mechanical orbitals.

The lowest-energy orbital in the quantum-mechanical model is the *1s orbital*. We specifiy it by the number 1 and the letter *s*. The number is the **principal quantum number** (n) and specifies the **principal shell** of the orbital. The higher the principal quantum number, the higher the energy of the orbital. The possible principal quantum numbers are $n = 1, 2, 3\ldots$, with energy increasing as n increases (◀ Figure 2.6). Because the 1*s* orbital has the lowest possible principal quantum number, it is in the lowest-energy shell and has the lowest possible energy.

The letter indicates the **subshell** of the orbital and specifies its shape. The possible letters are *s*, *p*, *d*, and *f*, and each letter corresponds to a different shape. For example, orbitals within the *s* subshell have a spherical shape. The 1*s* quantum-mechanical orbital is a three-dimensional probability map. We sometimes represent orbitals with dots (▼ Figure 2.7), where the dot density is proportional to the probability of finding the electron.

This analogy is purely hypothetical. It is impossible to photograph electrons in this way.

We can understand the dot representation of an orbital better with another analogy. Imagine taking a photograph of an electron in an atom every second for 10 or 15 minutes. One second the electron is very close to the nucleus; the next second it is farther away and so on. Each photo shows a dot representing the electron's position relative to the nucleus at that time. If we took hundreds of photos and superimposed

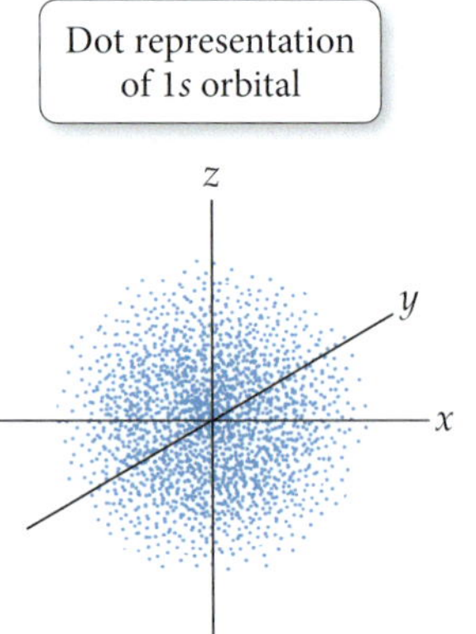

▲ **FIGURE 2.7 1s orbital** The dot density in this plot is proportional to the probability of finding the electron. The greater dot density near the middle indicates a higher probability of finding the electron near the nucleus.

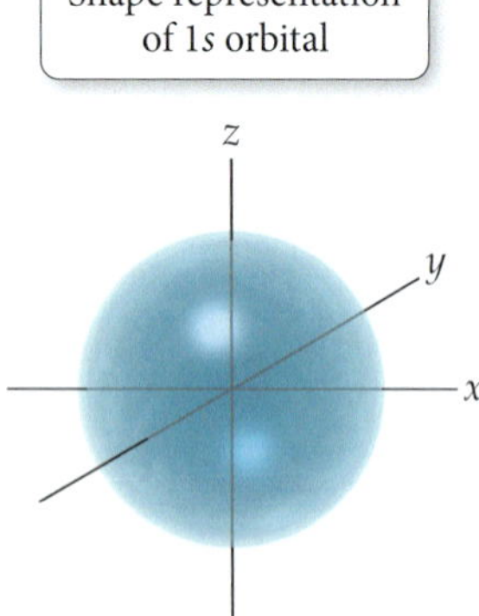

▲ **FIGURE 2.8 Shape representation of the 1s orbital** Because the distribution of electron density around the nucleus in Figure 2.7 is symmetrical—the same in all directions—we can represent the 1*s* orbital as a sphere.

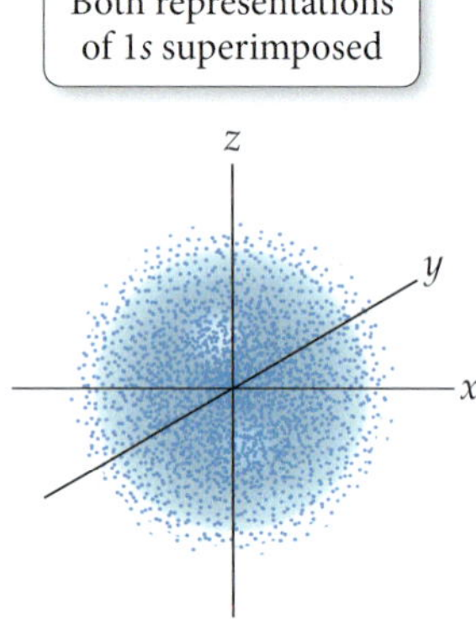

▲ **FIGURE 2.9 Orbital shape and dot representation for the 1s orbital** The shape representation of the 1*s* orbital superimposed on the dot density representation. We can see that when the electron is in the 1*s* orbital, it is most likely to be found within the sphere.

FIGURE 2.10 Subshells The number of subshells in a given principal shell is equal to the value of n.

Shell	Number of subshells	Letters specifying subshells			
$n = 4$	4	*s*	*p*	*d*	*f*
$n = 3$	3	*s*	*p*	*d*	
$n = 2$	2	*s*	*p*		
$n = 1$	1	*s*			

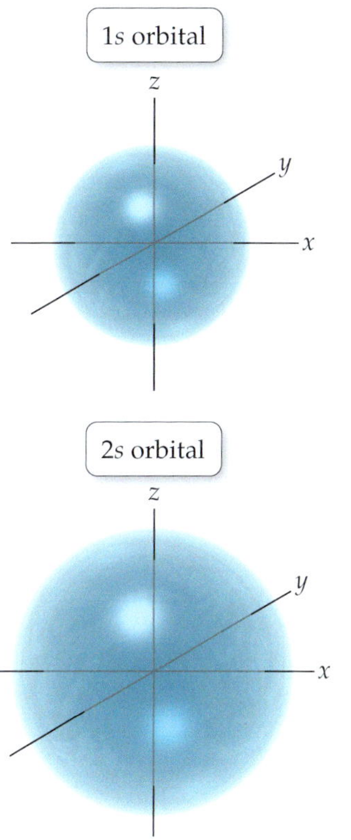

FIGURE 2.11 The 2*s* orbital The 2*s* orbital is similar to the 1*s* orbital, but larger in size.

all of them, we would have an image like Figure 2.7—a statistical representation of where the electron is found. Notice that the dot density for the 1*s* orbital is greatest near the nucleus and decreases farther away from the nucleus. This means that the electron is more likely to be found close to the nucleus than far away from it.

Orbitals can also be represented as geometric shapes that encompass most of the volume where the electron is likely to be found. For example, the 1*s* orbital can be represented as a sphere (Figure 2.8) that encompasses the volume within which the electron is found 90 % of the time. If we superimpose the dot representation of the 1*s* orbital on the shape representation (Figure 2.9), we can see that most of the dots are within the sphere, meaning that the electron is most likely to be found within the sphere when it is in the 1*s* orbital.

The single electron of an undisturbed hydrogen atom at room temperature is in the 1*s* orbital. This is the **ground state**, or lowest energy state, of the hydrogen atom. However, the quantum-mechanical model allows transitions to higher-energy orbitals upon the absorption of energy. What are these higher-energy orbitals? What do they look like?

The next orbitals in the quantum-mechanical model are those with principal quantum number $n = 2$ Unlike the $n = 1$ principal shell, which contains only one subshell (specified by *s*), the $n = 2$ principal shell contains two subshells, specified by *s* and *p*.

The number of subshells in a given principal shell is equal to the value of n. Therefore the $n = 1$ principal shell has one subshell, the $n = 2$ principal shell has two subshells, and so on (Figure 2.10). The *s* subshell contains the 2*s* orbital, higher in energy than the 1*s* orbital and slightly larger (Figure 2.11), but otherwise similar in shape. The *p* subshell contains three 2*p* orbitals (Figure 2.12), all with the same dumbbell-like shape but with different orientations.

The next principal shell, $n = 3$ contains three subshells specified by *s*, *p*, and *d*. The *s* and *p* subshells contain the 3*s* and 3*p* orbitals, similar in shape to the 2*s* and 2*p* orbitals, but slightly larger and higher in energy. The *d* subshell contains the five *d* orbitals shown in Figure 2.13. The next principal shell, $n = 4$ contains four subshells specified by *s*, *p*, *d*, and *f*. The *s*, *p*, and *d* subshells are similar to those in $n = 3$. The *f* subshell contains seven orbitals (called the 4*f* orbitals), whose shape we do not consider in this book.

As we have already discussed, hydrogen's single electron is usually in the 1*s* orbital because electrons generally occupy the lowest-energy orbital available. In

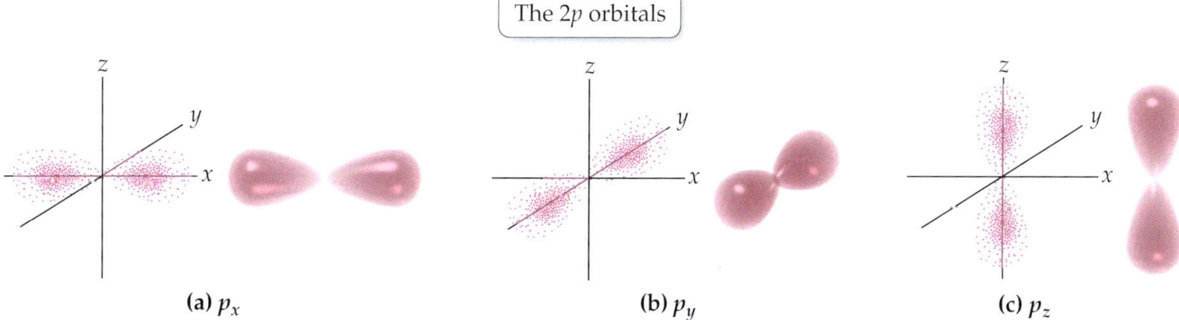

FIGURE 2.12 The 2*p* orbitals This figure shows both the dot representation (left) and shape representation (right) for each *p* orbital.

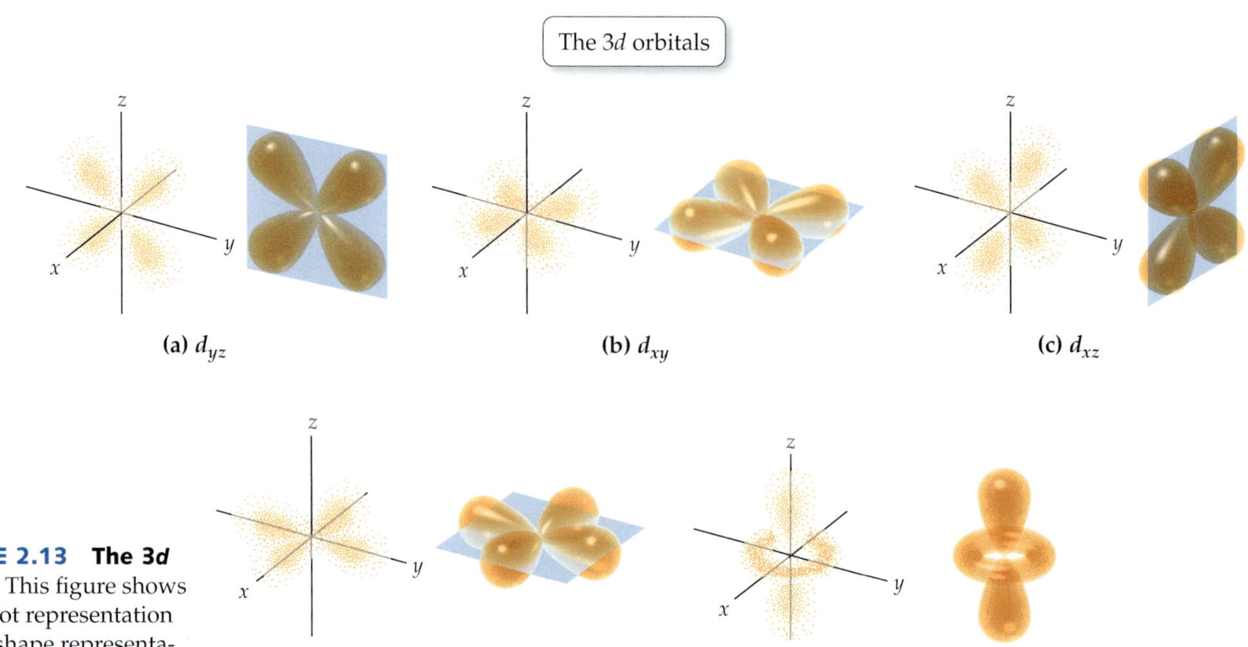

▶ **FIGURE 2.13 The 3*d* orbitals** This figure shows both the dot representation (left) and shape representation (right) for each *d* orbital.

hydrogen, the rest of the orbitals are normally empty. However, the absorption of energy by a hydrogen atom can cause the electron to jump (or make a transition) from the 1*s* orbital to a higher-energy orbital. When the electron is in a higher-energy orbital, we say that the hydrogen atom is in an **excited state**.

Because of their higher energy, excited states are unstable, and the electron will usually fall (or relax) back to a lower-energy orbital. In the process the electron emits energy, often in the form of light. The quantum-mechanical model predicts the different colours of light emitted in hydrogen, with one electron, as well as other elements with more than one electron.

Electron Configurations: How Electrons Occupy Orbitals

An **electron configuration** illustrates the occupation of orbitals by electrons for a particular atom. For example, the electron configuration for a ground-state (or lowest energy) hydrogen atom is:

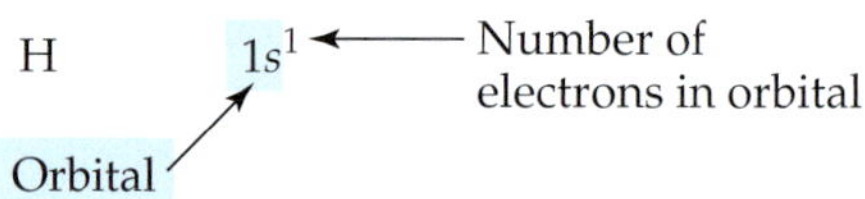

The electron configuration tells us that hydrogen's single electron is in the 1*s* orbital.

Another way to represent this information is with an **orbital diagram**, which gives similar information but shows the electrons as arrows in a box representing the orbital. The orbital diagram for a ground-state hydrogen atom is:

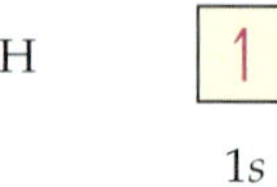

The box represents the 1*s* orbital, and the arrow within the box represents the electron in the 1*s* orbital. In orbital diagrams, the direction of the arrow (pointing up or pointing down) represents **electron spin**, a fundamental property of electrons. All electrons have spin. The **Pauli exclusion principle** states that *orbitals may hold no more than two electrons with opposing spins*. We symbolize this as two arrows pointing in opposite directions:

⥮

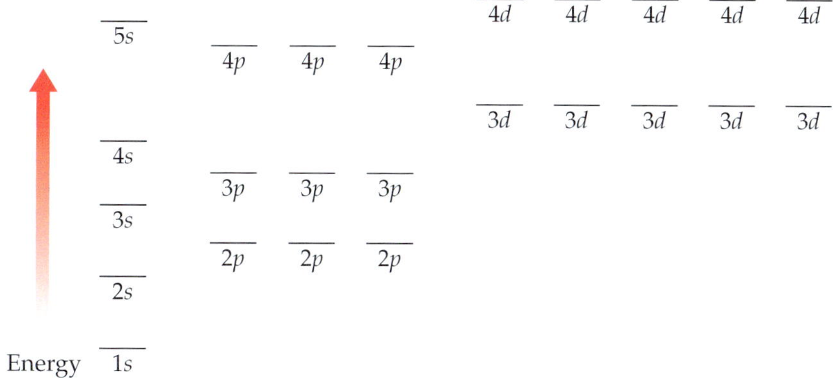

▶ **FIGURE 2.14 Energy ordering of orbitals for multi-electron atoms** Different subshells within the same principal shell have different energies.

A helium atom, for example, has two electrons. The electron configuration and orbital diagram for helium are:

	Electron configuration	Orbital diagram
He	$1s^2$	⇅ (1s)

Since we know that electrons occupy the lowest-energy orbitals available, and since we know that only two electrons (with opposing spins) are allowed in each orbital, we can continue to build ground-state electron configurations for the rest of the elements as long as we know the energy ordering of the orbitals. ▲ Figure 2.14 shows the energy ordering of a number of orbitals for multi-electron atoms.

In multi-electron atoms, the subshells within a principal shell do not have the same energy because of electron–electron interactions.

Notice that, for multi-electron atoms (in contrast to hydrogen which has only one electron), the subshells within a principal shell *do not* have the same energy. In elements other than hydrogen, the energy ordering is not determined by the principal quantum number alone. For example, in multi-electron atoms, the 4*s* subshell is lower in energy than the 3*d* subshell, even though its principal quantum number is higher. Using this relative energy ordering, we can write ground-state electron configurations and orbital diagrams for other elements. For lithium, which has three electrons, the electron configuration and orbital diagram are:

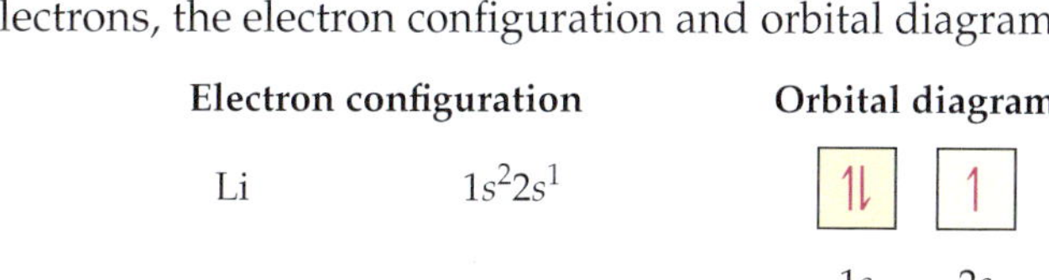

	Electron configuration	Orbital diagram
Li	$1s^2 2s^1$	⇅ (1s) ↑ (2s)

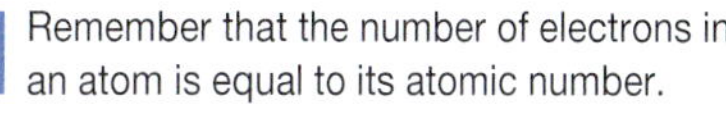

Remember that the number of electrons in an atom is equal to its atomic number.

For carbon, which has 6 electrons, the electron configuration and orbital diagram are:

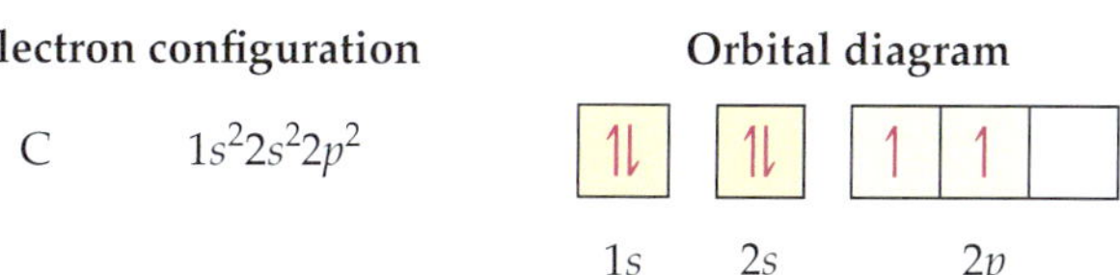

	Electron configuration	Orbital diagram
C	$1s^2 2s^2 2p^2$	⇅ (1s) ⇅ (2s) ↑ ↑ □ (2p)

Notice that the 2*p* electrons occupy the *p* orbitals (of equal energy) singly rather than pairing in one orbital. This is the result of **Hund's rule**, which states that *when filling orbitals of equal energy, electrons fill them singly first, with parallel spins.*

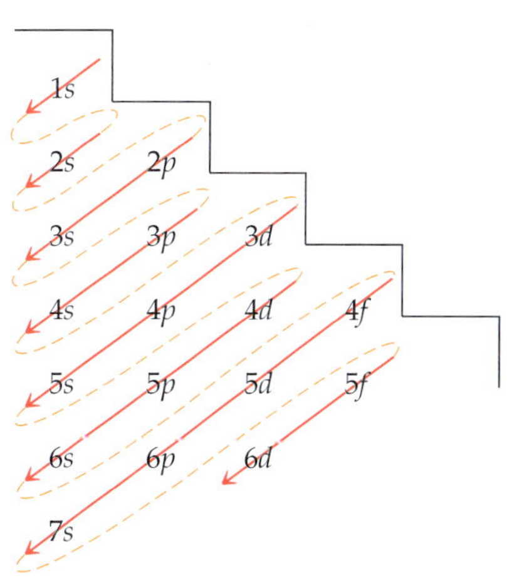

▲ **FIGURE 2.15 Orbital filling order** The arrows indicate the order in which orbitals fill.

Before we write electron configurations for other elements, let us summarize what we have learned so far:

- Electrons occupy orbitals so as to minimize the energy of the atom; therefore, lower-energy orbitals fill before higher-energy orbitals. Orbitals fill in the following order: 1*s* 2*s* 2*p* 3*s* 3*p* 4*s* 3*d* 4*p* 5*s* 4*d* 5*p* 6*s* (◀ Figure 2.15).

- Orbitals can hold no more than two electrons each. When two electrons occupy the same orbital, they must have opposing spins. This is known as the Pauli exclusion principle.
- When orbitals of identical energy are available, these are first occupied singly with parallel spins rather than in pairs. This is known as Hund's rule.

Consider the electron configurations and orbital diagrams for elements with atomic numbers 3 through 10:

Symbol ($\#e^-$)	Electron configuration	Orbital diagram 1s	2s	2p
Li (3)	$1s^22s^1$	[↑↓]	[↑]	
Be (4)	$1s^22s^2$	[↑↓]	[↑↓]	
B (5)	$1s^22s^22p^1$	[↑↓]	[↑↓]	[↑][ ][ ]
C (6)	$1s^22s^22p^2$	[↑↓]	[↑↓]	[↑][↑][ ]
N (7)	$1s^22s^22p^3$	[↑↓]	[↑↓]	[↑][↑][↑]
O (8)	$1s^22s^22p^4$	[↑↓]	[↑↓]	[↑↓][↑][↑]
F (9)	$1s^22s^22p^5$	[↑↓]	[↑↓]	[↑↓][↑↓][↑]
Ne (10)	$1s^22s^22p^6$	[↑↓]	[↑↓]	[↑↓][↑↓][↑↓]

Notice how the p orbitals fill. As a result of Hund's rule, the p orbitals fill with single electrons before they fill with paired electrons. The electron configuration of neon represents the complete filling of the $n = 2$ principal shell. When writing electron configurations for elements beyond neon—or beyond any other noble gas—we often abbreviate the electron configuration of the previous noble gas by the symbol for the noble gas in brackets. For example, the electron configuration of sodium is:

Na $1s^22s^22p^63s^1$

We can write this using the noble gas core notation as:

Na $[Ne]3s^1$

where [Ne] represents $1s^22s^22p^6$, the electron configuration for neon.

To write an electron configuration for an element, we first find its atomic number from the periodic table—this number equals the number of electrons in the neutral atom. Then we use the order of filling from Figure 2.14 to distribute the electrons in the appropriate orbitals. Remember that each orbital can hold a maximum of 2 electrons. Consequently:

- the *s* subshell has only 1 orbital and therefore can hold only 2 electrons.
- the *p* subshell has 3 orbitals and therefore can hold 6 electrons.
- the *d* subshell has 5 orbitals and therefore can hold 10 electrons.
- the *f* subshell has 7 orbitals and therefore can hold 14 electrons.

EXAMPLE 2.8 ELECTRON CONFIGURATIONS

Write electron configurations for each element.

(a) Mg **(b)** S **(c)** Ga

	SOLUTION
(a) Magnesium has 12 electrons. Distribute two of these into the 1*s* orbital, two into the 2*s* orbital, six into the 2*p* orbitals, and two into the 3*s* orbital. You can also write the electron configuration more compactly using the noble gas core notation. For magnesium, use [Ne] to represent $1s^22s^22p^6$.	Mg $1s^22s^22p^63s^2$ *or* Mg $[Ne]3s^2$
(b) Sulfur has 16 electrons. Distribute two of these into the 1*s* orbital, two into the 2*s* orbital, six into the 2*p* orbitals, two into the 3*s* orbital, and four into the 3*p* orbitals. You can write the electron configuration more compactly by using [Ne] to represent $1s^22s^22p^6$.	S $1s^22s^22p^63s^23p^4$ *or* S $[Ne]3s^23p^4$
(c) Gallium has 31 electrons. Distribute two of these into the 1*s* orbital, two into the 2*s* orbital, six into the 2*p* orbitals, two into the 3*s* orbital, six into the 3*p* orbitals, two into the 4*s* orbital, ten into the 3*d* orbitals, and one into the 4*p* orbitals. Notice that the *d* subshell has five orbitals and can therefore hold 10 electrons. You can write the electron configuration more compactly by using [Ar] to represent $1s^22s^22p^63s^23p^6$.	Ga $1s^22s^22p^63s^23p^64s^23d^{10}4p^1$ *or* Ga $[Ar]4s^23d^{10}4p^1$

▶ **SKILLBUILDER 2.8 | Electron Configurations**

Write electron configurations for each element.

(a) Al **(b)** Br **(c)** Sr

▶ **SKILLBUILDER PLUS** | Write electron configurations for each ion. (*Hint:* To determine the number of electrons to include in the electron configuration of an ion, add or subtract electrons as needed to account for the charge of the ion.)

(a) Al^{3+} **(b)** Cl^- **(c)** O^{2-}

▶ **FOR MORE PRACTICE** Problems 76, 77, 80, 81.

EXAMPLE 2.9 WRITING ORBITAL DIAGRAMS

Write an orbital diagram for silicon.

SOLUTION

Since silicon is atomic number 14, it has 14 electrons. Draw a box for each orbital, putting the lowest-energy orbital (1*s*) on the far left and proceeding to orbitals of higher energy to the right.

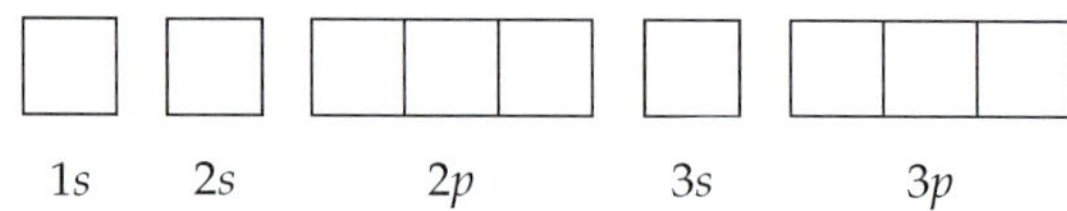

Distribute the 14 electrons into the orbitals, allowing a maximum of 2 electrons per orbital and remembering Hund's rule. The complete orbital diagram is:

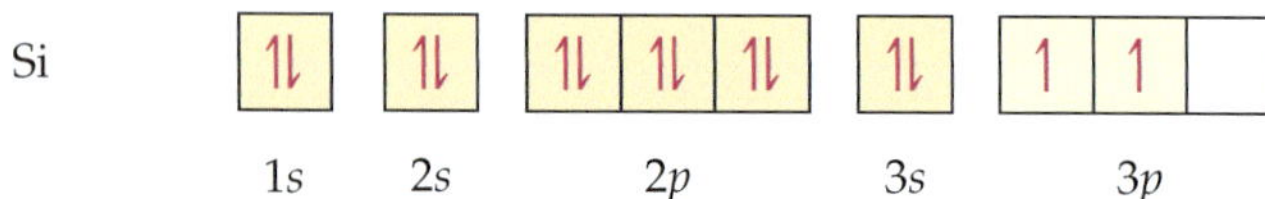

▶ **SKILLBUILDER 2.9 | Writing Orbital Diagrams**

Write an orbital diagram for argon.

▶ **FOR MORE PRACTICE** Example 2.22; Problems 78, 79.

CONCEPTUAL CHECKPOINT 2.7

Which pair of elements has the same *total* number of electrons in *p* orbitals?

(a) Na and K

(b) K and Kr

(c) P and N

(d) Ar and Ca

2.7 Looking for Patterns: The Periodic Law and the Periodic Table

LO: Use the periodic table to classify elements by group.

The organization of the periodic table has its origins in the work of Dmitri Mendeleev (1834–1907), a nineteenth-century Russian chemistry professor. In his time, about 65 different elements had been discovered. Thanks to the work of a number of chemists, much was known about each of these elements, including their relative masses, chemical activity, and some of their physical properties. However, there was no systematic way of organizing them.

▲ Dmitri Mendeleev, a Russian chemistry professor who arranged early versions of the periodic table. © Popova Olga/Fotolia.

1	2	3	4	5	6	7	8	9	10	11	12	13	14	15	16	17	18	19	20
H	He	Li	Be	B	C	N	O	F	Ne	Na	Mg	Al	Si	P	S	Cl	Ar	K	Ca

▲ **FIGURE 2.16 Recurring properties** These elements are listed in order of increasing atomic number (Mendeleev used relative mass, which is similar). The color of each element represents its properties. Notice that the properties (colors) of these elements form a repeating pattern.

In 1869, Mendeleev noticed that certain groups of elements had similar properties. He found that if he listed the elements in order of increasing relative mass, those similar properties recurred in a regular pattern (▲ Figure 2.16). Mendeleev summarized these observations in the **periodic law**:

| *Periodic* means "recurring regularly."

> When the elements are arranged in order of increasing relative mass, certain sets of properties recur periodically.

1 H							2 He
3 Li	4 Be	5 B	6 C	7 N	8 O	9 F	10 Ne
11 Na	12 Mg	13 Al	14 Si	15 P	16 S	17 Cl	18 Ar
19 K	20 Ca						

▲ **FIGURE 2.17 Making a periodic table** If we place the elements from Figure 2.16 in a table, we can arrange them in rows so that similar properties align in the same vertical columns. This is similar to Mendeleev's first periodic table.

Mendeleev organized all the known elements in a table in which relative mass increased from left to right and elements with similar properties were aligned in the same vertical columns (◀ Figure 2.17). Because many elements had not yet been discovered, Mendeleev's table contained some gaps, which allowed him to predict the existence of yet-undiscovered elements. For example, Mendeleev predicted the existence of an element he called *eka-silicon*, which fell below silicon on the table and between gallium and arsenic. In 1886, eka-silicon was discovered by German chemist Clemens Winkler (1838–1904) and was found to have almost exactly the properties that Mendeleev had anticipated. Winkler named the element germanium, after his home country.

Mendeleev's original listing has evolved into the modern **periodic table**. In the modern table, elements are listed in order of increasing atomic number rather than increasing relative mass. The modern periodic table also contains more elements than Mendeleev's original table because many more have been discovered since his time.

Mendeleev's periodic law was based on observation. Like all scientific laws, the periodic law summarized many observations but did not give the underlying reason for the observation—only theories do that. For now, we accept the periodic law as it is, but later in this module we will examine a powerful theory that explains the law and gives the underlying reasons for it.

We can broadly classify the elements in the periodic table as metals, nonmetals, and metalloids (▼ Figure 2.18). **Metals** occupy the left side of the periodic table and have similar properties: They are good conductors of heat and electricity; they can be pounded into flat sheets (malleability); they can be drawn into wires (ductility); they are often shiny; and they tend to lose electrons when they undergo chemical changes. Examples of metals are iron, magnesium, chromium, and sodium.

Nonmetals occupy the upper right side of the periodic table. The dividing line between metals and nonmetals is the zigzag diagonal line running from boron to astatine in Figure 2.17. Nonmetals have more varied properties—some are solids at room temperature, others are gases—but as a whole they tend to be poor

Metals

Nonmetals

Metalloids

	1A 1	2A 2	3B 3	4B 4	5B 5	6B 6	7B 7	8B 8	9	10	1B 11	2B 12	3A 13	4A 14	5A 15	6A 16	7A 17	8A 18
1	1 H																	2 He
2	3 Li	4 Be											5 B	6 C	7 N	8 O	9 F	10 Ne
3	11 Na	12 Mg											13 Al	14 Si	15 P	16 S	17 Cl	18 Ar
4	19 K	20 Ca	21 Sc	22 Ti	23 V	24 Cr	25 Mn	26 Fe	27 Co	28 Ni	29 Cu	30 Zn	31 Ga	32 Ge	33 As	34 Se	35 Br	36 Kr
5	37 Rb	38 Sr	39 Y	40 Zr	41 Nb	42 Mo	43 Tc	44 Ru	45 Rh	46 Pd	47 Ag	48 Cd	49 In	50 Sn	51 Sb	52 Te	53 I	54 Xe
6	55 Cs	56 Ba	57 La	72 Hf	73 Ta	74 W	75 Re	76 Os	77 Ir	78 Pt	79 Au	80 Hg	81 Tl	82 Pb	83 Bi	84 Po	85 At	86 Rn
7	87 Fr	88 Ra	89 Ac	104 Rf	105 Db	106 Sg	107 Bh	108 Hs	109 Mt	110 Ds	111 Rg	112 Cn	113 Nh	114 Fl	115 Mc	116 Lv	117 Ts	118 Og

Lanthanides	58 Ce	59 Pr	60 Nd	61 Pm	62 Sm	63 Eu	64 Gd	65 Tb	66 Dy	67 Ho	68 Er	69 Tm	70 Yb	71 Lu
Actinides	90 Th	91 Pa	92 U	93 Np	94 Pu	95 Am	96 Cm	97 Bk	98 Cf	99 Es	100 Fm	101 Md	102 No	103 Lr

▲ **FIGURE 2.18 Metals, nonmetals, and metalloids** The elements in the periodic table can be broadly classified as metals, nonmetals, or metalloids.

▲ Silicon is a metalloid used extensively in the computer and electronics industries. © Tom Grill/Corbis.

conductors of heat and electricity, and they all tend to gain electrons when they undergo chemical changes. Examples of nonmetals are oxygen, nitrogen, chlorine, and iodine.

Most of the elements that lie along the zigzag diagonal line dividing metals and nonmetals are **metalloids**, or semimetals, and display mixed properties. We also call metalloids **semiconductors** because of their intermediate electrical conductivity, which can be changed and controlled. This property makes semiconductors useful in the manufacture of the electronic devices that are central to computers, cell phones, and many other technological gadgets. Silicon, arsenic, and germanium are metalloids.

EXAMPLE 2.10 CLASSIFYING ELEMENTS AS METALS, NONMETALS, OR METALLOIDS

Classify each element as a metal, nonmetal, or metalloid.

(a) Ba **(b)** I **(c)** O **(d)** Te

SOLUTION

(a) Barium is on the left side of the periodic table; it is a metal.
(b) Iodine is on the right side of the periodic table; it is a nonmetal.
(c) Oxygen is on the right side of the periodic table; it is a nonmetal.
(d) Tellurium is in the middle-right section of the periodic table, along the line that divides the metals from the nonmetals; it is a metalloid.

▶SKILLBUILDER 2.10 | Classifying Elements as Metals, Nonmetals, or Metalloids

Classify each element as a metal, nonmetal, or metalloid.

(a) S **(b)** Cl **(c)** Ti **(d)** Sb

▶FOR MORE PRACTICE Problems 82, 83.

We can also broadly divide the periodic table into **main-group elements**, whose properties tend to be more predictable based on their position in the periodic table, and **transition elements** or **transition metals**, whose properties are less easily predictable based simply on their position in the periodic table (▼ Figure 2.19). Main-group elements are in columns labeled with a number and the letter A. Transition elements are in columns labeled with a number and the letter B. A competing numbering system does not use letters, but only the numbers 1–18. We show both numbering systems in the periodic table in the inside front cover of this book.

▶ **FIGURE 2.19 Main-group and transition elements** We can broadly divide the periodic table into main-group elements, whose properties we can generally predict based on their position, and transition elements, whose properties tend to be less predictable based on their position.

Main-group elements | Transition elements | Main-group elements

Group number

Periods	1A	2A	3B	4B	5B	6B	7B	8B	8B	8B	1B	2B	3A	4A	5A	6A	7A	8A
1	1 H																	2 He
2	3 Li	4 Be											5 B	6 C	7 N	8 O	9 F	10 Ne
3	11 Na	12 Mg											13 Al	14 Si	15 P	16 S	17 Cl	18 Ar
4	19 K	20 Ca	21 Sc	22 Ti	23 V	24 Cr	25 Mn	26 Fe	27 Co	28 Ni	29 Cu	30 Zn	31 Ga	32 Ge	33 As	34 Se	35 Br	36 Kr
5	37 Rb	38 Sr	39 Y	40 Zr	41 Nb	42 Mo	43 Tc	44 Ru	45 Rh	46 Pd	47 Ag	48 Cd	49 In	50 Sn	51 Sb	52 Te	53 I	54 Xe
6	55 Cs	56 Ba	57 La	72 Hf	73 Ta	74 W	75 Re	76 Os	77 Ir	78 Pt	79 Au	80 Hg	81 Tl	82 Pb	83 Bi	84 Po	85 At	86 Rn
7	87 Fr	88 Ra	89 Ac	104 Rf	105 Db	106 Sg	107 Bh	108 Hs	109 Mt	110 Ds	111 Rg	112 Cn	113 Nh	114 Fl	115 Mc	116 Lv	117 Ts	118 Og

✓ CONCEPTUAL CHECKPOINT 2.8

Which element is a main-group metal?

(a) O **(b)** Ag **(c)** P **(d)** Pb

The noble gases are inert (unreactive) compared to other elements. However, some noble gases, especially the heavier ones, form a limited number of compounds with other elements under special conditions.

Each column within the periodic table is a **family** or **group** of elements. The elements within a group of main-group elements usually have similar chemical properties, and some have a group name. For example, the Group 8A elements, the **noble gases**, are chemically inert gases. The most familiar noble gas is probably helium, used to fill balloons. Helium, like the other noble gases, is chemically stable—it won't combine with other elements to form compounds—and is therefore safe to put into balloons. Other noble gases include neon, often used in neon signs; argon, which makes up a small percentage of our atmosphere; krypton; and xenon. The Group 1A elements, the **alkali metals**, are all very reactive metals. A marble-sized piece of sodium can explode when dropped into water. Other alkali metals include lithium, potassium, and rubidium. The Group 2A elements, the **alkaline earth metals**, are also fairly reactive, although not quite as reactive as the alkali metals. Calcium, for example, reacts fairly vigorously when dropped into water but does not explode as readily as sodium. Other alkaline earth metals are magnesium, a common low-density structural metal; strontium; and barium. The Group 7A elements, the **halogens**, are very reactive nonmetals. Chlorine, a greenish-yellow gas with a pungent odor is probably the most familiar halogen. Because of its reactivity, people often use chlorine as a sterilizing and disinfecting agent (it reacts with and kills bacteria and other microscopic organisms). Other halogens include bromine, a red-brown liquid that readily evaporates into a gas; iodine, a purple solid; and fluorine, a pale yellow gas.

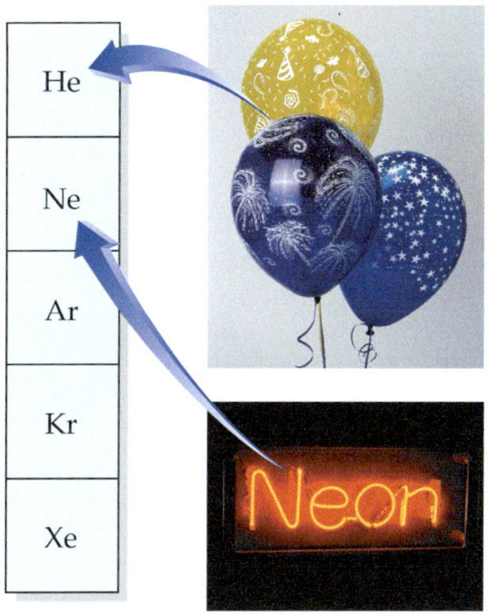

▲ The noble gases include helium (used in balloons), neon (used in neon signs), argon, krypton, and xenon. (top left) PhotoDisc/Getty Images. (bottom left) Graeme Dawes/Shutterstock. (top right) Richard Megna/Fundamental Photographs. (bottom right) Sciencephotos/Alamy.

▲ The alkali metals include lithium, sodium (shown in the second photo reacting with water), potassium, rubidium, and cesium.

▶ The periodic table with Groups 1A, 2A, 7A, and 8A highlighted.

1A	2A											3A	4A	5A	6A	7A	8A
1 H																	2 He
3 Li	4 Be											5 B	6 C	7 N	8 O	9 F	10 Ne
11 Na	12 Mg											13 Al	14 Si	15 P	16 S	17 Cl	18 Ar
19 K	20 Ca	21 Sc	22 Ti	23 V	24 Cr	25 Mn	26 Fe	27 Co	28 Ni	29 Cu	30 Zn	31 Ga	32 Ge	33 As	34 Se	35 Br	36 Kr
37 Rb	38 Sr	39 Y	40 Zr	41 Nb	42 Mo	43 Tc	44 Ru	45 Rh	46 Pd	47 Ag	48 Cd	49 In	50 Sn	51 Sb	52 Te	53 I	54 Xe
55 Cs	56 Ba	57 La	72 Hf	73 Ta	74 W	75 Re	76 Os	77 Ir	78 Pt	79 Au	80 Hg	81 Tl	82 Pb	83 Bi	84 Po	85 At	86 Rn
87 Fr	88 Ra	89 Ac	104 Rf	105 Db	106 Sg	107 Bh	108 Hs	109 Mt	110 Ds	111 Rg	112 Cn	113 Nh	114 Fl	115 Mc	116 Lv	117 Ts	118 Og

Lanthanides	58 Ce	59 Pr	60 Nd	61 Pm	62 Sm	63 Eu	64 Gd	65 Tb	66 Dy	67 Ho	68 Er	69 Tm	70 Yb	71 Lu
Actinides	90 Th	91 Pa	92 U	93 Np	94 Pu	95 Am	96 Cm	97 Bk	98 Cf	99 Es	100 Fm	101 Md	102 No	103 Lr

EXAMPLE 2.11 GROUPS OF ELEMENTS

To which group of elements does each element belong?

(a) Mg **(b)** N **(c)** K **(d)** Br

SOLUTION

(a) Mg is in Group 2A; it is an alkaline earth metal.
(b) N is in Group 5A.
(c) K is in Group 1A; it is an alkali metal.
(d) Br is in Group 7A; it is a halogen.

▶SKILLBUILDER 2.11 | Groups and Families of Elements

To which group of elements does each element belong?

(a) Li **(b)** B **(c)** I **(d)** Ar

▶FOR MORE PRACTICE Problems 88, 89, 90, 91, 92, 93, 94, 95.

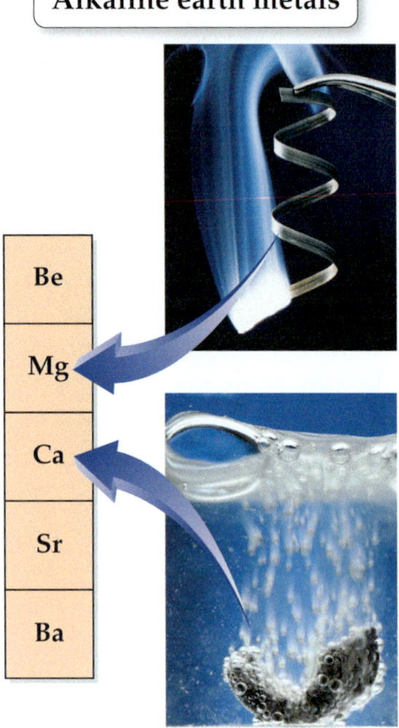

◀ The alkaline earth metals include beryllium, magnesium (shown burning in the first photo), calcium (shown reacting with water in the second photo), strontium, and barium. (top) Richard Megna/Fundamental Photographs. (bottom) Richard Megna/Fundamental Photographs.

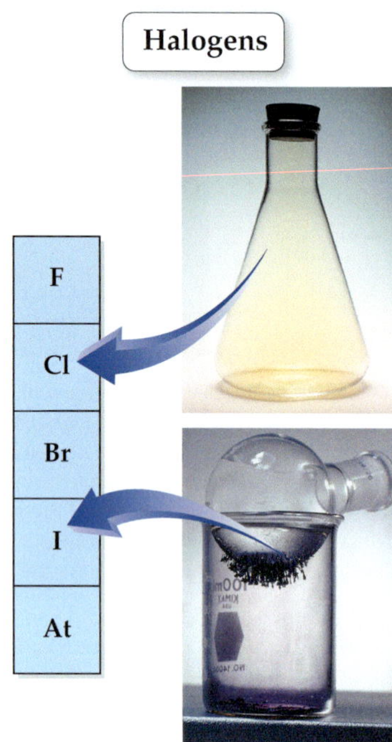

▶ The halogens include fluorine, chlorine, bromine, iodine, and astatine. © (top) Charles D. Winters/Science Source. (bottom) Tom Bochsler/Pearson.

CONCEPTUAL CHECKPOINT 2.9

Which statement is NEVER true?

(a) An element can be both a transition element and a metal.

(b) An element can be both a transition element and a metalloid.

(c) An element can be both a metalloid and a halogen.

(d) An element can be both a main-group element and a halogen.

2.8 Electron Configurations and the Periodic Table

LO: Identify valence electrons and core electrons.

LO: Write electron configurations for elements based on their positions in the periodic table.

Valence electrons are the electrons in the outermost principal shell (the principal shell with the highest principal quantum number, n). These electrons are important because, as we will see in the next module, they are held most loosely and are most easily lost or shared; therefore they are involved in chemical bonding. Electrons that are *not* in the outermost principal shell are **core electrons**. For example, silicon, with the electron configuration of $1s^22s^22p^63s^23p^2$, has 4 valence electrons (those in the $n = 3$ principal shell) and 10 core electrons.

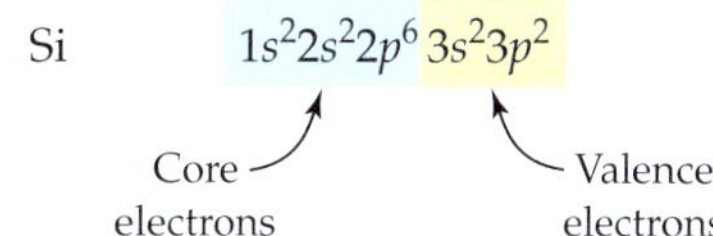

EXAMPLE 2.12 VALENCE ELECTRONS AND CORE ELECTRONS

Write an electron configuration for selenium and identify the valence electrons and the core electrons.

SOLUTION

Write the electron configuration for selenium by determining the total number of electrons from selenium's atomic number (34) and distributing them into the appropriate orbitals.

$$\text{Se} \quad 1s^22s^22p^63s^23p^64s^23d^{10}4p^4$$

The valence electrons are those in the outermost principal shell. For selenium, the outermost principal shell is the $n = 4$ shell, which contains 6 electrons (2 in the 4s orbital and 4 in the three 4p orbitals). All other electrons, including those in the 3d orbitals, are core electrons.

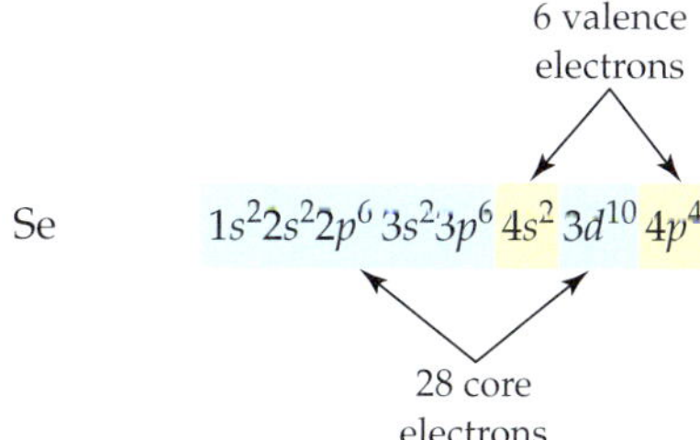

▶ **SKILLBUILDER 2.12 | Valence Electrons and Core Electrons**

Write an electron configuration for chlorine and identify the valence electrons and core electrons.

▶ **FOR MORE PRACTICE** Example 2.23; Problems 104, 105.

▶ **FIGURE 2.20 Outer electron configurations of the first 18 elements**

1A							8A
1 **H** $1s^1$	2A	3A	4A	5A	6A	7A	2 **He** $1s^2$
3 **Li** $2s^1$	4 **Be** $2s^2$	5 **B** $2s^22p^1$	6 **C** $2s^22p^2$	7 **N** $2s^22p^3$	8 **O** $2s^22p^4$	9 **F** $2s^22p^5$	10 **Ne** $2s^22p^6$
11 **Na** $3s^1$	12 **Mg** $3s^2$	13 **Al** $3s^23p^1$	14 **Si** $3s^23p^2$	15 **P** $3s^23p^3$	16 **S** $3s^23p^4$	17 **Cl** $3s^23p^5$	18 **Ar** $3s^23p^6$

▲ Figure 2.20 shows the first 18 elements in the periodic table with an outer electron configuration listed below each one. As we move across a row, the orbitals are simply filling in the correct order. As we move down a column, the highest principal quantum number increases, but the number of electrons in each subshell remains the same. Consequently, the elements within a column (or group) all have the same number of valence electrons and similar outer electron configurations.

A similar pattern exists for the entire periodic table (▼ Figure 2.21). Notice that, because of the filling order of orbitals, we can divide the periodic table into blocks representing the filling of particular subshells.

- The first two columns on the left side of the periodic table are the *s* block with outer electron configurations of ns^1 (first column) and ns^2 (second column).
- The six columns on the right side of the periodic table are the *p* block with outer electron configurations of: ns^2np^1, ns^2np^2, ns^2np^3, ns^2np^4, ns^2np^5 (halogens), and ns^2np^6 (noble gases).
- The transition metals are the *d* block.
- The lanthanides and actinides (also called the inner transition metals) are the *f* block.

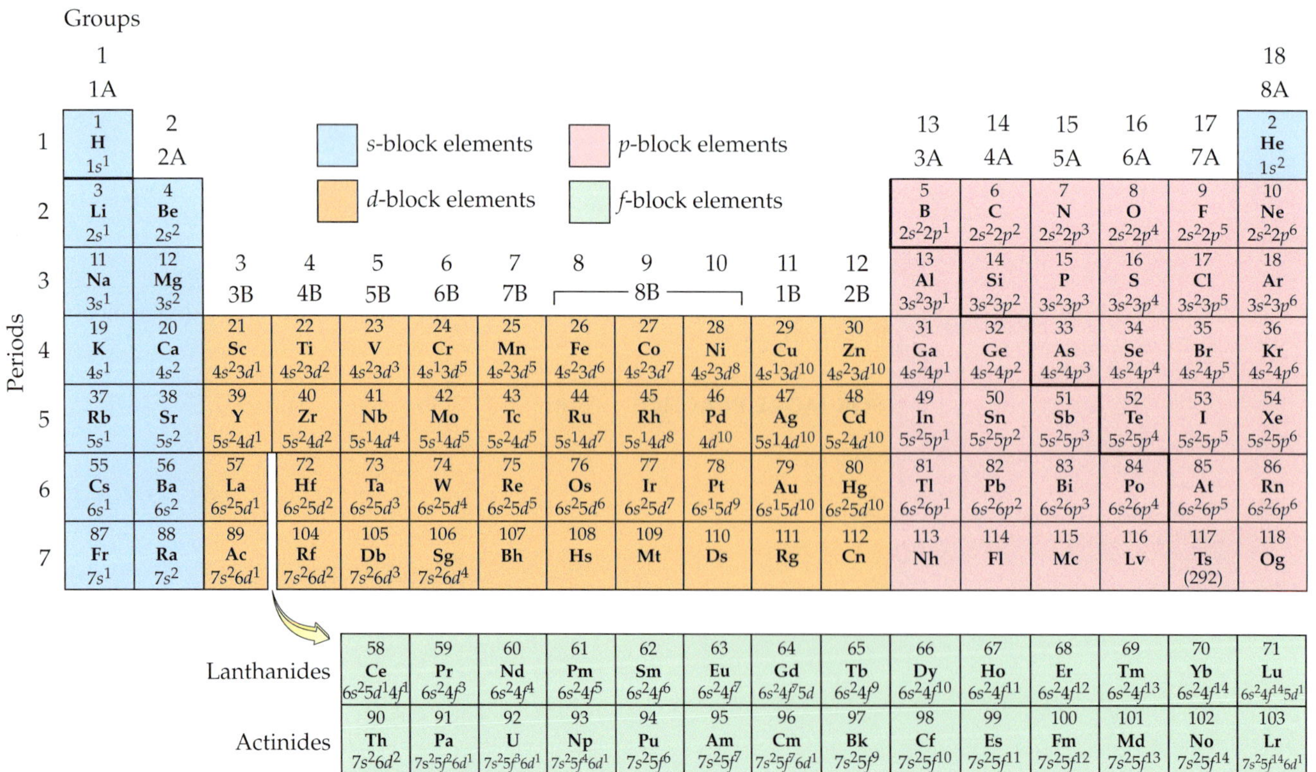

Period	1 1A	2 2A	3 3B	4 4B	5 5B	6 6B	7 7B	8 8B	9 8B	10 8B	11 1B	12 2B	13 3A	14 4A	15 5A	16 6A	17 7A	18 8A
1	1 **H** $1s^1$																	2 **He** $1s^2$
2	3 **Li** $2s^1$	4 **Be** $2s^2$											5 **B** $2s^22p^1$	6 **C** $2s^22p^2$	7 **N** $2s^22p^3$	8 **O** $2s^22p^4$	9 **F** $2s^22p^5$	10 **Ne** $2s^22p^6$
3	11 **Na** $3s^1$	12 **Mg** $3s^2$											13 **Al** $3s^23p^1$	14 **Si** $3s^23p^2$	15 **P** $3s^23p^3$	16 **S** $3s^23p^4$	17 **Cl** $3s^23p^5$	18 **Ar** $3s^23p^6$
4	19 **K** $4s^1$	20 **Ca** $4s^2$	21 **Sc** $4s^23d^1$	22 **Ti** $4s^23d^2$	23 **V** $4s^23d^3$	24 **Cr** $4s^13d^5$	25 **Mn** $4s^23d^5$	26 **Fe** $4s^23d^6$	27 **Co** $4s^23d^7$	28 **Ni** $4s^23d^8$	29 **Cu** $4s^13d^{10}$	30 **Zn** $4s^23d^{10}$	31 **Ga** $4s^24p^1$	32 **Ge** $4s^24p^2$	33 **As** $4s^24p^3$	34 **Se** $4s^24p^4$	35 **Br** $4s^24p^5$	36 **Kr** $4s^24p^6$
5	37 **Rb** $5s^1$	38 **Sr** $5s^2$	39 **Y** $5s^24d^1$	40 **Zr** $5s^24d^2$	41 **Nb** $5s^14d^4$	42 **Mo** $5s^14d^5$	43 **Tc** $5s^24d^5$	44 **Ru** $5s^14d^7$	45 **Rh** $5s^14d^8$	46 **Pd** $4d^{10}$	47 **Ag** $5s^14d^{10}$	48 **Cd** $5s^24d^{10}$	49 **In** $5s^25p^1$	50 **Sn** $5s^25p^2$	51 **Sb** $5s^25p^3$	52 **Te** $5s^25p^4$	53 **I** $5s^25p^5$	54 **Xe** $5s^25p^6$
6	55 **Cs** $6s^1$	56 **Ba** $6s^2$	57 **La** $6s^25d^1$	72 **Hf** $6s^25d^2$	73 **Ta** $6s^25d^3$	74 **W** $6s^25d^4$	75 **Re** $6s^25d^5$	76 **Os** $6s^25d^6$	77 **Ir** $6s^25d^7$	78 **Pt** $6s^15d^9$	79 **Au** $6s^15d^{10}$	80 **Hg** $6s^25d^{10}$	81 **Tl** $6s^26p^1$	82 **Pb** $6s^26p^2$	83 **Bi** $6s^26p^3$	84 **Po** $6s^26p^4$	85 **At** $6s^26p^5$	86 **Rn** $6s^26p^6$
7	87 **Fr** $7s^1$	88 **Ra** $7s^2$	89 **Ac** $7s^26d^1$	104 **Rf** $7s^26d^2$	105 **Db** $7s^26d^3$	106 **Sg** $7s^26d^4$	107 **Bh**	108 **Hs**	109 **Mt**	110 **Ds**	111 **Rg**	112 **Cn**	113 **Nh**	114 **Fl**	115 **Mc**	116 **Lv**	117 **Ts** (292)	118 **Og**

Lanthanides	58 **Ce** $6s^25d^14f^1$	59 **Pr** $6s^24f^3$	60 **Nd** $6s^24f^4$	61 **Pm** $6s^24f^5$	62 **Sm** $6s^24f^6$	63 **Eu** $6s^24f^7$	64 **Gd** $6s^24f^75d$	65 **Tb** $6s^24f^9$	66 **Dy** $6s^24f^{10}$	67 **Ho** $6s^24f^{11}$	68 **Er** $6s^24f^{12}$	69 **Tm** $6s^24f^{13}$	70 **Yb** $6s^24f^{14}$	71 **Lu** $6s^24f^{14}5d^1$
Actinides	90 **Th** $7s^26d^2$	91 **Pa** $7s^25f^26d^1$	92 **U** $7s^25f^36d^1$	93 **Np** $7s^25f^46d^1$	94 **Pu** $7s^25f^6$	95 **Am** $7s^25f^7$	96 **Cm** $7s^25f^76d^1$	97 **Bk** $7s^25f^9$	98 **Cf** $7s^25f^{10}$	99 **Es** $7s^25f^{11}$	100 **Fm** $7s^25f^{12}$	101 **Md** $7s^25f^{13}$	102 **No** $7s^25f^{14}$	103 **Lr** $7s^25f^{14}6d^1$

▲ **FIGURE 2.21 Outer electron configurations of the elements**

Remember that main-group elements are those in the two far–left columns (1A, 2A) and the six far–right columns (3A–8A) of the periodic table (see Section 2.5).

Notice that, except for helium, the number of valence electrons for any main-group element is equal to the group number of its column. For example, we can tell that chlorine has 7 valence electrons because it is in the column with group number 7A. The row number in the periodic table is equal to the number of the highest principal shell (n value). For example, since chlorine is in row 3, its highest principal shell is the $n = 3$ shell.

The transition metals have electron configurations with trends that differ somewhat from main-group elements. As we move across a row in the d block, the d orbitals are filling (▲ Figure 2.21). However, the principal quantum number of the d orbital being filled across each row in the transition series is equal to the row number minus one (in the fourth row, the $3d$ orbitals fill; in the fifth row, the $4d$ orbitals fill; and so on). For the first transition series, the outer configuration is $4s^2 3d^x$ (x = number of d electrons) with two exceptions: Cr is $4s^1 3d^5$ and Cu is $4s^1 3d^{10}$. These exceptions occur because a half-filled d subshell and a completely filled d subshell are particularly stable. Otherwise, the number of outershell electrons in a transition series does not change as we move across a period. In other words, *the transition series represents the filling of core orbitals, and the number of outershell electrons is mostly constant.*

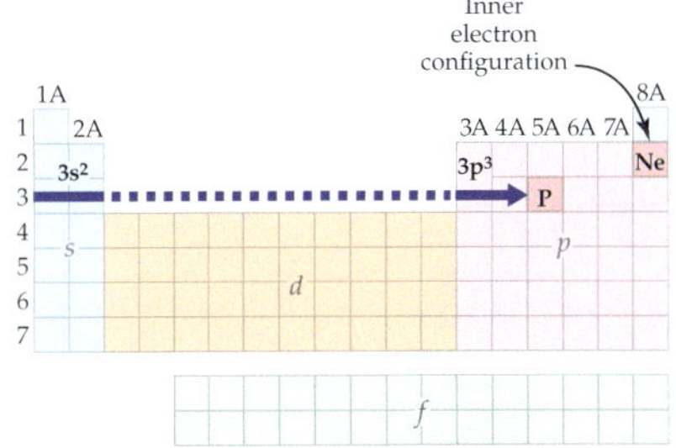

▲ FIGURE 2.22 Electron configuration of phosphorus Determining the electron configuration for P from its position in the periodic table.

We can now see that the organization of the periodic table allows us to write the electron configuration for any element based simply on its position in the periodic table. For example, suppose we want to write an electron configuration for P. The inner electrons of P are those of the noble gas that precedes P in the periodic table, Ne. So we can represent the inner electrons with [Ne]. We obtain the outer electron configuration by tracing the elements between Ne and P and assigning electrons to the appropriate orbitals (◀ Figure 2.22). Remember that the highest n value is given by the row number (3 for phosphorus). So we begin with [Ne], then add in the two $3s$ electrons as we trace across the s block, followed by three $3p$ electrons as we trace across the p block to P, which is in the third column of the p block. The electron configuration is:

$$\text{P} \qquad [\text{Ne}]3s^2 3p^3$$

Notice that P is in column 5A and therefore has 5 valence electrons and an outer electron configuration of $ns^2 np^3$.

To summarize writing an electron configuration for an element based on its position in the periodic table:

- The inner electron configuration for any element is the electron configuration of the noble gas that immediately precedes that element in the periodic table. We represent the inner configuration with the symbol for the noble gas in brackets.
- We can determine the outer electrons from the element's position within a particular block (s, p, d, or f) in the periodic table. We trace the elements between the preceding noble gas and the element of interest, and assign electrons to the appropriate orbitals.
- The highest principal quantum number (highest n value) is equal to the row number of the element in the periodic table.
- For any element containing d electrons, the principal quantum number (n value) of the outermost d electrons is equal to the row number of the element minus 1.

EXAMPLE 2.13 WRITING ELECTRON CONFIGURATIONS FROM THE PERIODIC TABLE

Write an electron configuration for arsenic based on its position in the periodic table.

SOLUTION

The noble gas that precedes arsenic in the periodic table is argon, so the inner electron configuration is [Ar]. Obtain the outer electron configuration by tracing the elements between Ar and As and assigning electrons to the appropriate orbitals.

Remember that the highest n value is given by the row number (4 for arsenic). So, begin with [Ar], then add in the two 4s electrons as you trace across the s block, followed by ten 3d electrons as you trace across the d block (the n value for d subshells is equal to the row number minus one), and finally the three 4p electrons as you trace across the p block to As, which is in the third column of the p block:

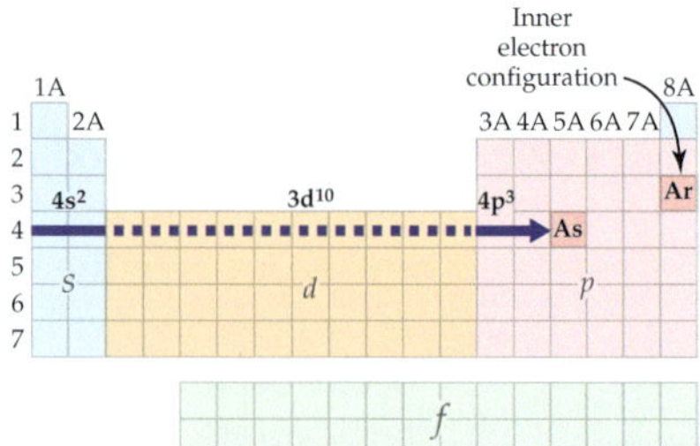

The electron configuration is:

$$\text{As} \qquad [\text{Ar}]4s^2 3d^{10} 4p^3$$

▶ **SKILLBUILDER 2.13 | Writing Electron Configurations from the Periodic Table**

Use the periodic table to determine the electron configuration for tin.

▶ **FOR MORE PRACTICE** Example 2.24; Problems 112, 113, 114, 115.

CONCEPTUAL CHECKPOINT 2.10

Which element has the *fewest* valence electrons?

(a) B **(b)** Ca **(c)** O

(d) K **(e)** Ga

2.9 The Explanatory Power of the Quantum-Mechanical Model

LO: Recognize that the chemical properties of elements are largely determined by the number of valence electrons they contain.

Noble gases 18 8A
2 **He** $1s^2$
10 **Ne** $2s^2 2p^6$
18 **Ar** $3s^2 3p^6$
36 **Kr** $4s^2 4p^6$
54 **Xe** $5s^2 5p^6$
86 **Rn** $6s^2 6p^6$

▶ **FIGURE 2.23 Electron configurations of the noble gases** The noble gases (except for helium) all have 8 valence electrons and completely full outer principal shells.

We can now see how *the chemical properties of elements are largely determined by the number of valence electrons they contain.* The properties of elements vary in a periodic fashion because the number of valence electrons is periodic.

Because elements within a group in the periodic table have the same number of valence electrons, they also have similar chemical properties. The noble gases, for example, all have 8 valence electrons, except for helium, which has 2 (◀ Figure 2.23). Although we don't get into the quantitative (or numerical) aspects of the quantum-mechanical model in this book, calculations show that atoms with 8 valence electrons (or 2 for helium) are particularly low in energy, and therefore stable. The noble gases are indeed chemically stable and thus relatively inert or nonreactive as accounted for by the quantum model.

Elements with electron configurations close to the noble gases are the most reactive because they can attain noble gas electron configurations by losing or gaining a small number of electrons. Alkali metals (Group 1) are among the most reactive metals since their outer electron configuration (ns^1) is 1 electron beyond a noble gas configuration (◀ Figure 2.24). If they can react to lose the ns^1 electron, they attain a noble gas configuration. This explains why—as we learned earlier in this module—the Group 1A metals tend to form 1+ cations. As an example, consider the electron configuration of sodium:

$$\text{Na} \quad 1s^2 2s^2 2p^6 3s^1$$

In reactions, sodium loses its 3s electron, forming a 1+ ion with the electron configuration of neon.

$$\text{Na}^+ \quad 1s^2 2s^2 2p^6$$

$$\text{Ne} \quad 1s^2 2s^2 2p^6$$

Similarly, alkaline earth metals, with an outer electron configuration of ns^2 also tend to be reactive metals, losing their ns^2 electrons to form 2+ cations (▼ Figure 2.25). For example, consider magnesium:

$$\text{Mg} \quad 1s^2 2s^2 2p^6 3s^2$$

In reactions, magnesium loses its two 3s electrons, forming a 2+ ion with the electron configuration of neon.

$$\text{Mg}^{2+} \quad 1s^2 2s^2 2p^6$$

On the other side of the periodic table, halogens are among the most reactive nonmetals because of their ns^2np^5 electron configurations (▼ Figure 2.26). They are only one electron away from a noble gas configuration and tend to react to gain that one electron, forming 1– ions. For example, consider fluorine:

$$\text{F} \quad 1s^2 2s^2 2p^5$$

In reactions, fluorine gains an additional 2p electron, forming a 1– ion with the electron configuration of neon.

$$\text{F}^- \quad 1s^2 2s^2 2p^6$$

The elements that form predictable ions are shown in ▶ Figure 2.27 (first introduced earlier in this module). Notice how the charge of these ions reflects their electron configurations—these elements form ions with noble gas electron configurations.

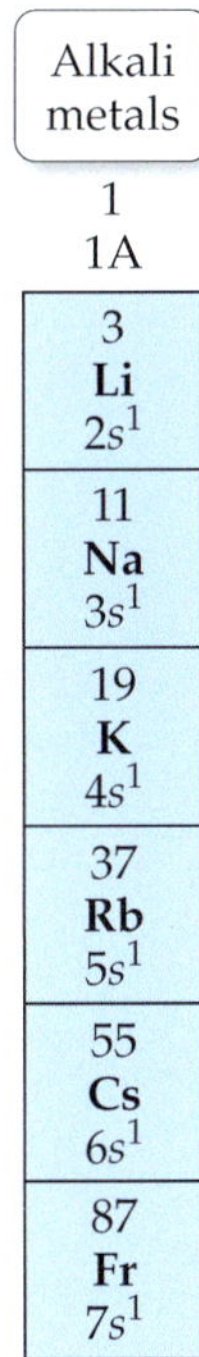

▲ **FIGURE 2.24 Electron configurations of the alkali metals** The alkali metals all have ns^1 electron configurations and are therefore 1 electron beyond a noble gas configuration. In their reactions, they tend to lose that electron, forming 1+ ions and attaining a noble gas configuration.

Atoms and/or ions that share the same electron configuration are termed isoelectronic.

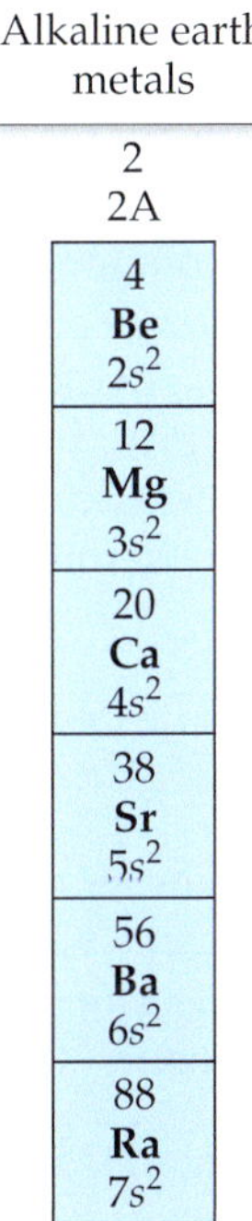

◀ **FIGURE 2.25 Electron configurations of the alkaline earth metals** The alkaline earth metals all have ns^2 electron configurations and are therefore 2 electrons beyond a noble gas configuration. In their reactions, they tend to lose 2 electrons, forming 2+ ions and attaining a noble gas configuration.

Halogens
17
7A
9
F
$2s^22p^5$
17
Cl
$3s^23p^5$
35
Br
$4s^24p^5$
53
I
$5s^25p^5$
85
At
$6s^26p^5$

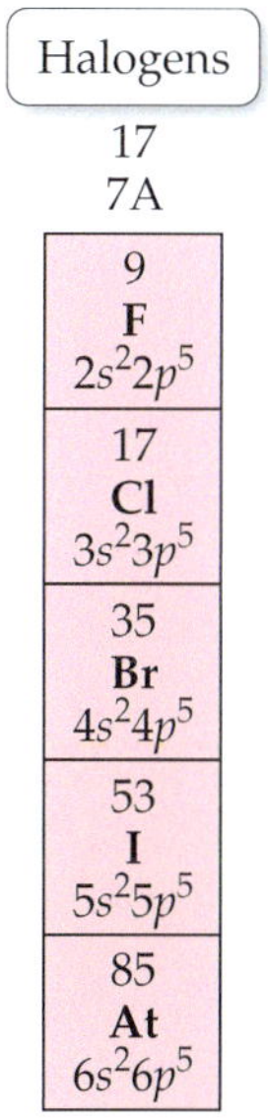

▶ **FIGURE 2.26 Electron configurations of the halogens** The halogens all have ns^2np^5 electron configurations and are therefore 1 electron short of a noble gas configuration. In their reactions, they tend to gain 1 electron, forming 1– ions and attaining a noble gas configuration.

1	2	Transition metals	3	4	5	6	7	8
Li^+	Be^{2+}				N^{3-}	O^{2-}	F^-	
Na^+	Mg^{2+}		Al^{3+}			S^{2-}	Cl^-	
K^+	Ca^{2+}	Most transition metals form cations with various charges	Ga^{3+}			Se^{2-}	Br^-	
Rb^+	Sr^{2+}		In^{3+}			Te^{2-}	I^-	
Cs^+	Ba^{2+}							

▲ **FIGURE 2.27 Elements that form predictable ions**

CONCEPTUAL CHECKPOINT 2.11

Below is the electron configuration of calcium:

$$\text{Ca} \quad 1s^2 2s^2 2p^6 3s^2 3p^6 4s^2$$

In its reactions, calcium tends to form the Ca^{2+} ion. Which electrons are lost upon ionization?

(a) the 4*s* electrons

(b) two of the 3*p* electrons

(c) the 3*s* electrons

(d) the 1*s* electrons

2.10 Periodic Trends: Atomic Size, Ionization Energy, and Metallic Character

LO: Identify and understand periodic trends such as atomic size, ionization energy, and metallic character.

The quantum-mechanical model also explains other periodic trends such as atomic size, ionization energy, and metallic character. We examine these trends individually in this section of the module.

Atomic Size

The **atomic size** of an atom is determined by the distance between the outermost electrons and the nucleus. As we move across a period in the periodic table, we know that electrons occupy orbitals with the same principal quantum number, n. Since the principal quantum number largely determines the size of an orbital, electrons are therefore filling orbitals of approximately the same size, and we might expect atomic size to remain constant across a period. However, with each step across a period, the number of protons in the nucleus also increases. This increase in the number of protons results in a greater pull on the electrons from the nucleus, causing atomic size to actually decrease. Therefore:

> As we move to the left across a period, or row, in the periodic table, atomic size increases, as shown in ▶Figure 2.28.

As we move down a column in the periodic table, the highest principal quantum number, n, increases. Because the size of an orbital increases with increasing principal quantum number, the electrons that occupy the outermost orbitals are farther from the nucleus as we move down a column. Therefore:

> As we move down a column, or group in the periodic table, atomic size increases, as shown in ▶ Figure 2.28.

▶ **FIGURE 2.28 Periodic properties: atomic size** Atomic size increases as we move to the left across a period and increases as we move down a column in the periodic table.

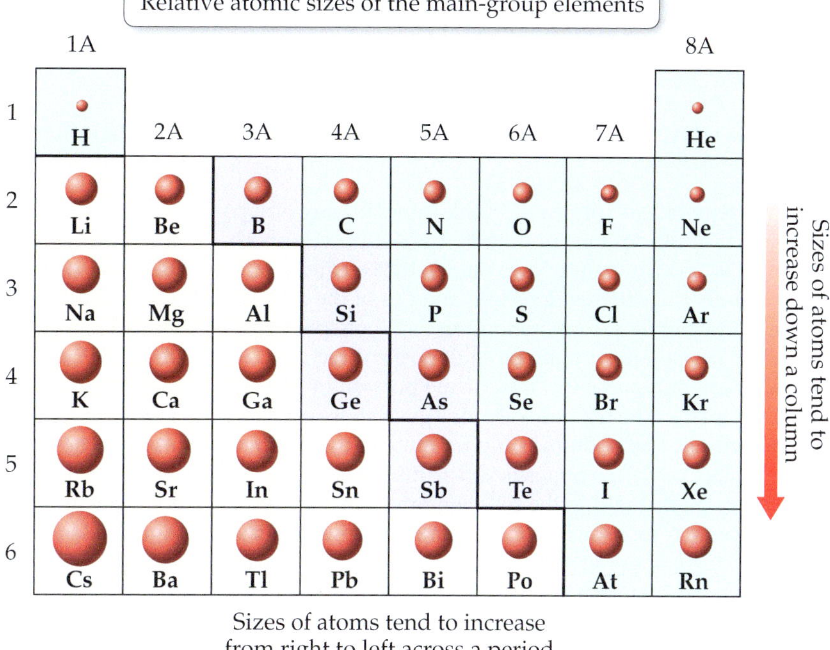

EXAMPLE 2.14 ATOMIC SIZE

Choose the larger atom in each pair.

(a) C or O **(b)** Li or K **(c)** C or Al **(d)** Se or I

SOLUTION

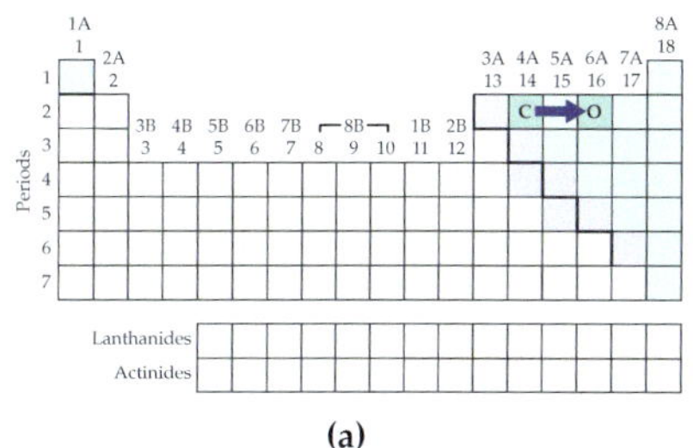

(a)

(a) C or O

Carbon atoms are larger than O atoms because, as you trace the path between C and O on the periodic table, you move to the right within the same period. Atomic size decreases as you go to the right.

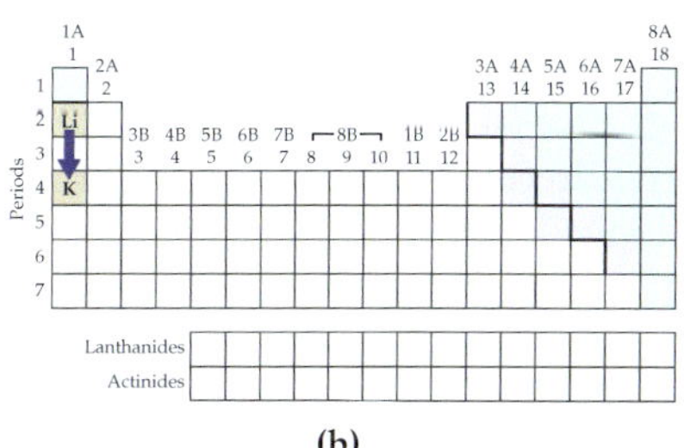

(b)

(b) Li or K

Potassium atoms are larger than Li atoms because, as you trace the path between Li and K on the periodic table, you move down a column. Atomic size increases as you go down a column.

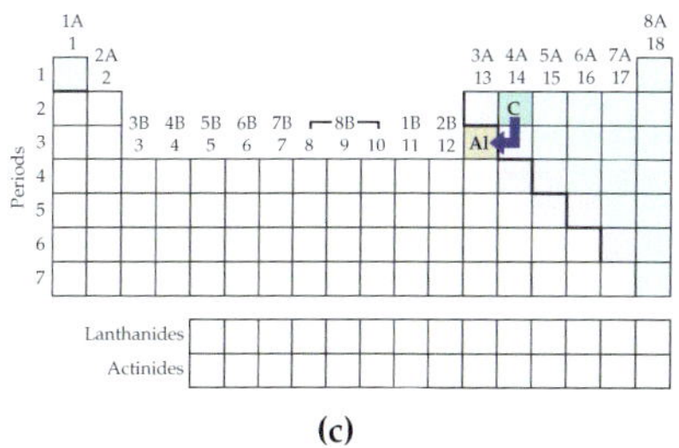

(c)

(c) C or Al

Aluminum atoms are larger than C atoms because, as you trace the path between C and Al on the periodic table you move down a column (atomic size increases) and then to the left across a period (atomic size increases). These effects add together for an overall increase.

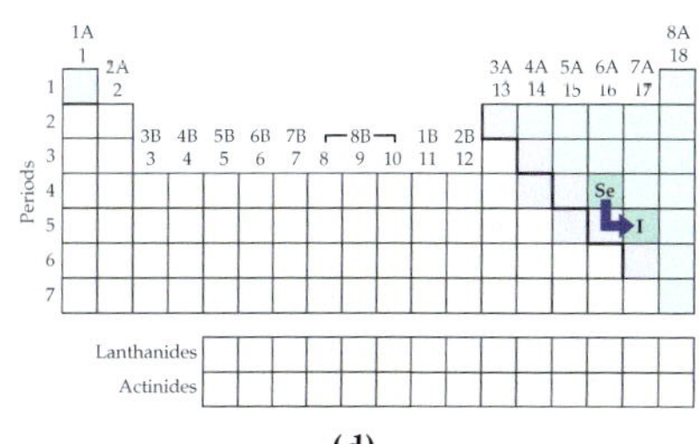

(d)

(d) Se or I

Based on periodic properties alone, you cannot tell which atom is larger because as you trace the path between Se and I, you go down a column (atomic size increases) and then to the right across a period (atomic size decreases). These effects tend to cancel one another.

▶ **SKILLBUILDER 2.14 | Atomic Size**

Choose the larger atom in each pair.

(a) Pb or Po **(b)** Rb or Na **(c)** Sn or Bi **(d)** F or Se

▶ **FOR MORE PRACTICE** Example 2.25a; Problems 136, 137, 138, 139.

CHEMISTRY AND HEALTH

▶ Pumping Ions: Atomic Size and Nerve Impulses

No matter what you are doing at this moment, tiny pumps in each of the trillions of cells that make up your body are hard at work. These pumps, located in the cell membrane, move a number of different ions into and out of the cell. The most important of these ions are sodium (Na^+) and potassium (K^+) which happen to be pumped in opposite directions. Sodium ions are pumped *out of cells*, while potassium ions are pumped *into cells*. The result is a *chemical gradient* for each ion: The concentration of sodium is higher outside the cell than within, while exactly the opposite is true for potassium.

The ion pumps within the cell membrane are analogous to water pumps in a high-rise building that pump water against the force of gravity to a tank on the roof. Other structures within the membrane, called ion channels, are like the building's faucets. When they open momentarily, bursts of sodium and potassium ions, driven by their concentration gradients, flow back across the membrane—sodium flowing in and potassium flowing out. These ion pulses are the basis for the transmission of nerve signals in the brain, heart, and throughout the body. Consequently, every move you make or every thought you have is mediated by the flow of these ions.

How do the pumps and channels differentiate between sodium and potassium ions? How do the ion pumps selectively move sodium out of the cell and potassium into the cell? To answer this question, we must examine the sodium and potassium ions more closely. In what ways do they differ? Both are cations of Group I metals. All Group I metals tend to lose one electron to form cations with 1+ charge, so the magnitude of the charge cannot be the decisive factor. But potassium (atomic number 19) lies directly below sodium in the periodic table (atomic number 11) and based on periodic properties is therefore larger than sodium. The potassium ion has a radius of 133 pm, while the sodium ion has a radius of 95 pm (1 pm = 10^{-12} m). The pumps and channels within cell membranes are so sensitive that they distinguish between the sizes of these two ions and selectively allow only one or the other to pass. The result is the transmission of nerve signals that allows you to read this page.

B2.2 CAN YOU ANSWER THIS? *Other ions, including calcium and magnesium, are also important to nerve signal transmission. Arrange these four ions in order of increasing size:* K^+, Na^+, Mg^{2+}, *and* Ca^{2+}.

Ionization Energy

The **ionization energy** of an atom is the energy required to remove n electron from the atom in the gaseous state. For example, the ionization of sodium can be represented with the equation:

$$Na + \text{Ionization energy} \rightarrow Na^+ + 1e^-$$

Based on what we know about electron configurations, what can we predict about ionization energy trends? Would it take more or less energy to remove an electron from Na than from Cl? The explanation is exactly in line with that for the trend in atomic radius. With each step across a period, the number of protons increases. Since this increase in the number of protons results in a greater pull on the electrons from the nucleus, the energy required to remove the outermost electron increases. It is easier to remove an electron from sodium than it is from chlorine. We can generalize this idea in this statement:

> As we move across a period, or row, to the right in the periodic table, ionization energy increases (▶Figure 2.29).

What happens to ionization energy as we move down a column? As we have learned, the principal quantum number, n, increases as we move down a column. Within a given subshell, orbitals with higher principal quantum numbers are larger than orbitals with smaller principal quantum numbers. Consequently, electrons in the outermost principal shell are farther away from the positively charged nucleus—and therefore are held less tightly—as we move down a column. This

▶ **FIGURE 2.29 Periodic properties: ionization energy** Ionization energy increases as we move to the right across a period and increases as we move up a column in the periodic table.

Ionization energy trends

Periods	1 1A	2 2A	3 3B	4 4B	5 5B	6 6B	7 7B	8 8B	9 8B	10 8B	11 1B	12 2B	13 3A	14 4A	15 5A	16 6A	17 7A	18 8A
1	1 H																	2 He
2	3 Li	4 Be											5 B	6 C	7 N	8 O	9 F	10 Ne
3	11 Na	12 Mg											13 Al	14 Si	15 P	16 S	17 Cl	18 Ar
4	19 K	20 Ca	21 Sc	22 Ti	23 V	24 Cr	25 Mn	26 Fe	27 Co	28 Ni	29 Cu	30 Zn	31 Ga	32 Ge	33 As	34 Se	35 Br	36 Kr
5	37 Rb	38 Sr	39 Y	40 Zr	41 Nb	42 Mo	43 Tc	44 Ru	45 Rh	46 Pd	47 Ag	48 Cd	49 In	50 Sn	51 Sb	52 Te	53 I	54 Xe
6	55 Cs	56 Ba	57 La	72 Hf	73 Ta	74 W	75 Re	76 Os	77 Ir	78 Pt	79 Au	80 Hg	81 Tl	82 Pb	83 Bi	84 Po	85 At	86 Rn
7	87 Fr	88 Ra	89 Ac	104 Rf	105 Db	106 Sg	107 Bh	108 Hs	109 Mt	110 Ds	111 Rg	112 Cn	113 Nh	114 Fl	115 Mc	116 Lv	117 Ts	118 Og

Lanthanides	58 Ce	59 Pr	60 Nd	61 Pm	62 Sm	63 Eu	64 Gd	65 Tb	66 Dy	67 Ho	68 Er	69 Tm	70 Yb	71 Lu
Actinides	90 Th	91 Pa	92 U	93 Np	94 Pu	95 Am	96 Cm	97 Bk	98 Cf	99 Es	100 Fm	101 Md	102 No	103 Lr

Ionization energy increases (across a period)

Ionization energy increases (up a column)

results in a lower ionization energy (if the electron is held less tightly, it is easier to pull away) as we move down a column. Therefore:

> As we move up a column (or group) in the periodic table, ionization energy increases (▲ Figure 2.29).

Notice that the trends in ionization energy are consistent with the trends in atomic size. Smaller atoms are more difficult to ionize because their electrons are held more tightly. Therefore, as we move across a period, atomic size decreases and ionization energy increases. Similarly, as we move down a column, atomic size increases and ionization energy decreases since electrons are farther from the nucleus and are therefore less tightly held.

EXAMPLE 2.15 IONIZATION ENERGY

Choose the element with the higher ionization energy from each pair.

(a) Mg or P
(b) As or Sb
(c) N or Si
(d) O or Cl

SOLUTION

(a) Mg or P

P has a higher ionization than Mg because, as you trace the path between Mg and P on the periodic table, you move to the right within the same period. Ionization energy increases as you go to the right.

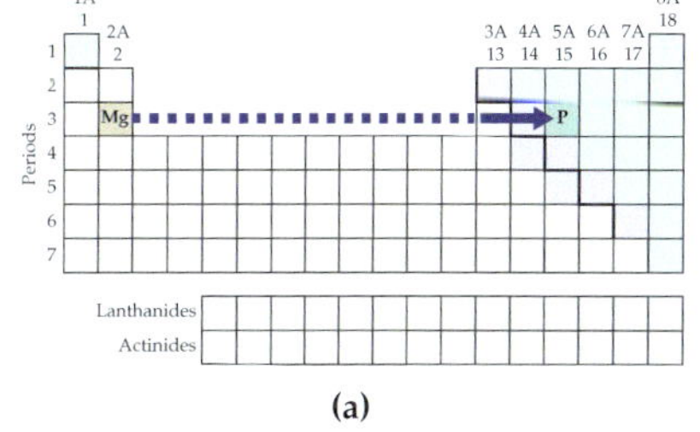

(a)

(b) As or Sb

As has a higher ionization energy than Sb because, as you trace the path between As and Sb on the periodic table, you move down a column. Ionization energy decreases as you go down a column.

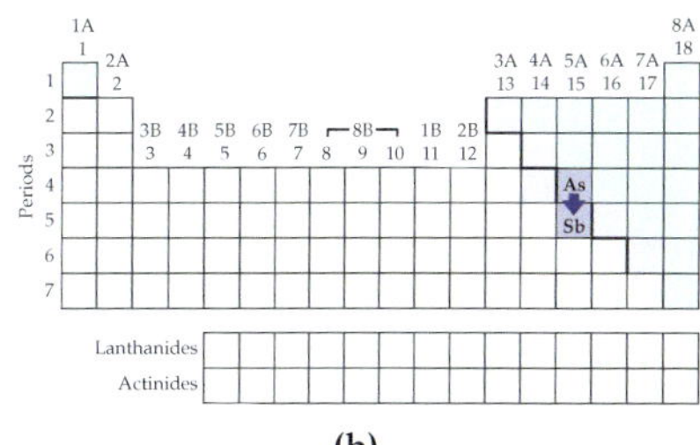

(b)

(c) N or Si

N has a higher ionization energy than Si because, as you trace the path between N and Si on the periodic table, you move down a column (ionization energy decreases) and then to the left across a period (ionization energy decreases). These effects sum together for an overall decrease.

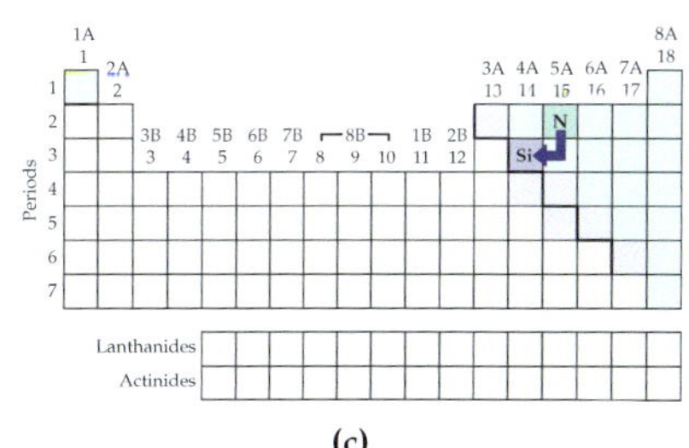

(c)

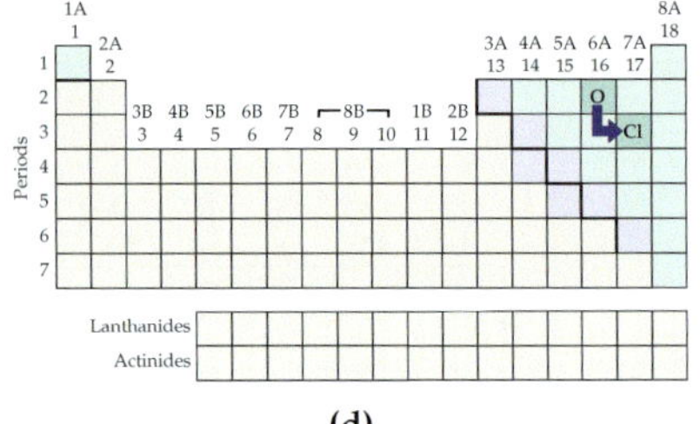

(d)

(d) O or Cl

Based on periodic properties alone, you cannot tell which has a higher ionization energy because, as you trace the path between O and Cl you move down a column (ionization energy decreases) and then to the right across a period (ionization energy increases). These effects tend to cancel.

▶ **SKILLBUILDER 2.15 | Ionization Energy**

Choose the element with the higher ionization energy from each pair.

(a) Mg or Sr
(b) In or Te
(c) C or P
(d) F or S

▶ **FOR MORE PRACTICE** Example 2.25b; Problems 132, 133, 134, 135.

Metallic Character

As we learned earlier in this module, metals tend to lose electrons in their chemical reactions, while nonmetals tend to gain electrons. As we move left across a period in the periodic table, ionization energy decreases, which means that electrons are more likely to be lost in chemical reactions. Consequently:

> As we move across a period, or row, to the left in the periodic table, **metallic character** increases (▼ Figure 2.30).

As we move down a column in the periodic table, ionization energy decreases, making electrons more likely to be lost in chemical reactions. Consequently:

> As we move down a column, or group in the periodic table, metallic character increases (▼ Figure 2.30).

These trends, based on the quantum-mechanical model, explain the distribution of metals and nonmetals that we were introduced to earlier in this module. Metals are found toward the left side of the periodic table and nonmetals (with the exception of hydrogen) toward the upper right.

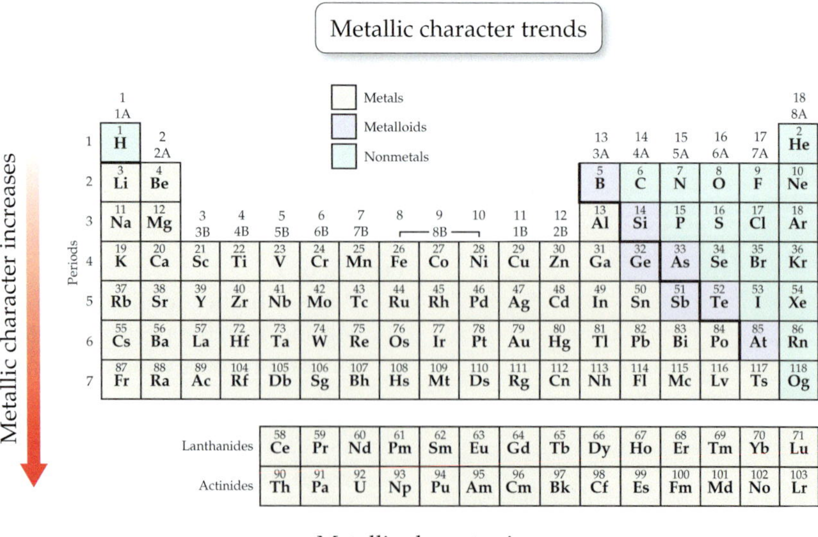

▲ **FIGURE 2.30 Periodic properties: metallic character** Metallic character decreases as we move to the right across a period and increases as we move down a column in the periodic table.

EXAMPLE 2.16 METALLIC CHARACTER

Choose the more metallic element from each pair.

(a) Sn or Te
(b) Si or Sn
(c) Br or Te
(d) Se or I

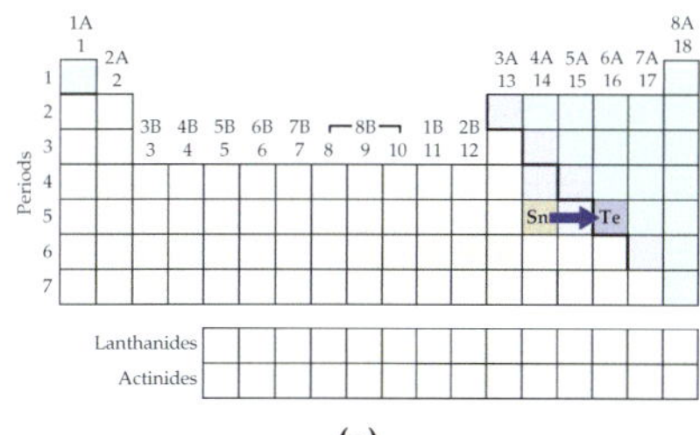

(a)

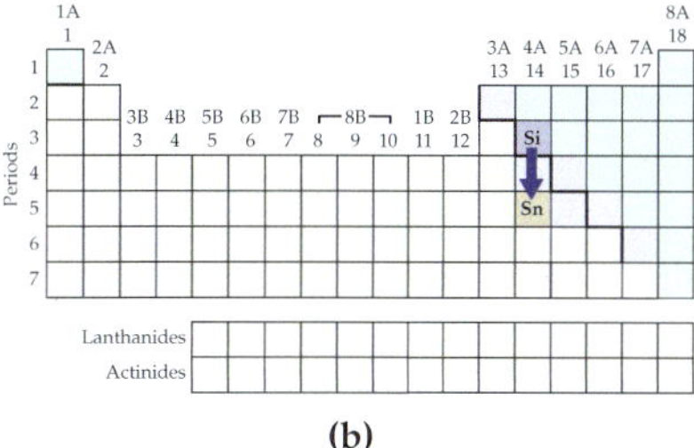

(b)

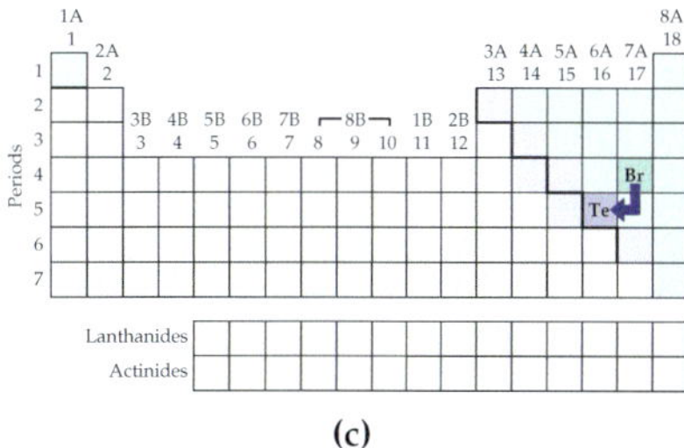

(c)

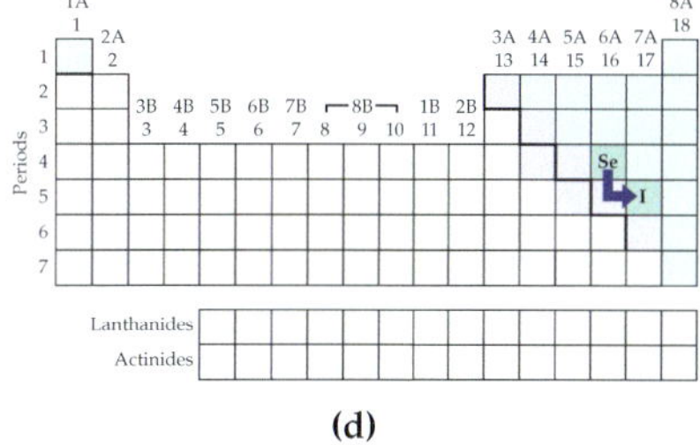

(d)

SOLUTION

(a) Sn or Te

Sn is more metallic than Te because, as you trace the path between Sn and Te on the periodic table, you move to the right within the same period. Metallic character decreases as you go to the right.

(b) Si or Sn

Sn is more metallic than Si because, as you trace the path between Si and Sn on the periodic table, you move down a column. Metallic character increases as you go down a column.

(c) Br or Te

Te is more metallic than Br because, as you trace the path between Br and Te on the periodic table, you move down a column (metallic character increases) and then to the left across a period (metallic character increases). These effects add together for an overall increase.

(d) Se or I

Based on periodic properties alone, you cannot tell which is more metallic because, as you trace the path between Se and I, you go down a column (metallic character increases) and then to the right across a period (metallic character decreases). These effects tend to cancel.

▶ SKILLBUILDER 2.16 | Metallic Character

Choose the more metallic element from each pair.

(a) Ge or In
(b) Ga or Sn
(c) P or Bi
(d) B or N

▶ FOR MORE PRACTICE Example 2.25c; Problems 140, 141, 142, 143.

CONCEPTUAL CHECKPOINT 2.12

Which property *increases* as you move from left to right across a row in the periodic table?

(a) Atomic size
(b) Ionization energy
(c) Metallic character

MODULE IN REVIEW

Self-Assessment Quiz

Q1. Which statement best summarizes the nuclear model of the atom?

(a) The atom is composed of a dense core that contains most of its mass and all of its positive charge, while low-mass negatively charged particles compose most of its volume.

(b) The atom is composed of a sphere of positive charge with many negatively charged particles within the sphere.

(c) Most of the mass of the atom is evenly distributed throughout its volume.

(d) All of the particles that compose an atom have exactly the same mass.

Q2. How many neutrons does the Fe-56 isotope contain?

(a) 26 **(b)** 30 **(c)** 56 **(d)** 112

Q3. Determine the number of protons, neutrons, and electrons in ${}^{32}_{16}S^{2-}$.

(a) 16 protons; 32 neutrons; and 18 electrons

(b) 16 protons; 16 neutrons; and 18 electrons

(c) 32 protons; 16 neutrons; and 2 electrons

(d) 16 protons; 48 neutrons; and 16 electrons

Q4. An element has four naturally occurring isotopes; the table lists the mass and natural abundance of each isotope. Find the atomic mass of the element.

Isotope	Mass (amu)	Natural abundance
A	203.9730	1.4 %
B	205.9744	24.1 %
C	206.9758	22.1 %
D	207.9766	52.4 %

(a) 207.2 amu

(b) 2.072×10^4 amu

(c) 206.2 amu

(d) 206.5 amu

Q5. An ion composed of which of these particles would have a mass of approximately 16 amu and a charge of 2–?

(a) 8 protons and 8 electrons

(b) 8 protons, 8 neutrons, and 10 electrons

(c) 8 protons, 8 neutrons, and 8 electrons

(d) 8 protons, 8 neutrons, and 6 electrons

Q6. What is the charge of the Cr ion that contains 21 electrons?

(a) 2– **(b)** 3– **(c)** 2+ **(d)** 3+

Q7. What is the electron configuration of arsenic (As)?

(a) $[Ar]4s^24p^3$ **(c)** $[Ar]4s^23d^64p^3$

(b) $[Ar]4s^24d^{10}4p^3$ **(d)** $[Ar]4s^23d^{10}4p^3$

Q8. Which orbital diagram corresponds to phosphorus (P)?

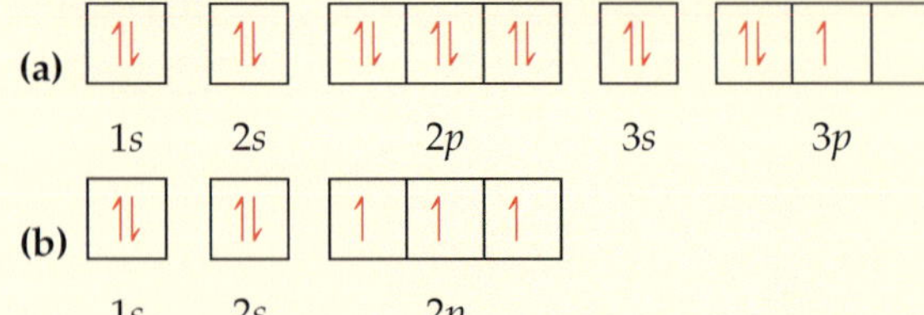

(c) 1s: ↑↓; 2s: ↑↓; 2p: ↑↓ ↑↓ ↑↓; 3s: ↑↓; 3p: ↑ ↑ ↑

(d) 1s: ↑↓; 2s: ↑↓; 2p: ↑↓ ↑↓ ↑↓; 3s: ↑↓; 3p: ↑↓ ↑↓ ↑

Q9. How many valence electrons does tellurium (Te) have?

(a) 5 **(b)** 6 **(c)** 16 **(d)** 52

Q10. Which element is a main-group metal with an even atomic number?

(a) K **(b)** Ca **(c)** Cr **(d)** Se

Q11. Which element is a halogen?

(a) Ne **(b)** O **(c)** Ca **(d)** I

Q12. Which pair of elements has the most similar chemical properties?

(a) Sr and Ba **(b)** S and Ar

(c) H and He **(d)** K and Se

Q13. Which element is a row 4 noble gas?

(a) Ne **(b)** Br **(c)** Zr **(d)** Kr

Q14. How many electrons does the predictable (most common) ion of fluorine contain?

(a) 1 **(b)** 4 **(c)** 9 **(d)** 10

Q15. The element sulfur forms an ion with what charge?

(a) 2– **(b)** 1– **(c)** 1+ **(d)** 2+

Q16. When aluminum forms an ion, it loses electrons. How many electrons does it lose, and which orbitals do the electrons come from?

(a) 1 electron from the 3*s* orbital

(b) 2 electrons; one from the 3*s* orbital and one from the 2*s* orbital

(c) 3 electrons; two from the 3*s* orbital and one from the 3*p* orbital

(d) 5 electrons from the 3*p* orbital

Q17. Order the elements Sr, Ca, and Se in order of decreasing atomic size.

(a) Se > Sr > Ca **(c)** Sr > Ca > Se

(b) Ca > Se > Sr **(d)** Se > Ca > Sr

Q18. Which element has the highest ionization energy?

(a) Sn **(b)** S **(c)** Si **(d)** F

Q19. Which element is most metallic?

(a) Al **(b)** N **(c)** P **(d)** O

Q20. Which property decreases as you move down a column in the periodic table?

(a) Atomic size

(b) Ionization energy

(c) Metallic character

(d) None of the above (all increase as you move down a column).

Answers: 1:a; 2:b; 3:b; 4:a; 5:b; 6:d; 7:d; 8:c; 9:b; 10:b; 11:d; 12:a; 13:d; 14:d; 15:a; 16:c; 17:c; 18:d; 19:a; 20:b

Chemical Principles	Relevance
The Nuclear Theory: According to this model, the atom is composed of protons and neutrons—which compose most of the atom's mass and are grouped together in a dense nucleus—and electrons, which compose most of the atom's volume. Protons and neutrons have similar masses (1 amu), while electrons have a much smaller mass (0.00055 amu).	**Discovery of the Atom's Nucleus:** We can understand why this is relevant by asking, what if it were otherwise? What if matter were *not* mostly empty space? While we cannot know for certain, it seems probable that such matter would not form the diversity of substances required for life—and then, of course, we would not be around to ask the question.
Charge: Protons and electrons both have electrical charge; the charge of the proton is 1+ and the charge of the electron is 1–. The neutron has no charge. When protons and electrons combine in atoms, their charges cancel.	**Charge:** Electrical charge is relevant to much of our modern world. Many of the machines and computers we depend on are powered by electricity, which is the movement of electrical charge.
The Periodic Table: The periodic table tabulates all known elements in order of increasing atomic number. The periodic table is arranged so that similar elements are grouped in columns. Columns of elements in the periodic table have similar properties and are called groups or families. Elements on the left side of the periodic table are metals and tend to lose electrons in their chemical changes. Elements on the upper right side of the periodic table are nonmetals and tend to gain electrons in their chemical changes. Elements between the two are metalloids.	**The Periodic Table:** The periodic table helps us organize the elements in ways that allow us to predict their properties. Helium, for example, is not toxic in small amounts because it is an inert gas—it does not react with anything. The gases in the column below it on the periodic table are also inert gases and form a group of elements called the noble gases. By tabulating the elements and grouping similar ones together, we begin to understand their properties.
Atomic Number: The characteristic that defines an element is the number of protons in the nuclei of its atoms; this number is the atomic number (Z).	**Atomic Number:** Elements are the fundamental building blocks from which all compounds are made.
Ions: When an atom gains or loses electrons, it becomes an ion. Positively charged ions are cations, and negatively charged ions are anions. Cations and anions occur together so that matter is ordinarily charge-neutral.	**Ions:** Ions occur in many common compounds, such as sodium chloride.
Isotopes: While all atoms of a given element have the same number of protons, they do not necessarily have the same number of neutrons. Atoms of the same element with different numbers of neutrons are isotopes. Isotopes are characterized by their mass number (A), the sum of the number of protons and the number of neutrons in the nucleus. Each naturally occurring sample of an element has the same percent natural abundance of each isotope. We can use these percentages, together with the mass of each isotope, to calculate the atomic mass of the element, a weighted average of the masses of the individual isotopes.	**Isotopes:** Isotopes are relevant because they influence atomic masses. To understand these masses, we must understand the presence and abundance of isotopes. In nuclear processes—processes in which the nuclei of atoms actually change—the presence of different isotopes becomes even more important. Some isotopes are not stable—they lose subatomic particles and transform into other elements. The emission of subatomic particles by unstable nuclei is called radioactive decay. In many situations, such as in diagnosing and treating certain diseases, nuclear radiation is extremely useful. In other situations, such as in the disposal of radioactive waste, it can pose environmental problems.

Arrangements of Electrons around the Nucleus: The quantum-mechanical model for the atom describes electron orbitals, which are electron probability maps that show the relative probability of finding an electron in various places surrounding the atomic nucleus. Orbitals are specified with a number (n), called the principal quantum number, and a letter. The principal quantum number ($n = 1, 2, 3 \ldots$) specifies the principal shell, and the letter (s, p, d, or f) specifies the subshell of the orbital. In the hydrogen atom, the energy of orbitals depends only on n. In multi-electron atoms, the energy ordering is $1s$ $2s$ $2p$ $3s$ $3p$ $4s$ $3d$ $4p$ $5s$ $4d$ $5p$ $6s$.

An electron configuration indicates which orbitals are occupied for a particular atom. Orbitals are filled in order of increasing energy and obey the Pauli exclusion principle (each orbital can hold a maximum of two electrons with opposing spins) and Hund's rule (electrons occupy orbitals of identical energy singly before pairing).

Arrangements of Electrons around the Nucleus: The quantum-mechanical model changed the way we view nature. Before the quantum-mechanical model, electrons were viewed as small particles, much like any other particle. Electrons were expected to follow the normal laws of motion, just as a baseball does. However, the electron, with its wavelike properties, does not follow these laws. Instead, electron motion is describable only through probabilistic predictions. Quantum theory single-handedly changed the predictability of nature at its most fundamental level.

The quantum-mechanical model of the atom predicts and explains many of the chemical properties we learned about in this module.

Period Trends: Elements within the same column of the periodic table have similar outer electron configurations and the same number of valence electrons (electrons in the outermost principal shell), and therefore similar chemical properties. We divide the periodic table into blocks (s block, p block, d block, and f block) in which particular sublevels are filled. As we move across a period to the right in the periodic table, atomic size decreases, ionization energy increases, and metallic character decreases. As we move down a column in the periodic table, atomic size increases, ionization energy decreases, and metallic character increases.

Period Trends: The periodic law exists because the number of valence electrons is periodic, and valence electrons determine chemical properties. Quantum theory also predicts that atoms with 8 outershell electrons (or 2 for helium) are particularly stable, thus explaining the inertness of the noble gases. Atoms without noble gas configurations undergo chemical reactions to attain them, explaining the reactivity of the alkali metals and the halogens as well as the tendency of several groups to form ions with certain charges.

Chemical Skills

LO: Determine atomic numbers, mass numbers, and isotope symbols for an isotope (Section 2.3).

- Refer to the periodic table or the alphabetical list of elements to find the atomic number of the element.
- The mass number (A) is equal to the atomic number plus the number of neutrons.
- Write the symbol for the isotope by writing the symbol for the element with the mass number in the upper left corner and the atomic number in the lower left corner.
- The other symbol for the isotope is simply the chemical symbol followed by a hyphen and the mass number.

LO: Determine number of protons and neutrons from isotope symbols (Section 2.3).

- The number of protons is equal to Z (lower left number).
- The number of neutrons is equal to

  A (upper left number) − Z (lower left number.)

Examples

EXAMPLE 2.17 DETERMINING ATOMIC NUMBERS, MASS NUMBERS, AND ISOTOPE SYMBOLS FOR AN ISOTOPE

What are the atomic number (Z), mass number (A), and symbols for the iron isotope with 30 neutrons?

SOLUTION

The atomic number of iron is 26.

$$A = 26 + 30 = 56$$

The mass number is 56.

$^{56}_{26}\text{Fe}$

Fe-56

EXAMPLE 2.18 DETERMINING NUMBER OF PROTONS AND NEUTRONS FROM ISOTOPE SYMBOLS

How many protons and neutrons are in $^{62}_{28}\text{Ni}$?

SOLUTION

28 protons

$$\#n = 62 - 28 = 34 \text{ neutrons}$$

LO: **Calculate atomic mass from percent natural abundances and isotopic masses (Section 2.4).**

- Convert the natural abundances from percent to decimal values by dividing by 100.
- Find the atomic mass by multiplying the fractions of each isotope by its respective mass and summing them.
- Check your work.

EXAMPLE 2.19 CALCULATING ATOMIC MASS FROM PERCENT NATURAL ABUNDANCES AND ISOTOPIC MASSES

Copper has two naturally occurring isotopes: Cu-63 with mass 62.9395 amu and a natural abundance of 69.17 %, and Cu-65 with mass 64.9278 amu and a natural abundance of 30.83 %. Calculate the atomic mass of copper.

SOLUTION

$$\text{Fraction Cu-63} = \frac{69.17}{100} = 0.6917$$

$$\text{Fraction Cu-65} = \frac{30.83}{100} = 0.3083$$

$$\begin{aligned}\text{Atomic mass} &= (0.6917 \times 62.9395\text{ amu}) + (0.3083 \times 64.9278)\\ &= 43.5353\text{ amu} + 20.0172\text{ amu}\\ &= 63.5525\text{ amu}\\ &= 63.55\text{ amu}\end{aligned}$$

LO: **Determine ion charge from numbers of protons and electrons (Section 2.5).**

- Refer to the periodic table or the alphabetical list of elements to find the atomic number of the element; this number is equal to the number of protons.
- Use the ion charge equation to calculate charge.

$$\text{Ion charge} = \#p^{+} - \#e^{-}$$

EXAMPLE 2.20 DETERMINING ION CHARGE FROM NUMBERS OF PROTONS AND ELECTRONS

Determine the charge of a selenium ion with 36 electrons.

SOLUTION

Selenium is atomic number 34; therefore, it has 34 protons.

$$\text{Ion charge} = 34 - 36 = 2-$$

LO: **Determine the number of protons and electrons in an ion (Section 2.5).**

- Refer to the periodic table or the alphabetical list of elements to find the atomic number of the element; this number is equal to the number of protons.
- Use the ion charge equation and substitute in the known values.

$$\text{Ion charge} = \#p^{+} - \#e^{-}$$

- Solve the equation for the number of electrons.

EXAMPLE 2.21 DETERMINING THE NUMBER OF PROTONS AND ELECTRONS IN AN ION

Find the number of protons and electrons in the O^{2-} ion.

SOLUTION

The atomic number of O is 8; therefore, it has 8 protons.

$$\begin{aligned}\text{Ion charge} &= \#p^{+} - \#e^{-}\\ 2- &= 8 - \#e^{-}\\ \#e^{-} &= 8 + 2 = 10\end{aligned}$$

The ion has 8 protons and 10 electrons.

LO: Write electron configurations and orbital diagrams for atoms (Section 2.6).

To write electron configurations, determine the number of electrons in the atom from the element's atomic number and then follow these rules:

- Electrons occupy orbitals so as to minimize the energy of the atom; therefore, lower-energy orbitals fill before higher-energy orbitals. Orbitals fill in the order: 1*s* 2*s* 2*p* 3*s* 3*p* 4*s* 3*d* 4*p* 5*s* 4*d* 5*p* 6*s* (Figure 2.15). The *s* subshells hold up to 2 electrons, *p* subshells hold up to 6, *d* subshells hold up to 10, and *f* subshells hold up to 14.
- Orbitals can hold no more than 2 electrons each. When 2 electrons occupy the same orbital, they must have opposing spins.
- When orbitals of identical energy are available, these are first occupied singly with parallel spins rather than in pairs.

EXAMPLE 2.22 WRITING ELECTRON CONFIGURATIONS AND ORBITAL DIAGRAMS

Write an electron configuration and orbital diagram (outer electrons only) for germanium.

SOLUTION

Germanium is atomic number 32; therefore, it has 32 electrons.

Electron Configuration

Ge $1s^22s^22p^63s^23p^64s^23d^{10}4p^2$

or

Ge $[Ar]4s^23d^{10}4p^2$

Orbital Diagram (Outer Electrons)

4*s* 3*d* 4*p*

LO: Identify valence electrons and core electrons (Section 2.8).

- Valence electrons are the electrons in the outermost principal energy shell (the principal shell with the highest principal quantum number).
- Core electrons are electrons that are not in the outermost principal shell.

EXAMPLE 2.23 IDENTIFYING VALENCE ELECTRONS AND CORE ELECTRONS

Identify the valence electrons and core electrons in the electron configuration of germanium (given in Example 2.13).

SOLUTION

Ge $1s^22s^22p^63s^23p^6\,4s^2\,3d^{10}\,4p^2$

28 core electrons 4 valence electrons

LO: Write electron configurations for elements based on their positions in the periodic table (Section 2.8).

- The inner electron configuration for any element is the electron configuration of the noble gas that immediately precedes that element in the periodic table. Represent the inner configuration with the symbol for the noble gas in brackets.
- The outer electrons can be determined from the element's position within a particular block (*s*, *p*, *d*, or *f*) in the periodic table. Trace the elements between the preceding noble gas and the element of interest and assign electrons to the appropriate orbitals. Figure 2.21 shows the outer electron configuration based on the position of an element in the periodic table.
- The highest principal quantum number (highest *n* value) is equal to the row number of the element in the periodic table.
- The principal quantum number (*n* value) of the outermost *d* electrons for any element containing *d* electrons is equal to the row number of the element minus 1.

EXAMPLE 2.24 WRITING AN ELECTRON CONFIGURATION FOR AN ELEMENT BASED ON ITS POSITION IN THE PERIODIC TABLE

Write an electron configuration for iodine based on its position in the periodic table.

SOLUTION

The inner configuration for I is [Kr].

Begin with the [Kr] inner electron configuration. As you trace from Kr to I, add two 5*s* electrons, ten 4*d* electrons, and five 5*p* electrons. The overall configuration is:

I $[Kr]5s^24d^{10}5p^5$

LO: Identify and understand periodic trends such as atomic size, ionization energy, and metallic character (Section 2.10).

On the periodic table:

- Atomic size decreases as you move to the right and increases as you move down.
- Ionization energy increases as you move to the right and decreases as you move down.
- Metallic character decreases as you move to the right and increases as you move down.

EXAMPLE 2.25 PERIODIC TRENDS: ATOMIC SIZE, IONIZATION ENERGY, AND METALLIC CHARACTER

Arrange Si, In, and S in order of **(a)** increasing atomic size, **(b)** increasing ionization energy, and **(c)** increasing metallic character.

SOLUTION

(a) S, Si, In
(b) In, Si, S
(c) S, Si, In

KEY TERMS

alkali metals **[2.7]**
alkaline earth metals **[2.7]**
anions **[2.5]**
atomic mass **[2.4]**
atomic mass unit (amu) **[2.1]**
atomic number (Z) **[2.2]**
atomic size **[2.10]**
cation **[2.5]**
charge **[2.1]**
chemical symbol **[2.2]**
core electrons **[2.6]**
electron **[2.1]**
electron configuration **[2.6]**
electron spin **[2.6]**
excited state **[2.6]**
ground state **[2.6]**
group (of elements) **[2.7]**
halogens **[2.7]**
Hund's rule **[2.6]**
ion **[2.5]**
ionization energy **[2.10]**
isotope **[2.3]**
main-group elements **[2.7]**
mass number (A) **[2.3]**
metallic character **[2.10]**
metalloids **[2.7]**
metals **[2.7]**
neutron **[2.1]**
noble gases **[2.7]**
nonmetals **[2.7]**
nuclear radiation **[2.4]**
nuclear theory of the atom **[2.1]**
nucleus **[2.1]**
orbital **[2.6]**
orbital diagram **[2.6]**
Pauli exclusion principle **[2.6]**
percent natural abundance **[2.3]**
periodic law **[2.7]**
periodic table **[2.7]**
principal quantum number **[2.6]**
principal shell **[2.6]**
radioactive **[2.4]**
semiconductor **[2.7]**
subshell **[2.6]**
transition elements **[2.7]**
transition metals **[2.7]**
valence electron **[2.6]**

EXERCISES

QUESTIONS

1. What are the main ideas in the nuclear theory of the atom?
2. List the three subatomic particles and their properties.
3. What is electrical charge?
4. Is matter usually charge-neutral? How would matter be different if it were not charge-neutral?
5. What does the atomic number of an element specify?
6. What is a chemical symbol?
7. List some examples of how elements were named.
8. What are isotopes?
9. What is the mass number of an isotope?
10. What notations are commonly used to specify isotopes? What do each of the numbers in these symbols mean?
11. What is the percent natural abundance of isotopes?
12. What is the atomic mass of an element?
13. What is an ion?
14. What is an anion? What is a cation?
15. What is the difference between the ground state of an atom and an excited state of an atom?
16. Explain how the motion of an electron is described by a probability map
17. Why do quantum-mechanical orbitals have "fuzzy" boundaries?
18. List the four possible subshells in the quantum-mechanical model, the number of orbitals in each subshell, and the maximum number of electrons that can be contained in each subshell.
19. List the quantum-mechanical orbitals through 5*s*, in the correct energy order for multi-electron atoms.
20. What is the Pauli exclusion principle? Why is it important when writing electron configurations?
21. What is Hund's rule? Why is it important when writing orbital diagrams?
22. Within an electron configuration, what do symbols such as [Ne] and [Kr] represent?
23. Explain the difference between valence electrons and core electrons.

24. What was Dmitri Mendeleev's main contribution to our modern understanding of chemistry?

25. What is the main idea in the periodic law?

26. How is the periodic table organized?

27. What are the properties of metals? Where are metals found on the periodic table?

28. What are the properties of nonmetals? Where are nonmetals found on the periodic table?

29. Where on the periodic table are metalloids found?

30. What is a group of elements?

31. Locate each group of elements on the periodic table and list its group number.
- **(a)** alkali metals
- **(b)** alkaline earth metals
- **(c)** halogens
- **(d)** noble gases

32. Identify each block in the blank periodic table.
- **(a)** *s* block
- **(b)** *p* block
- **(c)** *d* block
- **(d)** *f* block

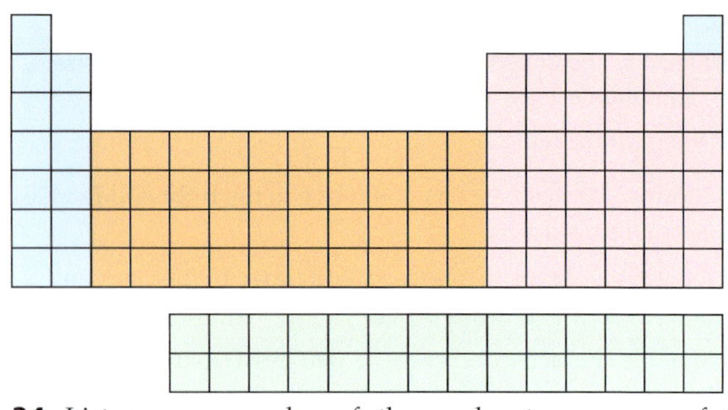

33. Locate each group on the periodic table and list the charge of the ions it tends to form.
- **(a)** Group 1A
- **(b)** Group 2A
- **(c)** Group 3A
- **(d)** Group 6A
- **(e)** Group 7A

34. List some examples of the explanatory power of the quantum-mechanical model.

35. Explain why Group 1 elements tend to form 1+ ions and Group 7 elements tend to form 1– ions.

36. Explain the periodic trends in each chemical property.
- **(a)** ionization energy
- **(b)** atomic size
- **(c)** metallic character

PROBLEMS

NUCLEAR THEORY OF THE ATOM

37. If atoms are mostly empty space and atoms compose all ordinary matter, why does solid matter seem to have no space within it?

38. Which statements about electrons are true?
- **(a)** Electrons repel each other.
- **(b)** Electrons are attracted to protons.
- **(c)** Some electrons have a charge of 1– and some have no charge.
- **(d)** Electrons are much lighter than neutrons.

39. Which statements about electrons are false?
- **(a)** Most atoms have more electrons than protons.
- **(b)** Electrons have a charge of 1–.
- **(c)** If an atom has an equal number of protons and electrons, it will be charge-neutral.
- **(d)** Electrons experience an attraction to protons.

40. Which statements about protons are true?
- **(a)** Protons have twice the mass of neutrons.
- **(b)** Protons have the same magnitude of charge as electrons but are opposite in sign.
- **(c)** Most atoms have more protons than electrons.
- **(d)** Protons have a charge of 1+.

41. Which statements about protons are false?
- **(a)** Protons have about the same mass as neutrons.
- **(b)** Protons have about the same mass as electrons.
- **(c)** Some atoms don't have any protons.
- **(d)** Protons have the same magnitude of charge as neutrons but are opposite in sign.

ELEMENTS, SYMBOLS, AND NAMES

42. Find the atomic number (Z) for each element.
- **(a)** Fr
- **(b)** Kr
- **(c)** Pa
- **(d)** Ge
- **(e)** Al

43. Find the atomic number (Z) for each element.
- **(a)** Si
- **(b)** W
- **(c)** Ni
- **(d)** Rn
- **(e)** Sr

44. How many protons are in the nucleus of an atom of each element?
- **(a)** Ar
- **(b)** Sn
- **(c)** Xe
- **(d)** O
- **(e)** Tl

45. How many protons are in the nucleus of an atom of each element?
- **(a)** Ti
- **(b)** Li
- **(c)** U
- **(d)** Br
- **(e)** F

46. List the symbol and atomic number of each element.
- **(a)** carbon
- **(b)** nitrogen
- **(c)** sodium
- **(d)** potassium
- **(e)** copper

47. List the symbol and atomic number of each element.
- **(a)** boron
- **(b)** neon
- **(c)** silver
- **(d)** mercury
- **(e)** curium

48. List the name and the atomic number of each element.
- **(a)** Mn
- **(b)** Ag
- **(c)** Au
- **(d)** Pb
- **(e)** S

49. List the name and the atomic number of each element.
- **(a)** Y
- **(b)** N
- **(c)** Ne
- **(d)** K
- **(e)** Mo

50. Fill in the blanks to complete the table.

Element Name	Element Symbol	Atomic Number
_____	Au	79
Tin	_____	_____
_____	As	_____
Copper	_____	29
_____	Fe	_____
_____	_____	80

51. Fill in the blanks to complete the table.

Element Name	Element Symbol	Atomic Number
_____	Al	13
Iodine	_____	_____
_____	Sb	_____
Sodium	_____	_____
_____	Rn	86
_____	_____	82

ISOTOPES

52. Determine the atomic number and mass number for each isotope.
- **(a)** the hydrogen isotope with 2 neutrons
- **(b)** the chromium isotope with 28 neutrons
- **(c)** the calcium isotope with 22 neutrons
- **(d)** the tantalum isotope with 109 neutrons

53. How many neutrons are in an atom with each atomic number and mass number?
- **(a)** Z = 28, A = 59
- **(b)** Z = 92, A = 235
- **(c)** Z = 21, A = 46
- **(d)** Z = 18, A = 42

54. Write isotopic symbols in the form $^{A}_{Z}X$ for each isotope.
- **(a)** the oxygen isotope with 8 neutrons
- **(b)** the fluorine isotope with 10 neutrons
- **(c)** the sodium isotope with 12 neutrons
- **(d)** the aluminum isotope with 14 neutrons

55. Write isotopic symbols in the form X-A (for example, C-13) for each isotope.
- **(a)** the iodine isotope with 74 neutrons
- **(b)** the phosphorus isotope with 16 neutrons
- **(c)** the uranium isotope with 142 neutrons
- **(d)** the argon isotope with 22 neutrons

56. Write the symbol for each isotope in the form ${}^{A}_{Z}X$.
(a) cobalt-60
(b) neon-22
(c) iodine-131
(d) plutonium-244

57. Write the symbol for each isotope in the form ${}^{A}_{Z}X$.
(a) U-235
(b) V-52
(c) P-32
(d) Xe-144

58. Determine the number of protons and neutrons in each isotope.
(a) ${}^{23}_{11}Na$
(b) ${}^{266}_{88}Ra$
(c) ${}^{208}_{82}Pb$
(d) ${}^{14}_{7}N$

59. Determine the number of protons and neutrons in each isotope.
(a) ${}^{33}_{15}P$
(b) ${}^{40}_{19}K$
(c) ${}^{222}_{86}Rn$
(d) ${}^{99}_{43}Tc$

60. Carbon-14, present within living organisms and substances derived from living organisms, is often used to establish the age of fossils and artifacts. Determine the number of protons and neutrons in a carbon-14 isotope and write its symbol in the form ${}^{A}_{Z}X$.

61. Plutonium-239 is used in nuclear bombs. Determine the number of protons and neutrons in plutonium-239 and write its symbol in the form ${}^{A}_{Z}X$.

ATOMIC MASS

62. Rubidium has two naturally occurring isotopes: Rb-85 with mass 84.9118 amu and a natural abundance of 72.17 %, and Rb-87 with mass 86.9092 amu and a natural abundance of 27.83 %. Calculate the atomic mass of rubidium.

63. Silicon has three naturally occurring isotopes: Si-28 with mass 27.9769 amu and a natural abundance of 92.21 %, Si-29 with mass 28.9765 amu and a natural abundance of 4.69 %, and Si-30 with mass 29.9737 amu and a natural abundance of 3.10 %. Calculate the atomic mass of silicon.

IONS

64. Complete each ionization equation.
(a) $Na \rightarrow Na^{+} + \_\_\_\_$
(b) $O + 2e^{-} \rightarrow \_\_\_\_$
(c) $Ca \rightarrow Ca^{2+} + \_\_\_\_$
(d) $Cl + e^{-} \rightarrow \_\_\_\_$

65. Complete each ionization equation.
(a) $Mg \rightarrow \_\_\_\_ + 2e^{-}$
(b) $Ba \rightarrow Ba^{2+} + \_\_\_\_$
(c) $I + e^{-} \rightarrow \_\_\_\_$
(d) $Al \rightarrow \_\_\_\_ + 3e^{-}$

66. Determine the charge of each ion.
(a) oxygen ion with 10 electrons
(b) aluminum ion with 10 electrons
(c) titanium ion with 18 electrons
(d) iodine ion with 54 electrons

67. Determine the charge of each ion.
(a) tungsten ion with 68 electrons
(b) tellurium ion with 54 electrons
(c) nitrogen ion with 10 electrons
(d) barium ion with 54 electrons

68. Determine the number of protons and electrons in each ion.
(a) Na^{+}
(b) Ba^{2+}
(c) O^{2-}
(d) Co^{3+}

69. Determine the number of protons and electrons in each ion.
(a) Al^{3+}
(b) S^{2-}
(c) I^{-}
(d) Ag^{+}

70. Determine whether each statement is true or false. If false, correct it.
(a) The Ti^{2+} ion contains 22 protons and 24 electrons.
(b) The I^{-} ion contains 53 protons and 54 electrons.
(c) The Mg^{2+} ion contains 14 protons and 12 electrons.
(d) The O^{2-} ion contains 8 protons and 10 electrons.

71. Determine whether each statement is true or false. If false, correct it.
(a) The Fe^{2+} ion contains 29 protons and 26 electrons.
(b) The Cs^{+} ion contains 55 protons and 56 electrons.
(c) The Se^{2-} ion contains 32 protons and 34 electrons.
(d) The Li^{+} ion contains 3 protons and 2 electrons.

ARRANGEMENTS OF ELECTRONS AROUND THE NUCLEUS

72. Sketch the 1*s* and 2*p* orbitals. How do the 2*s* and 3*p* orbitals differ from the 1*s* and 2*p* orbitals?

73. Sketch the 3*d* orbitals. How do the 4*d* orbitals differ from the 3*d* orbitals?

74. Which electron is, on average, closer to the nucleus: an electron in a 2*s* orbital or an electron in a 3*s* orbital?

75. Which electron is, on average, farther from the nucleus: an electron in a 3*p* orbital or an electron in a 4*p* orbital?

76. Write full electron configurations for each element.
(a) Sr
(b) Ge
(c) Li
(d) Kr

77. Write full electron configurations for each element.
(a) N
(b) Mg
(c) Ar
(d) Se

78. Write full orbital diagrams and indicate the number of unpaired electrons for each element.
(a) He
(b) B
(c) Li
(d) N

79. Write full orbital diagrams and indicate the number of unpaired electrons for each element.
(a) F
(b) C
(c) Ne
(d) Be

80. Write electron configurations for each element. Use the symbol of the previous noble gas in brackets to represent the core electrons.
(a) Ga
(b) As
(c) Rb
(d) Sn

81. Write electron configurations for each element. Use the symbol of the previous noble gas in brackets to represent the core electrons.
(a) Te
(b) Br
(c) I
(d) Cs

THE PERIODIC TABLE

82. Classify each element as a metal, nonmetal, or metalloid.
(a) Sr
(b) Mg
(c) F
(d) N
(e) As

83. Classify each element as a metal, nonmetal, or metalloid.
(a) Na
(b) Ge
(c) Si
(d) Br
(e) Ag

84. Which elements would you expect to lose electrons in chemical changes?
(a) potassium
(b) sulfur
(c) fluorine
(d) barium
(e) copper

85. Which elements would you expect to gain electrons in chemical changes?
(a) nitrogen
(b) iodine
(c) tungsten
(d) strontium
(e) gold

86. Which elements are main-group elements?
(a) Te
(b) K
(c) V
(d) Re
(e) Ag

87. Which elements are *not* main-group elements?
(a) Al
(b) Br
(c) Mo
(d) Cs
(e) Pb

88. Which elements are alkaline earth metals?
(a) sodium
(b) aluminum
(c) calcium
(d) barium
(e) lithium

89. Which elements are alkaline earth metals?
(a) rubidium
(b) tungsten
(c) magnesium
(d) cesium
(e) beryllium

90. Which elements are alkali metals?
(a) barium
(b) sodium
(c) gold
(d) tin
(e) rubidium

91. Which elements are alkali metals?
(a) scandium
(b) iron
(c) potassium
(d) lithium
(e) cobalt

92. Classify each element as a halogen, a noble gas, or neither.
(a) Cl
(b) Kr
(c) F
(d) Ga
(e) He

93. Classify each element as a halogen, a noble gas, or neither.
(a) Ne
(b) Br
(c) S
(d) Xe
(e) I

94. To what group number does each element belong?
(a) oxygen
(b) aluminum
(c) silicon
(d) tin
(e) phosphorus

95. To what group number does each element belong?
(a) germanium
(b) nitrogen
(c) sulfur
(d) carbon
(e) boron

96. Which element do you expect to be most like sulfur? Why?
(a) nitrogen
(b) oxygen
(c) fluorine
(d) lithium
(e) potassium

97. Which element do you expect to be most like magnesium? Why?
(a) potassium
(b) silver
(c) bromine
(d) calcium
(e) lead

98. Which pair of elements do you expect to be most similar? Why?
(a) Si and P
(b) Cl and F
(c) Na and Mg
(d) Mo and Sn
(e) N and Ni

99. Which pair of elements do you expect to be most similar? Why?
(a) Ti and Ga
(b) N and O
(c) Li and Na
(d) Ar and Br
(e) Ge and Ga

100. Which element is a main-group nonmetal?
(a) K
(b) Fe
(c) Sn
(d) S

101. Which element is a row 5 transition element?
(a) Sr
(b) Pd
(c) P
(d) V

102. Fill in the blanks to complete the table.

Chemical Symbol	Group Number	Group Name	Metal or Nonmetal
K	_____	_____	metal
Br	_____	halogens	_____
Sr	_____	_____	_____
He	8A	_____	_____
Ar	_____	_____	_____

103. Fill in the blanks to complete the table.

Chemical Symbol	Group Number	Group Name	Metal or Nonmetal
Cl	7A	_____	_____
Ca	_____	_____	metal
Xe	_____	_____	nonmetal
Na	_____	alkali metal	_____
F	_____	_____	_____

ELECTRON CONFIGURATIONS AND THE PERIODIC TABLE

104. Write full electron configurations and indicate the valence electrons and the core electrons for each element.
(a) Kr
(b) Ge
(c) Cl
(d) Sr

105. Write full electron configurations and indicate the valence electrons and the core electrons for each element.
(a) Sb
(b) N
(c) B
(d) K

106. Write orbital diagrams for the valence electrons and indicate the number of unpaired electrons for each element.
(a) Br
(b) Kr
(c) Na
(d) In

107. Write orbital diagrams for the valence electrons and indicate the number of unpaired electrons for each element.
(a) Ne
(b) I
(c) Sr
(d) Ge

108. How many valence electrons are in each element?
(a) O
(b) S
(c) Br
(d) Rb

109. How many valence electrons are in each element?
(a) Ba
(b) Al
(c) Be
(d) Se

110. List the outer electron configuration for each column in the periodic table.
(a) 1A
(b) 2A
(c) 5A
(d) 7A

111. List the outer electron configuration for each column in the periodic table.
(a) 3A
(b) 4A
(c) 6A
(d) 8A

112. Use the periodic table to write electron configurations for each element.
(a) Al
(b) Be
(c) In
(d) Zr

113. Use the periodic table to write electron configurations for each element.
(a) Tl
(b) Co
(c) Ba
(d) Sb

114. Use the periodic table to write electron configurations for each element.
(a) Sr
(b) Y
(c) Ti
(d) Te

115. Use the periodic table to write electron configurations for each element.
(a) Se
(b) Sn
(c) Pb
(d) Cd

116. How many 2*p* electrons are in an atom of each element?
(a) C
(b) N
(c) F
(d) P

117. How many 3*d* electrons are in an atom of each element?
(a) Fe
(b) Zn
(c) K
(d) As

118. List the number of elements in periods 1 and 2 of the periodic table. Why does each period have a different number of elements?

119. List the number of elements in periods 3 and 4 of the periodic table. Why does each period have a different number of elements?

120. Name the element in the third period (row) of the periodic table with:
(a) 3 valence electrons
(b) a total of four 3*p* electrons
(c) six 3*p* electrons
(d) two 3*s* electrons and no 3*p* electrons

121. Name the element in the fourth period of the periodic table with:
(a) 5 valence electrons
(b) a total of four 4*p* electrons
(c) a total of three 3*d* electrons
(d) a complete outer shell

122. Use the periodic table to identify the element with each electron configuration.
(a) $[Ne]3s^2 3p^5$
(b) $[Ar]4s^2 3d^{10} 4p^1$
(c) $[Ar]4s^2 3d^6$
(d) $[Kr]5s^1$

123. Use the periodic table to identify the element with each electron configuration.
(a) $[Ne]3s^1$
(b) $[Kr]5s^2 4d^{10}$
(c) $[Xe]6s^2$
(d) $[Kr]5s^2 4d^{10} 5p^3$

124. Write electron configurations for each transition metal.
(a) Zn
(b) Cu
(c) Zr
(d) Fe

125. Write electron configurations for each transition metal.
(a) Mn
(b) Ti
(c) Cd
(d) V

126. Predict the ion formed by each element.
(a) Rb
(b) K
(c) Al
(d) O

127. Predict the ion formed by each element.
(a) F
(b) N
(c) Mg
(d) Na

128. Predict how many electrons each element will most likely gain or lose.
(a) Ga
(b) Li
(c) Br
(d) S

129. Predict how many electrons each element will most likely gain or lose.
(a) I
(b) Ba
(c) Cs
(d) Se

130. Fill in the blanks to complete the table.

Symbol	Ion Commonly Formed	Number of Electrons in Ion	Number of Protons in Ion
Te	_____	54	_____
In	_____	_____	49
Sr	Sr^{2+}	_____	_____
_____	Mg^{2+}	_____	12
Cl	_____	_____	_____

131. Fill in the blanks to complete the table.

Symbol	Ion Commonly Formed	Number of Electrons in Ion	Number of Protons in Ion
F	_____	_____	9
_____	Be^{2+}	2	_____
Br	_____	36	_____
Al	_____	_____	13
O	_____	_____	_____

PERIODIC TRENDS

132. Choose the element with the higher ionization energy from each pair.
(a) As or Bi
(b) As or Br
(c) S or I
(d) S or Sb

133. Choose the element with the higher ionization energy from each pair.
(a) Al or In
(b) Cl or Sb
(c) K or Ge
(d) S or Se

134. Arrange the elements in order of increasing ionization energy: Te, Pb, Cl, S, Sn.

135. Arrange the elements in order of increasing ionization energy: Ga, In, F, Si, N.

136. Choose the element with the larger atoms from each pair.
(a) Al or In
(b) Si or N
(c) P or Pb
(d) C or F

137. Choose the element with the larger atoms from each pair.
(a) Sn or Si
(b) Br or Ga
(c) Sn or Bi
(d) Se or Sn

138. Arrange these elements in order of increasing atomic size: Ca, Rb, S, Si, Ge, F.

139. Arrange these elements in order of increasing atomic size: Cs, Sb, S, Pb, Se.

140. Choose the more metallic element from each pair.
(a) Sr or Sb
(b) As or Bi
(c) Cl or O
(d) S or As

141. Choose the more metallic element from each pair.
(a) Sb or Pb
(b) K or Ge
(c) Ge or Sb
(d) As or Sn

142. Arrange these elements in order of increasing metallic character: Fr, Sb, In, S, Ba, Se.

143. Arrange these elements in order of increasing metallic character: Sr, N, Si, P, Ga, Al.

CUMULATIVE PROBLEMS

144. Prepare a table like Table 2.2 for the four different isotopes of Sr that have the natural abundances and masses listed here.

Sr-84	0.56 %	83.9134 amu
Sr-86	9.86 %	85.9093 amu
Sr-87	7.00 %	86.9089 amu
Sr-88	82.58 %	87.9056 amu

Use your table and the listed atomic masses to calculate the atomic mass of strontium.

145. Determine the number of protons and neutrons in each isotope of chromium and use the listed natural abundances and masses to calculate its atomic mass.

Cr-50	4.345 %	49.9460 amu
Cr-52	83.79 %	51.9405 amu
Cr-53	9.50 %	52.9407 amu
Cr-54	2.365 %	53.9389 amu

146. The atomic mass of fluorine is 19.00 amu, and all fluorine atoms in a naturally occurring sample of fluorine have this mass. The atomic mass of chlorine is 35.45 amu, but no chlorine atoms in a naturally occurring sample of chlorine have this mass. Provide an explanation for the difference.

147. The atomic mass of germanium is 72.61 amu. Is it likely that any individual germanium atoms have a mass of 72.61 amu?

148. Fill in the blanks to complete the table.

Symbol	Z	A	Number of Protons	Number of Electrons	Number of Neutrons	Charge
Zn^{2+}	____	____	____	____	34	2+
____	25	55	____	22	____	____
____	____	____	15	15	16	____
O^{2-}	____	16	____	____	____	2–
____	____	____	16	18	18	____

149. Fill in the blanks to complete the table.

Symbol	Z	A	Number of Protons	Number of Electrons	Number of Neutrons	Charge
Mg^{2+}	____	25	____	____	13	2+
____	22	48	____	18	____	____
____	16	____	____	____	16	2–
Ga^{2+}	____	71	____	____	____	____
____	____	____	82	80	125	____

150. What is the maximum number of electrons that can occupy the $n = 3$ quantum shell?

151. Use the electron configurations of the alkaline earth metals to explain why they tend to form 2+ ions.

152. Use the electron configuration of oxygen to explain why it tends to form a 2– ion.

153. Write the electron configuration for each ion. What do all of the electron configurations have in common?
(a) Ca^{2+}
(b) K^{+}
(c) S^{2-}
(d) Br^{-}

154. Write the electron configuration for each ion. What do all of the electron configurations have in common?
(a) F^{-}
(b) P^{3-}
(c) Li^{+}
(d) Al^{3+}

155. Examine Figure 2.18, which shows the division of the periodic table into metals, nonmetals, and metalloids. Use what you know about electron configurations to explain these divisions.

156. Examine Figure 2.27, which shows the elements that form predictable ions. Use what you know about electron configurations to explain these trends.

157. Identify what is wrong with each electron configuration and write the correct ground state (or lowest energy) configuration based on the number of electrons.
(a) $1s^3 2s^3 2p^9$
(b) $1s^2 2s^2 2p^6 2d^4$
(c) $1s^2 1p^5$
(d) $1s^2 2s^2 2p^8 3s^2 3p^1$

158. Identify what is wrong with each electron configuration and write the correct ground state (or lowest energy) configuration based on the number of electrons.
(a) $1s^4 2s^4 2p^{12}$
(b) $1s^2 2s^2 2p^6 3s^2 3p^6 3d^{10}$
(c) $1s^2 2p^6 3s^2$
(d) $1s^2 2s^2 2p^6 3s^2 3p^6 4s^2 4d^{10} 4p^3$

159. Bromine is a highly reactive liquid, while krypton is an inert gas. Explain this difference based on their electron configurations.

160. Potassium is a highly reactive metal, while argon is an inert gas. Explain this difference based on their electron configurations.

HIGHLIGHT PROBLEMS

161. The figure shown here is a representation of 50 atoms of a fictitious element with the symbol Nt and atomic number 120. Nt has three isotopes represented by the following colors: Nt-304 (red), Nt-305 (blue), and Nt-306 (green).

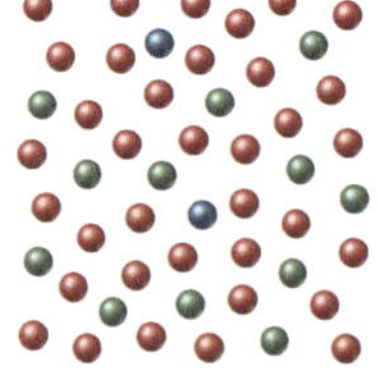

(a) Assuming that the figure is statistically representative of naturally occurring Nt, what is the percent natural abundance of each Nt isotope?

(b) Use the listed masses of each isotope to calculate the atomic mass of Nt. Then draw a box for the element similar to the boxes for each element shown in the periodic table in the inside front cover of this book. Make sure your box includes the atomic number, symbol, and atomic mass.

Nt-304	303.956 amu
Nt-305	304.962 amu
Nt-306	305.978 amu

Answers to Skillbuilder Exercises

Skillbuilder 2.1
- **(a)** sodium, 11
- **(b)** nickel, 28
- **(c)** phosphorus, 15
- **(d)** tantalum, 73

Skillbuilder 2.2 Z = 17, A = 35, Cl-35, and $^{35}_{17}Cl$

Skillbuilder 2.3 19 protons, 20 neutrons

Skillbuilder 2.4 24.31 amu

Skillbuilder 2.5
- **(a)** 2+
- **(b)** 1–
- **(c)** 3–

Skillbuilder 2.6 16 protons, 18 electrons

Skillbuilder 2.7 K^+ and Se^{2-}

Skillbuilder 2.8
- **(a)** Al $1s^22s^22p^63s^23p^1$ or $[Ne]3s^23p^1$
- **(b)** Br $1s^22s^22p^63s^23p^64s^23d^{10}4p^5$ or $[Ar]4s^23d^{10}4p^5$
- **(c)** Sr $1s^22s^22p^63s^23p^64s^23d^{10}4p^65s^2$ or $[Kr]5s^2$

Skillbuilder Plus, p. 37

Subtract 1 electron for each unit of positive charge. Add 1 electron for each unit of negative charge.
- **(a)** Al^{3+} $1s^22s^22p^6$
- **(b)** Cl^- $1s^22s^22p^63s^23p^6$
- **(c)** O^{2-} $1s^22s^22p^6$

Skillbuilder 2.9

Ar

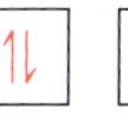

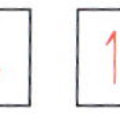

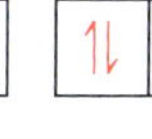
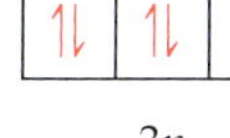

Skillbuilder 2.10
- **(a)** nonmetal
- **(b)** nonmetal
- **(c)** metal
- **(d)** metalloid

Skillbuilder 2.11 **(a)** alkali metal, Group 1A
- **(b)** Group 3A
- **(c)** halogen, Group 7A
- **(d)** noble gas, Group 8A

Skillbuilder 2.12

Cl

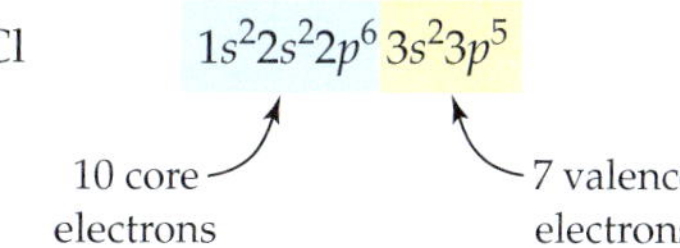

Skillbuilder 2.13 $[Kr]5s^24d^{10}5p^2$

Skillbuilder 2.14
- **(a)** Pb
- **(b)** Rb
- **(c)** cannot determine based on periodic properties
- **(d)** Se

Skillbuilder 2.15
- **(a)** Mg
- **(b)** Te
- **(c)** cannot determine based on periodic properties
- **(d)** F

Skillbuilder 2.16
- **(a)** In
- **(b)** cannot determine based on periodic properties
- **(c)** Bi
- **(d)** B

▸Answers to Conceptual Checkpoints

2.1 (c) The mass in amu is approximately equal to the number of protons plus the number of neutrons. In order to be charge-neutral, the number of protons must equal the number of electrons.

2.2

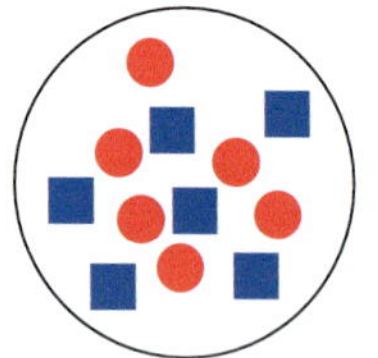

C-12 nucleus

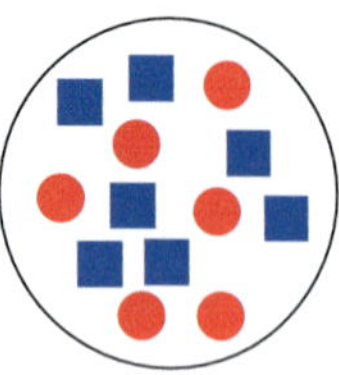

C-13 nucleus

2.3 (b) This atom must have (27 − 14) = 13 protons; the element with an atomic number of 13 is Al.

2.4 The isotopes C-12 and C-13 would not look different in this representation of atoms because the only difference between the two isotopes is that C-13 has an extra neutron in the nucleus. The illustration represents the whole atom and does not attempt to illustrate its nucleus. Because the nucleus of an atom is miniscule compared to the size of the atom itself, the extra neutron would not affect the size of the atom.

2.5 (b) The natural abundance of isotope B must be greater than the natural abundance of isotope A because the atomic mass is closer to the mass of isotope B than to the mass of isotope A.

2.6 (a) Both of these ions have 10 electrons.

2.7 (d) Both have 6 electrons in $2p$ orbitals and 6 electrons in $3p$ orbitals.

2.8 (d) Lead is a metal (see Figure 2.18) and a main-group element (see Figure 2.19).

2.9 (b) All of the metalloids are main-group elements (see Figures 2.18 and 2.19).

2.10 (d) The outermost principal shell for K is $n = 4$ which contains only a single valence electron, $4s^1$.

2.11 (a) Calcium loses its $4s$ electron and attains a noble gas configuration (that of Ar).

2.12 (b) Ionization energy increases as you move from left to right across a row in the periodic table. Both atomic size and metallic character decrease as you move from left to right across a row.

▲ Ordinary table sugar is a compound called sucrose. A sucrose molecule, such as the one shown here, contains carbon, hydrogen, and oxygen atoms. The properties of sucrose are, however, very different from those of carbon (also shown in the form of graphite), hydrogen, and oxygen. The properties of a compound are, in general, different from the properties of the elements that compose it.

Compounds 3

"Almost all aspects of life are engineered at the molecular level, and without understanding molecules, we can only have a very sketchy understanding of life itself."

—Francis Harry Compton Crick (1916–2004)

MODULE OUTLINE

3.1 Sugar and Salt

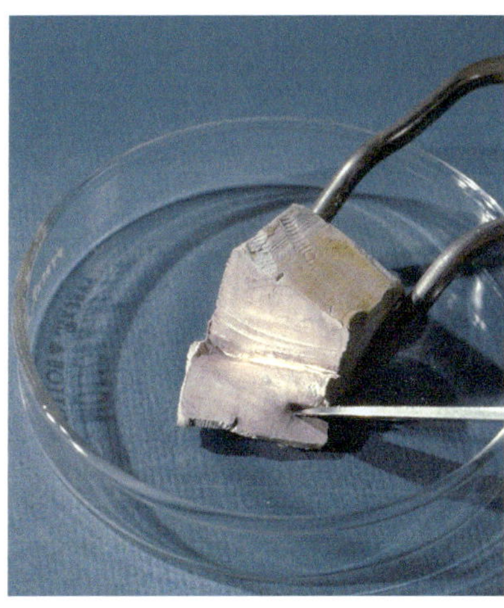

▲ **FIGURE 3.1 Elemental sodium** Sodium is an extremely reactive metal that dulls almost instantly upon exposure to air. © Tom Bochsler/Pearson Education/PH College.

Sodium, a shiny metal (◄ Figure 3.1) that dulls almost instantly upon exposure to air, is extremely reactive and poisonous. If you were to consume any appreciable amount of elemental sodium, you would need immediate medical help. Chlorine, a pale yellow gas (▼ Figure 3.2), is equally reactive and poisonous. Yet the compound formed from these two elements, sodium chloride, is the relatively harmless flavor enhancer that we call table salt (▼ Figure 3.3). When elements combine to form compounds, their properties completely change.

Consider also ordinary sugar. Sugar is a compound composed of carbon, hydrogen, and oxygen. Each of these elements has its own unique properties. Carbon is most familiar to us as the graphite found in pencils or as the diamonds in jewelry. Hydrogen is an extremely flammable gas used as a fuel for rocket engines, and oxygen is one of the gases that compose air. When these three elements combine to form sugar, however, a sweet, white, crystalline solid results.

► **FIGURE 3.2 Elemental chlorine** Chlorine is a yellow gas with a pungent odor. It is highly reactive and poisonous. © Charles D. Winters/Science Source.

▲ **FIGURE 3.3 Sodium chloride** The compound formed by sodium and chlorine is table salt. © Diane Diederich/diane39/iStockphoto/Getty Images.

In Module 2, you learned how protons, neutrons, and electrons combine to form different elements, each with its own properties and its own chemistry, each different from the other. In this module, you learn how these elements combine with each other to form different compounds, each with its own properties and its own chemistry, each different from all the others and different from the elements that compose it. This is the great wonder of nature: how from such simplicity—protons, neutrons, and electrons—we get such great complexity. It is exactly this complexity that makes life possible. Life could not exist with just 91 different elements if they did not combine to form compounds. It takes compounds in all of their diversity to make living organisms.

3.2 Compounds Display Constant Composition

LO: Restate and apply the law of constant composition.

Although some of the substances you encounter in everyday life are elements, most are not—they are compounds. Free atoms are rare in nature. As you learned in Module 1, a compound is different from a mixture of elements. In a compound, the elements combine in fixed, definite proportions, whereas in a mixture, they can have any proportions whatsoever. Consider the difference between a mixture of hydrogen and oxygen gas (▼ Figure 3.4) and the compound water (▼ Figure 3.5). A mixture of hydrogen and oxygen gas can contain any proportions of hydrogen and oxygen. Water, on the other hand, is composed of water molecules that consist of two hydrogen atoms bonded to one oxygen atom. Consequently, water has a definite proportion of hydrogen to oxygen.

The first chemist to formally state the idea that elements combine in fixed proportions to form compounds was Joseph Proust (1754–1826) in the **law of constant composition**, which states:

> All samples of a given compound have the same proportions of their constituent elements.

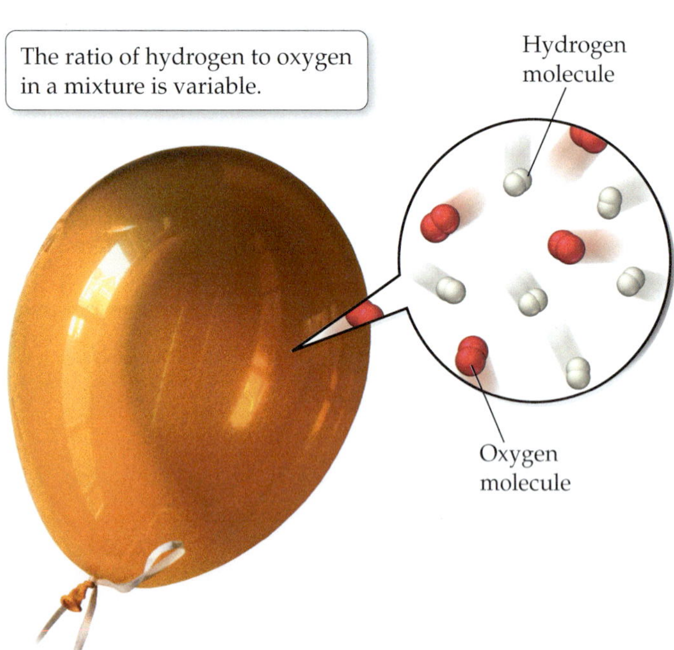

▲ **FIGURE 3.4 A mixture** This balloon is filled with a mixture of hydrogen and oxygen gas. The relative amounts of hydrogen and oxygen are variable. We could easily add either more hydrogen or more oxygen to the balloon. © Nancy R. Cohen/Photodisc/Getty Images.

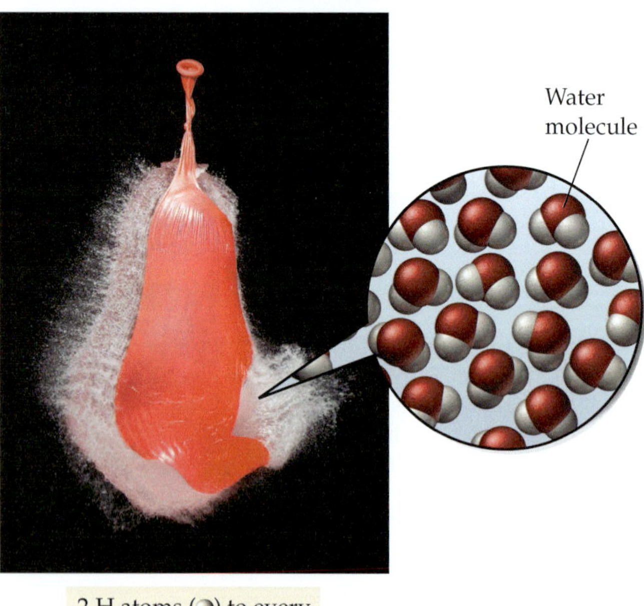

▲ **FIGURE 3.5 A chemical compound** This balloon is filled with water, composed of molecules that have a fixed ratio of hydrogen to oxygen. (*Source*: JoLynn E. Funk.) © Mike Kemp/RubberBall/Alamy.

For example, if we decompose an 18.0 g sample of water, we get 16.0 g of oxygen and 2.0 g of hydrogen, or an oxygen-to-hydrogen mass ratio of

Even though atoms combine in whole-number ratios, their mass ratios are not necessarily whole numbers.

$$\text{Mass ratio} = \frac{16.0 \text{ g O}}{2.0 \text{ g H}} = 8.0 \quad \text{or} \quad 8.0{:}1$$

This is true of any sample of pure water, no matter what its origin. The law of constant composition applies not only to water, but to every compound. If we decompose a 17.0 g sample of ammonia, a compound composed of nitrogen and hydrogen, we get 14.0 g of nitrogen and 3.0 g of hydrogen, or a nitrogen-to-hydrogen mass ratio of:

$$\text{Mass ratio} = \frac{14.0 \text{ g N}}{3.0 \text{ g H}} = 4.7 \quad \text{or} \quad 4.7{:}1$$

Again, this ratio is the same for every sample of ammonia—the composition of each compound is constant.

EXAMPLE 3.1 CONSTANT COMPOSITION OF COMPOUNDS

Two samples of carbon dioxide, obtained from different sources, are decomposed into their constituent elements. One sample produces 4.8 g of oxygen and 1.8 g of carbon, and the other sample produces 17.1 g of oxygen and 6.4 g of carbon. Show that these results are consistent with the law of constant composition.

	SOLUTION
Calculate the mass ratio of one element to the other by dividing the larger mass by the smaller one. For the first sample:	$\frac{\text{Mass oxygen}}{\text{Mass carbon}} = \frac{4.8 \text{ g}}{1.8 \text{ g}} = 2.7$
For the second sample:	$\frac{\text{Mass oxygen}}{\text{Mass carbon}} = \frac{17.1 \text{ g}}{6.4 \text{ g}} = 2.7$

Because the ratios are the same for the two samples, these results are consistent with the law of constant composition.

▶SKILLBUILDER 3.1 | Constant Composition of Compounds

Two samples of carbon monoxide, obtained from different sources, are decomposed into their constituent elements. One sample produces 4.3 g of oxygen and 3.2 g of carbon, and the other sample produces 7.5 g of oxygen and 5.6 g of carbon. Are these results consistent with the law of constant composition?

▶FOR MORE PRACTICE Example 3.26; Problems 46, 47.

CONCEPTUAL CHECKPOINT 3.1

A compound composed of two elements A and B has a ratio of $\frac{\text{Mass A}}{\text{Mass B}} = 3.0$. Decomposition of the compound produces 9.0 g of element A. What mass of element B is produced?

(a) 27.0 g B **(b)** 9.0 g B **(c)** 3.0 g B **(d)** 1.0 g B

3.3 Chemical Formulas: How to Represent Compounds

LO: Write chemical formulas.

LO: Determine the total number of atoms of each element in a chemical formula.

We represent a compound with a **chemical formula**, which indicates the elements present in the compound and the relative number of atoms of each element. For example, H_2O is the chemical formula for water; it indicates that water consists of hydrogen and oxygen atoms in a 2:1 ratio. (Note that the ratio in a chemical formula is a ratio of atoms, not a ratio of masses.) The formula contains the symbol

Compounds have constant composition with respect to mass (as you learned in the previous section) because they are composed of atoms in fixed ratios.

for each element, accompanied by a subscript indicating the number of atoms of that element. By convention, a subscript of 1 is omitted.

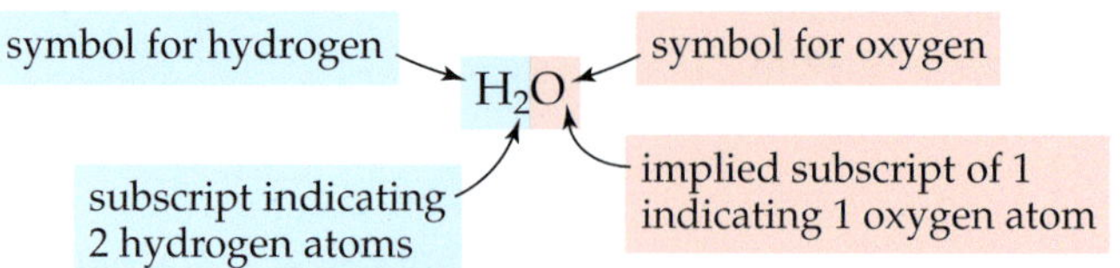

Other common chemical formulas include NaCl for table salt, indicating sodium and chlorine atoms in a 1:1 ratio; CO_2 for carbon dioxide, indicating carbon and oxygen atoms in a 1:2 ratio; and $C_{12}H_{22}O_{11}$ for table sugar (sucrose), indicating carbon, hydrogen, and oxygen atoms in a 12:22:11 ratio. The subscripts in a chemical formula are part of the compound's definition—if the subscripts change, the formula no longer specifies the same compound. For example, CO is the chemical formula for carbon monoxide, an air pollutant with adverse health effects on humans. When inhaled, carbon monoxide interferes with the blood's ability to carry oxygen, which can be fatal. CO is the primary substance responsible for the deaths of people who inhale too much automobile exhaust. If we change the subscript of the O in CO from 1 to 2, however, we get the formula for a totally different compound. CO_2 is the chemical formula for carbon dioxide, the relatively harmless product of combustion and human respiration. We breathe small amounts of CO_2 all the time with no harmful effects. So, remember that:

The subscripts in a chemical formula represent the relative numbers of atoms of each element in a chemical compound; they never change for a given compound.

Chemical formulas normally list the most metallic elements first. Therefore, the formula for table salt is NaCl, not ClNa. In compounds that do not include a metal, we list the more metal-like element first. Recall from Module 2 that metals occupy the left side of the periodic table and nonmetals the upper right side. Among nonmetals, those to the left in the periodic table are more metal-like than those to the right and are normally listed first. Therefore, we write CO_2 and NO, not O_2C and ON. Within a single column in the periodic table, elements toward the bottom are more metal-like than elements toward the top. So, we write SO_2, not O_2S. Table 3.1 lists the specific order for listing nonmetal elements in a chemical formula.

There are a few historical exceptions in which the most metallic element is not listed first, such as the hydroxide ion, which we write as OH^-.

TABLE 3.1 Order of Listing Nonmetal Elements in a Chemical Formula

C	P	N	H	S	I	Br	Cl	O	F

Elements on the left are generally listed before elements on the right.

EXAMPLE 3.2 WRITING CHEMICAL FORMULAS

Write a chemical formula for each compound.

(a) the compound containing two aluminum atoms to every three oxygen atoms
(b) the compound containing three oxygen atoms to every sulfur atom
(c) the compound containing four chlorine atoms to every carbon atom

	SOLUTION
Aluminum is the metal, so list it first.	**(a)** Al_2O_3
Sulfur is below oxygen on the periodic table and it occurs before oxygen in Table 3.1, so list it first.	**(b)** SO_3
Carbon is to the left of chlorine on the periodic table and it occurs before chlorine in Table 3.1, so list it first.	**(c)** CCl_4

▶SKILLBUILDER 3.2 | Writing Chemical Formulas

Write a chemical formula for each compound.

(a) the compound containing two silver atoms to every sulfur atom
(b) the compound containing two nitrogen atoms to every oxygen atom
(c) the compound containing two oxygen atoms to every titanium atom

▶FOR MORE PRACTICE Example 3.27; Problems 52, 53, 54, 55.

Polyatomic Ions in Chemical Formulas

Some chemical formulas contain groups of atoms that act as a unit. When more than one group of the same kind are present, we set their formula off in parentheses with a subscript to indicate the number of units of that group. Many of these groups of atoms have a charge associated with them and are called **polyatomic ions**. For example, NO_3^- is a polyatomic ion with a 1– charge. We describe polyatomic ions in more detail in Section 3.5.

To determine the total number of atoms of each element in a compound containing a group within parentheses, we multiply the subscript outside the parentheses by the subscript for each atom inside the parentheses. For example, $Mg(NO_3)_2$ indicates a compound containing one magnesium atom (present as the Mg^{2+}) and two NO_3^- groups.

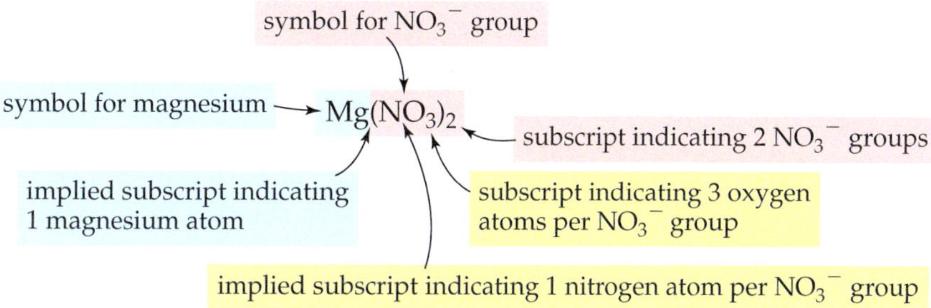

Therefore, the formula $Mg(NO_3)_2$ has the following numbers of atoms of each element.

Mg: 1 Mg
N: $1 \times 2 = 2$ N (implied 1 inside parentheses times 2 outside parentheses)
O: $3 \times 2 = 6$ O (3 inside parentheses times 2 outside parentheses)

EXAMPLE 3.3 TOTAL NUMBER OF ATOMS OF EACH ELEMENT IN A CHEMICAL FORMULA

Determine the number of atoms of each element in $Mg_3(PO_4)_2$.

SOLUTION

Mg: There are three Mg atoms (present as Mg^{2+} ions), as indicated by the subscript 3.

P: There are two P atoms. We determine this by multiplying the subscript outside the parentheses (2) by the subscript for P inside the parentheses, which is 1 (implied).

O: There are eight O atoms. We determine this by multiplying the subscript outside the parentheses (2) by the subscript for O inside the parentheses (4).

▶SKILLBUILDER 3.3 | Total Number of Atoms of Each Element in a Chemical Formula

Determine the number of atoms of each element in K_2SO_4.

▶SKILLBUILDER PLUS

Determine the number of atoms of each element in $Al_2(SO_4)_3$.

▶FOR MORE PRACTICE Example 3.28; Problems 56, 57, 58, 59.

CONCEPTUAL CHECKPOINT 3.2

Which formula represents the greatest total number of atoms?

(a) $Al(C_2H_3O_2)_3$ **(b)** $Al_2(Cr_2O_7)_3$ **(c)** $Pb(HSO_4)_4$
(d) $Pb_3(PO_4)_4$ **(e)** $(NH_4)_3PO_4$

Types of Chemical Formulas

We categorize chemical formulas into three types: empirical, molecular, and structural. An **empirical formula** is the simplest whole-number ratio of atoms of each element in a compound. A **molecular formula** is the *actual* number of atoms of each element in a molecule of the compound. For example, the molecular formula for hydrogen peroxide is H_2O_2, and its empirical formula is HO. The molecular formula is always a whole-number multiple of the empirical formula. For many compounds, the molecular and empirical formulas are the same. For example, the empirical and molecular formula for water is H_2O because water molecules contain two hydrogen atoms and one oxygen atom; no simpler whole-number ratio can express the number of hydrogen atoms relative to oxygen atoms.

Hydrogen
Carbon
Nitrogen
Oxygen
Fluorine
Phosphorus
Sulfur
Chlorine

A **structural formula** uses lines to represent chemical bonds and shows how the atoms in a molecule are connected to each other. The structural formula for hydrogen peroxide is H—O—O—H. We can also use **molecular models**—three-dimensional representations of molecules—to represent compounds. In this book, we use two types of molecular models: ball-and-stick and space-filling. In **ball-and-stick models**, we represent atoms as balls and chemical bonds as sticks. The balls and sticks are connected to represent the molecule's shape. The balls are color coded, and we assign each element a color as shown in the margin.

In **space-filling models**, atoms fill the space between each other to more closely represent our best idea for how a molecule might appear if we could scale it to a visible size. Consider the following ways to represent a molecule of methane, the main component of natural gas:

CH_4

Molecular formula

```
    H
    |
H — C — H
    |
    H
```

Structural formula

Ball-and-stick model

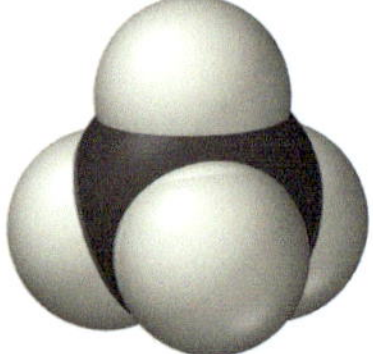

Space-filling model

Macroscopic
Molecular
H_2O
Symbolic

© malerapaso/iStockphoto/Getty Images.

The molecular formula of methane indicates that methane has one carbon atom and four hydrogen atoms. The structural formula shows how the atoms are connected: Each hydrogen atom is bonded to the central carbon atom. The ball-and-stick model and the space-filling model illustrate the *geometry* of the molecule: how the atoms are arranged in three dimensions.

Throughout this book, you have seen and will continue to see images that show the connection between the *macroscopic world* (what we see), the *atomic and molecular world* (the particles that compose matter), and the *symbolic way* that chemists represent the atomic and molecular world. For example, at left is a representation of water using this kind of image.

The main goal of these images is to help you visualize the main theme of this book: *the connection between the world around us and the world of atoms and molecules.*

3.4 A Molecular View of Elements and Compounds

LO: Classify elements as atomic or molecular.

LO: Classify compounds as ionic or molecular.

Recall from Module 1 that we can categorize pure substances as either elements or compounds. We can further subcategorize elements and compounds according to the basic units that compose them (▼ Figure 3.6). Pure substances may be elements, or they may be compounds. Elements may be either atomic or molecular. Compounds may be either molecular or ionic.

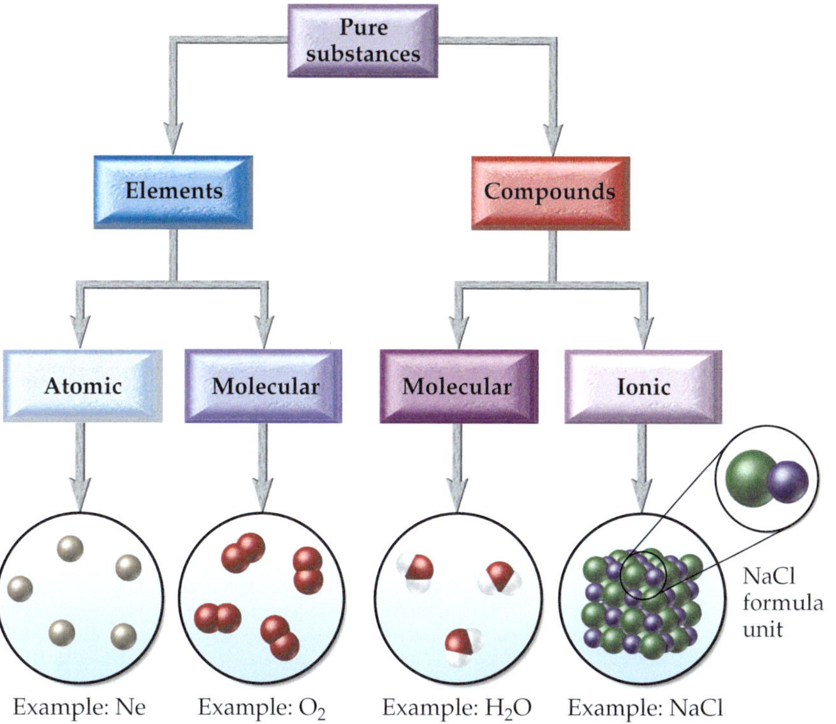

▲ **FIGURE 3.6 A molecular view of elements and compounds**

▲ **FIGURE 3.7 An atomic element** The basic units that compose mercury, an atomic element and a metal, are single mercury atoms. © Harry Taylor/Dorling Kindersley, Ltd.

A few molecular elements, such as S_8 and P_4, are composed of molecules containing several atoms.

Atomic Elements

Atomic elements have single atoms as their basic units. Most elements fall into this category. For example, helium is composed of helium atoms, copper is composed of copper atoms, and mercury of mercury atoms (▲ Figure 3.7).

Molecular Elements

Molecular elements do not normally exist in nature with single atoms as their basic units. Instead, these elements exist as *diatomic molecules*—two atoms of that element bonded together—as their basic units. For example, hydrogen is composed of H_2 molecules, oxygen is composed of O_2 molecules, and chlorine of Cl_2 molecules (◀ Figure 3.8). Table 3.2 and ▶ Figure 3.9 list elements that exist as diatomic molecules.

Molecular Compounds

Molecular compounds are composed of two or more nonmetals. The basic units of molecular compounds are molecules composed of the constituent atoms. For

◀ **FIGURE 3.8 A molecular element** The basic units that compose chlorine, a molecular element, are diatomic chlorine molecules; each molecule is composed of two chlorine atoms. © Charles D. Winters/Science Source.

TABLE 3.2 Elements That Occur as Diatomic Molecules

Name of Element	Formula of Basic Unit
hydrogen	H_2
nitrogen	N_2
oxygen	O_2
fluorine	F_2
chlorine	Cl_2
bromine	Br_2
iodine	I_2

Main groups — Transition metals — Main groups

Elements that exist as diatomic molecules (highlighted: H, N, O, F, Cl, Br, I)

Periods	1A 1	2A 2	3B 3	4B 4	5B 5	6B 6	7B 7	8B 8	8B 9	8B 10	1B 11	2B 12	3A 13	4A 14	5A 15	6A 16	7A 17	8A 18
1	1 H																	2 He
2	3 Li	4 Be											5 B	6 C	7 N	8 O	9 F	10 Ne
3	11 Na	12 Mg											13 Al	14 Si	15 P	16 S	17 Cl	18 Ar
4	19 K	20 Ca	21 Sc	22 Ti	23 V	24 Cr	25 Mn	26 Fe	27 Co	28 Ni	29 Cu	30 Zn	31 Ga	32 Ge	33 As	34 Se	35 Br	36 Kr
5	37 Rb	38 Sr	39 Y	40 Zr	41 Nb	42 Mo	43 Tc	44 Ru	45 Rh	46 Pd	47 Ag	48 Cd	49 In	50 Sn	51 Sb	52 Te	53 I	54 Xe
6	55 Cs	56 Ba	57 La	72 Hf	73 Ta	74 W	75 Re	76 Os	77 Ir	78 Pt	79 Au	80 Hg	81 Tl	82 Pb	83 Bi	84 Po	85 At	86 Rn
7	87 Fr	88 Ra	89 Ac	104 Rf	105 Db	106 Sg	107 Bh	108 Hs	109 Mt	110 Ds	111 Rg	112 Cn	113 Nh	114 Fl	115 Mc	116 Lv	117 Ts	118 Og

Lanthanides	58 Ce	59 Pr	60 Nd	61 Pm	62 Sm	63 Eu	64 Gd	65 Tb	66 Dy	67 Ho	68 Er	69 Tm	70 Yb	71 Lu
Actinides	90 Th	91 Pa	92 U	93 Np	94 Pu	95 Am	96 Cm	97 Bk	98 Cf	99 Es	100 Fm	101 Md	102 No	103 Lr

▲ **FIGURE 3.9 Elements that form diatomic molecules** Elements that normally exist as diatomic molecules are highlighted in yellow on this periodic table. Note that they are all nonmetals and include four of the halogens.

example, water is composed of H_2O molecules, dry ice is composed of CO_2 molecules (▼ Figure 3.10), and acetone (finger nail–polish remover) of C_3H_6O molecules.

▲ **FIGURE 3.10 A molecular compound** The basic units that compose dry ice, a molecular compound, are CO_2 molecules. © Universal Images Group Limited/Alamy.

Ionic Compounds

Ionic compounds are composed of one or more cations paired with one or more anions. In most cases, the cations are metals and the anions are nonmetals. When a metal, which has a tendency to lose electrons (see Section 2.5), combines with a nonmetal, which has a tendency to gain electrons, one or more electrons transfer from the metal to the nonmetal, creating positive and negative ions that are then attracted to each other. We can assume that a compound composed of a metal and a nonmetal is ionic. The basic unit of ionic compounds is the **formula unit**, the smallest electrically neutral collection of ions. Formula units are different from molecules in that they do not exist as discrete entities, but rather as part of a larger three

▲ **FIGURE 3.11 An ionic compound** The basic units that compose table salt, an ionic compound, are NaCl formula units. Unlike molecular compounds, ionic compounds do not contain individual molecules but rather sodium and chloride ions in an alternating three-dimensional array. © Paul Silverman/Fundamental Photographs.

dimensional array. For example, salt (NaCl) is composed of Na^+ and Cl^- ions in a 1:1 ratio. In table salt, Na^+ and Cl^- ions exist in an alternating three-dimensional array (◀ Figure 3.11). However, any one Na^+ ion does not pair with one specific Cl^- ion. Sometimes formula units are referred to as molecules but this is not strictly correct since ionic compounds do not contain distinct molecules.

EXAMPLE 3.4 CLASSIFYING SUBSTANCES AS ATOMIC ELEMENTS, MOLECULAR ELEMENTS, MOLECULAR COMPOUNDS, OR IONIC COMPOUNDS

Classify each substance as an atomic element, molecular element, molecular compound, or ionic compound.

(a) krypton **(b)** $CoCl_2$ **(c)** nitrogen **(d)** SO_2 **(e)** KNO_3

SOLUTION

(a) Krypton is not listed as diatomic in Table 3.2 because it is a noble gas (Group 8A) element; therefore, it is an atomic element.

(b) $CoCl_2$ is a compound composed of a metal (left side of periodic table) and nonmetal (right side of the periodic table); therefore, it is an ionic compound.

(c) Nitrogen is an element that is listed as diatomic in Table 3.2; therefore, it is a molecular element.

(d) SO_2 is a compound composed of two nonmetals; therefore, it is a molecular compound.

(e) KNO_3 is a compound composed of a metal and two nonmetals; therefore, it is an ionic compound.

▶**SKILLBUILDER 3.4 | Classifying Substances as Atomic Elements, Molecular Elements, Molecular Compounds, or Ionic Compounds**

Classify each substance as an atomic element, molecular element, molecular compound, or ionic compound.

(a) chlorine **(b)** NO **(c)** Au **(d)** Na_2O **(e)** $CrCl_3$

▶**FOR MORE PRACTICE** Example 3.29, Example 3.30(i); Problems 64, 65, 66, 67.

CONCEPTUAL CHECKPOINT 3.3

Which image represents a molecular compound?

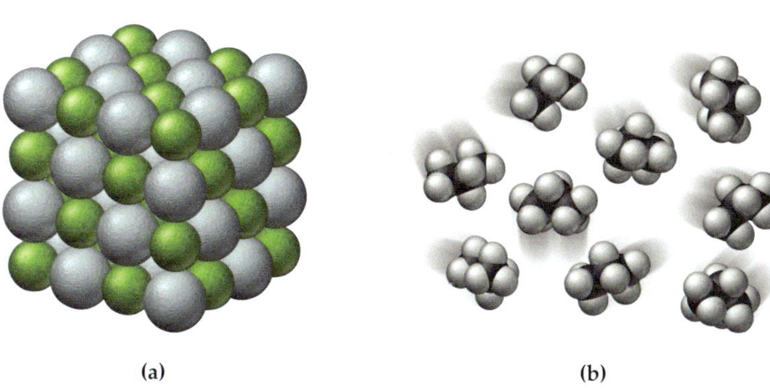

3.5 Writing Formulas for Ionic Compounds

LO: Write formulas for ionic compounds.

Because ionic compounds must be charge-neutral, and many elements form only one type of ion with a predictable charge, we can determine the formulas for many ionic compounds based on their constituent elements. For example, the formula for the ionic compound composed of sodium and chlorine must be NaCl and not anything else because in ionic compounds Na always forms 1+ cations and Cl always forms 1– anions. In order for the compound to be charge-neutral, it must contain one Na^+ cation to every Cl^- anion. The formula for the ionic compound composed of magnesium and chlorine, on the other hand, must be $MgCl_2$, because Mg always forms 2+ cations and Cl always forms 1– anions. In order for the compound to be charge-neutral, it must contain one Mg^{2+} cation to every two Cl^- anions. In general:

- Ionic compounds always contain positive and negative ions.
- In the chemical formula, the sum of the charges of the positive ions (cations) must always exactly cancel the sum of the charges of the negative ions (anions).

Writing Formulas for Ionic Compounds Containing Only Monoatomic Ions

Revisit Section 2.5 and Figure 2.5 to review the elements that form ions with a predictable charge.

To write the formula for an ionic compound containing monoatomic ions (no polyatomic ions), follow the procedure in the left column of the following examples. Two examples of how to apply the procedure are provided in the center and right columns.

WRITING FORMULAS FOR IONIC COMPOUNDS	EXAMPLE 3.5 Write a formula for the ionic compound that forms from aluminum and oxygen.	EXAMPLE 3.6 Write a formula for the ionic compound that forms from magnesium and oxygen.
1. Write the symbol for the metal and its charge followed by the symbol of the nonmetal and its charge. For many elements, you can determine these charges from their group number in the periodic table.	**SOLUTION** Al^{3+} O^{2-}	**SOLUTION** Mg^{2+} O^{2-}
2. Use the magnitude of the charge on each ion (without the sign) as the subscript for the other ion.	Al^{3+} O^{2-} Al_2O_3	Mg^{2+} O^{2-} Mg_2O_2
3. If possible, reduce the subscripts to give a ratio with the smallest whole numbers.	In this case, you cannot reduce the numbers any further; the correct formula is Al_2O_3.	To reduce the subscripts, divide both subscripts by 2. $Mg_2O_2 \div 2 = MgO$
4. Check to make sure that the sum of the charges of the cations exactly cancels the sum of the charges of the anions.	Cations: 2(3+) = 6+ Anions: 3(2–) = 6– The charges cancel.	Cations: 2+ Anions: 2– The charges cancel.
	▶**SKILLBUILDER 3.5** | Write a formula for the compound that forms from strontium and chlorine.	▶**SKILLBUILDER 3.6** | Write a formula for the compound that forms from aluminum and nitrogen.
		▶**FOR MORE PRACTICE** Example 3.30(ii); Problems 74, 75, 78.

Writing Formulas for Ionic Compounds Containing Polyatomic Ions

As noted previously, some ionic compounds contain **polyatomic ions** (ions that are themselves composed of a group of atoms with an overall charge). Table 3.3 lists the most common polyatomic ions. You need to be able to recognize polyatomic ions in a chemical formula, so it is good idea to become familiar with Table 3.3. To write a formula for ionic compounds containing polyatomic ions, use the formula and charge of the polyatomic ion as demonstrated in Example 3.7.

TABLE 3.3 Some Common Polyatomic Ions

Name	Formula	Name	Formula
acetate	$C_2H_3O_2^-$	hypochlorite	ClO^-
carbonate	CO_3^{2-}	chlorite	ClO_2^-
hydrogencarbonate (or bicarbonate)	HCO_3^-	chlorate	ClO_3^-
hydroxide	OH^-	perchlorate	ClO_4^-
nitrate	NO_3^-	permanganate	MnO_4^-
nitrite	NO_2^-	sulfate	SO_4^{2-}
chromate	CrO_4^{2-}	sulfite	SO_3^{2-}
dichromate	$Cr_2O_7^{2-}$	hydrogensulfite (or bisulfite)	HSO_3^-
phosphate	PO_4^{3-}	hydrogensulfate (or bisulfate)	HSO_4^-
hydrogenphosphate	HPO_4^{2-}	peroxide	O_2^{2-}
ammonium	NH_4^+	cyanide	CN^-

EXAMPLE 3.7 WRITING FORMULAS FOR IONIC COMPOUNDS CONTAINING POLYATOMIC IONS

Write a formula for the compound that forms from calcium and nitrate ions.

	SOLUTION
1. Write the symbol for the metal ion followed by the symbol for the polyatomic ion and their charges. You can deduce these charges from the group numbers in the periodic table. For polyatomic ions, look up the charges and names in Table 3.3.	Ca^{2+} NO_3^-
2. Use the magnitude of the charge on each ion as the subscript for the other ion.	$Ca_1(NO_3)_2$
3. Check to see if the subscripts can be reduced to simpler whole numbers. You should drop subscripts of 1 because they are implied.	In this case, you cannot further reduce the subscripts, but you can drop the subscript of 1. $Ca(NO_3)_2$
4. Confirm that the sum of the charges of the cations exactly cancels the sum of the charges of the anions.	Cations: 2+ Anions: 2(1−) = 2−

▶**SKILLBUILDER 3.7** | Write the formula for the compound that forms between aluminum and phosphate ions.

▶**SKILLBUILDER PLUS** | Write the formula for the compound that forms between sodium and sulfite ions.

▶**FOR MORE PRACTICE** Example 3.31; Problems 76, 77, 79.

CONCEPTUAL CHECKPOINT 3.4

Some metals can form ions of different charges in different compounds. Deduce the charge of the Cr ion in the compound $Cr(NO_3)_3$.

(a) 1+

(b) 2+

(c) 3+

(d) 1–

3.6 Nomenclature: Naming Compounds

LO: Distinguish between common and systematic names for compounds.

Because there are so many different compounds, chemists have developed systematic ways to name them. These naming rules can help you examine a compound's formula and determine its name, or vice versa. Many compounds also have a common name. For example, H_2O has the common name *water* and the systematic name *dihydrogen monoxide*. A common name is like a nickname for a compound, used by those who are familiar with it. Since water is such a familiar compound, everyone uses its common name and not its systematic name. In the sections that follow, you learn how to systematically name simple ionic and molecular compounds. Keep in mind, however, that some compounds also have common names that are often used instead of the systematic name. Common names can be learned only through familiarity.

3.7 Naming Ionic Compounds

LO: Name binary ionic compounds containing a metal that forms only one type of ion.

LO: Name binary ionic compounds containing a metal that forms more than one type of ion.

LO: Name ionic compounds containing a polyatomic ion.

The first step in naming an ionic compound is identifying it as one. Remember, any time we have a metal or ammonium and one or more nonmetals together in a chemical formula, we can assume the compound is ionic. Ionic compounds are categorized into two types (▼ Figure 3.12) depending on the metal in the compound. The first type (sometimes called Type I) contains a metal with an invariant charge—one that does not vary from one compound to another. Sodium, for instance, has a 1+ charge in all of its compounds. Table 3.4 lists examples of metals whose charge is invariant from one compound to another. The charge of most of these metals can be inferred from their group number in the periodic table.

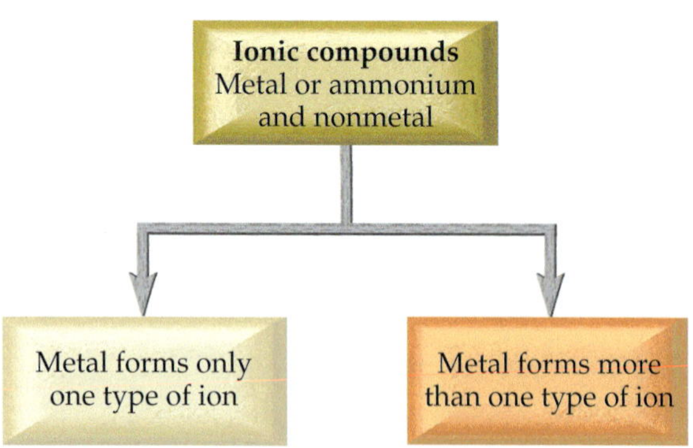

▲ **FIGURE 3.12 Classification of ionic compounds** We classify ionic compounds into two types, depending on the metal in the compound.

TABLE 3.4 Metals Whose Charge Is Invariant from One Compound to Another

Metal	Ion	Name	Group Number
Li	Li^+	lithium	1A
Na	Na^+	sodium	1A
K	K^+	potassium	1A
Rb	Rb^+	rubidium	1A
Cs	Cs^+	cesium	1A
Mg	Mg^{2+}	magnesium	2A
Ca	Ca^{2+}	calcium	2A
Sr	Sr^{2+}	strontium	2A
Ba	Ba^{2+}	barium	2A
Al	Al^{3+}	aluminum	3A
Zn	Zn^{2+}	zinc	*
Ag	Ag^+	silver	*

*The charge of these metals cannot be inferred from their group number. Zn and Ag can form more than one type of cation but they predominantly exist as 2+ and 1+ cations respectively.

TABLE 3.5 Some Metals That Form More Than One Type of Ion and Their Common Charges

Metal	Symbol Ion	Name	Older Name*
chromium	Cr^{2+}	chromium(II)	chromous
	Cr^{3+}	chromium(III)	chromic
iron	Fe^{2+}	iron(II)	ferrous
	Fe^{3+}	iron(III)	ferric
cobalt	Co^{2+}	cobalt(II)	cobaltous
	Co^{3+}	cobalt(III)	cobaltic
copper	Cu^{+}	copper(I)	cuprous
	Cu^{2+}	copper(II)	cupric
tin	Sn^{2+}	tin(II)	stannous
	Sn^{4+}	tin(IV)	stannic
mercury	$Hg_2{}^{2+}$	mercury(I)	mercurous
	Hg^{2+}	mercury(II)	mercuric
lead	Pb^{2+}	lead(II)	plumbous
	Pb^{4+}	lead(IV)	plumbic

* In an older naming system, the higher charge cation has a name ending in -ic, and the lower charge cation has a name ending in -ous. Under this system, chromium(II) oxide is named chromous oxide. In this text, we do *not* use this older system; the name of the metal and its charge will be used exclusively.

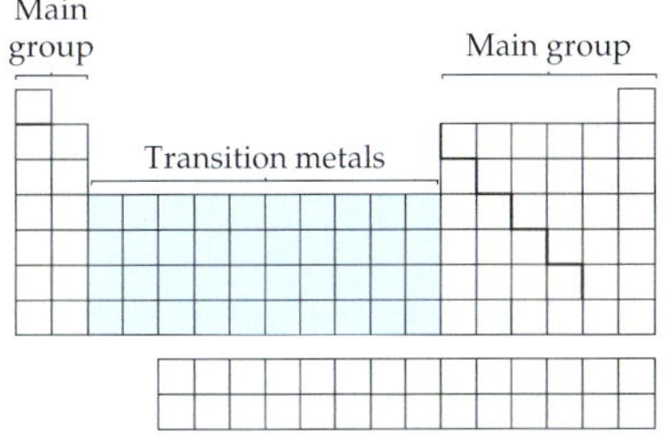

▲ **FIGURE 3.13 The transition metals** The metals that form more than one type of ion are usually (but not always) transition metals.

The second type of ionic compound (sometimes called Type II) contains a metal with a charge that can differ in different compounds. In other words, the metal in this second type of ionic compound can form more than one kind of cation (depending on the compound). Iron, for instance, has a 2+ charge in some of its compounds and a 3+ charge in others. Table 3.5 lists additional examples of metals that form more than one type of cation. Such metals are usually (but not always) found in the **transition metals** section of the periodic table (◀ Figure 3.13). However, some transition metals, such as Zn and Ag, form cations with the same charge in almost all of their compounds, and some main group metals, such as Pb and Sn, form more than one type of cation.

Naming Binary Ionic Compounds Containing a Metal That Forms Only One Type of Cation

Binary compounds contain only two different elements. The names for binary ionic compounds containing a metal that forms only one type of ion have the form:

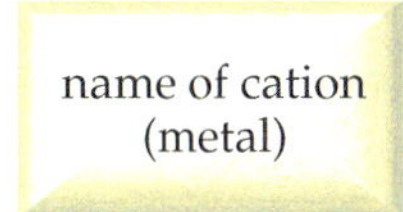

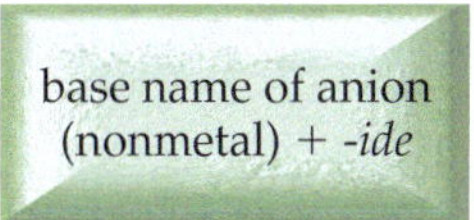

Because the charge of the metal is always the same for these types of compounds, we do not need to specify it in the compound's name. For example, the name for NaCl consists of the name of the cation, *sodium*, followed by the base name of the anion, *chlor*, with the ending *-ide*. The full name is *sodium chloride*.

The name of the cation in ionic compounds is the same as the name of the metal.

NaCl   sodium chloride

The name for MgO consists of the name of the cation, *magnesium*, followed by the base name of the anion, *ox*, with the ending *-ide*. The full name is *magnesium oxide*.

MgO   magnesium oxide

Table 3.6 contains the base names for various nonmetals and their most common charges in ionic compounds.

TABLE 3.6 Some Common Anions

Nonmetal	Symbol for Ion	Base Name	Anion Name
fluorine	F^-	fluor-	fluoride
chlorine	Cl^-	chlor-	chloride
bromine	Br^-	brom-	bromide
iodine	I^-	iod-	iodide
oxygen	O^{2-}	ox-	oxide
sulfur	S^{2-}	sulf-	sulfide
nitrogen	N^{3-}	nitr-	nitride

EXAMPLE 3.8 NAMING IONIC COMPOUNDS CONTAINING A METAL THAT FORMS ONLY ONE TYPE OF CATION

Name the compound MgF_2.

SOLUTION

The metal is magnesium that gives magnesium cation. The nonmetal is fluorine, which becomes *fluoride* anion. Its correct name is *magnesium fluoride*.

▶SKILLBUILDER 3.8 | Naming Ionic Compounds Containing a Metal That Forms Only One Type of Ion

Name the compound KBr.

▶SKILLBUILDER PLUS Name the compound Zn_3N_2.

▶FOR MORE PRACTICE Example 3.32; Problems 80, 81.

Naming Binary Ionic Compounds Containing a Metal That Forms More Than One Type of Cation

Because the charge of the metal cation in these types of compounds is not always the same, we must specify the charge in the metal's name. We specify the charge with a roman numeral (in parentheses) following the name of the metal. For example, we distinguish between Cu^+ and Cu^{2+} by writing a (I) to indicate the 1+ ion or a (II) to indicate the 2+ ion:

Cu^+ Copper(I)

Cu^{2+} Copper(II)

The full names for these types of compounds have the form:

name of cation (metal) (charge of cation (metal) in roman numerals in parentheses)

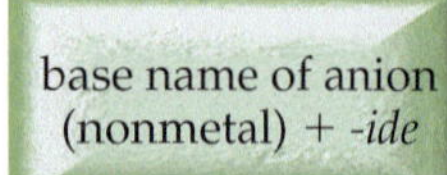

We can determine the charge of the metal from the chemical formula of the compound—remember that the sum of all the charges must be zero. For example, the charge of iron in $FeCl_3$ must be 3+ in order for the compound to be charge-neutral with the three Cl^- anions. The name for $FeCl_3$ is therefore the name of the cation, *iron*, followed by the charge of the cation in parentheses *(III)*, followed by the base name of the anion, *chlor*, with the ending *-ide*. The full name is *iron(III) chloride*.

$FeCl_3$ iron(III) chloride

Likewise, the name for CrO consists of the name of the cation, *chromium*, followed by the charge of the cation in parentheses *(II)*, followed by the base name of the anion, *ox-*, with the ending *-ide*. The full name is *chromium(II) oxide*.

CrO chromium(II) oxide

The charge of chromium must be 2+ in order for the compound to be charge-neutral with one O^{2-} anion.

EXAMPLE 3.9 NAMING IONIC COMPOUNDS CONTAINING A METAL THAT FORMS MORE THAN ONE TYPE OF CATION

Name the compound $PbCl_4$.

SOLUTION

The name for $PbCl_4$ consists of the name of the cation, *lead*, followed by the charge of the cation in parentheses *(IV)*, followed by the base name of the anion, *chlor-*, with the ending *-ide*. The full name is *lead(IV) chloride*. We know the charge on Pb is 4+ because the charge on Cl is 1–. Since there are 4 Cl^- anions, the Pb cation must be Pb^{4+}.

$PbCl_4$ lead(IV) chloride

▶SKILLBUILDER 3.9 | Naming Ionic Compounds Containing a Metal That Forms More Than One Type of Cation

Name the compound PbO.

▶FOR MORE PRACTICE Example 3.33; Problems 82, 83.

CONCEPTUAL CHECKPOINT 3.5

Explain why CaO is NOT named calcium(II) oxide.

Naming Ionic Compounds Containing a Polyatomic Ion

We name ionic compounds containing polyatomic ions using the same procedure we apply to other ionic compounds, except that we use the name of the poly atomic ion whenever it occurs (see Table 3.3). For example, we name KNO_3 using its cation, K^+, *potassium*, and its polyatomic anion, NO_3^-, *nitrate*. The full name is *potassium nitrate*.

KNO_3 potassium nitrate

We name $Fe(OH)_2$ according to its cation, *iron*, its charge *(II)*, and its polyatomic ion, *hydroxide*. Its full name is *iron(II) hydroxide*.

$Fe(OH)_2$ iron(II) hydroxide

If the compound contains both a polyatomic cation and a polyatomic anion, we use the names of both polyatomic ions. For example, NH_4NO_3 is *ammonium nitrate*.

NH_4NO_3 ammonium nitrate

Most polyatomic ions are **oxyanions**, anions containing oxygen. Notice in Table 3.3 that when a series of oxyanions contain different numbers of oxygen atoms, we name them systematically according to the number of oxygen atoms in the ion. If there are two ions in the series, we give the one with more oxygen atoms the ending *-ate* and we give the one with fewer oxygen atoms the ending *-ite*. For example, NO_3^- is *nitrate* and NO_2^- is *nitrite*.

NO_3^- nitrate

NO_2^- nitrite

If there are more than two ions in the series, then we use the prefixes *hypo-*, meaning "less than," and *per-*, meaning "more than." So we call ClO^- *hypochlorite*, meaning "less oxygen than chlorite," and we call ClO_4^- *perchlorate*, meaning "more oxygen than chlorate."

ClO^-	hypochlorite
ClO_2^-	chlorite
ClO_3^-	chlorate
ClO_4^-	perchlorate

EXAMPLE 3.10 NAMING IONIC COMPOUNDS CONTAINING A POLYATOMIC ION

Name the compound K_2CrO_4.

SOLUTION

The name for K_2CrO_4 consists of the name of the cation, *potassium*, followed by the name of the polyatomic ion, *chromate*.

K_2CrO_4 potassium chromate

▶SKILLBUILDER 3.10 | Naming Ionic Compounds Containing a Polyatomic Ion

Name the compound $Mn(NO_3)_2$.

▶FOR MORE PRACTICE Example 3.34; Problems 86, 87.

CONCEPTUAL CHECKPOINT 3.6

We just saw that the anion ClO_3^- is named chlorate. What is the name of the anion IO_3^-?

EVERYDAY CHEMISTRY

▶ Polyatomic Ions

A glance at the labels of household products reveals the importance of polyatomic ions in everyday compounds. For example, the active ingredient in household bleach is sodium hypochlorite, which acts to decompose color-causing molecules in clothes (bleaching action) and to kill bacteria (disinfection). A box of baking soda contains sodium bicarbonate (sodium hydrogencarbonate), which acts as an antacid when consumed in small quantities and as a source of carbon dioxide gas in baking. The pockets of carbon dioxide gas make baked goods fluffy rather than flat.

Calcium carbonate is the active ingredient in many antacids such as Tums™ and Alka-Mints™. It neutralizes stomach acids, relieving the symptoms of indigestion and heartburn. Too much calcium carbonate, however, can cause constipation, so Tums should not be overused. Sodium nitrite is a common food additive used to preserve packaged meats such as ham, hot dogs, and bologna. Sodium nitrite inhibits the growth of bacteria, especially those that cause botulism, an often fatal type of food poisoning.

▲ Compounds containing polyatomic ions are present in many consumer products. © Maxwell Art And Photo/ Pearson.

▲ The active ingredient in bleach is sodium hypochlorite. © Maxwell Art And Photo/Pearson.

B3.1 CAN YOU ANSWER THIS? *Write a formula for each of these compounds that contain polyatomic ions: sodium hypochlorite, sodium bicarbonate, calcium carbonate, sodium nitrite.*

3.8 Naming Molecular Compounds

LO: Name molecular compounds.

The first step in naming a molecular compound is identifying it as one. Remember, nearly all molecular compounds form from two or more nonmetals. In this section, we discuss how to name binary (two-element) molecular compounds. Their names have the form:

When writing the name of a molecular compound, as when writing the formula, the first element is the more metal-like one (see Table 3.1). The prefixes given to each element indicate the number of atoms present.

mono- 1	*hexa-* 6
di- 2	*hepta-* 7
tri- 3	*octa-* 8
tetra- 4	*nona-* 9
penta- 5	*deca-* 10

If there is only one atom of the *first element* in the formula, the prefix *mono-* is normally omitted. For example, the name for CO_2 begins with *carbon*, without a prefix because *mono-* is omitted for the first element, followed by the prefix *di-*, to indicate two oxygen atoms, followed by the base name of the second element, *ox*, with the ending *-ide*.

carbon   di- ox -ide

The full name is *carbon dioxide*.

CO_2 carbon dioxide

When the prefix ends with a vowel and the base name starts with a vowel, we sometimes drop the first vowel, especially in the case of mono oxide, which becomes monoxide.

The name for the compound N_2O, also called laughing gas, begins with the first element, *nitrogen*, with the prefix *di-*, to indicate that there are two of them, followed by the base name of the second element, *ox*, prefixed by *mono-*, to indicate one, and the suffix *-ide*. Because *mono-* ends with a vowel and *oxide* begins with one, we drop an *o* and combine the two as *monoxide*. The entire name is *dinitrogen monoxide*.

N_2O dinitrogen monoxide

EXAMPLE 3.11 NAMING MOLECULAR COMPOUNDS

Name each compound.

(a) CCl_4 **(b)** BCl_3 **(c)** SF_6

SOLUTION

(a) The name of the compound is the name of the first element, *carbon*, followed by the base name of the second element, *chlor*, prefixed by *tetra-* to indicate four, and the suffix *–ide*.

CCl_4 carbon tetrachloride

(b) The name of the compound is the name of the first element, *boron*, followed by the base name of the second element, *chlor*, prefixed by *tri-* to indicate three, and the suffix *-ide*.

BCl_3 boron trichloride

(c) The name of the compound is the name of the first element, *sulfur*, followed by the base name of the second element, *fluor*, prefixed by *hexa-* to indicate six, and the suffix *-ide*. The entire name is *sulfur hexafluoride*.

SF_6 sulfur hexafluoride

▶**SKILLBUILDER 3.11 | Naming Molecular Compounds**

Name the compound N_2O_4.

▶**FOR MORE PRACTICE** Example 3.35; Problems 92, 93.

3.9 Naming Acids

LO: Name binary acids.

LO: Name oxyacids containing an oxyanion ending in *-ate*.

LO: Name oxyacids containing an oxyanion ending in *-ite*.

Acids are molecular compounds that produce H^+ ions when dissolved in water. They are composed of hydrogen, which we usually write first in their formula, and one or more nonmetals, which we usually write second. Acids are characterized by their sour taste and their ability to dissolve some metals. For example, $HCl(aq)$ is an acid—the (aq) indicates that the compound is "aqueous" or "dissolved in water." $HCl(aq)$ has a characteristically sour taste. Since $HCl(aq)$ is present in stomach fluids, its sour taste becomes painfully obvious during vomiting. $HCl(aq)$ also dissolves some metals. If we drop a strip of zinc into a beaker of $HCl(aq)$, it will slowly disappear as the acid converts the zinc metal into dissolved Zn^{2+} cations.

$HCl(g)$ refers to HCl molecules in the gas state.

Acids are present in many foods, such as lemons and limes, and they are used in some household products such as toilet bowl cleaners and Lime-A-Way. In this section, we only learn how to name them, but in Module 8 we will learn more about the properties of acids. We categorize acids into two groups: **binary acids**, those containing only hydrogen and a nonmetal, and **oxyacids**, those containing hydrogen, a nonmetal, and oxygen (▼ Figure 3.14).

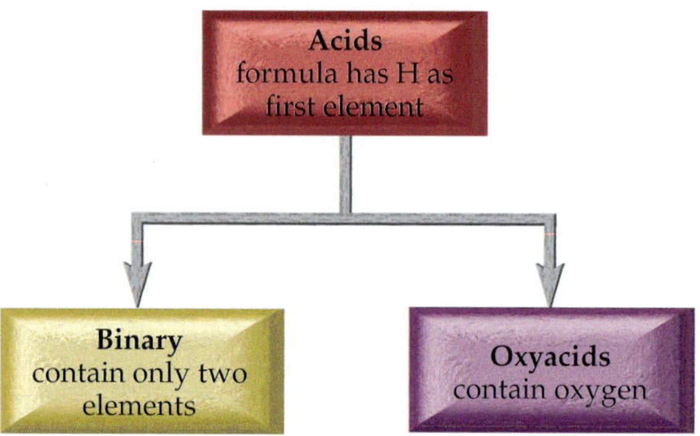

▶ **FIGURE 3.14 Classification of acids** We classify acids into two types, depending on the number of elements in the acid. If the acid contains only two elements, it is a binary acid. If it contains oxygen, it is an oxyacid.

Naming Binary Acids

Binary acids are composed of hydrogen and a nonmetal. The names for binary acids have the following form:

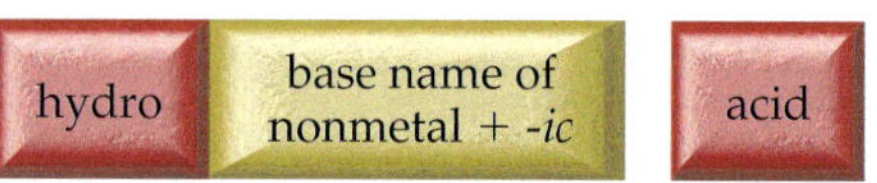

For example, $HCl(aq)$ is hydro*chlor*ic acid and $HBr(aq)$ is hydro*brom*ic acid.

$HCl(aq)$ hydrochloric acid $HBr(aq)$ hydrobromic acid

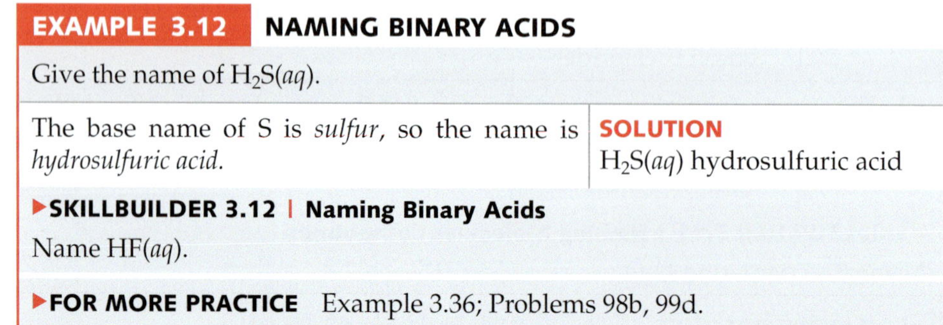

EXAMPLE 3.12 NAMING BINARY ACIDS

Give the name of $H_2S(aq)$.

The base name of S is *sulfur*, so the name is *hydrosulfuric acid*.

SOLUTION

$H_2S(aq)$ hydrosulfuric acid

▶**SKILLBUILDER 3.12 | Naming Binary Acids**

Name $HF(aq)$.

▶**FOR MORE PRACTICE** Example 3.36; Problems 98b, 99d.

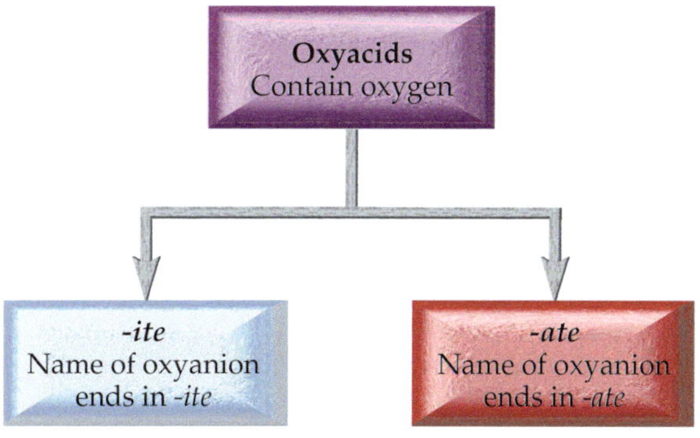

▲ **FIGURE 3.15 Classification of oxyacids** Oxyacids are classified into two types, depending on the endings of the oxyanions that they contain.

Naming Oxyacids

Oxyacids are acids that contain oxyanions, which are listed in the table of polyatomic ions (Table 3.3). For example, $HNO_3(aq)$ contains the nitrate (NO_3^-) ion, $H_2SO_3(aq)$ contains the sulfite (SO_3^{2-}) ion, and $H_2SO_4(aq)$ contains the sulfate (SO_4^{2-}) ion. All of these acids are a combination of one or more H^+ ions with an oxyanion. The number of H^+ ions depends on the charge of the oxyanion, so that the formula is always charge-neutral. The names of oxyacids depend on the ending of the oxyanion (◀ Figure 3.15).

The names of acids containing oxyanions ending with *-ite* take this form:

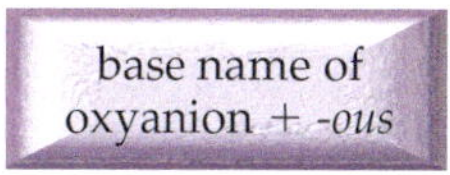

The names of acids containing oxyanions ending with *-ate* take this form:

You can use the saying, "*ic* I *ate* an acid" to remember the association of *-ic* with *-ate*.

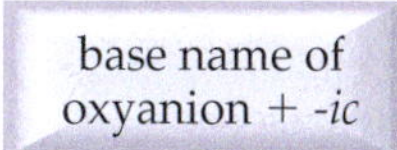

So H_2SO_3 is *sulfurous acid* (oxyanion is sulfite), and HNO_3 is *nitric acid* (oxyanion is nitrate).

$H_2SO_3(aq)$ sulfurous acid $HNO_3(aq)$ nitric acid

Table 3.7 lists some common oxyacids and their oxyanions.

EXAMPLE 3.13 NAMING OXYACIDS

Name $HC_2H_3O_2(aq)$.

The oxyanion is acetate, which ends in *-ate*; therefore, the name of the acid is *acetic acid*.

SOLUTION

$HC_2H_3O_2(aq)$ acetic acid

▶**SKILLBUILDER 3.13 | Naming Oxyacids**

Name $HNO_2(aq)$.

▶**FOR MORE PRACTICE** Examples 3.37, 3.38; Problems 98acd, 99abc.

TABLE 3.7 Names of Some Common Oxyacids and Their Oxyanions

Acid Formula	Acid Name	Oxyanion Name	Oxyanion Formula
HNO_2	nitrous acid	nitrite	NO_2^-
HNO_3	nitric acid	nitrate	NO_3^-
H_2SO_3	sulfurous acid	sulfite	SO_3^{2-}
H_2SO_4	sulfuric acid	sulfate	SO_4^{2-}
$HClO_2$	chlorous acid	chlorite	ClO_2^-
$HClO_3$	chloric acid	chlorate	ClO_3^-
$HC_2H_3O_2$	acetic acid	acetate	$C_2H_3O_2^-$
H_2CO_3	carbonic acid	carbonate	CO_3^{2-}

CHEMISTRY IN THE ENVIRONMENT

▶ Acid Rain

Acid rain occurs when rainwater mixes with air pollutants—such as NO, NO_2, and SO_2—that form acids. NO and NO_2, primarily from vehicular emission, combine with water to form $HNO_3(aq)$. SO_2, primarily from coal-powered electricity generation, combines with water and oxygen in air to form $H_2SO_4(aq)$. $HNO_3(aq)$ and $H_2SO_4(aq)$ both cause rainwater to become acidic. The problem is greatest in the northeastern United States, where pollutants from midwestern electrical power plants combine with rainwater to produce rain with acid levels that are 10 times higher than normal.

When acid rain falls or flows into lakes and streams, it makes them more acidic. Some species of aquatic animals—such as trout, bass, snails, salamanders, and clams—cannot tolerate the increased acidity and die. These deaths disturb the ecosystem of the lake, resulting in imbalances that may lead in turn to the deaths of other aquatic species. Acid rain also weakens trees by dissolving nutrients in the soil and by damaging leaves. Appalachian red spruce trees have been the hardest hit, with many forests showing significant acid rain damage.

Acid rain also damages building materials. Acids dissolve $CaCO_3$ (limestone), a component of marble and concrete, and iron, the main component of steel. Consequently, many statues, buildings, and bridges in the northeastern United States show significant deterioration, and some historical gravestones made of limestone are barely legible due to acid rain damage.

▲ A forest damaged by acid rain. © Photoshot Holdings Ltd/Alamy.

Although acid rain has been a problem for many years, innovative legislation has offered hope for change. In 1990, Congress passed several amendments to the Clean Air Act that included provisions requiring electrical utilities to reduce SO_2 emissions. Since then, SO_2 emissions have decreased, and rain in the northeastern United States has become somewhat less acidic. For example, according to the U.S. Environmental Protection Agency, average U.S. atmospheric SO_2 concentrations were 0.012 ppm in 1980, but have decreased to less than 0.003 ppm today. With time, and continued enforcement of the acid rain program, lakes, streams, and forests damaged by acid rain should recover. However, acid rain continues to worsen in countries such as China, where industrial growth is outpacing environmental controls. International cooperation is essential to solving environmental problems such as acid rain.

B3.2 CAN YOU ANSWER THIS? *Name each compound, given here as formulas:*

NO, NO_2, SO_2, $HNO_3(aq)$, $CaCO_3$

▲ Acid rain harms many materials, including the limestone often used for tombstones, buildings, and statues. © Nickos/iStockphoto/Getty Images.

3.10 Nomenclature Summary

LO: Recognize and name chemical compounds.

Acids are technically a subclass of molecular compounds; that is, they are molecular compounds that form H^+ ions when dissolved in water.

Naming compounds requires several steps. The flowchart in ▶ Figure 3.16 summarizes the different categories of compounds that we have covered in the module and how to identify and name them. The first step is to decide whether the compound is ionic, molecular, or an acid. We can recognize ionic compounds by the presence of a metal or ammonium and a nonmetal, molecular compounds by two or more nonmetals, and acids by the presence of hydrogen (written first) and one or more nonmetals.

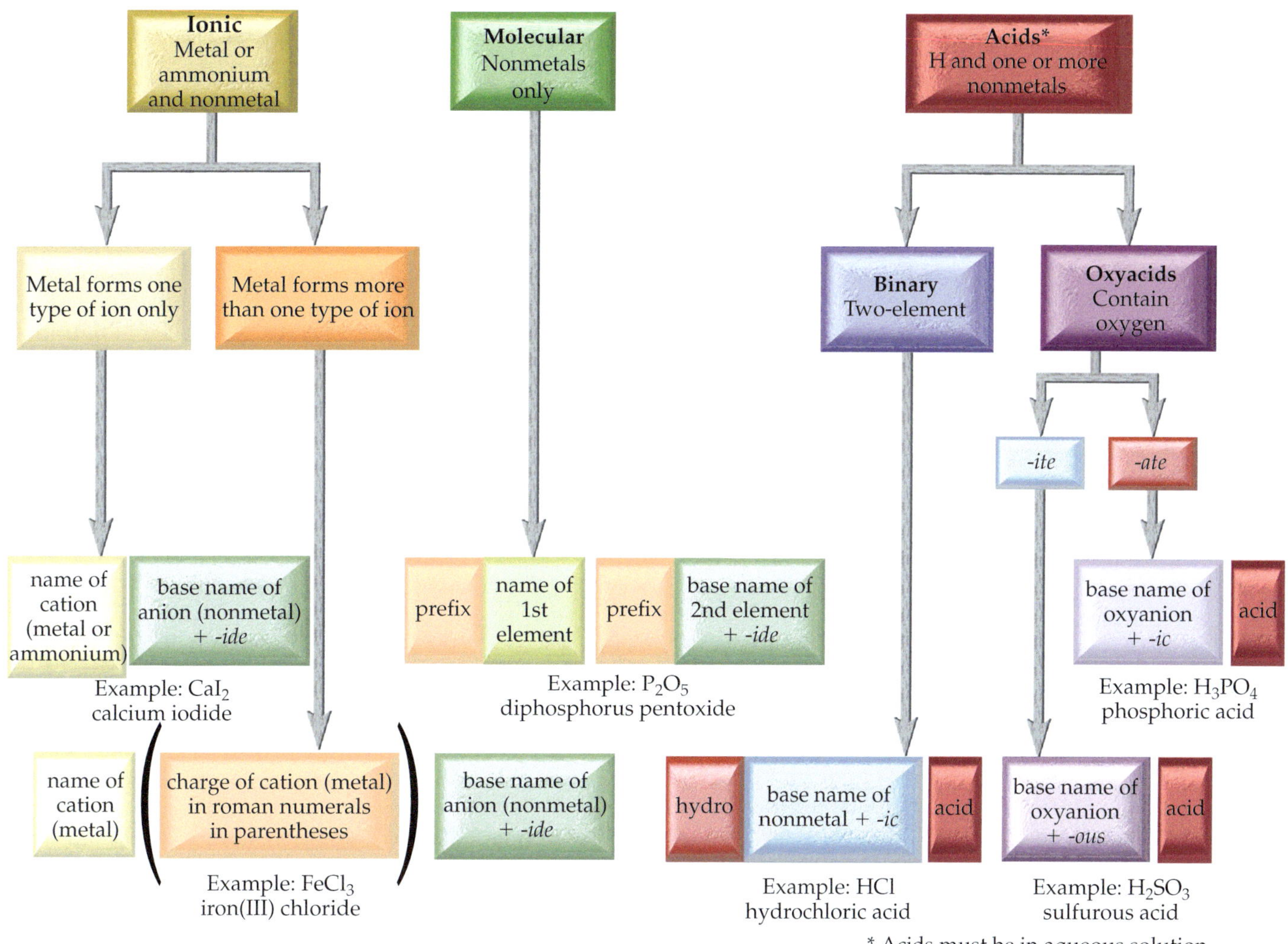

▲ **FIGURE 3.16 Nomenclature flowchart**

Ionic Compounds

Zinc, scandium, and silver predominantly exist as Zn^{2+}, Sc^{3+}, and Ag^{+} cations.

For an ionic compound, we must next decide whether the metal forms only one type of ion or more than one type of ion. Group 1A (alkali) metals, Group 2A (alkaline earth) metals, and aluminum always form only one type of ion (Figure 2.5). Most of the transition metals form more than one type of ion. Once we have identified the type of ionic compound, we name it according to the scheme in the chart. If the ionic compound contains a polyatomic ion—something we must learn to recognize by familiarity—we insert the name of the polyatomic ion in place of the metal (positive polyatomic ion) or the nonmetal (negative polyatomic ion).

Molecular Compounds

We have learned how to name only one type of molecular compound, the binary (two-element) compound. If we identify a compound as molecular, we name it according to the scheme in Figure 3.16.

Acids

To name an acid, we must first decide whether it is a binary (two-element) acid or an oxyacid (an acid containing oxygen). We name binary acids according to the scheme in Figure 3.16. We must further subdivide oxyacids based on the name of their corresponding oxyanion. If the oxyanion ends in *-ite*, we use one scheme; if it ends with *-ate*, we use the other.

EXAMPLE 3.14 NOMENCLATURE USING FIGURE 3.16

Name each compound: CO, CaF_2, HF(*aq*), $Fe(NO_3)_3$, $HClO_4$(*aq*). H_2SO_3(*aq*)

SOLUTION

The table below illustrates how to use Figure 3.16 to arrive at a name for each compound.

Formula	Flowchart Path	Name
CO	molecular	carbon monoxide
CaF_2	ionic → one type of ion →	calcium fluoride
HF(*aq*)	acid → binary →	hydrofluoric acid
$Fe(NO_3)_3$	ionic → more than one type of ion →	iron(III) nitrate
$HClO_4$(*aq*)	acid → oxyacid → *-ate* →	perchloric acid
H_2SO_3(*aq*)	acid → oxyacid → *-ite* →	sulfurous acid

▶**FOR MORE PRACTICE** Problems 180, 181.

3.11 Bonding Models and AIDS Drugs

In 1989, researchers discovered the structure of a molecule called HIV-protease. HIV-protease is a protein (a class of biological molecules) synthesized by the human immunodeficiency virus (HIV), which causes AIDS. HIV-protease is crucial to the virus's ability to replicate itself. Without HIV-protease, HIV could not spread in the human body because the virus could not copy itself, and AIDS would not develop.

With knowledge of the HIV-protease structure, drug companies set out to design a molecule that would disable the protease by sticking to the working part of the molecule (called the *active site*). To design such a molecule, researchers used **bonding theories**—models that predict how atoms bond together to form molecules—to simulate how potential drug molecules would interact with the protease molecule. By the early 1990s, these companies had developed several effective drug molecules. Since these molecules inhibit the action of HIV-protease, they are called *protease inhibitors*. In human trials, protease inhibitors in combination with other drugs have decreased the viral count in HIV-infected individuals to undetectable levels. Although these drugs do not cure AIDS, HIV-infected individuals who regularly take their medication can now expect nearly normal life spans.

Bonding theories are central to chemistry because they predict how atoms bond together to form compounds. They predict what combinations of atoms form compounds and what combinations do not. Bonding theories predict why salt is NaCl and not $NaCl_2$ and why water is H_2O and not H_3O. Bonding theories also explain the shapes of molecules, which in turn determine many of their physical and chemical properties. The bonding theory you will learn in this module is the **Lewis model**, named after G. N. Lewis (1875–1946), the American chemist who developed it. In this model, we represent electrons as dots and draw *dot structures* or *Lewis structures* to represent molecules. These structures, which are fairly simple to draw, have tremendous predictive power. It takes just a few minutes to apply the Lewis model to determine whether a particular set of atoms will form a stable molecule and what that molecule might look like. Although modern chemists also use more advanced bonding theories to better predict molecular properties, the Lewis model remains the simplest method for making quick, everyday predictions about molecules.

3.12 Representing Valence Electrons with Dots

LO: Write Lewis structures for elements.

As we discussed in Module 2, valence electrons are those electrons in the outermost principal shell. Since valence electrons are most important in bonding, the Lewis model focuses on these. In the Lewis model, the valence electrons of main-group elements are represented as dots surrounding the symbol of the element. The result is a **Lewis structure**, or **dot structure**. For example, the electron configuration of O is:

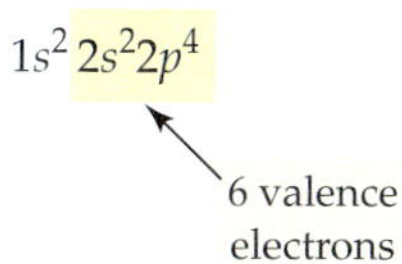

and its Lewis structure is:

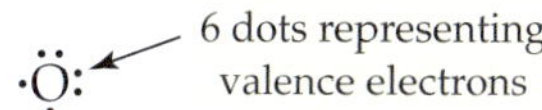

Remember, the number of valence electrons for any main-group element is equal to the group number of the element (except helium, which has 2 valence electrons but is in Group 8A).

Each dot represents a valence electron. We place the dots around the element's symbol with a maximum of two dots per side. Although the exact location of dots is not critical, in this book we fill in the dots singly first and then pair them (with the exception of helium, described shortly).

The Lewis structures for the period 2 elements are:

Lewis structures allow us to easily see the number of valence electrons in an atom. Atoms with 8 valence electrons—which are particularly stable—are easily identified because they have eight dots, an **octet**.

Helium is somewhat of an exception. Its electron configuration and Lewis structure are:

$1s^2$ He:

The Lewis structure of helium contains two paired dots (a **duet**). For helium, a duet represents a stable electron configuration.

In the Lewis model, a **chemical bond** involves the sharing or transfer of electrons to attain stable electron configurations for the bonding atoms. If the electrons are transferred, the bond is an **ionic bond**. If the electrons are shared, the bond is a **covalent bond**. In either case, the bonding atoms attain stable electron configurations. As we have seen, a stable configuration usually consists of eight electrons in the outermost or valence shell. This observation leads to the **octet rule**:

> In chemical bonding, atoms transfer or share electrons to obtain outer shells with eight electrons.

The octet rule generally applies to all main-group elements except hydrogen and lithium. Each of these elements achieves stability when it has two electrons (a duet) in its outermost shell.

EXAMPLE 3.15 WRITING LEWIS STRUCTURES FOR ELEMENTS

Write the Lewis structure of phosphorus.

	SOLUTION
Since phosphorus is in Group 5A in the periodic table, it has 5 valence electrons. Represent these as five dots surrounding the symbol for phosphorus.	$\cdot\dot{\underset{\cdot}{\text{P}}}:$

▶SKILLBUILDER 3.14 | Writing Lewis Structures for Elements

Write the Lewis structure of Mg.

▶FOR MORE PRACTICE Example 3.39; Problems 106, 107.

CONCEPTUAL CHECKPOINT 3.7

Which two elements have the most similar Lewis structures?

(a) C and Si **(b)** O and P **(c)** Li and F **(d)** S and Br

3.13 Lewis Structures of Ionic Compounds: Electrons Transferred

LO: Write Lewis structures for ionic compounds.

LO: Use the Lewis model to predict the chemical formula of an ionic compound.

Recall that when metals bond with nonmetals, electrons are transferred from the metal to the nonmetal. The metal becomes a cation and the nonmetal becomes an anion. The attraction between the cation and the anion results in an ionic compound. In the Lewis model, we represent this by moving electron dots from the metal to the nonmetal. For example, the Lewis structures for potassium and chlorine are:

$$\text{K}\cdot \quad :\ddot{\underset{\cdot}{\text{Cl}}}:$$

When potassium and chlorine bond, potassium transfers its valence electron to chlorine.

$$\text{K}\cdot \quad :\ddot{\underset{\cdot}{\text{Cl}}}: \longrightarrow \text{K}^+\ [:\ddot{\underset{\cdot\cdot}{\text{Cl}}}:]^-$$

Recall from Section 2.5 that atoms that lose electrons become positively charged and atoms that gain electrons become negatively charged.

The transfer of the electron gives chlorine an octet (shown as eight dots around chlorine) and leaves potassium with an octet in the previous principal shell, which is now the valence shell. Because the potassium lost an electron, it becomes positively charged, while the chlorine, which gained an electron, becomes negatively charged. The Lewis structure of an anion is usually written within brackets with the charge in the upper right corner (outside the brackets). The positive and negative charges attract one another, forming the compound KCl.

EXAMPLE 3.16 WRITING IONIC LEWIS STRUCTURES

Write the Lewis structure of the compound MgO.

	SOLUTION
Draw the Lewis structures of magnesium and oxygen by drawing two dots around the symbol for magnesium and six dots around the symbol for oxygen.	$\cdot\text{Mg}\cdot \quad \cdot\ddot{\underset{\cdot}{\text{O}}}:$
In MgO, magnesium loses its 2 valence electrons, resulting in a 2+ charge, and oxygen gains two electrons, attaining a 2– charge and an octet.	$\text{Mg}^{2+}\ [:\ddot{\underset{\cdot\cdot}{\text{O}}}:]^{2-}$

▶SKILLBUILDER 3.15 | Writing Ionic Lewis Structures

Write the Lewis structure of the compound NaBr.

▶FOR MORE PRACTICE Example 3.40; Problems 118, 119.

Recall from Section 3.4 that ionic compounds do not exist as distinct molecules, but rather as part of a large lattice of alternating cations and anions.

The Lewis model predicts the correct chemical formulas for ionic compounds. For the compound that forms between K and Cl, for example, the Lewis model predicts one potassium cation to every chlorine anion, KCl. As another example, consider the ionic compound formed between sodium and sulfur. The Lewis structures for sodium and sulfur are:

Na· ·S̤̈:

Notice that sodium must lose its one valence electron to obtain an octet (in the previous principal shell), while sulfur must gain two electrons to obtain an octet. Consequently, the compound that forms between sodium and sulfur requires two sodium atoms to every one sulfur atom. The Lewis structure is:

Na^+ [:S̤̈:]$^{2-}$ Na^+

The two sodium atoms each lose their single valence electron, while the sulfur atom gains two electrons and obtains an octet. The correct chemical formula is Na_2S.

EXAMPLE 3.17 USING THE LEWIS MODEL TO PREDICT THE CHEMICAL FORMULA OF AN IONIC COMPOUND

Use the Lewis model to predict the formula of the compound that forms between calcium and chlorine.

	SOLUTION
Draw the Lewis structures of calcium and chlorine by drawing two dots around the symbol for calcium and seven dots around the symbol for chlorine.	·Ca· :C̤̈l:
Calcium must lose its 2 valence electrons (to effectively attain an octet in its previous principal shell), while chlorine needs to gain only 1 electron to obtain an octet. Consequently, the compound that forms between Ca and Cl has two chlorine atoms to every one calcium atom.	[:C̤̈l:]$^-$ Ca^{2+} [:C̤̈l:]$^-$ The formula is therefore $CaCl_2$.

▶**SKILLBUILDER 3.16 | Using the Lewis Model to Predict the Chemical Formula of an Ionic Compound**

Use the Lewis model to predict the formula of the compound that forms between magnesium and nitrogen.

▶**FOR MORE PRACTICE** Example 3.41; Problems 120, 121, 122, 123.

CONCEPTUAL CHECKPOINT 3.8

Which nonmetal forms an ionic compound with aluminum that has the formula Al_2X_3 (where X represents the nonmetal)?

(a) Cl **(b)** S **(c)** N **(d)** C

3.14 Covalent Lewis Structures: Electrons Shared

LO: Write Lewis structures for covalent compounds.

Recall that when nonmetals bond with other nonmetals, a molecular compound results. Molecular compounds contain covalent bonds in which electrons are shared between atoms rather than transferred. In the Lewis model, we represent covalent bonding by allowing neighboring atoms to share some of their valence electrons in order to attain octets (or duets for hydrogen). For example, hydrogen and oxygen have the Lewis structures:

H· ·Ö:

In water, hydrogen and oxygen share their electrons so that each hydrogen atom gets a duet and the oxygen atom gets an octet.

H:Ö̤:H

The shared electrons—those that appear in the space between the two atoms—count toward the octets (or duets) of *both of the atoms*.

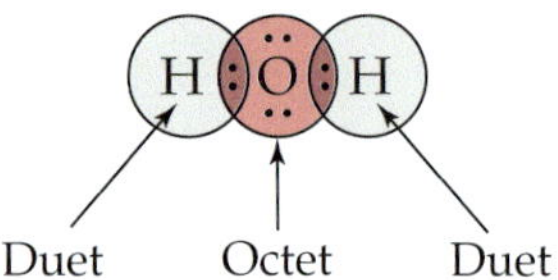

Electrons that are shared between two atoms are **bonding pair** electrons, while those that are only on one atom are **lone pair** (or nonbonding) electrons.

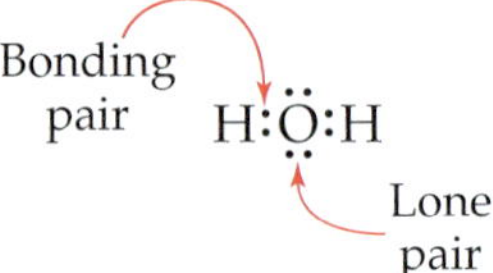

Bonding pair electrons are often represented by dashes to emphasize that they are a chemical bond.

Remember that each dash represents a *pair* of shared electrons.

H—Ö—H

The Lewis model also explains why the halogens form diatomic molecules. Consider the Lewis structure of chlorine:

If two Cl atoms pair, they can each attain an octet.

:C̈l:C̈l: *or* :C̈l—C̈l:

When we examine elemental chlorine, we find that it indeed exists as a diatomic molecule, just as the Lewis model predicts. The same is true for the other halogens.

Similarly, the Lewis model predicts that hydrogen, which has this Lewis structure:

H·

should exist as H_2. When two hydrogen atoms share their valence electrons, they each get a duet, a stable configuration for hydrogen.

H:H *or* H—H

Again, the Lewis model prediction is correct. In nature, elemental hydrogen exists as H_2 molecules.

Double and Triple Bonds

In the Lewis model, atoms can share more than one electron pair to attain an octet. For example, we know from Section 3.4 that oxygen exists as the diatomic molecule, O_2. The Lewis structure of an oxygen atom is:

·Ö:

If we pair two oxygen atoms together and then try to write the Lewis structure, we do not have enough electrons to give each O atom an octet.

:Ö:Ö:

However, we can convert a lone pair into an additional bonding pair by moving it into the bonding region.

:Ö:Ö: ⟶ :Ö::Ö: *or* :Ö=Ö:

Each oxygen atom now has an octet because the additional bonding pair counts toward the octet of both oxygen atoms.

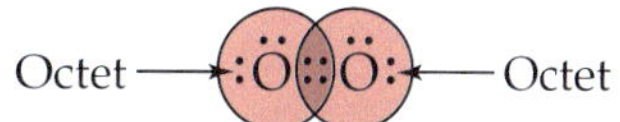

When two electron pairs are shared between two atoms, the resulting bond is a **double bond**. In general, double bonds are shorter and stronger than single bonds. For example, the distance between oxygen nuclei in an oxygen–oxygen double bond is 121 pm. In a single bond, it is 148 pm.

1 pm = 10^{-12} m

Atoms can also share three electron pairs. Consider the Lewis structure of N_2. Since each N atom has 5 valence electrons, the Lewis structure for N_2 has 10 electrons. A first attempt at writing the Lewis structure looks like this:

As with O_2, we do not have enough electrons to satisfy the octet rule for both N atoms. However, if we convert two additional lone pairs into bonding pairs, each nitrogen atom can get an octet.

:N:N: ⟶ :N:::N: *or* :N≡N:

The resulting bond is a **triple bond**. Triple bonds are even shorter and stronger than double bonds. The distance between nitrogen nuclei in a nitrogen–nitrogen triple bond is 110 pm. In a double bond, the distance is 124 pm. When we examine nitrogen in nature, we find that it exists as a diatomic molecule with a very strong short bond between the two nitrogen atoms. The bond is so strong that it is difficult to break, making N_2 a relatively unreactive molecule.

CONCEPTUAL CHECKPOINT 3.9

How many bonding electrons are in the Lewis structure of O_2?

(a) 2 **(b)** 4 **(c)** 6

3.15 Writing Lewis Structures for Covalent Compounds

LO: Write Lewis structures for covalent compounds.

When guessing at skeletal structures, put the less metallic elements in terminal positions and the more metallic elements in central positions. Halogens, which are among the least metallic elements, are nearly always terminal.

Nonterminal hydrogen atoms exist in some compounds. However, they are rare and beyond the scope of this text.

To write the Lewis structure for a covalent compound, follow these steps:

1. **Write the correct skeletal structure for the molecule.** The skeletal structure shows the relative positions of the atoms and does not include electrons, but it must have the atoms in the correct positions. For example, you could *not* write the Lewis structure for water if you started with the hydrogen atoms next to each other and the oxygen atom at the end (H H O). In nature, oxygen is the central atom, and the hydrogen atoms are **terminal atoms** (at the ends). The correct skeletal structure is H O H.

   The only way to absolutely know the correct skeletal structure for any molecule is to examine its structure in nature. However, you can write likely skeletal structures by remembering two guidelines. First, *hydrogen atoms are always terminal*. Since hydrogen requires only a duet, it is never a central atom because central atoms must be able to form at least two bonds and hydrogen can form only one. Second, *many molecules tend to be symmetrical*, so when a molecule contains several atoms of the same type, these tend to be in terminal positions. *This second guideline, however, has many exceptions*. In cases where the skeletal structure is unclear, this text provides you with the correct skeletal structure.

2. **Calculate the total number of electrons for the Lewis structure by summing the valence electrons of each atom in the molecule.** Remember that the number of valence electrons for any main-group element is equal to its group number

The central atom can have more than eight electrons. However the expanded octets and other exceptions to the Octet Rule are not covered in this book.

in the periodic table. **If you are writing a Lewis structure for a polyatomic ion, you must consider the charge of the ion when calculating the total number of electrons.** Add one electron for each negative charge and subtract one electron for each positive charge.

3. **Distribute the electrons among the atoms, giving octets (or duets for hydrogen) to as many atoms as possible.** Begin by placing two electrons between each pair of atoms. These are the minimal number of bonding electrons. Then distribute the remaining electrons, first to terminal atoms and then to the central atom, giving octets to as many atoms as possible.
4. **If any atoms lack an octet, form double or triple bonds as necessary to give them octets.** Do this by moving lone electron pairs from terminal atoms into the bonding region with the central atom.

A brief version of this procedure is presented in the left column. In the center and right columns, Examples 3.18 and 3.19 demonstrate the procedure.

WRITING LEWIS STRUCTURES FOR COVALENT COMPOUNDS	**EXAMPLE 3.18** Write the Lewis structure for CO_2.	**EXAMPLE 3.19** Write the Lewis structure for CCl_4.
1. Write the correct skeletal structure for the molecule.	**SOLUTION** Following the symmetry guideline, write: O C O	**SOLUTION** Following the symmetry guideline, write: Cl Cl C Cl Cl
2. Calculate the total number of electrons for the Lewis structure by summing the valence electrons of each atom in the molecule.	Total number of electrons for Lewis structure = $\begin{pmatrix}\#\text{ valence}\\ e^-\text{ for C}\end{pmatrix} + 2\begin{pmatrix}\#\text{ valence}\\ e^-\text{ for O}\end{pmatrix}$ $= 4 + 2(6)$ $= 16$	Total number of electrons for Lewis structure = $\begin{pmatrix}\#\text{ valence}\\ e^-\text{ for C}\end{pmatrix} + 4\begin{pmatrix}\#\text{ valence}\\ e^-\text{ for Cl}\end{pmatrix}$ $= 4 + 4(7)$ $= 32$
3. Distribute the electrons among the atoms, giving octets (or duets for hydrogen) to as many atoms as possible. Begin with the bonding electrons, and proceed to lone pairs on terminal atoms, and finally to lone pairs on the central atom.	Work with bonding electrons first. O:C:O (4 of 16 electrons used) Proceed to lone pairs on terminal atoms next. :Ö:C:Ö: (with lone pairs above and below each O) (16 of 16 electrons used)	Work with bonding electrons first. Cl Cl:C:Cl Cl (8 of 32 electrons used) Proceed to lone pairs on terminal atoms next. :Cl: :Cl:C:Cl: :Cl: (each Cl with four lone pairs) (32 of 32 electrons used)
4. If any atoms lack octets, form double or triple bonds as necessary to give them octets.	Move lone pairs from the oxygen atoms to bonding regions to form double bonds. :Ö:C:Ö: ⟶ :O::C::O: (each O with two lone pairs) ▶**SKILLBUILDER 3.17** | Write the Lewis structure for CO.	Since all of the atoms have octets, the Lewis structure is complete. ▶**SKILLBUILDER 3.18** | Write the Lewis structure for H_2CO. ▶**FOR MORE PRACTICE** Example 3.42; Problems 128, 129, 130, 131, 132, 133.

Writing Lewis Structures for Polyatomic Ions

We write Lewis structures for polyatomic ions by following the same procedure, but we pay special attention to the charge of the ion when calculating the number of electrons for the Lewis structure. We add one electron for each negative charge and subtract one electron for each positive charge. We normally show the Lewis structure for a polyatomic ion within brackets and write the charge of the ion in the upper right corner. For example, suppose we want to write the Lewis structure for the CN^- ion. We begin by writing the skeletal structure:

C N

Next we calculate the total number of electrons for the Lewis structure by summing the number of valence electrons for each atom and adding one for the negative charge.

Total number of electrons for Lewis structure = (# valence e^- in C) + (# valence e^- in N) + 1

= 4 + 5 + 1

= 10

Add one e^- to account for 1− charge of ion.

We then place 2 electrons between each pair of atoms

C:N (2 of 10 electrons used)

and distribute the remaining electrons.

:C̤̈:N̤̈: (10 of 10 electrons used)

Since neither of the atoms has octets, we move two lone pairs into the bonding region to form a triple bond, giving both atoms octets. We also enclose the Lewis structure in brackets and write the charge of the ion in the upper right corner.

$[:C:::N:]^-$ *or* $[:C{\equiv}N:]^-$

CONCEPTUAL CHECKPOINT 3.10

What is the total number of electrons in the Lewis structure of OH^-?

(a) 6 **(b)** 7 **(c)** 8 **(d)** 9

EXAMPLE 3.20 WRITING LEWIS STRUCTURES FOR POLYATOMIC IONS

Write the Lewis structure for the NH_4^+ ion.

	SOLUTION
Begin by writing the skeletal structure. Hydrogen atoms must be terminal, and following the guideline of symmetry, the nitrogen atom should be in the middle surrounded by four hydrogen atoms.	H H N H H
Calculate the total number of electrons for the Lewis structure by summing the number of valence electrons for each atom and subtracting 1 for the positive charge.	Total number of electrons for Lewis structure = 5 + 4 − 1 = 8 (5: # valence e^- in N; 4: 4 × (# valence e^- in H); −1: Subtract 1 e to account for 1+ charge of ion.)
Next, place 2 electrons between each pair of atoms.	H H:N̤̈:H (8 of 8 electrons used) H

Since the nitrogen atom has an octet and all of the hydrogen atoms have duets, the placement of electrons is complete. Write the entire Lewis structure in brackets indicating the charge of the ion in the upper right corner.

$$\left[\begin{array}{c} \text{H} \\ \text{H}:\ddot{\text{N}}:\text{H} \\ \text{H} \end{array}\right]^{+} \quad \textit{or} \quad \left[\begin{array}{c} \text{H} \\ | \\ \text{H}-\text{N}-\text{H} \\ | \\ \text{H} \end{array}\right]^{+}$$

▶SKILLBUILDER 3.19 | Writing Lewis Structures for Polyatomic Ions

Write the Lewis structure for the ClO^- ion.

▶FOR MORE PRACTICE Problems 136, 137.

CONCEPTUAL CHECKPOINT 3.11

Which two species have the same number of lone electron pairs in their Lewis structures?

(a) H_2O and H_3O^+

(b) NH_3 and H_3O^+

(c) NH_3 and CH_4

(d) NH_3 and NH_4^+

3.16 Resonance: Equivalent Lewis Structures for the Same Molecule

LO: Write resonance structures.

When writing Lewis structures, we may find that, for some molecules, we can write more than one good Lewis structure. For example, consider writing a Lewis structure for SO_2. We begin with the skeletal structure:

O S O

We then sum the valence electrons.

Total number of electrons for Lewis structure

$$= (\#\text{ valence } e^- \text{ in S}) + 2(\#\text{ valence } e^- \text{ in O})$$

$$= 6 + 2(6)$$

$$= 18$$

We next place 2 electrons between each pair of atoms

O:S:O (4 of 18 electrons used)

and distribute the remaining electrons, first to terminal atoms

$:\ddot{\underset{\cdot\cdot}{\text{O}}}:\text{S}:\ddot{\underset{\cdot\cdot}{\text{O}}}:$ (16 of 18 electrons used)

and finally to the central atom.

$:\ddot{\underset{\cdot\cdot}{\text{O}}}:\ddot{\text{S}}:\ddot{\underset{\cdot\cdot}{\text{O}}}:$ (18 of 18 electrons used)

Since the central atom lacks an octet, we move one lone pair from an oxygen atom into the bonding region to form a double bond, giving all of the atoms octets.

$:\ddot{\text{O}}::\ddot{\text{S}}:\ddot{\underset{\cdot\cdot}{\text{O}}}:$ *or* $:\ddot{\text{O}}=\ddot{\text{S}}-\ddot{\underset{\cdot\cdot}{\text{O}}}:$

However, we could have formed the double bond with the other oxygen atom.

$:\ddot{\underset{\cdot\cdot}{\text{O}}}-\ddot{\text{S}}=\ddot{\text{O}}:$

These two Lewis structures are equally correct. In cases such as this—where we can write two or more equivalent (or nearly equivalent) Lewis structures for the same molecule—we find that the molecule exists in nature as an average or intermediate between the two Lewis structures. Both of the two Lewis structures for SO_2 predict that SO_2 would contain two different kinds of bonds (one double bond and one single bond). However, when we examine SO_2 in nature, we find that both of the bonds are equivalent and intermediate in strength and length between a double bond and single bond. We address this in the Lewis model by representing the molecule with both structures, called **resonance structures**, with a double-headed arrow between them.

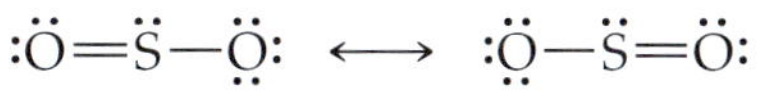

$$:\ddot{O}=\ddot{S}-\ddot{\underset{..}{O}}: \longleftrightarrow :\ddot{\underset{..}{O}}-\ddot{S}=\ddot{O}:$$

The true structure of SO_2 is intermediate between these two resonance structures.

EXAMPLE 3.21 WRITING RESONANCE STRUCTURES

Write the Lewis structure for the NO_3^- ion. Include resonance structures.

	SOLUTION
Begin by writing the skeletal structure. Applying the guideline of symmetry, make the three oxygen atoms terminal.	O (above N); O N O
Sum the valence electrons (adding one electron to account for the 1– charge) to determine the total number of electrons in the Lewis structure.	Total number of electrons for Lewis structure = 5 + 3(6) + 1 = 24 (5 = # valence e^- in N; 3(6) = 3 × (# valence e^- in O); 1 = Add one e^- to account for negative charge of ion.)
Place 2 electrons between each pair of atoms.	 O (above, bonded by : to N); O:N:O (6 of 24 electrons used)
Distribute the remaining electrons, first to terminal atoms.	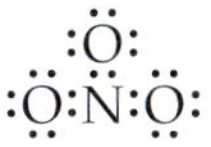 :Ö: (above N); :Ö:N:Ö: (24 of 24 electrons used)
Since there are no electrons remaining to complete the octet of the central atom, form a double bond by moving a lone pair from one of the oxygen atoms into the bonding region with nitrogen. Enclose the structure in brackets and write the charge at the upper right.	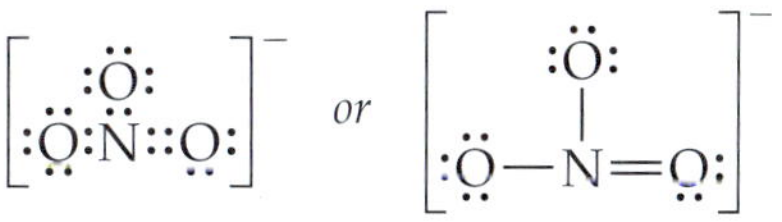 $[:\ddot{O}:N::O:]^-$ (with :Ö: above N) *or* $[:\ddot{O}-N=O:]^-$ (with :Ö: single-bonded above N)
Notice that you can form the double bond with either of the other two oxygen atoms as well.	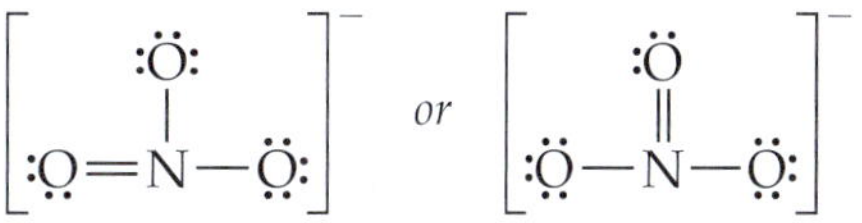 $[:O=N-\ddot{O}:]^-$ (with :Ö: single-bonded above N) *or* $[:\ddot{O}-N-\ddot{O}:]^-$ (with :Ö double-bonded above N)
Since the three Lewis structures are equally correct, write the three structures as resonance structures.	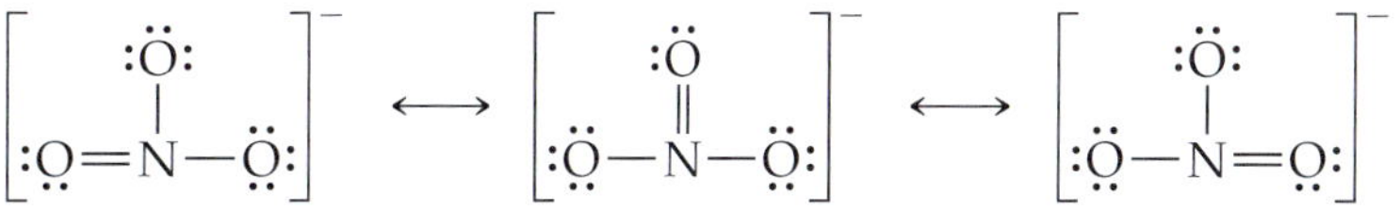 $[:O=N-\ddot{O}:]^-$ (:Ö: above, single bond) ⟷ $[:\ddot{O}-N-\ddot{O}:]^-$ (:Ö above, double bond) ⟷ $[:\ddot{O}-N=O:]^-$ (:Ö: above, single bond)

▶SKILLBUILDER 3.20 | Writing Resonance Structures

Write the Lewis structure for the NO_2^- ion. Include resonance structures.

▶FOR MORE PRACTICE Example 3.43; Problems 138, 139.

3.17 Predicting the Shapes of Molecules

LO: Predict the shapes of molecules.

We can use the Lewis model, in combination with **valence shell electron pair repulsion (VSEPR) theory**, to predict the shapes of molecules. VSEPR theory is based on the idea that **electron groups**—lone pairs, single bonds, or multiple bonds—repel each other. This repulsion between the negative charges of electron groups on the central atom determines the geometry of the molecule. For example, consider CO_2, which has the Lewis structure:

$$:\ddot{O}{=}C{=}\ddot{O}:$$

CHEMISTRY IN THE ENVIRONMENT

▸ The Lewis Structure of Ozone

Ozone is a form of oxygen in which three oxygen atoms bond together. The Lewis structure for O_3 consists of two resonance structures:

$$:\ddot{O}{=}\ddot{O}{-}\ddot{\underset{..}{O}}: \longleftrightarrow :\ddot{\underset{..}{O}}{-}\ddot{O}{=}\ddot{O}:$$

Compare the Lewis structure of ozone to the Lewis structure of O_2:

$$:\ddot{O}{=}\ddot{O}:$$

Which molecule, O_3 or O_2 do you think has the stronger oxygen–oxygen bond? If you deduced O_2 you are correct. O_2 has a stronger bond because it is a pure double bond. Ozone, on the other hand, has bonds that are intermediate between single and double, so O_3 has weaker bonds. The effects of this are significant. As discussed in the *Chemistry in the Environment* box in Module 5, O_3 shields us from harmful ultraviolet light entering Earth's atmosphere. O_3 is ideally suited to do this because photons at wavelengths of 280–320 nm (the most dangerous components of sunlight to humans) are just strong enough to break ozone's bonds. In the process, the photons are absorbed.

$$:\ddot{\underset{..}{O}}{-}\ddot{O}{=}\ddot{O}: + \text{UV light} \longrightarrow :\ddot{O}{=}\ddot{O}: + \cdot\ddot{\underset{..}{O}}\cdot$$

The same wavelengths of UV light, however, do not have sufficient energy to break the stronger double bond of O_2, which is therefore transparent to UV. Consequently, it is important that we continue, and even strengthen, the ban on ozone-depleting compounds.

B3.3 CAN YOU ANSWER THIS? *Why are these Lewis structures for ozone incorrect?*

$$:\ddot{\underset{..}{O}}{-}\ddot{\underset{..}{O}}{-}\ddot{\underset{..}{O}}: \qquad :\ddot{\underset{..}{O}}{=}O{=}\ddot{\underset{..}{O}}:$$

The geometry of CO_2 is determined by the repulsion between the two electron groups (the two double bonds) on the central carbon atom. These two electron groups get as far away from each other as possible, resulting in a bond angle of 180° and a **linear** geometry for CO_2.

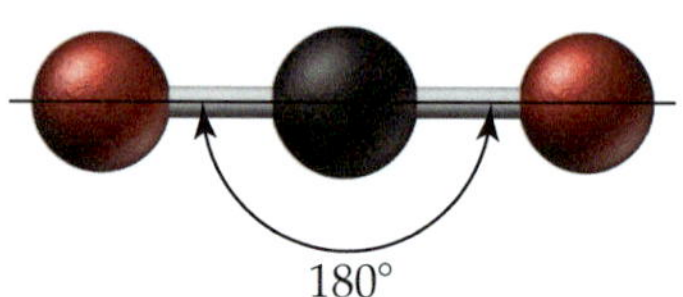

As another example, consider the molecule H_2CO. Its Lewis structure is:

$$\begin{array}{c} :\ddot{O}: \\ \| \\ H{-}C{-}H \end{array}$$

This molecule has three electron groups around the central atom. These three electron groups get as far away from each other as possible, resulting in a bond angle of 120° and a **trigonal planar** geometry.

~120° ~120° ~120°

The angles shown here for H_2CO are approximate. The C=O double bond contains more electron density than do C—H single bonds, resulting in a slightly greater repulsion; thus the HCH bond angle is actually 116°, and the HCO bond angles are actually 122°.

A tetrahedron is a geometric shape with four triangular faces.

CH_4 is shown here with both a ball-and-stick model (left) and a space-filling model (right). Although space-filling models more closely portray molecules, ball-and-stick models are often used to clearly illustrate molecular geometries.

If a molecule has four electron groups around the central atom, as CH_4 does, it has a **tetrahedral** geometry with bond angles of 109.5°.

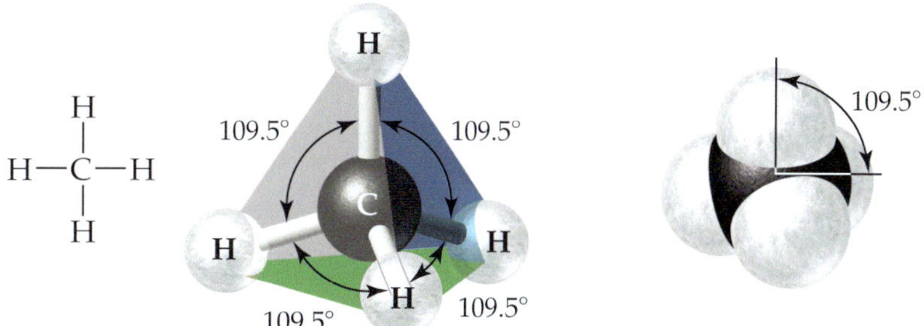

The mutual repulsion of the four electron groups causes the tetrahedral shape; the tetrahedron allows the maximum separation among the four groups. When we write the structure of CH_4 on paper, it may seem that the molecule should be square planar, with bond angles of 90°. However, in three dimensions the electron groups can get farther away from each other by forming the tetrahedral geometry.

Each of the molecules in the preceding examples has only bonding groups of electrons around the central atom. What happens in molecules with lone pairs around the central atom? These lone pairs also repel other electron groups. For example, consider the NH_3 molecule:

H
|
H—N̤—H

The four electron groups (one lone pair and three bonding pairs) get as far away from each other as possible. If we look only at the electrons, we find that the **electron geometry**—the geometrical arrangement of the electron groups—is tetrahedral.

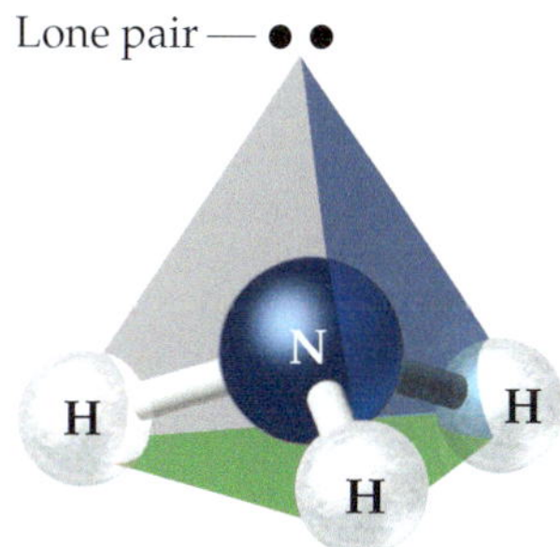

However, the **molecular geometry**—the geometrical arrangement of the atoms—is **trigonal pyramidal**.

Trigonal pyramidal structure

Notice that, although the electron geometry and the molecular geometry are different, the electron geometry is relevant to the molecular geometry. In other words, the lone pair exerts its influence on the bonding pairs.

Consider one last example, H_2O. Its Lewis structure is:

H—Ö̤—H

The bond angles in NH_3 and H_2O are actually a few degrees smaller than the ideal tetrahedral angles.

Since it has four electron groups, its electron geometry is also tetrahedral.

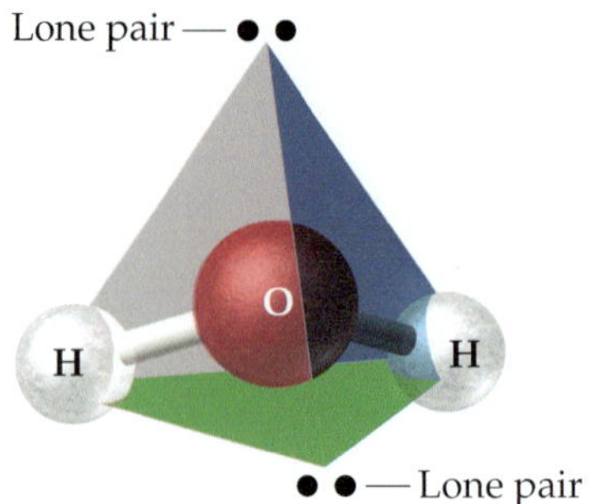

However, its molecular geometry is **bent**.

Bent structure

Table 3.8 summarizes the electron and molecular geometry of a molecule based on the total number of electron groups, the number of bonding groups, and the number of lone pairs.

To determine the geometry of any molecule, we use the procedure presented in the left column of Examples 3.22 and 3.23. As usual, the two examples of applying the steps are in the center and right columns.

TABLE 3.8 Electron and Molecular Geometries

Electron Groups*	Bonding Groups	Lone Pairs	Electron Geometry	Angle between Electron Groups**	Molecular Geometry	Example
2	2	0	linear	180°	linear	:Ö=C=Ö:
3	3	0	trigonal planar	120°	trigonal planar	Ö: ‖ H—C—H
3	2	1	trigonal planar	120°	bent	:Ö=S̈—Ö:
4	4	0	tetrahedral	109.5°	tetrahedral	H | H—C—H | H
4	3	1	tetrahedral	109.5°	trigonal pyramidal	H—N̈—H | H
4	2	2	tetrahedral	109.5°	bent	H—Ö—H

* Count only electron groups around the *central* atom. Each of the following is considered one electron group: a lone pair, a single bond, a double bond, and a triple bond.

** Angles listed here are idealized. Actual angles in specific molecules may vary by several degrees.

PREDICTING GEOMETRY USING VSEPR THEORY	EXAMPLE 3.22	EXAMPLE 3.23
	Predict the electron and molecular geometry of PCl_3.	Predict the electron and molecular geometry of the $[NO_3]^-$ ion.
1. Draw a Lewis structure for the molecule.	**SOLUTION** PCl_3 has 26 electrons. :C̈l: :C̈l:P̈:C̈l:	**SOLUTION** $[NO_3]^-$ has 24 electrons. [:Ö: :Ö:N::Ö:]⁻
2. Determine the total number of electron groups around the central atom. Lone pairs, single bonds, double bonds, and triple bonds each count as one group.	The central atom (P) has four electron groups.	The central atom (N) has three electron groups (the double bond counts as one group).
3. Determine the number of bonding groups and the number of lone pairs around the central atom. These should sum to the result from Step 2. Bonding groups include single bonds, double bonds, and triple bonds.	:C̈l: :C̈l:P̈:C̈l: Lone pair Three of the four electron groups around P are bonding groups, and one is a lone pair.	[:Ö: :Ö:N::Ö:]⁻ No lone pairs All three of the electron groups around N are bonding groups.
4. Refer to Table 3.8 to determine the electron geometry and molecular geometry.	The electron geometry is tetrahedral (four electron groups), and the molecular geometry—the shape of the molecule—is trigonal pyramidal (four electron groups, three bonding groups, and one lone pair). ▶**SKILLBUILDER 3.21** | Predict the molecular geometry of ClNO (N is the central atom).	The electron geometry is trigonal planar (three electron groups), and the molecular geometry—the shape of the molecule—is trigonal planar (three electron groups, three bonding groups, and no lone pairs). ▶**SKILLBUILDER 3.22** | Predict the molecular geometry of the SO_3^{2-} ion. ▶**FOR MORE PRACTICE** Example 3.44; Problems 144, 145, 148, 149, 154, 155.

CONCEPTUAL CHECKPOINT 3.12

Which condition necessarily leads to a molecular geometry that is identical to the electron geometry?

(a) The presence of a double bond between the central atom and a terminal atom.

(b) The presence of two or more identical terminal atoms bonded to the central atom.

(c) The presence of one or more lone pairs on the central atom.

(d) The absence of any lone pairs on the central atom.

Representing Molecular Geometries on Paper

Because molecular geometries are three-dimensional, they are often difficult to represent on two-dimensional paper. Many chemists use this notation for bonds to indicate three-dimensional structures on two-dimensional paper:

—	||||···	◀
Straight line	*Hashed lines*	*Wedge*
Bond in plane of paper	Bond projecting into the paper	Bond projecting out of the paper

CHEMISTRY AND HEALTH

▶Fooled by Molecular Shape

Artificial sweeteners, such as aspartame (Nutrasweet™), taste sweet but have few or no calories. Why? Because taste and caloric value are entirely separate properties of foods. The caloric value of a food depends on the amount of energy released when the food is metabolized. Sucrose (table sugar) is metabolized by oxidation to carbon dioxide and water:

$$C_{12}H_{22}O_{11} + 6\,O_2 \rightarrow 12\,CO_2 + 11\,H_2O \qquad \Delta H = -5644 \text{ kJ}$$

When your body metabolizes one mole of sucrose, it obtains 5644 kJ of energy. Some artificial sweeteners, such as saccharin, are not metabolized at all—they just pass through the body unchanged—and therefore have no caloric value. Other artificial sweeteners, such as aspartame, are metabolized but have a much lower caloric content (for a given amount of sweetness) than sucrose.

The *taste* of a food, however, is independent of its metabolism. The sensation of taste originates in the tongue, where specialized cells called taste cells act as highly sensitive and specific molecular detectors. These cells can distinguish the sugar molecules from the thousands of different types of molecules present in a mouthful of food. The main basis for this discrimination is the molecule's *shape*.

The surface of a taste cell contains specialized protein molecules called taste receptors. Each particular *tastant*—a molecule that you can taste—fits snugly into a special pocket on the taste receptor protein called the *active site*, just as a key fits into a lock. For example, a sugar molecule fits only into the active site of the sugar receptor protein called Tlr3. When the sugar molecule (the key) enters the active site (the lock), the different subunits of the Tlr3 protein split apart. This split causes a series of events that results in transmission of a nerve signal, which reaches the brain and registers a sweet taste.

Artificial sweeteners taste sweet because they fit into the receptor pocket that normally binds sucrose. In fact, both aspartame and saccharin bind to the active site in the Tlr3 protein more strongly than sugar does! For this reason, artificial sweeteners are "sweeter than sugar." It takes 200 times as much sucrose as aspartame to trigger the same amount of nerve signal transmission from taste cells.

This type of lock-and-key fit between the active site of a protein and a particular molecule is important not only to taste but to many other biological functions as well. For example, immune response, the sense of smell, and many types of drug action all depend on shape-specific interactions between molecules and proteins. The ability of scientists to determine the shapes of key biological molecules is largely responsible for the revolution in biology that has occurred over the last 50 years.

B3.4 CAN YOU ANSWER THIS? *Proteins are long-chain molecules in which each link is an amino acid. The simplest amino acid is glycine, which has this structure:*

```
     H   :O:
     |    ||
H—N̈—C—C—Ö—H
  |   |
  H   H
```

Determine the geometry around each interior atom in the glycine structure and make a three-dimensional sketch of the molecule.

The major molecular geometries used in this book are shown here using this notation:

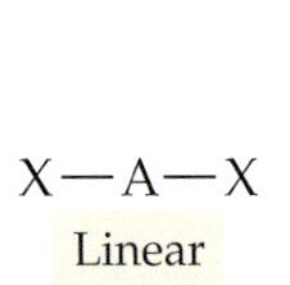

X—A—X
Linear

Trigonal planar

Bent

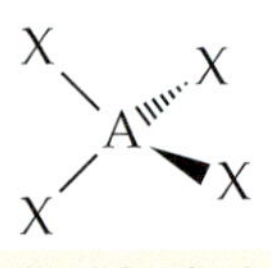

Tetrahedral

Trigonal pyramidal

3.18 Electronegativity and Polarity: Why Oil and Water Don't Mix

LO: Determine whether a molecule is polar.

▲ **FIGURE 3.17 Oil and water don't mix** Question: Why not?
© Kip Peticolas/Fundamental Photographs.

If we combine oil and water in a container, they separate into distinct regions (◀ Figure 3.17). Why? Something about water molecules causes them to bunch together into one region, expelling the oil molecules into a separate region. What is that something? We can begin to understand the answer by examining the Lewis structure of water.

$$H-\ddot{\underset{\cdot\cdot}{O}}-H$$

The two bonds between O and H each consist of an electron pair—2 electrons shared between the oxygen atom and the hydrogen atom. The oxygen and hydrogen atoms each donate one electron to this electron pair; however, like most children, they don't share them equally. The oxygen atom takes more than its fair share of the electron pair.

Electronegativity

The ability of an element to attract electrons within a covalent bond is called **electronegativity.** Oxygen is more electronegative than hydrogen, which means that, on average, the shared electrons are more likely to be found near the oxygen atom than near the hydrogen atom. Consider one of the two OH bonds:

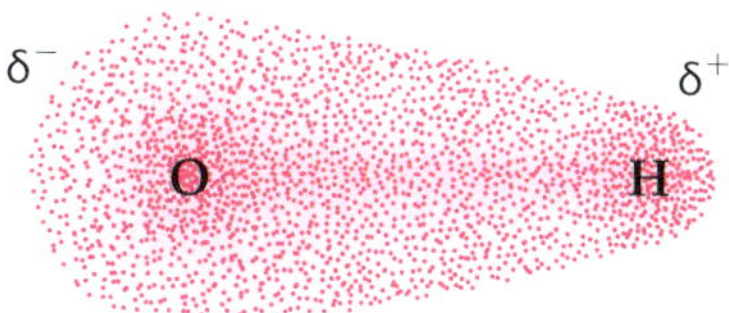

Dipole moment

Since the electron pair is unequally shared (with oxygen getting the larger share), the oxygen atom has a partial negative charge, symbolized by δ– (delta minus). The hydrogen atom (which gets the smaller share) has a partial positive charge, symbolized by δ+ (delta plus). The result of this uneven electron sharing is a **dipole moment**, a separation of charge within the bond. We call covalent bonds that have a dipole moment **polar covalent bonds**. The magnitude of the dipole moment, and therefore the degree of polarity of the bond, depend on the electronegativity difference between the two elements in the bond and the length of the bond. For a fixed bond length, the greater the electronegativity difference, the greater the dipole moment and the more polar the bond.

The value of electronegativity is assigned using a relative scale on which fluorine, the most electronegative element, has an electronegativity of 4.0. All other electronegativities are defined relative to fluorine.

▼ Figure 3.18 shows the relative electronegativities of the elements. Notice that electronegativity increases as we move toward the right across a period in the periodic table and decreases as we move down a column in the periodic

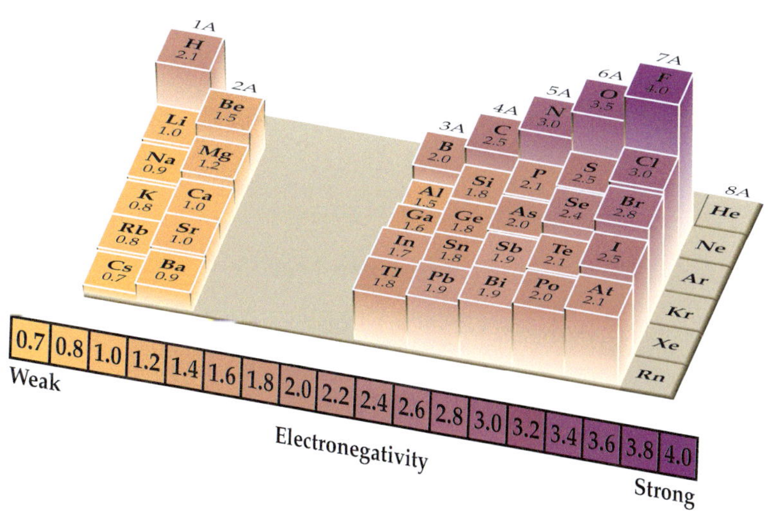

▶ **FIGURE 3.18 Electronegativity of the elements** Linus Pauling introduced the scale shown in this figure. He arbitrarily set the electronegativity of fluorine at 4.0 and calculated all other values relative to fluorine.

▲ **FIGURE 3.19 Pure covalent bonding** In Cl_2, the two Cl atoms share the electrons evenly. This is a pure covalent bond.

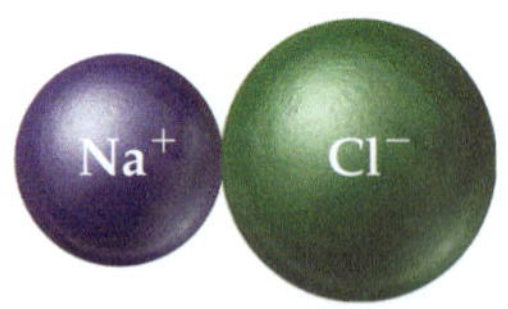

▲ **FIGURE 3.20 Ionic bonding** In NaCl, Na completely transfers an electron to Cl. This is an ionic bond.

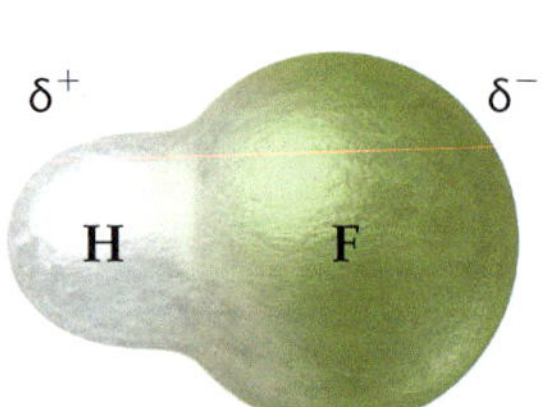

▲ **FIGURE 3.21 Polar covalent bonding** In HF, the electrons are shared, but the shared electrons are more likely to be found on F than on H. The bond is polar covalent.

table. If two atoms of the same element form a covalent bond, they share the electrons equally, and there is no dipole moment. For example, the chlorine molecule, composed of two chlorine atoms (which of course have identical electronegativities), has a pure covalent bond in which electrons are evenly shared (◀ Figure 3.19). The bond has no dipole moment, and the molecule is **nonpolar**.

If there is a large electronegativity difference between the two elements in a bond, such as normally occurs between a metal and a nonmetal, the electron is completely transferred and the bond is ionic. For example, sodium and chlorine form an ionic bond (◀ Figure 3.20).

If there is an intermediate electronegativity difference between the two elements, such as between two different nonmetals, the bond is polar covalent. For example, HF forms a polar covalent bond (◀ Figure 3.21). The larger the electronegativity difference between the two nonmetals, the more polar the bond is.

These concepts are summarized in ▼ Figure 3.22.

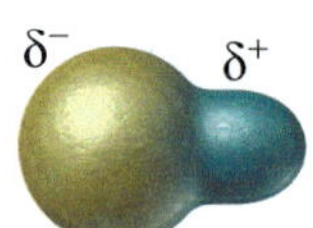

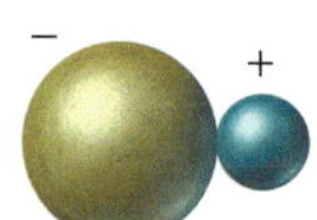

▲ **FIGURE 3.22 The continuum of bond types** The electronegativity difference between two bonded atoms determines the type of bond (pure covalent, polar covalent, or ionic).

EXAMPLE 3.24 CLASSIFYING BONDS AS PURE COVALENT, POLAR COVALENT, OR IONIC

Is the bond formed between each pair of atoms pure covalent, polar covalent, or ionic?

(a) Sr and F **(b)** N and O

SOLUTION

(a) Sr is a metal and F is a nonmetal so the bond between Sr and F is ionic.

(b) N and O are two different nonmetals so the bond between N and O is polar covalent.

▶**SKILLBUILDER 3.23 | Classifying Bonds as Pure Covalent, Polar Covalent, or Ionic**

Is the bond formed between each pair of atoms pure covalent, polar covalent, or ionic?

(a) I and I **(b)** Cs and Br **(c)** P and O

▶**FOR MORE PRACTICE** Problems 160, 161.

CONCEPTUAL CHECKPOINT 3.13

Which bond would you expect to be more polar: the bond in HCl or the bond in HBr?

Polar Bonds and Polar Molecules

Does the presence of one or more polar bonds in a molecule always result in a polar molecule? The answer is no. A **polar molecule** is one with polar bonds that add together—they do not cancel each other—to form a net dipole moment. For diatomic molecules, we can readily tell polar molecules from nonpolar ones. If a diatomic molecule contains a polar bond, then the molecule is polar. However, for molecules with more than two atoms, it is more difficult to tell polar molecules from nonpolar ones because two or more polar bonds may cancel one another. For example, consider carbon dioxide:

$$:\ddot{O}=C=\ddot{O}:$$

Each C=O *bond* is polar because the difference in electronegativity between oxygen and carbon is 1.0. However, since CO_2 has a linear geometry, the dipole moment of one bond completely cancels the dipole moment of the other and the *molecule* is nonpolar. We can understand this with an analogy. Imagine each polar bond to be a rope pulling on the central atom. In CO_2 we can see how the two ropes pulling in opposing directions cancel each other's effect:

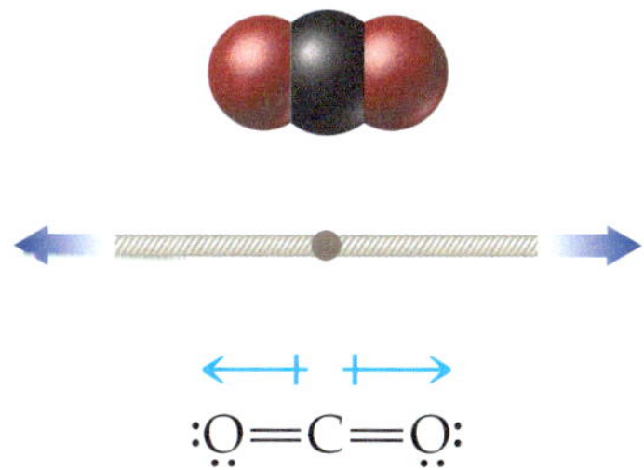

We can also represent polar bonds with arrows (or *vectors*) that point in the direction of the negative pole and have a plus sign at the positive pole (as we just saw for carbon dioxide). If the arrows (or vectors) point in exactly opposing directions as in carbon dioxide, the dipole moments cancel.

In the vector representation of a dipole moment, the vector points in the direction of the atom with the partial negative charge.

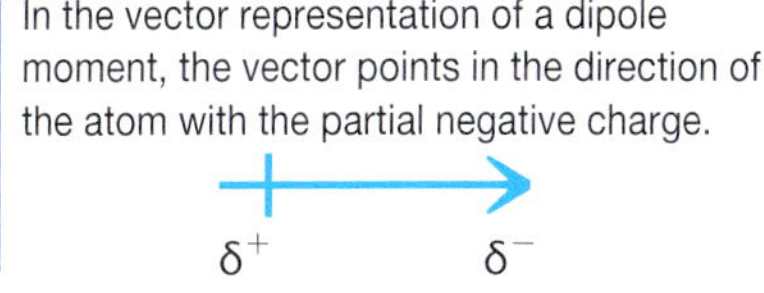

Water, on the other hand, has two dipole moments that do not cancel. If we imagine each bond as a rope pulling on oxygen, we see that, because of the angle between the bonds, the pulls of the two ropes do not cancel:

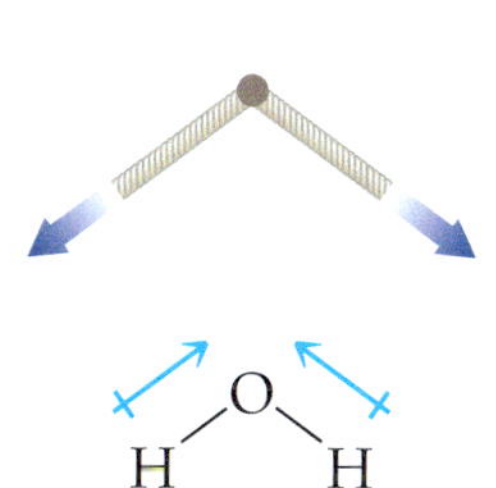

Consequently, water is a polar molecule. We can use symmetry as a guide to determine whether a molecule containing polar bonds is indeed polar. Highly symmetric molecules tend to be nonpolar even if they have polar bonds because the bond dipole moments (or the pulls of the ropes) tend to cancel. Asymmetric molecules that contain polar bonds tend to be polar because the bond dipole moments (or the pulls of the ropes) tend not to cancel. Table 3.9 summarizes some common cases.

In summary, to determine whether a molecule is polar:

- **We determine whether the molecule contains polar bonds.** A bond is polar if the two bonding atoms are from different elements. If there are no polar bonds, the molecule is nonpolar.
- **We determine whether the polar bonds add together to form a net dipole moment.** We must first use VSEPR to determine the geometry of the molecule. Then we visualize each bond as a rope pulling on the central atom. Is the molecule highly symmetrical? Do the pulls of the ropes cancel? If so, there is no net dipole moment and the molecule is nonpolar. If the molecule is asymmetrical and the pulls of the rope do not cancel, the molecule is polar.

TABLE 3.9 Common Cases of Adding Dipole Moments to Determine Whether a Molecule Is Polar

Nonpolar

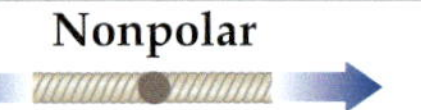

Two identical polar bonds pointing in opposite directions will cancel. The molecule is nonpolar.

Nonpolar

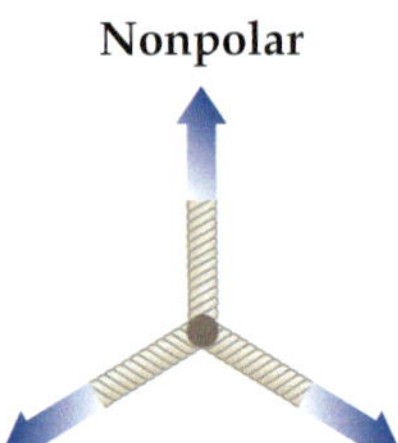

Three identical polar bonds at 120° from each other will cancel. The molecule is nonpolar.

Polar

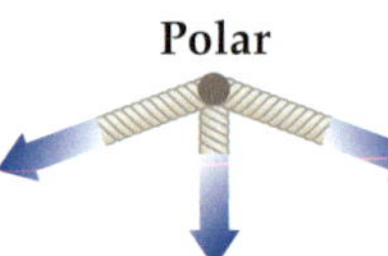

Three polar bonds in a trigonal pyramidal arrangement (109.5°) will not cancel. The molecule is polar.

Polar

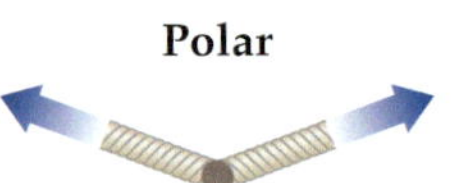

Two polar bonds with an angle of less than 180° between them will not cancel. The molecule is polar.

Nonpolar

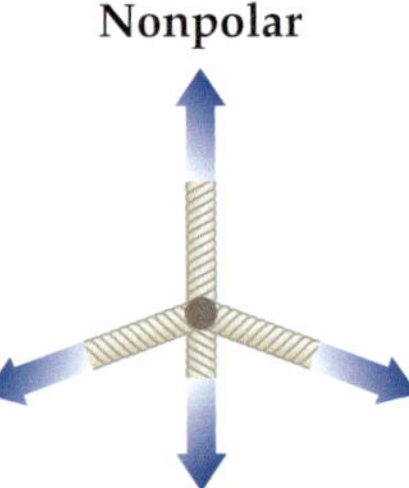

Four identical polar bonds in a tetrahedral arrangement (109.5° from each other) will cancel. The molecule is nonpolar.

Note: In all cases where the polar bonds cancel, the bonds are assumed to be identical. If one or more of the bonds are different than the other(s), the bonds will not cancel and the molecule is polar.

EXAMPLE 3.25 DETERMINING WHETHER A MOLECULE IS POLAR

Is NH_3 polar?

SOLUTION

Begin by drawing the Lewis structure of NH_3. Since N and H have different electronegativities, the bonds are polar.

H
|
H—N̤—H

The geometry of NH_3 is trigonal pyramidal (four electron groups, three bonding groups, one lone pair). Draw a three-dimensional picture of NH_3 and imagine each bond as a rope that is being pulled. The pulls of the ropes do not cancel and the molecule is polar.

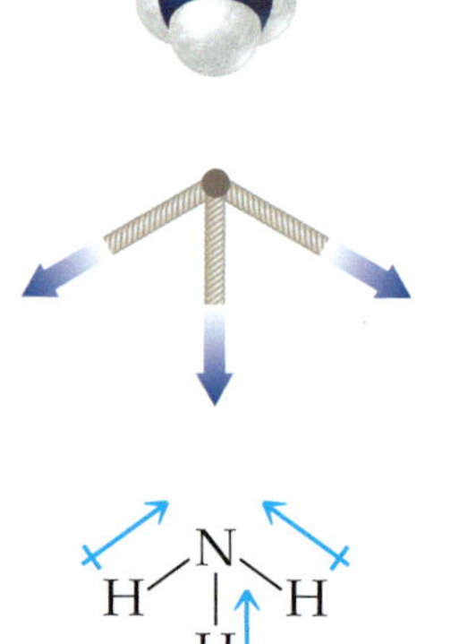

N
H H
H

NH_3 is polar

▶SKILLBUILDER 3.24 | Determining Whether a Molecule Is Polar

Determine whether CH_4 is polar.

▶FOR MORE PRACTICE Example 3.45; Problems 168, 169, 170, 171.

Polarity is important because polar molecules tend to behave differently than nonpolar molecules. Water and oil do not mix, for example, because water molecules are polar and the molecules that compose oil are generally nonpolar. Polar molecules interact strongly with other polar molecules because the positive end of one molecule is attracted to the negative end of another, just as the south pole of a

EVERYDAY CHEMISTRY

▶ How Soap Works

Imagine eating a greasy cheeseburger without flatware or napkins. By the end of the meal, your hands are coated with grease and oil. If you try to wash them with only water, they remain greasy. However, if you add a little soap, the grease washes away. Why? As we learned previously, water molecules are polar and the molecules that compose grease and oil are nonpolar. As a result, water and grease repel each other.

The molecules that compose soap, however, have a special structure that allows them to interact strongly with both water and grease. One end of a soap molecule is polar, while the other end is nonpolar.

Soap molecule

Polar head attracts water

Nonpolar tail attracts grease

The polar head of a soap molecule strongly attracts water molecules, while the nonpolar tail strongly attracts grease and oil molecules. Soap is a sort of molecular liaison, one end interacting with water and the other end interacting with grease. Soap therefore allows water and grease to mix, removing the grease from your hands and washing it down the drain.

B3.5 CAN YOU ANSWER THIS? *Consider this detergent molecule. Which end do you think is polar? Which end is nonpolar?*

$$CH_3(CH_2)_{11}OCH_2CH_2OH$$

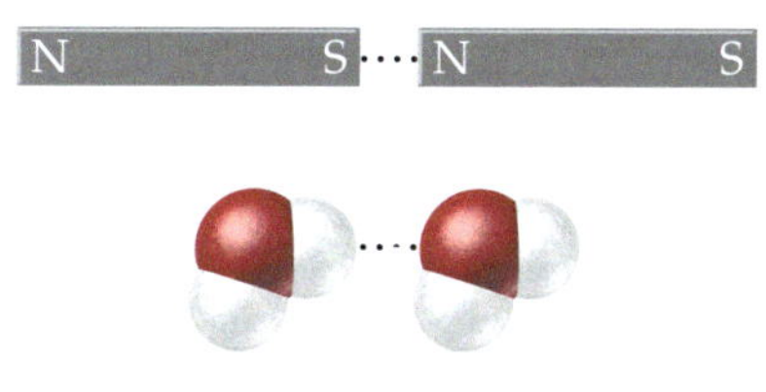

▶ **FIGURE 3.23 Dipole–dipole attraction** Just as the north pole of one magnet is attracted to the south pole of another, so the positive end of one molecule with a dipole is attracted to the negative end of another molecule with a dipole.

magnet is attracted to the north pole of another magnet (▲ Figure 3.23). A mixture of polar and nonpolar molecules is similar to a mixture of small magnetic and nonmagnetic particles. The magnetic particles clump together, excluding the nonmagnetic ones and separating into distinct regions (◀ Figure 3.24). Similarly, the polar water molecules attract one another, forming regions from which the nonpolar oil molecules are excluded (▼ Figure 3.25).

▲ **FIGURE 3.24 Magnetic and nonmagnetic particles** Magnetic particles (the colored marbles) attract one another, excluding nonmagnetic particles (the clear marbles). This behavior is analogous to that of polar and nonpolar molecules. © Richard Megna/Fundamental Photographs.

◀ **FIGURE 3.25 Polar and nonpolar molecules** A mixture of polar and nonpolar molecules, like a mixture of magnetic and nonmagnetic particles, separates into distinct regions because the polar molecules attract one another, excluding the nonpolar ones. Question: Can you think of some examples of this behavior? © Richard Megna/Fundamental Photographs.

MODULE IN REVIEW

Self-Assessment Quiz

Q1. Carbon tetrachloride has a chlorine-to-carbon mass ratio of 11.8:1. If a sample of carbon tetrachloride contains 35 g of chlorine, what mass of carbon does it contain?

(a) 0.34 g C

(b) 1.0 g C

(c) 3.0 g C

(d) 11.8 g C

Q2. Write a chemical formula for a compound that contains two chlorine atoms to every one oxygen atom.

(a) Cl_2O

(b) ClO_2

(c) $2ClO$

(d) $Cl(O_2)_2$

Q3. How many oxygen atoms are in the chemical formula $Fe_2(SO_4)_3$?

(a) 2 **(b)** 3 **(c)** 4 **(d)** 12

Q4. Which element is a molecular element?

(a) copper **(b)** iodine **(c)** krypton **(d)** potassium

Q5. Which compound is ionic?

(a) BrF_5 **(b)** HNO_3 **(c)** $MgSO_4$ **(d)** NI_3

Q6. Write a formula for the compound that forms between Sr and Br.

(a) SrBr **(b)** Sr_2Br **(c)** $SrBr_2$ **(d)** Sr_2Br_2

Q7. Write a formula for the compound that forms between sodium and chlorite ions.

(a) $NaClO_2$

(b) Na_2ClO_2

(c) $Na(ClO)_2$

(d) $NaClO_3$

Q8. Name the compound Li_3N.

(a) Trilithium mononitride

(b) Trilithium nitride

(c) Lithium(I) nitride

(d) Lithium nitride

Q9. Name the compound $CrCl_3$.

(a) Monochromium trichloride

(b) Chromium trichloride

(c) Chromium chloride

(d) Chromium(III) chloride

Q10. Name the compound $BaSO_4$.

(a) Barium sulfate

(b) Barium(II) sulfate

(c) Barium monosulfur tetraoxygen

(d) Barium tetrasulfate

Q11. Name the compound PF_5.

(a) Monophosphorus pentafluoride

(b) Phosphorus pentafluoride

(c) Phosphorus fluoride

(d) Phosphorus(III) fluoride

Q12. What is the formula for manganese(III) oxide?

(a) MnO

(b) Mn_3O

(c) Mn_2O_3

(d) MnO_3

Q13. Name the acid $H_3PO_4(aq)$.

(a) Hydrogen phosphate

(b) Phosphoric acid

(c) Phosphorus acid

(d) Hydrophosphic acid

Q14. What is the formula for hydrobromic acid?

(a) HBr

(b) HBrO

(c) $HBrO_2$

(d) $HBrO_3$

Q15. Which pair of elements has the most similar Lewis structures?

(a) N and S **(b)** F and Ar

(c) Cl and Ar **(d)** O and S

Q16. What is the Lewis structure for the compound that forms between K and S?

(a) K—S—K (S with lone pairs above and below)

(b) $K^+[:\ddot{S}:]^-$

(c) $K^+[:\ddot{S}:]^{2-}K^+$

(d) $[:\ddot{S}:]^{2-}K^+[:\ddot{S}:]^{2-}$

Q17. Use the Lewis model to predict the correct formula for the compound that forms between K and S.

(a) KS **(b)** K_2S **(c)** KS_2 **(d)** K_2S_2

Q18. What is the correct Lewis structure for H_2CS?

(a) H—H—C̤̈—S̤̈:

(b) H—C̤—H, with C single-bonded to :S̈: above

(c) H═C—H, with C single-bonded to :S̈: above

(d) H—C—H, with C double-bonded (‖) to :S̈ above

Q19. How many electron dots are in the Lewis structure of NO_2^-?

(a) 17 **(b)** 18 **(c)** 19 **(d)** 20

Q20. Which compound has two or more resonance structures?

(a) NO_2^- **(b)** CO_2 **(c)** NH_4^+ **(d)** CCl_4

Q21. What is the molecular geometry of PBr_3?

(a) Bent **(b)** Tetrahedral

(c) Trigonal pyramidal **(d)** Linear

Q22. What is the molecular geometry of N_2O (Nitrogen is the central atom.)?

(a) Bent

(b) Tetrahedral

(c) Trigonal pyramidal

(d) Linear

Q23. Which bond is most polar?

(a) A bond between C and S

(b) A bond between Br and Br

(c) A bond between C and O

(d) A bond between B and H

Q24. Which molecule is polar?

(a) SCl_2 **(b)** CS_2 **(c)** CF_4 **(d)** $SiCl_4$

Answers: 1:c, 2:a, 3:d, 4:b, 5:c, 6:c, 7:a, 8:d, 9:d, 10:a, 11:b, 12:c, 13:b, 14:a, 15:d, 16:c, 17:b, 18:d, 19:b, 20:a, 21:c, 22:d, 23:c, 24:a

Chemical Principles	Relevance
Compounds: Matter is ultimately composed of atoms, and those atoms are often combined in compounds. The most important characteristic of a compound is its constant composition. The elements that make up a particular compound are in fixed, definite proportions in all samples of the compound.	**Compounds:** Most of the matter you encounter is in the form of compounds. Water, salt, and carbon dioxide are all examples of common simple compounds. More complex compounds include caffeine, aspirin, acetone, and testosterone.
Chemical Formulas: Chemical formulas represent compounds. Formulas indicate the elements present in the compound and the relative number of atoms of each. These formulas represent the basic units that make up a compound. Pure substances can be categorized according to the basic units that compose them. Elements can be composed of atoms or molecules. Compounds can be molecular, in which case their basic units are molecules, or ionic, in which case their basic units are formula units (composed of cations and anions). We can write the formulas for many ionic compounds simply by knowing the elements in the compound.	**Chemical Formulas:** To understand compounds, you must understand their composition, which is represented by a chemical formula. The connection between the microscopic world and the macroscopic world hinges on the particles that compose matter. Since most matter is in the form of compounds, the properties of most matter depend on the molecules or ions that compose it. Molecular matter does what its molecules do; ionic matter does what its ions do. The world you see and experience is governed by what these particles are doing.
Chemical Nomenclature: We can write the names of simple ionic compounds, molecular compounds, and acids, by examining their chemical formulas. The nomenclature flowchart (Figure 3.16) shows the basic procedure for determining these names.	**Chemical Nomenclature:** Because there are so many compounds, there must be a systematic way to name them. By learning these few simple rules, you will be able to name thousands of different compounds. The next time you look at the label on a consumer product, try to identify as many of the compounds as you can by examining their names.
The Lewis Model: The Lewis model is a model for chemical bonding. According to the Lewis model, chemical bonds form when atoms transfer valence electrons (ionic bonding) or share valence electrons (covalent bonding) to attain noble gas electron configurations. In the Lewis model, we represent valence electrons as dots surrounding the symbol for an element. When two or more elements bond together, the dots are transferred or shared so that every atom attains eight dots (an octet), or two dots (a duet) in the case of hydrogen.	**The Lewis Model:** Bonding theories predict what combinations of elements will form stable compounds, and we can use them to predict the properties of those compounds. For example, pharmaceutical companies use bonding theories when they are designing drug molecules to interact with a specific part of a protein molecule.
Molecular Shapes: We can predict the shapes of molecules by combining the Lewis model with valence shell electron pair repulsion (VSEPR) theory. In this model, electron groups—lone pairs, single bonds, double bonds, and triple bonds—around the central atom repel one another and determine the geometry of the molecule.	**Molecular Shapes:** Molecular shapes determine many of the properties of compounds. Water's bent geometry, for example, causes it to be a liquid at room temperature instead of a gas. It is also the reason ice floats on water and snowflakes have hexagonal patterns.

Electronegativity and Polarity: Electronegativity refers to the relative ability of elements to attract electrons within a chemical bond. Electronegativity increases as we move to the right across a period in the periodic table and decreases as we move down a column. When two atoms of different nonmetals form a covalent bond, the electrons in the bond are not evenly shared and the bond is polar. In diatomic molecules, a polar bond results in a polar molecule. In molecules with more than two atoms, polar bonds may cancel, forming a nonpolar molecule, or they may sum, forming a polar molecule.

Electronegativity and Polarity: The polarity of a molecule influences many of its properties such as whether it is a solid, liquid, or gas at room temperature and whether it mixes with other compounds. Oil and water, for example, do not mix because water is polar while oil is nonpolar.

Chemical Skills

LO: Restate and apply the law of constant composition (Section 3.2).

The law of constant composition states that all samples of a given compound have the same ratio of their constituent elements.

To determine whether experimental data are consistent with the law of constant composition, calculate the ratios of the masses of each element in all samples. When calculating these ratios, it is most convenient to put the larger number in the numerator (top) and the smaller one in the denominator (bottom); that way, the ratio is greater than 1. If the ratios are the same, then the data are consistent with the law of constant composition.

LO: Write chemical formulas (Section 3.3).

Chemical formulas indicate the elements present in a compound and the relative number of atoms of each. When writing formulas, put the more metallic element first.

LO: Determine the total number of atoms of each element in a chemical formula (Section 3.3).

The numbers of atoms not enclosed in parentheses are given directly by their subscript.

Find the numbers of atoms within parentheses by multiplying their subscript within the parentheses by their subscript outside the parentheses.

Examples

EXAMPLE 3.26 CONSTANT COMPOSITION OF COMPOUNDS

Two samples said to be carbon disulfide (CS_2) are decomposed into their constituent elements. One sample produces 8.08 g S and 1.51 g C, while the other produces 31.3 g S and 3.85 g C. Are these results consistent with the law of constant composition?

SOLUTION

Sample 1

$$\frac{\text{Mass S}}{\text{Mass C}} = \frac{8.08 \text{ g}}{1.51 \text{ g}} = 5.35$$

Sample 2

$$\frac{\text{Mass S}}{\text{Mass C}} = \frac{31.3 \text{ g}}{3.85 \text{ g}} = 8.13$$

These results are not consistent with the law of constant composition, so the information that the two samples are the same substance must therefore be in error.

EXAMPLE 3.27 WRITING CHEMICAL FORMULAS

Write a chemical formula for the compound containing one nitrogen atom for every two oxygen atoms.

SOLUTION

NO_2

EXAMPLE 3.28 DETERMINING THE TOTAL NUMBER OF ATOMS OF EACH ELEMENT IN A CHEMICAL FORMULA

Determine the number of atoms of each element in $Pb(ClO_3)_2$.

SOLUTION

One Pb atom

Two Cl atoms

Six O atoms

LO: Classify elements as atomic or molecular (Section 3.4).

Most elements exist as atomic elements; their basic units in nature are individual atoms. However, several elements (H_2, N_2, O_2, F_2, Cl_2, Br_2, and I_2) exist as molecular elements; their basic units in nature are diatomic molecules.

EXAMPLE 3.29 CLASSIFYING ELEMENTS AS ATOMIC OR MOLECULAR

Classify each element as atomic or molecular: sodium, iodine, and nitrogen.

SOLUTION

sodium: atomic

iodine: molecular (I_2)

nitrogen: molecular (N_2)

LO: Classify compounds as ionic or molecular (Section 3.4).

Compounds containing a metal or ammonium and a nonmetal are ionic. If the metal is a transition metal, it is likely to form more than one type of ion (see exceptions in Tables 3.4 and 3.5). If the metal is not a transition metal, it is likely to form only one type of ion (see Table 3.4).

Compounds composed of nonmetals are molecular.

EXAMPLE 3.30 CLASSIFYING COMPOUNDS AS IONIC OR MOLECULAR

(i) Classify each compound as ionic or molecular.

(ii) If they are ionic, determine whether the metal forms only one type of ion or more than one type of ion.

$FeCl_3$, K_2SO_4, CCl_4

SOLUTION

$FeCl_3$: ionic, metal forms more than one type of ion

K_2SO_4: ionic, metal forms only one type of ion

CCl_4: molecular

LO: Write formulas for ionic compounds (Section 3.5).

1. Write the symbol for the metal ion followed by the symbol for the nonmetal ion (or polyatomic ion) and their charges. The charge of the non-metal ions can be deduced from the group numbers in the periodic table (see Figure 2.5). In the case of polyatomic ions, the charges come from Table 3.3.
2. Use the magnitude of the charge on each ion as the subscript for the other ion.
3. Check to see if you can reduce the subscripts to simpler whole numbers. Drop subscripts of 1; they are implied.
4. Confirm that the sum of the charges of the cations exactly cancels the sum of the charges of the anions.

EXAMPLE 3.31 WRITING FORMULAS FOR IONIC COMPOUNDS

Write a formula for the compound that forms from lithium and sulfate ions.

SOLUTION

Li^+ SO_4^{2-}

$Li_2(SO_4)$

In this case, the subscripts cannot be further reduced.

Li_2SO_4

Cations	Anions
2(1+) = 2+	2–

LO: Name binary ionic compounds containing a metal that forms only one type of ion (Section 3.7).

The name of the metal is unchanged. The name of the nonmetal is its base name with the ending *-ide*.

EXAMPLE 3.32 NAMING BINARY IONIC COMPOUNDS CONTAINING A METAL THAT FORMS ONLY ONE TYPE OF ION

Name the compound Al_2O_3.

SOLUTION

aluminum oxide

LO: Name binary ionic compounds containing a metal that forms more than one type of ion (Section 3.7).

Because the names of these compounds include the charge of the metal ion, first determine that charge by calculating the total charge of the nonmetal ions.

The total charge of the metal ions must equal the total charge of the nonmetal ions, but have the opposite sign.

The name of the compound is the name of the metal ion, followed by the charge of the metal ion, followed by the base name of the nonmetal + *-ide*.

EXAMPLE 3.33 NAMING BINARY IONIC COMPOUNDS CONTAINING A METAL THAT FORMS MORE THAN ONE TYPE OF ION

Name the compound Fe_2S_3.

SOLUTION

3 sulfide ions × (2−) = 6−

2 iron ions × (*ion charge*) = 6+

ion charge = 3+

Charge of each iron ion = 3+

iron(III) sulfide

LO: Name compounds containing a polyatomic ion (Section 3.7).

Name ionic compounds containing a polyatomic ion in the normal way, except substitute the name of the polyatomic ion (from Table 3.3) in place of the nonmetal.

Because the metal in this example forms more than one type of ion, you need to determine the charge on the metal ion. The charge of the metal ion must be equal in magnitude to the sum of the charges of the polyatomic ions but opposite in sign.

The name of the compound is the name of the metal ion, followed by the charge of the metal ion, followed by the name of the polyatomic ion.

EXAMPLE 3.34 NAMING COMPOUNDS CONTAINING A POLYATOMIC ION

Name the compound $Co(ClO_4)_2$.

SOLUTION

2 perchlorate ions × (1–) = 2–

Charge of cobalt ion = 2+

cobalt(II) perchlorate

LO: Name molecular compounds (Section 3.8).

The name consists of a prefix indicating the number of atoms of the first element, followed by the name of the first element, and a prefix for the number of atoms of the second element, followed by the base name of the second element plus the suffix *-ide*. The prefix *-mono* is normally dropped on the first element.

EXAMPLE 3.35 NAMING MOLECULAR COMPOUNDS

Name the compound NO_2.

SOLUTION

nitrogen dioxide

LO: Name binary acids (Section 3.9).

The name begins with *hydro-*, followed by the base name of the nonmetal, plus the suffix *–ic*, and the word *acid*.

EXAMPLE 3.36 NAMING BINARY ACIDS

Name the acid $HI(aq)$.

SOLUTION

hydroiodic acid

LO: Name oxyacids containing an oxyanion ending in *–ate* (Section 3.9).

The name is the base name of the oxyanion + *-ic*, followed by the word *acid* (sulfate violates the rule somewhat; in strict terms, the base name would be *sulf*).

EXAMPLE 3.37 NAMING OXYACIDS CONTAINING AN OXYANION ENDING IN *-ATE*

Name the acid $H_2SO_4(aq)$.

SOLUTION

The oxyanion is sulfate. The name of the acid is *sulfuric acid*.

LO: Name oxyacids containing an oxyanion ending in *-ite* (Section 3.9).

The name is the base name of the oxyanion + *-ous*, followed by the word *acid*.

EXAMPLE 3.38 NAMING OXYACIDS CONTAINING AN OXYANION ENDING IN *-ITE*

Name the acid $HClO_2(aq)$.

SOLUTION

The oxyanion is chlorite. The name of the acid is *chlorous acid*.

LO: Write Lewis structures for elements (Section 3.12).

The Lewis structure of any element is the symbol for the element with the valence electrons represented as dots drawn around the element. The number of valence electrons is equal to the group number of the element (for main-group elements).

EXAMPLE 3.39 LEWIS STRUCTURES FOR ELEMENTS

Draw the Lewis structure of sulfur.

SOLUTION

Since S is in Group 6A, it has 6 valence electrons. Draw these as dots surrounding its symbol, S.

LO: Write Lewis structures for ionic compounds (Section 3.13).

In an ionic Lewis structure, the metal loses all of its valence electrons to the nonmetal, which attains an octet. We place the nonmetal in brackets with the charge in the upper right corner.

EXAMPLE 3.40 WRITING LEWIS STRUCTURES OF IONIC COMPOUNDS

Write the Lewis structure for lithium bromide.

SOLUTION

$$Li^+ [:\ddot{\underset{..}{Br}}:]^-$$

LO: Use the Lewis model to predict the chemical formula of an ionic compound (Section 3.13).

To determine the chemical formula of an ionic compound, write the Lewis structures of each of the elements. Then choose the correct number of atoms of each element so that the metal atom(s) lose all of their valence electrons and the nonmetal atom(s) attain an octet.

EXAMPLE 3.41 USING THE LEWIS MODEL TO PREDICT THE CHEMICAL FORMULA OF AN IONIC COMPOUND

Use the Lewis model to predict the formula for the compound that forms between potassium and sulfur.

SOLUTION

The Lewis structures of K and S are:

$$K \qquad \cdot\ddot{\underset{\cdot}{S}}:$$

Potassium must lose 1 electron, and sulfur must gain 2. Consequently, there are two potassium atoms for every sulfur atom. The Lewis structure is:

$$K^+[:\ddot{\underset{..}{S}}:]^{2-}K^+$$

The correct formula is K_2S.

LO: Write Lewis structures for covalent compounds (Sections 3.14, 3.15).

To write covalent Lewis structures, follow these steps:

1. **Write the correct skeletal structure for the molecule.** Hydrogen atoms are always terminal, halogens are usually terminal, and many molecules tend to be symmetrical.
2. **Calculate the total number of electrons for the Lewis structure by summing the valence electrons of each atom in the molecule.** Remember that the number of valence electrons for any main-group element is equal to its group number in the periodic table. For polyatomic ions, add one electron for each negative charge and subtract one electron for each positive charge.

EXAMPLE 3.42 WRITING LEWIS STRUCTURES FOR COVALENT COMPOUNDS

Write the Lewis structure for CS_2.

SOLUTION

S C S

$$\begin{aligned}\text{Total e}^- &= 1 \times (\#\text{ valence e}^- \text{ in C})\\ &+ 2 \times (\#\text{ valence e}^- \text{ in S})\\ &= 4 + 2(6)\\ &= 16\end{aligned}$$

3. **Distribute the electrons among the atoms, giving octets (or duets for hydrogen) to as many atoms as possible.** Begin by placing 2 electrons between each pair of atoms. These are the bonding electrons. Then distribute the remaining electrons, first to terminal atoms and then to the central atom.

S:C:S (4 of 16 e^- used)

:S̤̈:C:S̤̈: (16 of 16 e^- used)

4. **If any atoms lack an octet, form double or triple bonds as necessary to give them octets.** Do this by moving lone electron pairs from terminal atoms into the bonding region with the central atom.

:S̈::C::S̈: or :S̈=C=S̈:

LO: Write resonance structures (Section 3.16).

When you can write two or more equivalent (or nearly equivalent) Lewis structures for a molecule, the true structure is an average between these. Represent this by writing all of the correct structures (called resonance structures) with double-headed arrows between them.

EXAMPLE 3.43 WRITING RESONANCE STRUCTURES

Write resonance structures for SeO_2.

SOLUTION

You can write the Lewis structure for SeO_2 by following the steps for writing covalent Lewis structures. You can write two equally correct structures, so draw them both as resonance structures.

:Ö̤—S̈e=Ö: ⟷ :Ö=S̈e—Ö̤:

LO: Predict the shapes of molecules (Section 3.17).

To determine the shape of a molecule, follow these steps:

1. **Draw the Lewis structure for the molecule.**

2. **Determine the total number of electron groups around the central atom.** Lone pairs, single bonds, double bonds, and triple bonds each count as one group.

3. **Determine the number of bonding groups and the number of lone pairs around the central atom.** These should sum to the result from Step 2. Bonding groups include single bonds, double bonds, and triple bonds.

4. Refer to Table 3.8 to determine the electron geometry and molecular geometry.

EXAMPLE 3.44 PREDICTING THE SHAPES OF MOLECULES

Predict the geometry of SeO_2.

SOLUTION

The Lewis structure for SeO_2 (as you determined in Example 3.43) is composed of the following two resonance structures.

:Ö̤—S̈e=Ö: ⟷ :Ö=S̈e—Ö̤:

Either of the resonance structures will give the same geometry.

Total number of electron groups = 3
Number of bonding groups = 2
Number of lone pairs = 1

Electron geometry = Trigonal planar
Molecular geometry = Bent

LO: Determine whether a molecule is polar (Section 3.18).

- **Determine whether the molecule contains polar bonds.** A bond is polar if the two bonding atoms are from different elements. If there are no polar bonds, the molecule is nonpolar.
- **Determine whether the polar bonds add together to form a net dipole moment.** Use VSEPR theory to determine the geometry of the molecule. Then visualize each bond as a rope pulling on the central atom. Is the molecule highly symmetrical? Do the pulls of the ropes cancel? If so, there is no net dipole moment and the molecule is nonpolar. If the molecule is asymmetrical and the pulls of the rope do not cancel, the molecule is polar.

EXAMPLE 3.45 DETERMINING WHETHER A MOLECULE IS POLAR

Determine whether SeO_2 is polar.

SOLUTION

Se and O are two different nonmetals. Therefore, the Se–O bonds are polar.

As you determined in Example 3.44, the geometry of SeO_2 is bent.

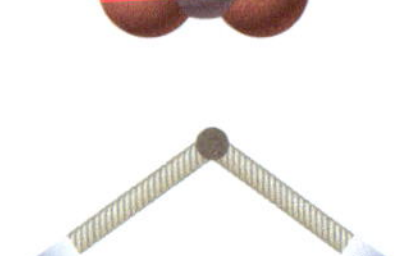

The polar bonds do not cancel but rather sum to give a net dipole moment. Therefore the molecule is polar.

KEY TERMS

acid **[3.9]**
atomic element **[3.4]**
ball-and-stick model **[3.3]**
bent **[3.17]**
binary acid **[3.9]**
binary compound **[3.7]**
bonding pair **[3.14]**
bonding theory **[3.11]**
chemical bond **[3.12]**
chemical formula **[3.3]**
covalent bond **[3.12]**
dipole moment **[3.18]**
dot structure **[3.12]**
double bond **[3.14]**
duet **[3.12]**
electron geometry **[3.17]**
electron group **[3.17]**
electronegativity **[3.18]**
empirical formula **[3.3]**
formula unit **[3.4]**
ionic bond **[3.12]**
ionic compound **[3.4]**
Lewis model **[3.11]**
Lewis structure **[3.12]**
law of constant composition **[3.2]**
linear **[3.17]**
lone pair **[3.14]**
molecular compound **[3.4]**
molecular element **[3.4]**
molecular formula **[3.3]**
molecular geometry **[3.17]**
molecular model **[3.3]**
nonpolar **[3.18]**
octet **[3.12]**
octet rule **[3.12]**
oxyacid **[3.9]**
oxyanion **[3.7]**
polar covalent bond **[3.18]**
polar molecule **[3.18]**
polyatomic ion **[3.3]**
resonance structures **[3.16]**
space-filling model **[3.3]**
structural formula **[3.3]**
terminal atom **[3.15]**
tetrahedral **[3.17]**
transition metals **[3.7]**
trigonal planar **[3.17]**
trigonal pyramidal **[3.17]**
triple bond **[3.14]**
valence shell electron pair repulsion (VSEPR) theory **[3.17]**

EXERCISES

QUESTIONS

1. Do the properties of an element change when it combines with another element to form a compound? Explain.
2. How might the world be different if elements did not combine to form compounds?
3. What is the law of constant composition? Who discovered it?
4. What is a chemical formula? List some examples.
5. In a chemical formula, which element is listed first?
6. In a chemical formula, how do you calculate the number of atoms of an element within parentheses? Provide an example.
7. Explain the difference between a molecular formula and an empirical formula.
8. What is a structural formula? What is the difference between a structural formula and a molecular model?
9. What is the difference between a molecular element and an atomic element? List the elements that occur as diatomic molecules.
10. What is the difference between an ionic compound and a molecular compound?
11. What is the difference between a common name for a compound and a systematic name?
12. List the metals that form only one type of ion (that is, metals whose charge is invariant from one compound to another). What are the group numbers of these metals?
13. Identify the block in the periodic table of metals that tend to form more than one type of ion.
14. What is the basic form for the names of ionic compounds containing a metal that forms only one type of ion?
15. What is the basic form for the names of ionic compounds containing a metal that forms more than one type of ion?
16. Why are roman numerals needed in the names of ionic compounds containing a metal that forms more than one type of ion?
17. How are compounds containing a polyatomic ion named?
18. Which polyatomic ions have a 2– charge? Which polyatomic ions have a 3– charge?
19. What is the basic form for the names of molecular compounds?
20. How many atoms does each prefix specify? *mono-*, *di-*, *tri-*, *tetra-*, *penta-*, *hexa-*.
21. What is the basic form for the names of binary acids?
22. What is the basic form for the name of oxyacids whose oxyanions end with *-ate*?
23. What is the basic form for the name of oxyacids whose oxyanions end with *-ite*?
24. Why are bonding theories important? Cite some examples of what bonding theories can predict.
25. Write the electron configurations for Ne and Ar. How many valence electrons does each element have?
26. In the Lewis model, what is an octet? What is a duet? What is a chemical bond?
27. What is the difference between ionic bonding and covalent bonding?
28. How can the Lewis model be used to determine the formula of ionic compounds? You may explain this with an example.
29. What is the difference between lone pair and bonding pair electrons?
30. How are double and triple bonds physically different from single bonds?
31. What is the procedure for writing a covalent Lewis structure?
32. How do you determine the number of electrons that go into the Lewis structure of a molecule?
33. How do you determine the number of electrons that go into the Lewis structure of a polyatomic ion?
34. Why does the octet rule have exceptions? List some examples.
35. What are resonance structures? Why are they necessary?
36. Explain how VSEPR theory predicts the shapes of molecules.
37. If all of the electron groups around a central atom are bonding groups (that is, there are no lone pairs), what is the molecular geometry for:
(a) two electron groups
(b) three electron groups
(c) four electron groups

38. Give the bond angles for each of the geometries in the preceding question.
39. What is the difference between electron geometry and molecular geometry in VSEPR theory?
40. What is electronegativity?
41. What is the most electronegative element on the periodic table?
42. What is a polar covalent bond?
43. What is a dipole moment?
44. What happens if you try to mix a polar liquid with a nonpolar one?
45. If a molecule has polar bonds, is the molecule itself polar? Why or why not?

PROBLEMS

CONSTANT COMPOSITION OF COMPOUNDS

46. Two samples of sodium chloride are decomposed into their constituent elements. One sample produces 4.65 g of sodium and 7.16 g of chlorine, and the other sample produces 7.45 g of sodium and 11.5 g of chlorine. Are these results consistent with the law of constant composition? Explain your answer.

47. Two samples of carbon tetrachloride are decomposed into their constituent elements. One sample produces 32.4 g of carbon and 373 g of chlorine, and the other sample produces 12.3 g of carbon and 112 g of chlorine. Are these results consistent with the law of constant composition? Explain your answer.

48. Upon decomposition, one sample of magnesium fluoride produced 1.65 kg of magnesium and 2.57 kg of fluorine. A second sample produced 1.32 kg of magnesium. How much fluorine (in grams) did the second sample produce?

49. The mass ratio of sodium to fluorine in sodium fluoride is 1.21:1. A sample of sodium fluoride produces 34.5 g of sodium upon decomposition. How much fluorine (in grams) forms?

50. Use the law of constant composition to complete the table summarizing the amounts of nitrogen and oxygen produced upon the decomposition of several samples of dinitrogen monoxide.

	Mass N_2O	Mass N	Mass O
Sample A	2.85 g	1.81 g	1.04 g
Sample B	4.55 g	____	____
Sample C	____	____	1.35 g
Sample D	____	1.11 g	____

51. Use the law of constant composition to complete the table summarizing the amounts of iron and chlorine produced upon the decomposition of several samples of iron(III) chloride.

	Mass $FeCl_3$	Mass Fe	Mass Cl
Sample A	3.785 g	1.303 g	2.482 g
Sample B	2.175 g	____	____
Sample C	____	2.012 g	____
Sample D	____	____	2.329 g

CHEMICAL FORMULAS

52. Write a chemical formula for the compound containing one nitrogen atom for every three iodine atoms.

53. Write a chemical formula for the compound containing one carbon atom for every four bromine atoms.

54. Write chemical formulas for compounds containing:
(a) three iron atoms for every four oxygen atoms
(b) one phosphorus atom for every three chlorine atoms
(c) one phosphorus atom for every five chlorine atoms
(d) two silver atoms for every oxygen atom

55. Write chemical formulas for compounds containing:
(a) one calcium atom for every two iodine atoms
(b) two nitrogen atoms for every four oxygen atoms
(c) one silicon atom for every two oxygen atoms
(d) one zinc atom for every two chlorine atoms

56. How many oxygen atoms are in each chemical formula?
(a) H_3PO_4
(b) Na_2HPO_4
(c) $Ca(HCO_3)_2$
(d) $Ba(C_2H_3O_2)_2$

57. How many hydrogen atoms are in each of the formulas in Question 56?

58. Determine the number of atoms of each element in each formula.
(a) $MgCl_2$
(b) $NaNO_3$
(c) $Ca(NO_2)_2$
(d) $Sr(OH)_2$

59. Determine the number of atoms of each element in each formula.
(a) NH_4Cl
(b) $Mg_3(PO_4)_2$
(c) NaCN
(d) $Ba(HCO_3)_2$

60. Complete the table.

Formula	Number of $C_2H_3O_2^-$ Units	Number of Carbon Atoms	Number of Hydrogen Atoms	Number of Oxygen Atoms	Number of Metal Atoms
$Mg(C_2H_3O_2)_2$	____	____	____	____	____
$NaC_2H_3O_2$	____	____	____	____	____
$Cr_2(C_2H_3O_2)_4$	____	____	____	____	____

61. Complete the table.

Formula	Number of SO_4^{2-} Units	Number of Sulfur Atoms	Number of Oxygen Atoms	Number of Metal Atoms
$CaSO_4$	____	____	____	____
$Al_2(SO_4)_3$	____	____	____	____
K_2SO_4	____	____	____	____

62. Give the empirical formula that corresponds to each molecular formula.
(a) C_2H_6
(b) N_2O_4
(c) $C_4H_6O_2$
(d) NH_3

63. Give the empirical formula that corresponds to each molecular formula.
(a) C_2H_6
(b) CO_2
(c) $C_6H_{12}O_6$
(d) B_2H_6

MOLECULAR VIEW OF ELEMENTS AND COMPOUNDS

64. Classify each element as atomic or molecular.
(a) chlorine
(b) argon
(c) cobalt
(d) hydrogen

65. Which elements have molecules as their basic units?
(a) helium
(b) oxygen
(c) iron
(d) bromine

66. Classify each compound as ionic or molecular.
(a) CS_2
(b) CuO
(c) KI
(d) PCl_3

67. Classify each compound as ionic or molecular.
(a) PtO_2
(b) CF_2Cl_2
(c) CO
(d) SO_3

68. Match the substances on the left with the basic units that compose them on the right.

helium	molecules
CCl_4	formula units
K_2SO_4	diatomic molecules
bromine	single atoms

69. Match the substances on the left with the basic units that compose them on the right.

NI_3	molecules
copper metal	single atoms
$CrCl_2$	diatomic molecules
nitrogen	formula units

70. What are the basic units—single atoms, molecules, or formula units—that compose each substance?
(a) $BaBr_2$
(b) Ne
(c) I_2
(d) CO

71. What are the basic units—single atoms, molecules, or formula units—that compose each substance?
(a) Rb_2O
(b) N_2
(c) $Fe(NO_3)_2$
(d) N_2F_4

72. Classify each compound as ionic or molecular. If it is ionic, determine whether the metal forms only one type of ion or more than one type of ion.
(a) KCl
(b) CBr_4
(c) NO_2
(d) $Sn(SO_4)_2$

73. Classify each compound as ionic or molecular. If it is ionic, determine whether the metal forms only one type of ion or more than one type of ion.
(a) $CoCl_2$
(b) CF_4
(c) $BaSO_4$
(d) NO

WRITING FORMULAS FOR IONIC COMPOUNDS

74. Write a formula for the ionic compound that forms from each pair of elements.
(a) sodium and sulfur
(b) strontium and oxygen
(c) aluminum and sulfur
(d) magnesium and chlorine

75. Write a formula for the ionic compound that forms from each pair of elements.
(a) aluminum and oxygen
(b) beryllium and iodine
(c) calcium and sulfur
(d) calcium and iodine

76. Write a formula for the compound that forms from potassium and
(a) acetate
(b) chromate
(c) phosphate
(d) cyanide

77. Write a formula for the compound that forms from calcium and
(a) hydroxide
(b) carbonate
(c) phosphate
(d) hydrogenphosphate

78. Write formulas for the compounds formed from the element on the left and each of the elements on the right.
(a) Li N, O, F
(b) Ba N, O, F
(c) Al N, O, F

79. Write formulas for the compounds formed from the element on the left and each polyatomic ion on the right.
(a) Rb NO_3^-, SO_4^{2-}, PO_4^{3-}
(b) Sr NO_3^-, SO_4^{2-}, PO_4^{3-}
(c) In NO_3^-, SO_4^{2-}, PO_4^{3-}
(Assume In charge is 3+.)

NAMING IONIC COMPOUNDS

80. Name each ionic compound. In each of these compounds, the metal forms only one type of ion.
(a) $CsCl$
(b) $SrBr_2$
(c) K_2O
(d) LiF

81. Name each ionic compound. In each of these compounds, the metal forms only one type of ion.
(a) LiI
(b) MgS
(c) BaF_2
(d) NaF

82. Name each ionic compound. In each of these compounds, the metal forms more than one type of ion.
(a) $CrCl_2$
(b) $CrCl_3$
(c) SnO_2
(d) PbI_2

83. Name each ionic compound. In each of these compounds, the metal forms more than one type of ion.
(a) $HgBr_2$
(b) Fe_2O_3
(c) CuI_2
(d) $SnCl_4$

84. Determine whether the metal in each ionic compound forms only one type of ion or more than one type of ion and name the compound accordingly.
(a) Cr_2O_3
(b) NaI
(c) $CaBr_2$
(d) SnO

85. Determine whether the metal in each ionic compound forms only one type of ion or more than one type of ion and name the compound accordingly.
(a) FeI_3
(b) $PbCl_4$
(c) SrI_2
(d) BaO

86. Name each ionic compound containing a polyatomic ion.
- **(a)** $Ba(NO_3)_2$
- **(b)** $Pb(C_2H_3O_2)_2$
- **(c)** NH_4I
- **(d)** $KClO_3$
- **(e)** $CoSO_4$
- **(f)** $NaClO_4$

87. Name each ionic compound containing a polyatomic ion.
- **(a)** $Ba(OH)_2$
- **(b)** $Fe(OH)_3$
- **(c)** $Cu(NO_2)_2$
- **(d)** $PbSO_4$
- **(e)** $KClO$
- **(f)** $Mg(C_2H_3O_2)_2$

88. Name each polyatomic ion.
- **(a)** BrO^-
- **(b)** BrO_2^-
- **(c)** BrO_3^-
- **(d)** BrO_4^-

89. Name each polyatomic ion.
- **(a)** IO^-
- **(b)** IO_2^-
- **(c)** IO_3^-
- **(d)** IO_4^-

90. Write a formula for each ionic compound.
- **(a)** copper(II) bromide
- **(b)** silver nitrate
- **(c)** potassium hydroxide
- **(d)** sodium sulfate
- **(e)** potassium hydrogensulfate
- **(f)** sodium hydrogencarbonate

91. Write a formula for each ionic compound.
- **(a)** copper(I) chlorate
- **(b)** potassium permanganate
- **(c)** lead(II) chromate
- **(d)** calcium fluoride
- **(e)** iron(II) phosphate
- **(f)** lithium hydrogensulfite

NAMING MOLECULAR COMPOUNDS

92. Name each molecular compound.
- **(a)** SO_2
- **(b)** NI_3
- **(c)** BrF_5
- **(d)** NO
- **(e)** N_4Se_4

93. Name each molecular compound.
- **(a)** XeF_4
- **(b)** PI_3
- **(c)** SO_3
- **(d)** $SiCl_4$
- **(e)** I_2O_5

94. Write a formula for each molecular compound.
- **(a)** carbon monoxide
- **(b)** disulfur tetrafluoride
- **(c)** dichlorine monoxide
- **(d)** phosphorus pentafluoride
- **(e)** boron tribromide
- **(f)** diphosphorus pentasulfide

95. Write a formula for each molecular compound.
- **(a)** chlorine monoxide
- **(b)** xenon tetroxide
- **(c)** xenon hexafluoride
- **(d)** carbon tetrabromide
- **(e)** diboron tetrachloride
- **(f)** tetraphosphorus triselenide

96. Determine whether the name shown for each molecular compound is correct. If not, provide the compound's correct name.
- **(a)** PBr_5 phosphorus (V) pentabromide
- **(b)** P_2O_3 phosphorus trioxide
- **(c)** SF_4 monosulfur hexafluoride
- **(d)** NF_3 nitrogen trifluoride

97. Determine whether the name shown for each molecular compound is correct. If not, provide the compound's correct name.
- **(a)** NCl_3 nitrogen chloride
- **(b)** CI_4 carbon(IV) iodide
- **(c)** CO carbon oxide
- **(d)** SCl_4 sulfur tetrachloride

NAMING ACIDS

98. Determine whether each acid is a binary acid or an oxyacid and name each acid. If the acid is an oxyacid, provide the name of the oxyanion.
(a) $HNO_2(aq)$
(b) $HI(aq)$
(c) $H_2SO_4(aq)$
(d) $HNO_3(aq)$

99. Determine whether each acid is a binary acid or an oxyacid and name each acid. If the acid is an oxyacid, provide the name of the oxyanion.
(a) $H_2CO_3\ (aq)$
(b) $HC_2H_3O_2(aq)$
(c) $H_3PO_4\ (aq)$
(d) $HCl(aq)$

100. Name each acid.
(a) $HClO$
(b) $HClO_2$
(c) $HClO_3$
(d) $HClO_4$

101. Name each acid. (*Hint*: The names of the oxyanions are analogous to the names of the oxyanions of chlorine.)
(a) $HBrO_3$
(b) HIO_3

102. Write a formula for each acid.
(a) phosphoric acid
(b) hydrobromic acid
(c) sulfurous acid

103. Write a formula for each acid.
(a) hydrofluoric acid
(b) hydrocyanic acid
(c) chlorous acid

WRITING LEWIS STRUCTURES FOR ELEMENTS

104. Write an electron configuration for each element and the corresponding Lewis structure. Indicate which electrons in the electron configuration are included in the Lewis structure.
(a) N
(b) C
(c) Cl
(d) Ar

105. Write an electron configuration for each element and the corresponding Lewis structure. Indicate which electrons in the electron configuration are included in the Lewis structure.
(a) Li
(b) P
(c) F
(d) Ne

106. Write the Lewis structure for each element.
(a) I
(b) S
(c) Ge
(d) Ca

107. Write the Lewis structure for each element.
(a) Kr
(b) P
(c) B
(d) Na

108. Write a generic Lewis structure for the halogens. Do the halogens tend to gain or lose electrons in chemical reactions? How many?

109. Write a generic Lewis structure for the alkali metals. Do the alkali metals tend to gain or lose electrons in chemical reactions? How many?

110. Write a generic Lewis structure for the alkaline earth metals. Do the alkaline earth metals tend to gain or lose electrons in chemical reactions? How many?

111. Write a generic Lewis structure for the elements in the oxygen family (Group 6A). Do the elements in the oxygen family tend to gain or lose electrons in chemical reactions? How many?

112. Write the Lewis structure for each ion.
(a) Al^{3+}
(b) Mg^{2+}
(c) Se^{2-}
(d) N^{3-}

113. Write the Lewis structure for each ion.
(a) Sr^{2+}
(b) S^{2-}
(c) Li^{+}
(d) Cl^{-}

114. Indicate the noble gas that has the same Lewis structure as each ion.
(a) Br^-
(b) O^{2-}
(c) Rb^+
(d) Ba^{2+}

115. Indicate the noble gas that has the same Lewis structure as each ion.
(a) Se^{2-}
(b) I^-
(c) Sr^{2+}
(d) F^-

LEWIS STRUCTURES FOR IONIC COMPOUNDS

116. Is each compound best represented by an ionic or a covalent Lewis structure?
(a) SF_6
(b) $MgCl_2$
(c) BrCl
(d) K_2S

117. Is each compound best represented by an ionic or a covalent Lewis structure?
(a) NO
(b) CO_2
(c) Rb_2O
(d) Al_2S_3

118. Write the Lewis structure for each ionic compound.
(a) NaF
(b) CaO
(c) $SrBr_2$
(d) K_2O

119. Write the Lewis structure for each ionic compound.
(a) SrO
(b) Li_2S
(c) CaI_2
(d) RbF

120. Use the Lewis model to determine the formula for the compound that forms from each pair of atoms.
(a) Ca and S
(b) Mg and Br
(c) Cs and I
(d) Ca and N

121. Use the Lewis model to determine the formula for the compound that forms from each pair of atoms.
(a) Al and S
(b) Na and S
(c) Sr and Se
(d) Ba and F

122. Draw the Lewis structure for the ionic compound that forms from Mg and each atom.
(a) F
(b) O
(c) N

123. Draw the Lewis structure for the ionic compound that forms from Al and each atom.
(a) F
(b) O
(c) N

124. Determine what is wrong with each ionic Lewis structure and write the correct structure.
(a) $[\mathrm{Cs}{:}]^+\ [{:}\ddot{\mathrm{Cl}}{:}]^-$
(b) $\mathrm{Ba}^+\ [{:}\underset{\cdot\cdot}{\overset{\cdot\cdot}{\mathrm{O}}}{:}]^-$
(c) $\mathrm{Ca}^{2+}\ [{:}\underset{\cdot\cdot}{\overset{\cdot\cdot}{\mathrm{I}}}{:}]^-$

125. Determine what is wrong with each ionic Lewis structure and write the correct structure.
(a) $[{:}\underset{\cdot\cdot}{\overset{\cdot\cdot}{\mathrm{O}}}{:}]^{2-}\ \mathrm{Na}^+[{:}\underset{\cdot\cdot}{\overset{\cdot\cdot}{\mathrm{O}}}{:}]^{2-}$
(b) $\mathrm{Mg}{:}\underset{\cdot\cdot}{\overset{\cdot\cdot}{\mathrm{O}}}{:}$
(c) $[\mathrm{Li}{:}]^+[{:}\underset{\cdot\cdot}{\overset{\cdot\cdot}{\mathrm{S}}}{:}]^-$

LEWIS STRUCTURES FOR COVALENT COMPOUNDS

126. Use the Lewis model to explain why each element exists as a diatomic molecule.
(a) hydrogen
(b) iodine
(c) nitrogen
(d) oxygen

127. Use the Lewis model to explain why the compound that forms between hydrogen and sulfur has the formula H_2S. Would you expect HS to be stable? H_3S?

128. Write the Lewis structure for each molecule.
(a) PH_3
(b) SCl_2
(c) F_2
(d) HI

129. Write the Lewis structure for each molecule.
(a) CH_4
(b) NF_3
(c) OF_2
(d) H_2O

130. Write the Lewis structure for each molecule.
(a) O_2
(b) CO
(c) HONO (N is central; H bonded to one of the O atoms.)
(d) SO_2

131. Write the Lewis structure for each molecule.
(a) N_2O (oxygen is terminal)
(b) SiH_4
(c) CI_4
(d) Cl_2CO (carbon is central)

132. Write the Lewis structure for each molecule.
(a) C_2H_2
(b) C_2H_4
(c) N_2H_2
(d) N_2H_4

133. Write the Lewis structure for each molecule.
(a) H_2CO (carbon is central)
(b) H_3COH (carbon and oxygen are both central)
(c) H_3COCH_3 (oxygen is between the two carbon atoms)
(d) H_2O_2

134. Determine what is wrong with each Lewis structure and write the correct structure.
(a) $:\ddot{N}=\ddot{N}:$
(b) $:\underset{..}{\ddot{S}}-Si-\underset{..}{\ddot{S}}:$
(c) $H-H-\underset{..}{\ddot{O}}:$
(d) $:\underset{..}{\ddot{I}}-\underset{\underset{:\underset{..}{I}:}{|}}{N}-\underset{..}{\ddot{I}}:$

135. Determine what is wrong with each Lewis structure and write the correct structure.
(a) $H-H-H-\underset{..}{\ddot{N}}:$
(b) $:\underset{..}{\ddot{Cl}}=O=\underset{..}{\ddot{Cl}}:$
(c) $H-\overset{\overset{:\ddot{O}:}{|}}{C}-\underset{..}{\ddot{O}}-H$
(d) $H=\underset{..}{\ddot{Br}}:$

136. Write the Lewis structure for each molecule or ion.
(a) SeO_2
(b) CO_3^{2-}
(c) ClO^-
(d) ClO_2^-

137. Write the Lewis structure for each molecule or ion.
(a) ClO_3^-
(b) ClO_4^-
(c) NO_3^-
(d) SO_3

138. Write the Lewis structure for each ion. Include resonance structures if necessary.
(a) PO_4^{3-}
(b) CN^-
(c) NO_2^-
(d) SO_3^{2-}

139. Write the Lewis structure for each ion. Include resonance structures if necessary.
(a) SO_4^{2-}
(b) HSO_4^- (S is central; H is attached to one of the O atoms)
(c) NH_4^+
(d) BrO_2^- (Br is central)

PREDICTING THE SHAPES OF MOLECULES

140. Determine the number of electron groups around the central atom for each molecule.
(a) OF_2
(b) NF_3
(c) CS_2
(d) CH_4

141. Determine the number of electron groups around the central atom for each molecule.
(a) CH_2Cl_2
(b) SBr_2
(c) H_2S
(d) PCl_3

142. Determine the number of bonding groups and the number of lone pairs for each of the molecules in Problem 140. The sum of these should equal your answers to Problem 140.

143. Determine the number of bonding groups and the number of lone pairs for each of the molecules in Problem 141. The sum of these should equal your answers to Problem 141.

144. Determine the molecular geometry of each molecule.
(a) CBr_4
(b) H_2CO
(c) CS_2

145. Determine the molecular geometry of each molecule.
(a) SiO_2
(b) $CFCl_3$ (carbon is central)
(c) H_2CS (carbon is central)

146. Determine the bond angles for each molecule in Problem 144.

147. Determine the bond angles for each molecule in Problem 145.

148. Determine the electron and molecular geometries of each molecule.
(a) N_2O (oxygen is terminal)
(b) SO_2
(c) H_2S
(d) PF_3

149. Determine the electron and molecular geometries of each molecule. (*Hint:* Determine the geometry around each of the two central atoms.)
(a) C_2H_2 (skeletal structure HCCH)
(b) C_2H_4 (skeletal structure H_2CCH_2)
(c) C_2H_6 (skeletal structure H_3CCH_3)

150. Determine the bond angles for each molecule in Problem 148.

151. Determine the bond angles for each molecule in Problem 149.

152. Determine the electron and molecular geometries of each molecule. For molecules with two central atoms, indicate the geometry about each central atom.
(a) N_2
(b) N_2H_2 (skeletal structure HNNH)
(c) N_2H_4 (skeletal structure H_2NNH_2)

153. Determine the electron and molecular geometries of each molecule. For molecules with more than one central atom, indicate the geometry about each central atom.
(a) CH_3OH (skeletal structure H_3COH)
(b) H_3COCH_3 (skeletal structure H_3COCH_3)
(c) H_2O_2 (skeletal structure HOOH)

154. Determine the molecular geometry of each polyatomic ion.
(a) CO_3^{2-}
(b) ClO_2^-
(c) NO_3^-
(d) NH_4^+

155. Determine the molecular geometry of each polyatomic ion.
(a) ClO_4^-
(b) BrO_2^-
(c) NO_2^-
(d) SO_4^{2-}

ELECTRONEGATIVITY AND POLARITY

156. Refer to Figure 3.18 to determine the electronegativity of each element.
(a) Mg
(b) Si
(c) Br

157. Refer to Figure 3.18 to determine the electronegativity of each element.
(a) F
(b) C
(c) S

158. List these elements in order of decreasing electronegativity: Rb, Si, Cl, Ca, Ga.

159. List these elements in order of increasing electronegativity: Ba, N, F, Si, Cs.

160. Classify the bond between each pair of elements as pure covalent, polar covalent, or ionic.
- **(a)** Mg and Br
- **(b)** Cr and F
- **(c)** Br and Br
- **(d)** Si and O

161. Classify the bond between each pair of elements as pure covalent, polar covalent, or ionic.
- **(a)** K and Cl
- **(b)** N and N
- **(c)** C and S
- **(d)** C and Cl

162. Arrange these diatomic molecules in order of increasing bond polarity: ICl, HBr, H_2, CO

163. Arrange these diatomic molecules in order of decreasing bond polarity: HCl, NO, F_2, HI

164. Classify each diatomic molecule as polar or nonpolar.
- **(a)** CO
- **(b)** O_2
- **(c)** F_2
- **(d)** HBr

165. Classify each diatomic molecule as polar or nonpolar.
- **(a)** I_2
- **(b)** HI
- **(c)** HCl
- **(d)** N_2

166. For each polar molecule in Problem 164 draw the molecule and indicate the positive and negative ends of the dipole moment.

167. For each polar molecule in Problem 165 draw the molecule and indicate the positive and negative ends of the dipole moment.

168. Classify each molecule as polar or nonpolar.
- **(a)** CS_2
- **(b)** SO_2
- **(c)** CH_4
- **(d)** CH_3Cl

169. Classify each molecule as polar or nonpolar.
- **(a)** H_2CO
- **(b)** CH_3OH
- **(c)** CH_2Cl_2
- **(d)** CO_2

170. Classify each molecule as polar or nonpolar.
- **(a)** $CHCl_3$
- **(b)** C_2H_2
- **(c)** NH_3

171. Classify each molecule as polar or nonpolar.
- **(a)** N_2H_2
- **(b)** H_2O_2
- **(c)** CF_4

CUMULATIVE PROBLEMS

172. Write a molecular formula for each molecular model. (White = hydrogen; red = oxygen; black = carbon; blue = nitrogen; yellow = sulfur)

(a) 

(b) 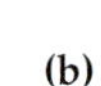 

(c) 

173. Write a molecular formula for each molecular model. (White = hydrogen; red = oxygen; black = carbon; blue = nitrogen; yellow = sulfur)

(a) 

(b) 

(c) 

174. How many chlorine atoms are in each set?
- **(a)** three carbon tetrachloride molecules
- **(b)** two calcium chloride formula units
- **(c)** four phosphorus trichloride molecules
- **(d)** seven sodium chloride formula units

175. How many oxygen atoms are in each set?
- **(a)** four dinitrogen monoxide molecules
- **(b)** two calcium carbonate formula units
- **(c)** three sulfur dioxide molecules
- **(d)** five perchlorate ions

176. Specify the number of hydrogen atoms (white) represented in each set of molecular models:

(a)

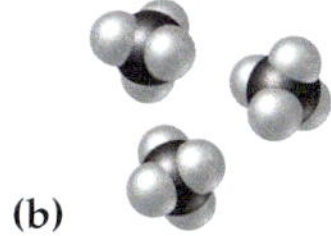
(b)

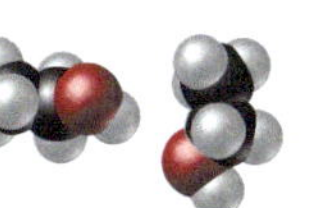
(c)

177. Specify the number of oxygen atoms (red) represented in each set of molecular models:

(a) (b)

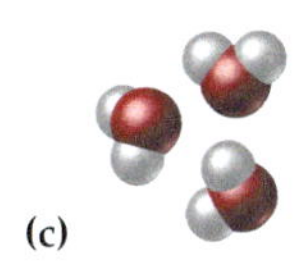
(c)

178. Complete the table:

Formula	Type of Compound (Ionic, Molecular, Acid)	Name
N_2H_4	molecular	________
________	________	potassium chloride
$H_2CrO_4(aq)$	________	________
________	________	cobalt(III) cyanide

179. Complete the table:

Formula	Type of Compound (Ionic, Molecular, Acid)	Name
$K_2Cr_2O_7$	ionic	________
$HBr(aq)$	________	hydrobromic acid
________	________	dinitrogen pentoxide
PbO_2	________	________

180. Is each name correct for the given formula? If not, provide the correct name.

(a) $Ca(NO_2)_2$ calcium nitrate
(b) K_2O dipotassium monoxide
(c) PCl_3 phosphorus chloride
(d) $PbCO_3$ lead (II) carbonate
(e) KIO_2 potassium hypoiodite

181. Is each name correct for the given formula? If not, provide the correct name.

(a) $HNO_3(aq)$ hydrogen nitrate
(b) NaClO sodium hypochlorite
(c) CaI_2 calcium diiodide
(d) $SnCrO_4$ tin chromate
(e) $NaBrO_3$ sodium bromite

182. Write electron configurations and Lewis structures for each element. Indicate which of the electrons in the electron configuration are shown in the Lewis structure.

(a) Ca
(b) Ga
(c) As
(d) I

183. Write electron configurations and Lewis structures for each element. Indicate which of the electrons in the electron configuration are shown in the Lewis structure.

(a) Rb
(b) Ge
(c) Kr
(d) Se

184. Determine whether each compound is ionic or covalent and write the appropriate Lewis structure.

(a) K_2S
(b) CHFO (carbon is central)
(c) MgSe
(d) PBr_3

185. Determine whether each compound is ionic or covalent and write the appropriate Lewis structure.

(a) HCN
(b) ClF
(c) MgI_2
(d) CaS

186. Write the Lewis structure for $OCCl_2$ (carbon is central) and determine whether the molecule is polar. Draw the three-dimensional structure of the molecule.

187. Write the Lewis structure for CH_3COH and determine whether the molecule is polar. Draw the three-dimensional structure of the molecule. The skeletal structure is:

```
    H  O
H   C  C  H
    H
```

188. Write the Lewis structure for acetic acid (a component of vinegar) CH_3COOH, and draw the three-dimensional sketch of the molecule. Its skeletal structure is:

```
    H  O

H   C  C  O  H

    H
```

189. Write the Lewis structure for benzene, C_6H_6, and draw a three-dimensional sketch of the molecule. The skeletal structure is the ring shown here. (*Hint:* The Lewis structure consists of two resonance structures.)

```
      H
      C
  HC     CH
  HC     CH
      C
      H
```

190. Each compound listed contains both ionic and covalent bonds. Write the ionic Lewis structure for each one including the covalent structure for the polyatomic ion. Write resonance structures if necessary.

(a) KOH
(b) KNO_3
(c) LiIO
(d) $BaCO_3$

191. Each of the compounds listed contains both ionic and covalent bonds. Write an ionic Lewis structure for each one, including the covalent structure for the polyatomic ion. Write resonance structures if necessary.

(a) $RbIO_2$
(b) $Ca(OH)_2$
(c) NH_4Cl
(d) $Sr(CN)_2$

HIGHLIGHT PROBLEMS

192. Examine each substance and the corresponding molecular view and classify it as an atomic element, a molecular element, a molecular compound, or an ionic compound.

(a)

(b)

(c)

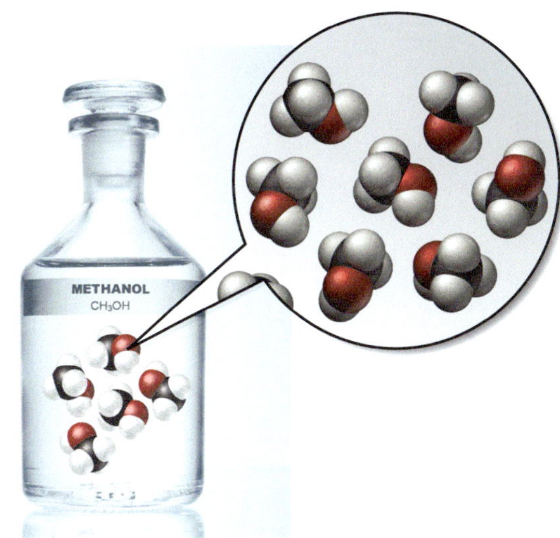

(d)

193. Examine each consumer product label. Write chemical formulas for as many of the compounds as possible based on what you have learned in this module.

(a)

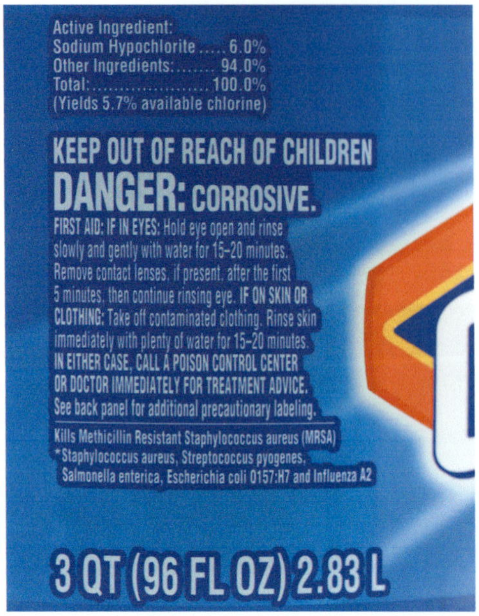

(b)

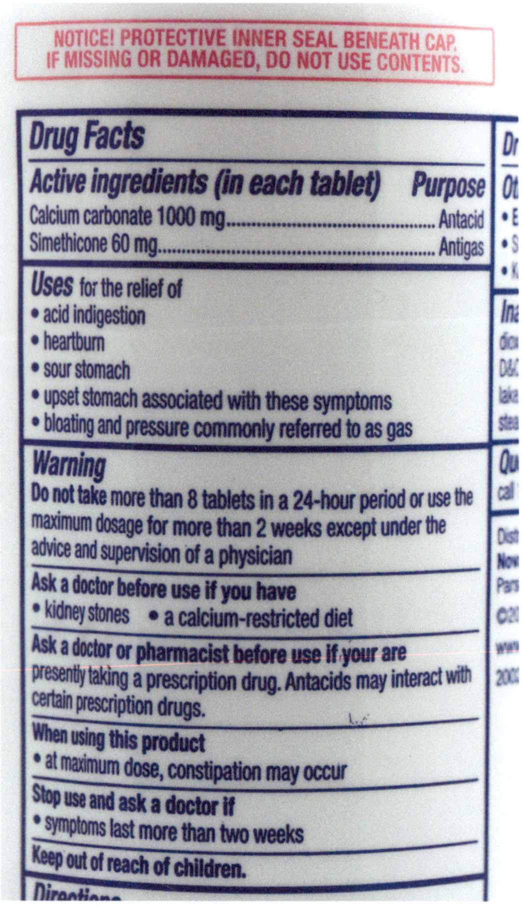

(c)

Drug Facts

Active ingredients (in each 5 mL teaspoon) — **Purposes**
Aluminum hydroxide (equivalent to dried gel, USP) 400 mg .... Antacid
Magnesium hydroxide 400 mg .... Antacid
Simethicone 40 mg .... Antigas

Use relieves: ■ heartburn ■ acid indigestion ■ sour stomach ■ upset stomach due to these symptoms ■ pressure and bloating commonly referred to as gas

Warnings
Ask a doctor before use if you have
■ kidney disease ■ a magnesium-restricted diet

Ask a doctor or pharmacist if you are taking a prescription drug. Antacids may interact with certain prescription drugs.

Stop use and ask a doctor if symptoms last more than 2 weeks.

Keep out of reach of children.

Directions ■ shake well ■ adults/children 12 years and older: take 2-4 teaspoonfuls between meals, at bedtime, or as directed by a doctor ■ do not take more than 12 teaspoonfuls in a 24-hour period, or use the maximum dosage for more than 2 weeks ■ children under 12 years: ask a doctor

Other information ■ each teaspoon contains: **magnesium 171 mg**
■ **do not use if breakaway band on plastic cap is broken or missing**
■ **does not meet USP requirements for preservative effectiveness**
■ do not freeze

Inactive ingredients butylparaben, carboxymethylcellulose sodium, flavors, hypromellose, microcrystalline cellulose, propylparaben, purified water, sodium saccharin, sorbitol

Questions or comments?
1-800-469-5268 (English) or
1-888-466-8746 (Spanish)

7 16837 62412

Johnson & Johnson · MERCK
Consumer Pharmaceuticals Co.
FORT WASHINGTON, PA 19034 USA

7844154

(d)

Nutrition Facts
Serving Size 1/8 tsp (0.6g)
Servings Per Container about 472

Amount Per Serving	
Calories 0	
	% Daily Value*
Total Fat 0g	**0%**
Sodium 65mg	**3%**
Total Carb. 0g	**0%**
Protein 0g	
Calcium 2%	

Not a significant source of calories from fat, saturated fat, trans fat, cholesterol, dietary fiber, sugars, vitamin A, vitamin C and iron.

*Percent Daily Values are based on a 2,000 calorie diet.

Ingredients: Cornstarch, Sodium Bicarbonate, Sodium Aluminum Sulfate, Monocalcium Phosphate.

CLABBER GIRL CORPORATION
TERRE HAUTE, IN 47808
davisbakingpowder.com
MADE IN USA

194. Examine the formulas and space-filling models of the molecules shown here. Determine whether the structure is correct. If the structure is incorrect, sketch the correct structure.

(a) H_2Se

(b) CSe_2

(c) PCl_3

(d) CF_2Cl_2

▸Answers to Skillbuilder Exercises

Skillbuilder 3.1 Yes, because in both cases

$$\frac{\text{Mass O}}{\text{Mass C}} = 1.3$$

Skillbuilder 3.2 (a) Ag_2S (b) N_2O (c) TiO_2

Skillbuilder 3.3 two K atoms, one S atom, four O atoms

Skillbuilder Plus, p. 77
two Al atoms, three S atoms, twelve O atoms

Skillbuilder 3.4
- **(a)** molecular element
- **(b)** molecular compound
- **(c)** atomic element
- **(d)** ionic compound
- **(e)** ionic compound

Skillbuilder 3.5 $SrCl_2$

Skillbuilder 3.6 AlN

Skillbuilder 3.7 $AlPO_4$

Skillbuilder Plus, p. 83 Na_2SO_3

Skillbuilder 3.8 potassium bromide

Skillbuilder Plus, p. 86 zinc nitride

Skillbuilder 3.9 lead(II) oxide

Skillbuilder 3.10 manganese(II) nitrate

Skillbuilder 3.11 dinitrogen tetroxide

Skillbuilder 3.12 hydrofluoric acid

Skillbuilder 3.13 nitrous acid

Skillbuilder 3.14 $\cdot\text{Mg}\cdot$

Skillbuilder 3.15 $\text{Na}^+\ [:\ddot{\underset{..}{\text{Br}}}:]^-$

Skillbuilder 3.16 Mg_3N_2

Skillbuilder 3.17 $:\text{C}\equiv\text{O}:$

Skillbuilder 3.18

$$\begin{array}{c} \ddot{\text{O}}: \\ \| \\ \text{H}-\text{C}-\text{H} \end{array}$$

Skillbuilder 3.19 $[:\ddot{\underset{..}{\text{Cl}}}-\ddot{\underset{..}{\text{O}}}:]^-$

Skillbuilder 3.20

$$[:\ddot{\text{O}}=\ddot{\text{N}}-\ddot{\underset{..}{\text{O}}}:]^- \longleftrightarrow [:\ddot{\underset{..}{\text{O}}}-\ddot{\text{N}}=\ddot{\text{O}}:]^-$$

Skillbuilder 3.21 bent

Skillbuilder 3.22 trigonal pyramidal

Skillbuilder 3.23
- **(a)** pure covalent
- **(b)** ionic
- **(c)** polar covalent

Skillbuilder 3.24 CH_4 is nonpolar

▸Answers to Conceptual Checkpoints

3.1 (c) The ratio of A/B is 3.0, and A is 9.0 g, so B must be 3.0 g.

3.2 (b) This formula represents 2 Al atoms + 3(2 Cr atoms + 7 O atoms) = 29 atoms.

3.3 (b) The figure represents a molecular compound because the compound exists as individual molecules. Figure (a) represents an ionic compound with formula units in a lattice structure.

3.4 (c) Because the nitrate ion has a charge of 1–, the three nitrate ions together have a charge of 3–. Because the compound must be charge-neutral, Cr must have a charge of 3+.

3.5 Because calcium forms only one type of ion (Ca^{2+}); therefore, the charge of the ion is not included in the name (it is always the same, 2+).

3.6 Iodate

3.7 (a) C and Si both have four dots in their Lewis structure because they are both in the same column in the periodic table.

3.8 (b) Aluminum must lose its 3 valence electrons to obtain an octet. Sulfur must gain 2 electrons to obtain an octet. Therefore, two Al atoms are required for every three S atoms.

3.9 (b) The Lewis structure of O_2 has one double bond that contains 4 electrons (all of them bonding electrons); therefore, the number of bonding electrons is four.

3.10 (c) The Lewis structure of OH^- has eight electrons: six from oxygen, one from hydrogen, and one from the negative charge.

3.11 (b) Both NH_3 and H_3O^+ have one lone electron pair.

3.12 (d) If there are no lone pairs on the central atom, all of its valence electrons are involved in bonds, so the molecular geometry must be the same as the electron geometry.

3.13 The bond in H—Cl is more polar than the bond in H—Br because Cl is more electronegative than Br, so the electronegativity difference between H and Cl is greater than the difference between H and Br.

▲ The graph in this image displays average global temperatures (relative to the mean) over the past 100 years.

Measurement 4

"The important thing in science is not so much to obtain new facts as to discover new ways of thinking about them."—Sir William Lawrence Bragg (1890–1971)

MODULE OUTLINE

4.1 Scientific Notation: Writing Large and Small Numbers

LO: Express very large and very small numbers using scientific notation.

Science constantly pushes the boundaries of the very large and the very small. We can, for example, now measure time periods as short as 0.000000000000001 second and distances as great as 14,000,000,000 light-years. Because the many zeros in these numbers are cumbersome to write, we use **scientific notation** to write them more compactly. In scientific notation, 0.000000000000001 is 1×10^{-15}, and 14,000,000,000 is 1.4×10^{10}. A number written in scientific notation consists of a **decimal part**, a number that is usually between 1 and 10, and an **exponential part**, 10 raised to an **exponent**, *n*.

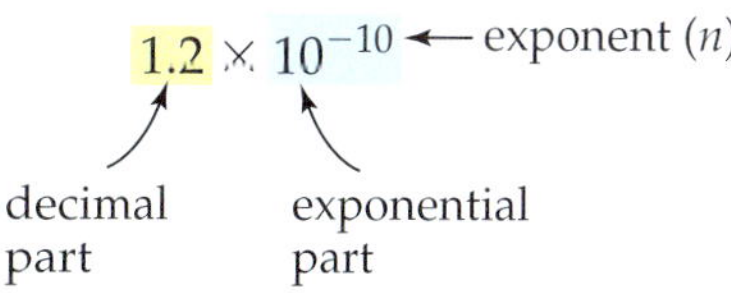

A positive exponent (*n*) means 1 multiplied by 10 *n* times.

$$10^0 = 1$$

$$10^1 = 1 \times 10 = 10$$

$$10^2 = 1 \times 10 \times 10 = 100$$

$$10^3 = 1 \times 10 \times 10 \times 10 = 1000$$

▲ Lasers such as this one can measure time periods as short as 1×10^{-15} s.

A negative exponent ($-n$) means 1 divided by 10 n times.

$$10^{-1} = \frac{1}{10} = 0.1$$

$$10^{-2} = \frac{1}{10 \times 10} = 0.01$$

$$10^{-3} = \frac{1}{10 \times 10 \times 10} = 0.001$$

To convert a number to scientific notation, we move the decimal point (either to the left or to the right, as needed) to obtain a number between 1 and 10 and then multiply that number (the decimal part) by 10 raised to the power that reflects the movement of the decimal point. For example, to write 5983 in scientific notation, we move the decimal point to the left three places to get 5.983 (a number between 1 and 10) and then multiply the decimal part by 1000 to compensate for moving the decimal point.

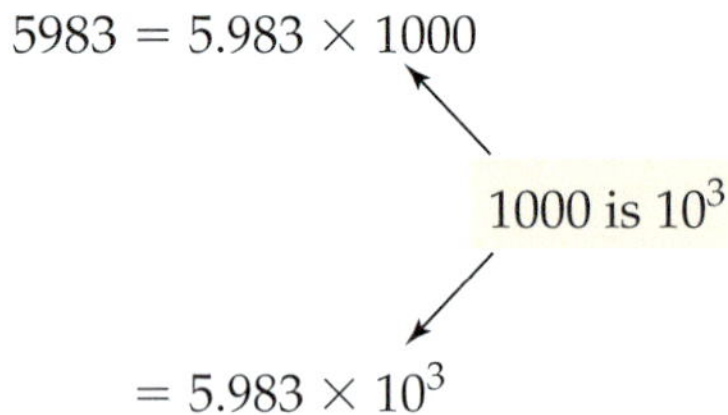

We can do this in one step by counting how many places we move the decimal point to obtain a number between 1 and 10 and then writing the decimal part multiplied by 10 raised to the number of places we moved the decimal point.

$$5983 = 5.983 \times 10^3$$

3 2 1

If the decimal point is moved to the left, as in the previous example, the exponent is positive. If the decimal is moved to the right, the exponent is negative.

The Mathematics Review Appendix (p. MR-2) includes a review of mathematical operations for numbers written in scientific notation.

$$0.00034 = 3.4 \times 10^{-4}$$

1 2 3 4

To express a number in scientific notation:

1. Move the decimal point to obtain a number between 1 and 10.
2. Write the result from Step 1 multiplied by 10 raised to the number of places you moved the decimal point.
   - The exponent is positive if you moved the decimal point to the left.
   - The exponent is negative if you moved the decimal point to the right.

Remember, large numbers have positive exponents and small numbers have negative exponents

EXAMPLE 4.1 SCIENTIFIC NOTATION

The 2013 U.S. population was estimated to be 315,000,000 people. Express this number in scientific notation.

To obtain a number between 1 and 10, move the decimal point to the left eight decimal places; the exponent is 8. Because you move the decimal point to the left, the sign of the exponent is positive.

SOLUTION

315,000,000 people = 3.15×10^8 people

▶**SKILLBUILDER 4.1 | Scientific Notation**

The total U.S national debt in 2013 was approximately $16,342,000,000,000. Express this number in scientific notation.

Note: The answers to all Skillbuilders appear at the end of the module.

▶**FOR MORE PRACTICE** Example 4.12; Problems 20, 21.

EXAMPLE 4.2 SCIENTIFIC NOTATION

The radius of a carbon atom is approximately 0.000000000070 m. Express this number in scientific notation.

To obtain a number between 1 and 10, move the decimal point to the right 11 decimal places; therefore, the exponent is 11. Because you moved the decimal point to the right, the sign of the exponent is negative.

SOLUTION

0.000000000070 m = 7.0×10^{-11} m

▶**SKILLBUILDER 4.2 | Scientific Notation**

Express the number 0.000038 in scientific notation.

▶**FOR MORE PRACTICE** Problems 22, 23.

CONCEPTUAL CHECKPOINT 4.1

The radius of a dust speck is 4.5×10^{-3} mm. What is the correct value of this number in decimal notation (i.e., express the number without using scientific notation)?

(a) 4500 mm

(b) 0.045 mm

(c) 0.0045 mm

(d) 0.00045 mm

Note: The answers to all Conceptual Checkpoints appear at the end of the module.

4.2 Significant Figures: Writing Numbers to Reflect Precision

LO: Report measured quantities to the right number of digits.

LO: Determine which digits in a number are significant.

▲ Because pennies come in whole numbers, seven pennies means 7.00000. . . pennies. This is an exact number and therefore never limits significant figures in calculations. © Maxwell Art And Photo/Pearson.

▲ Our knowledge of the amount of gold in a 10 g gold bar depends on how precisely it was measured. © Paul Silverman/Fundamental Photographs.

If we tell someone we have seven pennies, our meaning is clear. Pennies come in whole numbers, and seven pennies means seven whole pennies—it is unlikely that we would have 7.4 pennies. However, if we tell someone that we have a 10 g gold bar, the meaning is *unclear*. Our knowledge of the actual amount of gold in the bar depends on how precisely it was measured, which in turn depends on the scale or balance used to make the measurement. As we just learned, measured quantities are written to reflect the uncertainty in the measurement. If the gold measurement is rough, we could describe the bar as containing "10 g of gold." If a more precise balance is used, we can write the gold content as "10.0 g." We could report an even more precise measurement as "10.00 g."

We report scientific numbers so that every digit is certain except the last, which we estimate. For example, if a reported measurement is:

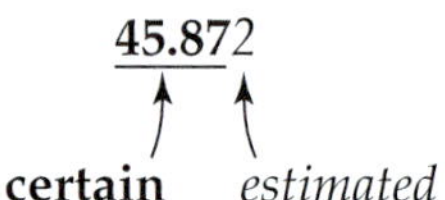

We know that the first four digits are certain; the last digit is estimated.

Suppose that we weigh an object on a balance with marks at every 1 g, and the pointer is between the 1 g mark and the 2 g mark (▼ Figure 4.1) but much closer to the 1 g mark. To record the measurement, we mentally divide the space between the 1 and 2 g marks into 10 equal spaces and estimate the position of the pointer. In this case, the pointer indicates about 1.2 g. We write the measurement as 1.2 g, indicating that we are sure of the "1" but have estimated the ".2."

If we measure an object using a balance with marks every tenth of a gram, we need to write the result with more digits. For example, suppose that on this more precise balance the pointer is between the 1.2 g mark and the 1.3 g mark (▼ Figure 4.2). We again divide the space between the two marks into 10 equal spaces and estimate the third digit. In the case of the nut shown in Figure 4.2, we report 1.26 g. Digital balances usually have readouts that report the mass to the correct number of digits.

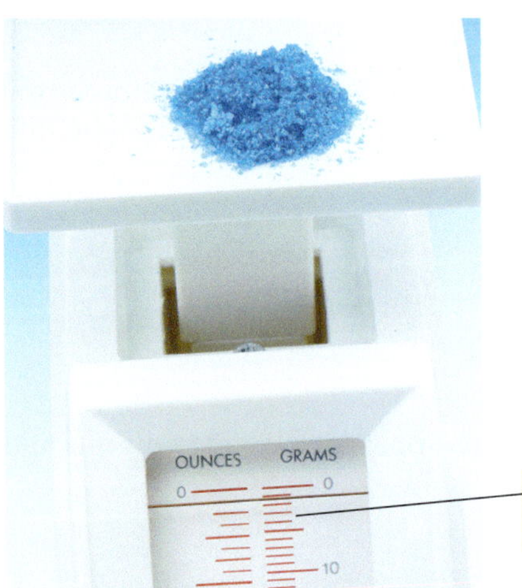

▲ **FIGURE 4.1 Estimating tenths of a gram** This balance has markings every 1 g, so we estimate to the tenths place. To estimate between markings, we mentally divide the space into 10 equal spaces and estimate the last digit. This reading is 1.2 g. © Richard Megna/Fundamental Photographs.

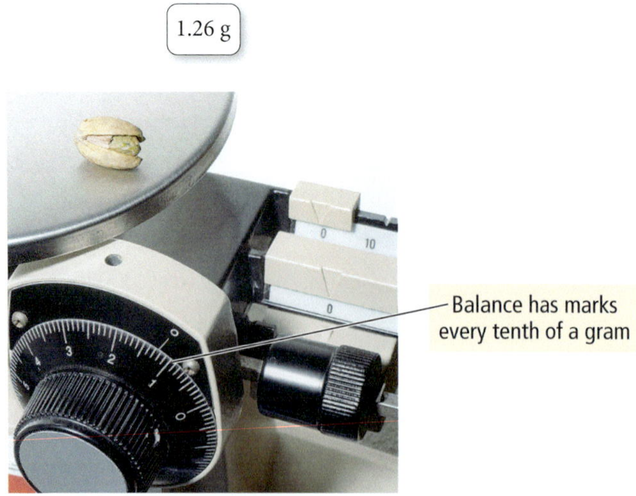

▲ **FIGURE 4.2 Estimating hundredths of a gram** Because this scale has markings every 0.1 g, we estimate to the hundredths place. The correct reading is 1.26 g. © Richard Megna/Fundamental Photographs.

EXAMPLE 4.3 REPORTING THE RIGHT NUMBER OF DIGITS

The bathroom scale in ▼ Figure 4.3 has markings at every 1 lb. Report the reading to the correct number of digits.

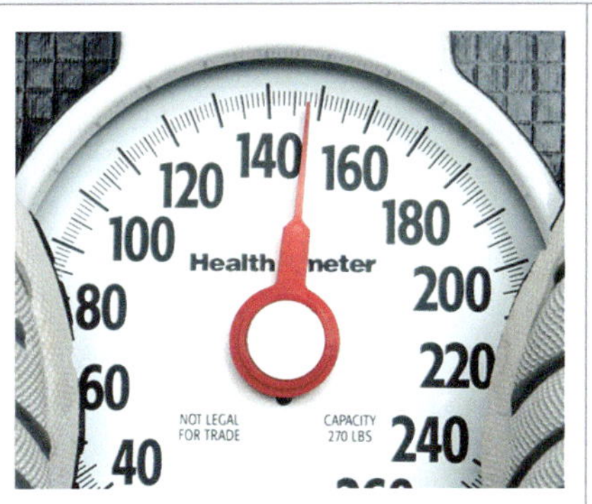

▲ **FIGURE 4.3 Reading a bathroom scale**

SOLUTION

Because the pointer is between the 147 and 148 lb markings, you mentally divide the space between the markings into 10 equal spaces and estimate the next digit. In this case, you should report the result as:

147.7 lb

What if you estimated a little differently and wrote 147.6 lb? In general, one unit of difference in the last digit is acceptable because the last digit is estimated and different people might estimate it slightly differently. However, if you wrote 147.2 lb, you would clearly be wrong.

▶**SKILLBUILDER 4.3 | Reporting the Right Number of Digits**

You use a thermometer to measure the temperature of a backyard hot tub, and you obtain the reading shown in ▼ Figure 4.4. Record the temperature reading to the correct number of digits.

◀ **FIGURE 4.4 Reading a thermometer**

▶**FOR MORE PRACTICE** Example 4.13; Problems 30, 31.

Counting Significant Figures

The non–place-holding digits in a measurement are **significant figures** (or **significant digits**). As we have seen, these significant figures represent the precision of a measured quantity—the greater the number of significant figures, the greater the precision of the measurement. We can determine the number of significant figures in a written number fairly easily; however, if the number contains zeros, we must distinguish between the zeros that are significant and those that simply mark the decimal place. In the number 0.002, for example, the leading zeros mark the decimal place; they *do not* add to the precision of the measurement. In the number 0.00200, however, the trailing zeros *do* add to the precision of the measurement.

To determine the number of significant figures in a number, we follow these rules:

1. All nonzero digits are significant.

   1.05 0.0110

2. Interior zeros (zeros between two numbers) are significant.

   4.0208 50.1

3. Trailing zeros (zeros to the right of a nonzero number) that fall after a decimal point are significant.

   5.10 3.00

When a number is expressed in scientific notation, all trailing zeros are significant.

4. Trailing zeros that fall before a decimal point are significant.

$$5\underline{0}.00 \qquad 17\underline{00}.24$$

5. Leading zeros (zeros to the left of the first nonzero number) are not significant. They only serve to locate the decimal point.

   For example, the number 0.0005 has only one significant digit.

6. Trailing zeros at the end of a number, but before an *implied* decimal point, are ambiguous and should be avoided by using scientific notation.

   For example, it is unclear if the number 350 has two or three significant figures. We can avoid confusion by writing the number as 3.5×10^2 to indicate two significant figures or as 3.50×10^2 to indicate three.

Exact Numbers

Some books put a decimal point after one or more trailing zeros if the zeros are to be considered significant. We avoid that practice in this book, but you should be aware of it.

Exact numbers have an unlimited number of significant figures. Exact numbers originate from three sources:

- Exact counting of discrete objects. For example, 3 atoms means 3.00000 … atoms.
- *Defined quantities*, such as the number of centimetres in 1 m. Because 100 cm is defined as 1 m,

$$100 \text{ cm} = 1 \text{ m means } 100.00000\ldots\text{cm} = 1.0000000\ldots\text{m}$$

- Integral numbers that are part of an equation. For example, in the equation, $radius = \frac{diameter}{2}$, the number 2 is exact and therefore has an unlimited number of significant figures.

EXAMPLE 4.4 DETERMINING THE NUMBER OF SIGNIFICANT FIGURES IN A NUMBER

How many significant figures are in each number?

(a) 0.0035
(b) 1.080
(c) 2371
(d) 2.97×10^5
(e) 1 dozen = 12
(f) 100.00
(g) 100,000

	SOLUTION
The 3 and the 5 are significant (rule 1). The leading zeros only mark the decimal place and are not significant (rule 5).	**(a)** 0.0035 two significant figures
The interior zero is significant (rule 2), and the trailing zero is significant (rule 3). The 1 and the 8 are also significant (rule 1).	**(b)** 1.080 four significant figures
All digits are significant (rule 1).	**(c)** 2371 four significant figures
All digits in the decimal part are significant (rule 1).	**(d)** 2.97×10^5 three significant figures
Defined numbers are exact and therefore have an unlimited number of significant figures.	**(e)** 1 dozen = 12 unlimited significant figures

The 1 is significant (rule 1), and the trailing zeros before the decimal point are significant (rule 4). The trailing zeros after the decimal point are also significant (rule 3).	**(f)** 100.00 five significant figures
This number is ambiguous. Write as 1×10^5 to indicate one significant figure or as 1.00000×10^5 to indicate six significant figures.	**(g)** 100,000 ambiguous

▶SKILLBUILDER 4.4 | Determining the Number of Significant Figures in a Number

How many significant figures are in each number?

(a) 58.31
(b) 0.00250
(c) 2.7×10^3
(d) 1 cm = 0.01 m
(e) 0.500
(f) 2100

▶FOR MORE PRACTICE Example 4.14; Problems 32, 33, 34, 35, 36, 37.

CHEMISTRY IN THE MEDIA

▶ The COBE Satellite and Very Precise Measurements That Illuminate Our Cosmic Past

Since the earliest times, humans have wondered about the origins of our planet. Scientists have probed this question and developed theories for how the universe and the Earth began. The most accepted theory today about the origin of the universe is the Big Bang theory. According to the Big Bang theory, the universe began in a tremendous expansion about 13.7 billion years ago and has been expanding ever since. A measurable prediction of this theory is the presence of a remnant "background radiation" from the expansion of the universe. That remnant radiation is characteristic of the current temperature of the universe. When the Big Bang occurred, the temperature of the universe was very hot and the associated radiation very bright. Today, 13.7 billion years later, the temperature of the universe is very cold and the background radiation very faint.

In the early 1960s, Robert H. Dicke, P. J. E. Peebles, and their colleagues at Princeton University began to build a device to measure this background radiation and by doing so take a direct look into the cosmological past and provide evidence for the Big Bang theory. At about the same time, quite by accident, Arno Penzias and Robert Wilson of Bell Telephone Laboratories measured excess radio noise on one of their communications satellites. As it turned out, this noise was the background radiation that the Princeton scientists were looking for. The two groups published papers together in 1965 reporting their findings along with the corresponding current temperature of the universe, about 3 degrees above absolute zero, or 3 K. We will define temperature measurement scales in Module 6. For now, know that 3 K is an extremely low temperature (–270 °C).

In 1989, NASA's Goddard Space Flight Center developed the Cosmic Background Explorer (COBE) satellite to measure the background radiation more precisely. The COBE satellite determined that the background radiation corresponded to a universe with a temperature of 2.735 K. (Notice the difference in significant figures from the previous measurement.) It went on to measure tiny fluctuations in the background radiation that amount to temperature differences of 1 part in 100,000. These fluctuations, though small, are an important prediction of the Big Bang theory. Scientists announced that the COBE satellite had produced the strongest evidence yet for the Big Bang theory of the creation of the universe. This is the way that science works. Measurement, and precision in measurement, are important to understanding the world—so important that we dedicate most of this module to the concept of measurement.

B4.1 CAN YOU ANSWER THIS? *How many significant figures are there in each of the preceding temperature measurements (3 K, 2.735 K)?*

◀ The COBE satellite, launched in 1989 to measure background radiation. Background radiation is a remnant of the Big Bang—the expansion that is believed to have formed the universe. © NASA Images.

✓ CONCEPTUAL CHECKPOINT 4.2

The Curiosity Rover on the surface of Mars recently measured a daily low temperature of –65.19 °C. What is implied range of the actual temperature?

(a) between –65.190 °C and –65.199 °C

(b) between –65.18 °C and –65.20 °C

(c) between –65.1 °C and –65.2 °C

(d) exactly –65.19 °C

4.3 Significant Figures in Calculations

LO: Round numbers to the correct number of significant figures.

LO: Determine the correct number of significant figures in the results of multiplication and division calculations.

LO: Determine the correct number of significant figures in the results of addition and subtraction calculations.

LO: Determine the correct number of significant figures in the results of calculations involving both addition/subtraction and multiplication/division.

When we use measured quantities in calculations, the results of the calculation must reflect the precision of the measured quantities. We should not lose or gain precision during mathematical operations.

Multiplication and Division

In multiplication or division, the result carries the same number of significant figures as the factor with the fewest significant figures.

For example:

$$\underset{\text{(3 sig. figures)}}{5.02} \times \underset{\text{(5 sig. figures)}}{89.665} \times \underset{\text{(2 sig. figures)}}{0.10} = 45.0118 = \underset{\text{(2 sig. figures)}}{45}$$

We round the intermediate result (in blue) to two significant figures to reflect the least precisely known factor (0.10), which has two significant figures.

In division, we follow the same rule.

$$\underset{\text{(4 sig. figures)}}{5.892} \div \underset{\text{(3 sig. figures)}}{6.10} = 0.96590 = \underset{\text{(3 sig. figures)}}{0.966}$$

We round the intermediate result (in blue) to three significant figures to reflect the least precisely known factor (6.10), which has three significant figures.

Rounding

When we round to the correct number of significant figures:

we round down if the last (or leftmost) digit dropped is 4 or less;
we round up if the last (or leftmost) digit dropped is 5 or more.

For example, when we round each of these numbers to two significant figures:

2.33 rounds to 2.3
2.37 rounds to 2.4
2.34 rounds to 2.3
2.35 rounds to 2.4

We consider only the *last (or leftmost) digit being dropped* when we decide in which direction to round—we ignore all digits to the right of it. For example, to round 2.349 to two significant figures, only the 4 in the hundredths place (2.349) determines which direction to round—the 9 is irrelevant.

2.349 rounds to 2.3

For calculations involving multiple steps, we round only the final answer—we do not round the intermediate results. This prevents small rounding errors from affecting the final answer.

EXAMPLE 4.5 SIGNIFICANT FIGURES IN MULTIPLICATION AND DIVISION

Perform each calculation to the correct number of significant figures.

(a) 1.01 × 0.12 × 53.51 ÷ 96
(b) 56.55 × 0.920 ÷ 34.2585

SOLUTION

Round the intermediate result (in blue) to two significant figures to reflect the two significant figures in the least precisely known quantities (0.12 and 96).

(a) 1.01 × 0.12 × 53.51 ÷ 96 = 0.067556 = 0.068

Round the intermediate result (in blue) to three significant figures to reflect the three significant figures in the least precisely known quantity (0.920).

(b) 56.55 × 0.920 ÷ 34.2585 = 1.51863 = 1.52

▶SKILLBUILDER 4.5 | Significant Figures in Multiplication and Division

Perform each calculation to the correct number of significant figures.

(a) 1.10 × 0.512 × 1.301 × 0.005 ÷ 3.4
(b) 4.562 × 3.99870 ÷ 89.5

▶FOR MORE PRACTICE Examples 4.15, 4.16; Problems 46, 47, 48, 49.

Addition and Subtraction

In addition or subtraction, the result has the same number of decimal places as the quantity with the fewest decimal places.

For example:

```
   5.74|
   0.82|3
 + 2.65|1
 -------
   9.21|4 = 9.21
```

It is sometimes helpful to draw a vertical line directly to the right of the number with the fewest decimal places. The line shows the number of decimal places that should be in the answer.

We round the intermediate answer (in blue) to two decimal places because the quantity with the fewest decimal places (5.74) has two decimal places.

For subtraction, we follow the same rule. For example:

```
   4.8|
 - 3.9|65
 --------
   0.8|35 = 0.8
```

We round the intermediate answer (in blue) to one decimal place because the quantity with the fewest decimal places (4.8) has one decimal place. Remember: *For multiplication and division, the quantity with the fewest* ***significant figures*** *determines the number of significant figures in the answer. For addition and subtraction, the quantity with the fewest* ***decimal places*** *determines the number of decimal places in the answer.* In multiplication and division we focus on significant figures, but in addition and subtraction we focus on decimal places. When a problem involves addition and subtraction, the answer may have a different number of significant figures than the initial quantities. For example:

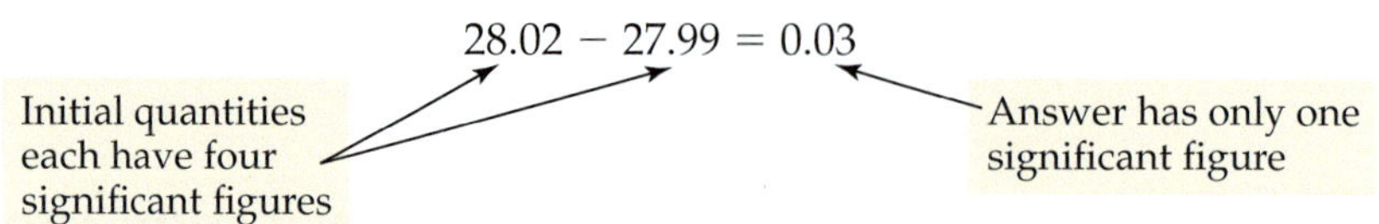

The answer has only one significant figure, even though the initial quantities each had four significant figures.

EXAMPLE 4.6 SIGNIFICANT FIGURES IN ADDITION AND SUBTRACTION

Perform the calculations to the correct number of significant figures.

(a)
$$\begin{array}{r} 0.987 \\ +125.1 \\ -1.22 \\ \hline \end{array}$$

(b)
$$\begin{array}{r} 0.765 \\ -3.449 \\ -5.98 \\ \hline \end{array}$$

SOLUTION

Round the intermediate answer (in blue) to one decimal place to reflect the quantity with the fewest decimal places (125.1). Notice that 125.1 is not the quantity with the fewest significant figures—it has four while the other quantities only have three—but because it has the fewest decimal places, it determines the number of decimal places in the answer.

(a)
$$\begin{array}{r} 0.987 \\ +125.1 \\ -1.22 \\ \hline 124.867 \end{array} = 124.9$$

Round the intermediate answer (in blue) to two decimal places to reflect the quantity with the fewest decimal places (5.98).

(b)
$$\begin{array}{r} 0.765 \\ -3.449 \\ -5.98 \\ \hline -8.664 \end{array} = -8.66$$

▶SKILLBUILDER 4.6 | Significant Figures in Addition and Subtraction

Perform the calculations to the correct number of significant figures.

(a)
$$\begin{array}{r} 2.18 \\ +5.621 \\ +1.5870 \\ -1.8 \\ \hline \end{array}$$

b)
$$\begin{array}{r} 7.876 \\ -0.56 \\ +123.792 \\ \hline \end{array}$$

▶FOR MORE PRACTICE Example 4.17; Problems 50, 51, 52, 53.

Calculations Involving Both Multiplication/Division and Addition/Subtraction

In calculations involving both multiplication/division and addition/subtraction, we do the steps in parentheses first; determine the correct number of significant figures in the intermediate answer; then complete the remaining steps.

For example:

$$3.489 \times (5.67 - 2.3)$$

We complete the subtraction step first.

$$5.67 - 2.3 = 3.37$$

We use the subtraction rule to determine that the intermediate answer (3.37) has only one significant decimal place. To avoid small errors, we do not round at this point; instead, we underline the least significant figure as a reminder.

$$= 3.489 \times 3.\underline{3}7$$

We then complete the multiplication step.

$$3.489 \times 3.\underline{3}7 = 11.758 = 12$$

The multiplication rule indicates that the intermediate answer (11.758) rounds to two significant figures (12) because it is limited by the two significant figures in $3.\underline{3}7$.

EXAMPLE 4.7 SIGNIFICANT FIGURES IN CALCULATIONS INVOLVING BOTH MULTIPLICATION/DIVISION AND ADDITION/SUBTRACTION

Perform the calculations to the correct number of significant figures.

(a) $6.78 \times 5.903 \times (5.489 - 5.01)$
(b) $19.667 - (5.4 \times 0.916)$

SOLUTION

Do the step in parentheses first. Use the subtraction rule to mark 0.479 to two decimal places because 5.01, the number in the parentheses with the least number of decimal places, has two.

Then perform the multiplication and round the answer to two significant figures because the number with the least number of significant figures has two.

(a)
$$6.78 \times 5.903 \times (5.489 - 5.01)$$
$$= 6.78 \times 5.903 \times (0.479)$$
$$= 6.78 \times 5.903 \times 0.4\underline{7}9$$
$$6.78 \times 5.903 \times 0.4\underline{7}9 = 19.1707$$
$$= 19$$

Do the step in parentheses first. The number with the least number of significant figures within the parentheses (5.4) has two, so mark the answer to two significant figures.

Then perform the subtraction and round the answer to one decimal place because the number with the least number of decimal places has one.

(b)
$$19.667 - (5.4 \times 0.916)$$
$$= 19.667 - (4.9464)$$
$$= 19.667 - 4.\underline{9}464$$
$$19.667 - 4.\underline{9}464 = 14.7206$$
$$= 14.7$$

▶**SKILLBUILDER 4.7 | Significant Figures in Calculations Involving Both Multiplication/Division and Addition/Subtraction**

Perform each calculation to the correct number of significant figures.

(a) $3.897 \times (782.3 - 451.88)$
(b) $(4.58 \div 1.239) - 0.578$

▶**FOR MORE PRACTICE** Example 4.18; Problems 54, 55, 56, 57.

CONCEPTUAL CHECKPOINT 4.3

Which calculation would have its result reported to the *greater* number of significant figures?

(a) $3 + (15/12)$
(b) $(3 + 15)/12$

4.4 The Basic Units of Measurement

LO: Recognize and work with the SI base units of measurement, prefix multipliers, and derived units.

By themselves, numbers have little meaning. Read this sentence: When my son was 7 he walked 3, and when he was 4 he could throw his baseball 8 and tell us that his school was 5 away. The sentence is confusing. We don't know what the numbers mean because the **units** are missing. The meaning becomes clear, however, when we add the missing units to the numbers: When my son was 7 *months old* he walked 3 *steps*, and when he was 4 *years old* he could throw his baseball 8 *feet* and tell us that his school was 5 *minutes* away. Units make all the difference. In chemistry, units are critical. Never write a number by itself; always use its associated units—otherwise your work will be as confusing as the initial sentence.

The two most common unit systems are the **English system**, used in the United States, and the **metric system**, used in most of the rest of the world. The English system uses units such as inches, yards, and pounds, while the metric system uses centimetres, metres, and kilograms. The most convenient system for science measurements is based on the metric system and is called the **International System** of units or **SI units**. SI units are a set of standard units agreed on by scientists throughout the world.

The abbreviation *SI* comes from the French *le Système International*.

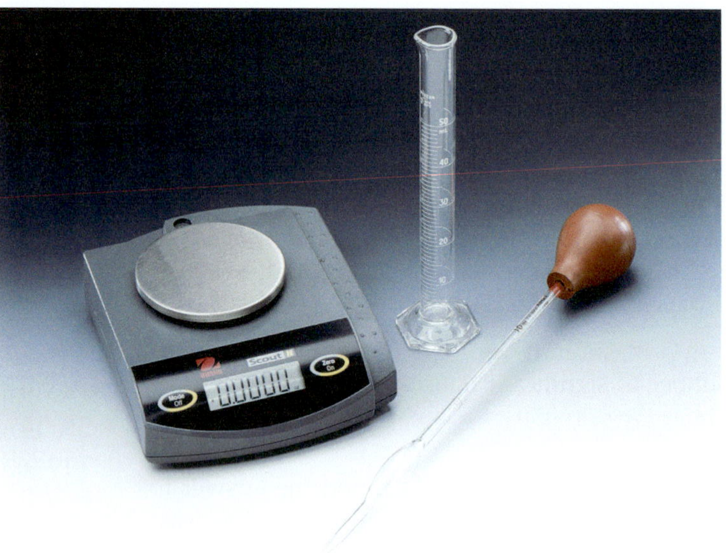

▲ Science uses instruments to make measurements. Every instrument is calibrated in a particular unit without which the measurements would be meaningless. © Richard Megna/Fundamental Photographs.

TABLE 4.1 Important SI Base Units

Quantity	Unit	Symbol
length	metre	m
mass	kilogram	kg
time	second	s
temperature*	kelvin	K

*Temperature units are discussed in Module 6.

The Base Units

Table 4.1 lists the common SI base units. They include the **metre (m)** as the base unit of length; the **kilogram (kg)** as the base unit of mass; and the **second (s)** as the base unit of time. Each of these base units is precisely defined. The metre is defined as the distance light travels in a certain period of time: 1/299,792,458 s (▼ Figure 4.5).

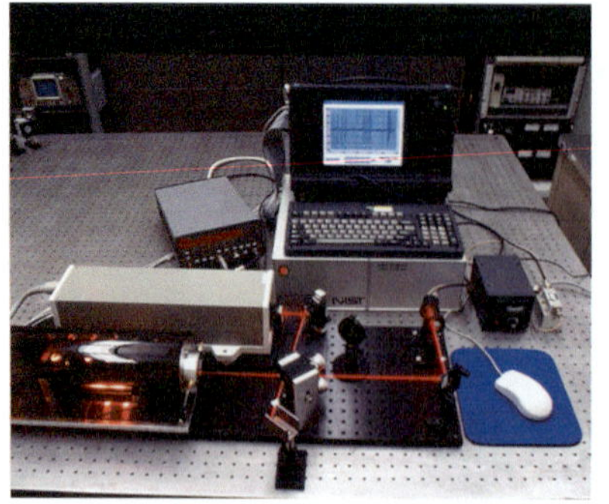

◀ **FIGURE 4.5 The base unit of length** The definition of a metre, established by international agreement in 1983, is the distance that light travels in vacuum in 1/299,792,458 s. © The National Institute of Standards and Technology Museum.
Question: Why is such a precise standard necessary?

▲ **FIGURE 4.6 The base unit of mass** A duplicate of the international standard kilogram, called kilogram 20, is kept at the National Institute of Standards and Technology near Washington, D.C. © BIPM/AFP/Getty Images/Newscom.

▲ **FIGURE 4.7 The base unit of time** The second is defined, using an atomic clock, as the duration of 9,192,631,770 periods of the radiation emitted from a certain transition in a cesium-133 atom. © The National Institute of Standards and Technology Museum.

Because the mass of the block of metal used to define a kilogram has changed slightly over the years, scientists are currently working on alternate ways to define the kilogram.

(The speed of light is 3.0×10^8 m/s.) The kilogram is defined as the mass of a block of metal kept at the International Bureau of Weights and Measures at Sèvres, France (▲ Figure 4.6). The second is defined using an atomic standard (▲ Figure 4.7). Most people are familiar with the SI base unit of time, the second.

The kilogram is a measure of mass, which is different from weight. The **mass** of an object is a measure of the quantity of matter within it, while the weight of an object is a measure of the gravitational pull on that matter. Consequently, weight depends on gravity while mass does not. If you were to weigh yourself on Mars, for example, the lower gravity would pull you toward the scale less than Earth's gravity would, resulting in a lower weight. A 68 kg person on Earth weighs only 26 kg on Mars. However, the person's mass, the quantity of matter in his or her body, remains the same. A second common unit of mass is the gram (g), defined as follows:

A nickel (5 cents) has a mass of about 5 g.

$$1000 \text{ g} = 10^3 \text{ g} = 1 \text{ kg}$$

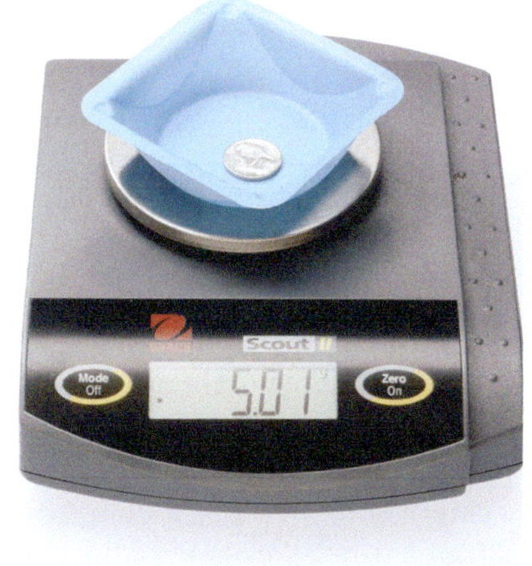

© Richard Megna/Fundamental Photographs.

Prefix Multipliers

The SI system employs **prefix multipliers** (Table 4.2) with the base units. These multipliers change the value of the unit by powers of 10. For example, the kilometre (km) features the prefix *kilo-*, meaning 1000 or 10^3. Therefore:

$$1 \text{ km} = 1000 \text{ m} = 10^3 \text{ m}$$

Similarly, the millisecond (ms) has the prefix *milli-*, meaning 0.001 or 10^{-3}.

$$1 \text{ ms} = 0.001 \text{ s} = 10^{-3} \text{ s}$$

The prefix multipliers allow us to express a wide range of measurements in units that are similar in size to the quantity we are measuring. We choose the prefix multiplier that is most convenient for a particular measurement. For example, to measure the diameter of a quarter, we use centimetres because a quarter has a diameter of about 2.43 cm. A centimetre is a common metric unit and is about equivalent to the width of a pinky finger (2.54 cm = 1 in.). The millimetre could also work to express the diameter of the quarter; the same quarter measures 24.3 mm.

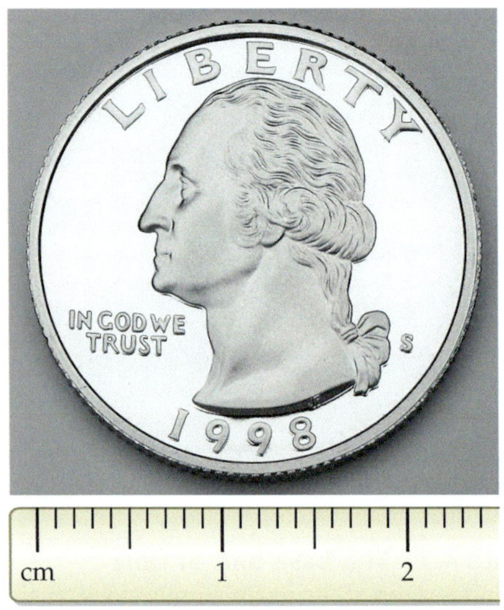

▲ The diameter of a quarter is about 2.43 cm.

Question: Why would we *not* use metres to measure a quarter?

TABLE 4.2 SI Prefix Multipliers

Prefix	Symbol	Meaning	Multiplier	
tera-	T	trillion	1,000,000,000,000	(10^{12})
giga-	G	billion	1,000,000,000	(10^{9})
mega-	M	million	1,000,000	(10^{6})
kilo-	k	thousand	1,000	(10^{3})
hecto-	h	hundred	100	10^{2}
deca-	da	ten	10	10^{1}
deci-	d	tenth	0.1	(10^{-1})
centi-	c	hundredth	0.01	(10^{-2})
milli-	m	thousandth	0.001	(10^{-3})
micro-	μ	millionth	0.000001	(10^{-6})
nano-	n	billionth	0.000000001	(10^{-9})
pico-	p	trillionth	0.000000000001	(10^{-12})
femto-	f	quadrillionth	0.000000000000001	(10^{-15})

The kilometre, however, would not work as well because, in that unit, the quarter's diameter is 0.0000243 km. We pick a unit similar in size to (or smaller than) the quantity we are measuring. Consider expressing the length of a short chemical bond, about 1.2×10^{-10} m. Which prefix multiplier should we use? The most convenient one is probably the picometre (pico = 10^{-12}). Chemical bonds measure about 120 pm.

CONCEPTUAL CHECKPOINT 4.4

Which is the most convenient unit to use to express the dimensions of a polio virus, which is about 2.8×10^{-8} m in diameter?

(a) Mm

(b) mm

(c) μm

(d) nm

Derived Units

A derived unit is formed from other units. For example, many units of **volume**, a measure of space, are derived units. Any unit of length, when cubed (raised to the third power), becomes a unit of volume. Therefore, cubic metres (m^3), cubic centimetres (cm^3), and cubic millimetres (mm^3) are all units of volume. A three-bedroom house has a volume of about 630 m^3, a can of soda pop has a volume of about 350 cm^3, and a rice grain has a volume of about 3 mm^3. We also use the **litre (L)** and millilitre (mL) to express volume (although these are not derived units). A millilitre is equivalent to 1 cm^3.

4.5 Problem Solving and Unit Conversion

Problem solving is one of the most important skills you will acquire in this course. Not only will this skill help you succeed in chemistry, but it will help you to learn how to think critically, which is important in every area of knowledge. Learning how to solve chemical problems helps you develop these kinds of skills. Although no simple formula applies to every problem, you can learn problem-solving strategies and begin to develop some chemical intuition. You can think of many of the problems in this book as *unit conversion problems*, where you are given one or more quantities and asked to convert them into different units. Other problems require the use of *specific equations* to get to the information you are trying to find. In the sections that follow, you will find strategies to help you solve both of these types of problems. Of course, many problems contain both conversions and equations, requiring the combination of these strategies, and some problems may require an altogether different approach, but the basic tools you learn here can be applied to those problems as well.

Units are critical in calculations. Knowing how to work with and manipulate units in calculations is a crucial part of problem solving. In calculations, units help determine correctness. You should always include units in calculations, and you can think of many calculations as converting from one unit to another. You can multiply, divide, and cancel units like any other algebraic quantity.

Remember:

1. Always write every number with its associated unit. Never ignore units; they are critical.
2. Always include units in your calculations, dividing them and multiplying them as if they were algebraic quantities. Do not let units magically appear or disappear in calculations. Units must flow logically from beginning to end.

General Problem-Solving Strategy

In this book, we use a standard problem-solving procedure that you can adapt to many of the problems encountered in chemistry and beyond. Solving any problem essentially requires that you assess the information given in the problem and devise a way to get to the requested information. In other words, you need to:

- Identify the starting point (the ***given*** information).
- Identify the end point (what you must ***find***).
- Devise a way to get from the starting point to the end point using what is given as well as what you already know or can look up. You can use a *solution map* to diagram the steps required to get from the starting point to the end point.

In graphic form, this progression looks like this:

Given → Solution Map → Find

Beginning students often have trouble knowing how to start solving a chemistry problem. Although no problem-solving procedure is applicable to all problems, the following four-step procedure can be helpful in working through many of the numerical problems you will encounter in this book.

1. **Sort**. Begin by sorting the information in the problem. *Given* information is the basic data provided by the problem—often one or more numbers with their associated units. The given information is the starting point for the problem. *Find* indicates what the problem is asking you to find (the end point of the problem).
2. **Strategize.** This is usually the hardest part of solving a problem. In this process, you must create a solution map—the series of steps that will get you from the given information to the information you are trying to find. Each arrow in a solution map represents a computational step. On the left side of the arrow is the quantity (or quantities) you had before the step; on the right side of the arrow is the quantity (or quantities) you will have after the step. Below the solution map is the information you will need—the relationship between the quantities or the equation for a calculation.

   In some cases, you may get stuck at the strategize step. If you cannot figure out how to get from the given information to the information you are asked to find, you might try working backwards. For example, you may want to look at the units of the quantity you are trying to find and look for relationships or equations to get to the units of the given quantity. You may even try a combination of strategies; work forward, backward, or some of both. If you persist, you will develop a strategy to solve the problem.
3. **Solve.** This is the most straightforward part of solving a problem. Once you set up the problem properly and devise a solution map, you follow the map to solve the problem. Carry out mathematical operations (paying attention to the rules for significant figures in calculations) and cancel units as needed.
4. **Check.** This step is often overlooked by beginning students. Experienced problem solvers always ask, Does this answer make physical sense? Are the units correct? Is the number of significant figures correct? When solving multistep problems, errors easily creep into the solution. You can catch most of these errors by simply checking the answer. For example, suppose you are calculating the number of atoms in a gold coin and end up with an answer of 1.1×10^{-6} atoms. Could the gold coin really be composed of one-millionth of one atom?

In Examples 4.8 and 4.9, you will find this problem-solving procedure applied to unit conversion problems. The procedure is summarized in the left column, and two examples of applying the procedure are shown in the middle and right columns. You will encounter this three-column format in selected examples throughout this text. It allows you to see how a particular procedure can be applied to two different problems. Work through one problem first (from top to bottom) and then examine how the same procedure is applied to the other problem. Recognizing the commonalities and differences between problems is a key part of problem solving.

PROBLEM-SOLVING PROCEDURE	EXAMPLE 4.8	EXAMPLE 4.9
	UNIT CONVERSION Convert 637 millilitres (mL) to L.	**UNIT CONVERSION** Convert 0.825 m to millimetres.
SORT Begin by sorting the information in the problem into *given* and *find*.	**GIVEN:** 637 mL **FIND:** L	**GIVEN:** 0.825 m **FIND:** mm
STRATEGIZE Draw a *solution map* for the problem. Begin with the *given* quantity and symbolize each step with an arrow. Below the arrow, write the relationship or equation for that step. The solution map ends at the *find* quantity.	**SOLUTION MAP** mL → L **RELATIONSHIPS USED** $1\ \text{mL} = 10^{-3}\ \text{L}\ (1000\ \text{mL/L})$	**SOLUTION MAP** m → mm **RELATIONSHIPS USED** 1 m = 1000 mm (1000 mm/m)
SOLVE Follow the *solution map* to solve the problem, using the appropriate relationship or equation. Round the answer to the correct number of significant figures.	**SOLUTION** $\frac{637\ \text{mL}}{1000\ \text{mL/L}} = 0.637\ \text{L}$ See how the mL in the numerator and denominator cancel to give the correct units for the answer. Leave the answer with three significant figures, because the given quantity has three significant figures.	**SOLUTION** $0.825\ \text{m} \times \frac{1000\ \text{mm}}{\text{m}} = 825\ \text{mm}$ See how the m and 1/m terms cancel to give the correct units for the answer. Leave the answer with three significant figures, because the quantity given has three significant figures.
CHECK Check your answer. Are the units correct? Does the answer make sense?	The units, L, are correct. The magnitude of the answer is reasonable. A litre is larger than a millilitre, so the value in litres should be smaller than the value in millilitres. ▶**SKILLBUILDER 4.8 | Unit Conversion** Convert 56 mL to L. ▶**FOR MORE PRACTICE** Example 4.19.	The units, mm, are correct and the magnitude is reasonable. A millimetre is shorter than a metre, so the value in millimetres should be larger than the value in metres. ▶**SKILLBUILDER 4.9 | Unit Conversion** Convert 5678 m to kilometres. ▶**FOR MORE PRACTICE** Problems 58, 59, 60, 61.

CHEMISTRY AND HEALTH

▶ Drug Dosage

The unit of choice in specifying drug dosage is the milligram (mg). A bottle of aspirin, Tylenol, or any other common drug lists the number of milligrams of the active ingredient contained in each tablet, as well as the number of tablets to take per dose. The following table shows the mass of the active ingredient per pill in several common pain relievers, all reported in milligrams. The remainder of each tablet is composed of inactive ingredients such as cellulose (or fiber) and starch.

The recommended adult dose for many of these pain relievers is one or two tablets every 4 to 8 hours (depending on the specific pain reliever). Notice that the extra-strength version of each pain reliever just contains a higher dose of the same compound found in the regular-strength version. For the pain relievers listed, three regular-strength tablets are the equivalent of two extra-strength tablets (and probably cost less).

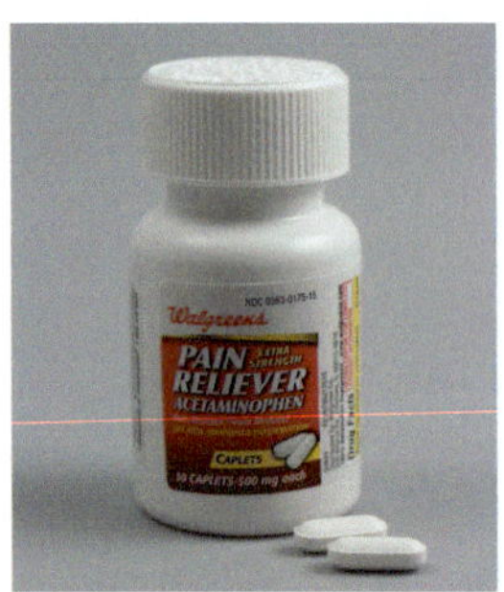

The dosages given in the table are fairly standard for each drug, regardless of the brand. On most drugstore shelves, there are many different brands of regular-strength ibuprofen, some sold under the generic name and others sold under their brand names (such as Advil). However, if you look closely at the labels, you will find that they all contain the same thing: 200 mg of the compound ibuprofen. There is no difference in the compound or in the amount of the compound. Yet these pain relievers will most likely all have different prices. Choose the least expensive. Why pay more for the same thing?

Drug Mass per Pill for Common Pain Relievers

Pain Reliever	Mass of Active Ingredient per Pill
aspirin	325 mg
aspirin, extra strength	500 mg
ibuprofen (Advil)	200 mg
ibuprofen, extra strength	300 mg
acetaminophen (Tylenol)	325 mg
acetaminophen, extra strength	500 mg

4.6 Density

LO: Calculate the density of a substance.

LO: Use density in calculations.

Why do some people pay over \$3,000 for a bicycle made of titanium? A steel frame would be just as strong for a fraction of the cost. The difference between the two bikes is their masses—the titanium bike is lighter. For a given volume of metal, titanium has less mass than steel. Titanium is said to be *less dense* than steel. The **density** of a substance is the ratio of its mass to its volume.

$$\text{Density} = \frac{\text{Mass}}{\text{Volume}} \quad \text{or} \quad d = \frac{m}{V}$$

Density is a fundamental property of a substance and differs from one substance to another. The units of density are those of mass divided by those of volume, most conveniently expressed in grams per cubic centimetre (g/cm^3) or grams per millilitre (g/mL). Table 4.3 lists the densities of some common substances. Aluminum is among the least dense structural metals with a density of 2.70 g/cm^3, while platinum is among the densest with a density of 21.4 g/cm^3. Titanium has a density of 4.50 g/cm^3.

TABLE 4.3 Densities of Some Common Substances

Substance	Density (g/cm^3)
charcoal, oak	0.57
ethanol	0.789
ice	0.92
water	1.0
glass	2.6
aluminum	2.7
titanium	4.50
iron	7.86
copper	8.96
lead	11.4
gold	19.3
platinum	21.4

Calculating Density

You can calculate the density of a substance by dividing the mass of a given amount of the substance by its volume. For example, a sample of liquid has a volume of 22.5 mL and a mass of 27.2 g. To find its density, use the equation $d = m/V$.

$$d = \frac{m}{V} = \frac{27.2\ \text{g}}{22.5\ \text{mL}} = 1.21\ \text{g/mL}$$

▲ Top-end bicycle frames are made of titanium because of titanium's low density and high relative strength. Titanium has a density of 4.50 g/cm^3, while iron, for example, has a density of 7.86 g/cm^3. © Rudy Umans/Shutterstock.

Remember that cubic centimetres and millilitres are equivalent units.

You can use a solution map for solving problems involving equations, but the solution map takes a slightly different form than one for a unit conversion problem. In a problem involving an equation, the solution map shows how the *equation* takes you from the *given* quantities to the *find* quantity. The solution map for this problem is:

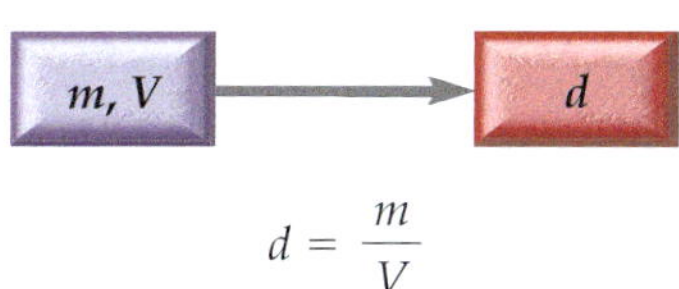

$$d = \frac{m}{V}$$

The solution map illustrates that the values of m and V, when substituted into the equation $d = \frac{m}{V}$, give the desired result, d.

EXAMPLE 4.10 CALCULATING DENSITY

A jeweler offers to sell a ring to a woman and tells her that it is made of platinum. Noting that the ring feels a little light, the woman decides to perform a test to determine the ring's density. She places the ring on a balance and finds that it has a mass of 5.84 g. She also finds that the ring *displaces* 0.556 cm^3 of water. Is the ring made of platinum? The density of platinum is 21.4 g/cm^3. (The displacement of water is a common way to measure the volume of irregularly shaped objects. To say that an object *displaces* 0.556 cm^3 of water means that when the object is submerged in a container of water filled to the brim, 0.556 cm^3 overflows. Therefore, the volume of the object is 0.556 cm^3.)

SORT
You are given the mass and volume of the ring and asked to find the density.

GIVEN: $m = 5.84$ g
$V = 0.556$ cm^3

FIND: density in g/cm^3

STRATEGIZE
If the ring is platinum, its density should match that of platinum. Build a solution map that represents how you get from the given quantities (mass and volume) to the find quantity (density). After the solution map is the information needed to solve the problem, the equation for density.

SOLUTION MAP

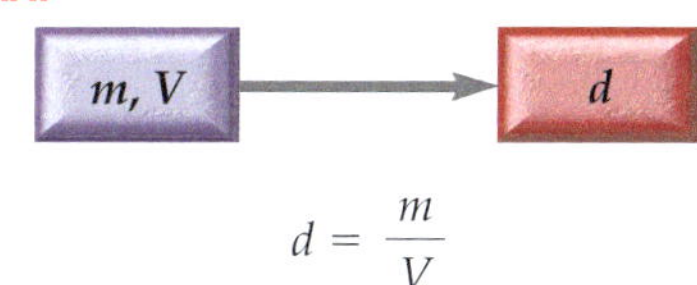

$$d = \frac{m}{V}$$

RELATIONSHIPS USED

$d = \frac{m}{V}$ (equation for density)

SOLVE
Follow the solution map. Substitute the given values into the density equation and calculate the density.

Round the answer to three significant figures to reflect the three significant figures in the given quantities.

SOLUTION

$$d = \frac{m}{V} = \frac{5.84\ \text{g}}{0.556\ \text{cm}^3} = 10.5\ \text{g/cm}^3$$

The density of the ring is much too low to be platinum; therefore the ring is a fake.

CHECK

Check your answer. Are the units correct? Does the answer make physical sense?

The units of the answer are correct, and the magnitude seems like it could be an actual density. As you can see from Table 4.3, the densities of liquids and solids range from below 1 g/cm^3 to just over 20 g/cm^3.

▶SKILLBUILDER 4.10 | Calculating Density

The woman takes the ring back to the jewelry shop, where she is met with endless apologies. The jeweler had accidentally made the ring out of silver rather than platinum. The jeweler gives her a new ring that she promises is platinum. This time when the customer checks the density, she finds the mass of the ring to be 9.67 g and its volume to be 0.452 cm^3. Is this ring genuine?

▶FOR MORE PRACTICE Example 4.20; Problems 66, 67, 68, 69, 70, 71.

Using Density in Calculations

Suppose you need 68.4 g of a liquid with a density of 1.32 g/cm^3 and want to measure the correct amount with a graduated cylinder (a piece of laboratory glassware used to measure volume). How much volume (in millilitres, mL) should you measure? This can be calculated by rearranging the equation for density $d = \frac{m}{V}$. Begin by sorting the information in the problem.

SOLUTION MAP

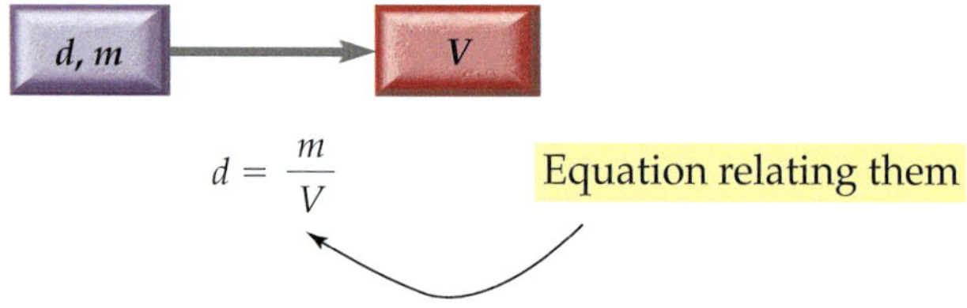

SOLUTION

Rearrange the equation for the unknown quantity, in this case it's volume (V).

$$V = \frac{m}{d} = \frac{68.4 \text{ g}}{1.32 \text{ g}/\text{cm}^3} = 51.8 \text{ cm}^3$$

Since 1 cm^3 is equivalent to 1 mL, you must measure 51.8 mL to obtain 68.4 g of the liquid.

EXAMPLE 4.11 USING DENSITY IN CALCULATIONS

The petrol in an automobile gas tank has a mass of 60.0 kg and a density of 0.752 g/cm^3. What is its volume in cm^3?

SORT You are given the mass in kilograms and asked to find the volume in cubic centimetres.	**GIVEN:** 60.0 kg Density = 0.752 g/cm^3 **FIND:** volume in cm^3
STRATEGIZE Build the solution map showing how the equation allows us to find the unknown from the known quantities. Note that the units of density are g/cm^3, so the mass given must first be converted from kg to g.	**SOLUTION MAP** 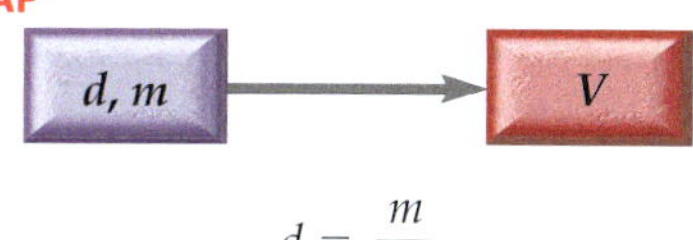 $d = \frac{m}{V}$ **RELATIONSHIPS USED** $d = \frac{m}{V}$ (equation for density) 1 kg = 1000 g
SOLVE Convert kg to g then follow the solution map to solve the problem. Round the answer to three significant figures to reflect the three significant figures in the given quantities.	**SOLUTION** 60.0 kg = 60 000 g $V = \frac{m}{d} = \frac{60\,000\,\cancel{g}}{0.752\,\cancel{g}/\text{cm}^3} = 7.98 \times 10^4\ \text{cm}^3$
CHECK Check your answer. Are the units correct? Does the answer make physical sense?	The units of the answer are those of volume, so they are correct. The magnitude seems reasonable because the density is somewhat less than 1 g/cm^3; therefore the volume of 60.0 kg should be somewhat more than 60.0 × 10^3 cm^3.

▶SKILLBUILDER 4.11 | Using Density in Calculations

A drop of acetone (nail polish remover) has a mass of 35 mg and a density of 0.788 g/cm^3. What is its volume in cubic centimetres?

▶SKILLBUILDER PLUS | A steel cylinder has a volume of 246 cm^3 and a density of 7.93 g/cm^3. What is its mass in kilograms?

▶FOR MORE PRACTICE Example 4.21; Problems 72, 73.

CHEMISTRY AND HEALTH

▶ Density, Cholesterol, and Heart Disease

Cholesterol is the fatty substance found in animal-derived foods such as beef, eggs, fish, poultry, and milk products. The body uses cholesterol for several purposes. However, excessive amounts of cholesterol in the blood—which can be caused by both genetic factors and diet—may result in the deposition of cholesterol in arterial walls, leading to a condition called atherosclerosis, or blocking of the arteries. These blockages are dangerous because they inhibit blood flow to important organs, causing heart attacks and strokes. The risk of stroke and heart attack increases with increasing blood cholesterol levels (Table 4.4). Cholesterol is carried in the bloodstream by a class of substances known as lipoproteins. Lipoproteins are often separated and classified according to their density.

The main carriers of blood cholesterol are low-density lipoproteins (LDLs). LDLs, also called bad cholesterol, have a density of 1.04 g/cm^3. They are bad because they tend to deposit cholesterol on arterial walls, increasing the risk of stroke and heart attack. Cholesterol is also carried by high-density lipoproteins (HDLs). HDLs, called good cholesterol, have a density of 1.13 g/cm^3. HDLs transport cholesterol to the liver for processing and excretion and therefore have a tendency to reduce cholesterol on arterial walls. Too low a level of HDLs (below 35 mg/100 mL) is considered a risk factor for heart disease. Exercise, along with a diet low in saturated fats, is believed to raise HDL levels in the blood while lowering LDL levels.

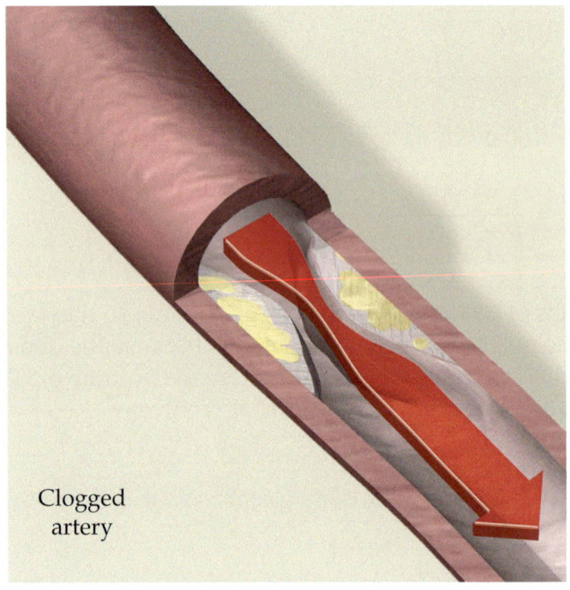

▲ Too many low-density lipoproteins in the blood can lead to the blocking of arteries.

B4.2 CAN YOU ANSWER THIS? *What mass of low-density lipoprotein is contained in a cylinder that is 1.25 cm long and 0.50 cm in diameter? (The volume of a cylinder, V, is given by $V = \pi r^2 \ell$, where r is the radius of the cylinder and ℓ is its length.)*

TABLE 4.4 Risk of Stroke and Heart Attack versus Blood Cholesterol Level

Risk Level	Total Blood Cholesterol (mg/100 mL)	LDL (mg/100 mL)
low	< 200	< 130
borderline	200–239	130–159
high	> 240	> 160

MODULE IN REVIEW

Self-Assessment Quiz

Q1. Express the number 0.000042 in scientific notation.

(a) 0.42×10^{-4}

(b) 4.2×10^{-5}

(c) 4.2×10^{-4}

(d) 4×10^{-5}

Q2. A graduated cylinder has markings every millilitre. Which measurement is accurately reported for this graduated cylinder?

(a) 21 mL

(b) 21.2 mL

(c) 21.23 mL

(d) 21.232 mL

Q3. How many significant figures are in the number 0.00620?

(a) 2 (b) 3 (c) 4 (d) 5

Q4. Round the number 89.04997 to three significant figures.

(a) 89.03 (b) 89.04 (c) 89.1 (d) 89.0

Q5. Perform this multiplication to the correct number of significant figures: $65.2 \times 0.0015 \times 12.02$

(a) 1.17 (b) 1.18 (c) 1.2 (d) 1.176

Q6. Perform this addition to the correct number of significant figures: 8.32 + 12.148 + 0.02

(a) 20.488 **(b)** 20.49 **(c)** 20.5 **(d)** 21

Q7. Perform this calculation to the correct number of significant figures: 78.222 × (12.02 – 11.52)

(a) 39 **(b)** 39.1 **(c)** 39.11 **(d)** 39.111

Q8. Convert 76.8 cm to m.

(a) 0.0768 m

(b) 7.68 m

(c) 0.768 m

(d) 7.68×10^{-2} m

Q9. Convert 2855 mg to kg.

(a) 2.855×10^{-3} kg

(b) 2.855 kg

(c) 0.02855 kg

(d) 3.503×10^{-4} kg

Q10. A cube measures 2.5 cm on each edge and has a mass of 66.9 g. Calculate the density of the material that composes the cube. (The volume of a cube is equal to the edge length cubed.)

(a) 10.7 g/cm^3

(b) 4.3 g/cm^3

(c) 0.234 g/cm^3

(d) 26.7 g/cm^3

Q11. What is the mass of 225 mL of a liquid that has a density of 0.880 g/mL?

(a) 198 g **(b)** 0.0039 g **(c)** 0.198 g **(d)** 0.256 g

Answers: 1:b, 2:b, 3:b, 4:d, 5:c, 6:b, 7:a, 8:c, 9:a, 10:b, 11:a

Chemical Principles

Relevance

Uncertainty: Scientists report measured quantities so that the number of digits reflects the certainty in the measurement. Write measured quantities so that every digit is certain except the last, which is estimated.

Uncertainty: Measurement is a hallmark of science, and you must communicate the precision of a measurement with the measurement so that others know how reliable the measurement is. When you write or manipulate measured quantities, you must show and retain the precision with which the original measurement was made.

Units: Measured quantities usually have units associated with them. The SI unit for length is the metre; for mass, the kilogram; and for time, the second. Prefix multipliers such as *kilo-* or *milli-* are often used in combination with these basic units. The SI units of volume are units of length raised to the third power; litres or millilitres are often used as well.

Units: The units in a measured quantity communicate what the quantity actually is. Without an agreed-on system of units, scientists could not communicate their measurements. Units are also important in calculations, and the tracking of units throughout a calculation is essential.

Density: The density of a substance is its mass divided by its volume, $d = m/V$, and is usually reported in units of grams per cubic centimetre or grams per millilitre. Density is a fundamental property of all substances and generally differs from one substance to another.

Density: The density of substances is an important consideration in choosing materials for manufacturing and production. Airplanes, for example, are made of low-density materials, while bridges are made of higher-density materials. Density can be used to calculate mass from volume and vice versa.

Chemical Skills

Examples

LO: Express very large and very small numbers using scientific notation (Section 4.1).

To express a number in scientific notation:

- Move the decimal point to obtain a number between 1 and 10.
- Write the decimal part multiplied by 10 raised to the number of places you moved the decimal point.
- The exponent is positive if you moved the decimal point to the left and negative if you moved the decimal point to the right.

EXAMPLE 4.12 SCIENTIFIC NOTATION

Express the number 45,000,000 in scientific notation.

45,000,000

7 6 5 4 3 2 1

4.5×10^7

LO: Report measured quantities to the right number of digits (Section 4.2).

Report measured quantities so that every digit is certain except the last, which is estimated.

EXAMPLE 4.13 REPORTING MEASURED QUANTITIES TO THE RIGHT NUMBER OF DIGITS

Record the volume of liquid in the graduated cylinder to the correct number of digits. Laboratory glassware is calibrated (and should therefore be read) from the bottom of the meniscus, the curved surface at the top of a column of liquid (see figure).

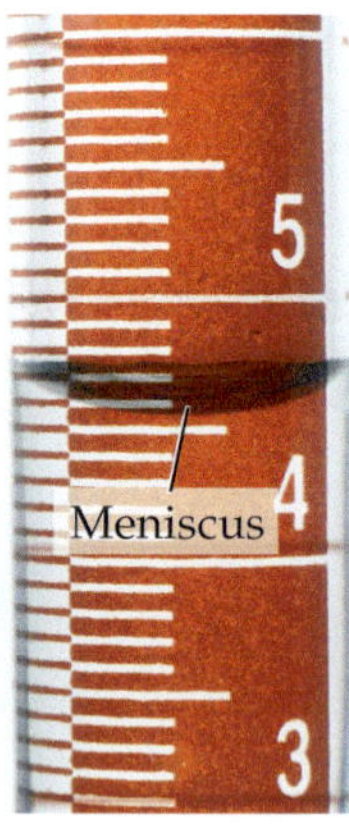

Because the graduated cylinder has markings every 0.1 mL, you should record the measurement to the nearest 0.01 mL. In this case, that is 4.57 mL.

LO: Determine which digits in a number are significant (Section 4.2).

Always count the following as significant:

- nonzero digits
- interior zeros
- trailing zeros after a decimal point
- trailing zeros before a decimal point but after a nonzero number

Never count the following digits as significant:

- zeros to the left of the first nonzero number

The following digits are ambiguous, and you should avoid them by using scientific notation:

- zeros at the end of a number but before a decimal point

EXAMPLE 4.14 COUNTING SIGNIFICANT DIGITS

How many significant figures are in the following numbers?

1.0050	five significant figures
0.00870	three significant figures
100.085	six significant digits
5400	It is not possible to tell in its current form.

The number must be written as 5.4×10^3, 5.40×10^3, or 5.400×10^3, depending on the number of significant figures intended.

LO: Round numbers to the correct number of significant figures (Section 4.3).

When rounding numbers to the correct number of significant figures, round down if the last digit dropped is 4 or less; round up if the last digit dropped is 5 or more.

EXAMPLE 4.15 ROUNDING

Round 6.442 and 6.456 to two significant figures each.

6.442 rounds to 6.4
6.456 rounds to 6.5

LO: Determine the correct number of significant figures in the results of multiplication and division calculations (Section 4.3).

The result of a multiplication or division should carry the same number of significant figures as the factor with the least number of significant figures.

EXAMPLE 4.16 SIGNIFICANT FIGURES IN MULTIPLICATION AND DIVISION

Perform the calculation and report the answer to the correct number of significant figures.

$$8.54 \times 3.589 \div 4.2 = 7.2976 = 7.3$$

Round the final result to two significant figures to reflect the two significant figures in the factor with the least number of significant figures (4.2).

LO: Determine the correct number of significant figures in the results of addition and subtraction calculations (Section 4.3).

The result of an addition or subtraction should carry the same number of decimal places as the quantity carrying the least number of decimal places.

EXAMPLE 4.17 SIGNIFICANT FIGURES IN ADDITION AND SUBTRACTION

Perform the operation and report the answer to the correct number of significant figures.

$$\begin{array}{r} 3.098 \\ +0.67 \\ -0.9452 \\ \hline 2.8228 = 2.82 \end{array}$$

Round the final result to two decimal places to reflect the two decimal places in the quantity with the least number of decimal places (0.67).

LO: Determine the correct number of significant figures in the results of calculations involving both addition/subtraction and multiplication/division (Section 4.3).

In calculations involving both addition/subtraction and multiplication/division, do the steps in parentheses first, keeping track of how many significant figures are in the answer by underlining the least significant figure, then proceeding with the remaining steps. Do not round off until the very end.

EXAMPLE 4.18 SIGNIFICANT FIGURES IN CALCULATIONS INVOLVING BOTH ADDITION/SUBTRACTION AND MULTIPLICATION/DIVISION

Perform the operation and report the answer to the correct number of significant figures.

$$8.16 \times (15.4323 - 5.4112) = 8.16 \times 0.02\underline{1}3 = 0.1738 = 0.17$$

LO: Convert between units (Section 4.5).

Solve unit conversion problems by following these steps.

1. **Sort** Write down the given quantity and its units and the quantity you are asked to find and its units.
2. **Strategize** Draw a solution map showing how to get from the given quantity to the quantity you are asked to find.
3. **Solve** Follow the solution map. Starting with the given quantity and its units, use the appropriate relationship to arrive at the answer. Round the final answer to the correct number of significant figures.
4. **Check** Are the units correct? Does the answer make physical sense?

EXAMPLE 4.19 UNIT CONVERSION

Convert 0.625 L to mL.

GIVEN: 0.625 L

FIND: mL

SOLUTION MAP

RELATIONSHIPS USED

1 L = 1000 mL (1000 mL/L)

SOLUTION

$$0.625\ \cancel{\text{L}} \times \frac{1000\ \text{mL}}{\cancel{\text{L}}}$$

$$= 625\ \text{mL}$$

LO: Calculate the density of a substance (Section 4.6).

The density of an object or substance is its mass divided by its volume.

$$d = \frac{m}{V}$$

1. **Sort** Write down the given quantity and its units and the quantity you are asked to find and its units.
2. **Strategize** Draw a solution map showing how to get from the given quantity to the quantity you are asked to find. Use the equation for density to take you from the mass and the volume to the density.
3. **Solve** Substitute the correct values into the equation for density.
4. **Check** Are the units correct? Does the answer make physical sense?

EXAMPLE 4.20 CALCULATING DENSITY

An object has a mass of 23.4 g and displaces 5.7 mL of water. Determine its density in grams per millilitre.

GIVEN: $m = 23.4$ g
$V = 5.7$ mL

FIND: density in g/mL

SOLUTION MAP

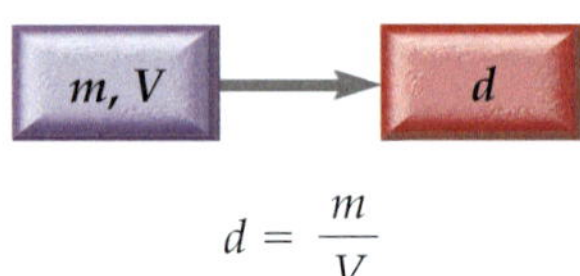

RELATIONSHIPS USED

$$d = \frac{m}{V} \quad \text{(equation for density)}$$

SOLUTION

$$d = \frac{m}{V}$$

$$= \frac{23.4\ \text{g}}{5.7\ \text{mL}}$$

$$= 4.11\ \text{g/mL}$$

$$= 4.1\ \text{g/mL}$$

The units (g/mL) are units of density. The answer is in the range of values for the densities of liquids and solids (see Table 4.3).

LO: Using density in calculations (Section 4.6).

You can use density to calculate mass from volume or volume from mass.

1. **Sort** Write down the given quantity and its units and the quantity you are asked to find and its units.

2. **Strategize** Draw a solution map showing how the equation allows us to find the unknown from the known quantities. Perform any necessary unit conversions.

3. **Solve** Follow the solution map to arrive at the answer. Rearrange the equation to find the unknown quantity. Round to the correct number of significant figures.

4. **Check** Are the units correct? Does the answer make physical sense?

EXAMPLE 4.21 USING DENSITY IN CALCULATIONS

What is the volume in litres of 321 g of a liquid with a density of 0.84 g/mL?

GIVEN: 321 g
0.84 g/mL

FIND: volume in L

SOLUTION MAP

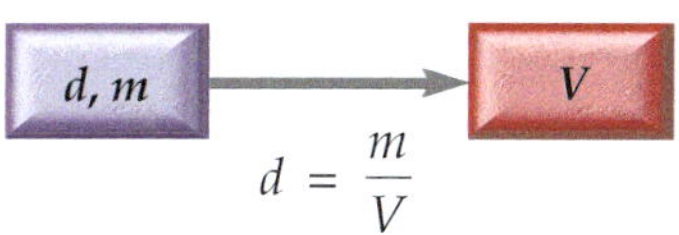

$d = \frac{m}{V}$

RELATIONSHIPS USED

$d = \frac{m}{V}$ (equation for density)

1 mL = 10^{-3} L (1000 mL/L)

SOLUTION

$$V = \frac{m}{d} = \frac{321\ \text{g}}{0.84\ \text{g/mL}} = 382\ \text{mL}$$

382 mL = 0.38 L

The answer is in the correct units. The magnitude seems right because the density is slightly less than 1; therefore the volume (382 mL) should be slightly greater than the mass (321 g).

KEY TERMS

decimal part **[4.1]**
density **[4.6]**
English system **[4.4]**
exponent **[4.1]**
exponential part **[4.1]**
International System **[4.4]**
kilogram (kg) **[4.4]**
litre (L) **[4.4]**
mass **[4.4]**
metre (m) **[4.4]**
metric system **[4.4]**
prefix multipliers **[4.4]**
scientific notation **[4.1]**
second (s) **[4.4]**
SI units **[4.4]**
significant figures (digits) **[4.2]**
solution map **[4.5]**
units **[4.4]**
volume **[4.4]**

EXERCISES

QUESTIONS

1. Why is it necessary to include units when reporting scientific measurements?
2. Why are the number of digits reported in scientific measurements important?
3. Why is scientific notation useful?
4. If a measured quantity is written correctly, which digits are certain? Which are uncertain?
5. When do zeros count as significant digits and when don't they count?
6. How many significant digits are there in exact numbers? What kinds of numbers are exact?
7. What limits the number of significant digits in a calculation involving only multiplication and division?
8. What limits the number of significant digits in a calculation involving only addition and subtraction?
9. How do we determine significant figures in calculations involving both addition/subtraction and multiplication/division?
10. What are the rules for rounding numbers?
11. What are the basic SI units of length, mass, and time?
12. List the common units of volume.
13. Suppose you are trying to measure the diametre of a Frisbee. What unit and prefix multiplier should you use?
14. What is the difference between mass and weight?
15. Explain why units are important in calculations.
16. How are units treated in a calculation?
17. Experienced problem solvers always consider both the value and units of their answer to a problem. Why?
18. What is density?
19. Explain how you would calculate the density of a substance.

PROBLEMS

SCIENTIFIC NOTATION

20. Express each number in scientific notation.
- **(a)** 37,692,000 (population of California)
- **(b)** 1,360,000 (population of Hawaii)
- **(c)** 19,306,000 (population of New York)
- **(d)** 568,000 (population of Wyoming)

21. Express each number in scientific notation.
- **(a)** 6,974,000,000 (population of the world)
- **(b)** 1,344,000,000 (population of China)
- **(c)** 11,254,000 (population of Cuba)
- **(d)** 4,487,000 (population of Ireland)

22. Express each number in scientific notation.
- **(a)** 0.00000000007461 m (length of a hydrogen–hydrogen chemical bond)
- **(b)** 0.0000158 mi (number of miles in an inch)
- **(c)** 0.000000632 m (wavelength of red light)
- **(d)** 0.000015 m (diameter of a human hair)

23. Express each number in scientific notation.
- **(a)** 0.000000001 s (time it takes light to travel 1 ft)
- **(b)** 0.143 s (time it takes light to travel around the world)
- **(c)** 0.000000000001 s (time it takes a chemical bond to undergo one vibration)
- **(d)** 0.000001 m (approximate size of a dust particle)

24. Express each number in decimal notation (i.e., express the number without using scientific notation).
- **(a)** 6.022×10^{23} (number of carbon atoms in 12.01 g of carbon)
- **(b)** 1.6×10^{-19} C (charge of a proton in coulombs)
- **(c)** 2.99×10^{8} m/s (speed of light)
- **(d)** 3.44×10^{2} m/s (speed of sound)

25. Express each number in decimal notation (i.e., express the number without using scientific notation).
- **(a)** 450×10^{-9} m (wavelength of blue light)
- **(b)** 13.7×10^{9} years (approximate age of the universe)
- **(c)** 5×10^{9} years (approximate age of Earth)
- **(d)** 5.0×10^{1} years (approximate age of this author)

26. Express each number in decimal notation (i.e., express the number without using scientific notation).
- **(a)** 3.22×10^{7}
- **(b)** 7.2×10^{-3}
- **(c)** 1.18×10^{11}
- **(d)** 9.43×10^{-6}

27. Express each number in decimal notation (i.e., express the number without using scientific notation).
- **(a)** 1.30×10^{6}
- **(b)** 1.1×10^{-4}
- **(c)** 1.9×10^{2}
- **(d)** 7.41×10^{-10}

28. Complete the table.

Decimal Notation	Scientific Notation
2,000,000,000	_______
_______	1.211×10^{9}
0.000874	_______
_______	3.2×10^{11}

29. Complete the table.

Decimal Notation	Scientific Notation
_______	4.2×10^{-3}
315,171,000	_______
_______	1.8×10^{-11}
1,232,000	_______

SIGNIFICANT FIGURES

30. Read each instrument to the correct number of significant figures. Laboratory glassware should always be read from the bottom of the *meniscus* (the curved surface at the top of the liquid column).

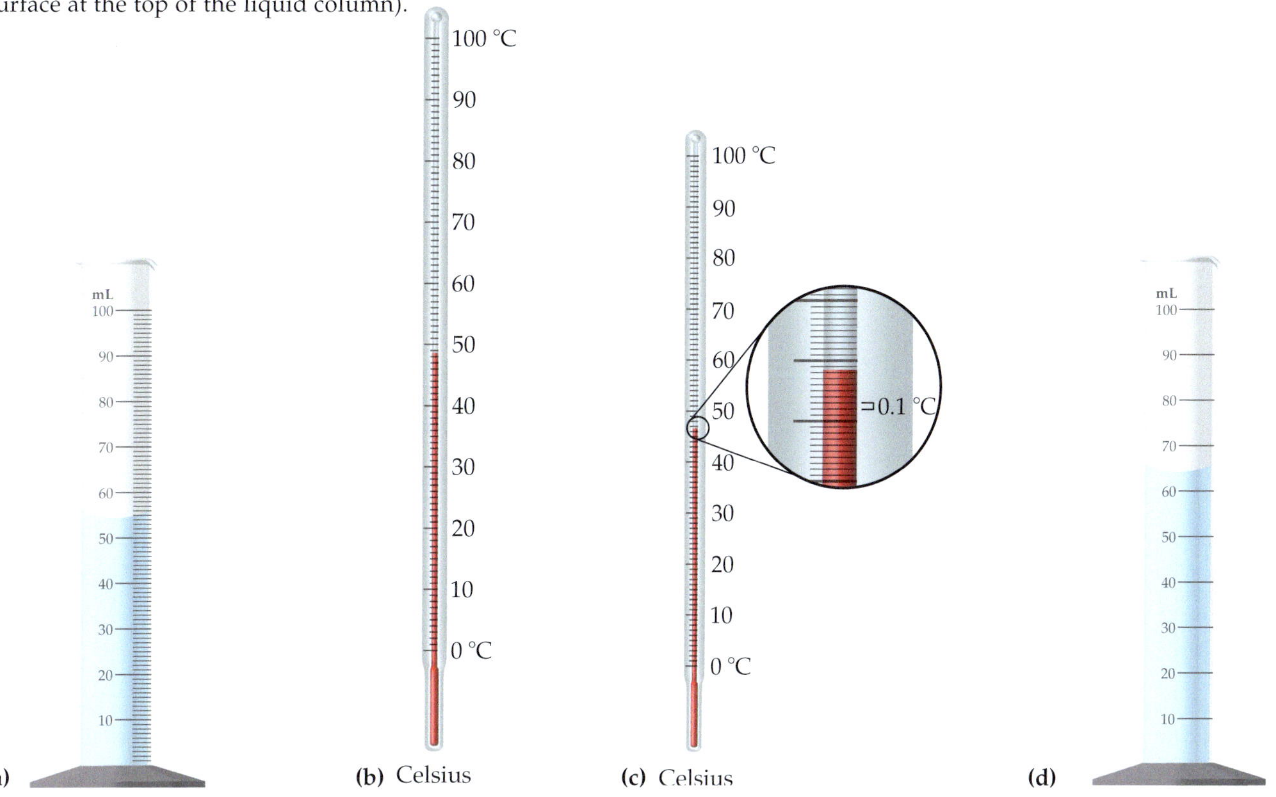

31. Read each instrument to the correct number of significant figures. Laboratory glassware should always be read from the bottom of the meniscus (the curved surface at the top of the liquid column).

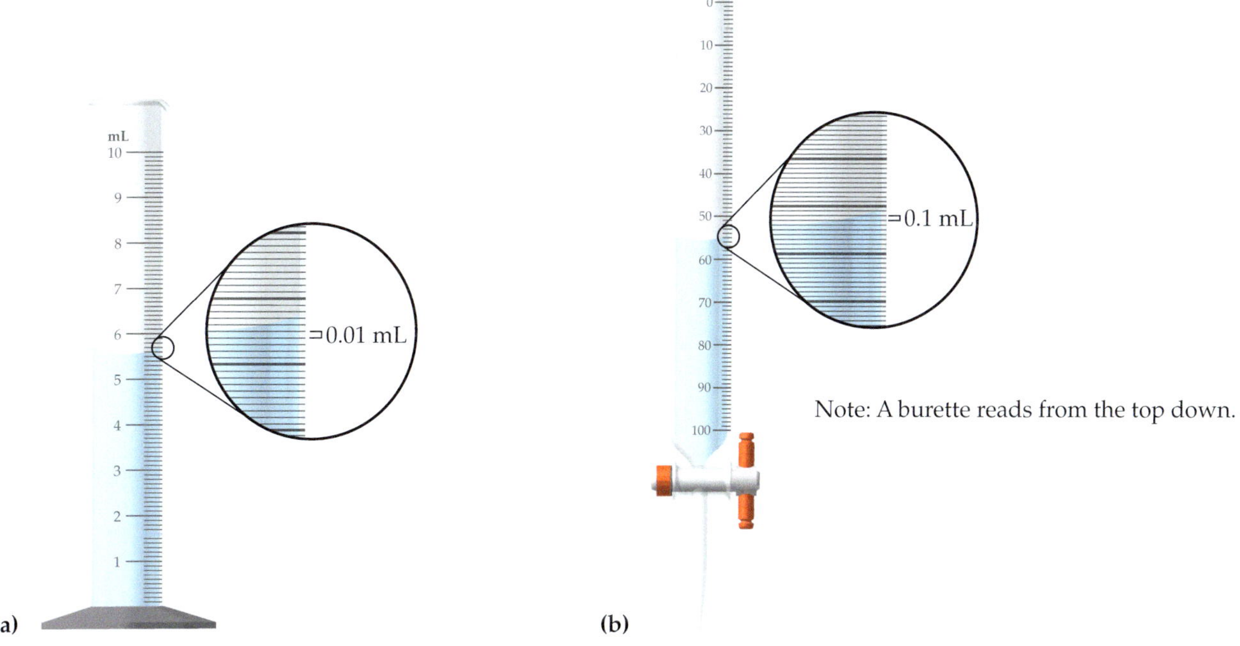

Note: A pipette reads from the top down.

(c)

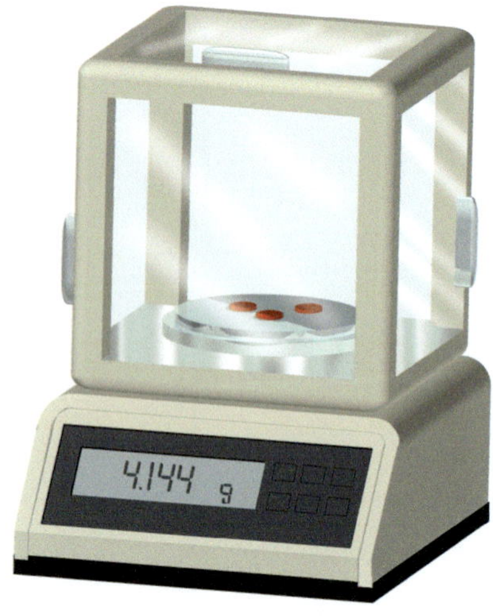

(d) Note: Digital balances normally display mass to the correct number of significant figures for that particular balance.

32. For each measured quantity, underline the zeros that are significant and draw an X through the zeros that are not.
- **(a)** 0.005050 m
- **(b)** 0.0000000000000060 s
- **(c)** 220,103 kg
- **(d)** 0.00108 in.

33. For each measured quantity, underline the zeros that are significant and draw an X through the zeros that are not.
- **(a)** 0.00010320 s
- **(b)** 1,322,600,324 kg
- **(c)** 0.0001240 in.
- **(d)** 0.02061 m

34. How many significant figures are in each measured quantity?
- **(a)** 0.001125 m
- **(b)** 0.1125 m
- **(c)** 1.12500×10^4 m
- **(d)** 11205 m

35. How many significant figures are in each measured quantity?
- **(a)** 13001 kg
- **(b)** 13111 kg
- **(c)** 1.30×10^4 kg
- **(d)** 0.00013 kg

36. Correct any entries in the table that are wrong.

Quantity	Significant Figures
(a) 895675 m	6
(b) 0.000869 kg	6
(c) 0.5672100 s	5
(d) 6.022×10^{23} atoms	4

37. Correct any entries in the table that are wrong.

Quantity	Significant Figures
(a) 24 days	2
(b) 5.6×10^{-12} s	3
(c) 3.14 m	3
(d) 0.00383 g	5

ROUNDING

38. Round each number to four significant figures.
- **(a)** 255.98612
- **(b)** 0.0004893222
- **(c)** 2.900856×10^{-4}
- **(d)** 2,231,479

39. Round each number to three significant figures.
- **(a)** 10,776.522
- **(b)** 4.999902×10^6
- **(c)** 1.3499999995
- **(d)** 0.0000344988

40. Round each number to two significant figures.
(a) 2.34
(b) 2.35
(c) 2.349
(d) 2.359

41. Round each number to three significant figures.
(a) 65.74
(b) 65.749
(c) 65.75
(d) 65.750

42. Each number is supposed to be rounded to three significant figures. Correct the ones that are incorrectly rounded.
(a) 42.3492 to 42.4
(b) 56.9971 to 57.0
(c) 231.904 to 232
(d) 0.04555 to 0.046

43. Each number is supposed to be rounded to two significant figures. Correct the ones that are incorrectly rounded.
(a) 1.249×10^3 to 1.3×10^3
(b) 3.999×10^2 to 40
(c) 56.21 to 56.2
(d) 0.009964 to 0.010

44. Round the number on the left to the number of significant figures indicated by the example in the first row. (Use scientific notation as needed to avoid ambiguity.)

Number	Rounded to 4 Significant Figures	Rounded to 2 Significant Figures	Rounded to 1 Significant Figure
1.45815	1.458	1.5	1
8.32466			
84.57225			
132.5512			

45. Round the number on the left to the number of significant figures indicated by the example in the first row. (Use scientific notation as needed to avoid ambiguity.)

Number	Rounded to 4 Significant Figures	Rounded to 2 Significant Figures	Rounded to 1 Significant Figure
94.52118	94.52	95	9×10^1
105.4545			
0.455981			
0.009999991			

SIGNIFICANT FIGURES IN CALCULATIONS

46. Perform each calculation to the correct number of significant figures.
(a) $4.5 \times 0.03060 \times 0.391$
(b) $5.55 \div 8.97$
(c) $(7.890 \times 10^{12}) \div (6.7 \times 10^4)$
(d) $67.8 \times 9.8 \div 100.04$

47. Perform each calculation to the correct number of significant figures.
(a) $89.3 \times 77.0 \times 0.08$
(b) $(5.01 \times 10^5) \div (7.8 \times 10^2)$
(c) $4.005 \times 74 \times 0.007$
(d) $453 \div 2.031$

48. Correct any answers that have the incorrect number of significant figures.
(a) $34.00 \times 567 \div 4.564 = 4.2239 \times 10^3$
(b) $79.3 \div 0.004 \times 35.4 = 7 \times 10^5$
(c) $89.763 \div 22.4581 = 3.997$
(d) $(4.32 \times 10^{12}) \div (3.1 \times 10^{-4}) = 1.4 \times 10^{16}$

49. Correct any answers that have the incorrect number of significant figures.
(a) $45.3254 \times 89.00205 = 4034.05$
(b) $0.00740 \times 45.0901 = 0.334$
(c) $49857 \div 904875 = 0.05510$
(d) $0.009090 \times 6007.2 = 54.605$

50. Perform each calculation to the correct number of significant figures.
(a) $87.6 + 9.888 + 2.3 + 10.77$
(b) $43.7 - 2.341$
(c) $89.6 + 98.33 - 4.674$
(d) $6.99 - 5.772$

51. Perform each calculation to the correct number of significant figures.
(a) $1459.3 + 9.77 + 4.32$
(b) $0.004 + 0.09879$
(c) $432 + 7.3 - 28.523$
(d) $2.4 + 1.777$

52. Correct any answers that have the incorrect number of significant figures.
(a) $(3.8 \times 10^5) - (8.45 \times 10^5) = -4.7 \times 10^5$
(b) $0.00456 + 1.0936 = 1.10$
(c) $8475.45 - 34.899 = 8440.55$
(d) $908.87 - 905.34095 = 3.5291$

53. Correct any answers that have the incorrect number of significant figures.
(a) $78.9 + 890.43 - 23 = 9.5 \times 10^2$
(b) $9354 - 3489.56 + 34.3 = 5898.74$
(c) $0.00407 + 0.0943 = 0.0984$
(d) $0.00896 - 0.007 = 0.00196$

54. Perform each calculation to the correct number of significant figures.
(a) $(78.4 - 44.889) \div 0.0087$
(b) $(34.6784 \times 5.38) + 445.56$
(c) $(78.7 \times 10^5 \div 88.529) + 356.99$
(d) $(892 \div 986.7) + 5.44$

55. Perform each calculation to the correct number of significant figures.
(a) $(1.7 \times 10^6 \div 2.63 \times 10^5) + 7.33$
(b) $(568.99 - 232.1) \div 5.3$
(c) $(9443 + 45 - 9.9) \times 8.1 \times 10^6$
(d) $(3.14 \times 2.4367) - 2.34$

56. Correct any answers that have the incorrect number of significant figures.
(a) $(78.56 - 9.44) \times 45.6 = 3152$
(b) $(8.9 \times 10^5 \div 2.348 \times 10^2) + 121 = 3.9 \times 10^3$
(c) $(45.8 \div 3.2) - 12.3 = 2$
(d) $(4.5 \times 10^3 - 1.53 \times 10^3) \div 34.5 = 86$

57. Correct any answers that have the incorrect number of significant figures.
(a) $(908.4 - 3.4) \div 3.52 \times 10^4 = 0.026$
(b) $(1206.7 - 0.904) \times 89 = 1.07 \times 10^5$
(c) $(876.90 + 98.1) \div 56.998 = 17.11$
(d) $(4.55 \div 4078.59) + 1.00098 = 1.00210$

UNIT CONVERSION

58. Perform each conversion.
(a) 3.55 kg to grams
(b) 8944 mm to metres
(c) 4598 mg to kilograms
(d) 0.0187 L to millilitres

59. Perform each conversion.
(a) 155.5 cm to metres
(b) 2491.6 g to kilograms
(c) 248 cm to millimetres
(d) 6781 mL to litres

60. Perform each conversion.
(a) 5.88 dL to litres
(b) 3.41×10^{-5} g to micrograms
(c) 1.01×10^{-8} s to nanoseconds
(d) 2.19 pm to metres

61. Perform each conversion.
(a) 1.08 Mm to kilometres
(b) 4.88 fs to picoseconds
(c) 7.39×10^{11} m to gigametres
(d) 1.15×10^{-10} m to picometres

62. Complete the table.

m	km	Mm	Gm	Tm
5.08×10^8 m	_____	508 Mm	_____	_____
_____	_____	27,976 Mm	_____	_____
_____	_____	_____	_____	1.77 Tm
_____	1.5×10^5 km	_____	_____	_____
_____	_____	_____	423 Gm	_____

63. Complete the table.

s	ms	μs	ns	ps
1.31×10^{-4} s	_____	131μs	_____	_____
_____	_____	_____	_____	12.6 ps
_____	_____	_____	155 ns	_____
_____	1.99×10^{-3} ms	_____	_____	_____
_____	_____	8.66×10^{-5} μs	_____	_____

64. Convert 2.255×10^{10} g to each unit.
(a) kg
(b) Mg
(c) mg
(d) metric tons (1 metric ton = 1000 kg)

65. Convert 1.88×10^{-6} g to each unit.
(a) mg
(b) cg
(c) ng
(d) μg

DENSITY

66. A sample of an unknown metal has a mass of 35.4 g and a volume of 3.11 cm^3. Calculate its density and identify the metal by comparison to Table 4.3.

67. A new penny has a mass of 2.49 g and a volume of 0.349 cm^3. Is the penny pure copper?

68. Glycerol is a syrupy liquid often used in cosmetics and soaps. A 2.50 L sample of pure glycerol has a mass of 3.15×10^3 g. What is the density of glycerol in grams per cubic centimetre?

69. An aluminum engine block has a volume of 4.77 L and a mass of 12.88 kg. What is the density of the aluminum in grams per cubic centimetre?

70. A supposedly gold crown is tested to determine its density. It displaces 10.7 mL of water and has a mass of 206 g. Could the crown be made of gold?

71. A vase is said to be solid platinum. It displaces 18.65 mL of water and has a mass of 157 g. Could the vase be solid platinum?

72. Ethylene glycol (antifreeze) has a density of 1.11 g/cm^3.
(a) What is the mass in grams of 387 mL of ethylene glycol?
(b) What is the volume in litres of 3.46 kg of ethylene glycol?

73. Acetone (fingernail-polish remover) has a density of 0.7857 g/cm^3.
(a) What is the mass in grams of 17.56 mL of acetone?
(b) What is the volume in millilitres of 7.22 g of acetone?

▸Answers to Skillbuilder Exercises

Skillbuilder 4.1 $1.6342 $\times 10^{13}$
Skillbuilder 4.2 3.8×10^{-5}
Skillbuilder 4.3 103.4 °F
Skillbuilder 4.4
(a) four significant figures
(b) three significant figures
(c) two significant figures
(d) unlimited significant figures
(e) three significant figures
(f) ambiguous
Skillbuilder 4.5
(a) 0.001 or 1×10^{-3}
(b) 0.204
Skillbuilder 4.6
(a) 7.6
(b) 131.11
Skillbuilder 4.7
(a) 1288
(b) 3.12
Skillbuilder 4.8 0.056 L.
Skillbuilder 4.9 5.678 km
Skillbuilder 4.10 Yes, the density is 21.4 g/cm^3 and matches that of platinum.
Skillbuilder 4.11 4.4×10^{-2} cm^3
Skillbuilder Plus, p. 159 1.95 kg

▸Answers to Conceptual Checkpoints

4.1 **(c)** Multiplying by 10^{-3} is equivalent to moving the decimal point three places to the left.
4.2 **(b)** The last digit is considered to be uncertain by ±1.
4.3 **(b)** The result of the calculation in **(a)** would be reported as 4; the result of the calculation in **(b)** would be reported as 1.5.
4.4 **(d)** The diameter would be expressed as 28 nm.

▲ Ordinary table salt is a compound called sodium chloride. The sodium within sodium chloride has been linked to high blood pressure. In this module, we learn how to determine how much sodium is in a given amount of sodium chloride.

Chemical Composition 5

"In science, you don't ask why, you ask how much." —Erwin Chargaff (1905–2002)

MODULE OUTLINE

5.1 How Much Sodium?

Sodium is an important dietary mineral that we eat in food, primarily as sodium chloride (table salt). Sodium is involved in the regulation of body fluids, and eating too much of it can lead to high blood pressure. High blood pressure, in turn, increases the risk of stroke and heart attack. Consequently, people with high blood pressure should limit their sodium intake. The National Health and Medical Research Council (NHMRC) recommends that a person consume between 460 and 920 mg of sodium per day. However, sodium is usually consumed as sodium chloride, so the mass of sodium that we eat is not the same as the mass of sodium chloride that we eat. How many grams of sodium chloride can we consume and still stay within the NHMRC recommendation for sodium?

▲ The mining of iron requires knowing how much iron is in a given amount of iron ore. © Richard Kittenberger/ Shutterstock.

To answer this question, we need to know the *chemical composition* of sodium chloride. From Module 3, we are familiar with its formula, NaCl, which indicates that there is one sodium ion to every chloride ion. However, because the masses of sodium and chlorine are different, the relationship between the mass of sodium and the mass of sodium chloride is not clear from the chemical formula alone. In this module, we learn how to use the information in a chemical formula, together with atomic and formula masses, to calculate the amount of a constituent element in a given amount of a compound (or vice versa).

▲ Estimating the threat of ozone depletion requires knowing the amount of chlorine in a given amount of a chlorofluorocarbon. © Joseph P. Sinnot/Fundamental Photographs.

Chemical composition is important not just for assessing dietary sodium intake, but for addressing many other questions as well. A company that mines iron, for example, wants to know how much iron it can extract from a given amount of iron ore; an organization interested in developing hydrogen as a potential fuel would want to know how much hydrogen it can extract from a given amount of water. Many environmental issues also require knowledge of chemical composition. An estimate of the threat of ozone depletion requires knowing how much chlorine is in a given amount of a particular chlorofluorocarbon such as freon-12.

5.2 Counting Atoms by the Gram

LO: Calculate moles from number of atoms and number of atoms from moles.

LO: Calculate mass from moles and moles from mass.

LO: Calculate the number of atoms from mass and mass from the number of atoms.

Determining the number of atoms in a sample with a certain mass is similar to determining the number of items in a sample with a certain weight. With nails for example, we can use a dozen as a convenient number to convert between mass and number of items, but a dozen is too small to use with atoms. We need a larger number because atoms are so small. The chemist's "dozen" is called the **mole (mol)** and has a value of 6.022×10^{23}.

$$1 \text{ mol} = 6.022 \times 10^{23}$$

This is **Avogadro's number**, named after Amedeo Avogadro (1776–1856). The unit of Avogadro's number is mol^{-1}

The first thing to understand about the mole is that it can specify Avogadro's number of anything. *One mole of anything is* 6.022×10^{23} *units of that thing*. For example, one mole of marbles corresponds to 6.022×10^{23} marbles, and one mole of sand grains corresponds to 6.022×10^{23} sand grains. One mole of atoms, ions, or molecules generally makes up objects of reasonable size. For example, eight \$1 coins contain approximately 1 mol of copper (Cu) atoms, and a couple of large helium balloons contain approximately 1 mol of helium (He) atoms.

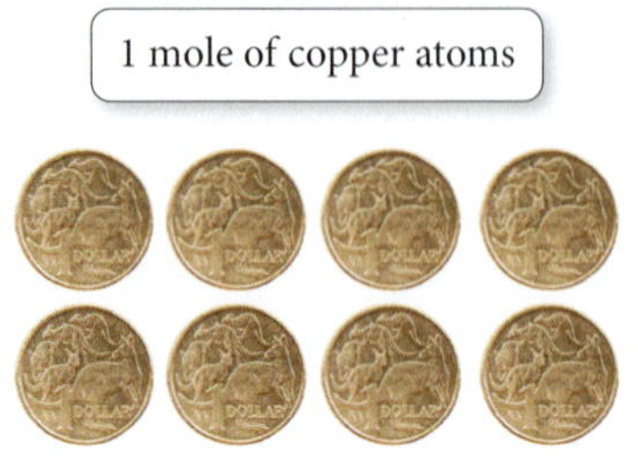

▲ Eight \$1 coins contain approximately 1 mol of copper atoms. The Australian \$1 coin consists of approximately 92 % by mass copper with the remainder being aluminium and nickel. © timhester.123rf.com

The second thing to understand about the mole is how it gets its specific value. *The numerical value of the mole is defined as being equal to the number of atoms in exactly 12 g of pure carbon-12.*

This definition of the mole establishes a relationship between mass (grams of carbon) and number of atoms (Avogadro's number). This relationship, as we will see shortly, allows us to count atoms by weighing them.

1 mole of helium atoms

▲ Two large helium balloons contain approximately 1 mol of helium atoms. © Nancy R. Cohen/Photodisc/Getty Images.

Calculations Involving Moles and Number of Atoms

Calculations involving moles and number of atoms are similar to how you would calculate the number of nails in several dozen nails. To calculate moles of atoms from number of atoms, and vice versa we use the equation:

Number of atoms = number of moles × Avogadro's number or :

$$N = n \times N_A$$

For example, suppose we want to calculate the number of helium atoms in 3.5 mol of helium. We set up the problem in the standard way.

GIVEN: $n = 3.5$ mol

FIND: N

SOLUTION MAP

We draw a solution map showing the calculation of He atoms from moles of He.

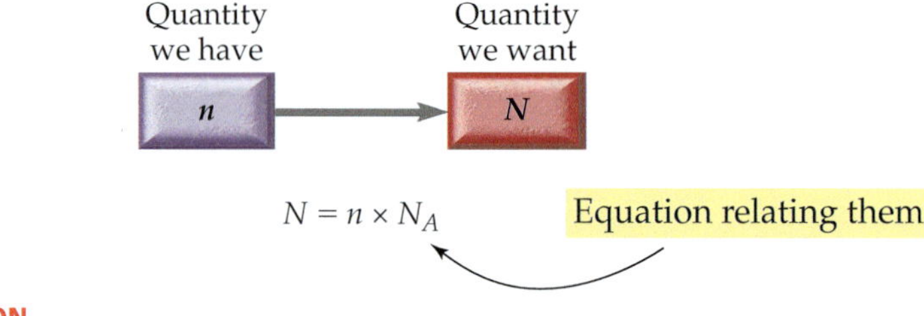

SOLUTION

$$N = n \times N_A = 3.5 \text{ mol} \times 6.022 \times 10^{23} \text{ mol}^{-1} = 2.1 \times 10^{24} \text{ He atoms}$$

EXAMPLE 5.1 **USING AVOGADRO'S NUMBER TO CALCULATE MOLES**

A silver ring contains 1.1×10^{22} silver atoms. How many moles of silver are in the ring?

SORT You are given the number of silver atoms and asked to find the number of moles.	**GIVEN:** $N = 1.1 \times 10^{22}$ Ag atoms **FIND:** n
STRATEGIZE Draw a solution map, beginning with silver atoms and ending at moles. What is the equation used to relate these two quantities?	**SOLUTION MAP** 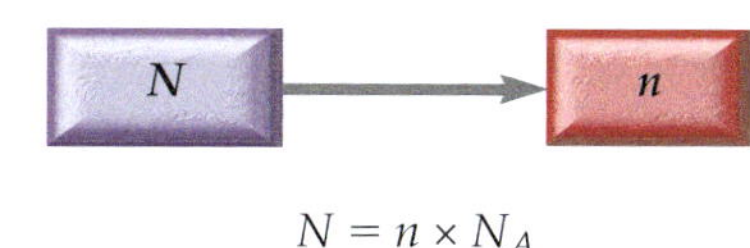 $N = n \times N_A$
SOLVE Follow the solution map to solve the problem. Beginning with 1.1×10^{22} Ag atoms, use the equation to calculate the moles of Ag.	**SOLUTION** Rearrange the equation for the unknown quantity, in this case, n. $N = n \times N_A$ $n = \frac{N}{N_A} = \frac{1.1 \times 10^{22}}{6.022 \times 10^{23}\,\text{mol}^{-1}} = 1.8 \times 10^{-2}$ mol Ag
CHECK Are the units correct? Does the answer make physical sense?	The units, mol Ag, are the desired units. The magnitude of the answer is orders of magnitude smaller than the given quantity because it takes many atoms to make a mole, so you expect the answer to be orders of magnitude smaller than the given quantity.

▶**SKILLBUILDER 5.1 | Calculate Number of Atoms from Moles**

How many gold atoms are in a pure gold ring containing 8.83×10^{-2} mol Au?

▶**FOR MORE PRACTICE** Example 5.14, Problems 17, 18, 19, 20.

Calculating the Mass of an Element

We just explained how to calculate moles from number of atoms, and vice versa, which is like converting between dozens and number of nails. We need one more equation to calculate the number of atoms in a sample from the mass of the sample. For nails, we used the weight of one dozen nails; for atoms, we use the mass of one mole of atoms.

> The mass of 1 mol of atoms of an element is its **molar mass**. The value of an element's molar mass in grams per mole is numerically equal to the element's atomic mass in atomic mass units.

Recall that Avogadro's number, the number of atoms in a mole, is defined as the number of atoms in exactly 12 g of carbon-12. The atomic mass unit is defined as one-twelfth of the mass of a carbon-12 atom, so it follows that the molar mass of any element—the mass of 1 mol of atoms in grams of that element—is equal to the atomic mass of that element expressed in atomic mass units. For example, copper has an atomic mass of 63.55 amu; therefore, 1 mol of copper atoms has a mass of 63.55 g, and the molar mass of copper is 63.55 g/mol. Just as the weight of 1 doz nails changes for different types of nails, so the mass of 1 mol of atoms changes for different elements: 1 mol of sulfur atoms (sulfur atoms are lighter than copper atoms) has a mass of 32.07 g; 1 mol of carbon atoms (lighter than sulfur) has a mass of 12.01 g; and 1 mol of lithium atoms (lighter yet) has a mass of 6.94 g.

$$32.07\text{ g sulfur} = 1\text{ mol sulfur} = 6.022 \times 10^{23}\text{ S atoms}$$

$$12.01\text{ g carbon} = 1\text{ mol carbon} = 6.022 \times 10^{23}\text{ C atoms}$$

$$6.94\text{ g lithium} = 1\text{ mol lithium} = 6.022 \times 10^{23}\text{ Li atoms}$$

The lighter the atom, the less mass in one mole of that atom (▼ Figure 5.1).

The molar mass of any element can be used to calculate the mass of an element from moles and the moles of an element from mass. The equation for this calculation is:

$$\text{Number of moles} = \frac{\text{mass}}{\text{molar mass}} \text{ or } n = \frac{m}{M}$$

1 dozen large nails

1 dozen small nails

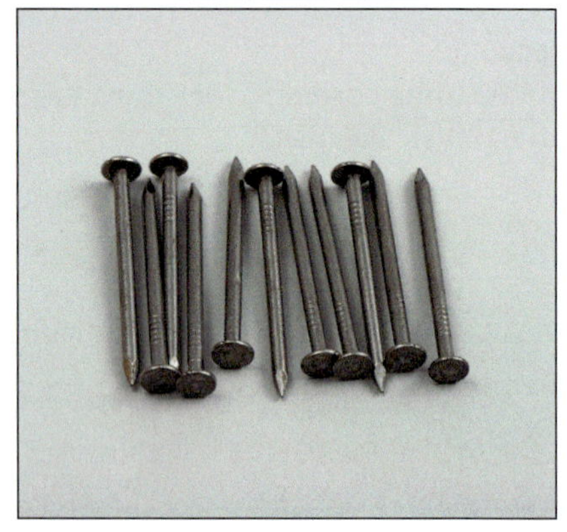

(a) © Maxwell Art And Photo/Pearson.

1 mole S (32.07 g)

1 mole C (12.01 g)

(b) © Richard Megna/Fundamental Photographs.

▶ **FIGURE 5.1 The mass of 1 mol** **(a)** Each of these pictures shows the same number of nails: 12. As you can see, 12 large nails have more weight and occupy more space than 12 small nails. The same is true for atoms. **(b)** Each of these samples has the same number of atoms: 6.022×10^{23}. Because sulfur atoms are more massive and larger than carbon atoms, 1 mol of S atoms is heavier and occupies more space than 1 mol of C atoms.

A 0.58 g diamond is about a three-carat diamond.

Suppose we want to calculate the number of moles of carbon in a 0.58 g diamond (pure carbon).

We first sort the information in the problem.

GIVEN: $m = 0.58 \text{ g C}$

Since we have a periodic table available, we have $M = 12.01 \text{ g/mol}$

FIND: mol C

SOLUTION MAP

We draw a solution map showing the calculation of moles of C from the mass.

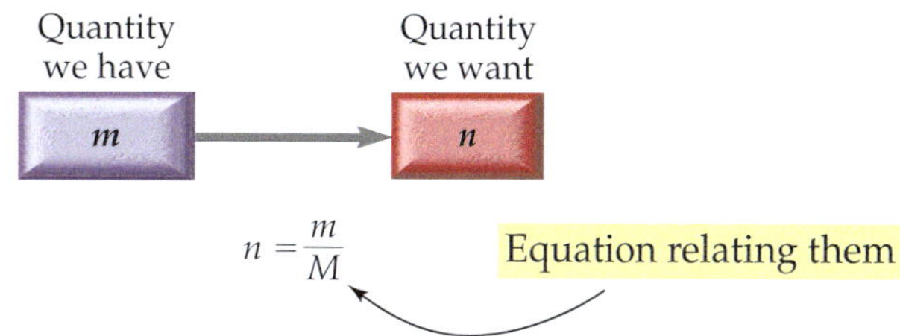

SOLUTION

Follow the solution map to calculate the moles of C.

$$n = \frac{m}{M} = \frac{0.58 \text{ g}}{12.01 \text{ g/mol}} = 4.8 \times 10^{-2} \text{ mol C}$$

EXAMPLE 5.2 THE MOLE CONCEPT—CALCULATING MOLES FROM MASS

Calculate the number of moles of sulfur in 57.8 g of sulfur.

SORT Begin by sorting the information in the question. You are given the mass of sulfur and asked to find the number of moles. Recall that you are always provided with a periodic table.	**GIVEN:** $m = 57.8 \text{ g S}$ $M = 32.06 \text{ g/mol}$ **FIND:** mol S
STRATEGIZE Draw a solution map showing the calculation of moles of S from the mass.	**SOLUTION MAP** 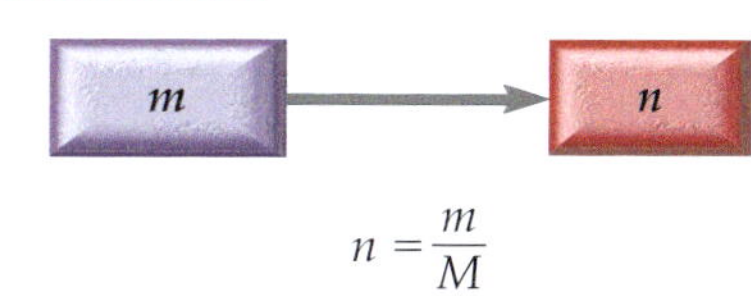
SOLVE Follow the solution map to solve the problem. Begin with 57.8 g S and use the mathematical equation to determine mol S.	**SOLUTION** $n = \frac{m}{M} = \frac{57.8 \text{ g}}{32.06 \text{ g/mol}} = 1.80 \text{ mol S}$
CHECK Check your answer. Are the units correct? Does the answer make physical sense?	The units (mol S) are correct. The magnitude of the answer makes sense because 1 mole of S has a mass of 32.06 g; therefore, 57.8 g of S should be close to 2 moles.

▶**SKILLBUILDER 5.2 | The Mole Concept—Calculating Mass from Moles**

Calculate the number of grams of sulfur in 2.78 mol of sulfur.

▶**FOR MORE PRACTICE** Example 5.15; Problems 25, 26, 27, 28, 29, 30.

Calculating the Number of Atoms of an Element from the Mass

Suppose we want to know the number of carbon *atoms* in the 0.58 g diamond. We can use the two equations we have introduced so far to determine this using the solution map below:

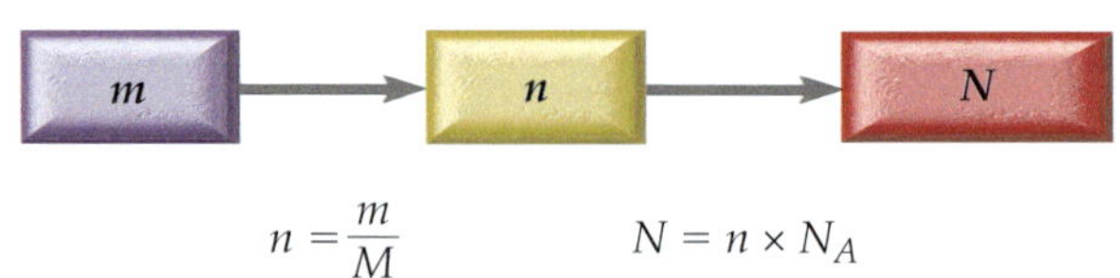

$$n = \frac{m}{M} \qquad N = n \times N_A$$

Beginning with 0.58 g carbon and using the solution map as a guide, we calculate to the number of carbon atoms.

We first calculate the number of moles of C in the sample:

$$n = \frac{m}{M} = \frac{0.58\ \cancel{g}}{12.01\ \cancel{g}/\text{mol}} = 4.8 \times 10^{-2}\ \text{mol C}$$

Once we have moles of C, we can use this to calculate the number of atoms of C:

$$N = n \times N_A = 4.8 \times 10^{-2}\ \cancel{\text{mol}} \times 6.022 \times 10^{23}\ \cancel{\text{mol}^{-1}} = 2.9 \times 10^{22}\ \text{C atoms}$$

EXAMPLE 5.3 THE MOLE CONCEPT—CONVERTING BETWEEN GRAMS AND NUMBER OF ATOMS

How many aluminum atoms are in an aluminum can with a mass of 16.2 g?

SORT Begin by sorting the information in the question. You are given the mass of aluminium and asked to find the number of moles. Recall that you are always provided with a periodic table.	**GIVEN:** $m = 16.2$ g Al $M = 26.98$ g/mol **FIND:** Al atoms
STRATEGIZE The solution map has two steps. In the first step, moles of Al are calculated from mass. In the second step, moles of Al are used to calculate the number of Al atoms. The molar mass of aluminum and the number of atoms in a mole are required information for the calculation.	**SOLUTION MAP** 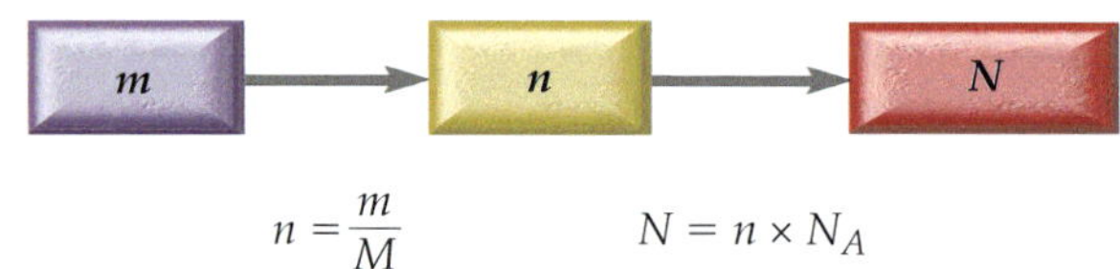 $n = \frac{m}{M}$ $\quad N = n \times N_A$
SOLVE Follow the solution map to solve the problem, using the two mathematical equations.	**SOLUTION** $n = \frac{m}{M} = \frac{16.2\ \cancel{g}}{26.98\ \cancel{g}/\text{mol}} = 6.00 \times 10^{-1}\ \text{mol Al}$ $N = n \times N_A = 6.00 \times 10^{-1}\ \cancel{\text{mol}} \times 6.022 \times 10^{23}\ \cancel{\text{mol}^{-1}} = 3.62 \times 10^{23}\ \text{Al atoms}$
CHECK Are the units correct? Does the answer make physical sense?	The units, Al atoms, are correct. The answer makes sense because the number of atoms in any macroscopic-sized sample of matter is very large.

▶**SKILLBUILDER 5.3 | The Mole Concept—Calculating the Mass of a sample from the Number of Atoms**

Calculate the mass of 1.23×10^{24} helium atoms.

▶**FOR MORE PRACTICE** Example 5.16; Problems 35, 36, 37, 38, 39, 40, 41, 42.

Before we move on, notice that numbers with large exponents, such as 6.022×10^{23}, are almost unimaginably large. Eight \$1 coins contain 6.022×10^{23} or 1 mol of copper atoms; 6.022×10^{23} coins would cover Earth's entire surface to a depth of over 2 km. Even objects that are small by everyday standards occupy a huge space when we have a mole of them. For example, one crystal of granulated sugar has a mass of less than 1 mg and a diameter of less than 0.1 mm, yet 1 mol of sugar crystals would cover the state of Queensland to a depth of approximately 1 m. For every increase of 1 in the exponent of a number, the number increases by 10. So a number with an exponent of 23 is incredibly large. A mole has to be a large number because atoms are so small.

CONCEPTUAL CHECKPOINT 5.1

Which statement is *always* true for samples of atomic elements, regardless of the type of element present in the samples?

(a) If two samples of different elements contain the same number of atoms, they contain the same number of moles.

(b) If two samples of different elements have the same mass, they contain the same number of moles.

(c) If two samples of different elements have the same mass, they contain the same number of atoms.

CONCEPTUAL CHECKPOINT 5.2

Without doing any calculations, determine which sample contains the most atoms.

(a) one gram of cobalt

(b) one gram of carbon

(c) one gram of lead

5.3 Counting Molecules by the Gram

LO: Calculate moles from mass and mass from moles of compounds.

LO: Calculate the number of molecules from mass and mass from the number of molecules.

The calculations we just performed for atoms can also be applied to molecules for covalent compounds or formula units for ionic compounds. We first calculate the moles of a compound from the mass, and then we calculate the number of molecules (or formula units) from moles.

Calculations Involving Grams and Moles of a Compound

Remember, ionic compounds do not contain individual molecules. In loose language, the smallest electrically neutral collection of ions is sometimes called a molecule but is a formula unit.

For elements, the molar mass is the mass of 1 mol of atoms of that element. For compounds, the molar mass is the mass of 1 mol of molecules or formula units of that compound. The molar mass of a compound in grams per mole is numerically equal to the formula mass of the compound in atomic mass units. For example, the formula mass of CO_2 is:

Remember, the formula mass for a compound is the sum of the atomic masses of all of the atoms in a chemical formula.

$$\begin{aligned} \text{Formula mass} &= 1(\text{Atomic mass of C}) + 2(\text{Atomic mass of O}) \\ &= 1(12.01\text{ amu}) + 2(16.00\text{ amu}) \\ &= 44.01\text{ amu} \end{aligned}$$

The molar mass of CO_2 is therefore:

$$\text{M}(CO_2) = 44.01\text{ g/mol}$$

Just as the molar mass of an element relates grams and moles of that element, the molar mass of a compound relates grams and moles of that compound. For example, suppose we want to find the number of moles in a 22.5 g sample of dry ice (solid CO_2). We begin by sorting the information.

GIVEN: 22.5 g CO_2

FIND: mol CO_2

SOLUTION MAP

We then strategize by drawing a solution map that shows how the molar mass can be used, along with the mass, to calculate moles of the compound. Recall that you can access the molar mass by using a periodic table and the formula of the compound.

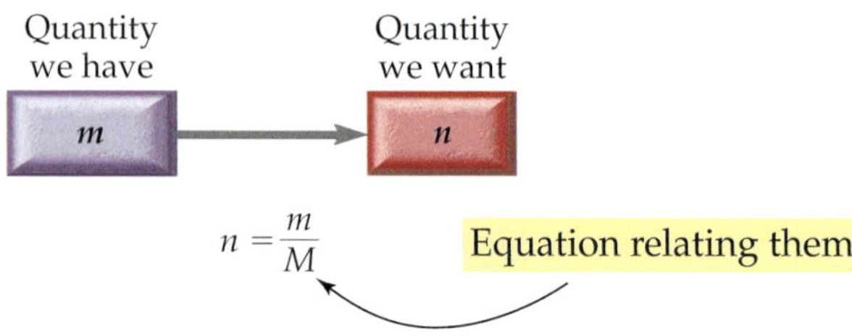

SOLUTION

Follow the solution map to solve the problem.

$$n = \frac{m}{M} = \frac{22.5\ \cancel{g}}{44.01\ \cancel{g}/\text{mol}} = 0.511 \text{ mol } CO_2$$

EXAMPLE 5.4 THE MOLE CONCEPT—CALCULATIONS INVOLVING GRAMS AND MOLES OF COMPOUNDS

Calculate the mass (in grams) of 1.75 mol of water.

SORT You are given moles of water and asked to find the mass.	**GIVEN:** 1.75 mol H_2O **FIND:** g H_2O
STRATEGIZE As per Example 5.2, we only need one step in the calculation as there is one equation relating moles to mass. We need the molar mass of water in order to calculate the mass of water present in 1.75 moles. Remembering that you always have access to a periodic table, M can be calculated from the atomic masses of each element in the formula for water in their abundance as shown.	**SOLUTION MAP** 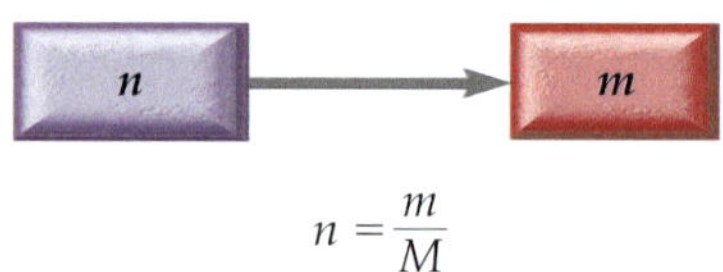 $n = \frac{m}{M}$ **RELATIONSHIPS USED** $M(H_2O) = 2(\text{atomic mass H}) + 1(\text{atomic mass O})$ $= 2(1.01) + 2(16.00)$ $= 18.02$ g/mol
SOLVE Follow the solution map to solve the problem. Begin with 1.75 mol of water and use the molar mass to calculate the mass of water.	**SOLUTION** Rearrange the equation for the unknown quantity, in this case, m. $n = \frac{m}{M}$ $m = n \times M = 1.75\ \cancel{\text{mol}} \times 18.02\ \text{g}/\cancel{\text{mol}} = 31.5 \text{ g } H_2O$
CHECK Check your answer. Are the units correct? Does the answer make physical sense?	The units (g H_2O) are the desired units. The magnitude of the answer makes sense because 1 mole of water has a mass of 18.02 g; therefore, 1.75 moles should have a mass that is slightly less than 36 g.

▶**SKILLBUILDER 5.4 | The Mole Concept—Calculations involving Grams and Moles of Compounds.**

Calculate the number of moles of NO_2 in 1.18 g of NO_2.

▶**FOR MORE PRACTICE** Problems 47, 48, 49, 50.

Calculating the Number of Molecules there are in a Mass of a Compound

Suppose that we want to find the *number of CO_2 molecules* in a sample of dry ice (solid CO_2) with a mass of 22.5 g. The solution map for the problem is:

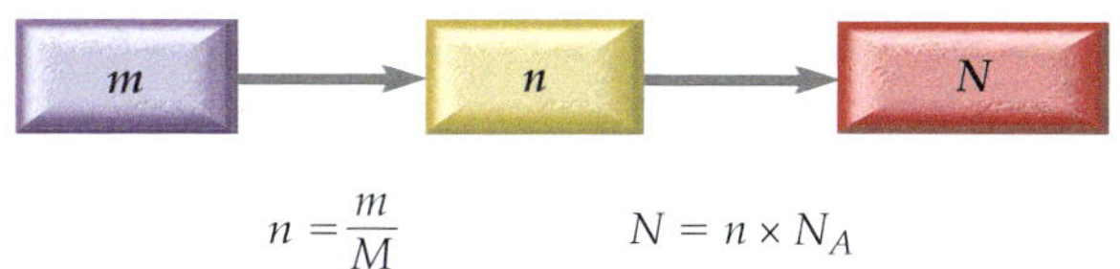

$$n = \frac{m}{M} \qquad N = n \times N_A$$

Notice that the first part of the solution map is identical to calculating the number of moles of CO_2 in 22.5 g of dry ice. The second part of the solution map shows the calculating the number of molecules from moles.

We first calculate the number of moles of CO_2 in the sample:

$$n = \frac{m}{M} = \frac{22.5\,\text{g}}{44.01\,\text{g/ mol}} = 0.511 \text{ mol } CO_2$$

Then we calculate number of molecules:

$$N = n \times N_A = 0.511 \text{ mol} \times 6.022 \times 10^{23} \text{ mol}^{-1} = 3.08 \times 10^{23} \text{ molecules of } CO_2$$

EXAMPLE 5.5 **THE MOLE CONCEPT—CALCULATIONS INVOLVING MASS OF A COMPOUND AND NUMBER OF MOLECULES**

What is the mass of 4.78×10^{24} NO_2 molecules?

SORT
You are given the number of NO_2 molecules and asked to find the mass.

GIVEN: 4.78×10^{24} NO_2 molecules

FIND: g NO_2

STRATEGIZE
Draw a solution map that relates the information given to that required, showing the mathematical equations that can be used for each step.

SOLUTION MAP

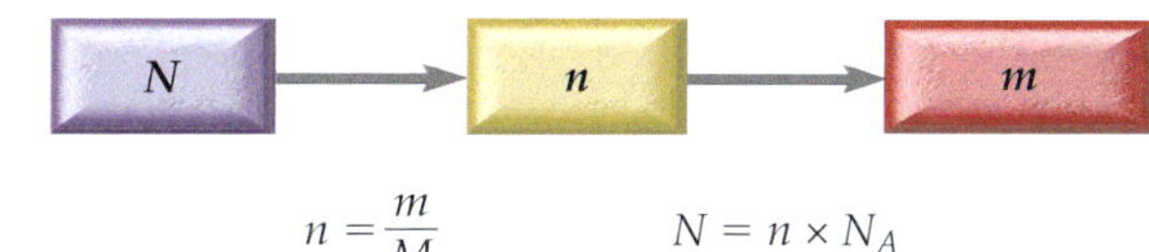

$$n = \frac{m}{M} \qquad N = n \times N_A$$

RELATIONSHIPS USED

M (NO_2) = 46.01 g/mol

SOLVE
We follow the solution map as a guide, begin with molecules of NO_2. Each equation needs to be rearranged to calculate the relevant unknown.

SOLUTION

Step 1: Calculate the number of moles of NO_2

$$N = n \times N_A$$

$$n = \frac{N}{N_A} = \frac{4.78 \times 10^{24}}{6.022 \times 10^{23} \text{mol}^{-1}} = 7.94 \text{ mol } NO_2$$

Step 2: Calculate the mass of NO_2

$$n = \frac{m}{M}$$

$$m = n \times M = 7.94 \text{ mol} \times 46.01 \text{ g/mol} = 365 \text{ g } NO_2$$

CHECK Check your answer. Are the units correct? Does the answer make physical sense?	The units, g NO_2, are correct. Because the number of NO_2 molecules is more than one mole, the answer should be more than the molar mass (more than 46.01 g), which it is; therefore, the magnitude of the answer is reasonable.

▶SKILLBUILDER 5.5 | The Mole Concept—Calculations involving Mass and Number of Molecules

How many H_2O molecules are in a sample of water with a mass of 3.64 g?

▶FOR MORE PRACTICE Problems 51, 52, 53, 54.

CONCEPTUAL CHECKPOINT 5.3

Compound A has a molar mass of 100 g/mol and Compound B has a molar mass of 200 g/mol. If you have samples of equal mass of both compounds, which sample contains the greater number of molecules?

5.4 Ratios of Atoms in Chemical Formulas

LO: Relate via ratios moles of a compound and moles of a constituent element.

LO: Relate via ratios grams of a compound and grams of a constituent element.

We are almost ready to address the sodium problem posed in Section 5.1. To determine how much of a particular element (such as sodium) is in a given amount of a particular compound (such as sodium chloride), we must understand the numerical relationships inherent in a chemical formula. We can understand these relationships with a straightforward analogy: Asking how much sodium is in a given amount of sodium chloride is similar to asking how many leaves are on a given number of clovers. For example, suppose we want to know the number of leaves on 14 clovers. We need to know the relationship between leaves and clovers. For clovers, the relationship comes from our everyday knowledge about them—we know that each clover has three leaves. We can express that relationship as a ratio between clovers and leaves.

3 leaves : 1 clover

▲ We know that each clover has three leaves. We can express that as a ratio: 3 leaves : 1 clover. © Siede Preis/Photodisc/Getty Images.

With this ratio, we can write a relationship to determine the number of leaves in 14 clovers. The solution map is:

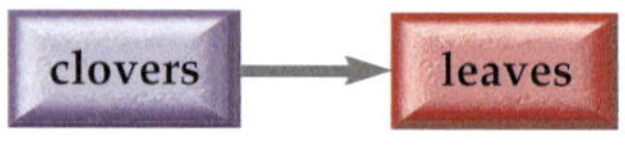

3 leaves : 1 clover

We solve the problem by beginning with clovers and calculating to leaves.

$$14 \text{ clovers} \times \frac{3 \text{ leaves}}{1 \text{ clover}} = 42 \text{ leaves}$$

Similarly, a chemical formula gives us ratios between elements and molecules for a particular compound. For example, the formula for carbon dioxide (CO_2) indicates that there are two O atoms per CO_2 molecule. We write this as:

2 O atoms : 1 CO_2 molecule

Just as 3 leaves : 1 clover can also be written as 3 dozen leaves: 1 dozen clovers, for molecules we can write:

2 doz O atoms : 1 doz CO_2 molecules

However, for atoms and molecules, we normally work in moles.

2 mol O : 1 mol CO_2

Chemical formulas are discussed in Module 3.

With ratios such as these—which come directly from the chemical formula—we can determine the amounts of the constituent elements present in a given amount of a compound.

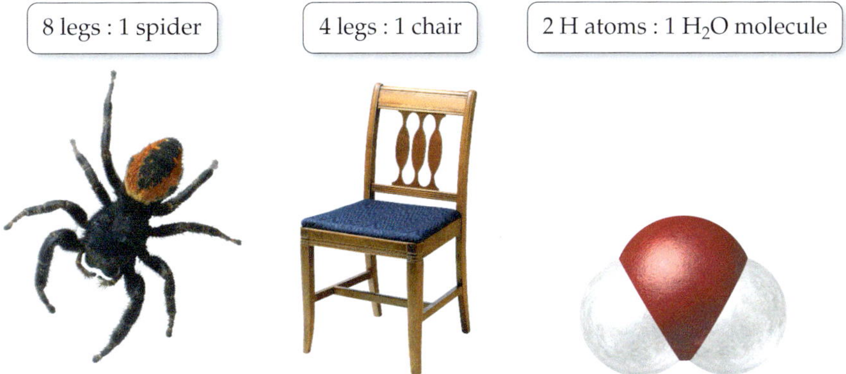

▲ Each of these shows a ratio. (Left) © GK Hart/Vikki Hart/Photodisc/Getty Images. (centre) © C Squared Studios/Photodisc/Getty Images.

Relating Moles of a Compound and Moles of a Constituent Element

Suppose we want to know the number of moles of O in 18 mol of CO_2. Our solution map is:

2 mol O : 1 mol CO_2

We can then calculate the moles of O.

$$18\ \text{mol}\ CO_2 \times \frac{2\ \text{mol O}}{1\ \text{mol}\ CO_2} = 36\ \text{mol O}$$

EXAMPLE 5.6 RATIOS OF ATOMS WITHIN CHEMICAL FORMULAS—RELATING MOLES OF A COMPOUND AND MOLES OF A CONSTITUENT ELEMENT

Determine the number of moles of O in 1.7 mol of $CaCO_3$.

SORT You are given the number of moles of $CaCO_3$ and asked to find the number of moles of O.	**GIVEN:** 1.7 mol $CaCO_3$ **FIND:** mol O
STRATEGIZE The solution map begins with moles of calcium carbonate and ends with moles of oxygen. Determine the ratio between O atoms and $CaCO_3$, which indicates three O atoms for every $CaCO_3$ unit.	**SOLUTION MAP** mol $CaCO_3$ → mol O n mol O : n mol $CaCO_3$ **RELATIONSHIPS USED** 3 mol O : 1 mol $CaCO_3$ (from chemical formula)
SOLVE Follow the solution map to solve the problem. The subscripts in a chemical formula are exact, so they never limit significant figures.	**SOLUTION** $1.7\ \text{mol}\ CaCO_3 \times \frac{3\ \text{mol O}}{1\ \text{mol}\ CaCO_3} = 5.1\ \text{mol O}$
CHECK Check your answer. Are the units correct? Does the answer make physical sense?	The units (mol O) are correct. The magnitude is reasonable as the number of moles of oxygen should be larger than the number of moles of $CaCO_3$ (because each $CaCO_3$ unit contains 3 O atoms).

▶SKILLBUILDER 5.6 | Ratios of Atoms within Chemical Formulas—Relating Moles of a Compound and Moles of a Constituent Element

Determine the number of moles of O in 1.4 mol of H_2SO_4.

▶FOR MORE PRACTICE Example 5.17; Problems 63, 64.

Relationship between Grams of a Compound and Grams of a Constituent Element

Now, we have the tools we need to solve our sodium problem from the beginning of the module. Suppose we want to know the mass of sodium in 15 g of NaCl. The chemical formula gives us the relationship between moles of Na and moles of NaCl:

$$1 \text{ mol Na} : 1 \text{ mol NaCl}$$

To use this relationship, we need *mol* NaCl, but we have *g* NaCl. We can use the *molar mass* of NaCl to calculate mol NaCl from g NaCl. Then we use the ratio from the chemical formula to calculate mol Na. Finally, we use the molar mass of Na to calculate g Na. The solution map is:

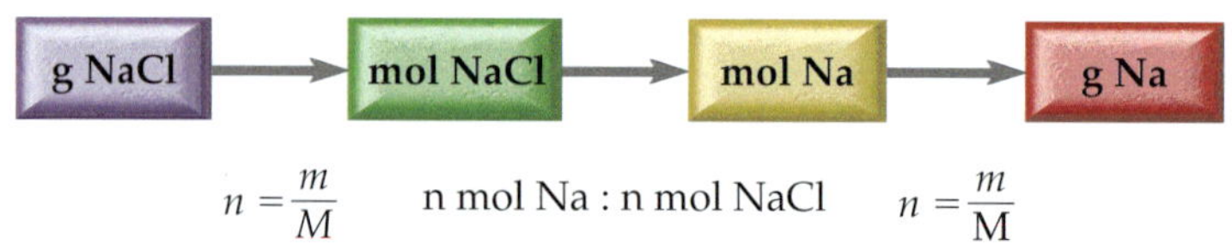

$n = \dfrac{m}{M}$ n mol Na : n mol NaCl $n = \dfrac{m}{M}$

Notice that we must calculate mol NaCl from g NaCl *before* we can use the ratio of Na to NaCl from the chemical formula.

The chemical formula gives us a relationship between moles of substances, not between grams.

We follow the solution map to solve the problem.

Step 1: Calculate moles of compound.

$$n = \frac{m}{M} = \frac{15 \cancel{g}}{58.44 \cancel{g}/\text{mol}} = 0.257 \text{ mol NaCl}$$

Step 2: Calculate moles of constituent element.

In this example, the ratio is 1 mol Na : 1 mol NaCl

$$n(\text{Na}) = n(\text{NaCl}) \times \frac{1 \cancel{\text{mol}} \text{ Na}}{1 \cancel{\text{mol}} \text{ NaCl}} = 0.257 \text{ mol Na}$$

Step 3: Calculate the mass of the element.

$$m = n \times M = 0.257 \cancel{\text{mol}} \times 22.99 \text{ g}/\cancel{\text{mol}} = 5.9 \text{ g Na}$$

The general form for solving problems where you are asked to find the mass of an element present in a given mass of a compound is:

Mass compound → **Moles** compound → **Moles** element → **Mass** element

Use the atomic or molar mass to calculate moles from mass and use the relationships inherent in the chemical formula to convert between moles and moles (▼Figure 5.2).

▶ FIGURE 5.2 Mole relationships from a chemical formula The relationships inherent in a chemical formula allow us to convert between moles of the compound and moles of a constituent element (or vice versa).

1 mol CCl_4 : 4 mol Cl

EXAMPLE 5.7 RATIOS OF ATOMS WITHIN CHEMICAL FORMULAS—RELATING MASS OF A COMPOUND AND MASS OF A CONSTITUENT ELEMENT

Carvone ($C_{10}H_{14}O$) is the main component of spearmint oil. It has a pleasant aroma and mint flavor. Carvone is added to chewing gum, liqueurs, soaps, and perfumes. Calculate the mass of carbon in 55.4 g of carvone.

SORT

You are given the mass of carvone and asked to find the mass of one of its constituent elements.

GIVEN: 55.4 g $C_{10}H_{14}O$

FIND: g C

STRATEGIZE

Base the solution map on:
Grams → Mole → Mole → Grams

SOLUTION MAP

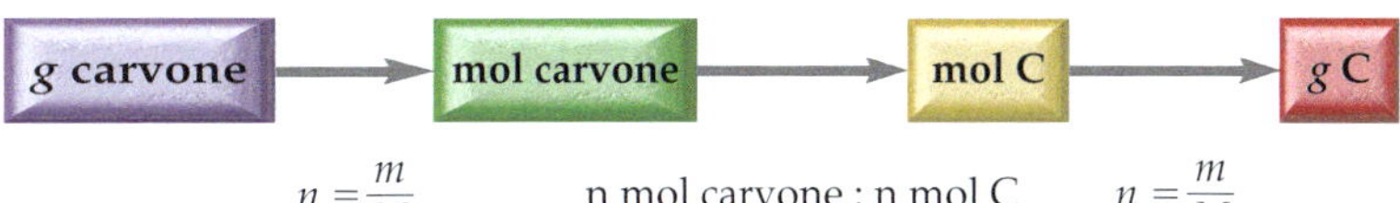

$n = \frac{m}{M}$ n mol carvone : n mol C $n = \frac{m}{M}$

You need the molar mass of carvone.

RELATIONSHIPS USED

$$\begin{aligned} M(\text{carvone}) &= 10(12.01) + 14(1.01) + 1(16.00) \\ &= 120.1 + 14.14 + 16.00 \\ &= 150.2 \text{ g/mol} \end{aligned}$$

The relationship between moles of carbon and moles of carvone can be found from the molecular formula.

10 mol C : 1 mol carvone (from chemical formula)

The molar mass of carbon is also required.

M(C) = 12.01 g/mol C (from periodic table)

SOLVE

Follow the solution map to solve the problem, beginning with g $C_{10}H_{14}O$ using the appropriate equations and ratios to arrive at g C.

SOLUTION

$$n = \frac{m}{M} = \frac{55.4 \cancel{\text{g}}}{150.2 \cancel{\text{g}}/\text{mol}} = 0.369 \text{ mol } C_{10}H_{14}O$$

$$n(C) = n(C_{10}H_{14}O) \times \frac{10 \cancel{\text{mol}} \text{ C}}{1 \cancel{\text{mol}} \ C_{10}H_{14}O} = 0.369 \times \frac{10}{1} = 3.69 \text{ mol C}$$

$$m = n \times M = 3.69 \cancel{\text{mol}} \times 12.01 \text{ g}/\cancel{\text{mol}} = 44.3 \text{ g C}$$

CHECK

Check your answer. Are the units correct? Does the answer make physical sense?

The units, g C, are correct. The magnitude of the answer is reasonable because the mass of carbon with the compound must be less than the mass of the compound itself. If you had arrived at a mass of carbon that was greater than the mass of the compound, you would know that you had made a mistake; the mass of a constituent element can never be greater than the mass of the compound itself.

▶SKILLBUILDER 5.7 | Ratios of Atoms within Chemical Formulas - Relating Mass of a Compound and Mass of a Constituent Element

Determine the mass of oxygen in a 5.8 g sample of sodium bicarbonate ($NaHCO_3$).

▶SKILLBUILDER PLUS Determine the mass of oxygen in a 7.20 g sample of $Al_2(SO_4)_3$.

▶FOR MORE PRACTICE Example 5.18; Problems 66, 67, 68, 69.

CONCEPTUAL CHECKPOINT 5.4

Without doing any detailed calculations, determine which sample contains the most fluorine atoms.

(a) 25 g of HF

(b) 1.5 mol of CH_3F

(c) 1.0 mol of F_2

CHEMISTRY IN THE ENVIRONMENT

▶ Chlorine in Chlorofluorocarbons

About 40 years ago, scientists began to suspect that synthetic compounds known as chlorofluorocarbons (CFCs) were destroying a vital compound called ozone (O_3) in Earth's upper atmosphere. Upper atmospheric ozone is important because it acts as a shield to protect life on Earth from harmful ultraviolet light (▶ Figure 5.3). CFCs are chemically inert molecules (they do not readily react with other substances) used primarily as refrigerants and industrial solvents. Their inertness has allowed them to leak into the atmosphere and stay there for many years. In the upper atmosphere, however, sunlight eventually breaks bonds within CFCs, resulting in the release of chlorine atoms. The chlorine atoms then react with ozone and destroy it by converting it from O_3 into O_2.

In 1985, scientists discovered a large hole in the ozone layer over Antarctica that has since been attributed to CFCs. The amount of ozone over Antarctica had depleted by a startling 50 %. The ozone hole is transient, existing only in the Antarctic spring, from late August to November. Examination of data from previous years showed that this gradually expanding ozone hole has formed each spring and had been growing bigger each year (▼ Figure 5.4). A similar hole has been observed during some years over the North Pole, and a smaller, but still significant, drop in ozone has been observed over more populated areas such as the northern United States and Canada. The thinning of ozone over these areas is dangerous because ultraviolet light can harm living things and induce skin cancer in humans. Based on this evidence, most developed nations banned the production of CFCs on January 1, 1996. The ozone hole continues to form today, but has reached a maximum and is starting to contract. The contraction of the ozone hole is attributed to the CFC ban.

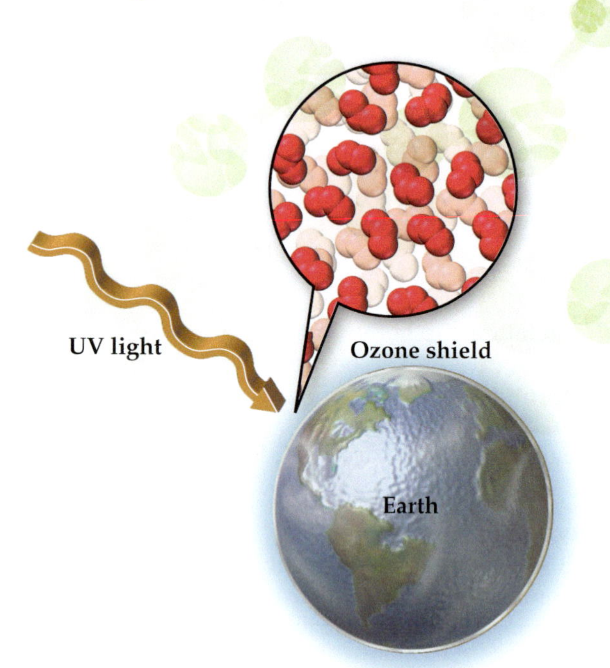

▲ **FIGURE 5.3 The ozone shield** Atmospheric ozone shields life on Earth from harmful ultraviolet light.

B5.1 CAN YOU ANSWER THIS? *Suppose a car air conditioner contains 2.5 kg of freon-12 (CCl_2F_2), a CFC. How many kilograms of Cl are contained within the freon?*

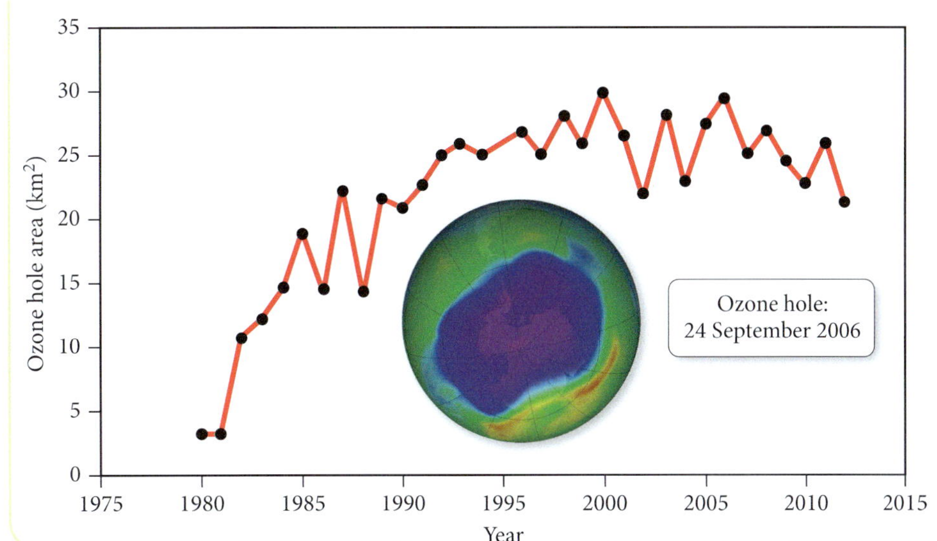

◀ **FIGURE 5.4 Growth of the ozone hole** Antarctic ozone levels from 1980 to 2012. © NASA/Tom Pantages. Page 183: Maxwell Art And Photo/Pearson.

5.5 Mass Percent Composition

LO: Determine mass percent composition from a chemical formula.

LO: Use mass percent composition to calculate the mass of elements in a sample of a compound.

The **mass percent composition** or simply **mass percent** of an element is the element's percentage of the total mass of the compound. We can calculate the mass percent of any element in a compound from the chemical formula for the compound. Based on the chemical formula, the mass percent of element *X* in a compound is:

$$\text{Mass percent of element } X = \frac{\text{Mass of element } X \text{ in 1 mol of compound}}{\text{Mass of 1 mol of compound}} \times 100\ \%$$

Suppose, for example, that we want to calculate the mass percent composition of Cl in the chlorofluorocarbon CCl_2F_2. The mass percent of Cl is given by:

$$CCl_2F_2$$

$$\text{Mass percent Cl} = \frac{2 \times \text{Molar mass Cl}}{\text{Molar mass } CCl_2F_2} \times 100\ \%$$

We must multiply the molar mass of Cl by 2 because the chemical formula has a subscript of 2 for Cl, meaning that 1 mol of CCl_2F_2 contains 2 mol of Cl atoms. We calculate the molar mass of CCl_2F_2 as follows:

$$\text{Molar mass} = 1(12.01) + 2(35.45) + 2(19.00) = 120.91\ \text{g/mol}$$

So the mass percent of Cl in CCl_2F_2 is:

$$\text{Mass percent Cl} = \frac{2 \times \text{Molar mass Cl}}{\text{Molar mass } CCl_2F_2} \times 100\ \% = \frac{2 \times 35.45\ \cancel{\text{g/mol}}}{120.91\ \cancel{\text{g/mol}}} \times 100\ \%$$
$$= 58.64\ \%$$

EXAMPLE 5.8 MASS PERCENT COMPOSITION

Calculate the mass percent of Cl in freon-114 ($C_2Cl_4F_2$).

SORT
You are given the molecular formula of freon-114 and asked to find the mass percent of Cl.

GIVEN: $C_2Cl_4F_2$

FIND: Mass % Cl

STRATEGIZE

Use the mass percent equation to obtain the mass percent Cl. The molar masses of the element and the compound are required.

SOLUTION MAP

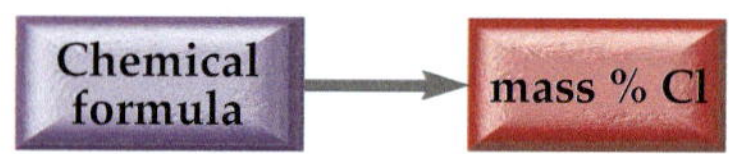

Mass percent of element X

$$= \frac{\text{Mass of element } X \text{ in 1 mol of compound}}{\text{Mass of 1 mol of compound}} \times 100\ \%$$

RELATIONSHIPS USED

$$\begin{aligned} M(\text{Cl}) &= 35.45\ \text{g/mol} \\ M(C_2Cl_4F_2) &= 2(12.01) + 4(35.45) + 2(19.00) \\ &= 24.02 + 141.8 + 38.00 \\ &= 203.8\ \text{g/mol} \end{aligned}$$

SOLVE

Substitute the values into the equation to find mass percent Cl.

SOLUTION

$$\begin{aligned} \text{Mass \% Cl} &= \frac{4 \times M(\text{Cl})}{M(C_2Cl_4F_2)} \times 100\ \% \\ &= \frac{4 \times 35.45\ \cancel{\text{g/mol}}}{203.8\ \cancel{\text{g/mol}}} \times 100\ \% \\ &= 69.58\ \% \end{aligned}$$

CHECK

Check your answer. Are the units correct? Does the answer make physical sense?

The units (%) are correct. The answer makes physical sense. Mass percent composition should never exceed 100 %. If your answer is greater than 100 %, you have made an error.

▶SKILLBUILDER 5.8 | Mass Percent Composition

Acetic acid ($HC_2H_3O_2$) is the active ingredient in vinegar. Calculate the mass percent composition of O in acetic acid.

▶FOR MORE PRACTICE Example 5.19; Problems 79, 80, 81.

The mass percent composition of compounds can be used to equate the mass of an element in a *sample* of compound not just 1 mole of the compound. In order to do this, we consider a sample of the compound rather than 1 mol of the compound and need to modify our equation as follows:

$$\text{Mass percent of element } X = \frac{\text{Mass of } X \text{ in a sample of the compound}}{\text{Mass of the sample of the compound}} \times 100\ \%$$

Suppose a 0.358 g sample of chromium reacts with oxygen to form 0.523 g of the metal oxide. Then the mass percent of chromium is:

$$\begin{aligned} \text{Mass percent Cr} &= \frac{\text{Mass Cr}}{\text{Mass metal oxide}} \times 100\ \% \\ &= \frac{0.358\ \cancel{\text{g}}}{0.523\ \cancel{\text{g}}} \times 100\ \% = 68.5\ \% \end{aligned}$$

We can use mass percent composition to equate grams of a constituent element and grams of the compound. We now have the tools to consider our sodium question from Section 5.1, as shown in Example 5.9.

EXAMPLE 5.9 USING MASS PERCENT COMPOSITION IN CALCULATIONS

The NHMRC recommends that adults consume up to 0.92 g of sodium per day. How many grams of sodium chloride can you consume and still be within the NHMRC guidelines? Sodium chloride is 39 % sodium by mass.

SORT

You are given the mass of sodium and the mass percent of sodium in sodium chloride. *Percent means per hundred*, so 39 % sodium indicates that there are 39 g Na per 100 g NaCl. You are asked to find the mass of sodium chloride that contains the given mass of sodium.

GIVEN: 0.92 g Na

39 % Na in NaCl by mass

FIND: g NaCl

STRATEGIZE

Draw a solution map that starts with the mass of sodium and uses the mass percent equation to calculate the mass of sodium chloride.

SOLUTION MAP

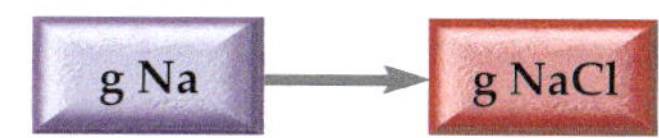

$$\text{Mass percent of element } X = \frac{\text{Mass of } X \text{ in a sample of the compound}}{\text{Mass of the sample of the compound}} \times 100\ \%$$

SOLVE

Follow the solution map to solve the problem. Begin by rearranging the equation. The amount of salt you can consume and still be within the NHMRC guideline is 2.4 g NaCl.

SOLUTION

$$\text{Mass percent of Na in NaCl} = \frac{\text{Mass of Na in a sample}}{\text{Mass of a sample of NaCl}} \times 100\ \%$$

$$\text{Mass of a sample of NaCl} = \frac{\text{Mass of Na in a sample}}{\text{Mass percent of Na in NaCl}} \times 100\ \%$$

$$= \frac{0.92\ \text{g}}{39\ \%} \times 100\ \% = 2.4\ \text{g NaCl}$$

CHECK

Check your answer. Are the units correct? Does the answer make physical sense?

The units, g NaCl, are correct. The answer makes physical sense because the mass of NaCl should be *larger* than the mass of Na. The mass of a compound containing a given mass of a particular element is always larger than the mass of the element itself.

▶SKILLBUILDER 5.9 | Using Mass Percent Composition in calculations

If a woman consumes 22 g of sodium chloride, how much sodium does she consume? Sodium chloride is 39 % sodium by mass.

▶FOR MORE PRACTICE Example 5.20; Problems 74, 75, 76, 77.

CONCEPTUAL CHECKPOINT 5.5

Which compound has the highest mass percent of O? (You should not have to perform any detailed calculations to answer this question.)

(a) CrO

(b) CrO_2

(c) Cr_2O_3

CHEMISTRY AND HEALTH

▶ Fluoridation of Drinking Water

In the early 1900s, scientists discovered that people whose drinking water naturally contained fluoride (F^-) ions had fewer cavities than people whose water did not. At appropriate levels, fluoride strengthens tooth enamel, which prevents tooth decay. In an effort to improve public health, fluoride can be artificially added to drinking water supplies. In Australia today, the majority of the population in most states drink artificially fluoridated drinking water.

The fluoridation of public drinking water, however, is often controversial. Some opponents argue that fluoride is available from other sources—such as toothpaste, mouthwash, drops, and pills—and therefore should not be added to drinking water. Anyone who wants fluoride can get it from these optional sources, they argue, and the government should not impose fluoride on the population. Other opponents argue that the risks associated with fluoridation are too great. Indeed, too much fluoride can cause teeth to become brown and spotted, a condition known as dental fluorosis. Extremely high levels can lead to skeletal fluorosis, a condition in which the bones become brittle and arthritic.

The scientific consensus is that, like many minerals, fluoride shows some health benefits at certain levels—about 1–4 mg/day for adults—but can have detrimental effects at higher levels. Consequently, most major cities fluoridate their drinking water up to 1 mg/L. Most adults drink between 1 and 2 L of water per day, so they receive beneficial amounts of fluoride from the water. Bottled water does not normally contain fluoride. Fluoridated bottled water can sometimes be found in the infant section of supermarkets.

B5.2 CAN YOU ANSWER THIS? *Fluoride is often added to water as sodium fluoride (NaF). What is the mass percent composition of F^- in NaF? How many grams of NaF must be added to 1500 L of water to fluoridate it at a level of 1.0 mg F^-/L?*

5.6 Calculating Empirical Formulas for Compounds

LO: Determine an empirical formula from experimental data.

LO: Calculate an empirical formula from reaction data.

In the previous section, we learned how to calculate mass percent composition from a chemical formula. But can we go the other way? Can we calculate a chemical formula from mass percent composition? This is important because laboratory analyses of compounds do not often give chemical formulas directly; rather, they give the relative masses of each element present in a compound. For example, if we decompose water into hydrogen and oxygen in the laboratory, we could measure the masses of hydrogen and oxygen produced. Can we determine the chemical formula for water from this kind of data?

▶ We just learned how to go from the chemical formula of a compound to its mass percent composition. Can we also go the other way?

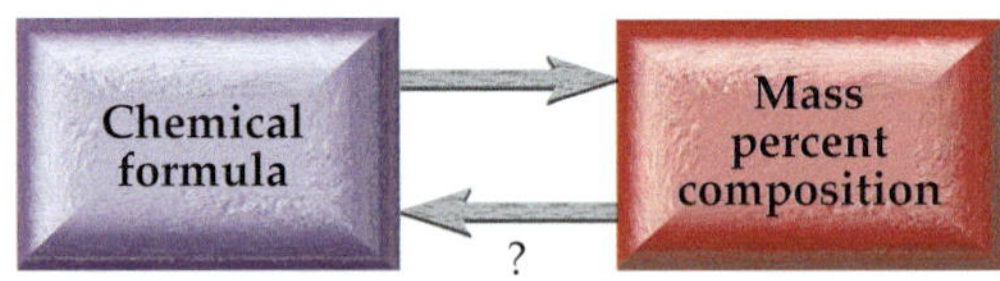

The answer is a qualified yes. We can determine a chemical formula, but it is the **empirical formula**, not the molecular formula. As we saw in Section 3.3, an empirical formula gives the smallest whole-number ratio of the atoms of each element in a compound, not the specific number of each type of atom in a molecule. Recall that the **molecular formula** is always a whole-number multiple of the empirical formula: Molecular formula = Empirical × n, where n = 1, 2, 3 . . .

A chemical formula represents a ratio of atoms or moles of atoms, not a ratio of masses.

For example, the molecular formula for hydrogen peroxide is H_2O_2, and its empirical formula is HO.

$$HO \times 2 \rightarrow H_2O_2$$

Calculating an Empirical Formula from Experimental Data

Suppose we decompose a sample of water in the laboratory and find that it produces 3.0 g of hydrogen atoms and 24 g of oxygen atoms. How do we determine an empirical formula from these data?

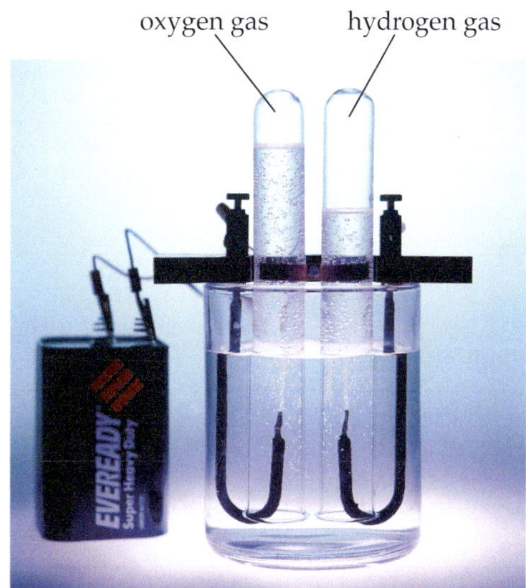

▲ Water can be decomposed by an electric current into hydrogen and oxygen. How can we find the empirical formula for water from the masses of its component elements? © Charles D. Winters/Science Source.

We know that an empirical formula represents a ratio of atoms or a ratio of moles of atoms, but it *does not* represent a ratio of masses. So the first thing we must do is calculate moles. How many moles of each element formed during the decomposition? To calculate moles, we substitute the mass and molar mass into the appropriate equation:

$$n(\text{H}) = \frac{m}{M} = \frac{3.0\,\cancel{\text{g}}}{1.01\,\cancel{\text{g}}/\text{mol}} = 3.0\text{ mol H}$$

$$n(\text{O}) = \frac{m}{M} = \frac{24\,\cancel{\text{g}}}{16.00\,\cancel{\text{g}}/\text{mol}} = 1.5\text{ mol O}$$

From these data, we know there are 3 mol of H for every 1.5 mol of O. We can now write a pseudoformula for water:

$$H_3O_{1.5}$$

To get whole-number subscripts in our formula, we divide all the subscripts by the smallest one, in this case 1.5.

$$H_{\frac{3}{1.5}}O_{\frac{1.5}{1.5}} = H_2O$$

Our empirical formula for water, which in this case also happens to be the molecular formula, is H_2O. The following procedure can be used to obtain the empirical formula of any compound from experimental data. The left column outlines the procedure, and the center and right columns contain two examples of how to apply the procedure.

OBTAINING AN EMPIRICAL FORMULA FROM EXPERIMENTAL DATA

EXAMPLE 5.10

You decompose a compound containing nitrogen and oxygen in the laboratory and produce 24.5 g of N and 70.0 g of O. Calculate the empirical formula of the compound.

EXAMPLE 5.11

A laboratory analysis of aspirin determines the following mass percent composition:

C 60.00 %
H 4.48 %
O 35.53 %

Find the empirical formula.

1. Write down (or calculate) the masses of each element present in a sample of the compound. If you are given mass percent composition, assume a 100 g sample and calculate the masses of each element from the given percentages.

Example 5.10:

GIVEN: 24.5 g N
70.0 g O

FIND: empirical formula

Example 5.11:

GIVEN: In a 100 g sample:
60.00 g C
4.48 g H
35.53 g O

FIND: empirical formula

2. Using the appropriate molar mass for each element, calculate the moles of each element.

Example 5.10:

SOLUTION

$$n(\mathrm{N}) = \frac{m}{M} = \frac{24.5\,\text{g}}{14.01\,\text{g / mol}} = 1.75\text{ mol N}$$

$$n(\mathrm{O}) = \frac{m}{M} = \frac{70.0\,\text{g}}{16.00\,\text{g / mol}} = 4.38\text{ mol O}$$

Example 5.11:

SOLUTION

$$n(\mathrm{C}) = \frac{m}{M} = \frac{60.00\,\text{g}}{12.01\,\text{g/mol}} = 4.996\text{ mol C}$$

$$n(\mathrm{H}) = \frac{m}{M} = \frac{4.48\,\text{g}}{1.01\,\text{g/mol}} = 4.44\text{ mol H}$$

$$n(\mathrm{O}) = \frac{m}{M} = \frac{35.53\,\text{g}}{16.00\,\text{g/mol}} = 2.221\text{ mol O}$$

3. Write down a pseudoformula for the compound, using the moles of each element (from Step 2) as subscripts.

Example 5.10: $N_{1.75}O_{4.38}$

Example 5.11: $C_{4.996}H_{4.44}O_{2.221}$

4. Divide all the subscripts in the formula by the smallest subscript.

Example 5.10: $N_{\frac{1.75}{1.75}}O_{\frac{4.38}{1.75}} \rightarrow N_1O_{2.5}$

Example 5.11: $C_{\frac{4.996}{2.221}}H_{\frac{4.44}{2.221}}O_{\frac{2.221}{2.221}} \rightarrow C_{2.25}H_2O_1$

5. If the subscripts are not whole numbers, multiply all the subscripts by a small whole number (see the following table for common examples) to arrive at whole-number subscripts.

Fractional Subscript	Multiply by This Number to Get Whole-Number Subscripts
_.10	10
_.20	5
_.25	4
_.33	3
_.50	2
_.66	3
_.75	4

Example 5.10: $N_1O_{2.5} \times 2 \rightarrow N_2O_5$

The correct empirical formula is N_2O_5.

Example 5.11: $C_{2.25}H_2O_1 \times 4 \rightarrow C_9H_8O_4$

The correct empirical formula is $C_9H_8O_4$.

▶SKILLBUILDER 5.10 | A sample of a compound is decomposed in the laboratory and produces 165 g of C, 27.8 g of H, and 220.2 g O. Calculate the empirical formula of the compound.

▶FOR MORE PRACTICE
Problems 86, 87, 88. 89.

▶SKILLBUILDER 5.11 | Ibuprofen, an aspirin substitute, has the mass percent composition: C 75.69 %; H 8.80 %; O 15.51 %. Calculate the empirical formula of the ibuprofen.

▶FOR MORE PRACTICE
Example 5.21; Problems 90, 91, 92, 93.

EXAMPLE 5.12 CALCULATING AN EMPIRICAL FORMULA FROM REACTION DATA

A 3.24 g sample of titanium reacts with oxygen to form 5.40 g of the metal oxide. What is the empirical formula of the metal oxide?

You are given the mass of titanium and the mass of the metal oxide that forms. You are asked to find the empirical formula. This is slightly different than the previous examples. In this case, you are given a total mass of titanium oxide, with an unknown ratio of Ti : O, but you know how much Ti metal was used to create it.	**GIVEN:** 3.24 g Ti 5.40 g metal oxide **FIND:** empirical formula
1. Write down (or calculate) the masses of each element present in a sample of the compound. In this case, you are given the mass of the initial Ti sample and the mass of its oxide after the sample reacts with oxygen. The mass of oxygen is the difference between the mass of the oxide and the mass of titanium.	**SOLUTION** 3.24 g Ti Mass O = Mass metal oxide – Mass titanium = 5.40 g – 3.24 g = 2.16 g O
2. Using the appropriate equation, calculate the moles of each element present in the sample of the metal oxide.	$n(\text{Ti}) = \frac{m}{M} = \frac{3.24\,\text{g}}{47.88\,\text{g/mol}} = 0.0677\ \text{mol Ti}$ $n(\text{O}) = \frac{m}{M} = \frac{2.16\,\text{g}}{16.00\,\text{g/mol}} = 0.135\ \text{mol O}$
3. Write down a pseudoformula for the compound, using the moles of each element obtained in Step 2 as subscripts.	$Ti_{0.0677}O_{0.135}$
4. Divide all the subscripts in the formula by the smallest subscript.	$Ti_{\frac{0.0677}{0.0677}}O_{\frac{0.135}{0.0677}} \rightarrow TiO_2$
5. If the subscripts are not whole numbers, multiply all the subscripts by a small whole number to arrive at whole-number subscripts.	As the subscripts are already whole numbers, this last step is unnecessary. The correct empirical formula is TiO_2.

▶SKILLBUILDER 5.12 | Calculating an Empirical Formula from Reaction Data

A 1.56 g sample of copper reacts with oxygen to form 1.95 g of the metal oxide. What is the formula of the metal oxide?

▶FOR MORE PRACTICE Problems 94, 95, 96, 97.

5.7 Calculating Molecular Formulas for Compounds

LO: Calculate a molecular formula from an empirical formula and molar mass.

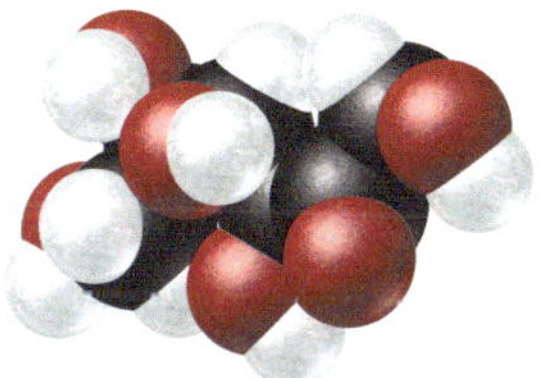

▲ Fructose, a sugar found in fruit.

We can determine the *molecular* formula of a compound from the empirical formula if we also know the molar mass of the compound. Recall from Section 3.3 that the molecular formula is always a whole-number multiple of the empirical formula.

$$\text{Molecular formula} = \text{Empirical formula} \times n, \text{ where } n = 1, 2, 3 \ldots$$

Suppose we want to find the molecular formula for fructose (a sugar found in fruit) from its empirical formula, CH_2O, and its molar mass, 180.2 g/mol. We know that the molecular formula is a whole-number multiple of CH_2O.

$$\text{Molecular formula} = CH_2O \times n$$

We also know that the molar mass is a whole-number multiple of the **empirical formula molar mass**, the sum of the masses of all the atoms in the empirical formula.

$$\text{Molar mass} = \text{Empirical formula molar mass} \times n$$

For a particular compound, the value of n in both cases is the same. Therefore, we can find n by calculating the ratio of the molar mass to the empirical formula molar mass.

$$n = \frac{\text{Molar mass}}{\text{Empirical formula molar mass}}$$

For fructose, the empirical formula molar mass is:

$$\text{Empirical formula molar mass} = 1(12.01) + 2(1.01) + 16.00 = 30.03\ \text{g/mol}$$

Therefore, n is:

$$n = \frac{180.2\ \cancel{\text{g/mol}}}{30.03\ \cancel{\text{g/mol}}} = 6$$

We can then use this value of n to find the molecular formula.

$$\text{Molecular formula} = CH_2O \times 6 = C_6H_{12}O_6$$

EXAMPLE 5.13 CALCULATING MOLECULAR FORMULA FROM EMPIRICAL FORMULA AND MOLAR MASS

Naphthalene is a compound containing carbon and hydrogen that is used in mothballs. Its empirical formula is C_5H_4 and its molar mass is 128.16 g/mol. What is its molecular formula?

SORT
You are given the empirical formula and the molar mass of a compound and asked to find its molecular formula.

GIVEN: empirical formula = C_5H_4
molar mass = 128.16 g/mol

FIND: molecular formula

STRATEGIZE
In the first step, use the molar mass (which is given) and the empirical formula molar mass (which you can calculate based on the empirical formula) to determine n (the integer by which you must multiply the empirical formula to determine the molecular formula).

In the second step, multiply the subscripts in the empirical formula by n to arrive at the molecular formula.

SOLUTION MAP

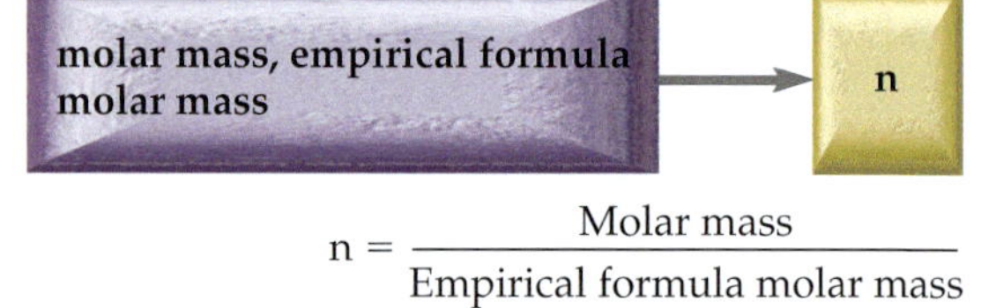

$$n = \frac{\text{Molar mass}}{\text{Empirical formula molar mass}}$$

Molecular formula = Empirical formula × n

SOLVE
First find the empirical formula molar mass.
Next follow the solution map. Find n by dividing the molar mass by the empirical formula molar mass (which you just calculated). Multiply the empirical formula by n to determine the molecular formula.

SOLUTION

$$\text{Empirical formula molar mass} = 5(12.01) + 4(1.01) = 64.09\ \text{g/mol}$$

$$n = \frac{\text{Molar mass}}{\text{Empirical formula mass}} = \frac{128.16\ \cancel{\text{g/mol}}}{64.09\ \cancel{\text{g/mol}}} = 2$$

$$\text{Molecular formula} = C_5H_4 \times 2 = C_{10}H_8$$

CHECK
Check your answer. Does the answer make physical sense?

The answer makes physical sense because it is a whole-number multiple of the empirical formula. Any answer containing fractional subscripts would be an error.

▶SKILLBUILDER 5.13 | Calculating Molecular Formula from Empirical Formula and Molar Mass

Butane is a compound containing carbon and hydrogen used as a fuel in butane lighters. Its empirical formula is C_2H_5, and its molar mass is 58.12 g/mol. Find its molecular formula.

▶SKILLBUILDER PLUS A compound with the following mass percent composition has a molar mass of 60.10 g/mol. Find its molecular formula.

C 39.97 % H 13.41 % N 46.62 %

▶FOR MORE PRACTICE Example 5.22; Problems 98, 99, 100, 101.

MODULE IN REVIEW

Self-Assessment Quiz

Q1. How many atoms are there in 5.8 mol helium?

(a) 23.2 atoms

(b) 9.6×10^{-24} atoms

(c) 5.8×10^{23} atoms

(d) 3.5×10^{24} atoms

Q2. A sample of pure silver has a mass of 155 g. How many moles of silver are in the sample?

(a) 1.44 mol

(b) 1.67×10^{4} mol

(c) 0.696 mol

(d) 155 mol

Q3. How many carbon atoms are there in a 12.5 kg sample of carbon?

(a) 6.27×10^{20} atoms

(b) 9.04×10^{28} atoms

(c) 6.27×10^{26} atoms

(d) 1.73×10^{-21} atoms

Q4. Which sample contains the greatest number of atoms?

(a) 15 g Ne

(b) 15 g Ar

(c) 15 g Kr

(d) None of the above (all contain the same number of atoms).

Q5. What is the average mass (in grams) of a single carbon dioxide molecule?

(a) 3.8×10^{-26} g

(b) 7.31×10^{-23} g

(c) 2.65×10^{25} g

(d) 44.01 g

Q6. How many moles of O are in 1.6 mol of $Ca(NO_3)_2$?

(a) 1.6 mol O

(b) 3.2 mol O

(c) 4.8 mol O

(d) 9.6 mol O

Q7. How many grams of Cl are in 25.8 g CF_2Cl_2?

(a) 7.56 g

(b) 3.78 g

(c) 15.1 g

(d) 0.427 g

Q8. Which sample contains the greatest number of F atoms?

(a) 2.0 mol HF

(b) 1.5 mol F_2

(c) 1.0 mol CF_4

(d) 0.5 mol CH_2F_2

Q9. The compound A_2X is 35.8 % A by mass. What mass of the compound contains 55.1 g A?

(a) 308 g

(b) 154 g

(c) 19.7 g

(d) 35.8 g

Q10. Which compound has the highest mass percent C?

(a) CO

(b) CO_2

(c) H_2CO_3

(d) H_2CO

Q11. What is the mass percent N in $C_2H_8N_2$?

(a) 23.3 % N

(b) 16.6 % N

(c) 215 % N

(d) 46.6 % N

Q12. A compound is 52.14 % C, 13.13 % H, and 34.73 % O by mass. What is the empirical formula of the compound?

(a) C_4HO_3

(b) C_2H_6O

(c) $C_2H_8O_3$

(d) C_3HO_6

Q13. A compound has the empirical formula CH_2O and a formula mass of 120.10 g/mol. What is the molecular formula of the compound?

(a) CH_2O

(b) $C_2H_4O_2$

(c) $C_3H_6O_3$

(d) $C_4H_8O_4$

Q14. A compound is decomposed in the laboratory and produces 1.40 g N and 0.20 g H. What is the empirical formula of the compound?

(a) NH

(b) N_2H

(c) NH_2

(d) N_7H

Answers: 1d; 2a; 3c; 4a; 5b; 6d; 7c; 8c; 9b; 10a; 11d; 12b; 13d; 14c

Chemical Principles

The Mole Concept: The mole is a specific number (6.022×10^{23}) that allows us to count atoms or molecules by weighing them. One mole of any element has a mass equivalent to its atomic mass in grams, and a mole of any compound has a mass equivalent to its formula mass in grams. The mass of 1 mol of an element or compound is its molar mass.

Chemical Formulas and Chemical Composition: Chemical formulas indicate the relative number of each kind of element in a compound. These numbers are based on atoms or moles. By using molar masses, we can use the information in a chemical formula to determine the relative masses of each kind of element in a compound. We can then relate the mass of a sample of a compound to the masses of the elements contained in the compound.

Empirical and Molecular Formulas from Laboratory Data: The relative masses of the elements within a compound allow us to determine the empirical formula of the compound. If we know the molar mass of the compound, we can also determine its molecular formula.

Relevance

The Mole Concept: The mole concept allows us to determine the number of atoms or molecules in a sample from its mass. Just as a hardware store customer wants to know the number of nails in a certain weight of nails, we want to know the number of atoms in a certain mass of atoms. Because atoms are too small to count, we use their mass.

Chemical Formulas and Chemical Composition: The chemical composition of compounds is important because it lets us determine how much of a particular element is contained within a particular compound. For example, to assess the threat to the Earth's ozone layer from chlorofluorocarbons (CFCs), we need to know how much chlorine is in a particular CFC.

Empirical and Molecular Formulas from Laboratory Data: The first thing we want to know about an unknown compound is its chemical formula, because the formula reveals the compound's composition. Chemists often arrive at formulas by analyzing compounds in the laboratory—either by decomposing them or by synthesizing them—to determine the relative masses of the elements they contain.

Chemical Skills

LO: Calculations involving moles and number of atoms (Section 5.2).

SORT
You are given moles of copper and asked to find the number of copper atoms.

STRATEGIZE
To calculate number of atoms from moles, use the equation relating N to n.

SOLVE
Follow the solution map to solve the problem.

CHECK
Check your answer. Are the units correct? Does the answer make physical sense?

Examples

EXAMPLE 5.14 CALCULATIONS INVOLVING MOLES AND NUMBER OF ATOMS

Calculate the number of atoms in 4.8 mol of copper.

GIVEN: 4.8 mol Cu

FIND: Cu atoms

SOLUTION MAP

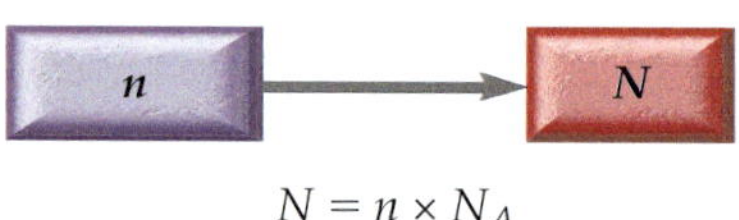

$$N = n \times N_A$$

SOLUTION

$$N = n \times N_A = 4.8\ \text{mol} \times 6.022 \times 10^{23}\ \text{mol}^{-1}$$
$$= 2.9 \times 10^{24}\ \text{Cu atoms}$$

The units, Cu atoms, are correct. The answer makes physical sense because the number is very large, as you would expect for nearly 5 moles of atoms.

LO: Calculations involving grams and moles (Section 5.2).

EXAMPLE 5.15 CALCULATIONS INVOLVING GRAMS AND MOLES

Calculate the mass of aluminum (in grams) of 6.73 moles of aluminum.

SORT
You are given the number of moles of aluminum and asked to find the mass of aluminum in grams.

GIVEN: 6.73 mol Al

M(Al) = 26.98 g/mol (from periodic table)

FIND: g Al

STRATEGIZE
Use the molar mass of aluminum to calculate mass using the equation relating mass to moles.

SOLUTION MAP

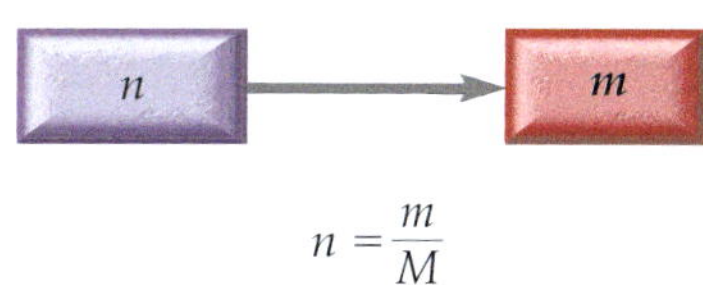

$$n = \frac{m}{M}$$

SOLVE
Follow the solution map to solve the problem.

SOLUTION

$$n = \frac{m}{M}$$

$$m = n \times M = 6.73\,\text{mol} \times 26.98\,\text{g/mol} = 182\,\text{g Al}$$

CHECK
Check your answer. Are the units correct? Does the answer make physical sense?

The units, g Al, are correct. The answer makes physical sense because each mole has a mass of about 27 g; therefore, nearly 7 moles should have a mass of nearly 190 g.

LO: Calculations involving grams and number of atoms or molecules (Section 5.2).

EXAMPLE 5.16 CALCULATIONS INVOLVING GRAMS AND NUMBER OF ATOMS OR MOLECULES

Determine the number of atoms in a 48.3 g sample of zinc.

SORT
You are given the mass of a zinc sample and asked to find the number of Zn atoms that it contains.

GIVEN: 48.3 g Zn

M(Zn) = 65.39 g/mol (from periodic table)

FIND: Zn atoms

STRATEGIZE
Use the molar mass of the element to calculate moles, and then use Avogadro's number to calculate number of atoms.

SOLUTION MAP

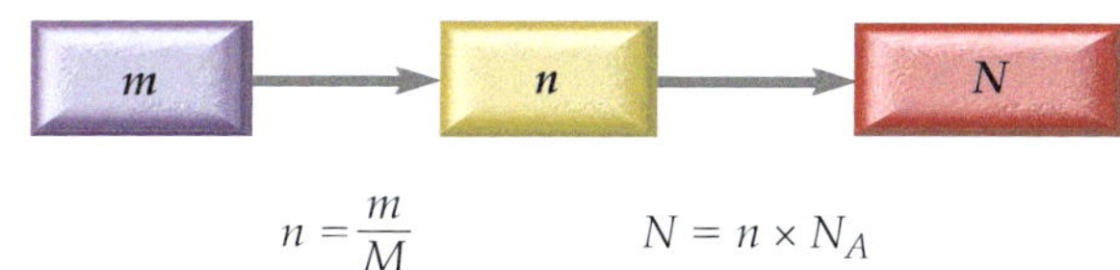

$$n = \frac{m}{M} \qquad N = n \times N_A$$

SOLVE
Follow the solution map to solve the problem.

SOLUTION

$$n = \frac{m}{M} = \frac{48.3\,\text{g}}{65.39\,\text{g/mol}} = 0.739\text{ mol Zn}$$

$$N = n \times N_A = 0.739\,\text{mol} \times 6.022 \times 10^{23}\,\text{mol}^{-1}$$
$$= 4.45 \times 10^{23}\text{ Zn atoms}$$

CHECK
Check your answer. Are the units correct? Does the answer make physical sense?

The units, Zn atoms, are correct. The answer makes physical sense because the number of atoms in any macroscopic-sized sample should be very large.

LO: Calculations involving moles of a compound and moles of a constituent element (Section 5.4).

EXAMPLE 5.17 **CALCULATIONS INVOLVING MOLES OF A COMPOUND AND MOLES OF A CONSTITUENT ELEMENT**

Determine the number of moles of oxygen in 7.20 mol of H_2SO_4.

SORT
You are given the number of moles of sulfuric acid and asked to find the number of moles of oxygen.

GIVEN: 7.20 mol H_2SO_4

FIND: mol O

STRATEGIZE
To calculate moles of a constituent element from moles of a compound, use the chemical formula of the compound to determine a ratio between the moles of the element and the moles of the compound.

SOLUTION MAP

n mol O : n mol H_2SO_4

RELATIONSHIPS USED

4 mol O : 1 mol H_2SO_4 (from chemical formula)

SOLVE
Follow the solution map to solve the problem.

SOLUTION

$$7.20 \text{ mol } H_2SO_4 \times \frac{4 \text{ mol O}}{1 \text{ mol } H_2SO_4} = 28.8 \text{ mol O}$$

CHECK
Check your answer. Are the units correct? Does the answer make physical sense?

The units, mol O, are correct. The answer makes physical sense because the number of moles of an element in a compound is equal to or greater than the number of moles of the compound itself.

LO: Calculations involving grams of a compound and grams of a constituent element (Section 5.4).

EXAMPLE 5.18 **CALCULATIONS INVOLVING GRAMS OF A COMPOUND AND GRAMS OF A CONSTITUENT ELEMENT**

Find the grams of iron in 79.2 g of Fe_2O_3.

SORT
You are given the mass of iron(III) oxide and asked to find the mass of iron contained within it.

GIVEN: 79.2 g Fe_2O_3

FIND: g Fe

STRATEGIZE
Use the molar mass of the compound to calculate moles of the compound from grams of the compound. Then use the chemical formula to calculate moles of the constituent element from moles of the compound. Finally, use the molar mass of the constituent element to calculate grams of the element from moles of the element.

SOLUTION MAP

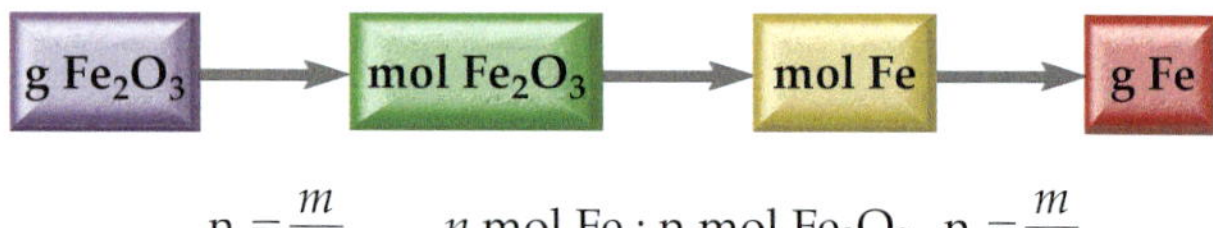

$n = \frac{m}{M}$ n mol Fe : n mol Fe_2O_3 $n = \frac{m}{M}$

RELATIONSHIPS USED

2 mol Fe : 1 mol Fe_2O_3 (from chemical formula)
$M(Fe_2O_3) = 159.70$ g/mol
$M(Fe) = 55.85$ g/mol

SOLVE
Follow the solution map to solve the problem.

SOLUTION

$$n = \frac{m}{M} = \frac{79.2 \text{ g}}{159.70 \text{ g/mol}} = 0.496 \text{ mol } Fe_2O_3$$

$$n(Fe) = n(Fe_2O_3) \times \frac{2 \text{ mol Fe}}{1 \text{ mol } Fe_2O_3} = 0.496 \times \frac{2}{1}$$

$$= 0.992 \text{ mol Fe}$$

$$m = n \times M = 0.992 \text{ mol} \times 55.85 \text{ g/mol} = 55.4 \text{ g Fe}$$

CHECK
Check your answer. Are the units correct? Does the answer make physical sense?

The units, g Fe, are correct. The answer makes physical sense because the mass of a constituent element within a compound should be less than the mass of the compound itself.

LO: Determine mass percent composition from a chemical formula (Section 5.5).

EXAMPLE 5.19 MASS PERCENT COMPOSITION

Calculate the mass percent composition of potassium in potassium oxide (K_2O).

SORT
You are given the formula of potassium oxide and asked to determine the mass percent of potassium within it.

GIVEN: K_2O

FIND: Mass % K

STRATEGIZE
The solution map shows how you can use mass percent equation to yield the mass percent of the element. The molar masses of the element and compound are required.

SOLUTION MAP

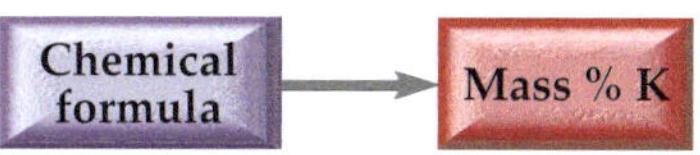

$$\text{Mass percent of element } X = \frac{\text{Mass of element } X \text{ in 1 mol of compound}}{\text{Mass of 1 mol of compound}} \times 100\ \%$$

RELATIONSHIPS USED

$$M(K) = 39.10\ g/mol$$

$$M(K_2O) = 2(39.10) + 16.00 = 94.20\ g/mol$$

SOLVE
Substitute the values into the equation to find mass percent.

SOLUTION

$$\text{Mass \% K} = \frac{2(39.10)\ g/\cancel{mol}}{94.20\ g/\cancel{mol}} \times 100\ \% = 83.01\ \%$$

CHECK
Check your answer. Are the units correct? Does the answer make physical sense?

The units, % K, are correct. The answer makes physical sense because it should be below 100 %.

LO: Use mass percent composition in calculations (Section 5.5).

EXAMPLE 5.20 USING MASS PERCENT COMPOSITION IN CALCULATIONS

Determine the mass of titanium in 57.2 g of titanium(IV) oxide. The mass percent of titanium in titanium(IV) oxide is 59.9 %.

SORT
You are given the mass of titanium(IV) oxide and the mass percent titanium in the oxide. You are asked to find the mass of titanium in the sample.

GIVEN: 57.2 g TiO_2

59.9 % Ti in TiO_2 by mass

FIND: g Ti

STRATEGIZE
Use the percent composition equation to relate grams of titanium(IV) oxide and grams of titanium.

SOLUTION MAP

$$\text{Mass percent of element } X = \frac{\text{Mass of } X \text{ in a sample of the compound}}{\text{Mass of the sample of the compound}} \times 100\ \%$$

SOLVE
Follow the solution map to solve the problem problem, rearranging the equation as appropriate.

SOLUTION

$$\text{Mass \% Ti in TiO}_2 = \frac{\text{Mass Ti in sample}}{\text{Mass of TiO}_2\text{ sample}} \times 100\ \%$$

$$\text{Mass Ti in sample} = \frac{\text{Mass \% Ti in TiO}_2}{100\ \%} \times \text{Mass of TiO}_2\text{ sample}$$

$$= \frac{59.9\ \%}{100\ \%} \times 57.2\text{ g} = 34.3\text{ g Ti}$$

CHECK
check your answer. are the units correct? does the answer make physical sense?

The units, g Ti, are correct. The answer makes physical sense because the mass of an element within a compound should be less than the mass of the compound itself.

LO: Determine an empirical formula from experimental data (Section 5.6).

EXAMPLE 5.21 DETERMINING AN EMPIRICAL FORMULA FROM EXPERIMENTAL DATA

A laboratory analysis of vanillin, the flavoring agent in vanilla, determined the mass percent composition: C, 63.15 %; H, 5.30 %; O, 31.55 %. Determine the empirical formula of vanillin.

You need to recognize this problem as one requiring a special procedure. Follow these steps to solve the problem.

GIVEN: 63.15 % C, 5.30 % H, and 31.55 % O.

FIND: empirical formula

SOLUTION

1. Write down (or calculate) the masses of each element present in a sample of the compound. If you are given mass percent composition, assume a 100 g sample and calculate the masses of each element from the given percentages.

In a 100 g sample:

63.15 g C

5.30 g H

31.55 g O

2. Calculate the moles of each element using the appropriate molar mass for each element.

$$n(\text{C}) = \frac{m}{M} = \frac{63.15\text{ g}}{12.01\text{ g/mol}} = 5.258\ \text{mol C}$$

$$n(\text{H}) = \frac{m}{M} = \frac{5.30\text{ g}}{1.01\text{ g/mol}} = 5.25\ \text{mol H}$$

$$n(\text{O}) = \frac{m}{M} = \frac{31.55\text{ g}}{16.00\text{ g/mol}} = 1.972\ \text{mol O}$$

3. Write down a pseudoformula for the compound using the moles of each element (from Step 2) as subscripts.

$C_{5.258}H_{5.25}O_{1.972}$

4. Divide all the subscripts in the formula by the smallest subscript.

$$C_{\frac{5.258}{1.972}}H_{\frac{5.25}{1.972}}O_{\frac{1.972}{1.972}} \rightarrow C_{2.67}H_{2.66}O_1$$

5. If the subscripts are not whole numbers, multiply all the subscripts by a small whole number to arrive at whole-number subscripts.

$$C_{2.67}H_{2.66}O_1 \times 3 \rightarrow C_8H_8O_3$$

The correct empirical formula is $C_8H_8O_3$.

LO: Calculate a molecular formula from an empirical formula and molar mass (Section 5.7).

EXAMPLE 5.22 CALCULATING A MOLECULAR FORMULA FROM AN EMPIRICAL FORMULA AND MOLAR MASS

Acetylene, a gas used in welding torches, has the empirical formula CH and a molar mass of 26.04 g/mol. Find its molecular formula.

SORT

You are given the empirical formula and molar mass of acetylene and asked to find the molecular formula.

GIVEN: empirical formula = CH

molar mass = 26.04 g/mol

FIND: molecular formula

STRATEGIZE

In the first step, use the molar mass (which is given) and the empirical formula molar mass (which you can calculate based on the empirical formula) to determine n (the integer by which you must multiply the empirical formula to arrive at the molecular formula).

In the second step, multiply the coefficients in the empirical formula by n to arrive at the molecular formula.

SOLUTION MAP

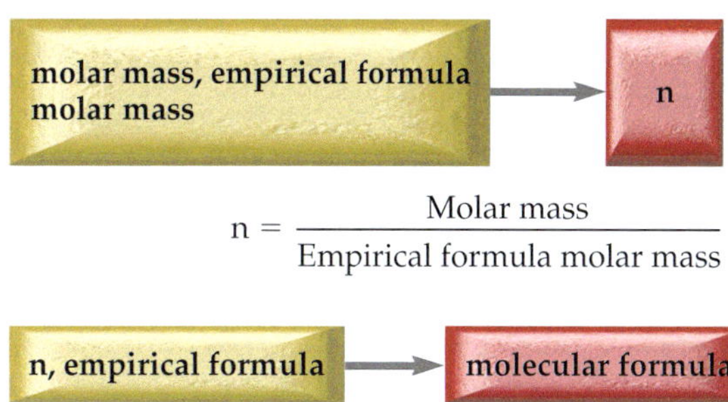

$$n = \frac{\text{Molar mass}}{\text{Empirical formula molar mass}}$$

$$\text{Molecular formula} = \text{Empirical formula} \times n$$

SOLVE

Follow the solution map to solve the problem. Calculate the empirical formula molar mass, which is the sum of the masses of all the atoms in the empirical formula.

SOLUTION

Empirical formula molar mass

$$= 12.01 + 1.01$$

$$= 13.02 \text{ g/mol}$$

Next, find n, the ratio of the molar mass to empirical mass.

$$n = \frac{\text{Molar mass}}{\text{Empirical formula molar mass}}$$

$$= \frac{26.04 \text{ g/mol}}{13.02 \text{ g/mol}} = 2$$

Finally, multiply the empirical formula by n to get the molecular formula.

$$\text{Molecular formula} = CH \times 2 \rightarrow C_2H_2$$

CHECK

Check your answer. Does the answer make physical sense?

The answer makes physical sense because the formula subscripts are all integers. Any answer with fractional integers is incorrrect.

KEY TERMS

Avogadro's number **[5.2]**
empirical formula **[5.7]**
empirical formula molar mass **[5.8]**
mass percent (composition) **[5.5]**
molar mass **[5.2]**
mole (mol) **[5.2]**
molecular formula **[5.7]**

EXERCISES

QUESTIONS

1. Why is chemical composition important?

2. How can you efficiently determine the number of atoms in a sample of an element? Why is counting them not an option?

3. How many atoms are in 1 mol of atoms?

4. How many molecules are in 1 mol of molecules?

5. What is the mass of 1 mol of atoms for an element?

6. What is the mass of 1 mol of molecules for a compound?

7. What is the mass of 1 mol of atoms of each element?
(a) P **(b)** Pt **(c)** C **(d)** Cr

8. What is the mass of 1 mol of molecules of each compound?
(a) CO_2 **(b)** CH_2Cl_2 **(c)** $C_{12}H_{22}O_{11}$ **(d)** SO_2

9. The subscripts in a chemical formula give relationships between moles of the constituent elements and moles of the compound. Explain why these subscripts *do not* give relationships between grams of the constituent elements and grams of the compound.

10. Write the ratios between moles of each constituent element and moles of the compound for $C_{12}H_{22}O_{11}$.

11. What is the mathematical equation for calculating mass percent composition from a chemical formula?

12. How are the empirical formula and the molecular formula of a compound related?

13. Why is it important to be able to calculate an empirical formula from experimental data?

14. What is the empirical formula mass of a compound?

15. How are the molar mass and empirical formula mass for a compound related?

PROBLEMS

THE MOLE CONCEPT

16. How many mercury atoms are in 5.8 mol of mercury?

17. How many moles of gold atoms do 3.45×10^{24} gold atoms constitute?

18. How many atoms are in each elemental sample?
(a) 3.4 mol Cu **(b)** 9.7×10^{-3} mol C
(c) 22.9 mol Hg **(d)** 0.215 mol Na

19. How many moles of atoms are in each elemental sample?
(a) 4.6×10^{24} Pb atoms **(b)** 2.87×10^{22} He atoms
(c) 7.91×10^{23} K atoms **(d)** 4.41×10^{21} Ca atoms

20. Complete the table.

Element	Moles	Number of Atoms
Ne	0.552	_____
Ar	_____	3.25×10^{24}
Xe	1.78	_____
He	_____	1.08×10^{20}

21. Complete the table.

Element	Moles	Number of Atoms
Cr	_____	9.61×10^{23}
Fe	1.52×10^{-5}	_____
Ti	0.0365	_____
Hg	_____	1.09×10^{23}

22. Consider these definitions.
1 doz = 12 1 gross = 144
1 ream = 500 1 mol = 6.022×10^{23}

Suppose you have 872 sheets of paper. How many ______ of paper sheets do you have?
(a) dozens **(b)** gross **(c)** reams **(d)** moles

23. An Australian $1 coin contains approximately 7.4×10^{22} copper atoms. Use the definitions in the previous problem to determine how many ______ of copper atoms are in a $1 coin.
(a) dozens **(b)** gross **(c)** reams **(d)** moles

24. How many moles of tin atoms are in a pure tin cup with a mass of 38.1 g?

25. A lead fishing weight contains 0.12 mol of lead atoms. What is its mass?

26. A pure gold coin contains 0.145 mol of gold. What is its mass?

27. A helium balloon contains 0.46 g of helium. How many moles of helium does it contain?

28. How many moles of atoms are in each elemental sample?
(a) 1.34 g Zn **(b)** 24.9 g Ar
(c) 72.5 g Ta **(d)** 0.0223 g Li

29. What is the mass in grams of each elemental sample?
(a) 6.64 mol W **(b)** 0.581 mol Ba
(c) 68.1 mol Xe **(d)** 1.57 mol S

30. Complete the table.

Element	Moles	Mass
Ne	_____	22.5 g
Ar	0.117	_____
Xe	_____	1.00 kg
He	1.44×10^{-4}	_____

31. Complete the table.

Element	Moles	Mass
Cr	0.00442	_____
Fe	_____	73.5 mg
Ti	1.009×10^{-3}	_____
Hg	_____	1.78 kg

32. A pure silver ring contains 0.0134 mmol (millimol) Ag. How many silver atoms does it contain?

33. A pure gold ring contains 0.0102 mmol (millimol) Au. How many gold atoms does it contain?

34. How many aluminum atoms are in 3.78 g of aluminum?

35. What is the mass of 4.91×10^{21} platinum atoms?

36. How many atoms are in each elemental sample?
(a) 16.9 g Sr
(b) 26.1 g Fe
(c) 8.55 g Bi
(d) 38.2 g P

37. Calculate the mass in grams of each elemental sample.
(a) 1.32×10^{20} uranium atoms
(b) 2.55×10^{22} zinc atoms
(c) 4.11×10^{23} lead atoms
(d) 6.59×10^{24} silicon atoms

38. How many carbon atoms are in a diamond (pure carbon) with a mass of 38 mg?

39. How many helium atoms are in a helium blimp containing 495 kg of helium?

40. How many titanium atoms are in a pure titanium bicycle frame with a mass of 1.28 kg?

41. How many copper atoms are in a pure copper statue with a mass of 133 kg?

42. Complete the table.

Element	Mass	Moles	Number of Atoms
Na	38.5 mg	____	____
C	____	1.12	____
V	____	____	214
Hg	1.44 kg	____	____

43. Complete the table.

Element	Mass	Moles	Number of Atoms
Pt	____	0.0449	____
Fe	____	____	1.14×10^{25}
Ti	23.8 mg	____	____
Hg	____	2.05	____

44. Which sample contains the greatest number of atoms?
(a) 27.2 g Cr **(b)** 55.1 g Ti **(c)** 205 g Pb

45. Which sample contains the greatest number of atoms?
(a) 10.0 g He **(b)** 25.0 g Ne **(c)** 115 g Xe

46. Determine the number of moles of molecules (or formula units) in each sample.
(a) 38.2 g sodium chloride
(b) 36.5 g nitrogen monoxide
(c) 4.25 kg carbon dioxide
(d) 2.71 mg carbon tetrachloride

47. Determine the mass of each sample.
(a) 1.32 mol carbon tetrafluoride
(b) 0.555 mol magnesium fluoride
(c) 1.29 mmol carbon disulfide
(d) 1.89 kmol sulfur trioxide

48. Complete the table.

Compound	Mass	Moles	Number of Molecules
H_2O	112 kg	____	____
N_2O	6.33 g	____	____
SO_2	____	2.44	____
CH_2Cl_2	____	0.0643	____

49. Complete the table.

Compound	Mass	Moles	Number of Molecules
CO_2	____	0.0153	____
CO	____	0.0150	____
BrI	23.8 mg	____	____
CF_2Cl_2	1.02 kg	____	____

50. A mothball, composed of naphthalene ($C_{10}H_8$), has a mass of 1.32 g. How many naphthalene molecules does it contain?

51. Calculate the mass in grams of a single water molecule.

52. How many molecules are in each sample?
(a) 3.5 g H_2O
(b) 56.1 g N_2
(c) 89 g CCl_4
(d) 19 g $C_6H_{12}O_6$

53. Calculate the mass in grams of each sample.
(a) 5.94×10^{20} H_2O_2 molecules
(b) 2.8×10^{22} SO_2 molecules
(c) 4.5×10^{25} O_3 molecules
(d) 9.85×10^{19} CH_4 molecules

54. A sugar crystal contains approximately 1.8×10^{17} sucrose ($C_{12}H_{22}O_{11}$) molecules. What is its mass in milligrams?

55. A salt crystal has a mass of 0.12 mg. How many NaCl formula units does it contain?

56. How much money, in dollars, does one mole of cents represent? If this amount of money were evenly distributed to the entire world's population (about 7.1 billion people), how much would each person get? Would each person be a millionaire? Billionaire? Trillionaire?

57. A typical dust particle has a diameter of about 10.0 μm. If 1.0 mol of dust particles were laid end to end along the equator, how many times would they encircle the planet? The circumference of the Earth at the equator is 40,076 km.

RATIOS OF ATOMS IN CHEMICAL FORMULAS

58. Determine the number of moles of Cl in 2.7 mol $CaCl_2$.

59. How many moles of O are in 12.4 mol $Fe(NO_3)_3$?

60. Which sample contains the greatest number of moles of O?
(a) 2.3 mol H_2O **(b)** 1.2 mol H_2O_2
(c) 0.9 mol $NaNO_3$ **(d)** 0.5 mol $Ca(NO_3)_2$

61. Which sample contains the greatest number of moles of Cl?
(a) 3.8 mol HCl **(b)** 1.7 mol CH_2Cl_2
(c) 4.2 mol $NaClO_3$ **(d)** 2.2 mol $Mg(ClO_4)_2$

62. Determine the number of moles of C in each sample.
(a) 2.5 mol CH_4 **(b)** 0.115 mol C_2H_6
(c) 5.67 mol C_4H_{10} **(d)** 25.1 mol C_8H_{18}

63. Determine the number of moles of H in each sample.
(a) 4.67 mol H_2O **(b)** 8.39 mol NH_3
(c) 0.117 mol N_2H_4 **(d)** 35.8 mol $C_{10}H_{22}$

64. For each set of molecular models, write a relationship between moles of hydrogen and moles of molecules. Then determine the total number of hydrogen atoms present. (H—white; O—red; C—black; N—blue)

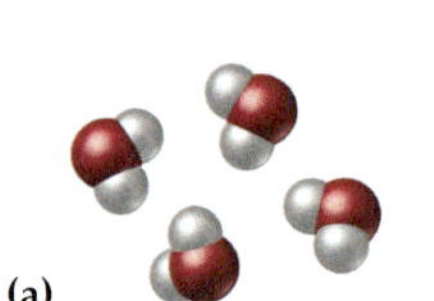
(a)

(b)

(c)

65. For each set of molecular models, write a relationship between moles of oxygen and moles of molecules. Then determine the total number of oxygen atoms present. (H—white; O—red; C—black; S—yellow)

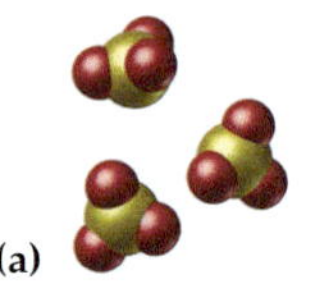
(a)

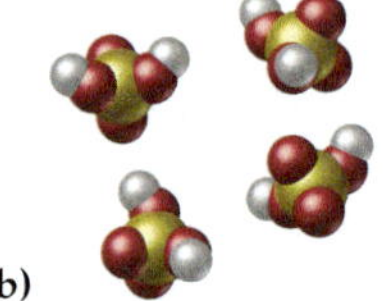
(b)

(c)

66. How many grams of Cl are in 38.0 g of each sample of chlorofluorocarbons (CFCs)?
(a) CF_2Cl_2
(b) $CFCl_3$
(c) $C_2F_3Cl_3$
(d) CF_3Cl

67. Calculate the number of grams of sodium in 1.00 g of each sodium-containing food additive.
(a) NaCl (table salt)
(b) Na_3PO_4 (sodium phosphate)
(c) $NaC_7H_5O_2$ (sodium benzoate)
(d) $Na_2C_6H_6O_7$ (sodium hydrogen citrate)

68. Iron is found in Earth's crust as several different iron compounds. Calculate the mass (in kg) of each compound that contains 1.0×10^3 kg of iron.
(a) Fe_2O_3 (hematite)
(b) Fe_3O_4 (magnetite)
(c) $FeCO_3$ (siderite)

69. Lead is found in Earth's crust as several lead compounds. Calculate the mass (in kg) of each compound that contains 1.0×10^3 kg of lead.
(a) PbS (galena)
(b) $PbCO_3$ (cerussite)
(c) $PbSO_4$ (anglesite)

MASS PERCENT COMPOSITION

70. A 2.45 g sample of strontium completely reacts with oxygen to form 2.89 g of strontium oxide. Use this data to calculate the mass percent composition of strontium in strontium oxide.

71. A 4.78 g sample of aluminum completely reacts with oxygen to form 6.67 g of aluminum oxide. Use this data to calculate the mass percent composition of aluminum in aluminum oxide.

72. A 1.912 g sample of calcium chloride is decomposed into its constituent elements and found to contain 0.690 g Ca and 1.222 g Cl. Calculate the mass percent composition of Ca and Cl in calcium chloride.

73. A 0.45 g sample of aspirin is decomposed into its constituent elements and found to contain 0.27 g C, 0.020 g H, and 0.16 g O. Calculate the mass percent composition of C, H, and O in aspirin.

74. Copper(II) fluoride contains 37.42 % F by mass. Use this percentage to calculate the mass of fluorine in grams contained in 28.5 g of copper(II) fluoride.

75. Silver chloride, used in silver plating, contains 75.27 % Ag. Calculate the mass of silver chloride in grams required to make 4.8 g of silver plating.

76. In small amounts, the fluoride ion (often consumed as NaF) prevents tooth decay. It is recommended that an adult female should consume 3.0 mg of fluorine per day. Calculate the amount of sodium fluoride (45.24 % F) that a woman should consume to get the recommended amount of fluorine.

77. The iodide ion, usually consumed as potassium iodide, is a dietary mineral essential to good nutrition. In countries where potassium iodide is added to salt, iodine deficiency or goiter has been almost completely eliminated. The recommended daily allowance (RDA) for iodine is 150 μg/day. How much potassium iodide (76.45 % I) should you consume to meet the RDA?

78. Calculate the mass percent composition of nitrogen in each compound.
(a) N_2O **(b)** NO **(c)** NO_2 **(d)** N_2O_5

79. Calculate the mass percent composition of carbon in each compound.
(a) C_2H_2 **(b)** C_3H_6 **(c)** C_2H_6 **(d)** C_2H_6O

80. Calculate the mass percent composition of each element in each compound.
(a) $C_2H_4O_2$ **(b)** CH_2O_2 **(c)** C_3H_9N **(d)** $C_4H_{12}N_2$

81. Calculate the mass percent composition of each element in each compound.
(a) $FeCl_3$ **(b)** TiO_2 **(c)** H_3PO_4 **(d)** HNO_3

82. Calculate the mass percent composition of O in each compound.
(a) calcium nitrate **(b)** iron(II) sulfate
(c) carbon dioxide

83. Calculate the mass percent composition of Cl in each compound.
(a) carbon tetrachloride **(b)** calcium hypochlorite
(c) perchloric acid

84. Various iron ores have different amounts of iron per kilogram of ore. Calculate the mass percent composition of iron for each iron ore: Fe_2O_3 (hematite), Fe_3O_4 (magnetite), $FeCO_3$ (siderite). Which ore has the highest iron content?

85. Plants need nitrogen to grow, so many fertilizers consist of nitrogen-containing compounds. Calculate the mass percent composition of nitrogen in each fertilizer: NH_3, $CO(NH_2)_2$, NH_4NO_3, $(NH_4)_2SO_4$. Which fertilizer has the highest nitrogen content?

CALCULATING EMPIRICAL FORMULAS

86. A compound containing nitrogen and oxygen is decomposed in the laboratory and produces 1.78 g of N and 4.05 g of O. Calculate the empirical formula of the compound.

87. A compound containing selenium and fluorine is decomposed in the laboratory and produces 2.231 g of Se and 3.221 g of F. Calculate the empirical formula of the compound.

88. Samples of several compounds are decomposed, and the masses of their constituent elements are measured. Calculate the empirical formula for each compound.
(a) 1.245 g Ni, 5.381 g I **(b)** 1.443 g Se, 5.841 g Br
(c) 2.128 g Be, 7.557 g S, 15.107 g O

89. Samples of several compounds are decomposed, and the masses of their constituent elements are measured. Calculate the empirical formula for each compound.
(a) 2.677 g Ba, 3.115 g Br **(b)** 1.651 g Ag, 0.1224 g O
(c) 0.672 g Co, 0.569 g As, 0.486 g O

90. The rotten smell of a decaying animal carcass is partially due to a nitrogen-containing compound called putrescine. Elemental analysis of putrescine indicates that it consists of 54.50 % C, 13.73 % H, and 31.77 % N by mass. Calculate the empirical formula of putrescine.

91. Citric acid, the compound responsible for the sour taste of lemons, has the elemental composition: C, 37.51 %; H, 4.20 %; O, 58.29 % by mass. Calculate the empirical formula of citric acid.

92. These compounds are found in many natural flavors and scents. Calculate the empirical formula for each compound. given the mass percent composition.
(a) ethyl butyrate (pineapple oil): C, 62.04 %; H, 10.41 %; O, 27.55 %
(b) methyl butyrate (apple flavor): C, 58.80 %; H, 9.87 %; O, 31.33 %
(c) benzyl acetate (oil of jasmine): C, 71.98 %; H, 6.71 %; O, 21.31 %

93. Calculate the empirical formula for each over-the-counter pain reliever given the mass percent composition.
(a) acetaminophen (Tylenol): C, 63.56 %; H, 6.00 %, N, 9.27 %; O, 21.17 %
(b) naproxen (Aleve): C, 73.03 %; H, 6.13 %; O, 20.84 %

94. A 1.45 g sample of phosphorus burns in air and forms 2.57 g of a phosphorus oxide. Calculate the empirical formula of the oxide.

95. A 2.241 g sample of nickel reacts with oxygen to form 2.852 g of the metal oxide. Calculate the empirical formula of the oxide.

96. A 0.77 mg sample of nitrogen reacts with chlorine to form 6.61 mg of the chloride. What is the empirical formula of the nitrogen chloride?

97. A 45.2 mg sample of phosphorus reacts with selenium to form 131.6 mg of the selenide. What is the empirical formula of the phosphorus selenide?

CALCULATING MOLECULAR FORMULAS

98. A compound containing carbon and hydrogen has a molar mass of 56.11 g/mol and an empirical formula of CH_2. Determine its molecular formula.

99. A compound containing phosphorus and oxygen has a molar mass of 219.9 g/mol and an empirical formula of P_2O_3. Determine its molecular formula.

100. The molar masses and empirical formulas of several compounds containing carbon and chlorine are listed here. Find the molecular formula of each compound.
(a) 284.77 g/mol, CCl **(b)** 131.39 g/mol, C_2HCl_3
(c) 181.44 g/mol, C_2HCl

101. The molar masses and empirical formulas of several compounds containing carbon and nitrogen are listed here. Find the molecular formula of each compound.
(a) 163.26 g/mol, $C_{11}H_{17}N$ **(b)** 186.24 g/mol, C_6H_7N
(c) 312.29 g/mol, C_3H_2N

CUMULATIVE PROBLEMS

102. Complete the table.

Substance	Mass	Moles	Number of Particles (atoms or molecules)
Ar	_____	4.5×10^{-4}	_____
NO_2	_____	_____	1.09×10^{20}
K	22.4 mg	_____	_____
C_8H_{18}	3.76 kg	_____	_____

103. Complete the table.

Substance	Mass	Moles	Number of Particles (atoms or molecules)
$C_6H_{12}O_6$	15.8 g	_____	_____
Pb	_____	_____	9.04×10^{21}
CF_4	22.5 mg	_____	_____
C	_____	0.0388	_____

104. Determine the chemical formula of each compound and refer to the formula to calculate the mass percent composition of each constituent element.
(a) copper(II) iodide **(b)** sodium nitrate
(c) lead(II) sulfate **(d)** calcium fluoride

105. Determine the chemical formula of each compound and refer to the formula to calculate the mass percent composition of each constituent element.
(a) nitrogen triiodide **(b)** xenon tetrafluoride
(c) phosphorus trichloride **(d)** carbon monoxide

106. The rock in a particular iron ore deposit contains 78 % Fe_2O_3 by mass. How many kilograms of the rock must a mining company process to obtain 1.0×10^3 kg of iron?

107. The rock in a lead ore deposit contains 84 % PbS by mass. How many kilograms of the rock must a mining company process to obtain 1.0×10^3 kg of Pb?

108. A leak in the air conditioning system of an office building releases 12 kg of CHF_2Cl per month. If the leak continues, how many kilograms of Cl are emitted into the atmosphere each year?

109. A leak in the air conditioning system of an older car releases 55 g of CF_2Cl_2 per month. How much Cl is emitted into the atmosphere each year by this car?

110. Hydrogen is a possible future fuel. However, elemental hydrogen is rare, so it must be obtained from a hydrogen-containing compound such as water. If hydrogen were obtained from water, how much hydrogen, in grams, could be obtained from 1.0 kg of water

111. Hydrogen, a possible future fuel mentioned in Problem 110, can also be obtained from ethanol. Ethanol can be made from the fermentation of crops such as corn. How much hydrogen, in grams, could be obtained from 1.0 kg of ethanol (C_2H_5OH)?

112. Complete the table of compounds that contain only carbon and hydrogen.

Formula	Molar Mass	% C (by mass)	% H (by mass)
C_2H_4	_____	_____	_____
_____	58.12	82.66 %	_____
C_4H_8	_____	_____	_____
_____	44.09	_____	18.29 %

113. Complete the table of compounds that contain only chromium and oxygen.

Formula	Name	Molar Mass	% Cr (by mass)	% O (by mass)
_____	Chromium (III) oxide	_____	_____	_____
_____	_____	84.00	61.90 %	_____
_____	_____	100.00	_____	48.00 %

114. Butanedione, a component of butter and body odor, has a cheesy smell. Elemental analysis of butanedione gave the mass percent composition: C, 55.80 %; H, 7.03 %; O, 37.17 %. The molar mass of butanedione is 86.09 g/mol. Determine the molecular formula of butanedione.

115. Caffeine, a stimulant found in coffee and soda, has the mass percent composition: C, 49.48 %; H, 5.19 %; N, 28.85 %; O, 16.48 %. The molar mass of caffeine is 194.19 g/mol. Find the molecular formula of caffeine.

116. Nicotine, a stimulant found in tobacco, has the mass percent composition: C, 74.03 %; H, 8.70 %; N, 17.27 %. The molar mass of nicotine is 162.26 g/mol. Find the molecular formula of nicotine.

117. Estradiol is a female sexual hormone that causes maturation and maintenance of the female reproductive system. Elemental analysis of estradiol gave the mass percent composition: C, 79.37 %; H, 8.88 %; O, 11.75 %. The molar mass of estradiol is 272.37 g/mol. Find the molecular formula of estradiol.

HIGHLIGHT PROBLEMS

118. Because of increasing evidence of damage to the ozone layer, chlorofluorocarbon (CFC) production was banned in 1996. However, there are about 100 million auto air conditioners that still use CFC-12 (CF_2Cl_2). These air conditioners are recharged from stockpiled supplies of CFC-12. If each of the 100 million automobiles contains 1.1 kg of CFC-12 and leaks 25 % of its CFC-12 into the atmosphere per year, how much Cl in kilograms is added to the atmosphere each year by auto air conditioners? (Assume two significant figures in your calculations.)

119. In 1996, the media reported that possible evidence of life on Mars was found on a meteorite called Allan Hills 84001 (AH 84001). The meteorite was discovered in Antarctica in 1984 and is believed to have originated on Mars. Elemental analysis of substances within its crevices revealed carbon-containing compounds that normally derive only from living organisms. Suppose that one of those compounds had a molar mass of 202.23 g/mol and the mass percent composition: C, 95.02 %; H, 4.98 %. What is the molecular formula for the carbon-containing compound?

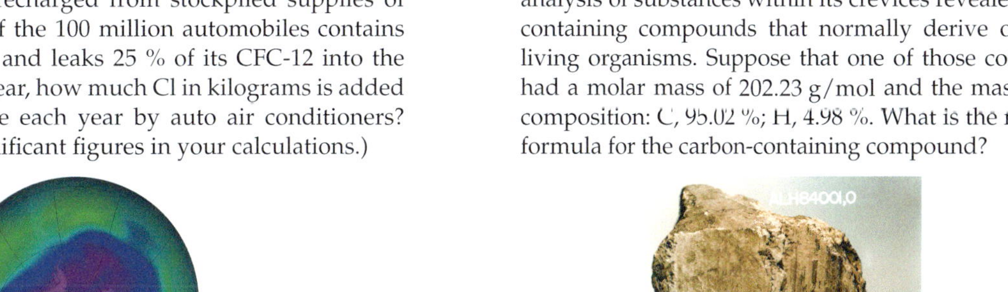

▲ The ozone hole over Antarctica on September 24, 2009. The dark blue and purple areas over the South Pole represent depressed ozone concentrations. © NASA Johnson Space Center.

▲ The Allan Hills 84001 meteorite. Elemental analysis of the substances within the crevices of this meteorite revealed carbon-containing compounds that normally originate from living organisms. © NASA Johnson Space Center.

▸Answers to Skillbuilder Exercises

Skillbuilder 5.1	5.32×10^{22} Au atoms	**Skillbuilder 5.8**	53.28 % O
Skillbuilder 5.2	89.1 g S	**Skillbuilder 5.9**	8.6 g Na
Skillbuilder 5.3	8.17 g He	**Skillbuilder 5.10**	CH_2O
Skillbuilder 5.4	2.56×10^{-2} mol NO_2	**Skillbuilder 5.11**	$C_{13}H_{18}O_2$
Skillbuilder 5.5	1.22×10^{23} H_2O molecules	**Skillbuilder 5.12**	CuO
Skillbuilder 5.6	5.6 mol O	**Skillbuilder 5.13**	C_4H_{10}
Skillbuilder 5.7	3.3 g O	**Skillbuilder Plus, p. 194**	$C_2H_8N_2$
Skillbuilder Plus, p. 185	4.04 g O		

▸Answers to Conceptual Checkpoints

5.1 (a) The mole is a counting unit; it represents a definite number (Avogadro's number, 6.022×10^{23}). Therefore, a given number of atoms always represents a precise number of moles, regardless of what atom is involved. Atoms of different elements have different masses. So if samples of different elements have the same mass, they *cannot* contain the same number of atoms or moles.

5.2 (b) Because carbon has lower molar mass than cobalt or lead, a one-gram sample of carbon contains more atoms than one gram of cobalt or lead.

5.3 Sample A would have the greatest number of molecules. Sample A has a lower molar mass than sample B, so a given mass of sample A has more moles and therefore more molecules than the same mass of sample B.

5.4 (c) 1.0 mole of F_2 contains 2.0 mol of F atoms. The other two options each contain less than two moles of F atoms.

5.5 (b) This compound has the highest ratio of oxygen atoms to chromium atoms and therefore has the greatest mass percent of oxygen.

▲ When we drink from a straw, we remove some of the molecules from inside the straw. This creates a pressure difference between the inside of the straw and the outside of the straw that results in the liquid being pushed up the straw. The pushing is done by molecules in the atmosphere—primarily nitrogen and oxygen—as shown here.

States of Matter 1 6

"The generality of men are so accustomed to judge of things by their senses that, because the air is invisible, they ascribe but little to it, and think it but one removed from nothing."—Robert Boyle (1627–1691)

MODULE OUTLINE

6.1 Kinetic Molecular Theory and Properties of Gases

LO: Describe how the main properties of a gas are predicted by kinetic molecular theory.

In prior modules we have seen the importance of models or theories in understanding nature. A simple model for understanding the behavior of gases is the **kinetic molecular theory**. This model predicts the correct behavior for most gases under many conditions. Like other models, the kinetic molecular theory is not perfect and breaks down under certain conditions. In this book, however, we focus on conditions where it works well.

Kinetic molecular theory makes the following assumptions (▶ Figure 6.1):

1. A gas is a collection of particles (molecules or atoms) in constant, straight-line motion.
2. Gas particles do not attract or repel each other—they do not interact. The particles collide with each other and with the surfaces around them, but they bounce back from these collisions like idealized billiard balls.
3. There is a lot of space between gas particles compared with the size of the particles themselves.
4. The average kinetic energy—energy due to motion—of gas particles is proportional to the temperature of the gas in kelvins. This means that as the temperature increases, the particles move faster and therefore have more energy.

Kinetic molecular theory is consistent with, and indeed predicts, the properties of gases. Recall from Section 1.2 that gases:

- are compressible
- assume the shape and volume of their container
- have low densities in comparison with liquids and solids

▶ **FIGURE 6.1 Simplified representation of an ideal gas** In reality, the spaces between the gas molecules would be much larger in relation to the size of the molecules than shown here.

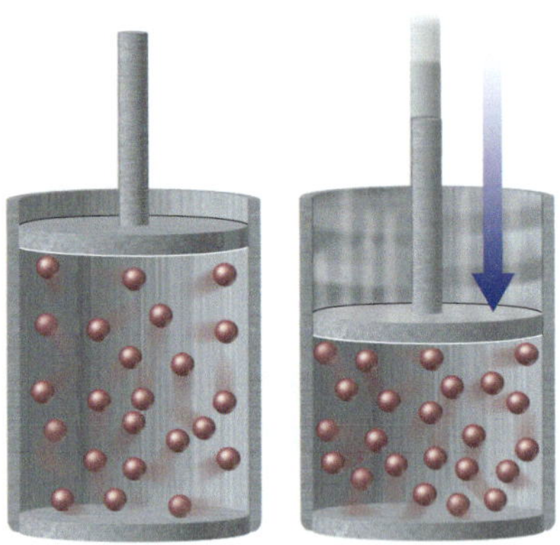

▲ **FIGURE 6.2 Compressibility of gases** Gases are compressible because there is so much empty space between gas particles.

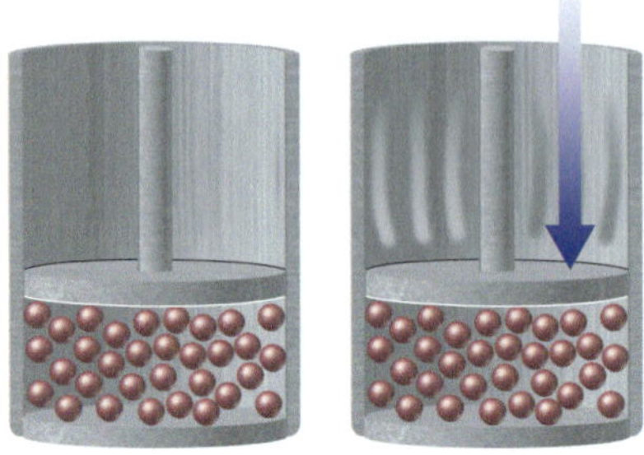

▲ **FIGURE 6.3 Incompressibility of liquids** Liquids are not compressible because there is so little space between the liquid particles.

Gases are compressible because the atoms or molecules that compose them have a lot of space between them. By applying external pressure to a gas sample, the atoms or molecules are forced closer together, compressing the gas. Liquids and solids, in contrast, are not compressible because the atoms or molecules composing them are already in close contact—they cannot be forced any closer together. We can witness the compressibility of a gas, for example, by pushing a piston into a cylinder containing a gas. The piston goes down (◀ Figure 6.2) in response to the external pressure. If the cylinder were filled with a liquid or a solid, the piston would not move when pushed (◀ Figure 6.3).

Gases assume the shape and volume of their container because gaseous atoms or molecules are in constant, straight-line motion. In contrast to a solid or liquid, whose atoms or molecules interact with one another, the atoms or molecules in a gas do not interact with one another (or more precisely, their interactions are negligible). They simply move in straight lines, colliding with each other and with the walls of their container. As a result, they fill the entire container, collectively assuming its shape (◀ Figure 6.4).

Gases have a low density in comparison with solids and liquids because there is so much empty space between the atoms or molecules in a gas. For example, if the water in a 350 mL can of soda were converted to steam (gaseous water), the steam would occupy a volume of 595 L (the equivalent of 1700 soda cans).

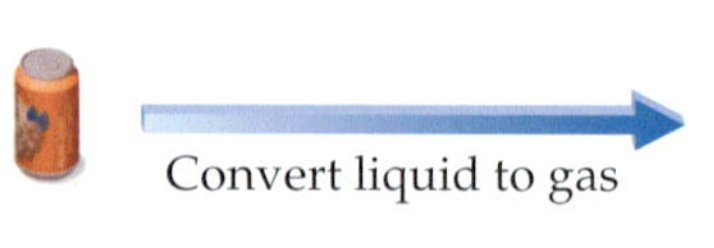

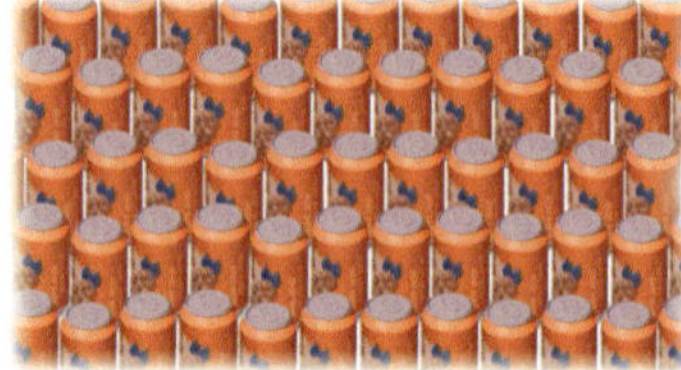

▲ If all of the water in a 350 mL can of orange soda were converted to gaseous steam (at 1 atm pressure and 100 °C), the steam would occupy a volume equal to 1700 soda cans.

◀ **FIGURE 6.4 A gas assumes the shape of its container** Since the attractions between molecules in a gas are negligible, and since the particles are in constant motion, a gas expands to fill the volume of its container.

6.2 Gas Pressure

LO: Identify and understand the relationship between pressure, force, and area.

LO: Convert among pressure units.

A prediction of kinetic molecular theory—is the very existence of pressure. **Pressure** is the result of the constant collisions between the atoms or molecules in a gas and the surfaces around them. Because of pressure, we can drink from straws, inflate basketballs, and move air into and out of our lungs. Variation in pressure in Earth's atmosphere creates wind, and changes in pressure help predict weather. Pressure is all around us and even inside us. The pressure exerted by a gas sample is defined as the force per unit area that results from the collisions of gas particles with surrounding surfaces.

$$\text{Pressure} = \frac{\text{Force}}{\text{Area}}$$

The pressure exerted by a gas depends on several factors, including the number of gas particles in a given volume (◀ Figure 6.5). The fewer the gas particles, the lower the pressure. Pressure decreases, for example, with increasing altitude. As we climb a mountain or ascend in an airplane, there are fewer molecules per unit volume in air and the pressure consequently drops. For this reason, most airplane cabins are artificially pressurized (see the *Everyday Chemistry* box on page 241.)

You may feel the effect of a drop in pressure as a pain in your ears. This pain is caused by air-containing cavities within your ear (▼ Figure 6.6). When you climb a mountain, for example, the external pressure (that pressure that surrounds you) drops while the pressure within your ear cavities (the internal pressure) remains the same. This creates an imbalance—the lower external pressure causes your eardrum to bulge outward, causing pain. With time and a yawn or two, the excess air within your ears' cavities escapes, equalizing the internal and external pressure and relieving the pain.

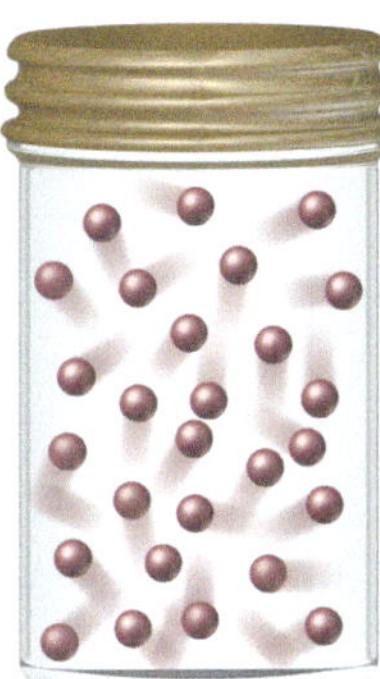

▲ **FIGURE 6.5 Pressure** Since pressure is a result of collisions between gas particles and the surfaces around them, the amount of pressure increases when the number of particles in a given volume increases (assuming constant temperature).

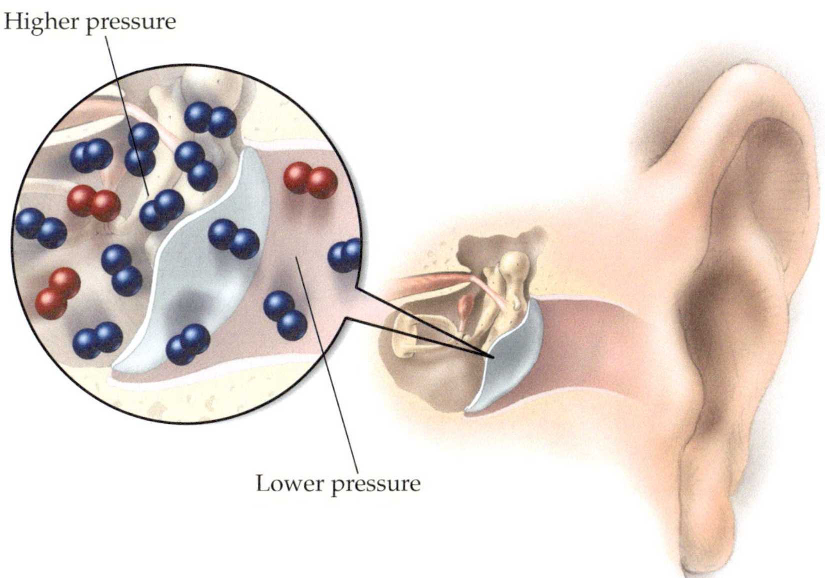

▲ **FIGURE 6.6 Pressure imbalance** The pain you feel in your ears upon climbing a mountain or ascending in an airplane is caused by an imbalance of pressure between the cavities inside your ear and the outside air.

Vacuum
Glass tube
760 mm (29.92 in.)
Atmospheric pressure
Mercury

▲ **FIGURE 6.7 The mercury barometer** Average atmospheric pressure at sea level pushes a column of mercury to a height of 760 mm. **Question: What happens to the height of the mercury column if the external pressure decreases? increases?**

Since mercury is 13.5 times as dense as water, it is pushed up 1/13.5 times as high as water by atmospheric pressure.

CONCEPTUAL CHECKPOINT 6.1

Which sample of an ideal gas has the *lowest* pressure? Assume that all of the particles are identical and that the three samples are at the same temperature.

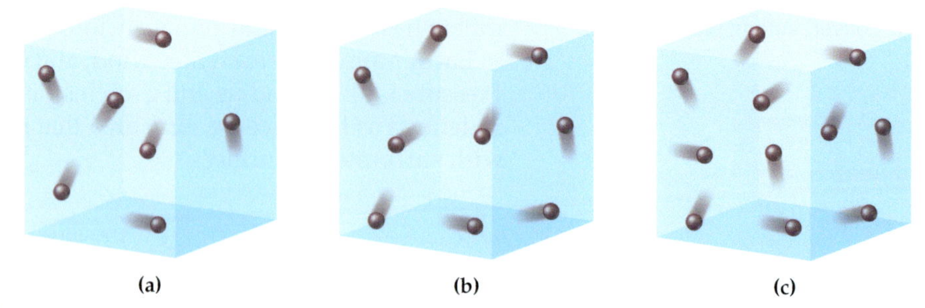

Pressure Units

The simplest unit of pressure is the **atmosphere (atm)**, the average pressure at sea level. The SI unit of pressure is the **pascal (Pa)**, defined as 1 newton (N) per square metre.

$$1 \text{ Pa} = 1 \text{ N/m}^2$$

The pascal is a much smaller unit of pressure; 1 atm is equal to 101,325 Pa.

$$1 \text{ atm} = 101{,}325 \text{ Pa}$$

The pressure in a fully inflated mountain bike tire is about 6 atm, and the pressure on top of Mount Everest is about 0.31 atm.

A third unit of pressure, the **millimetre of mercury (mm Hg)**, originates from how pressure is measured with a barometer (◄ Figure 6.7). A barometer is an evacuated glass tube whose tip is submerged in a pool of mercury. Liquid is pushed up an evacuated tube by the atmospheric pressure on the liquid's surface. We learned that water is pushed up to a height of 10.3 m by the average pressure at sea level. Mercury, however, with its higher density, is pushed up to a height of only 0.760 m, or 760 mm, by the average pressure at sea level. This shorter length—0.760 m instead of 10.3 m—makes a column of mercury a convenient way to measure pressure.

In a barometer, the mercury column rises or falls with changes in atmospheric pressure. If the pressure increases, the level of mercury within the column rises. If the pressure decreases, the level of mercury within the column falls. Since 1 atm of pressure pushes a column of mercury to a height of 760 mm, 1 atm and 760 mm Hg are equal.

$$1 \text{ atm} = 760 \text{ mm Hg}$$

A millimetre of mercury is also called a **torr**, after Italian physicist Evangelista Torricelli (1608–1647), who invented the barometer.

$$1 \text{ mm Hg} = 1 \text{ torr}$$

Table 6.1 lists the common units of pressure.

TABLE 6.1 Common Units of Pressure

Unit	Average Air Pressure at Sea Level
pascal (Pa)	101,325 Pa
atmosphere (atm)	1 atm
millimetre of mercury (mm Hg)	760 mm Hg (exact)
torr (torr)	760 torr (exact)

CONCEPTUAL CHECKPOINT 6.2

A liquid that is about twice as dense as water is used in a barometre. With this barometer, normal atmospheric pressure would be about:

(a) 0.38 m **(b)** 1.52 m **(c)** 5.15 m **(d)** 20.6 m

Pressure Unit Conversion

See Section 4.5 for a review on converting between units.

We convert one pressure unit to another in the same way that we converted between other units in Module 4. For example, suppose we want to convert 0.311 atm (the approximate average pressure at the top of Mount Everest) to millimetres of mercury. We begin by sorting the information in the problem statement.

GIVEN: 0.311 atm

FIND: mm Hg

SOLUTION MAP

We then strategize by building a solution map that shows how to convert from atm to mm Hg.

RELATIONSHIPS USED

1 atm = 760 mm Hg (from Table 6.1)

SOLUTION

The solution begins with the given value (0.311 atm) and converts it to mm Hg.

$$0.311\ \text{atm} \times \frac{760\ \text{mm Hg}}{1\ \text{atm}} = 236\ \text{mm Hg}$$

EXAMPLE 6.1 CONVERTING BETWEEN PRESSURE UNITS

A high-performance road bicycle tire is inflated to a total pressure of 8.50 atm. What is this pressure in millimetres of mercury?

SORT You are given a pressure in atm and asked to convert it to mm Hg.	**GIVEN:** 8.50 atm **FIND:** mm Hg
STRATEGIZE Begin the solution map with the given units of atm. Use the conversion factor to convert atm to mm Hg.	**SOLUTION MAP** atm → mm Hg **RELATIONSHIPS USED** 760 mm Hg = 1 atm (Table 6.1)
SOLVE Follow the solution map to solve the problem.	**SOLUTION** $8.50\ \text{atm} \times \frac{760\ \text{mm Hg}}{1\ \text{atm}} = 6.46 \times 10^3\ \text{mm Hg}$

CHECK Check your answer. Are the units correct? Does the answer make physical sense?	The answer has the correct units, mm Hg. The answer is reasonable because the mm Hg is a smaller unit than atm; therefore the value of the pressure in mm Hg should be greater than the value of the same pressure in atm.

▶**FOR MORE PRACTICE** Example 6.14, Problems 36, 37.

6.3 Boyle's Law: Pressure and Volume

LO: Restate and apply Boyle's law.

The pressure of a gas sample depends, in part, on its volume. If the temperature and the amount of gas are constant, the pressure of a gas sample *increases* for a *decrease* in volume and *decreases* for an *increase* in volume. A simple hand pump, for example, works on this principle. A hand pump is basically a cylinder equipped with a moveable piston (▼ Figure 6.8). The volume in the cylinder increases when you pull the handle up (the upstroke) and decreases when you push the handle down (the downstroke). On the upstroke, the *increasing* volume causes a *decrease* in the internal pressure (the pressure within the pump's cylinder). This, in turn, draws air into the pump's cylinder through a one-way valve. On the downstroke, the *decreasing* volume causes an *increase* in the internal pressure. This increase forces the air out of the pump, through a different one-way valve, and into the tire or whatever else is being inflated.

The relationships between gas properties—such as the relationship between pressure and volume—are described by gas laws. These laws show how a change in one of these properties affects one or more of the others. The relationship between volume and pressure was discovered by Robert Boyle (1627–1691) and is called **Boyle's law**.

Boyle's law assumes constant temperature and a constant number of gas particles.

Boyle's law: The volume of a gas and its pressure are inversely proportional.

$$V \propto \frac{1}{P}$$

$\propto$ means "proportional to"

If two quantities are inversely proportional, increasing one decreases the other (▶ Figure 6.9). As we saw for the hand pump, when the volume of a gas sample is decreased, its pressure increases and vice versa. Kinetic molecular theory explains the observed change in pressure. If the volume of a gas sample is decreased, the same number of gas particles is crowded into a smaller volume, causing more collisions with the walls of the container and therefore increasing the pressure (▶ Figure 6.10).

Scuba divers learn about Boyle's law during certification courses because it explains why ascending too quickly toward the surface is dangerous. For every 10 m of depth that a diver descends in water, he experiences an additional 1 atm of pressure due to the weight of the water above him. The pressure regulator used in scuba diving delivers air at a pressure that matches the external pressure;

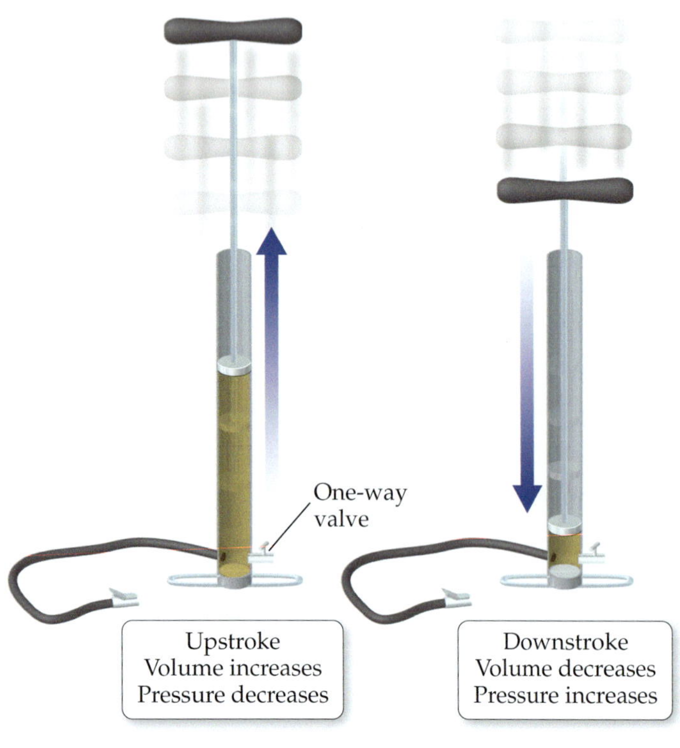

◀ **FIGURE 6.8 Operation of a hand pump**

EVERYDAY CHEMISTRY

▶ Airplane Cabin Pressurization

Most commercial airplanes fly at elevations between 7700 m and 15,000 m. At these elevations, atmospheric pressure is below 0.50 atm, much less than the normal atmospheric pressure to which our bodies are accustomed. The physiological effects of these lowered pressures—and the correspondingly lowered oxygen levels (see Section 6.9)—include dizziness, headache, shortness of breath, and even unconsciousness. Consequently, commercial airplanes pressurize the air in their cabins. If, for some reason, an airplane cabin should lose its pressurization, passengers are directed to breathe oxygen through an oxygen mask.

Cabin air pressurization is accomplished as part of the cabin's overall air circulation system. As air flows into the plane's jet engines, the large turbines at the front of the engines compress it. Most of this compressed (or pressurized) air exits out the back of the engines, creating the thrust that drives the plane forward. However, some of the pressurized air is directed into the cabin, where it is cooled and mixed with existing cabin air. This air is then circulated through the cabin through the overhead vents. The air leaves the cabin through ducts that direct it into the lower portion of the airplane. About half of this exiting air is mixed with incoming, pressurized air to circulate again. The other half is vented out of the plane through an outflow valve. This valve is adjusted to maintain the desired cabin pressure. Federal regulations require that cabin pressure in commercial airliners be greater than the equivalent of outside air pressure at 2500 m. Some newer jets, such as the Boeing 787 Dreamliner, pressurize their cabins at an equivalent pressure of 1800 m for greater passenger comfort.

B6.1 CAN YOU ANSWER THIS? *Atmospheric pressure at elevations of 2400 m averages about 0.72 atm. Convert this pressure to millimetres of mercury. Would a cabin pressurized at 500 mm Hg meet federal standards?*

◀ Commercial airplane cabins must be pressurized to a pressure greater than the equivalent atmospheric pressure at an elevation of 2500 m. © DriendL/Photodisc/Getty Images.

▶ **FIGURE 6.9 Volume versus pressure** **(a)** We can use a J-tube to measure the volume of a gas at different pressures. Adding mercury to the J-tube causes the pressure on the gas sample to increase and its volume to decrease. **(b)** A plot of the volume of a gas as a function of pressure.

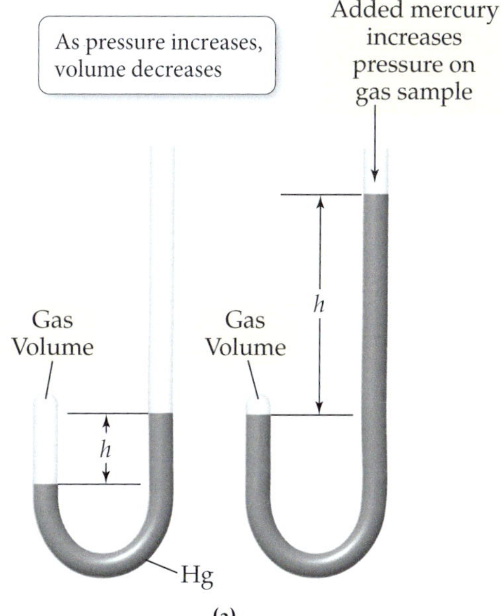

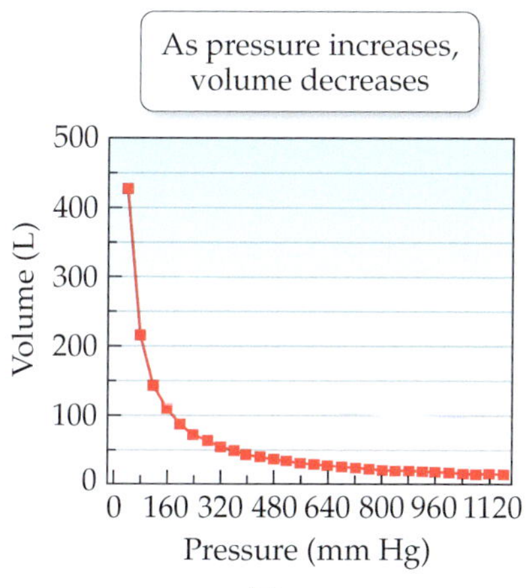

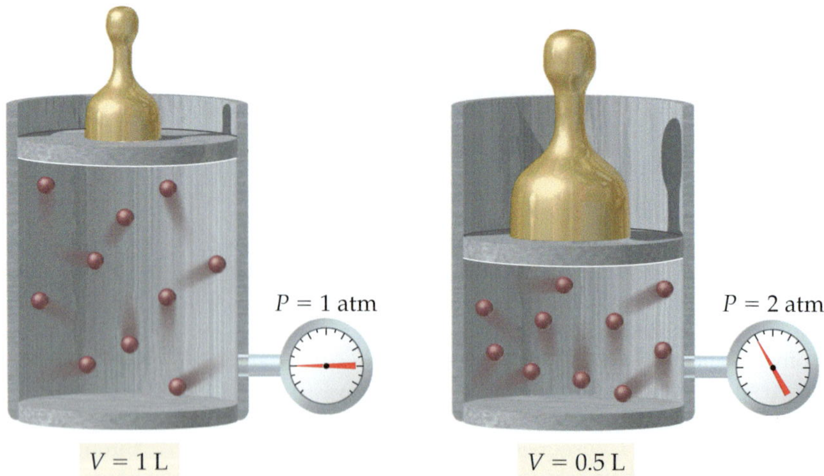

▶ **FIGURE 6.10 Volume versus pressure: a molecular view** As the volume of a sample of gas is decreased, the number of collisions between the gas molecules and each square metre of the container increases. This raises the pressure exerted by the gas.

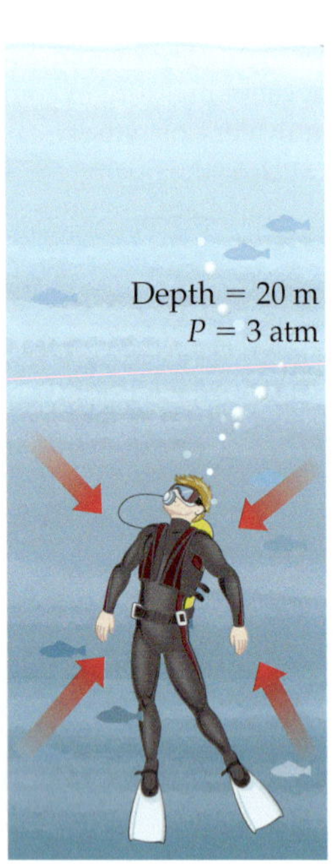

◀ **FIGURE 6.11 Pressure at depth** For every 10 m of depth, a diver experiences an additional 1 atm of pressure due to the weight of the water surrounding him. At 20 m, the diver experiences a total pressure of 3 atm (1 atm from atmospheric pressure plus an additional 2 atm from the weight of the water).

otherwise the diver could not inhale the air (see the *Everyday Chemistry* box on page 221). For example, when a diver is at 20 m of depth, the regulator delivers air at a pressure of 3 atm to match the 3 atm of pressure around the diver—1 atm due to normal atmospheric pressure and 2 additional atmospheres due to the weight of the water at 20 m (◀ Figure 6.11).

Suppose that a diver inhales a lungful of 3 atm air and swims quickly to the surface (where the pressure drops to 1 atm) while holding his breath. What happens to the volume of air in his lungs? Since the pressure decreases by a factor of 3, the volume of the air in his lungs increases by a factor of 3, severely damaging his lungs and possibly killing him (▶ Figure 6.12). The volume increase in the diver's lungs would be so great that the diver would not be able to hold his breath all the way to the surface—the air would force itself out of his mouth. So the most important rule in diving is *never hold your breath.* Divers must ascend slowly and breathe continuously, allowing the regulator to bring the air pressure in their lungs back to 1 atm by the time they reach the surface.

We can use Boyle's law to calculate the volume of a gas following a pressure change or the pressure of a gas following a volume change *as long as the temperature and the amount of gas remain constant.* For these calculations, we write Boyle's law in a slightly different way.

$$\text{Since } V \propto \frac{1}{P}\text{, then } V = \frac{\text{Constant}}{P}$$

If we multiply both sides by P, we get:

$$PV = \text{Constant}$$

This relationship is true because if the pressure increases, the volume decreases, but the product $P \times V$ is always equal to the same constant. For two different sets of conditions, we can say that

$$P_1V_1 = \text{Constant} = P_2V_2\text{, or}$$

$$P_1V_1 = P_2V_2$$

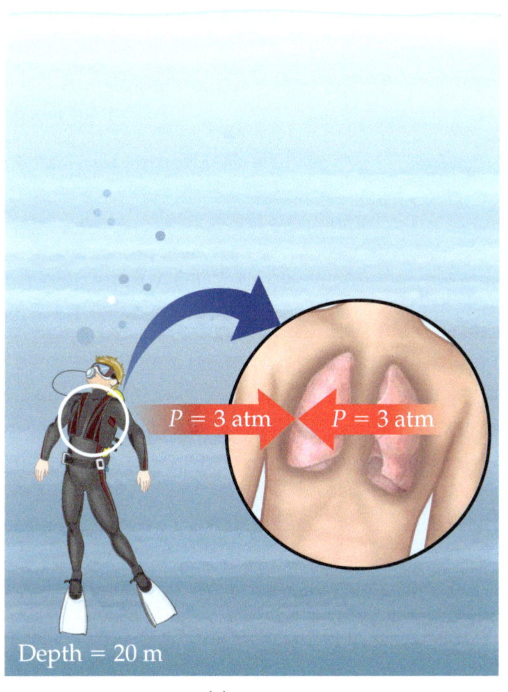

(a)

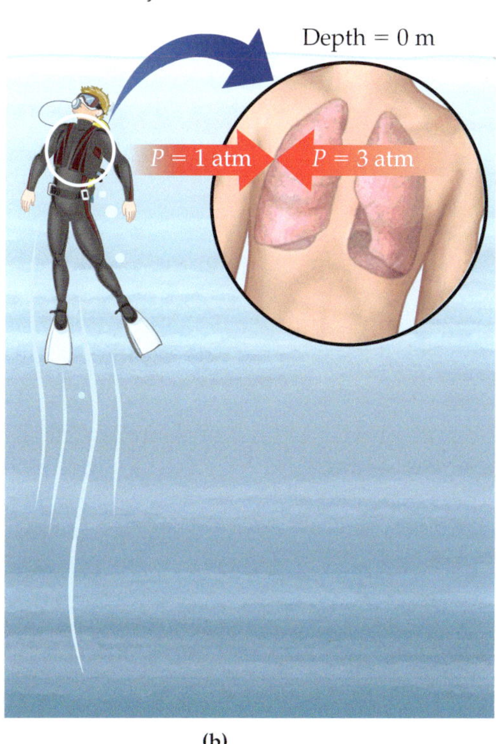

(b)

▶ **FIGURE 6.12**
The dangers of decompression **(a)** A diver at 20 m experiences an external pressure of 3 atm and breathes air pressurized at 3 atm. **(b)** If the diver shoots toward the surface with lungs full of 3 atm air, his lungs will expand as the external pressure drops to 1 atm.

where P_1 and V_1 are the initial pressure and volume of the gas, and P_2 and V_2 are the final volume and pressure. For example, suppose we want to calculate the pressure of a gas that was initially at 765 mm Hg and 1.78 L and later compressed to 1.25 L. We first sort the information in the problem.

Based on Boyle's law, and before doing any calculations, do you expect P_2 to be greater than or less than P_1?

GIVEN: $P_1 = 765$ mm Hg
$V_1 = 1.78$ L
$V_2 = 1.25$ L

FIND: P_2

SOLUTION MAP

We then draw a solution map showing how the equation takes us from the given quantities (what we have) to the find quantity (what we want to find).

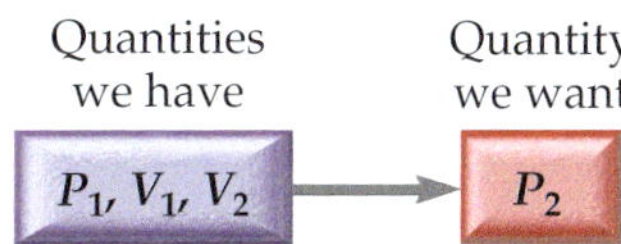

RELATIONSHIPS USED

$P_1V_1 = P_2V_2$ (Boyle's law, presented in this section)

SOLUTION

We then solve the equation for the quantity we are trying to find (P_2).

$$P_1V_1 = P_2V_2$$

$$P_2 = \frac{P_1V_1}{V_2}$$

Lastly, we substitute the numerical values into the equation and calculate the answer.

$$P_2 = \frac{P_1V_1}{V_2} = \frac{765 \text{ mm Hg} \times 1.78 \cancel{\text{L}}}{1.25 \cancel{\text{L}}}$$
$$= 1.09 \times 10^3 \text{ mm Hg}$$

EXAMPLE 6.2 BOYLE'S LAW

A cylinder equipped with a moveable piston has an applied pressure of 4.0 atm and a volume of 6.0 L. What is the volume of the cylinder if the applied pressure is decreased to 1.0 atm?

SORT
You are given an initial pressure, an initial volume, and a final pressure. You are asked to find the final volume.

GIVEN: $P_1 = 4.0$ atm
$V_1 = 6.0$ L
$P_2 = 1.0$ atm

FIND: V_2

STRATEGIZE
Draw a solution map beginning with the given quantities. Boyle's law shows the relationship necessary to get to the find quantity.

SOLUTION MAP

RELATIONSHIPS USED
$P_1V_1 = P_2V_2$ (Boyle's law, presented in this section)

SOLVE
Solve the equation for the quantity you are trying to find (V_2), and substitute the numerical quantities into the equation to calculate the answer.

SOLUTION
$$P_1V_1 = P_2V_2$$
$$V_2 = \frac{P_1V_1}{P_2}$$
$$= \frac{4.0 \cancel{\text{atm}} \times 6.0 \text{ L}}{1.0 \cancel{\text{atm}}}$$
$$= 24 \text{ L}$$

CHECK
Check your answer. Are the units correct? Does the answer make physical sense?

The answer has a unit of volume (L) as expected. The answer is reasonable because you expect the volume to increase as the pressure decreases.

▶SKILLBUILDER 6.1 | Boyle's Law

A snorkeler takes a syringe filled with 16 mL of air from the surface, where the pressure is 1.0 atm, to an unknown depth. The volume of the air in the syringe at this depth is 7.5 mL. What is the pressure at this depth? If pressure increases by an additional 1 atm for every 10 m of depth, how deep is the snorkeler?

▶FOR MORE PRACTICE Example 6.15; Problems 42, 43, 44, 45.

CONCEPTUAL CHECKPOINT 6.3

A flask contains a gas sample at pressure x. If the volume of the container triples at constant temperature and constant amount of gas, the pressure becomes:

(a) $3x$ **(b)** $\frac{1}{3}x$ **(c)** $9x$

EVERYDAY CHEMISTRY

▶ Extra-long Snorkels

Several episodes of *The Flintstones* featured Fred Flintstone and Barney Rubble snorkeling. Their snorkels, however, were not the modern kind, but long reeds that stretched from the surface of the water down to many metres of depth. Fred and Barney swam around in deep water while breathing air provided to them by these extra-long snorkels. Would this work? Why do people bother with scuba diving equipment if they could simply use 10 m snorkels like Fred and Barney did?

When we breathe, we expand the volume of our lungs, lowering the pressure within them (Boyle's law). Air from outside our lungs then flows into them. Extra-long snorkels, such as those used by Fred and Barney, do not work because of the pressure caused by water at depth. A diver at 10 m experiences a pressure of 2 atm that compresses the air in his lungs to a pressure of 2 atm. If the diver had a snorkel that went to the surface—where the air pressure is 1 atm—air would flow out of his lungs, not into them. It would be impossible to breathe.

B6.2 CAN YOU ANSWER THIS? *Suppose a diver takes a balloon with a volume of 2.5 L from the surface, where the pressure is 1.0 atm, to a depth of 20 m, where the pressure is 3.0 atm. What would happen to the volume of the balloon? What if the end of the balloon was on a long tube that went to the surface and was attached to another balloon, as shown in the drawing below? Which way would air flow as the diver descends?*

◀ Fred and Barney used reeds to breathe air from the surface, even when they were at depth. This would not work because the pressure at depth would push air out of their lungs, preventing them from breathing.

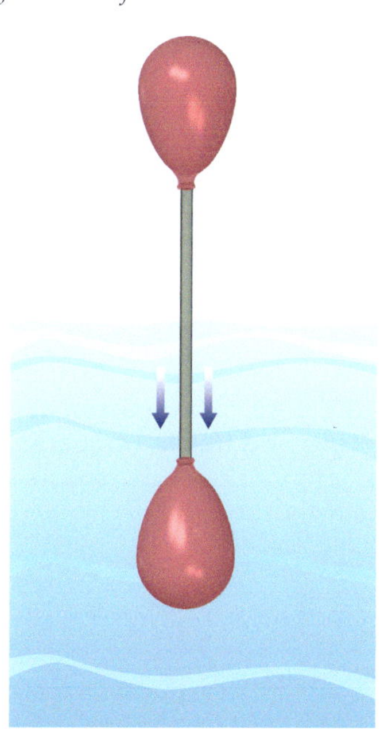

▶ If one end of a long tube with balloons tied on both ends is submerged in water, in which direction does air flow?

6.4 Temperature and the Kelvin Scale

LO: Convert between Celsius, and Kelvin temperature scales.

The atoms and molecules that compose matter are in constant random motion—they contain *thermal energy*. The **temperature** of a substance is a measure of its thermal energy. The hotter an object, the greater the random motion of the atoms and molecules that compose it, and the higher its temperature. We must be careful to not confuse *temperature* with *heat*. **Heat**, which has units of energy, is the *transfer* or *exchange* of thermal energy caused by a temperature difference. For example, when a cold ice cube is dropped into a warm cup of water, energy is transferred as heat from the water to the ice, resulting in the cooling of the water. Temperature, by contrast, is a *measure* of the thermal energy of matter (not the exchange of thermal energy).

Two different temperature scales are in common use. The scale scientists use is the **Celsius (°C) scale**. On this scale, water freezes at 0 °C and boils at 100 °C. Room temperature is approximately 22 °C.

The Celsius scale contains negative temperatures. A second temperature scale, called the **Kelvin (K) scale**, avoids negative temperatures by assigning 0 K to the coldest temperature possible, absolute zero. Absolute zero (–273.15 °C) is the temperature at which molecular motion virtually stops. There is no lower temperature. The Kelvin degree, or Kelvin (K), is the same size as the Celsius degree—the only difference is the temperature that each scale designates as zero (▼ Figure 6.13).

We can convert between these temperature scales using the following formulas.

$$K = °C + 273.15$$

For example, suppose we want to convert 212 K to Celsius. Following the procedure for solving numerical problems (Section 4.5), we first sort the information in the problem statement:

The degree symbol is used with the Celsius scale, but not with the Kelvin scale.

GIVEN: 212 K

FIND: °C

We then strategize by building a solution map.

In a solution map involving a formula, the formula establishes the relationship between the variables. However, the formula under the arrow is not necessarily solved for the correct variable until later, as is the case here.

SOLUTION MAP

K → °C

▶ **FIGURE 6.13 Comparison of the Celsius and Kelvin temperature scales** The Celsius degree and the Kelvin degree are the same size.

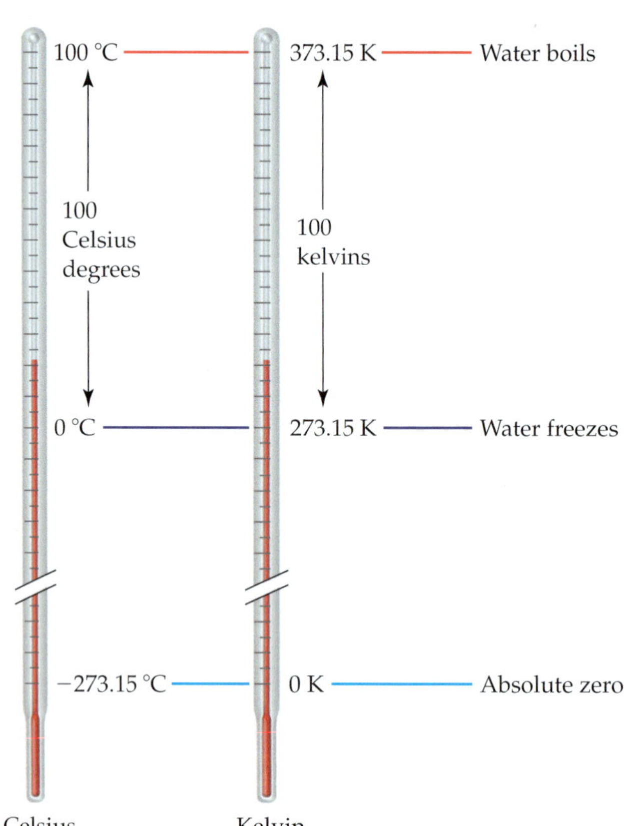

RELATIONSHIPS USED

K = °C + 273.15 (This equation relates the *given* quantity (K) to the *find* quantity (°C) and is given in this section.)

SOLUTION

Finally, we follow the solution map to solve the problem. The equation given shows the relationship between K and °C, but it is not solved for the correct variable. Before using the equation, we must solve it for °C.

$$K = °C + 273.15$$

$$°C = K - 273.15$$

We can now substitute the given value for K and calculate the answer to the correct number of significant figures (see Section 4.3).

$$°C = 212 - 273.15$$

$$= -61\ °C$$

EXAMPLE 6.3 CONVERTING BETWEEN CELSIUS AND KELVIN TEMPERATURE SCALES

Convert –25 °C to kelvins.

SORT You are given a temperature in degrees Celsius and asked to find the value of the temperature in kelvins.	**GIVEN:** –25 °C **FIND:** K
STRATEGIZE Draw a solution map. Use the equation that relates the temperature in kelvins to the temperature in Celsius to convert from the given quantity to the quantity you want to find.	**SOLUTION MAP** °C → K **RELATIONSHIPS USED** K = °C + 273.15 (presented in this section)
SOLVE Follow the solution map by substituting the correct value for °C and calculating the answer to the correct number of significant figures (see Section 4.3).	**SOLUTION** K = °C + 273.15 K = –25 °C + 273.15 = 248 K The significant figures are limited to the ones place because the given temperature (–25 °C) is only given to the ones place. Remember that for addition and subtraction, the number of *decimal places* in the least precisely known quantity determines the number of *decimal places* in the result.
CHECK Check your answer. Are the units correct? Does the answer make physical sense?	The unit (K) is correct. The answer makes sense because the value in kelvins should be a more positive number than the value in degrees Celsius.

▶SKILLBUILDER 6.2 | Converting between Celsius and Kelvin Temperature Scales

Convert 358 K to Celsius.

▶FOR MORE PRACTICE Example 6.16; Problems 38, 39.

6.5 Charles's Law: Volume and Temperature

LO: Restate and apply Charles's law.

Recall from Section 4.6 that Density = Mass/Volume. If the volume increases and the mass remains constant, the density must decrease.

Have you ever noticed that hot air rises? You may have walked upstairs in your house and noticed it getting warmer. Or you may have witnessed a hot-air balloon take flight. The air that fills a hot-air balloon is warmed with a burner, which causes the balloon to rise in the cooler air around it. Why does hot air rise? Hot air rises because the volume of a gas sample at constant pressure increases with increasing temperature. As long as the amount of gas (and therefore its mass) remains constant, warming decreases its density because density is mass divided by volume. A lower-density gas floats in a higher-density gas just as wood floats in water.

Suppose we keep the pressure of a gas sample constant and measure its volume at a number of different temperatures. The results of a number of such measurements are shown in ▼ Figure 6.14. The plot reveals the relationship between volume and temperature: The volume of a gas increases with increasing temperature. Note also that temperature and volume are *linearly related.* If two variables are linearly related, plotting one against the other produces a straight line.

▲ Heating the air in a balloon makes it expand (Charles's law). As the volume occupied by the hot air increases, its density decreases, allowing the balloon to float in the cooler, denser air that surrounds it. © Larry Brownstein/Photodisc/Getty Images.

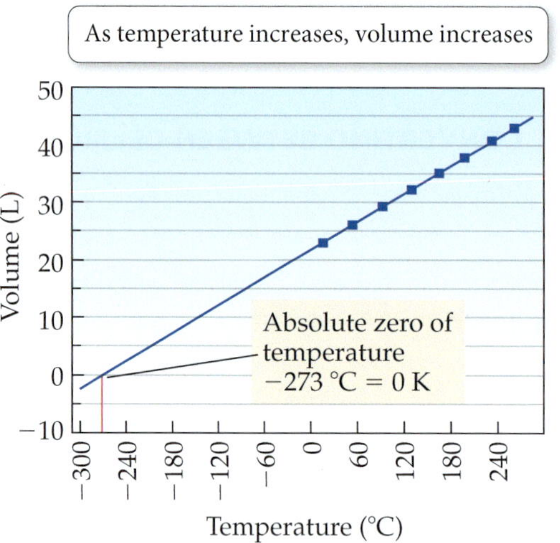

▲ **FIGURE 6.14 Volume versus temperature** The volume of a gas increases linearly with increasing temperature. **Question: How does this graph demonstrate that –273 °C is the coldest possible temperature?**

The extrapolated line cannot be measured experimentally because all gases condense into liquids before –273 °C is reached.

Section 6.4 summarizes the two different temperature scales.

We can predict an important property of matter by extending the line on our plot backward from the lowest measured point—a process called *extrapolation.* Our extrapolated line shows that the gas should have a zero volume at –273 °C. Recall that –273 °C corresponds to 0 K, the coldest possible temperature. Our extrapolated line shows that below –273 °C, our gas would have a negative volume, which is physically impossible. For this reason, we refer to 0 K as **absolute zero**—colder temperatures do not exist.

The first person to carefully quantify the relationship between the volume of a gas and its temperature was J. A. C. Charles (1746–1823), a French mathematician and physicist. Charles was interested in gases and was among the first people to ascend in a hydrogen-filled balloon. The law he formulated is called **Charles's law**.

Charles's law assumes constant pressure and a constant amount of gas.

Charles's law: The volume (V) of a gas and its Kelvin temperature (T) are directly proportional.

$$V \propto T$$

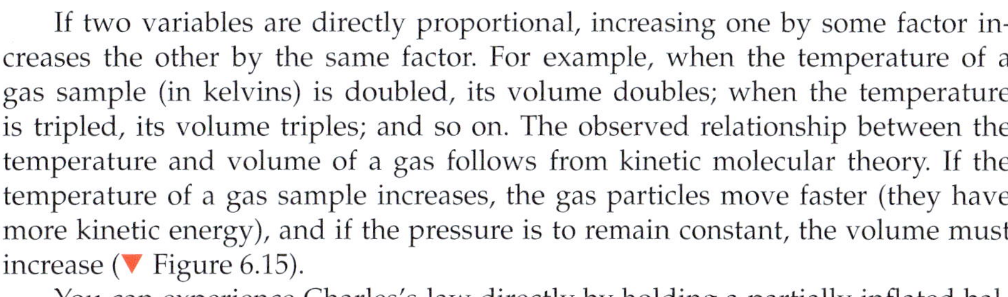

If two variables are directly proportional, increasing one by some factor increases the other by the same factor. For example, when the temperature of a gas sample (in kelvins) is doubled, its volume doubles; when the temperature is tripled, its volume triples; and so on. The observed relationship between the temperature and volume of a gas follows from kinetic molecular theory. If the temperature of a gas sample increases, the gas particles move faster (they have more kinetic energy), and if the pressure is to remain constant, the volume must increase (▼ Figure 6.15).

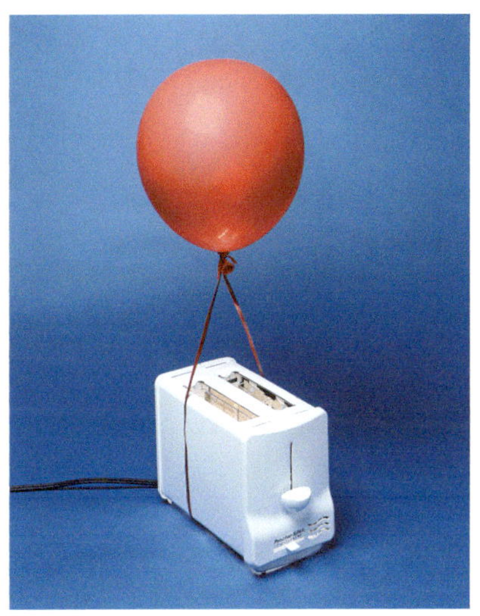

▲ If you hold a partially inflated balloon over a warm toaster, the balloon will expand as the air within the balloon warms. © Richard Megna/Fundamental Photographs.

You can experience Charles's law directly by holding a partially inflated balloon over a warm toaster. As the air in the balloon warms, you can feel the balloon expanding. Alternatively, you can put an inflated balloon in the freezer or take it outside on a very cold day (below freezing) and see that it becomes smaller as it cools.

We can use Charles's law to calculate the volume of a gas following a temperature change or the temperature of a gas following a volume change as *long as the pressure and the amount of gas are constant.* For these calculations, we express Charles's law in a different way.

$$\text{Since } V \propto T\text{, then } V = \text{Constant} \times T$$

If we divide both sides by T, we get:

$$\frac{V}{T} = \text{Constant}$$

If the temperature increases, the volume increases in direct proportion so that the quotient, V/T, is always equal to the same constant. So, for two different measurements, we can say that

$$\frac{V_1}{T_1} = \text{Constant} = \frac{V_2}{T_2}\text{, or}$$

$$\frac{V_1}{T_1} = \frac{V_2}{T_2}$$

As temperature increases, the volume of the balloon increases

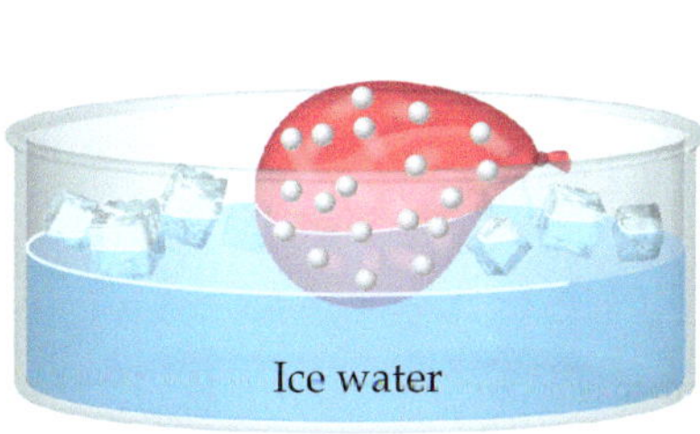

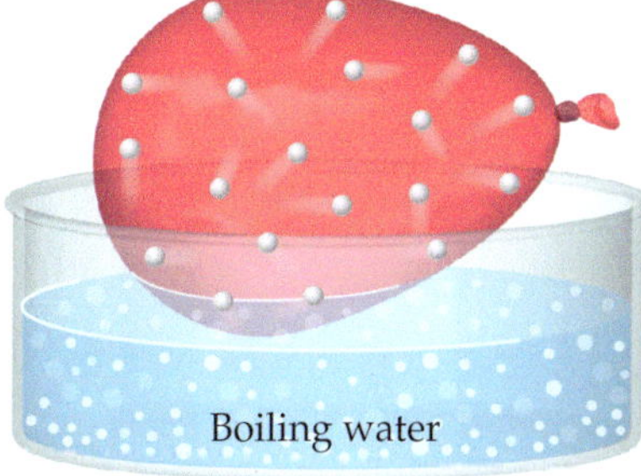

▲ **FIGURE 6.15 Volume versus temperature: a molecular view** If a balloon is moved from an ice-water bath into a boiling-water bath, the gas molecules inside it move faster (they have more kinetic energy) due to the increased temperature. If the external pressure remains constant, the molecules will expand the balloon and collectively occupy a larger volume.

where V_1 and T_1 are the initial volume and temperature of the gas and V_2 and T_2 are the final volume and temperature. *All temperatures must be expressed in kelvins.*

For example, suppose we have a 2.37 L sample of a gas at 298 K that is then heated to 354 K with no change in pressure. To determine the final volume of the gas, we begin by sorting the information in the problem statement.

Based on Charles's law, and before doing any calculations, do you expect V_2 to be greater than or less than V_1?

GIVEN: $T_1 = 298\text{ K}$

$V_1 = 2.37\text{ L}$

$T_2 = 354\text{ K}$

FIND: V_2

SOLUTION MAP

We then strategize by building a solution map that shows how the equation takes us from the given quantities to the unknown quantity.

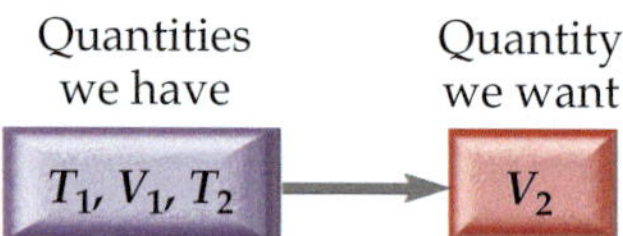

RELATIONSHIPS USED

$$\frac{V_1}{T_1} = \frac{V_2}{T_2} \text{ (Charles's law, presented in this section)}$$

SOLUTION

We then solve the equation for the quantity we are trying to find (V_2).

$$\frac{V_1}{T_1} = \frac{V_2}{T_2}$$

$$V_2 = \frac{V_1}{T_1}T_2$$

Lastly, we substitute the numerical values into the equation and calculate the answer.

$$V_2 = \frac{V_1}{T_1}T_2$$

$$= \frac{2.37\text{ L}}{298\text{ K}} \times 354\text{ K}$$

$$= 2.82\text{ L}$$

EXAMPLE 6.4 CHARLES'S LAW

A sample of gas has a volume of 2.80 L at an unknown temperature. When the sample is submerged in ice water at $t = 0$ °C, its volume decreases to 2.57 L. What was its initial temperature (in kelvins and in Celsius)? Assume a constant pressure. (To distinguish between the two temperature scales, use t for temperature in °C and T for temperature in K.)

SORT
You are given an initial volume, a final volume, and a final temperature. You are asked to find the intitial temperature in both kelvins (T_1) and degrees Celsius (t_1).

GIVEN: $V_1 = 2.80$ L

$V_2 = 2.57$ L

$t_2 = 0$ °C

FIND: T_1 and t_1

STRATEGIZE
Draw a solution map beginning with the given quantities. Charles's law shows the relationship necessary to get to the find quantity.

SOLUTION MAP

RELATIONSHIPS USED

$$\frac{V_1}{T_1} = \frac{V_2}{T_2}$$ (Charles's law, presented in this section)

SOLVE
Solve the equation for the quantity you are trying to find (T_1).
Before you substitute in the numerical values, you need to convert the temperature to kelvins. *Remember, you must always work gas law problems using Kelvin temperatures.* Once you have converted the temperature to kelvins, substitute into the equation to find T_1. Convert the temperature to degrees Celsius to find t_1.

SOLUTION

$$\frac{V_1}{T_1} = \frac{V_2}{T_2}$$

$$T_1 = \frac{V_1}{V_2} T_2$$

$$T_2 = 0 + 273 = 273 \text{ K}$$ (To simplify the calculation, 273 can be used instead of 273.15)

$$T_1 = \frac{V_1}{V_2} T_2 = \frac{2.80 \cancel{\text{L}}}{2.57 \cancel{\text{L}}} \times 273 \text{ K} = 297 \text{ K}$$

$$t_1 = 297 - 273 = 24 \text{ °C}$$

CHECK
Check your answer. Are the units correct? Does the answer make physical sense?

The answers have the correct units, K and °C. The answer is reasonable because the initial volume was larger than the final volume; therefore the initial temperature must be higher than the final temperature.

▶SKILLBUILDER 6.3 | Charles's Law

A gas in a cylinder with a moveable piston has an initial volume of 88.2 mL and is heated from 35 °C to 155 °C. What is the final volume of the gas in millilitres?

▶FOR MORE PRACTICE Problems 48, 49, 50, 51.

CONCEPTUAL CHECKPOINT 6.4

A volume of gas is confined to a cylinder with a freely moveable piston at one end. If you apply enough heat to double the Kelvin temperature of the gas, what happens? (Assume constant pressure.)

(a) the volume doubles

(b) the volume remains the same

(c) the volume falls to half of the initial volume

6.6 The Combined Gas Law: Pressure, Volume, and Temperature

LO: Restate and apply the combined gas law.

Boyle's law shows how P and V are related at constant temperature, and Charles's law shows how V and T are related at constant pressure. But what if two of these variables change at once? For example, what happens to the volume of a gas if both its pressure and its temperature are changed?

Since volume is inversely proportional to pressure ($V \propto 1/P$) and directly proportional to temperature ($V \propto T$), we can write:

$$V \propto \frac{T}{P} \quad \text{or} \quad \frac{PV}{T} = \text{Constant}$$

For a sample of gas under two different sets of conditions we use the **combined gas law**.

The combined gas law encompasses both Boyle's law and Charles's law, and we can use it in place of them. If one physical property (P, V, or T) is constant, it cancels out of our calculations when we use the combined gas law.

The combined gas law: $$\frac{P_1V_1}{T_1} = \frac{P_2V_2}{T_2}$$

The combined gas law applies only when the *amount* of gas is constant. We must express the temperature (as with Charles's law) in kelvins.

Suppose a person carries a cylinder with a moveable piston that has an initial volume of 3.65 L up a mountain. The pressure at the bottom of the mountain is 755 mm Hg, and the temperature is 302 K. The pressure at the top of the mountain is 687 mm Hg, and the temperature is 291 K. What is the volume of the cylinder at the top of the mountain? We begin by sorting the information in the problem statement.

GIVEN: $P_1 = 755$ mm Hg $\quad T_2 = 291$ K

$V_1 = 3.65$ L $\quad P_2 = 687$ mm Hg

$T_1 = 302$ K

FIND: V_2

SOLUTION MAP

We strategize by building a solution map that shows how the combined gas law equation takes us from the given quantities to the find quantity.

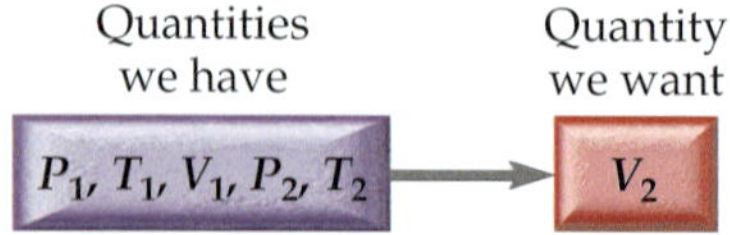

RELATIONSHIPS USED

$$\frac{P_1V_1}{T_1} = \frac{P_2V_2}{T_2}$$ (combined gas law, presented in this section)

SOLUTION

We then solve the equation for the quantity we are trying to find (V_2).

$$\frac{P_1V_1}{T_1} = \frac{P_2V_2}{T_2}$$

$$V_2 = \frac{P_1V_1T_2}{T_1P_2}$$

Lastly, we substitute in the appropriate values and calculate the answer.

$$V_2 = \frac{P_1V_1T_2}{T_1P_2}$$

$$= \frac{755 \text{ mm Hg} \times 3.65 \text{ L} \times 291 \text{ K}}{302 \text{ K} \times 687 \text{ mm Hg}}$$

$$= 3.87 \text{ L}$$

EXAMPLE 6.5 THE COMBINED GAS LAW

A sample of gas has an initial volume of 158 mL at a pressure of 735 mm Hg and a temperature of 34 °C. If the gas is compressed to a volume of 108 mL and heated to a temperature of 85 °C, what is its final pressure in millimetres of mercury?

SORT

You are given an initial pressure, temperature, and volume as well as a final temperature and volume. You are asked to find the final pressure.

GIVEN: $P_1 = 735$ mm Hg

$t_1 = 34$ °C $t_2 = 85$ °C

$V_1 = 158$ mL $V_2 = 108$ mL

FIND: P_2

STRATEGIZE

Draw a solution map beginning with the given quantities. The combined gas law shows the relationship necessary to get to the find quantity.

SOLUTION MAP

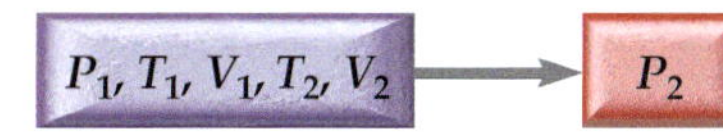

RELATIONSHIPS USED

$$\frac{P_1V_1}{T_1} = \frac{P_2V_2}{T_2}$$ (combined gas law, presented in this section)

SOLVE

Solve the equation for the quantity you are trying to find (P_2).

SOLUTION

$$\frac{P_1V_1}{T_1} = \frac{P_2V_2}{T_2}$$

$$P_2 = \frac{P_1V_1T_2}{T_1V_2}$$

Before you substitute in the numerical values, you must convert the temperatures to kelvins. Once you have converted the temperature to kelvins, substitute into the equation to find P_2.

$$T_1 = 34 + 273 = 307 \text{ K}$$

$$T_2 = 85 + 273 = 358 \text{ K}$$

$$P_2 = \frac{735 \text{ mm Hg} \times 158 \text{ mL} \times 358 \text{ K}}{307 \text{ K} \times 108 \text{ mL}}$$

$$= 1.25 \times 10^3 \text{ mm Hg}$$

CHECK

Check your answer. Are the units correct? Does the answer make physical sense?

The answer has the correct unit, mm Hg. The answer is reasonable because the decrease in volume and the increase in temperature should result in a pressure that is higher than the initial pressure.

▶SKILLBUILDER 6.4 | The Combined Gas Law

A balloon has a volume of 3.7 L at a pressure of 1.1 atm and a temperature of 30 °C. If the balloon is submerged in water to a depth where the pressure is 4.7 atm and the temperature is 15 °C, what will its volume be (assume that any changes in pressure caused by the skin of the balloon are negligible)?

▶FOR MORE PRACTICE Example 6.17; Problems 60, 61, 62, 63, 64, 65.

CONCEPTUAL CHECKPOINT 6.5

A volume of gas is confined to a container. If you apply enough heat to double the Kelvin temperature of the gas and expand the size of the container to double its initial volume, what happens to the pressure?

(a) the pressure doubles

(b) the pressure falls to half of its initial value

(c) the pressure is the same as its initial value

6.7 Avogadro's Law: Volume and Moles

LO: Restate and apply Avogadro's law.

So far, we have learned how V, P, and T are interrelated, but we have considered only a constant amount of a gas. What happens when the amount of gas changes? If we make several measurements of the volume of a gas sample (at constant temperature and pressure) while varying the number of moles in the sample, we get results similar to those shown in ▼ Figure 6.16. We can see that the relationship between volume and number of moles is linear. An extrapolation to zero moles shows a zero volume, as we might expect. This relationship was first stated formally by Amedeo Avogadro (1776–1856) and is called **Avogadro's law**.

Avogadro's law assumes constant temperature and pressure.

Avogadro's law: The volume of a gas and the amount of the gas in moles (n) are directly proportional.

$$V \propto n$$

Since $V \propto n$, then V/n = Constant. If the number of moles increases, then the volume increases in direct proportion so that the quotient, V/n, is always equal to the same constant. Thus, for two different measurements, we can say that $\frac{V_1}{n_1} = \text{Constant} = \frac{V_2}{n_2}$ or $\frac{V_1}{n_1} = \frac{V_2}{n_2}$.

When the amount of gas in a sample increases, its volume increases in direct proportion, which is yet another prediction of kinetic molecular theory. If the number of gas particles increases at constant pressure and temperature, the particles must occupy more volume.

We experience Avogadro's law when we inflate a balloon, for example. With each exhaled breath, we add more gas particles to the inside of the balloon, increasing its volume (▼ Figure 6.17). We can use Avogadro's law to calculate the volume of a gas following a change in the amount of the gas *as long as the pressure and temperature of the gas are constant.* For these calculations, Avogadro's law is expressed as

$$\frac{V_1}{n_1} = \frac{V_2}{n_2}$$

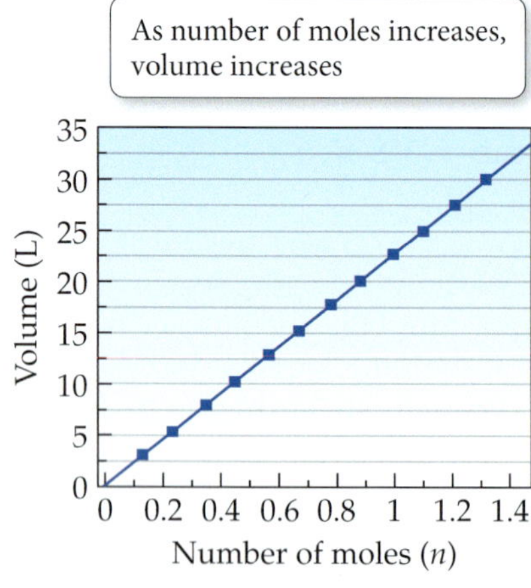

◀ **FIGURE 6.16 Volume versus number of moles** The volume of a gas sample increases linearly with the number of moles in the sample.

▶ **FIGURE 6.17 Blow-up** As you exhale into a balloon, you add gas molecules to the inside of the balloon, increasing its volume. © matha_Wariatka/Shutterstock.

where V_1 and n_1 are the initial volume and number of moles of the gas and V_2 and n_2 are the final volume and number of moles. In calculations, we use Avogadro's law in a manner similar to the other gas laws, as shown in Example 6.6.

EXAMPLE 6.6 AVOGADRO'S LAW

A 4.8 L sample of helium gas contains 0.22 mol of helium. How many additional moles of helium gas should you add to the sample to obtain a volume of 6.4 L? Assume constant temperature and pressure.

SORT
You are given an initial volume, an initial number of moles, and a final volume. You are (essentially) asked to find the final number of moles.

GIVEN: $V_1 = 4.8\ \text{L}$

$n_1 = 0.22\ \text{mol}$

$V_2 = 6.4\ \text{L}$

FIND: n_2

STRATEGIZE
Draw a solution map beginning with the given quantities. Avogadro's law addresses the relationship necessary to get to the find quantity.

SOLUTION MAP

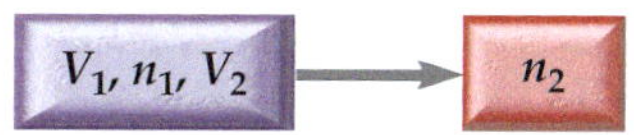

RELATIONSHIPS USED

$$\frac{V_1}{n_1} = \frac{V_2}{n_2}$$ (Avogadro's law, presented in this section)

SOLVE
Solve the equation for the quantity you are trying to find (n_2) and substitute the appropriate quantities to calculate n_2.

Since the balloon already contains 0.22 mol, subtract this quantity from the final number of moles to determine how much you must add.

SOLUTION

$$\frac{V_1}{n_1} = \frac{V_2}{n_2}$$

$$n_2 = \frac{V_2}{V_1} n_1$$

$$= \frac{6.4\ \cancel{\text{L}}}{4.8\ \cancel{\text{L}}} \times 0.22\ \text{mol}$$

$$= 0.29\ \text{mol}$$

$$\text{mol to add} = 0.29\ \text{mol} - 0.22\ \text{mol} = 0.07\ \text{mol}$$

CHECK
Check your answer. Are the units correct? Does the answer make physical sense?

The answer has the correct unit, mol. The answer is reasonable because the increase in the number of moles is proportional to the given increase in the volume.

▶SKILLBUILDER 6.5 | Avogadro's Law

A chemical reaction occurring in a cylinder equipped with a moveable piston produces 0.58 mol of a gaseous product. The cylinder also contained 0.11 mol of another gas that did not participate in the chemical reaction. This 'spectator' gas occupied a volume of 2.1 L. What was the volume of both gases in the cylinder after the reaction?

▶FOR MORE PRACTICE Problems 54, 55, 56, 57.

CONCEPTUAL CHECKPOINT 6.6

If each gas sample has the same temperature and pressure, which has the greatest volume?

(a) 1 g O_2

(b) 1 g Ar

(c) 1 g H_2

6.8 The Ideal Gas Law: Pressure, Volume, Temperature, and Moles

LO: Restate and apply the ideal gas law.

The relationships covered so far can be combined into a single law that encompasses all of them. So far, we know that:

$$V \propto \frac{1}{P} \text{ (Boyle's law)}$$
$$V \propto T \text{ (Charles's law)}$$
$$V \propto n \text{ (Avogadro's law)}$$

Combining these three expressions, we arrive at:

$$V \propto \frac{nT}{P}$$

The volume of a gas is directly proportional to the number of moles of gas and the temperature of the gas and is inversely proportional to the pressure of the gas. We can replace the proportional sign with an equal sign by adding R, a proportionality constant called the **ideal gas constant**.

$$V = \frac{RnT}{P}$$

Rearranging, we get the **ideal gas law**:

The ideal gas law: $PV = nRT$

The value of R, the ideal gas constant, is:

R can also be expressed in other units, but its numerical value will be different.

$$R = 0.0821 \frac{\text{L} \cdot \text{atm}}{\text{mol} \cdot \text{K}}$$

The ideal gas law contains within it the simple gas laws. For example, recall that Boyle's law states that $V \propto 1/P$ when the amount of gas (n) and the temperature of the gas (T) are kept constant. To derive Boyle's law, we can rearrange the ideal gas law as follows:

$$PV = nRT$$

First, divide both sides by P.

$$V = \frac{nRT}{P}$$

Then put the variables that are constant in parentheses.

$$V = (nRT)\frac{1}{P}$$

Since n and T are constant in this case and since R is always a constant,

$$V = (\text{Constant}) \times \frac{1}{P}$$

which gives us Boyle's law $\left(V \propto \frac{1}{P}\right)$.

The ideal gas law also shows how other pairs of variables are related. For example, from Charles's law we know that volume is proportional to temperature at constant pressure and a constant number of moles. But what if we heat a sample of gas at constant *volume* and a constant number of moles? This question applies to the warning labels on aerosol cans such as hair spray or deodorants. These labels warn the user against excessive heating or incineration of the can, even after the contents are used up. Why? An aerosol can that appears empty actually contains a fixed amount of gas trapped in a fixed volume. What would happen if we heated the can? Let's rearrange the ideal gas law to clearly see the relationship between pressure and temperature at constant volume and a constant number of moles.

$$PV = nRT$$

If we divide both sides by V, we get:

$$P = \frac{nRT}{V}$$

$$P = \left(\frac{nR}{V}\right)T$$

The relationship between pressure and temperature is also known as Gay-Lussac's law.

Since n and V are constant and since R is always a constant:

$$P = \text{Constant} \times T$$

As the temperature of a fixed amount of gas in a fixed volume increases, the pressure increases. In an aerosol can, this pressure increase can cause the can to explode, which is why aerosol cans should not be heated or incinerated. Table 6.2 summarizes the relationships between all of the simple gas laws and the ideal gas law.

TABLE 6.2 Relationships between Simple Gas Laws and Ideal Gas Law

Variable Quantities	Constant Quantities	Ideal Gas Law in Form of Variables-Constant	Simple Gas Law	Name of Simple Law
V and P	n and T	$PV = nRT$	$P_1V_1 = P_2V_2$	Boyle's law
V and T	n and P	$\frac{V}{T} = \frac{nR}{P}$	$\frac{V_1}{T_1} = \frac{V_2}{T_2}$	Charles's law
P and T	n and V	$\frac{P}{T} = \frac{nR}{V}$	$\frac{P_1}{T_1} = \frac{P_2}{T_2}$	Gay-Lussac's law
P and n	V and T	$\frac{P}{n} = \frac{RT}{V}$	$\frac{P_1}{n_1} = \frac{P_2}{n_2}$	
V and n	T and P	$\frac{V}{n} = \frac{RT}{P}$	$\frac{V_1}{n_1} = \frac{V_2}{n_2}$	Avogadro's law

We can use the ideal gas law to determine the value of any one of the four variables (P, V, n, or T) given the other three. However, each of the quantities in the ideal gas law *must be expressed* in the units within R.

- Pressure ***(P)*** must be expressed in atmospheres.
- Volume ***(V)*** must be expressed in litres.
- Amount of gas ***(n)*** must be expressed in moles.
- Temperature ***(T)*** must be expressed in kelvins.

For example, suppose we want to know the pressure of 0.18 mol of a gas in a 1.2 L flask at 298 K. We begin by sorting the information in the problem statement.

GIVEN: $n = 0.18$ mol
$V = 1.2$ L
$T = 298$ K

FIND: P

SOLUTION MAP

We then strategize by drawing a solution map that shows how the ideal gas law takes us from the given quantities to the find quantity.

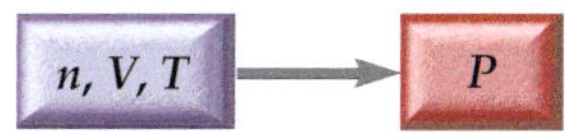

RELATIONSHIPS USED

$PV = nRT$ (ideal gas law, presented in this section)

SOLUTION

We then solve the equation for the quantity we are trying to find (in this case, P).

$$PV = nRT$$

$$P = \frac{nRT}{V}$$

Next we substitute in the numerical values and calculate the answer.

$$P = \frac{0.18\ \cancel{\text{mol}} \times 0.0821 \frac{\cancel{\text{L}} \cdot \text{atm}}{\cancel{\text{mol}} \cdot \cancel{\text{K}}} \times 298\ \cancel{\text{K}}}{1.2\ \cancel{\text{L}}}$$

$$= 3.7\ \text{atm}$$

Notice that all units cancel except the unit of the quantity we need (atm).

EXAMPLE 6.7 THE IDEAL GAS LAW

 Calculate the volume occupied by 0.845 mol of nitrogen gas at a pressure of 1.37 atm and a temperature of 315 K.

SORT You are given the number of moles, the pressure, and the temperature of a gas sample. You are asked to find the volume.	**GIVEN:** $n = 0.845$ mol $P = 1.37$ atm $T = 315$ K **FIND:** V
STRATEGIZE Draw a solution map beginning with the given quantities. The ideal gas law shows the relationship necessary to get to the find quantity.	**SOLUTION MAP**  **RELATIONSHIPS USED** $PV = nRT$ (ideal gas law, presented in this section)
SOLVE Solve the equation for the quantity you are trying to find (V) and substitute the appropriate quantities to calculate V.	**SOLUTION** $PV = nRT$ $V = \frac{nRT}{P}$ $V = \frac{0.845\ \cancel{\text{mol}} \times 0.0821 \frac{\text{L} \cdot \cancel{\text{atm}}}{\cancel{\text{mol}} \cdot \cancel{\text{K}}} \times 315\ \cancel{\text{K}}}{1.37\ \cancel{\text{atm}}} = 16.0\ \text{L}$
CHECK Check your answer. Are the units correct? Does the answer make physical sense?	The answer has the correct unit for volume, litres. The *value* of the answer is a bit more difficult to judge. However, at standard temperature and pressure ($t = 0$ °C or $T = 273.15$ K and $P = 1$ atm), 1 mol gas occupies 22.4 L (this is explained later in this section). Therefore your answer of 16.0 L seems reasonable for the volume of 0.85 mol of gas under conditions that are not too far from standard temperature and pressure.

▶SKILLBUILDER 6.6 | The Ideal Gas Law

An 8.5 L tire is filled with 0.55 mol of gas at a temperature of 305 K. What is the pressure of the gas in the tire?

▶FOR MORE PRACTICE Example 6.18; Problems 68, 69, 70, 71.

If the units given in an ideal gas law problem are different from those of the ideal gas constant (atm, L, mol, and K), we must convert to the correct units before we substitute into the ideal gas equation, as demonstrated in Example 6.8.

EXAMPLE 6.8 THE IDEAL GAS LAW REQUIRING UNIT CONVERSION

Calculate the number of moles of gas in a basketball inflated to a total pressure of 1251 torr with a volume of 3.2 L at 25 °C.

SORT

You are given the pressure, the volume, and the temperature of a gas sample. You are asked to find the number of moles.

GIVEN: $P = 1251$ torr

$V = 3.2$ L

$t = 25$ °C

FIND: n

STRATEGIZE

Draw a solution map beginning with the given quantities. The ideal gas law shows the relationship necessary to get to the find quantity.

SOLUTION MAP

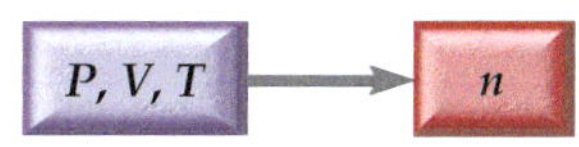

RELATIONSHIPS USED

$PV = nRT$ (ideal gas law, presented in this section)

SOLVE

Solve the equation for the quantity you are trying to find (n).

SOLUTION

$$PV = nRT$$

$$n = \frac{PV}{RT}$$

Before substituting into the equation, convert P and t into the correct units. (Since 1.6461 atm is an intermediate answer, mark the least significant digit, but don't round until the end.)

$$P = 1251 \text{ torr} \times \frac{1 \text{ atm}}{760 \text{ torr}} = 1.6\underline{4}61 \text{ atm}$$

$$T = t + 273$$

$$= 25 + 273 = 298 \text{ K}$$

Finally, substitute into the equation to calculate n.

$$n = \frac{1.6\underline{4}61 \cancel{\text{atm}} \times 3.2 \cancel{\text{L}}}{0.0821 \frac{\cancel{\text{L}} \cdot \cancel{\text{atm}}}{\text{mol} \cdot \cancel{\text{K}}} \times 298 \cancel{\text{K}}}$$

$$= 0.22 \text{ mol}$$

CHECK

Check your answer. Are the units correct? Does the answer make physical sense?

The answer has the correct unit, mol. The *value* of the answer is a bit more difficult to judge. Again, it is helpful to know that at standard temperature and pressure ($t = 0$ °C or $T = 273.15$ K and $P = 1$ atm), 1 mol of gas occupies 22.4 L (see Check step in Example 6.7). A 3.2 L sample of gas at standard temperature and pressure (STP) would contain about 0.15 mol; therefore at a greater pressure, the sample should contain a bit more than 0.15 mol, which is consistent with the answer.

▶SKILLBUILDER 6.7 | The Ideal Gas Law Requiring Unit Conversion

How much volume does 0.556 mol of gas occupy when its pressure is 715 mm Hg and its temperature is 58 °C?

▶SKILLBUILDER PLUS

Find the pressure in millimetres of mercury of a 0.133 g sample of helium gas at 32 °C contained in a 648 mL container.

▶FOR MORE PRACTICE Problems 72, 73.

▲ **FIGURE 6.18 Conditions for ideal gas behavior** At high temperatures and low pressures, the assumptions of the kinetic molecular theory apply.

Although a complete derivation is beyond the scope of this book, the ideal gas law follows directly from the kinetic molecular theory of gases. Consequently, the ideal gas law holds only under conditions where the kinetic molecular theory holds. The ideal gas law works exactly only for gases that are acting ideally (◀ Figure 6.18), which means that (a) the volume of the gas particles is small compared to the space between them and (b) the forces between the gas particles are not significant. These assumptions break down (▼ Figure 6.19) under conditions of high pressure (because the space between particles is no longer much larger than the size of the particles themselves) or low temperatures (because the gas particles move so slowly that their interactions become significant). For all of the problems encountered in this book, you may assume ideal gas behavior.

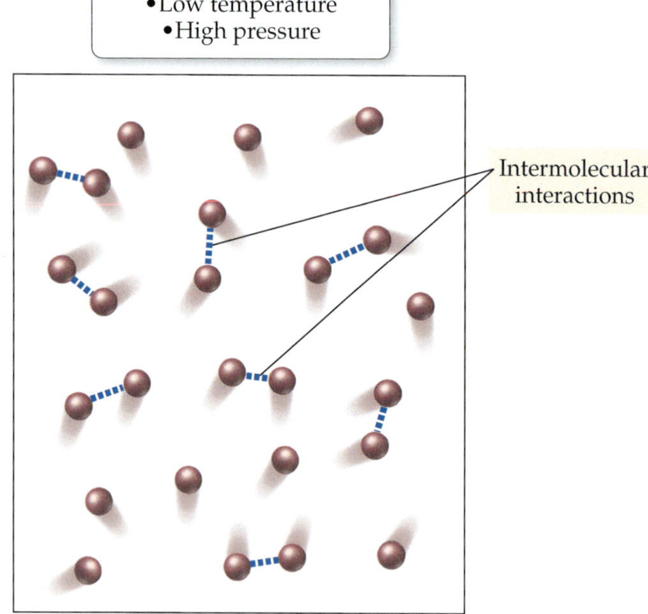

▶ **FIGURE 6.19 Conditions for nonideal gas behavior** At low temperatures and high pressures, the assumptions of the kinetic molecular theory do not apply.

Molar Volume at Standard Temperature and Pressure

Recall from the Check step of Example 6.7 that the volume occupied by 1 mol of gas at 0 °C (273.15 K) and 1 atm is 22.4 L. These conditions are called **standard temperature and pressure (STP)**, and the volume occupied by 1 mol of gas under these conditions is called the **molar volume** of an ideal gas at STP. Using the ideal gas law, we can confirm that the molar volume at STP is 22.4 L.

$$\begin{aligned} V &= \frac{nRT}{P} \\ &= \frac{1.00\ \cancel{\text{mol}} \times 0.0821\ \frac{\text{L}\cdot\cancel{\text{atm}}}{\cancel{\text{mol}}\cdot\cancel{\text{K}}} \times 273\ \cancel{\text{K}}}{1.00\ \cancel{\text{atm}}} \\ &= 22.4\ \text{L} \end{aligned}$$

Under standard conditions, therefore, we can use this ratio to calculate the volume of gas for a given number of moles of gas.

The molar volume of 22.4 L applies only at STP.

$$1\ \text{mol} : 22.4\ \text{L}$$

One mole of any gas at standard temperature and pressure (STP) occupies 22.4 L.

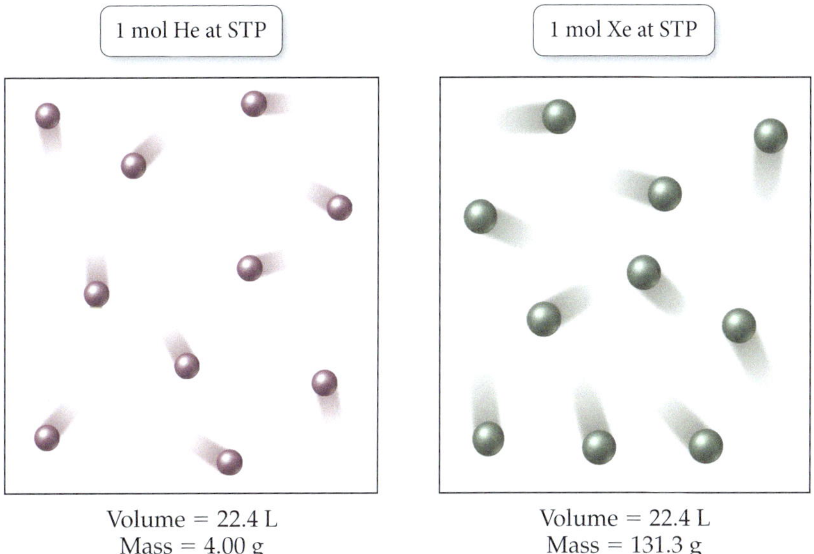

CHEMISTRY IN THE ENVIRONMENT

▶ Air Pollution

All major cities in the world have polluted air. This pollution comes from a number of sources, including electricity generation, motor vehicles, and industrial waste. While there are many different kinds of air pollutants, the major gaseous air pollutants include the following:

Sulfur dioxide (SO_2)—Sulfur dioxide is primarily a by-product of electricity generation and industrial metal refining. SO_2 is a lung and eye irritant that affects the respiratory system. SO_2 is also one of the main precursors of acid rain.

Carbon monoxide (CO)—Carbon monoxide is formed by the incomplete combustion of fossil fuels (petroleum, natural gas, and coal). It is emitted mainly by motor vehicles. CO displaces oxygen in the blood and causes the heart and lungs to work harder. At high levels, CO can cause sensory impairment, decreased thinking ability, unconsciousness, and even death.

Ozone (O_3)—Ozone in the upper atmosphere is a normal part of our environment. Upper atmospheric ozone filters out part of the harmful UV light contained in sunlight. Lower-atmospheric or *ground-level ozone*, on the other hand, is a pollutant that results from the action of sunlight on motor vehicle emissions. Ground-level ozone is an eye and lung irritant. Prolonged exposure to ozone has been shown to permanently damage the lungs.

Nitrogen dioxide (NO_2)—Nitrogen dioxide is emitted by motor vehicles and by electricity generation plants. It is an orange-brown gas that causes the dark haze often seen over polluted cities. NO_2 is an eye and lung irritant and a precursor of acid rain.

In the United States, the U.S. Environmental Protection Agency (EPA) has set standards for these pollutants. Beginning in the 1970s, the U.S. Congress passed the Clean Air Act and its amendments, requiring U.S. cities to reduce their pollution and maintain levels below the limits set by the EPA. As a result of this legislation, pollutant levels in U.S. cities have decreased significantly over the last 30 years, even as the number of vehicles has increased. For example, according to the EPA, the levels of all four of the previously mentioned pollutants in major U.S cities decreased during 1980–2010. Table 6.3 lists the amounts of these decreases.

TABLE 6.3 CHANGES IN POLLUTANT LEVELS FOR MAJOR U.S. CITIES, 1980–2010

Pollutant	Change, 1980–2010
SO_2	–76 %
CO	–82 %
O_3	–28 %
NO_2	–52 %

(*Source:* U.S. EPA)

Although the levels of pollutants (especially ozone) in many cities are still above what the EPA considers safe, much progress has been made. These trends demonstrate that good legislation can clean up our environment.

B6.3 CAN YOU ANSWER THIS? *Calculate the amount (in grams) of SO_2 emitted when 1.0 kg of coal containing 4.0 % S by mass is completely burned. Under standard conditions, what volume in litres does this SO_2 occupy?*

▲ Air pollution plagues most large cities. © Latitude Stock/Alamy.

6.9 Dalton's Law: Mixture of Gases

LO: Restate and apply Dalton's law of partial pressures.

TABLE 6.4 Composition of Dry Air

Gas	Percent by Volume (%)
nitrogen (N_2)	78
oxygen (O_2)	21
argon (Ar)	0.9
carbon dioxide (CO_2)	0.04

The fractional composition is the percent composition divided by 100.

Gas mixture (80 % He, 20 % Ne)
$P_{tot} = 1.0$ atm
$P_{He} = 0.80$ atm
$P_{Ne} = 0.20$ atm

▲ **FIGURE 6.20 Partial pressures** A gas mixture at a total pressure of 1.0 atm consisting of 80 % helium and 20 % neon has a helium partial pressure of 0.80 atm and a neon partial pressure of 0.20 atm.

Many gas samples are not pure but consist of mixtures of gases. The air in our atmosphere, for example, is a mixture containing 78 % nitrogen, 21 % oxygen, 0.9 % argon, 0.04 % carbon dioxide (Table 6.4), and a few other gases in smaller amounts.

According to the kinetic molecular theory, each of the components in a gas mixture acts independently of the others. For example, the nitrogen molecules in air exert a certain pressure—78 % of the total pressure—that is independent of the presence of the other gases in the mixture. Likewise, the oxygen molecules in air exert a certain pressure—21 % of the total pressure—that is also independent of the presence of the other gases in the mixture. The pressure due to any individual component in a gas mixture is called the **partial pressure** of that component. The partial pressure of any component is that component's fractional composition times the total pressure of the mixture (◀ Figure 6.20).

Partial pressure of component:

$$= \text{Fractional composition of component} \times \text{Total pressure}$$

For example, the partial pressure of nitrogen (P_{N_2}) in air at 1.0 atm is:

$$P_{N_2} = 0.78 \times 1.0 \text{ atm}$$
$$= 0.78 \text{ atm}$$

Similarly, the partial pressure of oxygen in air at 1.0 atm is:

$$P_{O_2} = 0.21 \times 1.0 \text{ atm}$$
$$= 0.21 \text{ atm}$$

The sum of the partial pressures of each of the components in a gas mixture must equal the total pressure, as expressed by **Dalton's law of partial pressures**:

Dalton's law of partial pressures:

$$P_{tot} = P_a + P_b + P_c + \ldots$$

where P_{tot} is the total pressure and P_a, P_b, P_c, ... are the partial pressures of the components.

For 1 atm air:

$$P_{tot} = P_{N_2} + P_{O_2} + P_{Ar}$$
$$P_{tot} = 0.78 \text{ atm} + 0.21 \text{ atm} + 0.01 \text{ atm}$$
$$= 1.00 \text{ atm}$$

EXAMPLE 6.9 TOTAL PRESSURE AND PARTIAL PRESSURE

A mixture of helium, neon, and argon has a total pressure of 558 mm Hg. If the partial pressure of helium is 341 mm Hg and the partial pressure of neon is 112 mm Hg, what is the partial pressure of argon?

SORT
You are given the total pressure of a gas mixture and the partial pressures of two (of its three) components. You are asked to find the partial pressure of the third component.

GIVEN: $P_{tot} = 558$ mm Hg
$P_{He} = 341$ mm Hg
$P_{Ne} = 112$ mm Hg

FIND: P_{Ar}

SOLVE
To solve this problem, solve Dalton's law for the partial pressure of argon and substitute the correct values to calculate it.

SOLUTION

$$P_{tot} = P_{He} + P_{Ne} + P_{Ar}$$
$$P_{Ar} = P_{tot} - P_{He} - P_{Ne}$$
$$= 558 \text{ mm Hg} - 341 \text{ mm Hg} - 112 \text{ mm Hg}$$
$$= 105 \text{ mm Hg}$$

▶ **SKILLBUILDER 6.8 | Total Pressure and Partial Pressure**

A sample of hydrogen gas is mixed with water vapor. The mixture has a total pressure of 745 torr, and the water vapor has a partial pressure of 24 torr. What is the partial pressure of the hydrogen gas?

▶ **FOR MORE PRACTICE** Example 6.19; Problems 82, 83, 84, 85.

▲ Mountain climbers on Mount Everest require oxygen because the pressure is so low that the lack of oxygen causes hypoxia, a condition that in severe cases can be fatal. © AP Photo/Pasang Geljen Sherpa.

Deep-Sea Diving and Partial Pressure

Our lungs have evolved to breathe oxygen at a partial pressure of $P_{O_2} = 0.21$ atm. If the total pressure decreases—as happens when we climb a mountain, for example—the partial pressure of oxygen also decreases. For example, on top of Mount Everest, where the total pressure is 0.311 atm, the partial pressure of oxygen is only 0.065 atm. As we learned earlier, low oxygen levels can have negative physiological effects, a condition called **hypoxia**, or oxygen starvation. Mild hypoxia causes dizziness, headache, and shortness of breath. Severe hypoxia, which occurs when P_{O_2} drops below 0.1 atm, may cause unconsciousness or even death. For this reason, climbers hoping to make the summit of Mount Everest usually carry oxygen to breathe.

High oxygen levels can also have negative physiological effects. Scuba divers, as we have learned, breathe pressurized air. At 30 m, a scuba diver breathes air at a total pressure of 4.0 atm, making P_{O_2} about 0.84 atm. This increased partial pressure of oxygen causes a higher density of oxygen molecules in the lungs (▶ Figure 6.21), which results in a higher concentration of oxygen in body tissues. When P_{O_2} increases beyond 1.4 atm, the increased oxygen concentration in body tissues causes a condition called **oxygen toxicity**, which is characterized by muscle twitching, tunnel vision, and convulsions (▶ Figure 6.22). Divers who venture too deep without proper precautions have drowned because of oxygen toxicity.

A second problem associated with breathing pressurized air is the increase in nitrogen in the lungs. At 30 m a scuba diver breathes nitrogen at $P_{N_2} = 3.1$ atm, which causes an increase in nitrogen concentration in bodily tissues and fluids. When P_{N_2} increases beyond about 4 atm, a condition called **nitrogen narcosis**, which is referred to as *rapture of the deep*, results. Divers describe this condition as a feeling of being tipsy. A diver breathing compressed air at 60 m feels as if he has had too much wine.

To avoid oxygen toxicity and nitrogen narcosis, deep-sea divers—those venturing beyond 50 m—breathe specialized mixtures of gases. One common mixture is called heliox, a mixture of helium and oxygen. These mixtures usually contain a smaller percentage of oxygen than would be found in air, thereby lowering the risk of oxygen toxicity. Heliox also contains helium instead of nitrogen, eliminating the risk of nitrogen narcosis.

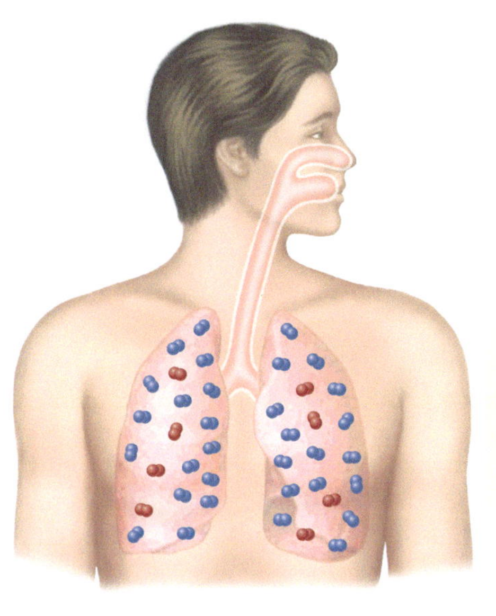

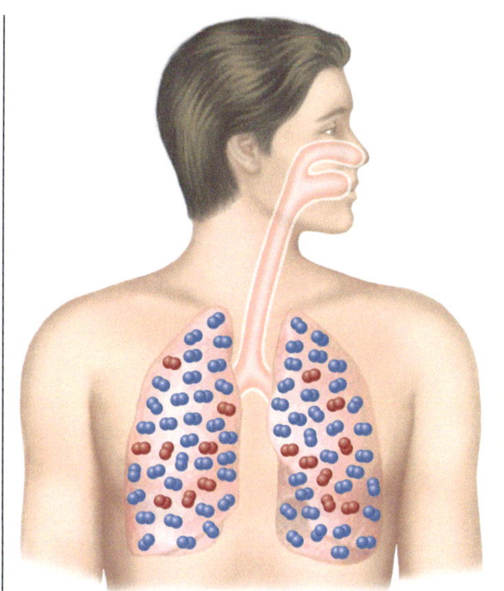

▲ **FIGURE 6.21 Too much of a good thing** When a person is breathing compressed air, there is a larger partial pressure of oxygen in the lungs. A large oxygen partial pressure in the lungs results in a larger amount of oxygen in bodily tissues. When the oxygen partial pressure increases beyond 1.4 atm, oxygen toxicity results. (In this figure, the red molecules are oxygen and the blue ones are nitrogen.)

▶ **FIGURE 6.22 Oxygen partial pressure limits** The partial pressure of oxygen in air at sea level is 0.21 atm. If this pressure drops by 50 %, fatal hypoxia can result. High oxygen levels can also be harmful, but only if the partial pressure of oxygen increases by a factor of 7 or more.

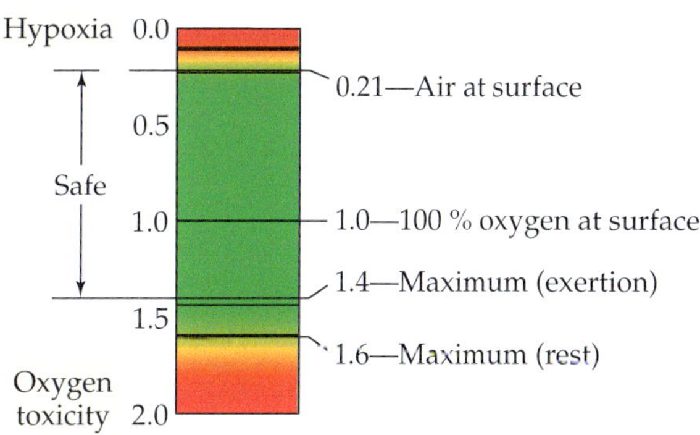

EXAMPLE 6.10 PARTIAL PRESSURE, TOTAL PRESSURE, AND PERCENT COMPOSITION

Calculate the partial pressure of oxygen that a diver breathes with a heliox mixture containing 2.0 % oxygen at a depth of 100 m where the total pressure is 11.0 atm.

SORT
You are given the percent oxygen in the mixture and the total pressure. You are asked to find the partial pressure of oxygen.

GIVEN: O_2 percent = 2.0 %
P_{tot} = 11.0 atm

FIND: P_{O_2}

SOLVE
The partial pressure of a component in a gas mixture is equal to the fractional composition of the component multiplied by the total pressure. Calculate the fractional composition of O_2 by dividing the percent composition by 100. Calculate the partial pressure of O_2 by multiplying the fractional composition by the total pressure.

SOLUTION
Partial pressure of component
= Fractional composition of component × Total pressure

$$\text{Fractional composition of } O_2 = \frac{2.0}{100} = 0.020$$

$$P_{O_2} = 0.020 \times 11.0 \text{ atm} = 0.22 \text{ atm}$$

▶SKILLBUILDER 6.9 | Partial Pressure, Total Pressure, and Percent Composition

A diver breathing heliox with an oxygen composition of 5.0 % wants to adjust the total pressure so that $P_{O_2} = 0.21$ atm. What must the total pressure be?

▶FOR MORE PRACTICE Problems 88, 89, 90, 91.

6.10 Interactions Between Molecules

Bite into a candy bar and taste its sweetness. Drink a cup of strong coffee and experience its bitterness. What causes these flavors? Most tastes originate from interactions between molecules. Certain molecules in coffee, for example, interact with molecular receptors on the surface of specialized cells on the tongue. The receptors are highly specific, recognizing only certain types of molecules. The interaction between the molecule and the receptor triggers a signal that goes to the brain, which we interpret as a bitter taste. Bitter tastes are usually unpleasant because many of the molecules that cause them are poisons. Our response to the bitterness of these molecules is probably an evolutionary adaptation that helps us avoid these poisons.

The interaction between the molecules in coffee that taste bitter and the taste receptors on the tongue is caused by **intermolecular forces**—attractive forces that exist *between* molecules. Living organisms depend on intermolecular forces not only for taste but also for many other physiological processes. Intermolecular forces help determine the shapes of protein molecules—the workhorse molecules in living organisms. Later in this module—in the *Chemistry and Health* box in Section 6.13—we learn how intermolecular forces are central to DNA, the inheritable molecules that serve as blueprints for life.

The interactions between bitter molecules in coffee and molecular receptors on the tongue are highly specific. However, less specific intermolecular forces exist between all molecules and atoms. These intermolecular forces are responsible for the very existence of liquids and solids. The state of a sample of matter—solid, liquid, or gas—depends on the magnitude of intermolecular forces relative to the amount of thermal energy in the sample. Recall from earlier in this module, that the molecules and atoms that compose matter are in constant random motion that increases with increasing temperature. The energy associated with this motion is called **thermal energy**. The weaker the intermolecular forces relative to thermal energy, the more likely the sample will be gaseous. The stronger the intermolecular forces relative to thermal energy, the more likely the sample will be liquid or solid.

6.11 Three States of Matter at the Molecular Level

LO: Describe the properties of solids and liquids and relate them to their constituent atoms and molecules.

We are all familiar with solids and liquids. Water, gasoline, rubbing alcohol, and fingernail-polish remover are all common liquids. Ice, dry ice, and diamond are familiar solids. In contrast to gases—in which molecules or atoms are separated by large distances—the molecules or atoms that compose liquids and solids are in close contact with one another (▶ Figure 6.23).

The difference between solids and liquids is in the freedom of movement of the constituent molecules or atoms. In liquids, even though the atoms or molecules are in close contact, they are still free to move around each other. In solids, the atoms or molecules are fixed in their positions, although thermal energy causes them to vibrate about a fixed point. These molecular properties of solids and liquids result in characteristic macroscopic properties.

Properties of Liquids

- High densities in comparison to gases.
- Indefinite shape; liquids assume the shape of their container.
- Definite volume; liquids are not easily compressed.

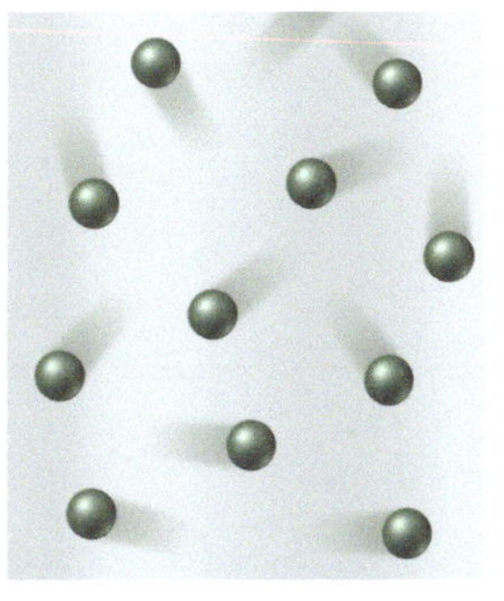

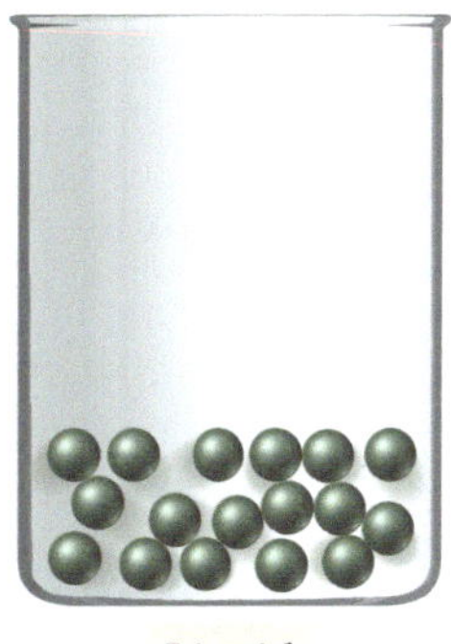

▶ **FIGURE 6.23 Gas, liquid, and solid states**

Properties of Solids

- High densities in comparison to gases.
- Definite shape; solids do not assume the shape of their container.
- Definite volume; solids are not easily compressed.
- May be crystalline (ordered) or amorphous (disordered).

Table 6.5 summarizes these properties, as well as the properties of gases for comparison.

TABLE 6.5 Properties of the States of Matter

Phase	Density	Shape	Volume	Strength of Intermolecular Forces[a]	Example
gas	low	indefinite	indefinite	weak	carbon dioxide gas (CO_2)
liquid	high	indefinite	definite	moderate	liquid water (H_2O)
solid	high	definite	definite	strong	sugar ($C_{12}H_{22}O_{11}$)

[a]Relative to thermal energy.

As we will see in Section 7.2, ice is less dense than liquid water because water expands when it freezes due to its unique crystalline structure.

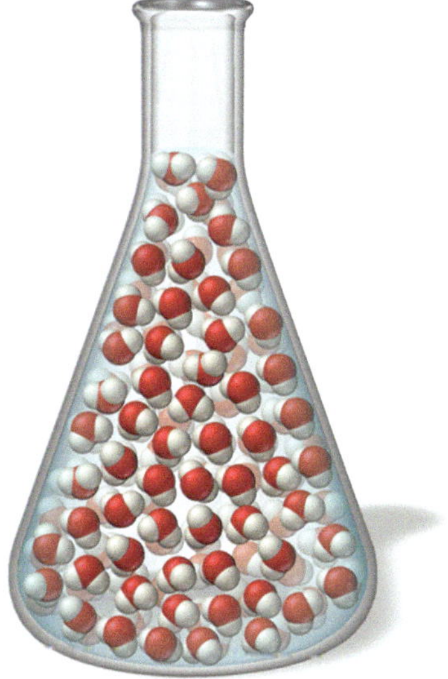

▲ **FIGURE 6.24 A liquid assumes the shape of its container** Because the molecules in liquid water are free to move around each other, they flow and assume the shape of their container.

Liquids have high densities in comparison to gases because the atoms or molecules that compose liquids are much closer together. The density of liquid water, for example, is 1.0 g/cm^3 (at 25 °C), while the density of gaseous water at 100 °C and 1 atm is 0.59 g/L, or 5.9×10^{-4} g/cm^3. Liquids assume the shape of their containers because the atoms or molecules that compose them are free to flow. When we pour water into a flask, the water flows and assumes the shape of the flask (◀ Figure 6.24). Liquids are not easily compressed because the molecules or atoms that compose them are in close contact—they cannot be pushed closer together.

Like liquids, solids have high densities in comparison to gases because the atoms or molecules that compose solids are also close together. The densities of solids are usually just slightly greater than those of the corresponding liquids. A major exception is water, whose solid (ice) is slightly less dense than liquid water. Solids have a definite shape because, in contrast to liquids or gases, the molecules or atoms that compose solids are fixed in place (▶ Figure 6.25). Each molecule or atom in a solid only vibrates about a fixed point. Like liquids, solids have a definite volume and cannot be compressed because the molecules or atoms composing them are in close contact. Solids may be *crystalline*, in which case the atoms or molecules that compose them arrange themselves in a well-ordered, three-dimensional array, or they may be *amorphous*, in which case the atoms or molecules that compose them have no long-range order.

CONCEPTUAL CHECKPOINT 6.7

A substance has a definite shape and definite volume. What is the state of the substance?

(a) Solid **(b)** Liquid **(c)** Gas

▲ **FIGURE 6.25 Solids have a definite shape** In a solid such as ice, the molecules are fixed in place. However, they vibrate about fixed points. © Tom Brakefield/Photodisc/Getty Images.

6.12 Changes of State and Associated Energy Requirements

LO: Describe and explain the processes of evaporation and condensation.

LO: Describe and explain the processes of melting, freezing, and sublimation.

Leave a glass of water in the open for several days and the water level within the glass slowly drops. Why? The first reason is that water molecules at the surface of the liquid—which experience fewer attractions to neighboring molecules and are therefore held less tightly—can break away from the rest of the liquid. The second reason is that all of the molecules in the liquid have a *distribution of kinetic energy* at any given temperature (▼ Figure 6.26). At any given moment, some molecules in the liquid are moving faster than the average (higher energy), and others are moving more slowly (lower energy). Some of the molecules that are moving faster have enough energy to break free from the surface, resulting in **evaporation** or **vaporization**, a physical change in which a substance converts from its liquid state to its gaseous state (▶ Figure 6.27).

In evaporation or vaporization, a substance is converted from its liquid state into its gaseous state.

If we spill the same amount of water (as was in the glass) on a table, it evaporates more quickly, probably within a few hours. Why? The surface area of the spilled water is greater, leaving more molecules susceptible to evaporation. If we warm the glass of water, it also evaporates more quickly because the greater

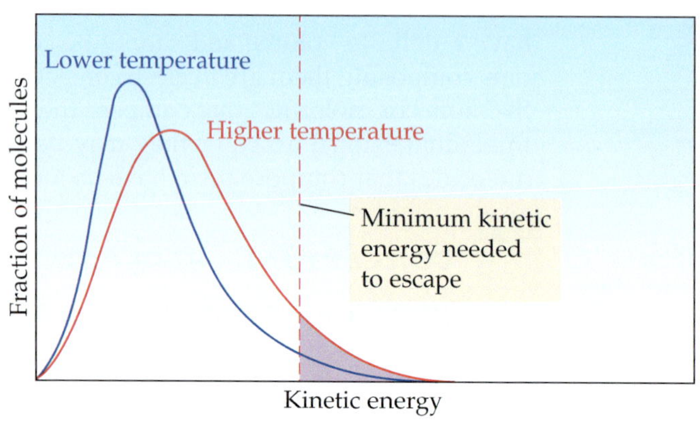

▶ **FIGURE 6.26 Energy distribution** At a given temperature, a sample of molecules or atoms will have a distribution of kinetic energies, as shown here. Only a small fraction of molecules has enough energy to escape. At a higher temperature, the fraction of molecules with enough energy to escape increases.

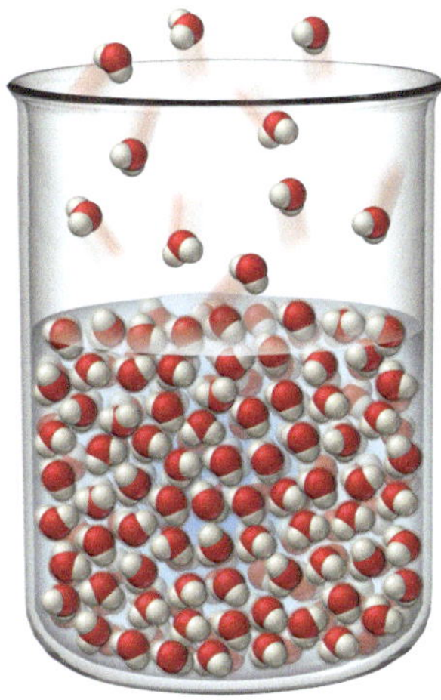

▲ **FIGURE 6.27 Evaporation** Because molecules on the surface of a liquid are held less tightly than those in the interior, the most energetic among them can break away into the gas state in the process called *evaporation*.

thermal energy causes a greater fraction of molecules to have enough energy to break away from the surface (see Figure 6.27). If we fill the glass with rubbing alcohol instead of water, the liquid again evaporates more quickly because the intermolecular forces between the alcohol molecules are weaker than the intermolecular forces between water molecules. In general, the rate of vaporization increases with:

- Increasing surface area
- Increasing temperature
- Decreasing strength of intermolecular forces

Liquids that evaporate easily are termed **volatile**, while those that do not vaporize easily are termed **nonvolatile**. Rubbing alcohol, for example, is more volatile than water. Motor oil at room temperature is virtually nonvolatile.

If we leave water in a *closed* container, its level remains constant because the molecules that leave the liquid are trapped in the air space above the water. These gaseous molecules bounce off the walls of the container and eventually hit the surface of the water again and recondense. **Condensation** is a physical change in which a substance converts from its gaseous state to its liquid state.

Dynamic equilibrium is so named because both condensation and evaporation of individual molecules continue, but at the same rate.

Evaporation and condensation are opposites: Evaporation is a liquid turning into a gas, and condensation is a gas turning into a liquid. When we initially put liquid water into a closed container, more evaporation happens than condensation because there are so few gaseous water molecules in the space above the water (▼ Figure 6.28(a). However, as the number of gaseous water molecules increases, the rate of condensation also increases (▼ Figure 6.28(b). At the point where the rates of condensation and evaporation become equal (▼ Figure 6.28(c), **dynamic equilibrium** is reached and the number of gaseous water molecules above the liquid remains constant. The **vapor pressure** of a liquid is the partial pressure of its vapor in dynamic equilibrium with its liquid. At 25 °C, water's vapor pressure is 23.8 mm Hg. Vapor pressure increases with:

- Increasing temperature
- Decreasing strength of intermolecular forces

Vapor pressure is independent of surface area because an increase in surface area at equilibrium equally affects the rate of evaporation and the rate of condensation.

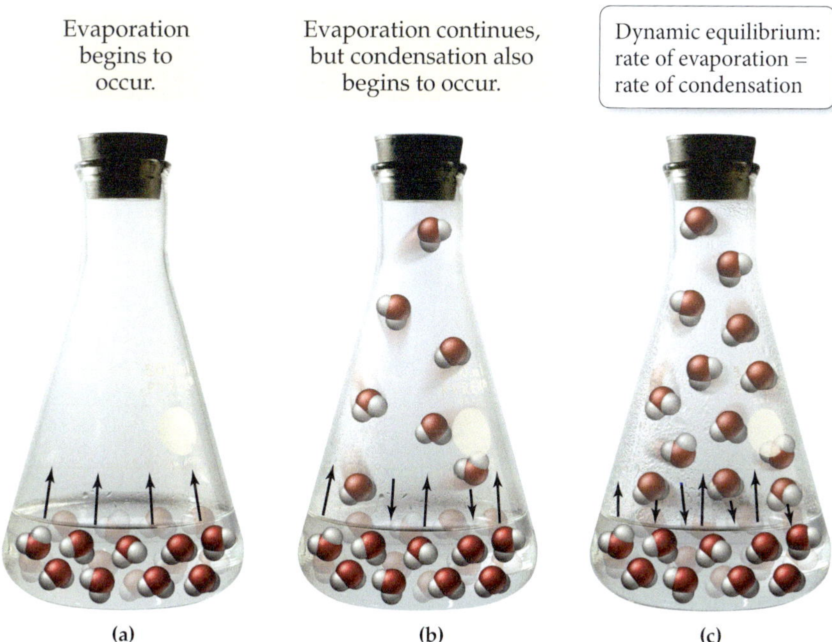

▶ **FIGURE 6.28 Evaporation and condensation** **(a)** When water is first put into a closed container, water molecules begin to evaporate. **(b)** As the number of gaseous molecules increases, some of the molecules begin to collide with the liquid and are recaptured—that is, they recondense into liquid. **(c)** When the rate of evaporation equals the rate of condensation, dynamic equilibrium occurs, and the number of gaseous molecules remains constant. © Richard Megna/Fundamental Photographs.

Boiling

Sometimes you see bubbles begin to form in hot water below 100 °C. These bubbles are dissolved air—not gaseous water—leaving the liquid. Dissolved air comes out of water as you heat it because the solubility of a gas in a liquid decreases with increasing temperature (Section 7.6).

As you increase the temperature of water in an open container, the increasing thermal energy causes molecules to leave the surface and vaporize at a faster and faster rate. At the **boiling point**—the temperature at which the vapor pressure of a liquid is equal to the pressure above it—the thermal energy is enough for molecules within the *interior* of the liquid (not just those at the surface) to break free into the gas phase (▼ Figure 6.29). Water's **normal boiling point**—its boiling point at a pressure of 1 atmosphere—is 100 °C. When a sample of water reaches 100 °C, you can see bubbles form within the liquid. These bubbles are pockets of gaseous water. The bubbles quickly rise to the surface of the liquid, and the water molecules that were in the bubble leave as gaseous water, or steam. Once the boiling point of a liquid is reached, additional heating only causes more rapid boiling; it does not raise the temperature of the liquid above its boiling point (▼ Figure 6.30). A mixture of boiling water *and* steam always has a temperature of 100 °C (at 1 atm pressure). Only after all the water has been converted to steam can the temperature of the steam rise beyond 100 °C.

▶ **FIGURE 6.29 Boiling** During boiling, thermal energy is enough to cause water molecules in the interior of the liquid to become gaseous, forming bubbles containing gaseous water molecules. © Magnascan/iStockphoto/Getty Images.

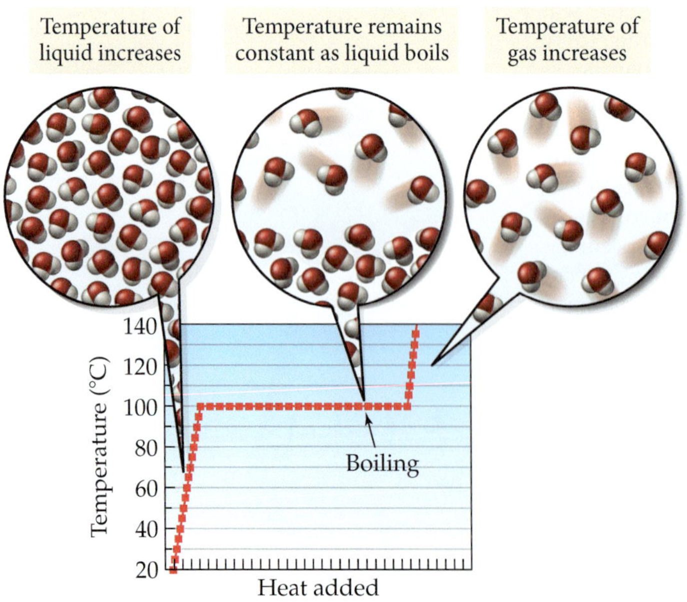

▶ **FIGURE 6.30 Heating curve during boiling** The temperature of water as it is heated from room temperature through boiling. During boiling, the temperature remains at 100 °C until all the liquid is evaporated.

Energetics of Evaporation and Condensation

In an endothermic process, heat is absorbed; in an exothermic process, heat is released.

Evaporation is *endothermic*—when a liquid is converted into a gas, it absorbs heat because energy is required to break molecules away from the rest of the liquid. Imagine a collection of water molecules in the liquid state. As the water evaporates, it cools—typical of endothermic processes—because only the fastest-moving molecules break away, which leaves the slower-moving molecules behind. Under ordinary conditions, the slight decrease in temperature of water as it evaporates is counteracted by thermal energy transfer from the surroundings, which warms the water back up. However, if the evaporating water were thermally isolated from the surroundings, it would continue to cool down as it evaporated.

You can observe the endothermic nature of evaporation by turning off the heat beneath a boiling pot of water; it quickly stops boiling as the heat lost due to vaporization causes the water to cool below its boiling point. Our bodies use the endothermic nature of evaporation for cooling. When we overheat, we sweat, causing our skin to be covered with liquid water. As this water evaporates it absorbs heat from our bodies, cooling us down. A fan intensifies the cooling effect because it blows newly vaporized water away from the skin, allowing more sweat to vaporize and cause even more cooling. High humidity, however, slows down evaporation, preventing cooling. When the air already contains high amounts of water vapor, sweat does not evaporate as easily, making our cooling system less efficient.

Condensation, the opposite of evaporation, is *exothermic*—heat is released when a gas condenses to a liquid. If you have ever accidentally put your hand above a steaming kettle, you may have experienced a *steam burn*. As the steam condenses to a liquid on your skin, it releases heat, causing a severe burn. The exothermic nature of condensation is also the reason that winter overnight temperatures in coastal cities, which tend to have water vapor in the air, do not get as low as in deserts, which tend to have dry air. As the air temperature in a coastal city drops, water condenses out of the air, releasing heat and preventing the temperature from dropping further. In deserts, there is little moisture in the air to condense, so the temperature drop is greater.

As the temperature of a solid increases, thermal energy causes the molecules and atoms composing the solid to vibrate faster. At the **melting point**, atoms and molecules have enough thermal energy to overcome the intermolecular forces that hold them at their stationary points, and the solid turns into a liquid. The melting point of ice, for example, is 0 °C. Once the melting point of a solid is reached, additional heating only causes more rapid melting; it does not raise the temperature of the solid above its melting point (◀ Figure 6.31). Only after all of the ice has melted does additional heating raise the temperature of the liquid water past 0 °C. A mixture of water *and* ice always has a temperature of 0 °C (at 1 atm pressure).

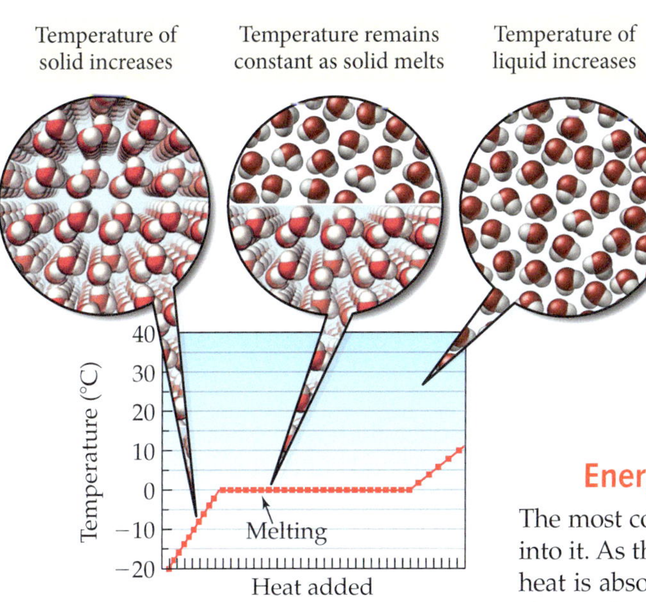

▲ **FIGURE 6.31 Heating curve during melting** A graph of the temperature of ice as it is heated from −20 °C to 35 °C. During melting, the temperature of the solid and the liquid remains at 0 °C until the entire solid is melted.

Energetics of Melting and Freezing

The most common way to cool down a drink is to drop several ice cubes into it. As the ice melts, the drink cools because melting is endothermic—heat is absorbed when a solid is converted into a liquid. The melting ice absorbs heat from the liquid in the drink and cools the liquid. Melting is endothermic because energy is required to partially overcome the attractions between molecules in the solid and free them into the liquid state.

Freezing, the opposite of melting, is exothermic—heat is released when a liquid freezes into a solid. For example, as water in your freezer turns into ice, it releases heat, which must be removed by the refrigeration system of the freezer. If the refrigeration system did not remove the heat, the water would not completely

▶ When ice melts, water molecules break free from the solid structure and become liquid. As long as ice and water are both present, the temperature is 0.0 °C. © Richard Megna/Fundamental Photographs.

▲ Dry ice is solid carbon dioxide. The solid does not melt but rather sublimes. It transforms directly from solid carbon dioxide to gaseous carbon dioxide. © Charles D Winters/Science Source.

freeze into ice. The heat released as it began to freeze would warm the freezer, preventing further freezing.

Sublimation

Sublimation is a physical change in which a substance changes from its solid state directly to its gaseous state. When a substance sublimes, molecules leave the surface of the solid, where they are held less tightly than in the interior, and become gaseous. For example, dry ice, which is solid carbon dioxide, does not melt under atmospheric pressure (at any temperature). At −78 °C, the CO_2 molecules have enough energy to leave the surface of the dry ice and become gaseous. Regular ice slowly sublimes at temperatures below 0 °C. You can observe the sublimation of ice in cold climates; ice or snow laying on the ground gradually disappears, even if the temperature remains below 0 °C. Similarly, ice cubes left in the freezer for a long time slowly become smaller, even though the freezer is always below 0 °C. In both cases, the ice is subliming, turning directly into water vapor.

Ice also sublimes out of frozen foods. You can clearly see this in food that is frozen in an airtight plastic bag for a long time. The ice crystals that form in the bag are water that has sublimed out of the food and redeposited on the surface of the bag. For this reason, food that remains frozen for too long becomes dried out. This can be avoided to some degree by freezing foods to colder temperatures (further below 0 °C), a process called deep-freezing. The colder temperature lowers the rate of sublimation and preserves the food longer.

CONCEPTUAL CHECKPOINT 6.8

1. The gas over a rapidly boiling pot of water is sampled and analyzed. Which substance composes a large fraction of the gas sample?

   (a) $H_2(g)$ **(b)** $H_2O(g)$ **(c)** $O_2(g)$ **(d)** $H_2O_2(g)$

2. Solid carbon dioxide (dry ice) can be depicted as follows:

© Photo Researchers/Getty Images.

Which image best represents the dry ice after it has sublimed?

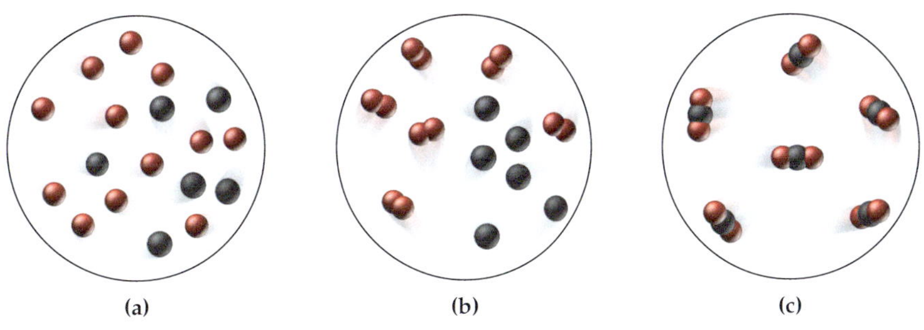

6.13 Types of Intermolecular Forces

LO: Compare and contrast four types of intermolecular forces: dispersion, dipole–dipole, hydrogen bonds, and ion–dipole.

LO: Determine the types of intermolecular forces in compounds.

LO: Use intermolecular forces to determine relative melting and/or boiling points.

LO: Use intermolecular forces to predict the miscibility of different substances.

The strength of the intermolecular forces between the molecules or atoms that compose a substance determines the state—solid, liquid, or gas—of the substance at room temperature. Strong intermolecular forces tend to result in liquids and solids (with high melting and boiling points). Weak intermolecular forces tend to result in gases (with low melting and boiling points). In this book, we focus on four fundamental types of intermolecular forces. In order of increasing strength, they are the dispersion force, the dipole–dipole force, the hydrogen bond, and the ion–dipole force.

Dispersion Force

The default intermolecular force, present in all molecules and atoms, is the **dispersion force** (also called the *London force*). Dispersion forces are caused by fluctuations in the electron distribution within molecules or atoms. Since all atoms and molecules have electrons, they all have dispersion forces. The electrons in an atom or a molecule may, at any one instant, be unevenly distributed. For example, imagine a frame-by-frame movie of a helium atom in which each "frame" captures the position of the helium atom's two electrons (▼ Figure 6.32). In any one frame, the electrons are not symmetrically arranged around the nucleus. In Frame 3, for example, helium's two electrons are on the left side of the helium atom. The left side then acquires a slightly negative charge (δ^-). The right side of the atom, which is void of electrons, acquires a slightly positive charge (δ^+).

The nature of dispersion forces was first recognized by Fritz W. London (1900–1954), a German-American physicist.

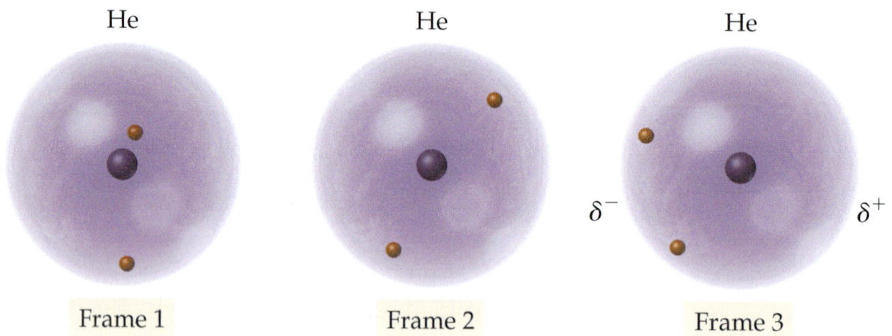

▶ **FIGURE 6.32 Instantaneous dipoles** Random fluctuations in the electron distribution of a helium atom cause instantaneous dipoles to form.

This fleeting charge separation is called an **instantaneous dipole** (or *temporary dipole*). An instantaneous dipole on one helium atom induces an instantaneous dipole on its neighboring atoms because the positive end of the instantaneous dipole attracts electrons in the neighboring atoms (▼ Figure 6.33). The dispersion force occurs as neighboring atoms attract one another—the positive end of one instantaneous dipole attracts the negative end of another. The dipoles responsible for the dispersion force are transient, constantly appearing and disappearing in response to fluctuations in electron clouds.

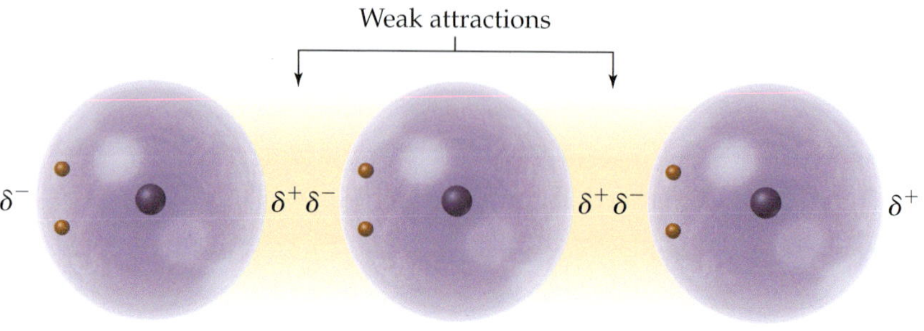

▶ **FIGURE 6.33 Dispersion force** An instantaneous dipole on any one helium atom induces instantaneous dipoles on neighboring atoms. The neighboring atoms then attract one another. This attraction is called the dispersion force.

To *polarize* means to form a dipole moment.

TABLE 6.6 Noble Gas Boiling Points

Noble Gas	Molar Mass (g/mol)	Boiling Point (K)
He	4.00	4.2
Ne	20.18	27
Ar	39.95	87
Kr	83.80	120
Xe	131.29	165

The magnitude of the dispersion force depends on how easily the electrons in the atom or molecule move or *polarize* in response to an instantaneous dipole, which in turn depends on the size of the electron cloud. A larger electron cloud results in a greater dispersion force because the electrons are held less tightly by the nucleus and therefore can polarize more easily. If all other variables are constant, the dispersion force increases with increasing molar mass. For example, consider the boiling points of the noble gases displayed in Table 6.6. As the molar mass of the noble gases increases, their boiling points increase. While molar mass alone does not determine the magnitude of the dispersion force, it can be useful as a guide when comparing dispersion forces within a family of similar elements or compounds.

EXAMPLE 6.11 DISPERSION FORCES

Which halogen, Cl_2 or I_2, has the higher boiling point?

SOLUTION

The molar mass of Cl_2 is 70.90 g/mol, and the molar mass of I_2 is 253.80 g/mol. Since I_2 has the higher molar mass, it has stronger dispersion forces and therefore the higher boiling point.

▶SKILLBUILDER 6.10 | Dispersion Forces

Which hydrocarbon, CH_4 or C_2H_6, has the higher boiling point?

▶FOR MORE PRACTICE Problems 102, 103.

Dipole–Dipole Force

See Module 3.18 to review how to determine whether a molecule is polar.

The **dipole–dipole force** exists in all polar molecules. Polar molecules have **permanent dipoles** (Module 3) that interact with the permanent dipoles of neighboring molecules (◀ Figure 6.34). The positive end of one permanent dipole is attracted to the negative end of another; this attraction is the dipole–dipole force (◀ Figure 6.35). Polar molecules, therefore, have higher melting and boiling points than nonpolar molecules of similar molar mass. Remember that all molecules (including polar ones) have dispersion forces. In addition, polar molecules have dipole–dipole forces. These additional attractive forces raise their melting and boiling points relative to nonpolar molecules of similar molar mass. For example, consider the compounds formaldehyde and ethane:

▲ **FIGURE 6.34 A permanent dipole** Molecules such as formaldehyde are polar and therefore have a permanent dipole.

Name	Formula	Molar mass (g/mol)	Structure	bp (°C)	mp (°C)
Formaldehyde	CH_2O	30.0	O ‖ H—C—H	−19.5	−92
Ethane	C_2H_6	30.1	H₃C—CH₃ (H—C(H)(H)—C(H)(H)—H)	−88	−172

Formaldehyde is polar and therefore has a higher melting point and boiling point than nonpolar ethane, even though the two compounds have the same molar mass.

The polarity of molecules composing liquids is also important in determining a liquid's **miscibility**—its ability to mix without separating into two phases. In general, polar liquids are miscible with other polar liquids but are not miscible with nonpolar liquids. For example, water, a polar liquid, is not miscible with pentane (C_5H_{12}), a nonpolar liquid (▶ Figure 6.36). Similarly, water and oil (also nonpolar) do not mix. Consequently, oily hands or oily stains on clothes cannot be washed away with plain water (see Module 3, *Everyday Chemistry: How Soap Works*).

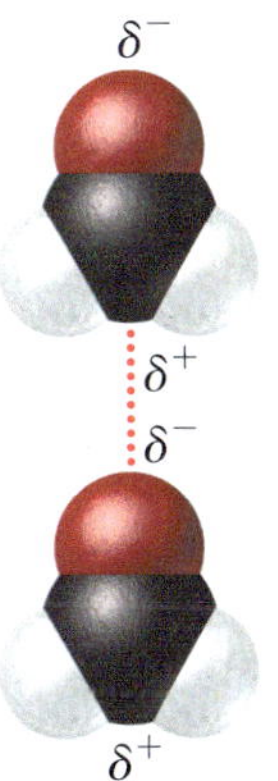

▲ **FIGURE 6.35 Dipole–dipole attraction** The positive end of a polar molecule is attracted to the negative end of its neighbor, giving rise to the dipole–dipole force.

(a)
© Richard Megna/ Fundamental Photographs.

(b)
© Richard Haynes/Prentice Hall School Division/Pearson.

(c)
© RGB Ventures LLC dba SuperStock/Alamy.

◀ **FIGURE 6.36 Polar and nonpolar compounds (a)** Pentane, a nonpolar compound, does not mix with water, a polar compound. **(b)** For the same reason, the oil and vinegar (vinegar is largely a water solution of acetic acid) in salad dressing tend to separate into distinct layers. **(c)** An oil spill from a tanker demonstrates dramatically that petroleum and seawater are not miscible.

EXAMPLE 6.12 DIPOLE–DIPOLE FORCES

Determine whether each molecule has dipole–dipole forces.

(a) CO_2 **(b)** CH_2Cl_2 **(c)** CH_4

SOLUTION

A molecule has dipole–dipole forces if it is polar. To find out whether a molecule is polar, you must:

1. determine whether the molecule contains polar bonds, and
2. determine whether the polar bonds add together to form a net dipole moment (Section 3.18).

(a) C and O are different nonmetals so the C — O bonds are polar (see Section 3.18). The geometry of CO_2 is linear. So, the polarity of the bonds cancels. Consequently, the molecule is not polar and does not have dipole–dipole forces.

O=C=O

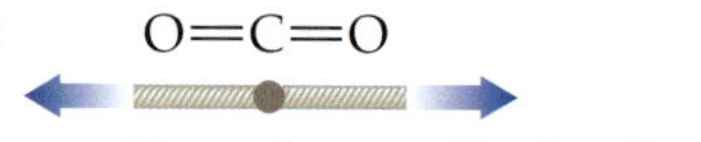

Nonpolar; no dipole–dipole forces

(b) C, H and Cl are different nonmetals, so CH_2Cl_2 has two polar bonds (C — Cl) and two polar bonds (C — H). The geometry of CH_2Cl_2 is tetrahedral. Since the C — Cl bonds and the C — H bonds are different, they do not cancel, but sum to a net dipole moment. Therefore the molecule is polar and has dipole–dipole forces.

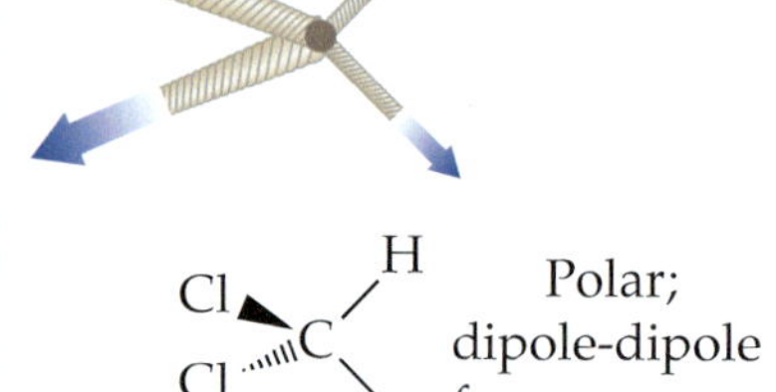

Polar; dipole-dipole forces

(c) C and H are different nonmetals, so the C — H bonds are polar. In addition, because the geometry of the molecule is tetrahedral, any polarities that the bonds have will cancel. CH_4 is therefore nonpolar and does not have dipole–dipole forces.

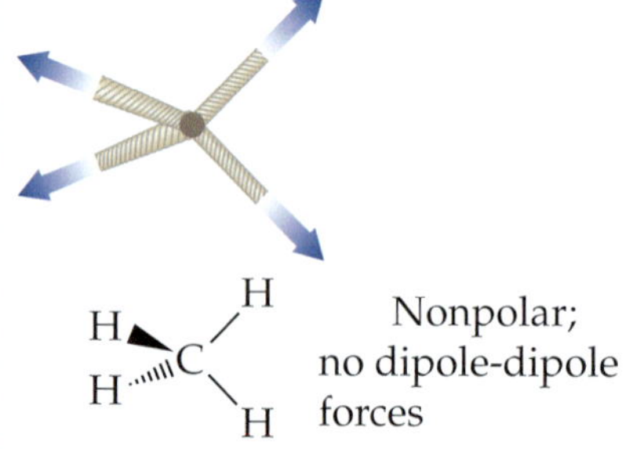

Nonpolar; no dipole-dipole forces

▶**SKILLBUILDER 6.11 | Dipole–Dipole Forces**

Determine whether or not each molecule has dipole–dipole forces.

(a) CI_4 **(b)** CH_3Cl **(c)** HCl

▶**FOR MORE PRACTICE** Problems 98, 99.

Hydrogen Bonding

Polar molecules containing hydrogen atoms bonded directly to fluorine, oxygen, or nitrogen exhibit an additional intermolecular force called a **hydrogen bond**. HF, NH_3, and H_2O, for example, all undergo hydrogen bonding. A hydrogen bond is a sort of *super* dipole–dipole force. The large electronegativity difference between hydrogen and these electronegative elements, as well as the small size of these atoms (which allows neighboring molecules to get very close to each other), gives rise to a strong attraction between the hydrogen in each of these molecules and the F, O, or N on neighboring molecules. This attraction between a hydrogen atom and an electronegative atom on a neighboring molecule is the hydrogen bond. For example, in HF the hydrogen is strongly attracted to the fluorine on neighboring molecules (◀ Figure 6.37).

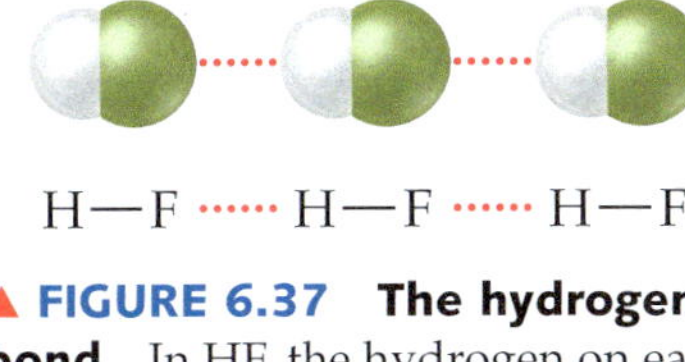

▲ **FIGURE 6.37 The hydrogen bond** In HF, the hydrogen on each molecule is strongly attracted to the fluorine on its neighbor. The intermolecular attraction of a hydrogen atom to an electronegative atom on a neighboring molecule is called a *hydrogen bond*.

Do not confuse hydrogen bonds with chemical bonds. Chemical bonds occur between *individual atoms within a molecule* and are generally much stronger than hydrogen bonds. A hydrogen bond has only 2 to 5 % the strength of a typical covalent chemical bond. Hydrogen bonds—like dispersion forces and dipole–dipole forces—are intermolecular forces that occur *between molecules*. In liquid water, for example, the hydrogen bonds are transient, constantly forming, breaking, and re-forming as water molecules move within the liquid. Hydrogen bonds are, however, strong intermolecular forces. Substances composed of molecules that form hydrogen bonds have much higher melting and boiling points than you would predict based on molar mass. For example, consider the two compounds, methanol and ethane:

Name	Formula	Molar mass (g/mol)	Structure	bp (°C)	mp (°C)
Methanol	CH_3OH	32.0	H—C(H)(H)—O—H	64.7	−97.8
Ethane	C_2H_6	30.1	H—C(H)(H)—C(H)(H)—H	−88	−172

Since methanol contains hydrogen directly bonded to oxygen, its molecules have hydrogen bonding as an intermolecular force. The hydrogen that is directly bonded to oxygen is strongly attracted to the oxygen on neighboring molecules (◀ Figure 6.38). This strong attraction makes the boiling point of methanol 64.7 °C. Consequently, methanol is a liquid at room temperature. Water is another good example of a molecule with hydrogen bonding as an intermolecular force (▼ Figure 6.39). The boiling point of water (100 °C) is remarkably high for a molecule with such a low molar mass (18.02 g/mol). Hydrogen bonding is important in biological molecules. The shapes of proteins and nucleic acids are largely influenced by hydrogen bonding; for example, the two halves of DNA are held together by hydrogen bonds (see the *Chemistry and Health* box later in this section).

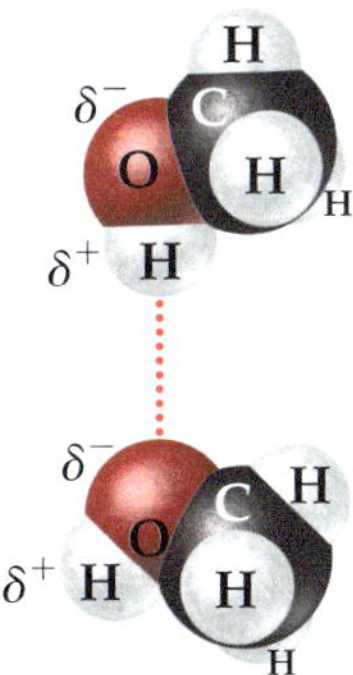

▲ **FIGURE 6.38 Hydrogen bonding in methanol** Since methanol contains hydrogen atoms directly bonded to oxygen, methanol molecules form hydrogen bonds to one another. The hydrogen atom on each methanol molecule is attracted to the oxygen atom of its neighbor.

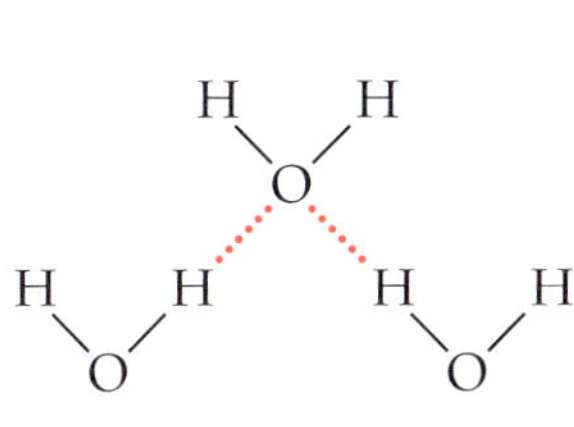

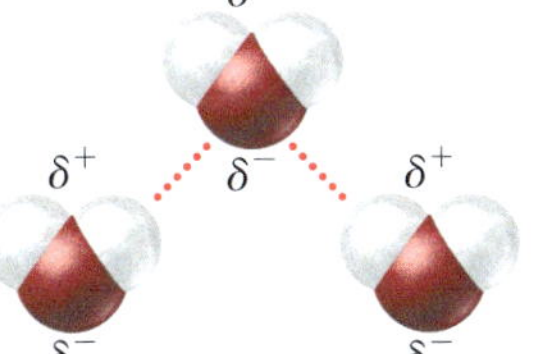

◀ **FIGURE 6.39 Hydrogen bonding in water** Water molecules form strong hydrogen bonds with one another.

EXAMPLE 6.13 HYDROGEN BONDING

One of these compounds is a liquid at room temperature. Which one and why?

O ‖ H—C—H	H | H—C—F | H	H—O—O—H
Formaldehyde	Fluoromethane	Hydrogen peroxide

SOLUTION

The three compounds have similar molar masses.

formaldehyde	30.03 g/mol
fluoromethane	34.04 g/mol
hydrogen peroxide	34.02 g/mol

In some cases, hydrogen bonding can occur between one molecule in which H is directly bonded to F, O, or N and another molecule containing an electronegative atom. (See the box, *Chemistry and Health*, in this section for an example.)

Therefore, the strengths of their dispersion forces are similar. All three compounds are also polar, so they have dipole–dipole forces. Hydrogen peroxide, however, is the only compound that also contains H bonded directly to F, O, or N. Therefore it also has hydrogen bonding and is most likely to have the highest boiling point of the three. Since the problem stated that only one of the compounds was a liquid, we can safely assume that hydrogen peroxide is the liquid. Note that although fluoromethane *contains* both H and F, H is not *directly bonded* to F, so fluoromethane does not have hydrogen bonding as an intermolecular force. Similarly, although formaldehyde *contains* both H and O, H is not *directly bonded* to O, so formaldehyde does not have hydrogen bonding either.

▶SKILLBUILDER 6.12 | Hydrogen Bonding

Which has the higher boiling point, HF or HCl? Why?

▶FOR MORE PRACTICE Examples 6.20, 6.21; Problems 104, 105, 106, 107, 108, 109.

CONCEPTUAL CHECKPOINT 6.9

Three molecular compounds A, B, and C have nearly identical molar masses. Substance A is nonpolar, substance B is polar, and substance C undergoes hydrogen bonding. What is most likely to be the relative order of their boiling points?

(a) $A < B < C$

(b) $C < B < A$

(c) $B < C < A$

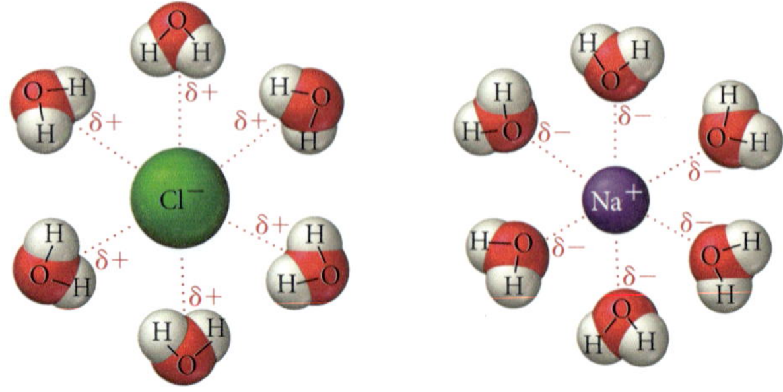

▲ **FIGURE 6.40 Ion–dipole forces.** Ion–dipole forces exist between Na^+ and the negative ends of H_2O molecules and between Cl^- and the positive ends of H_2O molecules.

Ion–Dipole Force

The **ion–dipole force** occurs in mixtures of ionic compounds and polar compounds; it is especially important in aqueous solutions of ionic compounds. For example, when sodium chloride is mixed with water, the sodium and chloride ions interact with water molecules via ion–dipole forces, as shown in ◀ Figure 6.40. The positive sodium ions interact with the negative poles of water molecules, while the negative chloride ions interact with the positive poles. Ion–dipole forces are the strongest of the four types of intermolecular forces discussed and are responsible for the ability of ionic substances to form solutions with water. We will discuss aqueous solutions more thoroughly in Module 7.

Table 6.7 summarizes the different types of intermolecular forces. Remember that dispersion forces, the weakest kind of intermolecular force, are present in all molecules and atoms and increase with increasing molar mass. These forces are weak in small molecules, but they become substantial in molecules with high molar masses. Dipole–dipole forces are present only in polar molecules. Hydrogen bonds are present in molecules containing hydrogen bonded directly to fluorine, oxygen, or nitrogen, and ion–dipole forces occur in mixtures of ionic compounds and polar compounds.

TABLE 6.7 Types of Intermolecular Forces

Type of Force	Relative Strength	Present in	Example
dispersion force (or London force)	weak, but increases with increasing molar mass	all atoms and molecules	H_2 H_2
dipole–dipole force	moderate	only polar molecules	δ^+ HCl δ^- δ^+ HCl δ^-
hydrogen bond	strong	molecules containing H bonded directly to F, O, or N	δ^+ HF δ^- δ^+ HF δ^-
ion–dipole	very strong	mixtures of ionic compounds and polar compounds	+ δ^- δ^- δ^- δ^- δ^- δ^-

CHEMISTRY AND HEALTH

▶ Hydrogen Bonding in DNA

DNA is a long chainlike molecule that acts as a blueprint for living organisms. Copies of DNA are passed from parent to offspring, which is why we inherit traits from our parents. A DNA molecule is composed of thousands of repeating units called *nucleotides* (▶ Figure 6.41). Each nucleotide contains one of four different bases: adenine, thymine, cytosine, and guanine (abbreviated A, T, C, and G, respectively). The order of these bases in DNA encodes the instructions that specify how proteins—the workhorse molecules in living organisms—are made in each cell of the body. Proteins determine virtually all human characteristics, including how we look, how we fight infections, and even how we behave. Consequently, human DNA is a blueprint for how humans are made.

Each time a human cell divides, it must copy the blueprint—which means replicating its DNA. The replicating mechanism is related to the structure of DNA, discovered in 1953 by James Watson and Francis Crick. DNA consists of two complementary strands wrapped around each other in the now famous double helix. Each strand is held to the other by hydrogen bonds that occur between the bases on each strand. DNA replicates because each base (A, T, C, and

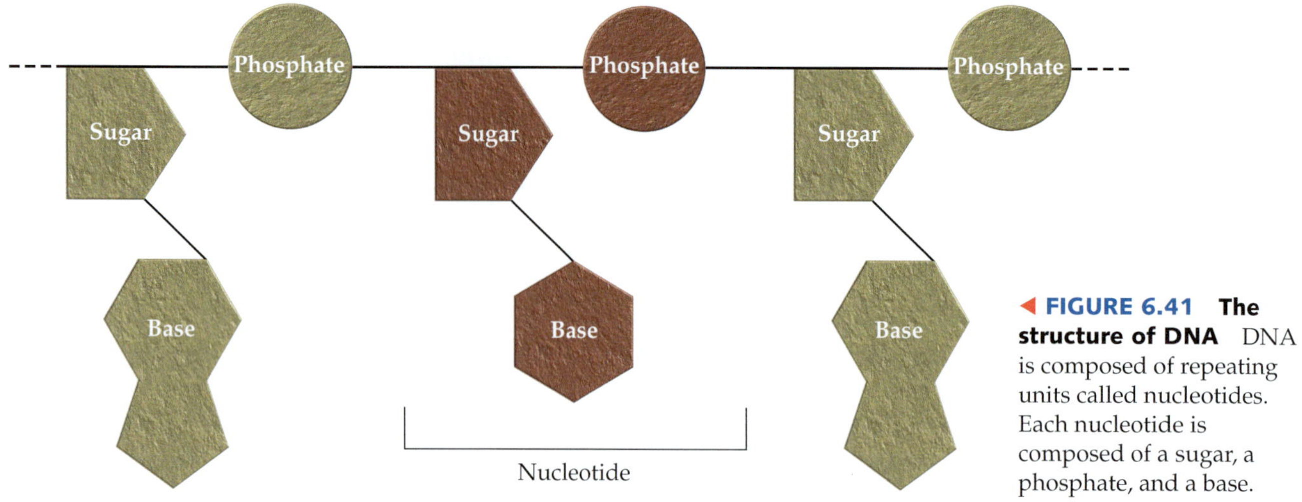

◀ **FIGURE 6.41 The structure of DNA** DNA is composed of repeating units called nucleotides. Each nucleotide is composed of a sugar, a phosphate, and a base.

G) has a complementary partner with which it hydrogen-bonds (▼ Figure 6.42). Adenine (A) hydrogen-bonds with thymine (T), and cytosine (C) hydrogen-bonds with guanine (G). The hydrogen bonds are so specific that each base will pair only with its complementary partner. When a cell is going to divide, the DNA unzips across the hydrogen bonds that run along its length. Then new nucleotides, containing bases complementary to the bases in each half, add along each of the halves, forming hydrogen bonds with their complement. The result is two identical copies of the original DNA.

B6.4 CAN YOU ANSWER THIS? *Why would dispersion forces not work as a way to hold the two halves of DNA together? Why would covalent bonds not work?*

▶ **FIGURE 6.42 Hydrogen bonding in DNA** The two halves of the DNA double helix are held together by hydrogen bonds.

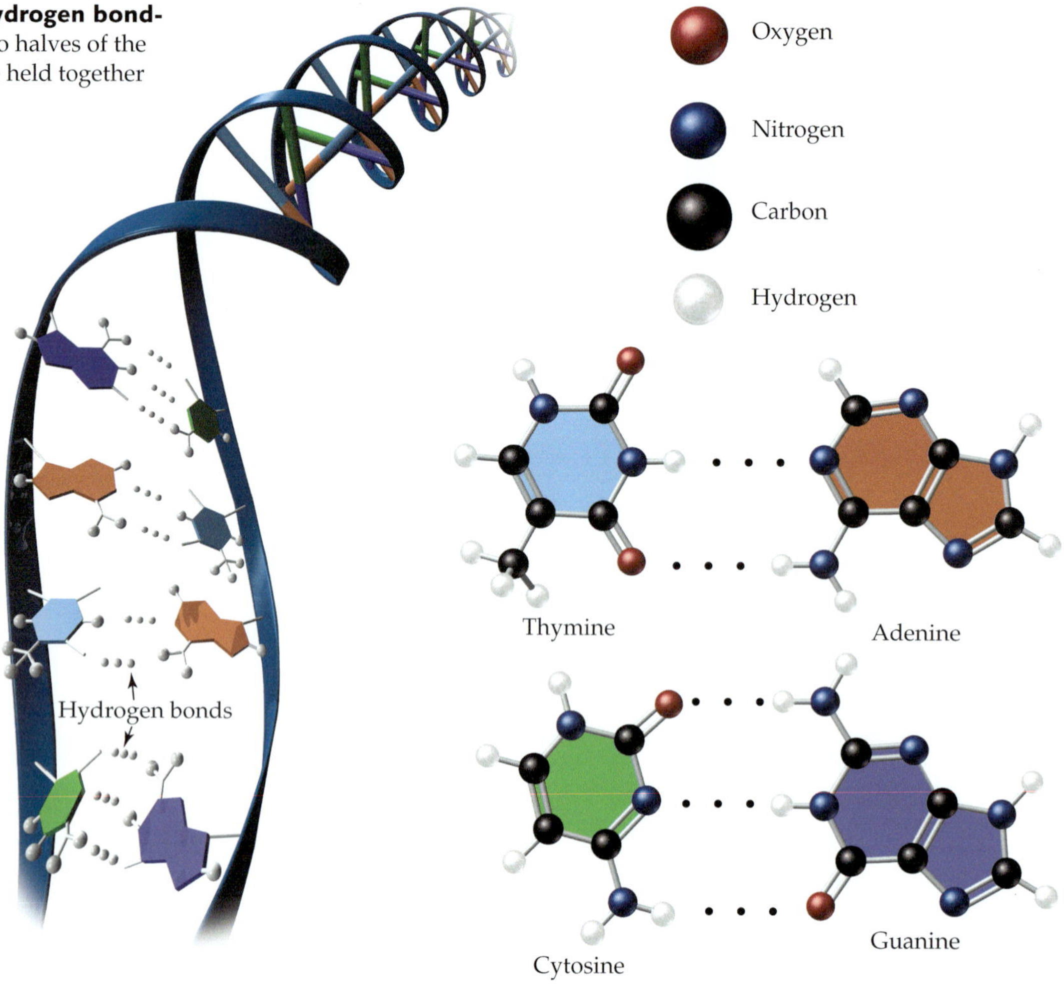

CONCEPTUAL CHECKPOINT 6.10

When dry ice sublimes, what forces are overcome?

(a) chemical bonds between carbon atoms and oxygen atoms

(b) hydrogen bonds between carbon dioxide molecules

(c) dispersion forces between carbon dioxide molecules

(d) dipole–dipole forces between carbon dioxide molecules

MODULE IN REVIEW

Self-Assessment Quiz

Q1. According to kinetic molecular theory, the particles that compose a gas:

(a) attract one another strongly

(b) are packed closely together—nearly touching one another

(c) have an average kinetic energy proportional to the temperature in kelvins

(d) All of the above

Q2. Convert 558 mm Hg to atm.

(a) 0.734 atm **(c)** 38.0 atm

(b) 4.24×10^5 atm **(d)** 18.6 atm

Q3. If you double the volume of a constant amount of gas at a constant temperature, what happens to the pressure?

(a) The pressure doubles.

(b) The pressure quadruples.

(c) The pressure does not change.

(d) The pressure is one-half the initial pressure.

Q4. A 2.55 L gas sample in a cylinder with a freely moveable piston is initially at 25.0 °C. What is the volume of the gas when it is heated to 145 °C (at constant pressure)?

(a) 1.82 L **(b)** 3.58 L **(c)** 0.440 L **(d)** 14.8 L

Q5. A gas sample has an initial volume of 55.2 mL, an initial temperature of 35.0 °C, and an initial pressure of 735 mm Hg. The volume is decreased to 48.8 mL and the temperature is increased to 72.5 °C. What is the final pressure?

(a) 933 mm Hg **(c)** 1.72×10^3 mm Hg

(b) 186 mm Hg **(d)** 401 mm Hg

Q6. Each gas sample has the same temperature and pressure. Which sample occupies the greatest volume?

(a) 4.0 g He **(c)** 40.0 g Ar

(b) 40.0 g Ne **(d)** 84.0 g Kr

Q7. Find the volume occupied by 22.0 g of helium gas at 25.0 °C and 1.18 atm of pressure.

(a) 123 L **(c)** 114 L

(b) 9.57 L **(d)** 159 L

Q8. A mixture of three gases containing an equal number of moles of each gas has a total pressure of 9.0 atm. What is the partial pressure of each gas?

(a) 1.0 atm **(c)** 9.0 atm

(b) 3.0 atm **(d)** 27.0 atm

Q9. A gas mixture containing only helium and neon is 22.1 % neon (by volume) and has a total pressure of 748 mm Hg. What is the partial pressure of neon?

(a) 165 mm Hg **(c)** 583 mm Hg

(b) 1.69×10^4 mm Hg **(d)** 22.1 mm Hg

Q10. Convert the boiling point of water (100.00 °C) to K.

(a) −173.15 K **(c)** 100.00 K

(b) 0 K **(d)** 373.15 K

Q11. A gas sample at STP contains 1.15 g oxygen and 1.55 g nitrogen. What is the volume of the gas sample?

(a) 1.26 L **(c)** 4.09 L

(b) 2.04 L **(d)** 60.5 L

Q12. The first diagram shown here represents liquid water. Which of the diagrams that follow best represents the water after it has boiled?

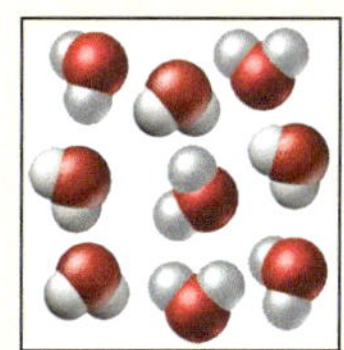

(a)

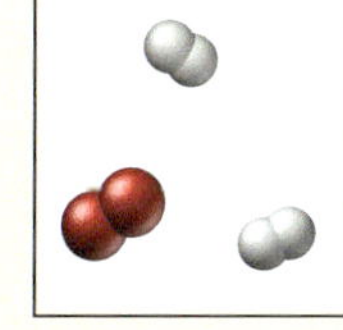

(b)

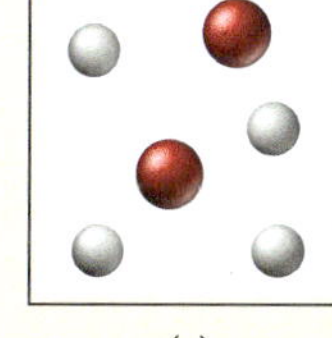

(c)

Q13. Which change affects the vapor pressure of a liquid?

(a) Pouring the liquid into a different container

(b) Increasing the surface area of the liquid

(c) Increasing the temperature of the liquid

(d) All of the above

Q14. A sample of ice is heated past its melting point, and its temperature is monitored. The graph below shows the results. What is the first point on the graph where the sample contains no solid ice?

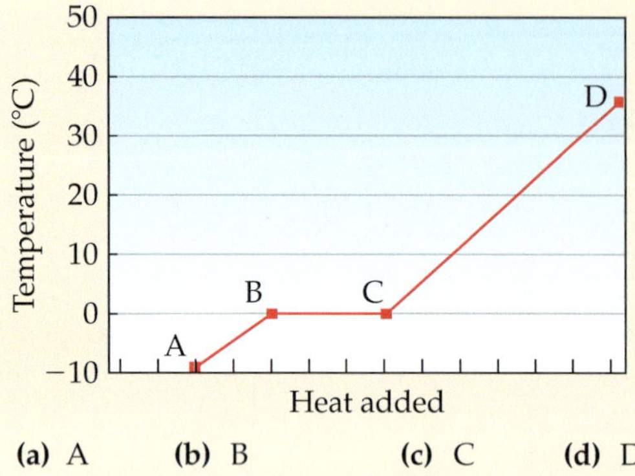

(a) A **(b)** B **(c)** C **(d)** D

Q15. Which halogen has the highest boiling point?

(a) F_2 **(b)** Cl_2 **(c)** Br_2 **(d)** I_2

Q16. Which substance has dipole–dipole forces?

(a) OF_2 **(b)** CBr_4 **(c)** CS_2 **(d)** Br_2

Q17. Which substance exhibits hydrogen bonding?

(a) CF_4 **(b)** CH_3OH **(c)** H_2 **(d)** PH_3

Q18. Which substance would you expect to have the highest melting point?

(a) CH_3CH_3 **(b)** CH_3CH_2Cl

(c) CH_4 **(d)** CH_3CH_2OH

Answers: 1:c, 2:a, 3:d, 4:b, 5:a, 6:b, 7:c,8:b, 9:a, 10:d, 11:b, 12:a, 13:c, 14:c, 15:d, 16:a, 17:b, 18:d

Chemical Principles

Kinetic Molecular Theory: The kinetic molecular theory is a model for gases. In this model, gases are composed of widely spaced, noninteracting particles whose average kinetic energy depends on temperature.

Pressure: Pressure is the force per unit area that results from the collision of gas particles with surfaces. The SI unit of pressure is the pascal, but we often express pressure in other units such as atmospheres, millimetres of mercury, and torr.

Simple Gas Laws: The simple gas laws show how one of the properties of a gas varies with another property of the same gas.

Volume (V) and Pressure (P)

$$V \propto \frac{1}{P} \quad \text{(Boyle's law)}$$

Volume (V) and Temperature (T)

$$V \propto T \quad \text{(Charles's law)}$$

Volume (V) and Moles (n)

$$V \propto n \quad \text{(Avogadro's law)}$$

Relevance

Kinetic Molecular Theory: The kinetic molecular theory predicts many of the properties of gases, including their low density in comparison to solids, their compressibility, and their tendency to assume the shape and volume of their container. The kinetic molecular theory also predicts the ideal gas law.

Pressure: Pressure is a fundamental property of a gas. It allows tires to be inflated and makes it possible to drink from straws.

Simple Gas Laws: Each of the simple gas laws allows us to see how two properties of a gas are interrelated. They are also useful in calculating how one of the properties of a gas changes when another does. We can use Boyle's law, for example, to calculate how the volume of a gas will change in response to a change in pressure, or vice versa.

Temperature: The temperature of matter is related to the random motions of the molecules and atoms that compose it—the greater the motion, the higher the temperature. To measure temperature we use two scales: Celsius (°C), and Kelvin (K).

Temperature: The temperature of matter and its measurement are relevant to many everyday phenomena. Humans are understandably interested in the weather, and air temperature is a fundamental part of weather. We use body temperature as one measure of human health and global temperature as one measure of the planet's health.

The Combined Gas Law: The combined gas law joins Boyle's law and Charles's law.

$$\frac{P_1V_1}{T_1} = \frac{P_2V_2}{T_2}$$

The Combined Gas Law: We use the combined gas law to calculate how a property of a gas (pressure, volume, or temperature) changes when two other properties change at the same time.

The Ideal Gas Law: The ideal gas law combines the four properties of a gas—pressure, volume, temperature, and number of moles—in a single equation showing their interrelatedness.

$$PV = nRT$$

The Ideal Gas Law: We can use the ideal gas law to find any one of the four properties of a gas if we know the other three.

Mixtures of Gases: The pressure due to an individual component in a mixture of gases is its partial pressure, and we define it as the fractional composition of the component multiplied by the total pressure:

Partial pressure of component = Fractional composition of component × Total pressure

Dalton's law states that the total pressure of a mixture of gases is equal to the sum of the partial pressures of its components.

$$P_{\text{tot}} = P_a + P_b + P_c + \ldots$$

Mixtures of Gases: Since many gases are not pure but mixtures of several components, it is useful to know how each component contributes to the properties of the entire mixture. The concepts of partial pressure are relevant to deep-sea diving, for example, and to collecting gases over water, where water vapor mixes with the gas being collected.

Properties of Liquids:

- High densities in comparison to gases.
- Indefinite shape; they assume the shape of their container.
- Definite volume; they are not easily compressed.

Properties of Liquids: Common liquids include water, acetone (fingernail-polish remover), and rubbing alcohol. Water is the most common and most important liquid on Earth. It is difficult to imagine life without water.

Properties of Solids:

- High densities in comparison to gases.
- Definite shape; they do not assume the shape of their container.
- Definite volume; they are not easily compressed.
- May be crystalline (ordered) or amorphous (disordered).

Properties of Solids: Much of the matter we encounter is solid. Common solids include ice, dry ice, and diamond. Understanding the properties of solids involves understanding the particles that compose them and how those particles interact.

Evaporation and Condensation: Evaporation or vaporization—an endothermic physical change—is the conversion of a liquid to a gas. Condensation—an exothermic physical change—is the conversion of a gas to a liquid. When the rates of evaporation and condensation in a liquid/gas sample are equal, dynamic equilibrium is reached and the partial pressure of the gas at that point is its vapor pressure. When the vapor pressure equals the external pressure, the boiling point is reached. At the boiling point, thermal energy causes molecules in the interior of the liquid, as well as those at the surface, to convert to gas, resulting in the bubbling.

Evaporation and Condensation: Evaporation is the body's natural cooling system. When we get overheated, we sweat; the sweat then evaporates and cools us. Evaporation and condensation both play roles in moderating climate. Humid areas, for example, cool less at night because as the temperature drops, water condenses out of the air, releasing heat and preventing a further temperature drop.

Melting and Freezing: Melting—an endothermic physical change—is the conversion of a solid to a liquid, and freezing—an exothermic physical change—is the conversion of a liquid to a solid.

Melting and Freezing: We use the melting of solid ice, for example, to cool drinks when we place ice cubes in them. Since melting is endothermic, it absorbs heat from the liquid and cools it.

Types of Intermolecular Forces: The four main types of intermolecular forces are:

Dispersion forces—Dispersion forces occur between all molecules and atoms due to instantaneous fluctuations in electron charge distribution. The strength of the dispersion force increases with increasing molar mass.

Dipole–dipole forces—Dipole–dipole forces exist between molecules that are polar. Consequently, polar molecules have higher melting and boiling points than nonpolar molecules of similar molar mass.

Hydrogen bonding—Hydrogen bonding exists between molecules that have H bonded directly to F, O, or N. Hydrogen bonds are stronger than dispersion forces or dipole–dipole forces.

Ion–dipole forces—The ion–dipole force occurs in mixtures of ionic compound and polar compounds.

Types of Intermolecular Forces: The type of intermolecular force present in a substance determines many of the properties of the substance. The stronger the intermolecular force, for example, the greater the melting and boiling points of the substance. In addition, the miscibility of liquids—their ability to mix without separating—depends on the relative kinds of intermolecular forces present within them. In general, polar liquids are miscible with other polar liquids, but not with nonpolar liquids. Hydrogen bonding is important in many biological molecules such as proteins and DNA. Ion–dipole forces are important in mixtures of ionic compounds and water.

Chemical Skills

LO: Convert among pressure units (Section 6.2).

SORT
You are given a pressure in atm and asked to convert the units to mm Hg.

STRATEGIZE
Begin with the quantity you are given and multiply by the appropriate conversion factor(s) to determine the quantity you are trying to find.

SOLVE
Follow the solution map to solve the problem.

CHECK
Are the units correct? Does the answer make physical sense?

Examples

EXAMPLE 6.14 PRESSURE UNIT CONVERSION

Convert 0.89 atm to mm Hg.

GIVEN: 0.89 atm

FIND: mm Hg

SOLUTION MAP

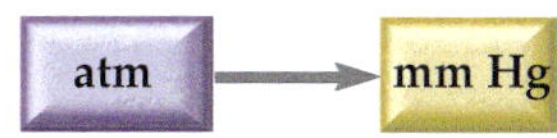

RELATIONSHIPS USED

1 atm = 760 mm Hg (Table 6.1)

SOLUTION

$$0.89\ \text{atm} \times \frac{760\ \text{mm Hg}}{1\ \text{atm}} = 676\ \text{mm Hg}$$

The unit (mm Hg) is correct. The value of the answer is reasonable because 0.89 atm is less than 1 atm. Therefore the answer should be less than 760 mm Hg.

LO: Restate and apply Boyle's law (Section 6.3).

SORT
You are given the initial and final pressures and the initial volume. You are asked to find the final volume.

STRATEGIZE
Calculations involving the simple gas laws usually consist of finding one of the initial or final conditions given the other initial and final conditions. In this case, use Boyle's law to find V_2 given P_1, V_1, and V_2.

SOLVE
Solve Boyle's law for V_2 and substitute the correct variables to calculate its value.

CHECK
Are the units correct? Does the answer make physical sense?

EXAMPLE 6.15 SIMPLE GAS LAWS

A gas has a volume of 5.7 L at a pressure of 3.2 atm. What is its volume at 4.7 atm? (Assume constant temperature.)

GIVEN: $P_1 = 3.2$ atm
$V_1 = 5.7$ L
$P_2 = 4.7$ atm

FIND: V_2

SOLUTION MAP

RELATIONSHIPS USED

$P_1V_1 = P_2V_2$ (Boyle's law, Section 6.3)

SOLUTION

$$P_1V_1 = P_2V_2$$

$$V_2 = \frac{P_1}{P_2}V_1$$

$$= \frac{3.2\ \text{atm}}{4.7\ \text{atm}} \times 5.7\ \text{L}$$

$$= 3.9\ \text{L}$$

The unit (L) is correct. The value is reasonable because as pressure increases volume should decrease.

LO: Convert between Celsius, and Kelvin temperature Scales (Section 6.4).

EXAMPLE 6.16 CONVERTING BETWEEN CELSIUS AND KELVIN TEMPERATURE SCALES

Convert 257 K to Celsius.

SORT
You are given the temperature in kelvins and asked to convert it to degrees Celsius.

GIVEN: 257 K

FIND: °C

STRATEGIZE
Draw a solution map. Use the equation that relates the *given* quantity to the *find* quantity.

SOLUTION MAP

RELATIONSHIPS USED

$$K = °C + 273.15 \text{ (Section 6.4)}$$

SOLVE
Solve the equation for the *find* quantity (°C) and substitute the temperature in K into the equation. Calculate the answer to the correct number of significant figures.

SOLUTION

$$K = °C + 273.15$$
$$°C = K - 273.15$$
$$°C = 257 - 273.15 = -16\ °C$$

CHECK
Are the units correct? Does the answer make physical sense?

The answer has the correct unit, and its magnitude seems correct.

LO: Restate and apply the combined gas law (Section 6.6).

EXAMPLE 6.17 THE COMBINED GAS LAW

A sample of gas has an initial volume of 2.4 L at a pressure of 855 mm Hg and a temperature of 298 K. If the gas is heated to a temperature of 387 K and expanded to a volume of 4.1 L, what is its final pressure in millimetres of mercury?

SORT
You are given the initial and final volume and temperature, and you are given the initial pressure. You are asked to find the final pressure.

GIVEN: $P_1 = 855$ mm Hg
$V_1 = 2.4$ L
$T_1 = 298$ K
$V_2 = 4.1$ L
$T_2 = 387$ K

FIND: P_2

STRATEGIZE
Problems involving the combined gas law usually consist of finding one of the initial or final conditions given the other initial and final conditions. In this case, you can use the combined gas law to find the unknown quantity, P_2.

SOLUTION MAP

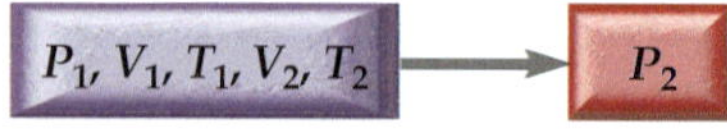

RELATIONSHIPS USED

$$\frac{P_1V_1}{T_1} = \frac{P_2V_2}{T_2} \text{ (combined gas law, Section 6.6)}$$

SOLVE

Solve the combined gas law for the quantity you are trying to find, in this case P_2, and substitute the known quantities to calculate the value of P_2.

SOLUTION

$$\frac{P_1V_1}{T_1} = \frac{P_2V_2}{T_2}$$

$$P_2 = \frac{P_1V_1T_2}{T_1V_2}$$

$$= \frac{855 \text{ mm Hg} \times 2.4 \text{ L} \times 387 \text{ K}}{298 \text{ K} \times 4.1 \text{ L}}$$

$$= 6.5 \times 10^2 \text{ mm Hg}$$

CHECK

Are the units correct? Does the answer make physical sense?

The unit (mm Hg) is correct. The value of the answer makes sense because the volume increase was proportionally more than the temperature decrease; therefore you would expect the pressure to decrease.

LO: **Restate and apply the ideal gas law (Section 6.8).**

EXAMPLE 6.18 THE IDEAL GAS LAW

Calculate the pressure exerted by 1.2 mol of gas in a volume of 28.2 L and at a temperature of 334 K.

SORT

You are given the number of moles of a gas, its volume, and its temperature. You are asked to find its pressure.

GIVEN: $n = 1.2$ mol

$V = 28.2$ L

$T = 334$ K

FIND: P

STRATEGIZE

Calculations involving the ideal gas law often involve finding one of the four variables (P, V, n, or T) given the other three. In this case, you are asked to find P. Use the given variables and the ideal gas law to arrive at P.

SOLUTION MAP

RELATIONSHIPS USED

$PV = nRT$ (ideal gas law, Section 6.8)

SOLVE

Solve the ideal gas law equation for P and substitute the given variables to calculate P.

SOLUTION

$$PV = nRT$$

$$P = \frac{nRT}{V}$$

$$= \frac{1.2 \text{ mol} \times 0.0821 \frac{\text{L} \cdot \text{atm}}{\text{mol} \cdot \text{K}} \times 334 \text{ K}}{28.2 \text{ L}}$$

$$= 1.2 \text{ atm}$$

CHECK

Are the units correct? Does the answer make physical sense?

The unit (atm) is correct for pressure. The *value* of the answer is a bit more difficult to judge. However, knowing that at standard temperature and pressure ($t = 0$ °C or $T = 273.15$ K and $P = 1$ atm) 1 mol of gas occupies 22.4 L, you can see that the answer is reasonable because you have a bit more than one mole of a gas at a temperature not too far from standard temperature. The volume of the gas is a bit higher than 22.4 L; therefore you might expect the pressure to be a bit higher than 1 atm.

LO: Restate and apply Dalton's law of partial pressures (Section 6.9).

EXAMPLE 6.19 TOTAL PRESSURE AND PARTIAL PRESSURE

A mixture of three gases has the partial pressures:

$$P_{CO_2} = 289 \text{ mm Hg}$$
$$P_{O_2} = 342 \text{ mm Hg}$$
$$P_{N_2} = 122 \text{ mm Hg}$$

What is the total pressure of the mixture?

SORT

You are given the partial pressures of three gases in a mixture and asked to find the total pressure.

GIVEN: $P_{CO_2} = 289$ mm Hg
$P_{O_2} = 342$ mm Hg
$P_{N_2} = 122$ mm Hg

FIND: P_{tot}

SOLVE

Use Dalton's law of partial pressures ($P_{tot} = P_a + P_b + P_c + \ldots$) to solve the problem. Sum the partial pressures to obtain the total pressure.

SOLUTION

$$\begin{aligned} P_{tot} &= P_{CO_2} + P_{O_2} + P_{N_2} \\ &= 289 \text{ mm Hg} + 342 \text{ mm Hg} + 122 \text{ mm Hg} \\ &= 753 \text{ mm Hg} \end{aligned}$$

LO: Determine the types of intermolecular forces in compounds (Section 6.13).

EXAMPLE 6.20 DETERMINING THE TYPE OF INTERMOLECULAR FORCES IN A COMPOUND

Determine the types of intermolecular forces present in each substance.

(a) N_2
(b) CO
(c) NH_3

All substances exhibit dispersion forces. Polar substances—those whose molecules have polar bonds that add to net dipole moment—also exhibit dipole–dipole forces. Substances whose molecules contain H bonded directly to F, O, or N exhibit hydrogen bonding as well.

SOLUTION

(a) N_2 is nonpolar and therefore has only dispersion forces.
(b) CO is polar and therefore has dipole–dipole forces (in addition to dispersion forces).
(c) NH_3 has hydrogen bonding (in addition to dispersion forces and dipole–dipole forces).

LO: Use intermolecular forces to determine melting and/or boiling points (Section 6.13).

EXAMPLE 6.21 USING INTERMOLECULAR FORCES TO DETERMINE MELTING AND/OR BOILING POINTS

Arrange each group of compounds in order of increasing boiling point.

(a) F_2, Cl_2, Br_2
(b) HF, HCl, HBr

To determine relative boiling points and melting points among compounds, you must evaluate the types of intermolecular forces that each compound exhibits. Dispersion forces are the weakest kind of intermolecular force, but they increase with increasing molar mass. Dipole–dipole forces are stronger than dispersion forces. If two compounds have similar molar mass, but one is polar, it will have higher melting and boiling points. Hydrogen bonds are a stronger type of intermolecular force. Substances that exhibit hydrogen bonding will have much higher boiling and melting points than substances without hydrogen bonding, even if the substance without hydrogen bonding is of higher molar mass.

SOLUTION

(a) Since these all have only dispersion forces, and since they are similar substances (all halogens), the strength of the dispersion force will increase with increasing molar mass. Therefore, the correct order is $F_2 < Cl_2 < Br_2$.
(b) Since HF has hydrogen bonding, it has the highest boiling point. Between HCl and HBr, HBr (because of its higher molar mass) has a higher boiling point. Therefore the correct order is HCl < HBr < HF.

KEY TERMS

absolute zero **[6.4]**
atmosphere (atm) **[6.2]**
Avogadro's law **[6.7]**
boiling point **[6.12]**
Boyle's law **[6.3]**
Charles's law **[6.5]**
combined gas law **[6.6]**
condensation **[6.12]**
Dalton's law of partial pressures **[6.9]**
dipole–dipole force **[6.13]**
dispersion force **[6.13]**
dynamic equilibrium **[6.12]**
evaporation **[6.12]**
heat **[6.4]**
hydrogen bond **[6.13]**
hypoxia **[6.9]**
ideal gas constant (*R*) **[6.8]**
ideal gas law **[6.8]**
instantaneous (temporary) dipole **[6.13]**
intermolecular forces **[6.10]**
ion–dipole force **[6.13]**
Kelvin (K) scale **[6.4]**
kinetic molecular theory **[6.1]**
melting point **[6.12]**
millimetre of mercury (mm Hg) **[6.2]**
miscibility **[6.13]**
nitrogen narcosis **[6.9]**
nonvolatile **[6.12]**
normal boiling point **[6.12]**
oxygen toxicity **[6.9]**
partial pressure **[6.9]**
pascal (Pa) **[6.2]**
permanent dipole **[6.13]**
pressure **[6.2]**
sublimation **[6.12]**
temperature **[6.4]**
thermal energy **[6.4, 6.10-6.12]**
torr **[6.2]**
vaporization **[6.12]**
vapor pressure **[6.12]**
volatile **[6.12]**

EXERCISES

QUESTIONS

1. What are the main assumptions of kinetic molecular theory?
2. Describe the main properties of a gas. How are these predicted by kinetic molecular theory?
3. What units are used to measure pressure?
4. What is Boyle's law? Explain Boyle's law from the perspective of kinetic molecular theory.
5. Explain the difference between heat and temperature.
6. How do the two temperature scales differ?
7. What is Charles's law? Explain Charles's law from the perspective of kinetic molecular theory.
8. What is the combined gas law? When is it useful?
9. What is Avogadro's law? Explain Avogadro's law from the perspective of kinetic molecular theory.
10. What is the ideal gas law? When is it useful?
11. Under what conditions is the ideal gas law most accurate? Under what conditions does the ideal gas law break down? Why?
12. What is partial pressure?
13. What is Dalton's law?
14. What is standard temperature and pressure (STP)? What is the molar volume of a gas at STP?
15. What is evaporation? Condensation?
16. Why does a glass of water evaporate more slowly in the glass than if you spilled the same amount of water on a table?
17. Explain the difference between evaporation below the boiling point of a liquid and evaporation at the boiling point of a liquid.
18. What is the boiling point of a liquid? What is the normal boiling point?
19. Acetone evaporates more quickly than water at room temperature. What can you say about the relative strength of the intermolecular forces in the two compounds? Which substance is more volatile?
20. Explain condensation and dynamic equilibrium.
21. What is the vapor pressure of a substance? How does it depend on the temperature and strength of intermolecular forces?
22. Explain how sweat cools the body.
23. Explain why a steam burn from gaseous water at 100 °C is worse than a water burn involving the same amount of liquid water at 100 °C.
24. Explain what happens when a liquid boils.
25. Explain why the water in a cup placed in a small ice chest (without a refrigeration mechanism) initially at –5 °C does *not* freeze.
26. Explain how ice cubes cool down beverages.
27. What are dispersion forces? How does the strength of dispersion forces relate to molar mass?
28. What are dipole–dipole forces? How can you tell whether a compound has dipole–dipole forces?
29. What is hydrogen bonding? How can you tell whether a compound has hydrogen bonding?
30. What are ion–dipole forces? What kinds of substances contain ion–dipole forces?
31. List the four types of intermolecular forces discussed in this chapter in order of increasing relative strength.

PROBLEMS

CONVERTING BETWEEN PRESSURE AND TEMPERATURE UNITS

32. Convert each measurement to atm.
(a) 1277 mm Hg
(b) 2.38×10^5 Pa
(c) 455 torr

33. Convert each measurement to atm.
(a) 921 torr
(b) 4.8×10^4 Pa
(c) 3422 mm Hg

34. Perform each conversion.
(a) 2.3 atm to torr
(b) 4.7×10^{-2} atm to millimetres of mercury

35. Perform each conversion.
(a) 1.06 atm to millimetres of mercury
(b) 95,422 Pa to millimetres of mercury

36. Complete the table.

Pascals	Atmospheres	Millimetres of Mercury	Torr
882 Pa	______	6.62 mm Hg	______
______	0.558 atm	______	______
______	______	______	764 torr
______	______	249 mm Hg	______

37. Complete the table.

Pascals	Atmospheres	Millimetres of Mercury	Torr
______	1.91 atm	______	1.45×10^3 torr
1.15×10^4 Pa	______	______	______
______	______	______	721 torr
______	______	109 mm Hg	______

38. Vodka does not freeze in the freezer because it contains a high percentage of ethanol. The freezing point of pure ethanol is –114 °C. Convert that temperature to degrees Kelvin.

39. Liquid helium boils at 4.2 K. Convert this temperature to degrees Celsius.

40. Complete the table.

Kelvin	Celsius
0.0 K	–273.0 °C
______	8.5 °C

41. Complete the table.

Kelvin	Celsius
______	0.0 °C
385 K	______

SIMPLE GAS LAWS

42. A sample of gas has an initial volume of 3.95 L at a pressure of 705 mm Hg. If the volume of the gas is increased to 5.38 L, what is the pressure? (Assume constant temperature.)

43. A sample of gas has an initial volume of 22.8 L at a pressure of 1.65 atm. If the sample is compressed to a volume of 10.7 L, what is its pressure? (Assume constant temperature.)

44. A snorkeler with a lung capacity of 6.3 L inhales a lungful of air at the surface, where the pressure is 1.0 atm. The snorkeler then descends to a depth of 25 m, where the pressure increases to 3.5 atm. What is the capacity of the snorkeler's lungs at this depth? (Assume constant temperature.)

45. A scuba diver with a lung capacity of 5.2 L inhales at a depth of 45 m and a pressure of 5.5 atm. If the diver were to ascend to the surface (where the pressure is 1.0 atm) while holding her breath, to what volume would the air in her lungs expand? (Assume constant temperature.)

46. Use Boyle's law to complete the table (assume temperature and number of moles of gas to be constant).

P_1	V_1	P_2	V_2
755 mm Hg	2.85 L	885 mm Hg	______
______	1.33 L	4.32 atm	2.88 L
192 mm Hg	382 mL	______	482 mL
2.11 atm	______	3.82 atm	125 mL

47. Use Boyle's law to complete the table (assume temperature and number of moles of gas to be constant).

P_1	V_1	P_2	V_2
______	1.90 L	4.19 atm	1.09 L
755 mm Hg	118 mL	709 mm Hg	______
2.75 atm	6.75 mL	______	49.8 mL
343 torr	______	683 torr	8.79 L

48. A balloon with an initial volume of 3.2 L at a temperature of 299 K is warmed to 376 K. What is its volume at 376 K?

49. A dramatic classroom demonstration involves cooling a balloon from room temperature (298 K) to liquid nitrogen temperature (77 K). If the initial volume of the balloon is 2.7 L, what is its volume after it cools? (Ignore the effect of any gases that might liquefy upon cooling.)

50. A 48.3 mL sample of gas in a cylinder equipped with a piston is warmed from 22 °C to 87 °C. What is its volume at the final temperature?

51. A syringe containing 1.55 mL of oxygen gas is cooled from 95.3 °C to 0.0 °C. What is the final volume of oxygen gas?

52. Use Charles's law to complete the table (assume pressure and number of moles of gas to be constant).

V_1	T_1	V_2	T_2
1.08 L	25.4 °C	1.33 L	______
______	77 K	228 mL	298 K
115 cm^3	______	119 cm^3	22.4 °C
232 L	18.5 °C	______	96.2 °C

53. Use Charles's law to complete the table (assume pressure and number of moles of gas to be constant).

V_1	T_1	V_2	T_2
119 L	10.5 °C	______	112.3 °C
______	135 K	176 mL	315 K
2.11 L	15.4 °C	2.33 L	______
15.4 cm^3	______	19.2 cm^3	10.4 °C

54. A 0.12 mole sample of nitrogen gas occupies a volume of 2.55 L. What is the volume of 0.32 mol of nitrogen gas under the same conditions?

55. A 0.48 mole sample of helium gas occupies a volume of 11.7 L. What is the volume of 0.72 mol of helium gas under the same conditions?

56. A balloon contains 0.128 mol of gas and has a volume of 2.76 L. If an additional 0.073 mol of gas is added to the balloon, what is its final volume?

57. A cylinder with a moveable piston contains 0.87 mol of gas and has a volume of 334 mL. What will its volume be if an additional 0.22 mol of gas is added to the cylinder?

58. Use Avogadro's law to complete the table (assume pressure and temperature to be constant).

V_1	n_1	V_2	n_2
38.5 mL	1.55×10^{-3} mol	49.4 mL	______
______	1.37 mol	26.8 L	4.57 mol
11.2 L	0.628 mol	______	0.881 mol
422 mL	______	671 mL	0.0174 mol

59. Use Avogadro's law to complete the table (assume pressure and temperature to be constant).

V_1	n_1	V_2	n_2
25.2 L	5.05 mol	______	3.03 mol
______	1.10 mol	414 mL	0.913 mol
8.63 L	0.0018 mol	10.9 L	______
53 mL	______	13 mL	2.61×10^{-4} mol

THE COMBINED GAS LAW

60. A sample of gas with an initial volume of 28.4 L at a pressure of 725 mm Hg and a temperature of 305 K is compressed to a volume of 14.8 L and warmed to a temperature of 375 K. What is the final pressure of the gas?

61. A cylinder with a moveable piston contains 218 mL of nitrogen gas at a pressure of 1.32 atm and a temperature of 298 K. What must the final volume be for the pressure of the gas to be 1.55 atm at a temperature of 335 K?

62. A scuba diver takes a 2.8 L balloon from the surface, where the pressure is 1.0 atm and the temperature is 34 °C, to a depth of 25 m, where the pressure is 3.5 atm and the temperature is 18 °C. What is the volume of the balloon at this depth?

63. A bag of potato chips contains 585 mL of air at 25 °C and a pressure of 765 mm Hg. Assuming the bag does not break, what will be its volume at the top of a mountain where the pressure is 442 mm Hg and the temperature is 5.0 °C?

64. A gas sample with a volume of 5.3 L has a pressure of 735 mm Hg at 28 °C. What is the pressure of the sample if the volume remains at 5.3 L but the temperature rises to 86 °C?

65. The total pressure in a 11.7 L automobile tire is 3.00 atm at 11 °C. How much does the pressure in the tire rise if its temperature increases to 37 °C and the volume remains at 11.7 L?

66. Use the combined gas law to complete the table (assume the number of moles of gas to be constant).

P_1	V_1	T_1	P_2	V_2	T_2
1.21 atm	1.58 L	12.2 °C	1.54 atm	______	32.3 °C
721 torr	141 mL	135 K	801 torr	152 mL	______
5.51 atm	0.879 L	22.1 °C	______	1.05 L	38.3 °C

67. Use the combined gas law to complete the table (assume the number of moles of gas to be constant).

P_1	V_1	T_1	P_2	V_2	T_2
1.01 atm	______	2.7 °C	0.54 atm	0.58 L	42.3 °C
123 torr	41.5 mL	______	626 torr	36.5 mL	205 K
______	1.879 L	20.8 °C	0.412 atm	2.05 L	48.1 °C

THE IDEAL GAS LAW

68. What is the volume occupied by 0.255 mol of helium gas at 1.25 atm and 305 K?

69. What is the pressure in a 20.0 L cylinder filled with 0.683 mol of nitrogen gas at 325 K?

70. A cylinder contains 28.5 L of oxygen gas at a pressure of 1.8 atm and a temperature of 298 K. How many moles of gas are in the cylinder?

71. What is the temperature of 0.52 mol of gas at a pressure of 1.3 atm and a volume of 11.8 L?

72. A cylinder contains 11.8 L of air at a total pressure of 2.94 atm and a temperature of 25 °C. How many moles of gas does the cylinder contain?

73. What is the pressure in millimetres of mercury of 0.0115 mol of helium gas with a volume of 214 mL at 45 °C?

74. Use the ideal gas law to complete the table.

P	*V*	*n*	*T*
1.05 atm	1.19 L	0.112 mol	______
112 torr	______	0.241 mol	304 K
______	28.5 mL	1.74×10^{-3} mol	25.4 °C
0.559 atm	0.439 L	______	255 K

75. Use the ideal gas law to complete the table.

P	*V*	*n*	*T*
2.39 atm	1.21 L	______	205 K
512 torr	______	0.741 mol	298 K
0.433 atm	0.192 L	0.0131 mol	______
______	20.2 mL	5.71×10^{-3} mol	20.4 °C

MOLAR VOLUME

76. Calculate the volume of each gas sample at STP.
(a) 22.5 mol Cl_2
(b) 3.6 mol nitrogen
(c) 2.2 mol helium
(d) 27 mol CH_4

77. Calculate the volume of each gas sample at STP.
(a) 21.2 mol N_2O
(b) 0.215 mol CO
(c) 0.364 mol CO_2
(d) 8.6 mol C_2H_6

78. Calculate the volume of each gas sample at STP.
(a) 73.9 g N_2
(b) 42.9 g O_2
(c) 148 g NO_2
(d) 245 mg CO_2

79. Calculate the volume of each gas sample at STP.
(a) 48.9 g He
(b) 45.2 g Xe
(c) 48.2 mg Cl_2
(d) 3.83 kg SO_2

80. Calculate the mass of each gas sample at STP.
(a) 178 mL CO_2
(b) 155 mL O_2
(c) 1.25 L SF_6

81. Calculate the mass of each gas sample at STP.
(a) 5.82 L NO
(b) 0.324 L N_2
(c) 139 cm^3 Ar

PARTIAL PRESSURE

82. A gas mixture contains each gas at the indicated partial pressure.

N_2	217 torr
O_2	106 torr
He	248 torr

What is the total pressure of the mixture?

83. A gas mixture contains each gas at the indicated partial pressure.

CO_2	422 mm Hg
Ar	102 mm Hg
O_2	165 mm Hg
H_2	52 mm Hg

What is the total pressure of the mixture?

84. A heliox deep-sea diving mixture delivers an oxygen partial pressure of 0.30 atm when the total pressure is 11.0 atm. What is the partial pressure of helium in this mixture?

85. A mixture of helium, nitrogen, and oxygen has a total pressure of 752 mm Hg. The partial pressures of helium and nitrogen are 234 mm Hg and 197 mm Hg, respectively. What is the partial pressure of oxygen in the mixture?

86. The hydrogen gas formed in a chemical reaction is collected over water at 30 °C at a total pressure of 732 mm Hg. Given the partial pressure of water is 31.8 mm Hg, what is the partial pressure of the hydrogen gas collected in this way?

87. The oxygen gas emitted from an aquatic plant during photosynthesis is collected over water at 25 °C and a total pressure of 753 torr. Given the partial pressure of water is 23.8 mm Hg, what is the partial pressure of the oxygen gas?

88. A gas mixture contains 78 % nitrogen and 22 % oxygen. If the total pressure is 1.12 atm, what are the partial pressures of each component?

89. An air sample contains 0.038 % CO_2. If the total pressure is 758 mm Hg, what is the partial pressure of CO_2?

90. A heliox deep-sea diving mixture contains 4.0 % oxygen and 96.0 % helium. What is the partial pressure of oxygen when this mixture is delivered at a total pressure of 8.5 atm?

91. A scuba diver breathing normal air descends to 100 m of depth, where the total pressure is 11 atm. What is the partial pressure of oxygen that the diver experiences at this depth? Is the diver in danger of experiencing oxygen toxicity?

EVAPORATION, CONDENSATION, MELTING, AND FREEZING

92. Which evaporates more quickly: 55 mL of water in a beaker with a diameter of 4.5 cm or 55 mL of water in a dish with a diameter of 12 cm? Why?

93. Two samples of pure water of equal volume are put into separate dishes and kept at room temperature for several days. The water in the first dish is completely vaporized after 2.8 days, while the water in the second dish takes 8.3 days to completely evaporate. What can you conclude about the two dishes?

94. Several ice cubes are placed in a beaker on a lab bench, and their temperature, initially at –5.0 °C, is monitored. Explain what happens to the temperature as a function of time. Make a sketch of how the temperature might change with time. (Assume that the lab is at 25 °C.)

95. Water is put into a beaker and heated with a Bunsen burner. The temperature of the water, initially at 25 °C, is monitored. Explain what happens to the temperature as a function of time. Make a sketch of how the temperature might change with time. (Assume that the Bunsen burner is hot enough to heat the water to its boiling point.)

INTERMOLECULAR FORCES

96. What kinds of intermolecular forces are present in each substance?
(a) Kr
(b) N_2
(c) CO
(d) HF

97. What kinds of intermolecular forces are present in each substance?
(a) HCl
(b) H_2O
(c) Br_2
(d) He

98. What kinds of intermolecular forces are present in each substance?
(a) NCl_3 (trigonal pyramidal)
(b) NH_3 (trigonal pyramidal)
(c) SiH_4 (tetrahedral)
(d) CCl_4 (tetrahedral)

99. What kinds of intermolecular forces are present in each substance?
(a) O_3
(b) HBr
(c) CH_3OH
(d) I_2

100. What kinds of intermolecular forces are present in a mixture of potassium chloride and water?

101. What kinds of intermolecular forces are present in a mixture of calcium bromide and water?

102. Which substance has the highest boiling point? Why? *Hint:* They are all nonpolar.
(a) CH_4
(b) CH_3CH_3
(c) $CH_3CH_2CH_3$
(d) $CH_3CH_2CH_2CH_3$

103. Which noble gas has the highest boiling point? Why?
(a) Kr
(b) Xe
(c) Rn

104. One of these two substances is a liquid at room temperature and the other one is a gas. Which one is the liquid and why?

CH_3OH CH_3SH

105. One of these two substances is a liquid at room temperature and the other one is a gas. Which one is the liquid and why?

CH_3OCH_3 CH_3CH_2OH

106. A flask containing a mixture of $NH_3(g)$ and $CH_4(g)$ is cooled. At –33.3 °C a liquid begins to form in the flask. What is the liquid?

107. Explain why CS_2 is a liquid at room temperature while CO_2 is a gas.

108. Are $CH_3CH_2CH_2CH_2CH_3$ and H_2O miscible?

109. Are CH_3OH and H_2O miscible?

110. Determine whether a homogeneous solution forms when each pair of substances is mixed.
(a) CCl_4 and H_2O
(b) Br_2 and CCl_4
(c) CH_3CH_2OH and H_2O

111. Determine whether a homogeneous solution forms when each pair of substances is mixed.
(a) $CH_3CH_2CH_2CH_2CH_3$ and $CH_3CH_2CH_2CH_2CH_2CH_3$
(b) CBr_4 and H_2O
(c) Cl_2 and H_2O

CUMULATIVE PROBLEMS

112. Use the ideal gas law to show that the molar volume of a gas at STP is 22.4 L.

113. Use the ideal gas law to show that 28.0 g of nitrogen gas and 4.00 g of helium gas occupy the same volume at any temperature and pressure.

114. A mixture containing 235 mg of helium and 325 mg of neon has a total pressure of 453 torr. What is the partial pressure of helium in the mixture?

115. A mixture containing 4.33 g of CO_2 and 3.11 g of CH_4 has a total pressure of 1.09 atm. What is the partial pressure of CO_2 in the mixture?

116. Draw a Lewis structure for each molecule and determine its molecular geometry. What kind of intermolecular forces are present in each substance?
(a) H_2Se
(b) SO_2
(c) $CHCl_3$
(d) CO_2

117. Draw a Lewis structure for each molecule and determine its molecular geometry. What kind of intermolecular forces are present in each substance?
(a) HCOH (carbon is central; each H and O bonded directly to C)
(b) CS_2
(c) NCl_3

118. Explain the observed trend in the melting points of the hydrogen halides. Why is HF atypical?

Compound	Melting Point
HI	–50.8 °C
HBr	–88.5 °C
HCl	–114.8 °C
HF	–83.1 °C

119. Explain the observed trend in the boiling points of the compounds listed. Why is H_2O atypical?

Compound	Boiling Point
H_2Te	–2 °C
H_2Se	–41.5 °C
H_2S	–60.7 °C
H_2O	+100 °C

HIGHLIGHT PROBLEMS

120. Which gas sample has the greatest pressure? Assume they are all at the same temperature. Explain.

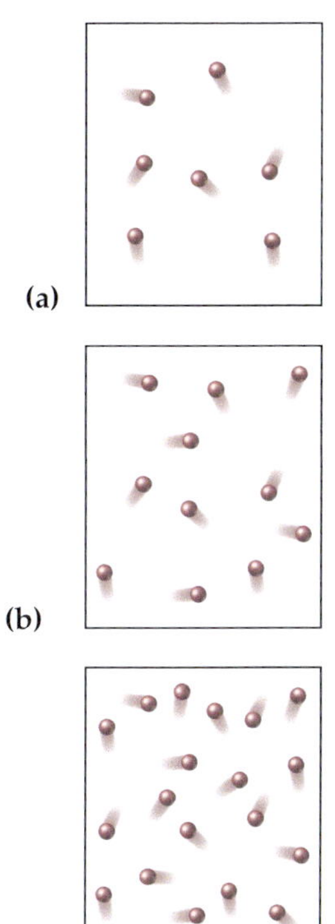

121. This image represents a sample of gas at a pressure of 1 atm, a volume of 1 L, and a temperature of 25 °C. Draw a similar picture showing what happens if the volume is reduced to 0.5 L and the temperature is increased to 250 °C. What happens to the pressure?

$V = 1.0$ L
$T = 25$ °C
$P = 1.0$ atm

122. In a common classroom demonstration, a balloon is filled with air and submerged into liquid nitrogen. The balloon contracts as the gases within the balloon cool. Suppose the balloon initially contains 2.95 L of air at 25.0 °C and a pressure of 0.998 atm. Calculate the expected volume of the balloon upon cooling to −196 °C (the boiling point of liquid nitrogen). When the demonstration is carried out, the actual volume of the balloon decreases to 0.61 L. How well does the observed volume of the balloon compare to your calculated value? Can you explain the difference?

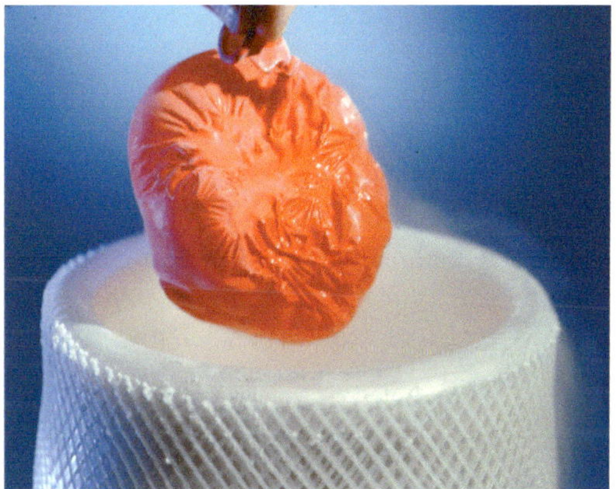

123. Aerosol cans carry clear warnings against incineration because of the high pressures that can develop upon heating. Suppose a can contains a residual amount of gas at a pressure of 755 mm Hg and 25 °C. What would the pressure be if the can were heated to 1155 °C?

124. Consider the molecular view of water shown here. Pick a molecule in the interior and draw a line to each of its direct neighbors. Pick a molecule near the edge (analogous to a molecule on the surface in three dimensions) and do the same. Which molecule has the most neighbors? Which molecule is more likely to evaporate?

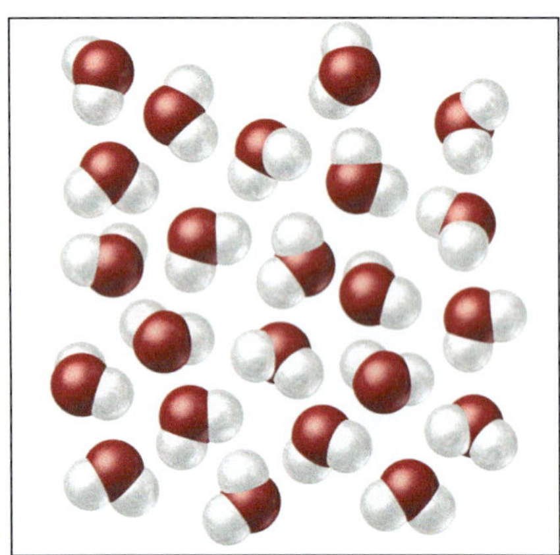

▶ Answers to Skillbuilder Exercises

Skillbuilder 6.1	$P_2 = 2.1$ atm; depth is approximately 11 m
Skillbuilder 6.2	85 °C
Skillbuilder 6.3	123 mL
Skillbuilder 6.4	0.82 L
Skillbuilder 6.5	13 L
Skillbuilder 6.6	1.6 atm
Skillbuilder 6.7	16.1 L
Skillbuilder Plus, p. 235	977 mm Hg
Skillbuilder 6.8	721 torr
Skillbuilder 6.9	$P_{tot} = 4.2$ atm
Skillbuilder 6.10	C_2H_6

Skillbuilder 6.11

(a) no dipole–dipole forces
(b) yes, it has dipole–dipole forces
(c) yes, it has dipole–dipole forces

Skillbuilder 6.12 HF, because it has hydrogen bonding as an intermolecular force

▶ Answers to Conceptual Checkpoints

6.1 (a) Since all the particles are identical and since (a) has the smallest number of particles per unit volume, it will have the lowest pressure.

6.2 (c) Atmospheric pressure will support a column of water 10.3 m in height. If the liquid in a barometer were twice as dense as water, a column of it would be twice as heavy and the pressure it exerted at its base would be twice as great. Therefore, atmospheric pressure would be able to support a column only half as high.

6.3 (b) Since the volume triples, and since according to Boyle's law the volume and pressure are inversely proportional, the pressure will fall by a factor of 3.

6.4 (a) At constant pressure, the volume of the gas will be proportional to the temperature—if the Kelvin temperature doubles, the volume will double.

6.5 (c) Doubling the temperature in kelvins doubles the pressure, but doubling the volume halves the pressure. The net result is that the pressure is the same as its initial value.

6.6 (c) Since hydrogen gas has the lowest molar mass of the set, 1 g will have the greatest number of moles and therefore the greatest volume.

6.7 (a) The substance has definite volume *and* a definite shape, so it must be a solid.

6.8 1 (b) Since boiling is a physical rather than a chemical change, the water molecules (H_2O) undergo no chemical alteration—they merely change from the liquid to the gaseous state.

6.8 2 (c) Since sublimation is a physical change, the carbon dioxide molecules do not decompose into other molecules or atoms; they simply change state from the solid to the gaseous state.

6.9 (a) All three compounds have nearly identical molar masses, so the strength of the dispersion forces should be similar in all three. A is nonpolar; it has only dispersion forces and will have the lowest boiling point. B is polar; it has dipole–dipole forces in addition and therefore has the next highest boiling point. C has hydrogen bonding in addition; it therefore has the highest boiling point.

6.10 (c) The chemical bonds between carbon and oxygen atoms are not broken by changes of state such as sublimation. Because carbon dioxide contains no hydrogen atoms, it cannot undergo hydrogen bonding, and because the molecule is nonpolar, it does not experience dipole–dipole interactions.

▲ Flavors are caused by the interactions of molecules in foods or drinks with molecular receptors on the surface of the tongue. This image shows a caffeine molecule, one of the substances responsible for the sometimes bitter flavors in coffee.

States of Matter 2 7

"It will be found that everything depends on the composition of the forces with which the particles of matter act upon one another; and from these forces ... all phenomena of nature take their origin." —Roger Joseph Boscovich (1711–1787)

MODULE OUTLINE

7.1 Liquids, Surface Tension, and Viscosity

LO: Describe how surface tension and viscosity are manifestations of the intermolecular forces in liquids.

▲ **FIGURE 7.1 Floating flies** Even though they are denser than water, fly-fishing lures float on the surface of a stream or lake because of surface tension. © Anthony West/Corbis.

The most important manifestation of intermolecular forces is the very existence of molecular liquids and solids. Without intermolecular forces, molecular solids and liquids would not exist (they would be gaseous). In liquids, we can observe several other manifestations of intermolecular forces including surface tension and viscosity.

Surface Tension

A fly fisherman delicately casts a small metal hook (with a few feathers and strings attached to make it look like an insect) onto the surface of a moving stream. The hook floats on the surface of the water and attracts trout (◀ Figure 7.1). The hook floats because of **surface tension**, the tendency of liquids to minimize their surface area. This tendency causes liquids to have a sort of "skin" that resists penetration. For the fisherman's hook to sink into the water, the water's surface area would have to increase slightly. The increase is resisted because molecules at the surface interact with fewer neighbors than those in the interior of the liquid (▶ Figure 7.2). Since molecules at the surface have fewer interactions with other molecules, they are inherently less stable than those in the interior; consequently, liquids have a tendency to minimize the number of the molecules at the surface, which results in surface tension. You can observe surface tension by carefully placing a paper clip on the surface of water (▶ Figure 7.3). The paper clip, even though it is denser than water, will float on the surface of the water. A slight tap on the clip overcomes surface tension and causes the clip to sink. Surface tension increases with increasing intermolecular forces. You can't float a paper clip on gasoline, for example, because the intermolecular forces among the molecules composing gasoline are weaker than the intermolecular forces among water molecules; they are not under as much tension, so they do not form a "skin."

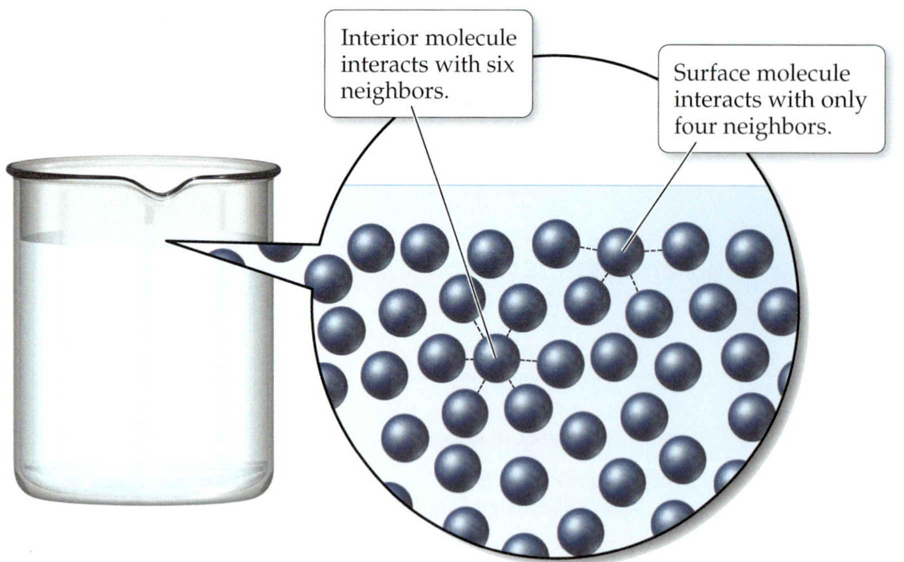

▲ **FIGURE 7.2 Origin of surface tension** Molecules at the surface of a liquid interact with fewer molecules than those in the interior; the lower number of interactions makes the surface molecules less stable.

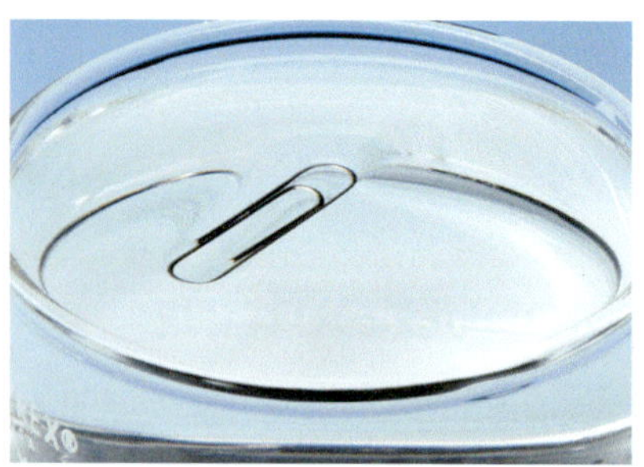

▲ **FIGURE 7.3 Surface tension at work** A paper clip will float on water if it is carefully placed on the surface of the water. It is held up by surface tension. © Fundamental Photographs.

Viscosity

Another manifestation of intermolecular forces is **viscosity**, the resistance of a liquid to flow. Liquids that are viscous flow more slowly than liquids that are not viscous. For example, motor oil is more viscous than gasoline, and maple syrup is more viscous than water (◀ Figure 7.4). Viscosity is greater in substances with stronger intermolecular forces because molecules cannot move around each other as freely, hindering flow. Long molecules, such as the hydrocarbons in motor oil, also tend to form viscous liquids because of molecular entanglement (the long chainlike molecules get tangled together).

◀ **FIGURE 7.4 Viscosity** Maple syrup is more viscous than water because its molecules interact strongly and so cannot flow past one another easily. © Vetta Collection/iStockphoto/Getty Images.

7.2 Remarkable Water

LO: Describe and explain the properties that make water unique among molecules.

Water is easily the most common and important liquid on Earth. It fills our oceans, lakes, and streams. In its solid form, it covers nearly an entire continent (Antarctica), as well as large regions around the North Pole, and caps our tallest mountains. In its gaseous form, it humidifies our air. We drink water, we sweat water, and we excrete bodily wastes dissolved in water. Indeed, the majority of our body mass *is* water. Life is impossible without water, and in most places on Earth where liquid water exists, life exists. Evidence of water on Mars—that existed either in the past or exists in the present—has fueled hopes of finding life or evidence of life there. Water is remarkable.

Among liquids, water is unique. It has a low molar mass (18.02 g/mol), yet is a liquid at room temperature. No other compound of similar molar mass even comes close to being a liquid at room temperature. For example, nitrogen (28.02 g/mol) and carbon dioxide (44.01 g/mol) are both gases at room temperature. Water's relatively high boiling point (for its low molar mass) can be understood by

EVERYDAY CHEMISTRY

▶ Why Are Water Drops Spherical?

Have you ever seen a close-up photograph of tiny water droplets (▼ Figure 7.5) or carefully watched water in free fall? In both cases, the distorting effects of gravity are diminished, and the water forms nearly perfect spheres. On the space shuttle, the complete absence of gravity results in floating spheres of water (▼ Figure 7.6). Why? Water drops are spherical because of the surface tension caused by the attractive forces between water molecules. Just as gravity pulls matter within a planet or star inward to form a sphere, so intermolecular forces pull water molecules inward to form a sphere. The sphere minimizes the surface-area-to-volume ratio, thereby minimizing the number of molecules at the surface.

▲ **FIGURE 7.5 An almost perfect sphere** If a water droplet is small enough, it will largely be free of the distorting effects of gravity and be almost perfectly spherical. © lostbear/iStockphoto/Getty Images.

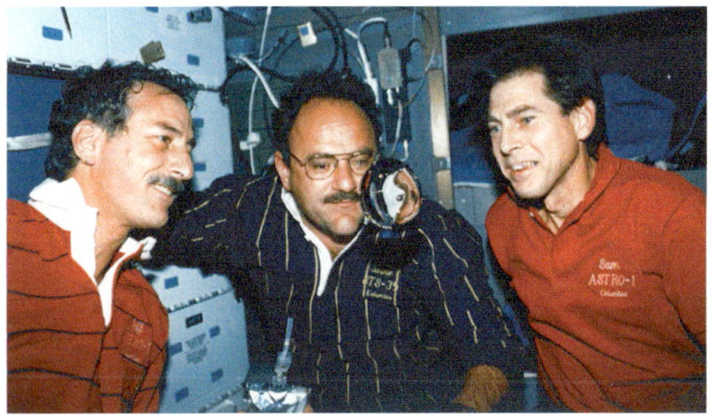

▲ **FIGURE 7.6 A perfect sphere** In the absence of gravity, as in this picture taken on the space shuttle, water assumes the shape of a sphere. © NASA.

A collection of magnetic marbles provides a good physical model of a water drop. Each magnetic marble is like a water molecule, attracted to the marbles around it. If you agitate these marbles slightly, so that they can find their preferred configuration, they tend toward a spherical shape (▼ Figure 7.7) because the attractions between the marbles cause them to minimize the number of marbles at the surface.

▲ **FIGURE 7.7 An analogy for surface tension** Magnetic marbles tend to arrange themselves in a spherical shape. © Fundamental Photographs.

B7.1 CAN YOU ANSWER THIS? *How does the tendency of a liquid to form spherical drops depend on the strength of intermolecular forces? Do liquids with weaker intermolecular forces have a higher or lower tendency to form spherical drops?*

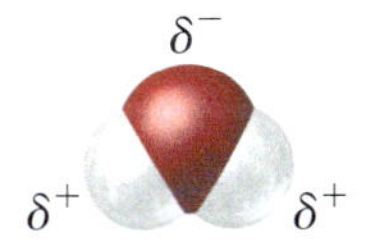

▲ **FIGURE 7.8 The water molecule**

examining the structure of the water molecule (◀ Figure 7.8). The bent geometry of the water molecule and the highly polar nature of the O — H bonds result in a molecule with a significant dipole moment. Water's two O — H bonds (hydrogen directly bonded to oxygen) allow water molecules to form strong hydrogen bonds with other water molecules, resulting in a relatively high boiling point. Water's high polarity also allows it to dissolve many other polar and ionic compounds. Consequently, water is the main solvent of living organisms, transporting nutrients and other important compounds throughout the body.

The way water freezes is also unique. Unlike other substances, which contract upon freezing, water expands upon freezing. This seemingly trivial property has

Water reaches its maximum density at 4.0 °C.

significant consequences. Because liquid water expands when it freezes, ice is less dense than liquid water. Consequently, ice cubes and icebergs both float. The frozen layer of ice at the surface of a winter lake insulates the water in the lake from further freezing. If this ice layer were to sink, it would kill bottom-dwelling aquatic life and possibly allow the lake to freeze solid, eliminating virtually all aquatic life in the lake.

The expansion of water upon freezing, however, is one reason that most organisms do not survive freezing. When the water within a cell freezes, it expands and often ruptures the cell, just as water freezing within a pipe bursts the pipe. Many foods, especially those with high water content, do not survive freezing very well either. Have you ever tried, for example, to freeze a vegetable? Try putting lettuce or spinach in the freezer. When you defrost it, it will be limp and damaged. The frozen food industry gets around this problem by *flash-freezing* vegetables and other foods. In this process, foods are frozen instantaneously, preventing water molecules from settling into their preferred crystalline structure. Consequently, the water does not expand very much, and the food remains largely undamaged.

▲ Lettuce does not survive freezing because the expansion of water upon freezing ruptures the cells within the lettuce leaf. © Kristen Brochmann/Fundamental Photographs.

CHEMISTRY IN THE ENVIRONMENT

▶ Water Pollution

Water quality is critical to human health. Many human diseases—especially in developing nations—are caused by poor water quality. Several kinds of pollutants, including biological and chemical contaminants, can get into water supplies. Biological contaminants are microorganisms that cause diseases such as hepatitis, cholera, dysentery, and typhoid. Drinking water in developed nations is usually treated to kill microorganisms. Most biological contaminants can be eliminated from untreated water by boiling. Water containing biological contaminants poses an immediate danger to human health and should not be consumed.

Chemical contaminants get into drinking water from sources such as industrial dumping, pesticide and fertilizer use, and household dumping. These contaminants include organic compounds, such as carbon tetrachloride and dioxin, and inorganic elements and compounds such as mercury, lead, and nitrates. Since many chemical contaminants are neither volatile nor alive like biological contaminants, they are *not* eliminated through boiling.

The Environmental Protection Agency (EPA), under the Safe Drinking Water Act of 1974 and its amendments, sets standards that specify the maximum contamination level (MCL) for nearly 100 biological and chemical contaminants in water. Water providers that serve more than 25 people must periodically test the water they deliver to their consumers for these contaminants. If levels exceed the standards set by the EPA, the water provider must notify the consumer and take appropriate measures to remove the contaminant from the water. According to the EPA, if water comes from a provider that serves more than 25 people, it should be safe to consume over a lifetime. If it is not safe to drink for a short period of time, consumers are notified.

▲ Safe drinking water has a major effect on public health and the spread of disease. In many parts of the world, the water supply is unsafe to drink. In the United States the Environmental Protection Agency (EPA) is charged with maintaining water safety. © Shutterstock.

B7.2 CAN YOU ANSWER THIS? *Suppose a sample of water is contaminated by a nonvolatile contaminant such as lead. Why doesn't boiling eliminate the contaminant?*

7.3 Types and Properties of Solids

LO: Identify types of crystalline solids.

As we learned in Module 1, solids may be crystalline (showing a well-ordered array of atoms or molecules) or amorphous (having no long-range order). We divide crystalline solids into three categories—molecular, ionic, and atomic—based on the individual units that compose the solid (▼ Figure 7.9).

Molecular Solids

Molecular solids are solids whose composite units are *molecules*. Ice (solid H_2O) and dry ice (solid CO_2) are examples of molecular solids. Molecular solids are held together by the kinds of intermolecular forces—dispersion forces, dipole–dipole forces, and hydrogen bonding—that we discussed in Section 6.13. For example, ice is held together by hydrogen bonds, and dry ice is held together by dispersion forces. Molecular solids as a whole tend to have low to moderately low melting points; ice melts at 0 °C and dry ice sublimes at –78.5 °C.

Ionic Solids

Ionic solids are solids whose composite units are *formula units*, the smallest electrically neutral collection of cations and anions that compose the compound. Table salt (NaCl) and calcium fluoride (CaF_2) are good examples of ionic solids. Ionic solids are held together by electrostatic attractions between cations and anions. For example, in NaCl, the attraction between the Na^+ cation and the Cl^- anion holds the solid lattice together because the lattice is composed of alternating Na^+ cations and Cl^- anions in a three-dimensional array. In other words, the forces that hold ionic solids together are actual ionic bonds. Since ionic bonds are much stronger than any of the intermolecular forces discussed previously, ionic solids tend to have much higher melting points than molecular solids. For example, sodium chloride melts at 801 °C, while carbon disulfide CS_2—a molecular solid with a higher molar mass—melts at –110 °C.

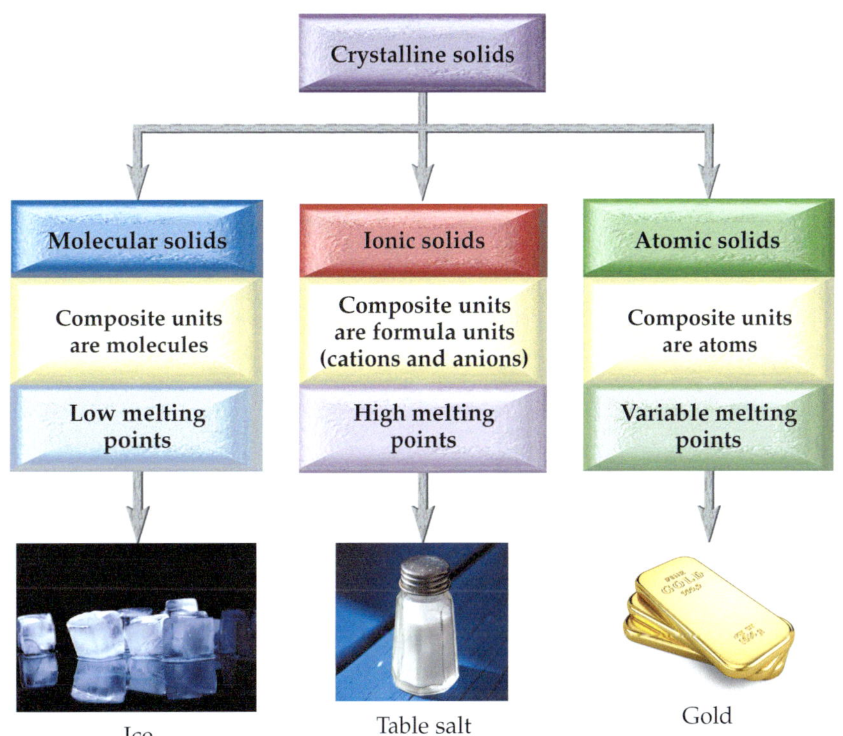

▲ **FIGURE 7.9 A classification scheme for crystalline solids** © (left) Matejay/iStockphoto/Getty Images. (centre): Sashkin/Shutterstock.

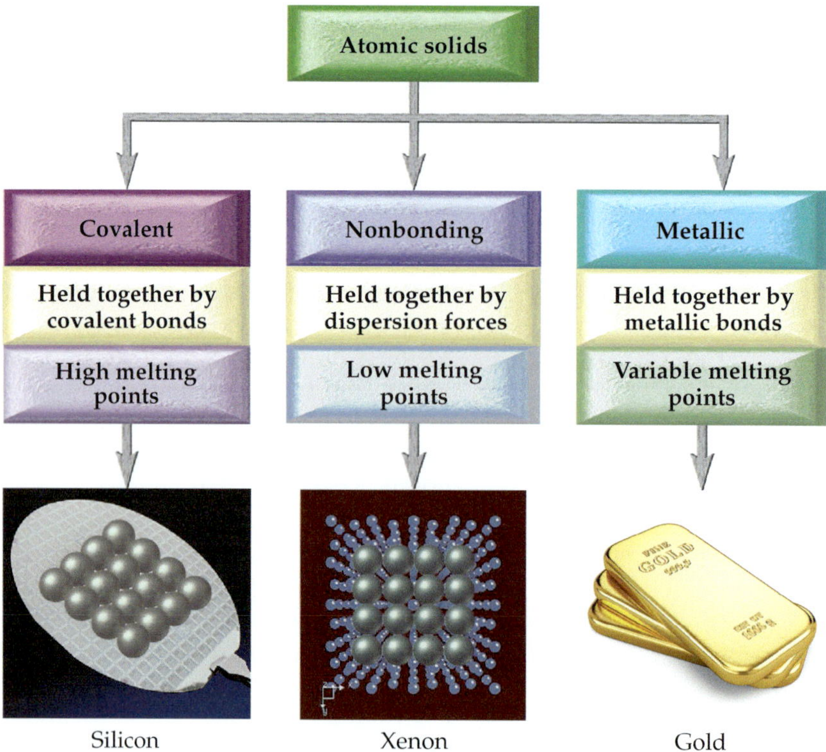

▲ **FIGURE 7.10 A classification scheme for atomic solids** © (left) GeoStock/Photodisc/Getty Images. (right) Sashkin/Shutterstock.

Atomic Solids

Atomic solids are solids whose composite units are *individual atoms*. Diamond (C), iron (Fe), and solid xenon (Xe) are good examples of atomic solids. We can divide atomic solids into three categories—**covalent atomic solids**, **nonbonding atomic solids**, and **metallic atomic solids**—each held together by a different kind of force (▲ Figure 7.10).

Covalent atomic solids, such as diamond, are held together by covalent bonds. In diamond (◀ Figure 7.11), each carbon atom forms four covalent bonds to four other carbon atoms in a tetrahedral geometry. This structure extends throughout the entire crystal, so that a diamond crystal can be thought of as a giant molecule held together by these covalent bonds. Since covalent bonds are very strong, covalent atomic solids have high melting points. Diamond is estimated to melt at about 3800 °C.

Nonbonding atomic solids, such as solid xenon, are held together by relatively weak dispersion forces. Xenon atoms have stable electron configurations and therefore do not form covalent bonds with each other. Consequently, solid xenon, like other nonbonding atomic solids, has a very low melting point (about −112 °C).

Metallic atomic solids, such as iron, have variable melting points. Metals are held together by metallic bonds that, in the simplest model, consist of positively charged ions in a sea of electrons (▶ Figure 7.12). Metallic bonds are of varying strengths, with some metals, such as mercury, having melting points below room temperature, and other metals, such as iron, having relatively high melting points (iron melts at 1809 °C).

▲ **FIGURE 7.11 Diamond: a covalent atomic solid** In diamond, carbon atoms form covalent bonds in a three-dimensional cubic structure. © Maciej Toporowicz, NYC/Flickr/Getty Images.

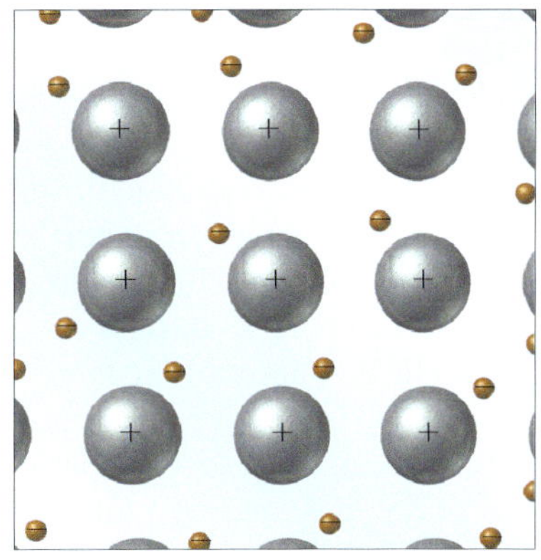

▲ **FIGURE 7.12 Structure of a metallic atomic solid** In the simplest model of a metal, each atom donates one or more electrons to an "electron sea." The metal consists of the metal cations in a negatively charged electron sea.

EXAMPLE 7.1 IDENTIFYING TYPES OF CRYSTALLINE SOLIDS

Identify each solid as molecular, ionic, or atomic.

(a) $CaCl_2(s)$ **(b)** $Co(s)$ **(c)** $CS_2(s)$

SOLUTION

(a) $CaCl_2$ is an ionic compound (metal and nonmetal) and therefore forms an ionic solid ($CaCl_2$ melts at 772 °C).
(b) Co is a metal and therefore forms a metallic atomic solid (Co melts at 1768 °C).
(c) CS_2 is a molecular compound (nonmetal bonded to a nonmetal) and therefore forms a molecular solid (CS_2 melts at –110 °C).

▶**SKILLBUILDER 7.1 | Identifying Types of Crystalline Solids**

Identify each solid as molecular, ionic, or atomic.

(a) $NH_3(s)$ **(b)** $CaO(s)$ **(c)** $Kr(s)$

▶**FOR MORE PRACTICE** Problems 28, 29, 30, 31.

7.4 Solutions

LO: Define solution, solute, and solvent.

On August 22, 1986, most people living near Lake Nyos in Cameroon, West Africa, began their day in an ordinary way. Unfortunately, the day ended in tragedy. On that evening, a large cloud of carbon dioxide gas, burped up from the depths of Lake Nyos, killed more than 1700 people and about 3000 head of cattle. Survivors tell of smelling rotten eggs, feeling a warm sensation, and then losing consciousness. Two years before that, a similar tragedy occurred in Lake Monoun, just 60 miles away, killing 37 people. Today, scientists have taken steps to prevent these lakes from burping again.

▲ Cameroon is in West Africa.

Carbon dioxide, a colorless and odorless gas, displaced the air in low-lying regions surrounding Lake Nyos, leaving no oxygen for the inhabitants to breathe. The rotten-egg smell was an indication of the presence of additional sulfur-containing gases.

Lake Nyos is a water-filled volcanic crater. About 80 km beneath the surface of the lake, molten volcanic rock (magma) produces carbon dioxide gas that seeps into the lake through the volcano's plumbing system. The carbon dioxide then mixes with the lake water. However, as we will see later in this module, the concentration of a gas (such as carbon dioxide) that can build up in water increases with increasing pressure. The great pressure at the bottom of the deep lake allows the concentration of carbon dioxide to become very high (just as the pressure in a soda can allows the concentration of carbon dioxide in soda to be very high). Over time, the carbon dioxide and water mixture at the bottom of the lake became so concentrated that—either because of the high concentration itself or because of some other natural trigger, such as a landslide—some gaseous carbon dioxide escaped. The rising bubbles disrupted the lake water, causing the highly concentrated carbon dioxide and water mixture at the bottom of the lake to rise, which lowered the pressure on it. The drop in pressure on the mixture released more carbon dioxide bubbles just as the drop in pressure upon opening a soda can releases carbon dioxide bubbles. This in turn caused further churning and more carbon dioxide release. Since carbon dioxide is more dense than air, once freed from the lake, it traveled down the sides of the volcano and into the nearby valley, displacing air and asphyxiating many of the local residents.

In efforts to prevent these events from occurring again—by 2001, carbon dioxide concentrations had already returned to dangerously high levels—scientists built a piping system to slowly vent carbon dioxide from the lake bottom. Since

▶ Engineers watch as the carbon dioxide vented from the bottom of Lake Nyos creates a geyser. The controlled release of carbon dioxide from the lake bed is designed to prevent future catastrophes like the one that killed more than 1700 people in 1986.

Aqueous comes from the Latin word *aqua*, which means "water."

In a solid/liquid solution, the liquid is usually considered the solvent, regardless of the relative proportions of the components.

2001, this system has gradually been releasing the carbon dioxide into the atmosphere, preventing a repeat of the tragedy.

The carbon dioxide and water mixture at the bottom of Lake Nyos is an example of a **solution**, a homogeneous mixture of two or more substances. Solutions are common—most of the liquids that we encounter every day are actually solutions. When most people think of a solution, they think of a solid dissolved in water. The ocean, for example, is a solution of various salts dissolved in water. Blood plasma (blood that has had blood cells removed from it) is a solution of several solids (as well as some gases) dissolved in water. In addition to these, many other kinds of solutions exist.

The most common solutions are those containing a solid, a liquid, or a gas and water. These are *aqueous solutions*—they are critical to life and are the main focus of this module. Common examples of aqueous solutions include sugar water and salt water, both solutions of solids and water.

A solution has at least two components. The majority component is usually called the **solvent**, and the minority component is called the **solute**. In our carbon-dioxide-and-water solution, carbon dioxide is the solute and water is the solvent. In a salt-and-water solution, salt is the solute and water is the solvent. Because water is so abundant on Earth, it is a common solvent. However, other solvents are often used in the laboratory, in industry, and even in the home, especially to form solutions with nonpolar solutes. For example, you may use paint thinner, a nonpolar solvent, to remove grease from a dirty bicycle chain or from ball bearings. The paint thinner dissolves (or forms a solution with) the grease, removing it from the metal.

In general, polar solvents dissolve polar or ionic solutes, and nonpolar solvents dissolve nonpolar solutes. This tendency is described by the rule *like dissolves like*. Thus, similar kinds of solvents dissolve similar kinds of solutes. Table 7.1 lists some common polar and nonpolar laboratory solvents.

Solutions that have water as the solvent are called aqueous solutions; otherwise they are nonaqueous solutions.

TABLE 7.1 Common Laboratory Solvents

Common Polar Solvents	Common Nonpolar Solvents
water (H_2O)	hexane (C_6H_{14})
acetone (CH_3COCH_3)	diethyl ether ($CH_3CH_2OCH_2CH_3$)
methyl alcohol (CH_3OH)	toluene (C_7H_8)

CONCEPTUAL CHECKPOINT 7.1

Which compound would you expect to be *least* soluble in water?

(a) CCl_4

(b) CH_3Cl

(c) NH_3

(d) KF

7.5 Solids in Aqueous Solution

LO: Relate the solubility of solids in water to temperature.

We have already seen several examples of solutions of a solid dissolved in water. The ocean, for example, is a solution of several salts dissolved in water. A sweetened cup of coffee is a solution of sugar and other solids dissolved in water. Blood plasma is a solution of several solids (and some gases) dissolved in water. Not all solids, however, dissolve in water. We already know that nonpolar solids—such as lard and shortening—do not dissolve in water.

When a solid is put into water, there is competition between the attractive forces that hold the solid together (the solute–solute interactions) and the attractive forces occurring between the water molecules and the particles that compose the solid (the solvent–solute interactions). The solvent–solute interactions are usually intermolecular forces of the type discussed in Module 6. For example, when sodium chloride is put into water, there is competition between the mutual attraction of Na^+ cations and Cl^- anions and the ion–dipole forces between Na^+ or Cl^- and water molecules as shown in the margin. For sodium ions, the attraction is between the positive charge of the sodium ion and the negative side of water's dipole moment as shown in ▼ Figure 7.13 (see Section 3.18 to review dipole moment). For chloride ions, the attraction is between the negative charge of the chloride ion and the positive side of water's dipole moment. In the case of NaCl, the attraction to water wins, and sodium chloride dissolves (▶ Figure 7.14). However, in the case of calcium carbonate ($CaCO_3$), the attraction between Ca^{2+} ions and CO_3^{2-} ions is very strong, so only a very small amount of calcium carbonate dissolves in water.

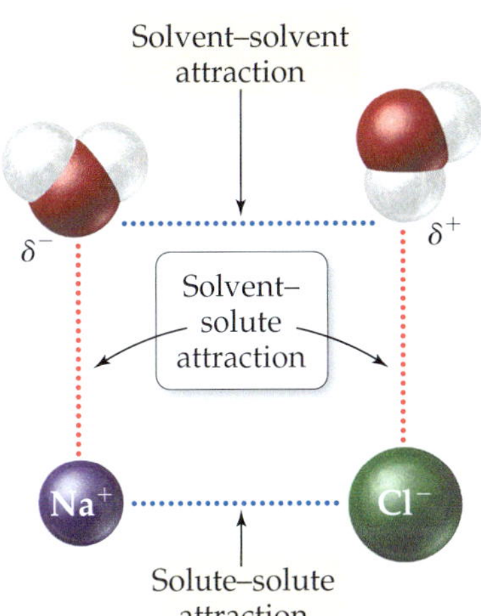

▲ When NaCl is put into water, the attraction between water molecules and Na^+ and Cl^- ions (solvent–solute attraction) overcomes the attraction between Na^+ and Cl^- ions (solute-solute attraction) and between water molecules (solvent-solvent extraction).

Solubility and Saturation

The **solubility** of a compound is defined as the amount of the compound, usually in grams, that dissolves in a certain amount of liquid. For example, the solubility of sodium chloride in water at 25 °C is 36 g NaCl per 100 g water, while the solubility of calcium carbonate in water is close to zero. A solution that contains 36 g of NaCl per 100 g water is a *saturated* sodium chloride solution. A **saturated solution** holds the maximum amount of solute under the solution conditions. If additional solute is added to a saturated solution, it will not dissolve. An **unsaturated solution** is holding less than the maximum amount of solute. If additional solute is added to an unsaturated solution, it will dissolve. A **supersaturated solution** is one holding more than the normal maximum amount of solute. The solute will normally *precipitate* from (or come out of) a supersaturated solution. As the carbon dioxide and water solution rose from the bottom of Lake Nyos, for example, it became supersaturated because of the drop in pressure. The excess gas came out of the solution and rose to the surface of the lake, where it was emitted into the surrounding air.

Supersaturated solutions can form under special circumstances, such as the sudden release in pressure that occurs in a soda can when it is opened.

Molecular solids may be soluble in water depending on whether the solid is polar. Table sugar $C_{12}H_{22}O_{11}$ for example, is polar and soluble in water. Nonpolar solids, such as lard and vegetable shortening, are usually insoluble in water.

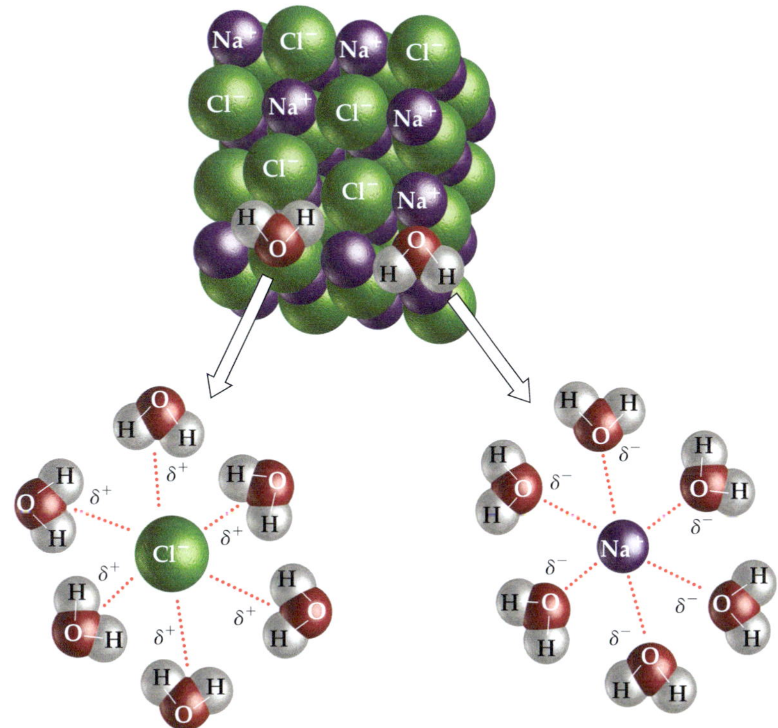

▶ **FIGURE 7.13 How a solid dissolves in water** The positive ends of the water dipoles are attracted to the negatively charged Cl^- ions, and the negative ends of the water dipoles are attracted to the positively charged Na^+ ions via ion–dipole forces. The water molecules surround the ions of NaCl and disperse them in the solution.

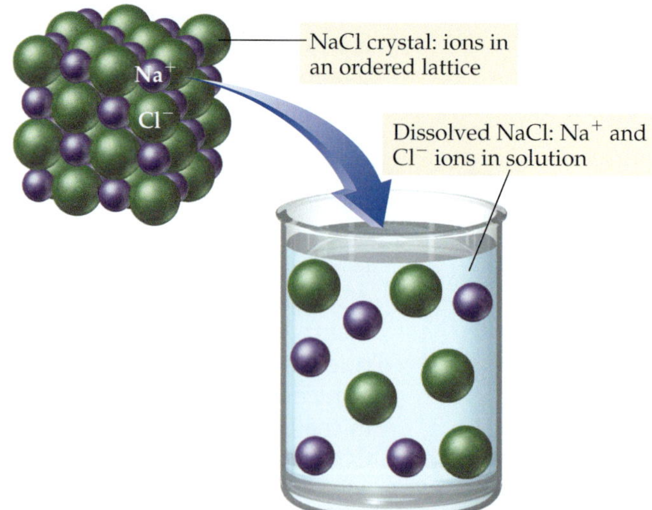

▶ **FIGURE 7.14 A sodium chloride solution** In a solution of NaCl, the Na^+ and Cl^- ions are dispersed in the water.

Electrolyte Solutions: Dissolved Ionic Solids

When ionic compounds such as NaCl dissolve in water, they usually dissociate into their component ions. A sodium chloride solution, represented as NaCl(*aq*), does not contain any NaCl units; only dissolved Na^+ ions and Cl^- ions are present.

We know that NaCl is present as independent sodium and chloride ions in solution because sodium chloride solutions conduct electricity, which requires the presence of freely moving charged particles. Substances (such as NaCl) that completely dissociate into ions in solution are called *strong electrolytes*, and the resultant solutions are called **strong electrolyte solutions** (▶ Figure 7.15). Similarly, a silver nitrate solution, represented as $AgNO_3(aq)$, does not contain any $AgNO_3$ units, but only dissolved Ag^+ ions and NO_3^- ions. It, too, is a strong electrolyte solution. When compounds containing polyatomic ions such as NO_3^- dissolve, the polyatomic ions dissolve as intact units.

Not all ionic compounds are highly soluble in water. AgCl, for example is sparingly soluble in water. If we add AgCl to water, most of it remains as solid AgCl and appears as a white solid at the bottom of the beaker and only a very small amount of AgCl dissociates into dissolved Ag^+ ions and Cl^- ions.

A sugar solution (containing a molecular solid) and a salt solution (containing an ionic solid) are very different, as shown in ▶ Figure 7.16. In a salt solution the dissolved particles are ions, while in a sugar solution the dissolved particles are molecules. The ions in the salt solution are mobile charged particles and can therefore conduct electricity. As described above, a solution containing a solute that dissociates into ions is an **electrolyte solution**. The sugar solution contains dissolved sugar molecules and cannot conduct electricity; it is a **nonelectrolyte solution**. In general, soluble ionic solids form electrolyte solutions, while soluble molecular solids form nonelectrolyte solutions.

Weak electrolyte solutions are covered in Module 8.

(a)

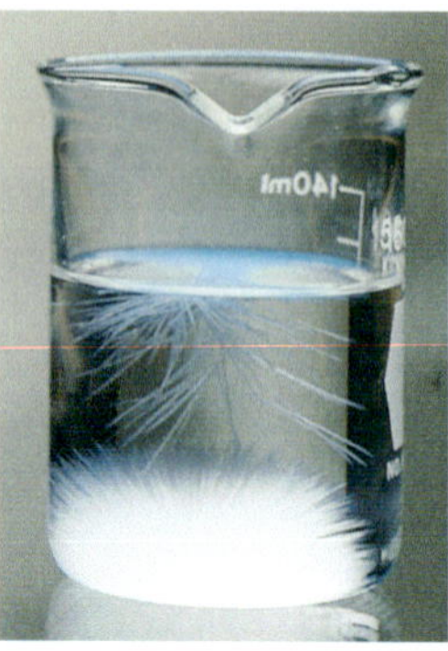

(b)

(c)

▶ A supersaturated solution holds more than the normal maximum amount of solute. In some cases, such as the sodium acetate solution pictured here, a supersaturated solution may be temporarily stable. Any disturbance, however, such as dropping in a small piece of solid sodium acetate **(a)**, causes the solid to come out of solution **(b, c)**.

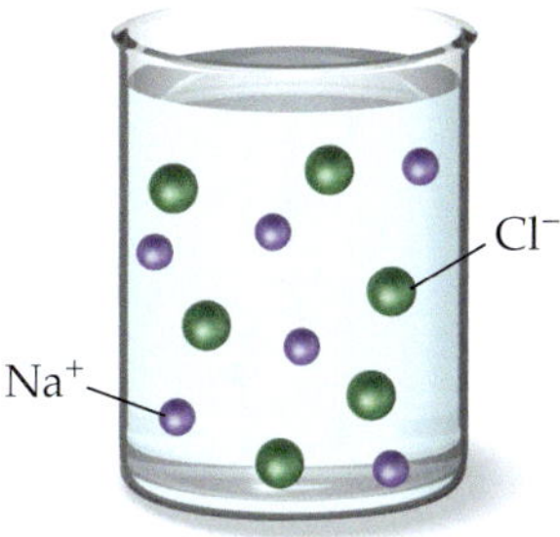

A sodium chloride solution contains independent $\mathbf{Na^+}$ and $\mathbf{Cl^-}$ ions.

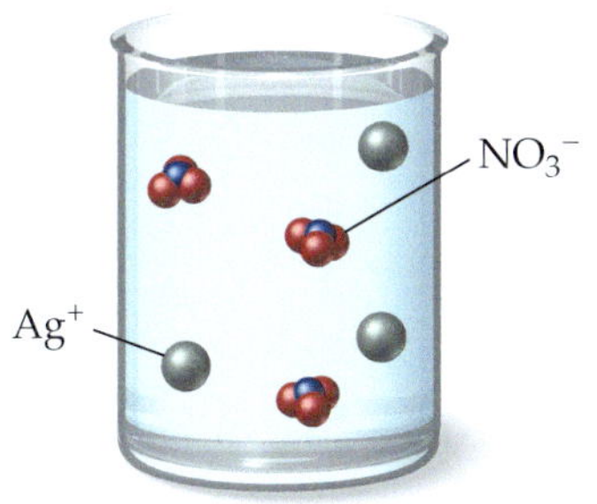

A silver nitrate solution contains independent $\mathbf{Ag^+}$ and $\mathbf{NO_3^-}$ ions.

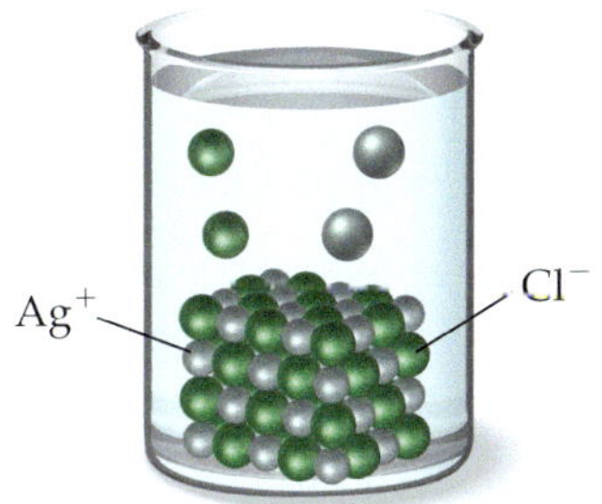

When silver chloride is added to water, most of it remains as solid AgCl—only a very small amount dissolves into independent Ag^+ and Cl^- ions.

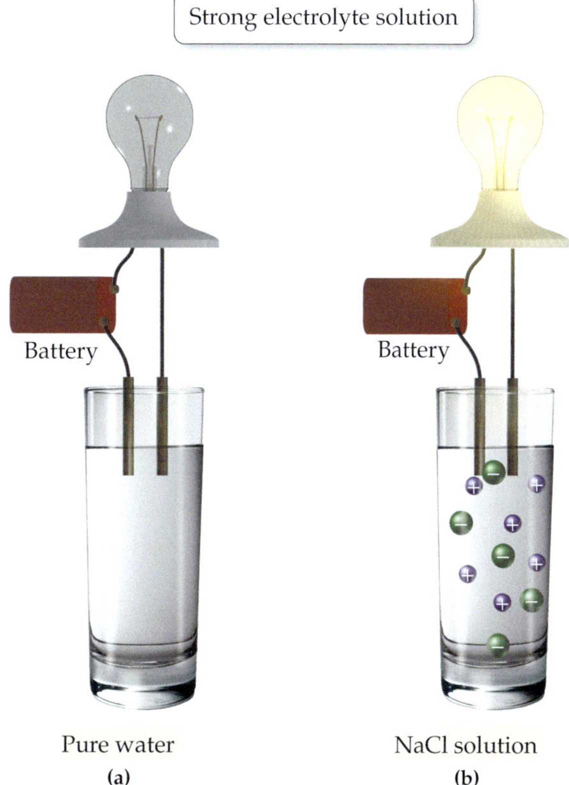

▲ **FIGURE 7.15 Ions as conductors** **(a)** Pure water does not conduct electricity. **(b)** Ions in a sodium chloride solution conduct electricity, causing the bulb to light. Solutions such as NaCl are called strong electrolyte solutions.

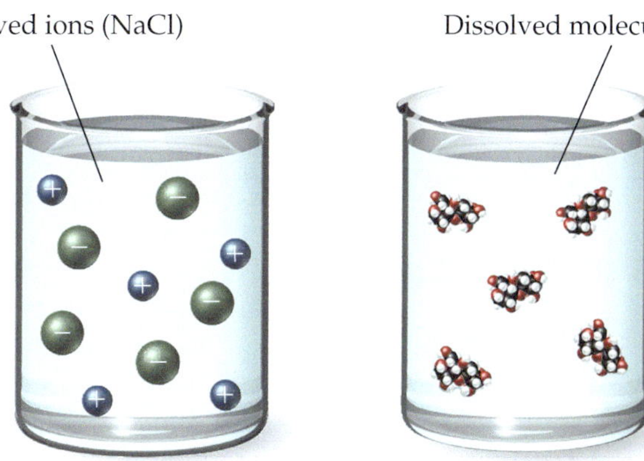

▲ **FIGURE 7.16 Electrolyte and nonelectrolyte solutions** Electrolyte solutions contain dissolved ions (charged particles) and therefore conduct electricity. Nonelectrolyte solutions contain dissolved molecules (neutral particles) and therefore do not conduct electricity.

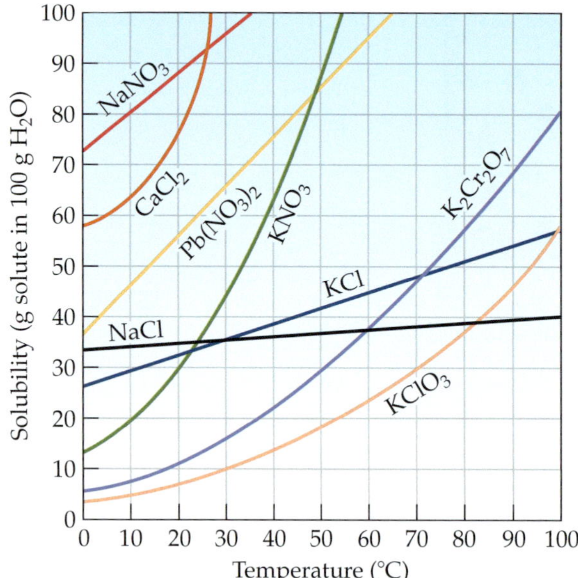

▶ **FIGURE 7.17 Solubility of some ionic solids as a function of temperature**

How Solubility Varies with Temperature

Have you ever noticed how much easier it is to dissolve sugar in hot tea than in cold tea? The solubility of solids in water can be highly dependent on temperature. In general, the solubility of *solids* in water increases with increasing temperature (▲ Figure 7.17). For example, the solubility of potassium nitrate (KNO_3) at 20 °C is about 30 g KNO_3 per 100 g of water. However, at 50 °C, the solubility rises to 88 g KNO_3 per 100 g of water. A common way to purify a solid is **recrystallization**. Recrystallization involves putting the solid into water (or some other solvent) at an elevated temperature. Enough solid is added to the solvent to create a saturated solution at the elevated temperature. As the solution cools, the solubility decreases, causing some of the solid to precipitate from solution. If the solution cools slowly, the solid will form crystals as it comes out. The crystalline structure tends to reject impurities, resulting in a purer solid.

▲ Rock candy is composed of sugar crystals that form through recrystallization.

Rock Candy

We can use recrystallization to make rock candy. To make rock candy, we prepare a saturated sucrose (table sugar) solution at an elevated temperature. We dangle a string in the solution, and leave it to cool and stand for several days. As the solution cools, it becomes supersaturated and sugar crystals grow on the string. After several days, beautiful and sweet crystals, or "rocks," of sugar cover the string, ready to be admired and eaten.

7.6 Gases in Aqueous Solution

LO: Relate the solubility of gases in liquids to temperature and pressure.

The water at the bottom of Lake Nyos and a can of soda pop are both examples of solutions in which a gas (carbon dioxide) is dissolved in a liquid (water). Most liquids exposed to air contain some dissolved gases. Lake water and seawater, for example, contain dissolved oxygen necessary for the survival of fish. Our blood contains dissolved nitrogen, oxygen, and carbon dioxide. Even tap water contains dissolved atmospheric gases.

We can see the dissolved gases in ordinary tap water when we heat it it on a stove. Before the water reaches its boiling point, small bubbles develop in the water. These bubbles are dissolved air (mostly nitrogen and oxygen) coming out of solution. Once the water boils, the bubbling becomes more vigorous—these larger bubbles are composed of water vapor. The dissolved air comes out of solution upon heating because—unlike solids, whose solubility *increases* with increasing temperature—the solubility of gases in water *decreases* with increasing temperature. As the temperature

▲ Warm soda pop fizzes more than cold soda pop because the solubility of the dissolved carbon dioxide decreases with increasing temperature.

of the water rises, the solubility of the dissolved nitrogen and oxygen decreases and these gases come out of solution, forming small bubbles around the bottom of the pot.

The decrease in the solubility of gases with increasing temperature is the reason that warm soda pop bubbles more than cold soda pop and also the reason that warm soda goes flat faster than cold soda. The carbon dioxide comes out of solution faster (bubbles more) at room temperature than at cooler temperature because the gas is less soluble at room temperature.

The solubility of gases also depends on pressure. The higher the pressure above a liquid, the more soluble the gas is in the liquid (▼ Figure 7.18), a relationship known as **Henry's law**.

In a can of soda pop and at the bottom of Lake Nyos, high pressure maintains the carbon dioxide in solution. In soda pop, the pressure is provided by a large amount of carbon dioxide gas that is pumped into the can before sealing it. When we open the can, we release the pressure and the solubility of carbon dioxide decreases, resulting in bubbling (▼ Figure 7.19). The bubbles are formed by the carbon dioxide gas as it escapes. In Lake Nyos, the mass of the lake water provides the pressure by pushing down on the carbon-dioxide–rich water at the bottom of the lake. When the stratification (or layering) of the lake is disturbed, the pressure

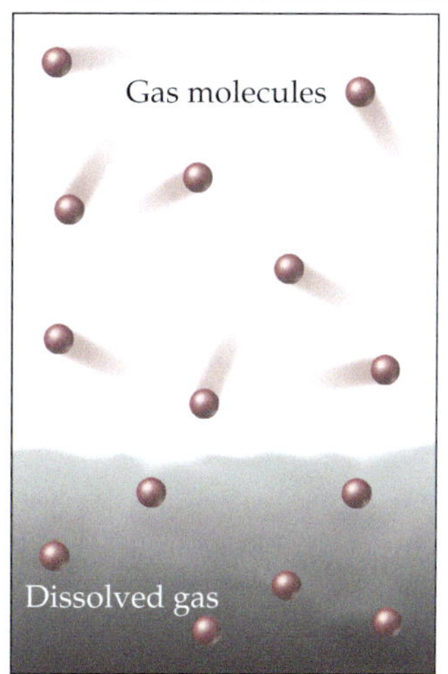

Gas at low pressure over a liquid

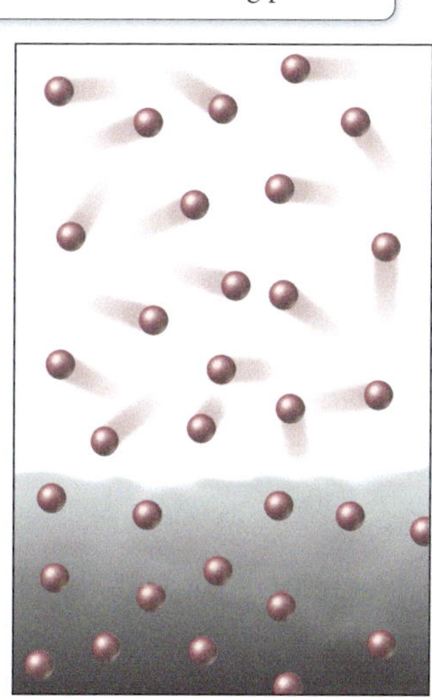

Gas at high pressure over a liquid

▶ **FIGURE 7.18 Pressure and solubility** The higher the pressure above a liquid, the more soluble the gas is in the liquid.

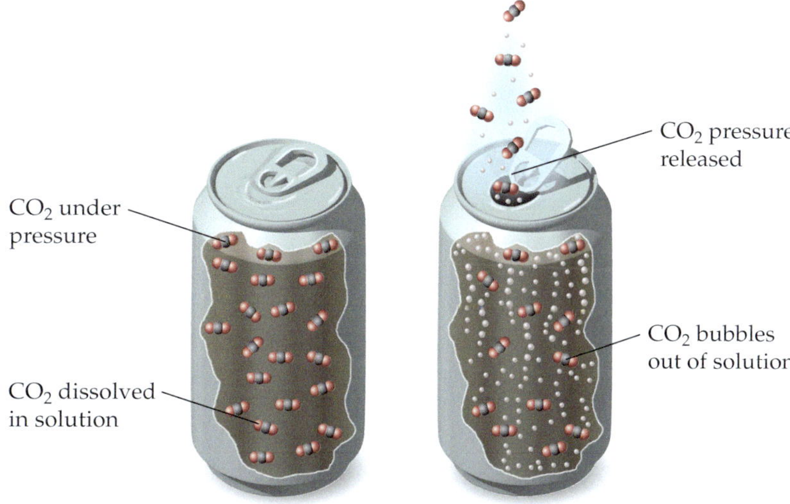

▶ **FIGURE 7.19 Pop! Fizz!** A can of soda pop is pressurized with carbon dioxide. When the can is opened, the pressure is released, lowering the solubility of carbon dioxide in the solution and causing it to come out of solution as bubbles.

on the carbon dioxide solution is lowered and the solubility of carbon dioxide decreases, resulting in the release of excess carbon dioxide gas.

CONCEPTUAL CHECKPOINT 7.2

A solution is saturated in both nitrogen gas (N_2) and potassium chloride (KCl) at 75 °C. What happens when the solution is cooled to room temperature?

(a) Some nitrogen gas bubbles out of solution.

(b) Some potassium chloride precipitates out of solution.

(c) Both (a) and (b).

(d) Nothing happens.

7.7 Specifying Solution Composition: Mass Percent

LO: Calculate mass percent.
LO: Use mass percent in calculations.

As we have seen, the amount of solute in a solution is an important property of the solution. For example, the amount of carbon dioxide in the water at the bottom of Lake Nyos is an important predictor of when the deadly event may repeat itself. A **dilute solution** is one containing small amounts of solute relative to solvent. If the water at the bottom of Lake Nyos were a dilute carbon dioxide solution, it would pose little threat. A **concentrated solution** is one containing large amounts of solute relative to solvent. If the carbon dioxide in the water at the bottom of Lake Nyos becomes concentrated (through the continual feeding of carbon dioxide from magma into the lake), it becomes a threat. A common method of reporting solution concentration is *mass percent*.

Mass Percent

Also in common use are *parts per million* (ppm), the number of grams of solute per 1 million (or 1 Mg) of solution, and *parts per billion* (ppb), the number of grams of solute per 1 billion (or 1 Gg) of solution.

Mass percent is the number of grams of solute per 100 g of solution. A solution with a concentration of 14 % by mass, for example, contains 14 g of solute per 100 g of solution. To calculate mass percent, we divide the mass of the solute by the mass of the solution (solute *and* solvent) and multiply by 100 %.

Note that the denominator is the mass of , not the mass of solvent.

$$\text{Mass percent} = \frac{\text{Mass solute}}{\text{Mass solute} + \text{Mass solvent}} \times 100\ \%$$

Suppose we want to calculate the mass percent of NaCl in a solution containing 15.3 g of NaCl and 155.0 g of water. We begin by sorting the information in the problem statement.

GIVEN: 15.3 g NaCl
155.0 g H_2O

FIND: mass percent

SOLUTION

To solve this problem, we substitute the correct values into the mass percent equation just presented.

$$\begin{aligned}\text{Mass percent} &= \frac{\text{Mass solute}}{\text{Mass solute} + \text{Mass solvent}} \times 100\ \% \\ &= \frac{15.3\ \text{g}}{15.3\ \text{g} + 155.0\ \text{g}} \times 100\ \% \\ &= \frac{15.3\ \cancel{\text{g}}}{170.3\ \cancel{\text{g}}} \times 100\ \% \\ &= 8.98\ \%\end{aligned}$$

The solution is 8.98 % NaCl by mass.

EXAMPLE 7.2 CALCULATING MASS PERCENT

Calculate the mass percent of a solution containing 27.5 g of ethanol (C_2H_6O) and 175 g of H_2O.

SORT

You are given the mass of ethanol and the mass of water. You are asked to find the mass percent of the solution.

GIVEN: 27.5 g C_2H_6O (mass of solute)
175 g H_2O (mass of solvent)

FIND: mass percent

STRATEGIZE

To find the mass percent, substitute the correct quantities into the equation for mass percent and calculate the mass percent.

RELATIONSHIP USED

$$\text{Mass percent} = \frac{\text{Mass solute}}{\text{Mass solute} + \text{Mass solvent}} \times 100\ \%$$

SOLUTION

$$\text{Mass percent} = \frac{27.5\ \text{g}}{27.5\ \text{g} + 175\ \text{g}}$$

$$= \frac{27.5\ \text{g}}{202.5\ \text{g}} \times 100\ \%$$

$$= 13.6\ \%$$

▶SKILLBUILDER 7.2 | Calculating Mass Percent

Calculate the mass percent of a sucrose solution containing 11.3 g of sucrose and 412.1 g of water.

▶FOR MORE PRACTICE Example 7.8; Problems 54, 55, 56, 57, 58, 59.

Using Mass Percent in Calculations

For example, let's consider a water sample from the bottom of Lake Nyos containing 8.5 % carbon dioxide by mass. We can determine how much carbon dioxide in grams is contained in 2.95×10^4 g of the water solution. We begin by sorting the information in the problem statement.

GIVEN: 8.5 % CO_2 by mass
2.95×10^4 g mass of solution

FIND: g CO_2

SOLUTION MAP

We strategize by drawing a solution map that begins with g solution and mass %. Then we proceed to g CO_2, using the mass percent equation.

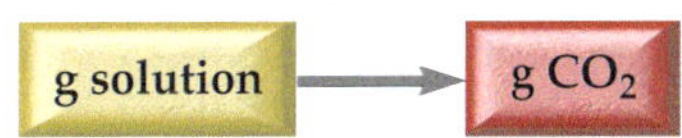

RELATIONSHIPS USED

$$\text{mass}\ \% = \frac{\text{mass solute}}{\text{mass solution}} \times 100\ \%$$

Rearrange the equation in terms of mass solute.

$$\text{mass solute} = \frac{\text{mass}\ \% \times \text{mass solution}}{100\ \%}$$

SOLUTION

$$\text{mass solute} = \frac{8.5\ \% \times 2.95 \times 10^4\ \text{g}}{100\ \%} = 2.5 \times 10^3\ \text{g}$$

In this example, we used mass percent to calculate the amount of solute present in the solution from a given amount of solution. In Example 7.3, you use mass percent to calculate the amount of solution containing a solute from a given amount of that solute.

EXAMPLE 7.3 USING MASS PERCENT IN CALCULATIONS

A soft drink contains 11.5 % sucrose ($C_{12}H_{22}O_{11}$) by mass. What mass of the soft drink solution contains 85.2 g of sucrose?

SORT
You are given the concentration (mass percent) of sucrose in a soft drink and a mass of sucrose. You are asked to find the mass of the soft drink that contains the given mass of sucrose.

GIVEN: 11.5 % $C_{12}H_{22}O_{11}$ by mass
85.2 g $C_{12}H_{22}O_{11}$

FIND: g solution (soft drink)

STRATEGIZE
Draw a solution map to calculate from g solute to g solution using the equation for mass percent.

SOLUTION MAP

RELATIONSHIPS USED:

$$\text{mass \%} = \frac{\text{mass solute} \times 100\ \%}{\text{mass solution}}$$

Rearrange the equation above for mass solution to get the equation below.

$$\text{mass solution} = \frac{\text{mass solute} \times 100\ \%}{\text{mass \%}}$$

SOLVE
Follow the solution map to solve the problem.

SOLUTION

$$\text{mass solution} = \frac{85.2\ \text{g} \times 100\ \%}{11.5\ \%} = 741\ \text{g}$$

CHECK
Check your answer. Are the units correct? Does the answer make physical sense?

The unit is correct. The magnitude of the answer makes sense because each 100 g of solution contains 11.5 g sucrose; therefore 741 g should contain a bit more than 77 g, which is close to the given amount of 85.2 g.

▶SKILLBUILDER 7.3 | Using Mass Percent in Calculations

How much sucrose ($C_{12}H_{22}O_{11}$) in grams is contained in 355 g of the soft drink in Example 7.3?

▶FOR MORE PRACTICE Example 7.9; Problems 60, 61, 62, 63, 64, 65.

7.8 Specifying Solution Composition: Molarity

LO: Calculate molarity.
LO: Use molarity in calculations.
LO: Calculate ion concentration.

Note that molarity is sometimes abbreviated with a capital *M* (which can be confused with M for molar mass).
The units of molarity are mol/L which sometimes is abbreviated as M.
Both *M* for molarity and M for mol/L are not used in this textbook.

A second way to express solution concentration (c) is **molarity** defined as the number of moles of solute (n) per litre of solution (V). Molarity is also known as molar concentration. We calculate the molarity of a solution as follows:

$$c = \frac{n}{V} \text{ that means } \textbf{Molarity} = \frac{\text{Moles solute}}{\text{Litres solution}}$$

Note that molarity is moles of solute per litre (mol/L or mol L^{-1}) of *solution*, not per litre of solvent. To make a solution of a specified molarity, we usually put the solute into a flask and then add water to the desired volume of solution. For example, to make 1.00 L of a 1.00 mol/L NaCl solution, we add 1.00 mol of NaCl to a flask and then add water to make 1.00 L of solution (▼ Figure 7.20). We *do not* combine 1.00 mol of NaCl with 1.00 L of water because that would result in a total volume exceeding 1.00 L and therefore a molarity of less than 1.00 mol/L.

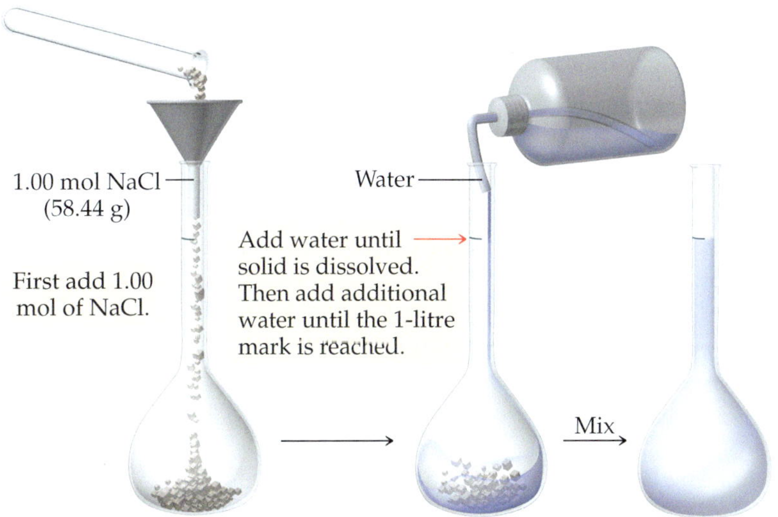

▶ **FIGURE 7.20 Making a solution of specific molarity** To make 1.00 L of a 1.00 mol/L NaCl solution, we add 1.00 mol (58.44 g) of sodium chloride to a flask and then dilute to 1.00 L of total volume.
Question: What would happen if we added 1 L of water to 1 mol of sodium chloride? Would the resulting solution be 1 mol/L?

CHEMISTRY IN THE ENVIRONMENT

The Dirty Dozen

A number of potentially harmful chemicals—such as DDT, dioxin, and polychlorinated biphenyls (PCBs)—make their way into water sources from industrial dumping, atmospheric emissions, agriculture, and household dumping. Since crops, livestock, and fish all rely on water, they too can accumulate these chemicals from water. Human consumption of food or water contaminated with these chemicals leads to a number of diseases and adverse health effects such as increased cancer risk, liver damage, or central nervous system damage. Governments around the world have joined forces to ban a number of these chemicals—called persistent organic pollutants or POP—from production. The original treaty (known as the Stockholm Convention on Persistent Organic Pollutants) targeted 12 such substances called the dirty dozen (Table 7.2).

TABLE 7.2 The Dirty Dozen

1. aldrin (insecticide)
2. chlordane (insecticide by-product)
3. DDT (insecticide)
4. dieldrin (insecticide)
5. dioxin (industrial by-product)
6. eldrin (insecticide)
7. furan (industrial by-product)
8. heptachlor (insecticide)
9. hexachlorobenzene (fungicide, industrial by-product)
10. mirex (insecticide, fire retardant)
11. polychlorinated biphenyls (PCBs) (electrical insulators)
12. toxaphene (insecticide)

A difficult problem posed by these chemicals is their persistence. Once they get into the environment, they stay there for a long time. A second problem is their tendency to undergo *bioamplification*. Because these chemicals are nonpolar, they are stored and concentrated in the fatty tissues of the organisms that consume them. As larger organisms eat smaller ones that have consumed the chemical, the larger organisms consume even more of the stored chemicals. The result is an increase in the concentrations of these chemicals as they move up the food chain.

Under the treaty, nearly all intentional production of these chemicals is banned. In the United States, the presence of these contaminants in water supplies is monitored under supervision of the Environmental Protection Agency (EPA). The EPA has set limits, called maximum contaminant levels (MCLs), for each of the dirty dozen in food and drinking water. Some MCLs for selected compounds in water supplies are listed in Table 7.3.

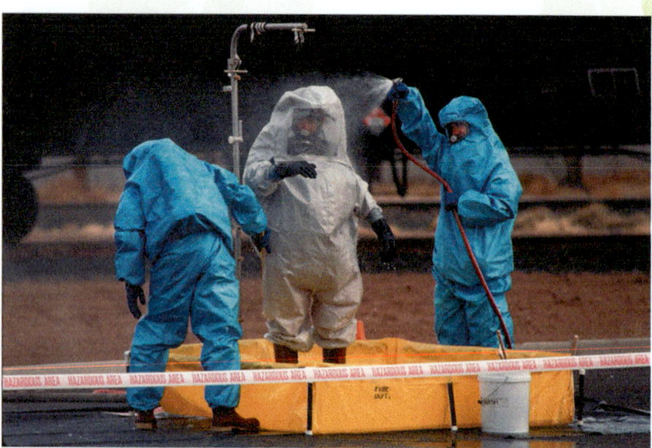

▲ Potentially dangerous chemicals can leak into the environment and contaminate water and food supplies. © Don B. Stevenson/Alamy.

TABLE 7.3 EPA Maximum Contaminant Level (MCL) for Several "Dirty Dozen" Chemicals

chlordane	0.002 mg/L
dioxin	0.00000003 mg/L
heptachlor	0.0004 mg/L
hexachlorobenzene	0.001 mg/L

Notice the units that the EPA uses to express the concentration of the contaminants: milligrams per litre. This unit shows the mass in milligrams of the pollutant in each litre of water consumed. According to the EPA, as long as the contaminant concentrations are below these levels, the water is safe to drink.

B7.3 CAN YOU ANSWER THIS? *Calculate how much of each of the chemicals in Table 7.3 (at their MCL) is present in 715 L of water, the approximate amount of water consumed by an adult in one year.*

To calculate molarity, we divide the number of moles of the solute by the volume of the solution (solute *and* solvent) in litres, this means $c = n/V$. For example, to calculate the molarity of a sucrose ($C_{12}H_{22}O_{11}$) solution made with 1.58 mol of sucrose diluted to a total volume of 5.0 L of solution, we begin by sorting the information in the problem statement.

GIVEN: 1.58 mol $C_{12}H_{22}O_{11}$
5.0 L solution

FIND: molarity

SOLUTION

We substitute the correct values into the equation for molarity and calculate the answer.

$$c = \frac{n}{V}$$

$$= \frac{1.58 \text{ mol } C_{12}H_{22}O_{11}}{5.0 \text{ L solution}}$$

$$= 0.32 \text{ mol/L}$$

EXAMPLE 7.4 **CALCULATING MOLARITY**

Calculate the molarity of a solution made by putting 15.5 g NaCl (molar mass 58.44 g/mol) into a beaker and adding water to make 1.50 L of NaCl solution.

SORT You are given the mass and the molar mass of NaCl (the solute) and the volume of solution. You are asked to find the molarity of the solution.	**GIVEN:** 15.5 g NaCl 58.44 g/mol NaCl 1.50 L solution **FIND:** molarity
STRATEGIZE To calculate molarity, substitute the correct values into the equation and calculate the answer. You must first calculate moles of NaCl from grams using the molar mass of NaCl.	**SOLUTION** $n = \frac{m}{M} = \frac{15.5 \text{ g}}{58.44 \text{ g/mol}} = 0.26\underline{5}2 \text{ mol}$ $c = \frac{n}{V} = \frac{0.2652 \text{ mol}}{1.50 \text{ L}} = 0.177 \text{ mol/L}$

▶**SKILLBUILDER 7.4** | **Calculating Molarity**

Calculate the molarity of a solution made by putting 55.8 g of $NaNO_3$ into a beaker and diluting to 2.50 L.

▶**FOR MORE PRACTICE** Example 7.10; Problems 66, 67, 68, 69, 70, 71.

Using Molarity in Calculations

If the molarity of a solution is known, (c) it can be used in calculations to determine the volume of the solution (V) or the amount of the solute (n) in the solution. This is done by rearranging the equation $c = n/V$.

For example, to determine how many grams of sucrose ($C_{12}H_{22}O_{11}$, molar mass 342.34 g/mol) are contained in 1.72 L of 0.758 mol/L sucrose solution, we begin by sorting the information in the problem statement.

GIVEN: 0.758 mol/L $C_{12}H_{22}O_{11}$
1.72 L solution
molar mass of sucrose = 342.34 g/mol

FIND: g $C_{12}H_{22}O_{11}$

SOLUTION MAP

We strategize by drawing a solution map that begins with L solution and shows the calculation of moles of sucrose using the molarity, and then the calculation of mass of sucrose using the molar mass.

RELATIONSHIPS USED

$$c = \frac{n}{V} \qquad n = \frac{m}{M}$$

Rearrange the equations above for n and m respectively to get the equations below.

$$n = cV \qquad m = nM$$

SOLUTION

$n = cV = 0.758 \text{ mol}/\text{L} \times 1.72 \text{ L} = 1.30 \text{ mol}$

$m = nM = 1.30 \text{ mol} \times 342.34 \text{ g}/\text{mol} = 446 \text{ g}$

In this example, we used molarity (c) to calculate the amount of solute (n) in a solution from a given amount of solution (V). In Example 7.5, we use the molarity to calculate the amount of a solution from a given amount of the solute in that solution.

EXAMPLE 7.5 USING MOLARITY IN CALCULATIONS

How many litres of a 0.114 mol/L NaOH solution contains 1.24 mol of NaOH?

SORT

You are given the molarity of an NaOH solution and the number of moles of NaOH. You are asked to find the volume of solution that contains the given number of moles.

GIVEN: 0.114 mol/L NaOH
1.24 mol NaOH

FIND: L solution

STRATEGIZE

The solution map begins with mol NaOH and shows the calculation of the volume in litres of solution using the equation for molarity.

SOLUTION MAP

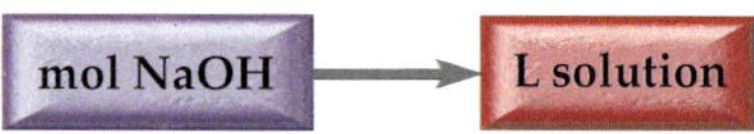

RELATIONSHIP USED:

$$c = \frac{n}{V}$$

Rearrange the equation above for V to get the equation below.

$$V = \frac{n}{c}$$

SOLVE

Solve the problem by following the solution map.

SOLUTION

$$V = \frac{1.24 \text{ mol}}{0.114 \text{ mol}/\text{L}} = 10.9 \text{ L}$$

CHECK Check your answer. Are the units correct? Does the answer make physical sense?	The unit (L) is correct. The magnitude of the answer makes sense because each L of solution contains a little more than 0.10 moles; therefore about 10 L contains a little more than 1 mole.

▶SKILLBUILDER 7.5 | Using Molarity in Calculations

How much of a 0.225 mol/L KCl solution contains 55.8 g of KCl?

▶FOR MORE PRACTICE Example 7.11; Problems 72, 73, 74, 75, 76, 77.

Ion Concentrations

When an ionic compound dissolves in solution, some of the cations and anions may pair up, so that the actual concentrations of the ions are lower than what you would expect if you assume complete dissociation occurred.

The reported concentration of a solution containing a *molecular* compound usually reflects the concentration of the solute as it actually exists in solution. For example, a 1.0 mol/L glucose ($C_6H_{12}O_6$) solution indicates that the solution contains 1.0 mol of $C_6H_{12}O_6$ per litre of solution. However, the reported concentration of solution containing an *ionic* compound reflects the concentration of the solute *before it is dissolved in solution*. For example, a 1.0 mol/L $CaCl_2$ solution contains 1.0 mol of Ca^{2+} per litre and 2.0 mol of Cl^- per litre. The concentration of the individual ions present in a solution containing an ionic compound can usually be approximated from the overall concentration as shown by the following example.

EXAMPLE 7.6 CALCULATING ION CONCENTRATION

Determine the molar concentrations of Na^+ and PO_4^{3-} in a 1.50 mol/L Na_3PO_4 solution.

SORT You are given the molarity of an ionic solution and asked to find the molar concentrations of the component ions.	**GIVEN:** 1.50 mol/L Na_3PO_4 **FIND:** molarity (mol/L) of Na^+ and PO_4^{3-}
STRATEGIZE A formula unit of Na_3PO_4 contains 3 Na^+ ions (as indicated by the subscript), so the concentration of Na^+ is three times the concentration of Na_3PO_4. Since the same formula unit contains one PO_4^{3-} ion, the concentration of PO_4^{3-} is equal to the concentration of Na_3PO_4.	**SOLUTION** molarity of $Na^+ = 3 \times 1.50$ mol/L $= 4.50$ mol/L; molarity of $PO_4^{3-} = 1.50$ mol/L

▶SKILLBUILDER 7.6 | Ion Concentration

Determine the molar concentrations of Ca^{2+} and Cl^- in a 0.75 mol/L $CaCl_2$ solution.

▶FOR MORE PRACTICE Problems 84, 85, 86, 87.

CONCEPTUAL CHECKPOINT 7.3

A solution is 0.15 mol/L in K_2SO_4. What is the concentration of K^+ in solution?

(a) 0.075 mol/L

(b) 0.15 mol/L

(c) 0.30 mol/L

(d) 0.45 mol/L

7.9 Solution Dilution

LO: Use the dilution equation in calculations.

When diluting acids, always add the concentrated acid to the water. *Never add water to concentrated acid solutions.*

To save space in laboratory storerooms, solutions are often stored in concentrated forms called **stock solutions**. For example, hydrochloric acid is typically stored as a 12 mol/L stock solution. However, many lab procedures call for much less concentrated hydrochloric acid solutions, so chemists must dilute the stock solution to the required concentration. This is normally done by diluting a certain amount of the stock solution with water. How do we determine how much of the stock solution to use? The easiest way to solve this problem is to use the dilution equation:

$$c_1V_1 = c_2V_2$$

where c_1 and V_1 are the molarity and volume of the initial concentrated solution and c_2 and V_2 are the molarity and volume of the final diluted solution. This equation works because the molarity multiplied by the volume gives the number of moles of solute ($c \times V$ = mol), which is the same in both solutions. For example, suppose a laboratory procedure calls for 5.00 L of a 1.50 mol/L HCl solution. How should we prepare this solution from a 10.0 mol/L stock solution? We begin by sorting the information in the problem statement.

The equation $c_1V_1 = c_2V_2$ applies only to solution dilution, NOT to stoichiometry.

GIVEN: $c_1 = 10.0$ mol/L
$c_2 = 1.50$ mol/L
$V_2 = 5.00$ L

FIND: V_1

SOLUTION

We solve the solution dilution equation for V_1 (the volume of the stock solution required for the dilution) and substitute in the correct values to calculate it.

$$\begin{aligned} c_1V_1 &= c_2V_2 \\ V_1 &= \frac{c_2V_2}{c_1} \\ &= \frac{1.50\,\frac{\text{mol}}{\cancel{\text{L}}} \times 5.00\text{ L}}{10.0\,\frac{\text{mol}}{\cancel{\text{L}}}} \\ &= 0.750\text{ L} \end{aligned}$$

We can therefore make the solution by diluting 0.750 L of the stock solution to a total volume of 5.00 L (V_2). The resulting solution is 1.50 mol/L in HCl (▶ Figure 7.21).

EXAMPLE 7.7 SOLUTION DILUTION

To what volume should you dilute 0.100 L of a 15 mol/L NaOH solution to obtain a 1.0 mol/L NaOH solution?

SORT

You are given the initial volume and molar concentration of an NaOH solution and a final molar concentration. You are asked to find the volume required to dilute the initial solution to the given final concentration.

GIVEN: $V_1 = 0.100$ L
$c_1 = 15$ mol/L
$c_2 = 1.0$ mol/L

FIND: V_2

STRATEGIZE

Solve the solution dilution equation for V_2 (the volume of the final solution) and substitute the required quantities to calculate V_2.

You can make the solution by diluting 0.100 L of the stock solution to a total volume of 1.5 L (V_2). The resulting solution has a concentration of 1.0 mol/L.

SOLUTION

$$\begin{aligned} c_1V_1 &= c_2V_2 \\ V_2 &= \frac{c_1V_1}{c_2} \\ &= \frac{15\,\frac{\text{mol}}{\cancel{\text{L}}} \times 0.100\text{ L}}{1.0\,\frac{\text{mol}}{\cancel{\text{L}}}} \\ &= 1.5\text{ L} \end{aligned}$$

▶SKILLBUILDER 7.7 | Solution Dilution

How much 6.0 mol/L $NaNO_3$ solution should you use to make 0.585 L of a 1.2 mol/L $NaNO_3$ solution?

▶FOR MORE PRACTICE Example 7.12; Problems 88, 89, 90, 91, 92, 94, 94, 95.

$$c_1V_1 = c_2V_2$$

$$\frac{10.0\text{ mol}}{\cancel{\text{L}}} \times 0.750\cancel{\text{L}} = \frac{1.50\text{ mol}}{\cancel{\text{L}}} \times 5.00\cancel{\text{L}}$$

$$7.50\text{ mol} = 7.50\text{ mol}$$

▶ **FIGURE 7.21 Making a solution by dilution of a more concentrated solution**

CONCEPTUAL CHECKPOINT 7.4

What is the molarity of a solution in which 100.0 mL of 1.0 mol/L KCl is diluted to 1.0 L?

(a) 0.10 mol/L

(b) 1.0 mol/L

(c) 10.0 mol/L

MODULE IN REVIEW

Self-Assessment Quiz

Q1. Which compound forms an electrolyte solution when dissolved in water?

(a) KBr

(b) Br_2

(c) CH_3OH

(d) $C_6H_{12}O_6$ (glucose)

Q2. A solution is saturated in O_2 gas and KNO_3 at room temperature. What happens if the solution is warmed to 75 °C?

(a) Solid KNO_3 precipitates out of solution.

(b) Gaseous O_2 bubbles out of solution.

(c) Solid KNO_3 precipitates out of solution *and* gaseous O_2 bubbles out of solution.

(d) Nothing happens (both O_2 and KNO_3 remain in solution).

Q3. What is the mass percent concentration of a solution containing 25.0 g NaCl and 155.0 g of water?

(a) 2.76 %

(b) 6.20 %

(c) 13.9 %

(d) 16.1 %

Q4. A glucose solution is 3.25 % glucose by mass. What mass of glucose is contained in 59.9 g of this solution?

(a) 18.4 g **(b)** 184 g

(c) 195 g **(d)** 1.95 g

Q5. What is the molarity of a solution containing 11.2 g $Ca(NO_3)_2$ in 175 mL of solution?

(a) 64.0 mol/L **(b)** 0.390 mol/L

(c) 0.064 mol/L **(d)** 1.96×10^3 mol/L

Q6. What mass of glucose ($C_6H_{12}O_6$) is contained in 75.0 mL of a 1.75 mol/L glucose solution?

(a) 23.6 g **(b)** 7.72 g

(c) 0.131 g **(d)** 75.0 g

Q7. What is the molar concentration of potassium ions in a 0.250 mol/L K_2SO_4 solution?

(a) 0.125 mol/L **(b)** 0.250 mol/L

(c) 0.500 mol/L **(d)** 0.750 mol/L

Q8. What volume of a 3.50 mol/L Na_3PO_4 solution should you use to make 1.50 L of a 2.55 mol/L Na_3PO_4 solution?

(a) 0.917 L **(b)** 13.4 L

(c) 2.06 L **(d)** 1.09 L

Answers: 1:a, 2:b, 3:c, 4:d, 5:b, 6:a, 7:c, 8:d

Chemical Principles

Manifestations of Intermolecular Forces: Surface tension—the tendency for liquids to minimize their surface area—is a direct result of intermolecular forces. Viscosity—the resistance of liquids to flow—is another result of intermolecular forces. Both surface tension and viscosity increase with greater intermolecular forces.

Water: Water is a unique molecule. Because of its strong hydrogen bonding, water is a liquid at room temperature. Unlike most liquids, water expands when it freezes. In addition, water is highly polar, making it a good solvent for many polar substances.

Types of Crystalline Solids: We divide crystalline solids into three categories based on the individual units composing the solid:

Molecular solids—Molecules are the composite units of molecular solids, which are held together by dispersion forces, dipole–dipole forces, or hydrogen bonding.

Ionic solids—Formula units (the smallest electrically neutral collection of cations and anions) are the composite units of ionic solids. They are held together by the electrostatic attractions that occur between cations and anions.

Atomic solids—Atoms are the composite units of atomic solids, which are held together by different forces depending on the particular solid.

Solutions: A solution is a homogeneous mixture with two or more components. Usually, the solvent is the majority component, and the solute is the minority component. Water is the solvent in aqueous solutions.

Solid-and-Liquid Solutions: The solubility—the amount of solute that dissolves in a certain amount of solvent—of solids in liquids often increases with increasing temperature. Recrystallization involves dissolving a solid into hot solvent to saturation and then allowing it to cool. As the solution cools, it becomes supersaturated and the solid crystallizes.

Relevance

Manifestations of Intermolecular Forces: Many insects can walk on water due to surface tension. The viscosity of a liquid is one of its defining properties and is important in applications such as automobile lubrication; the viscosity of a motor oil must be high enough to coat an engine's surfaces, but not so high that it can't flow to remote parts of the engine.

Water: Water is critical to life. On Earth, wherever there is water, there is life. Water acts as a solvent and transport medium, and virtually all the chemical reactions on which life depends take place in aqueous solution. The expansion of water upon freezing allows life within frozen lakes to survive the winter. The ice on top of the lake acts as insulation, protecting the rest of the lake (and the life within it) from freezing.

Types of Crystalline Solids: Solids have different properties depending on their individual units and the forces that hold those units together. Molecular solids tend to have low melting points. Ionic solids tend to have intermediate to high melting points. Atomic solids have varied melting points, depending on the particular solid.

Solutions: Solutions are all around us—most of the fluids that we encounter every day are solutions. Common solutions include seawater (solid and liquid), soda pop (gas and liquid), and blood (solid, gas, and liquid).

Solid-and-Liquid Solutions: Solutions of solids dissolved in liquids, such as seawater, coffee, and sugar water, are important both in chemistry and in everyday life. Recrystallization is used extensively in the laboratory to purify solids.

Gas-and-Liquid Solutions: The solubility of gases in liquids decreases with increasing temperature but increases with increasing pressure.

Gas-and-Liquid Solutions: The temperature and pressure dependence of gas solubility is the reason that soda pop fizzes when opened and the reason that warm soda goes flat.

Solution Concentration: We use solution concentration to specify how much of the solute is present in a given amount of solution. Two common ways to express solution concentration are mass percent and molarity (also known as molar concentration).

$$\text{Mass percent} = \frac{\text{Mass solute}}{\text{Mass solute} + \text{Mass solvent}} \times 100\ \%$$

$$c = \frac{n}{V}$$

Solution Concentration: Solution concentration is useful in calculations involving amounts of solute and solution. Mass percent and molarity are the most common concentration units.

Solution Dilution: We can solve solution dilution problems using the following equation:

$$c_1V_1 = c_2V_2$$

Solution Dilution: Since many solutions are stored in concentrated form, it is often necessary to dilute them to a desired concentration.

Chemical Skills

LO: Calculate mass percent (Section 7.7).

SORT
You are given the mass of the solute and the solvent and asked to find the concentration of the solution in mass percent.

STRATEGIZE
To calculate mass percent concentration, divide the mass of the solute by the mass of the solution (solute and solvent) and multiply by 100 %.

Examples

EXAMPLE 7.8 CALCULATING MASS PERCENT

Find the mass percent concentration of a solution containing 19 g of solute and 158 g of solvent.

GIVEN: 19 g solute
158 g solvent

FIND: mass percent

SOLUTION

$$\text{Mass percent} = \frac{\text{mass solute}}{\text{mass solute} + \text{mass solvent}} \times 100\ \%$$

$$\text{Mass percent} = \frac{19\text{ g}}{19\text{ g} + 158\text{ g}} \times 100\ \%$$

$$\text{Mass percent} = \frac{19\text{ g}}{177\text{ g}} \times 100\ \%$$

$$= 11\ \%$$

LO: Use mass percent in calculations (Section 7.7).

EXAMPLE 7.9 USING MASS PERCENT IN CALCULATIONS

How much KCl in grams is in 354 g of a 5.80 % mass percent KCl solution?

SORT

You are given the mass of a KCl solution and its mass percent concentration. You are asked to find the mass of potassium chloride.

GIVEN: 5.80 % KCl by mass
354 g solution

FIND: g KCl

STRATEGIZE

Draw a solution map. Begin with the given mass of the solution in g and use the equation for the mass percent to get to the mass of KCl.

SOLUTION MAP

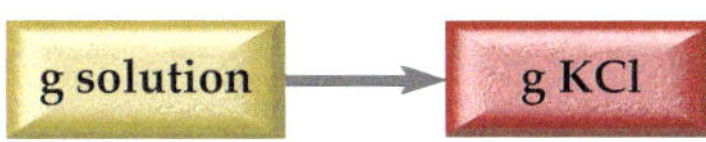

RELATIONSHIPS USED

$$\text{mass \%} = \frac{\text{mass solute}}{\text{mass solution}} \times 100\,\%$$

Rearrange the equation above for mass solute to get the equation shown below.

$$\text{mass solute} = \frac{\text{mass \%} \times \text{mass solution}}{100\,\%}$$

SOLVE

Follow your solution map to calculate the answer.

SOLUTION

$$\text{mass solute} = \frac{5.80\,\% \times 354\text{ g}}{100\,\%} = 20.5\text{ g}$$

CHECK

Check your answer. Are the units correct? Does the answer make physical sense?

The unit is correct. The magnitude of the answer makes sense because 354 g is a bit more than 300 g of solution. Each 100 g of solution contains about 6 g KCl. Therefore the answer should be a bit more than 18 g.

LO: Calculate molarity (Section 7.8).

EXAMPLE 7.10 CALCULATING MOLARITY

Calculate the molarity of a KCl solution containing 0.22 mol of KCl in 0.455 L of solution.

SORT

You are given the number of moles of potassium chloride and the volume of solution. You are asked to find the molarity.

GIVEN: 0.22 mol KCl
0.455 L solution

FIND: molarity

STRATEGIZE

To calculate molarity, divide the number of moles of solute by the volume of the solution in litres.

SOLUTION

$$c = \frac{n}{V} = \frac{0.22\text{ mol}}{0.455\text{ L}} = 0.48\text{ mol/L}$$

LO: Use molarity in calculations (Section 7.8).

EXAMPLE 7.11 USING MOLARITY IN CALCULATIONS

How much KCl (molar mass 74.55 g/mol) in grams is contained in 0.488 L of 1.25 mol/L KCl solution?

SORT
You are given the molar mass of KCl, the volume and molarity of a KCl solution and asked to find the mass of KCl contained in the solution.

GIVEN: 1.25 mol/L KCl
74.55 g/mol KCl
0.488 L solution

FIND: g KCl

STRATEGIZE
Draw a solution map beginning with litres of solution and calculating moles of solute using the molarity. Then calculate mass using the molar mass.

SOLUTION MAP

RELATIONSHIPS USED

$$c = \frac{n}{V} \qquad n = \frac{m}{M}$$

Rearrange the equations above for n and m respectively to get the equations below.

$$n = cV \qquad m = nM$$

SOLVE
Follow your solution map to calculate the answer.

SOLUTION

$$n = cV = 1.25\ \text{mol}/\cancel{\text{L}} \times 0.488\ \cancel{\text{L}} = 0.610\ \text{mol}$$

$$m = nM = 0.610\ \cancel{\text{mol}} \times 74.55\ \text{g}/\cancel{\text{mol}} = 45.5\ \text{g}$$

CHECK
Check your answer. Are the units correct? Does the answer make physical sense?

The unit (g) is correct. The magnitude of the answer makes sense because if each litre of solution contains 1.25 mol, then the given amount of solution (which is about 0.5 L) should contain a bit more than 0.5 mol, which would have a mass that is bit more than about 37 g.

LO: Use the dilution equation in calculations (Section 7.9).

EXAMPLE 7.12 SOLUTION DILUTION

How much of an 8.0 mol/L HCl solution should be used to make 0.400 L of a 2.7 mol/L HCl solution?

SORT
You are given the initial molarity and final molarity of a solution as well as the final volume. You are asked to find the initial volume.

GIVEN: $c_1 = 8.0$ mol/L
$c_2 = 2.7$ mol/L
$V_2 = 0.400$ L

FIND: V_1

STRATEGIZE
Most solution dilution problems will use equation $c_1V_1 = c_2V_2$.

Solve the equation for the quantity you are trying to find (in this case, V_1) and substitute in the correct values to calculate it.

SOLUTION

$$c_1V_1 = c_2V_2$$

$$V_1 = \frac{c_2V_2}{c_1}$$

$$= \frac{2.7\ \frac{\cancel{\text{mol}}}{\cancel{\text{L}}} \times 0.400\ \text{L}}{8.0\ \frac{\cancel{\text{mol}}}{\cancel{\text{L}}}}$$

$$= 0.14\ \text{L}$$

KEY TERMS

atomic solid **[7.3]**
covalent atomic solid **[7.3]**
electrolyte solution **[7.5]**
Henry's law **[7.6]**
ionic solid **[7.3]**
mass percent **[7.7]**
metallic atomic solid **[7.3]**
molarity **[7.8]**
molecular solid **[7.3]**
nonbonding atomic solid **[7.3]**
nonelectrolyte solution **[7.5]**
recrystallization **[7.5]**
saturated solution **[7.5]**
solubility **[7.5]**
solute **[7.4]**
solution **[7.4]**
solvent **[7.4]**
stock solution **[7.9]**
supersaturated solution **[7.5]**
surface tension **[7.1]**
unsaturated solution **[7.5]**
viscosity **[7.1]**

EXERCISES

QUESTIONS

1. What is surface tension? How does it depend on intermolecular forces?
2. What is viscosity? How does it depend on intermolecular forces?
3. In what ways is water unique?
4. How would ice be different if it were denser than water? How would that affect aquatic life in cold-climate lakes?
5. What is a molecular solid? What kinds of forces hold molecular solids together?
6. How do the melting points of molecular solids relate to those of other types of solids?
7. What is an ionic solid? What kinds of forces hold ionic solids together?
8. How do the melting points of ionic solids relate to those of other types of solids?
9. What is an atomic solid? What are the properties of atomic solids?
10. What is a solution? List some examples.
11. What is an aqueous solution?
12. In a solution, what is the solvent? What is the solute? List some examples.
13. Explain what "like dissolves like" means.
14. What is solubility?
15. Describe what happens when additional solute is added to:
(a) a saturated solution
(b) an unsaturated solution
(c) a supersaturated solution
16. Explain what happens to an ionic substance when it dissolves in water.
17. Do polyatomic ions dissociate when they dissolve in water, or do they remain intact?
18. What is a strong electrolyte solution?
19. Explain the difference between a strong electrolyte solution and a nonelectrolyte solution. What kinds of solutes form strong electrolyte solutions?
20. How does gas solubility depend on temperature?
21. Explain recrystallization.
22. When you heat water on a stove, bubbles form on the bottom of the pot *before* the water boils. What are these bubbles? Why do they form?
23. Explain why warm soda pop goes flat faster than cold soda pop.
24. How does gas solubility depend on pressure? How does this relationship explain why a can of soda pop fizzes when opened?
25. What is the difference between a dilute solution and a concentrated solution?
26. Define the concentration units mass percent and molarity.
27. What is a stock solution?

PROBLEMS

TYPES OF SOLIDS

28. Identify each solid as molecular, ionic, or atomic.
(a) $Ar(s)$
(b) $H_2O(s)$
(c) $K_2O(s)$
(d) $Fe(s)$

29. Identify each solid as molecular, ionic, or atomic.
(a) $CaCl_2(s)$
(b) $CO_2(s)$
(c) $Ni(s)$
(d) $I_2(s)$

30. Identify each solid as molecular, ionic, or atomic.
(a) $H_2S(s)$
(b) $KCl(s)$
(c) $N_2(s)$
(d) $NI_3(s)$

31. Identify each solid as molecular, ionic, or atomic.
(a) $SF_6(s)$
(b) $C(s)$
(c) $MgCl_2(s)$
(d) $Ti(s)$

32. Which solid has the highest melting point? Why?
- **(a)** Ar(s)
- **(b)** $CCl_4(s)$
- **(c)** LiCl(s)
- **(d)** $CH_3OH(s)$

33. Which solid has the highest melting point? Why?
- **(a)** C (s, diamond)
- **(b)** Kr(s)
- **(c)** NaCl(s)
- **(d)** $H_2O(s)$

34. For each pair of solids, determine which solid has the higher melting point and explain why.
- **(a)** Ti(s) and Ne(s)
- **(b)** $H_2O(s)$ and $H_2S(s)$
- **(c)** Kr(s) and Xe(s)
- **(d)** NaCl(s) and $CH_4(s)$

35. For each pair of solids, determine which solid has the higher melting point and explain why.
- **(a)** Fe(s) and $CCl_4(s)$
- **(b)** KCl(s) or HCl(s)
- **(c)** $TiO_2(s)$ or HOOH(s)

36. List these substances in order of increasing boiling point:

H_2O, Ne, NH_3, NaF, SO_2

37. List these substances in order of decreasing boiling point:

CO_2, Ne, CH_3OH, KF

SOLUTIONS

38. Determine whether or not each mixture is a solution.
- **(a)** sand and water mixture
- **(b)** oil and water mixture
- **(c)** salt and water mixture
- **(d)** carbon dioxide and water mixture

39. Identify the solute and solvent in each solution.
- **(a)** salt water
- **(b)** sugar water
- **(c)** soda water
- **(d)** oxygenated water

40. Pick an appropriate solvent from Table 7.1 to dissolve:
- **(a)** motor oil (nonpolar)
- **(b)** sugar (polar)
- **(c)** lard (nonpolar)
- **(d)** potassium chloride (ionic)

41. Pick an appropriate solvent from Table 7.1 to dissolve:
- **(a)** glucose (polar)
- **(b)** salt (ionic)
- **(c)** vegetable oil (nonpolar)
- **(d)** sodium nitrate (ionic)

SOLIDS DISSOLVED IN WATER

42. What are the dissolved particles in a solution containing an ionic solute? What is the name for this kind of solution?

43. What are the dissolved particles in a solution containing a molecular solute? What is the name for this kind of solution?

44. A solution contains 30 g of NaCl per 100 g of water at 25 °C. Is the solution unsaturated, saturated, or supersaturated? (See Figure 7.17.)

45. A solution contains 28 g of KNO_3 per 100 g of water at 25 °C. Is the solution unsaturated, saturated, or supersaturated? (See Figure 7.17.)

46. A KNO_3 solution containing 45 g of KNO_3 per 100 g of water is cooled from 40 °C to 0 °C. What happens during cooling? (See Figure 7.17.)

47. A KCl solution containing 42 g of KCl per 100 g of water is cooled from 60 °C to 0 °C. What happens during cooling? (See Figure 7.17.)

48. Refer to Figure 7.17 to determine whether each of the given amounts of solid will completely dissolve in the given amount of water at the indicated temperature.
- **(a)** 30.0 g $KClO_3$ in 85.0 g of water at 35 °C
- **(b)** 65.0 g $NaNO_3$ in 125 g of water at 15 °C
- **(c)** 32.0 g KCl in 70.0 g of water at 82 °C

49. Refer to Figure 7.17 to determine whether each of the given amounts of solid will completely dissolve in the given amount of water at the indicated temperature.
- **(a)** 45.0 g $CaCl_2$ in 105 g of water at 5 °C
- **(b)** 15.0 g $KClO_3$ in 115 g of water at 25 °C
- **(c)** 50.0 g $Pb(NO_3)_2$ in 95.0 g of water at 10 °C

GASES DISSOLVED IN WATER

50. Some laboratory procedures involving oxygen-sensitive reactants or products call for using preboiled (and then cooled) water. Explain why this is so.

51. A person preparing a fish tank uses preboiled (and then cooled) water to fill it. When the fish is put into the tank, it dies. Explain.

52. Scuba divers breathing air at increased pressure can suffer from nitrogen narcosis—a condition resembling drunkenness—when the partial pressure of nitrogen exceeds about 4 atm. What property of gas/water solutions causes this to happen? How could the diver reverse this effect?

53. Scuba divers breathing air at increased pressure can suffer from oxygen toxicity—too much oxygen in the bloodstream—when the partial pressure of oxygen exceeds about 1.4 atm. What happens to the amount of oxygen in a diver's bloodstream when he or she breathes oxygen at elevated pressures? How can this be reversed?

MASS PERCENT

54. Calculate the concentration of each solution in mass percent.
(a) 41.2 g $C_{12}H_{22}O_{11}$ in 498 g H_2O
(b) 178 mg $C_6H_{12}O_6$ in 4.91 g H_2O
(c) 7.55 g NaCl in 155 g H_2O

55. Calculate the concentration of each solution in mass percent.
(a) 132 g KCl in 598 g H_2O
(b) 22.3 mg KNO_3 in 2.84 g H_2O
(c) 8.72 g C_2H_6O in 76.1 g H_2O

56. A soft drink contains 42 g of sugar in 311 g of H_2O. What is the concentration of sugar in the soft drink in mass percent?

57. A soft drink contains 32 mg of sodium in 309 g of H_2O. What is the concentration of sodium in the soft drink in mass percent?

58. Complete the table.

Mass Solute	Mass Solvent	Mass Solution	Mass Percent
15.5 g	238.1 g	_____	_____
22.8 g	_____	_____	12.0 %
_____	183.3 g	212.1 g	_____
_____	315.2 g	_____	15.3 %

59. Complete the table.

Mass Solute	Mass Solvent	Mass Solution	Mass Percent
2.55 g	25.0 g	_____	_____
_____	45.8 g	_____	3.8 %
1.38 g	_____	27.2 g	_____
23.7 g	_____	_____	5.8 %

60. Ocean water contains 3.5 % NaCl by mass. How much salt can be obtained from 254 g of seawater?

61. A saline solution contains 1.1 % NaCl by mass. How much NaCl is present in 96.3 g of this solution?

62. Determine the amount of sucrose in each solution.
(a) 48 g of a solution containing 3.7 % sucrose by mass
(b) 103 mg of a solution containing 10.2 % sucrose by mass
(c) 3.2 kg of a solution containing 14.3 % sucrose by mass

63. Determine the amount of potassium chloride in each solution.
(a) 19.7 g of a solution containing 1.08 % KCl by mass
(b) 23.2 kg of a solution containing 18.7 % KCl by mass
(c) 38 mg of a solution containing 12 % KCl by mass

64. Determine the mass (in g) of each NaCl solution that contains 1.5 g of NaCl.
(a) 0.058 % NaCl by mass
(b) 1.46 % NaCl by mass
(c) 8.44 % NaCl by mass

65. Determine the mass (in g) of each sucrose solution that contains 12 g of sucrose.
(a) 4.1 % sucrose by mass
(b) 3.2 % sucrose by mass
(c) 12.5 % sucrose by mass

MOLARITY

66. Calculate the molarity of each solution.
(a) 0.127 mol of sucrose in 655 mL of solution
(b) 0.205 mol of KNO_3 in 0.875 L of solution
(c) 1.1 mol of KCl in 2.7 L of solution

67. Calculate the molarity of each solution.
(a) 1.54 mol of LiCl in 22.2 L of solution
(b) 0.101 mol of $LiNO_3$ in 6.4 L of solution
(c) 0.0323 mol of glucose in 76.2 mL of solution

68. Calculate the molarity of each solution.
(a) 22.6 g of $C_{12}H_{22}O_{11}$ in 0.442 L of solution
(b) 42.6 g of NaCl in 1.58 L of solution
(c) 315 mg of $C_6H_{12}O_6$ in 58.2 mL of solution

69. Calculate the molarity of each solution.
(a) 33.2 g of KCl in 0.895 L of solution
(b) 61.3 g of C_2H_6O in 3.4 L of solution
(c) 38.2 mg of KI in 112 mL of solution

70. A 205 mL sample of ocean water contains 6.8 g of NaCl. What is the molarity of the solution with respect to NaCl?

71. A 355 mL can of soda pop contains 41 g of sucrose ($C_{12}H_{22}O_{11}$). What is the molarity of the solution with respect to sucrose?

72. How many moles of NaCl are contained in each solution?
(a) 1.5 L of a 1.2 mol/L NaCl solution
(b) 0.448 L of a 0.85 mol/L NaCl solution
(c) 144 mL of a 1.65 mol/L NaCl solution

73. How many moles of sucrose are contained in each solution?
(a) 3.4 L of a 0.100 mol/L sucrose solution
(b) 0.952 L of a 1.88 mol/L sucrose solution
(c) 21.5 mL of a 0.528 mol/L sucrose solution

74. What volume of each solution contains 0.15 mol of KCl?
(a) 0.255 mol/L KCl
(b) 1.8 mol/L KCl
(c) 0.995 mol/L KCl

75. What volume of each solution contains 0.325 mol of NaI?
(a) 0.152 mol/L NaI
(b) 0.982 mol/L NaI
(c) 1.76 mol/L NaI

76. Complete the table.

Solute	Solute Mass	Mol Solute	Volume Solution	Molarity
KNO_3	22.5 g	____	125.0 mL	____
$NaHCO_3$	____	____	250.0 mL	0.100 mol/L
$C_{12}H_{22}O_{11}$	55.38 g	____	____	0.150 mol/L

77. Complete the table.

Solute	Solute Mass	Mol Solute	Volume Solution	Molarity
$MgSO_4$	0.588 g	____	25.0 mL	____
NaOH	____	____	100.0 mL	1.75 mol/L
CH_3OH	12.5 g	____	____	0.500 mol/L

78. Calculate the mass of NaCl in a 35 mL sample of a 1.3 mol/L NaCl solution.

79. Calculate the mass of glucose ($C_6H_{12}O_6$) in a 105 mL sample of a 1.02 mol/L glucose solution.

80. A chemist wants to make 2.5 L of a 0.100 mol/L KCl solution. How much KCl in grams should the chemist use?

81. A laboratory procedure calls for making 500.0 mL of a 1.4 mol/L KNO_3 solution. How much KNO_3 in grams is needed?

82. How many litres of a 0.500 mol/L sucrose ($C_{12}H_{22}O_{11}$) solution contain 1.5 kg of sucrose?

83. What volume of a 0.35 mol/L $Mg(NO_3)_2$ solution contains 87 g of $Mg(NO_3)_2$?

84. Determine the concentration of Cl^- in each aqueous solution. (Assume complete dissociation of each compound.)
(a) 0.15 mol/L NaCl
(b) 0.15 mol/L $CuCl_2$
(c) 0.15 mol/L $AlCl_3$

85. Determine the concentration of NO_3^- in each aqueous solution. (Assume complete dissociation of each compound.)
(a) 0.10 mol/L KNO_3
(b) 0.10 mol/L $Ca(NO_3)_2$
(c) 0.10 mol/L $Cr(NO_3)_3$

86. Determine the concentration of the cation and anion in each aqueous solution. (Assume complete dissociation of each compound.)
(a) 0.12 mol/L Na_2SO_4
(b) 0.25 mol/L K_2CO_3
(c) 0.11 mol/L RbBr

87. Determine the concentration of the cation and anion in each aqueous solution. (Assume complete dissociation of each compound.)
(a) 0.20 mol/L $SrSO_4$
(b) 0.15 mol/L $Cr_2(SO_4)_3$
(c) 0.12 mol/L SrI_2

SOLUTION DILUTION

88. A 122 mL sample of a 1.2 mol/L sucrose solution is diluted to 500.0 mL. What is the molarity of the diluted solution?

89. A 3.5 L sample of a 5.8 mol/L NaCl solution is diluted to 55 L. What is the molarity of the diluted solution?

90. Describe how you would make 2.5 L of a 0.100 mol/L KCl solution from a 5.5 mol/L stock KCl solution.

91. Describe how you would make 500.0 mL of a 0.200 mol/L NaOH solution from a 15.0 mol/L stock NaOH solution.

92. To what volume should you dilute 25 mL of a 12 mol/L stock HCl solution to obtain a 0.500 mol/L HCl solution?

93. To what volume should you dilute 75 mL of a 10.0 mol/L H_2SO_4 solution to obtain a 1.75 mol/L H_2SO_4 solution?

94. How much of a 12.0 mol/L HNO_3 solution should you use to make 850.0 mL of a 0.250 mol/L HNO_3 solution?

95. How much of a 5.0 mol/L sucrose solution should you use to make 85.0 mL of a 0.040 mol/L solution?

CUMULATIVE PROBLEMS

96. A 125 mL sample of an 8.5 mol/L NaCl solution is diluted to 2.5 L. What volume of the diluted solution contains 10.8 g of NaCl?

97. A 45.8 mL sample of a 5.8 mol/L KNO_3 solution is diluted to 1.00 L. What volume of the diluted solution contains 15.0 g of KNO_3?

98. To what final volume should you dilute 50.0 mL of a 5.00 mol/L KI solution so that 25.0 mL of the diluted solution contains 3.25 g of KI?

99. To what volume should you dilute 125 mL of an 8.00 mol/L $CuCl_2$ solution so that 50.0 mL of the diluted solution contains 5.9 g $CuCl_2$?

100. What is wrong with this molecular view of a sodium chloride solution? What would make the picture correct?

▸Answers to Skillbuilder Exercises

Skillbuilder 7.1
(a) molecular
(b) ionic
(c) atomic

Skillbuilder 7.2 2.67 %

Skillbuilder 7.3 40.8 g sucrose

Skillbuilder 7.4 0.263 mol/L

Skillbuilder 7.5 3.33 L

Skillbuilder 7.6 0.75 mol/L Ca^{2+} and 1.5 mol/L Cl^-

Skillbuilder 7.7 0.12 L

▸Answers to Conceptual Checkpoints

7.1 (a) CH_3Cl and NH_3 are both polar compounds, and KF is ionic. All three would therefore interact more strongly with water molecules (which are polar) than CCl_4, which is nonpolar.

7.2 (b) Some potassium chloride precipitates out of solution. The solubility of most solids decreases with decreasing temperature. However, the solubility of gases increases with decreasing temperature. Therefore, the nitrogen becomes more soluble and will not bubble out of solution.

7.3 (c) The solution is 0.30 mol/L in K^+ because the compound K_2SO_4 forms two moles of K^+ in solution for each mole of K_2SO_4 that dissolves.

7.4 (a) The original 100.0 mL solution contains 0.10 mol, which is present in 1.0 L after the dilution. Therefore the molarity of the diluted solution is 0.10 mol/L.

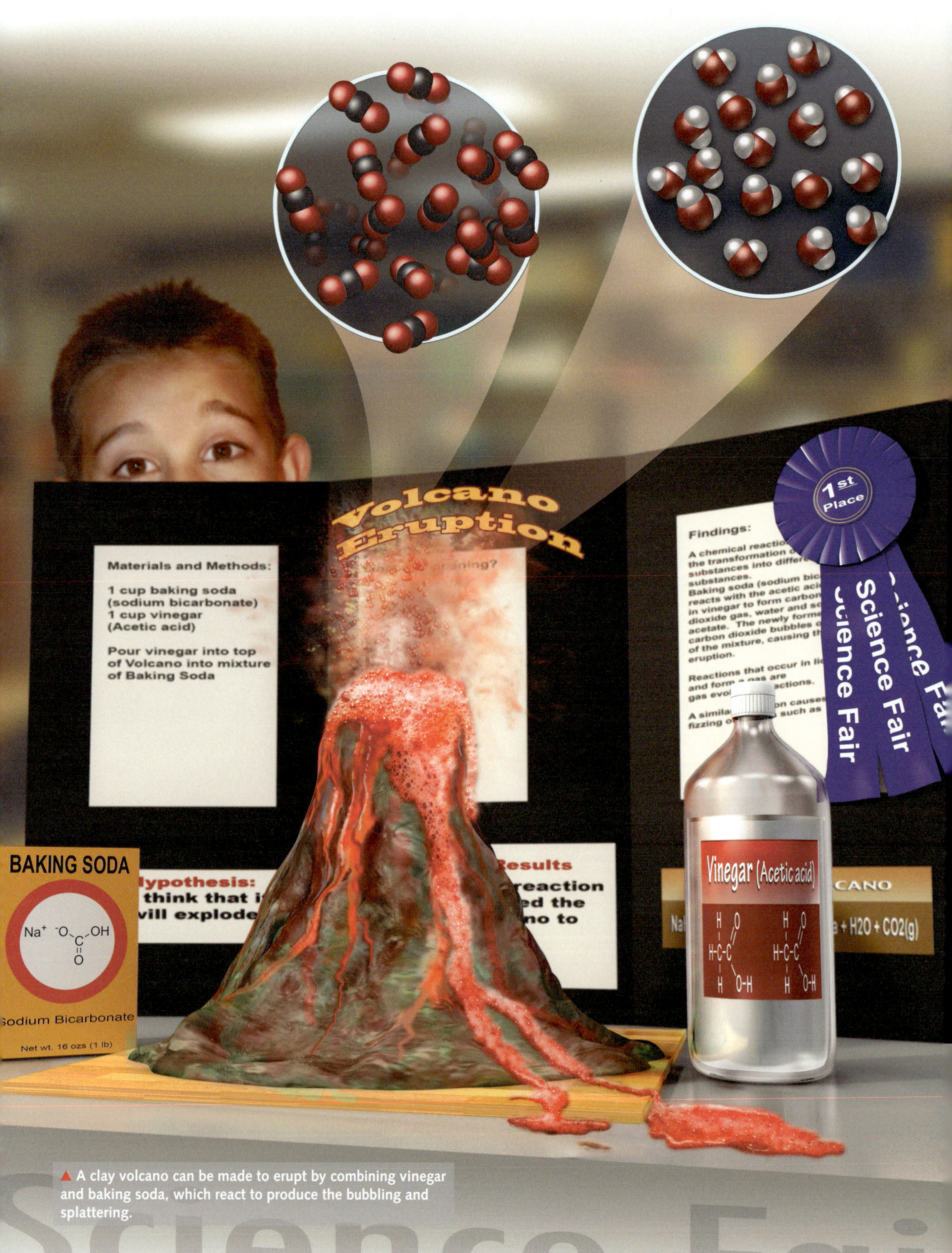

▲ A clay volcano can be made to erupt by combining vinegar and baking soda, which react to produce the bubbling and splattering.

Chemical Reactions 1 8

"Chemistry . . . is one of the broadest branches of science if for no other reason than, when we think about it, everything is chemistry."—Luciano Caglioti (1933–)

MODULE OUTLINE

8.1 Chemical Reactions

Did you ever make a clay volcano in primary school that erupted when filled with vinegar and baking soda? Have you pushed the gas pedal of a car and felt the acceleration as the car moved forward? Have you wondered why laundry detergents work better than hand soap to clean your clothes? Each of these processes involves a *chemical reaction*—the transformation of one or more substances into different substances.

In the classic primary school volcano, the baking soda (which is sodium bicarbonate/hydrogencarbonate) reacts with acetic acid in the vinegar to form carbon dioxide gas, water, and sodium acetate. The newly formed carbon dioxide bubbles out of the mixture, causing the eruption. Reactions that occur in liquids and form a gas are *gas evolution reactions*. A similar reaction causes the fizzing of antacids such as Alka-Seltzer™.

When you drive a car, hydrocarbons such as octane (in gasoline) react with oxygen from the air to form carbon dioxide gas and water (▼ Figure 8.1). This reaction

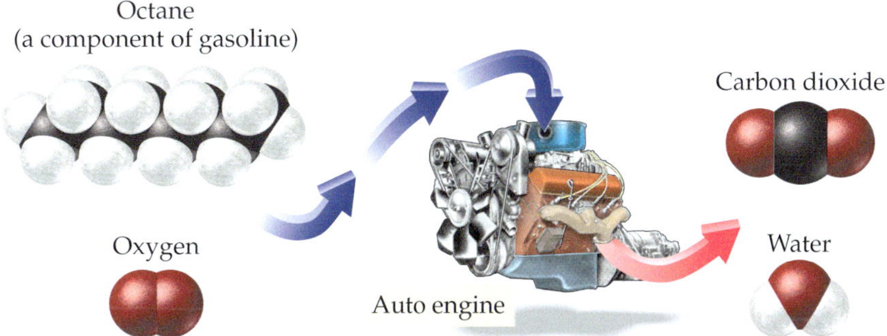

▲ **FIGURE 8.1 A combustion reaction** In an automobile engine, hydrocarbons such as octane (C_8H_{18}) from gasoline combine with oxygen from the air and react to form carbon dioxide and water.

▲ **FIGURE 8.2 Soap and water** Soap forms suds with pure water (left), but reacts with the ions in hard water (right) to form a gray residue that adheres to clothes. © Maxwell Art And Photo.

produces heat, which expands the gases in the car's cylinders, accelerating it forward. Reactions such as this one—in which a substance reacts with oxygen, emitting heat and forming one or more oxygen-containing compounds—are *combustion reactions*. Combustion reactions are a subcategory of *oxidation–reduction reactions*, in which electrons are transferred from one substance to another. The formation of rust and the dulling of automobile paint are other examples of oxidation–reduction reactions.

Laundry detergent works better than hand soap to wash clothes because it contains substances that soften hard water. Hard water contains dissolved calcium (Ca^{2+}) and magnesium (Mg^{2+}) ions. These ions interfere with the cleaning action of soap by reacting with it to form a gray, slimy substance called *curd* or *soap scum* (◀ Figure 8.2). If you have ever washed your clothes in ordinary soap, you may have noticed gray soap scum residue on your clothes.

Laundry detergents inhibit curd formation because they contain substances such as sodium carbonate (Na_2CO_3) that remove calcium and magnesium ions from the water. When sodium carbonate dissolves in water, it *dissociates*, or separates into sodium ions (Na^+) and carbonate ions (CO_3^{2-}). The dissolved carbonate ions react with calcium and magnesium ions in the hard water to form solid calcium carbonate ($CaCO_3$) and solid magnesium carbonate ($MgCO_3$). These solids simply settle to the bottom of the laundry mixture, resulting in the removal of the ions from the water. In other words, laundry detergents contain substances that react with the ions in hard water to immobilize them. Reactions such as these—that form solid substances in water—are *precipitation reactions*. Precipitation reactions are also used to remove dissolved toxic metals in industrial wastes.

Chemical reactions take place all around us and even inside us. They are involved in many of the products we use daily and in many of our experiences. Chemical reactions can be relatively simple, like the combination of hydrogen and oxygen to form water, or they can be complex, like the synthesis of a protein molecule from thousands of simpler molecules. In some cases, such as the neutralization reaction that occurs in a swimming pool when acid is added to adjust the water's acidity level, chemical reactions are not noticeable to the naked eye. In other cases, such as the combustion reaction that produces a pillar of smoke and fire under a rocket during liftoff, chemical reactions are very obvious. In all cases, however, chemical reactions produce changes in the arrangements of the molecules and atoms that compose matter. Often, these molecular changes cause macroscopic changes that we can directly experience.

8.2 Evidence of a Chemical Reaction

LO: Identify a chemical reaction.

If we could see the atoms and molecules that compose matter, we could easily identify a chemical reaction. Do atoms combine with other atoms to form compounds? Do new molecules form? Do the original molecules decompose? Do atoms in one molecule change places with atoms in another? If the answer to one or more of these questions is yes, a chemical reaction has occurred.

Although we can't see atoms, many chemical reactions do produce easily detectable changes as they occur. For example, when the color-causing molecules in a brightly colored shirt decompose with repeated exposure to sunlight, the color of the shirt fades. Similarly, when the molecules embedded in the plastic of a child's temperature-sensitive spoon transform upon warming, the color of the spoon changes. These *color changes* are evidence that a chemical reaction has occurred.

Other changes that identify chemical reactions include the *formation of a solid* (▶ Figure 8.3) or *the formation of a gas* (▶ Figure 8.4). Dropping Alka-Seltzer tablets into water or combining baking soda and vinegar (as in our opening example of the grade school volcano) are both good examples of chemical reactions that produce a gas—the gas is visible as bubbles in the liquid.

Solid formation

▲ **FIGURE 8.3 A precipitation reaction** The formation of a solid in a previously clear solution is evidence of a chemical reaction. © Richard Megna/Fundamental Photographs.

Gas formation

▲ **FIGURE 8.4 A gas evolution reaction** The formation of a gas is evidence of a chemical reaction. © Charles D. Winters/Science Source.

A reaction that emits heat is an *exothermic* reaction and one that absorbs heat is an *endothermic* reaction.

Heat absorption or *emission*, as well as *light emission*, are also evidence of reactions. For example, a natural gas flame produces heat and light. A chemical cold pack becomes cold when the plastic barrier separating two substances is broken. Both of these changes suggest that a chemical reaction is occurring.

Color change

▲ A child's temperature-sensitive spoon changes color upon warming due to a reaction induced by the higher temperature. © Maxwell Art And Photo/Pearson.

Heat absorption

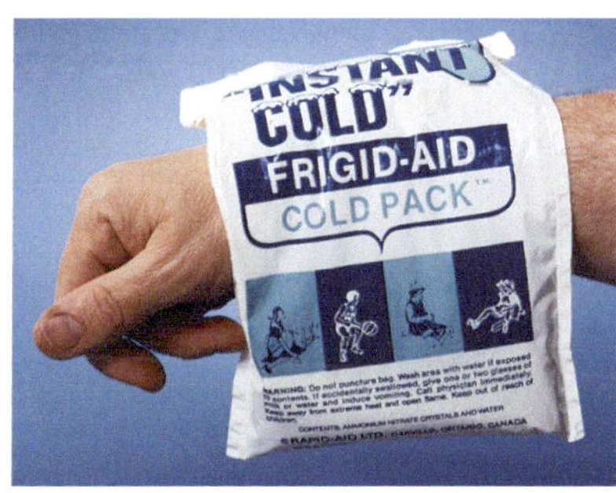

▲ A change in temperature due to absorption or emission of heat is evidence of a chemical reaction. This chemical cold pack becomes cold when the barrier separating two substances is broken.© Tom Bochsler/Pearson Education/PH College.

▲ **FIGURE 8.5 Boiling: A physical change** When water boils, bubbles are formed and a gas is evolved. However, no chemical change has occurred because the gas, like the liquid water, is also composed of water molecules. © Roman Sigaev/iStockphoto/Getty Images.

While these changes provide evidence of a chemical reaction, they are not *definitive* evidence. Only chemical analysis showing that the initial substances have changed into other substances conclusively proves that a chemical reaction has occurred. We can be fooled. For example, when water boils, bubbles form but no chemical reaction has occurred. Boiling water forms gaseous steam, but both water and steam are composed of water molecules—no chemical change has occurred (◀ Figure 8.5). On the other hand, chemical reactions may occur without any obvious signs, yet chemical analysis may show that a reaction has indeed occurred. The changes occurring at the atomic and molecular level determine whether a chemical reaction has taken place.

In summary, each of the following provides evidence of a chemical reaction.

- a *color change*

© Richard Megna/Fundamental Photographs.

- the *emission of light*

© Jeff J. Daly/Fundamental Photographs.

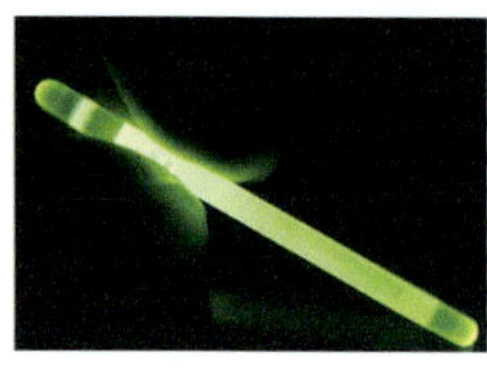

- the *formation of a solid* in a previously clear (unclouded) solution

© Richard Megna/Fundamental Photographs.

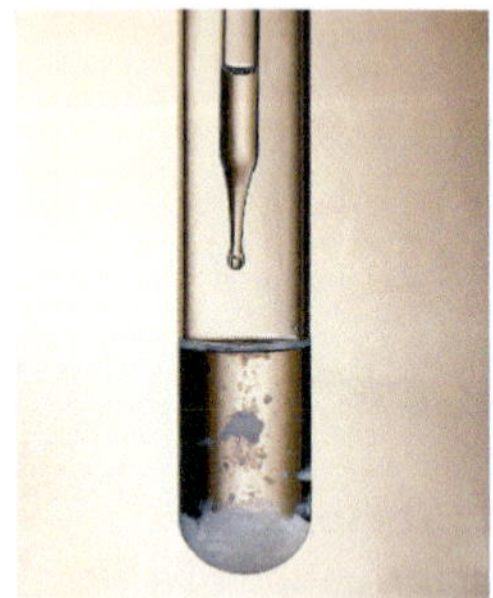

- the *emission* or *absorption of heat*

© Ayazad/Shutterstock.

- the *formation of a gas* when we add a substance to a solution

© Richard Megna/Fundamental Photographs.

EXAMPLE 8.1 EVIDENCE OF A CHEMICAL REACTION

Which changes involve a chemical reaction? Explain your answers.

(a) ice melting upon warming
(b) an electric current passing through water, resulting in the formation of hydrogen and oxygen gas that appears as bubbles rising in the water
(c) iron rusting
(d) bubbles forming when a soda can is opened

SOLUTION

(a) not a chemical reaction; melting ice forms water, but both the ice and water are composed of water molecules.
(b) chemical reaction; water decomposes into hydrogen and oxygen, as evidenced by the bubbling.
(c) chemical reaction; iron changes into iron oxide, changing color in the process.
(d) not a chemical reaction; even though there is bubbling, it is just carbon dioxide coming out of the liquid.

▶SKILLBUILDER 8.1 | Evidence of a Chemical Reaction

Which changes involve a chemical reaction? Explain your answers.

(a) butane burning in a butane lighter
(b) butane evaporating out of a butane lighter
(c) wood burning
(d) dry ice subliming

▶FOR MORE PRACTICE Example 8.25; Problems 42, 43, 44, 45.

CONCEPTUAL CHECKPOINT 8.1

These images portray molecular views of various substances before and after a change. Determine whether a chemical reaction has occurred in each case.

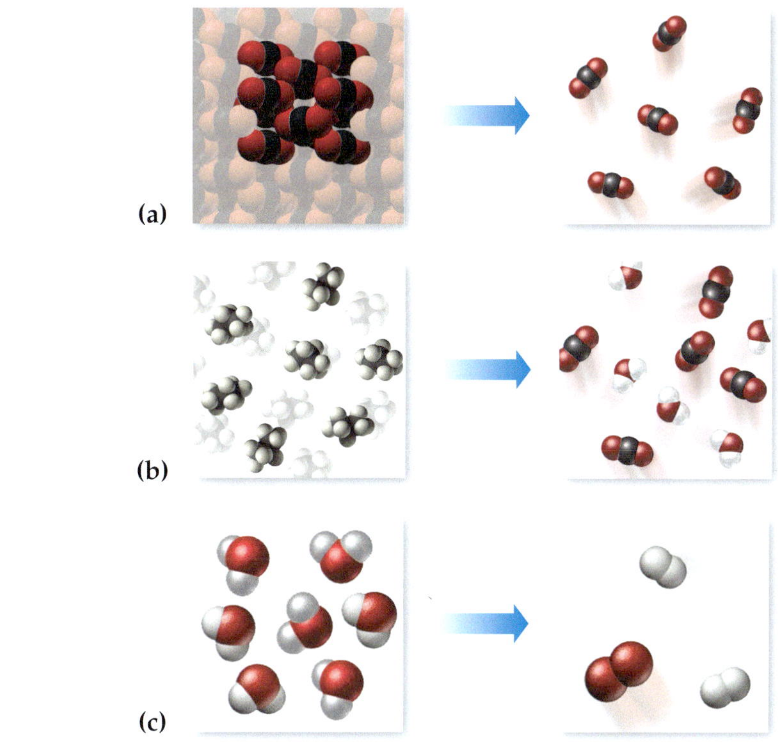

8.3 The Chemical Equation

LO: Identify balanced chemical equations.

We represent chemical reactions with *chemical equations*. For example, the reaction occurring in a natural-gas flame, such as the flame on a kitchen stove, is methane (CH_4) reacting with oxygen (O_2) to form carbon dioxide (CO_2) and water (H_2O). We represent this reaction with the equation:

$$\underbrace{CH_4 + O_2}_{\text{reactants}} \rightarrow \underbrace{CO_2 + H_2O}_{\text{products}}$$

The substances on the left side of the equation are the *reactants*, and the substances on the right side are the *products*. We often specify the state of each reactant or product in parentheses next to the formula. If we add states to our equation, it becomes:

$$CH_4(g) + O_2(g) \rightarrow CO_2(g) + H_2O(g)$$

The (g) indicates that these substances are gases in the reaction. Table 8.1 summarizes the common states of reactants and products and the symbols used in chemical reactions.

Let's look more closely at the equation for the burning of natural gas. How many oxygen atoms are on each side of the equation?

TABLE 8.1 Abbreviations Indicating the States of Reactants and Products in Chemical Equations

Abbreviation	State
(g)	gas
(l)	liquid
(s)	solid
(aq)	aqueous (water solution)*

*The (aq) designation stands for *aqueous*, which indicates that a substance is dissolved in water. When a substance dissolves in water, the mixture is called a *solution* (see Module 7).

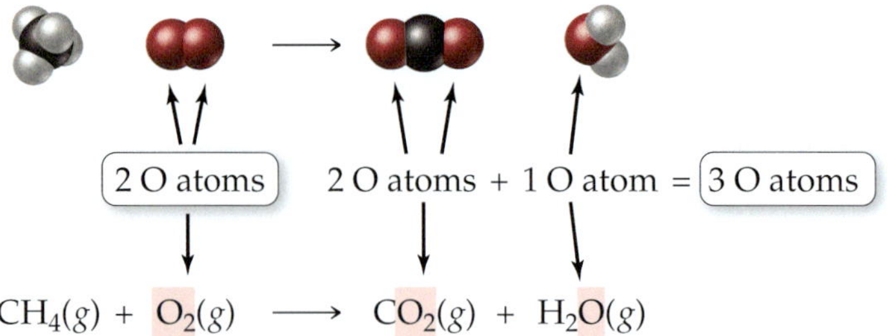

In chemical equations, atoms cannot change from one element to another—hydrogen atoms cannot change into oxygen atoms, for example nor can atoms disappear.

The left side of the equation has two oxygen atoms, and the right side has three. Since chemical equations represent real chemical reactions, atoms cannot simply appear or disappear in chemical equations because, as we know, atoms don't simply appear or disappear in nature. We must account for the atoms on both sides of the equation. Notice that the left side of the equation has four hydrogen atoms and the right side only two.

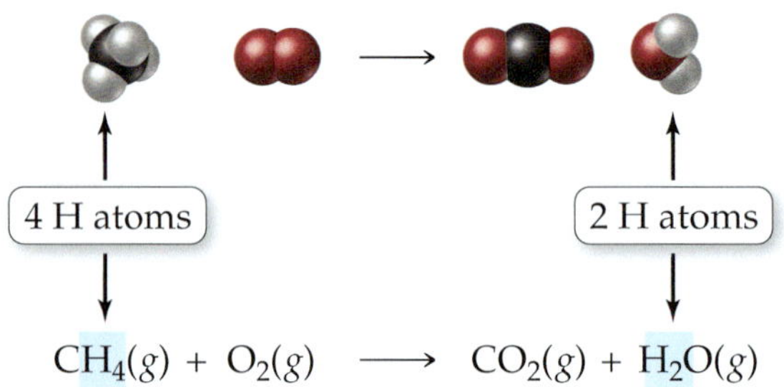

To correct these problems, we must create a **balanced equation**, one in which the numbers of atoms of each element on both sides of the equation are equal. To balance an equation, we insert coefficients—not subscripts—in front of the chemical formulas as needed to make the number of each type of atom in the reactants equal to the number of each type of atom in the products. New atoms do not form during a reaction, nor do atoms vanish—matter must be conserved.

When we balance chemical equations by inserting coefficients as needed in front of the formulas of the reactants and products, it changes the number of molecules in the equation, but it does not change the *kinds* of molecules. To balance the preceding equation, for example, we put the coefficient 2 before O_2 in the reactants, and the coefficient 2 before H_2O in the products.

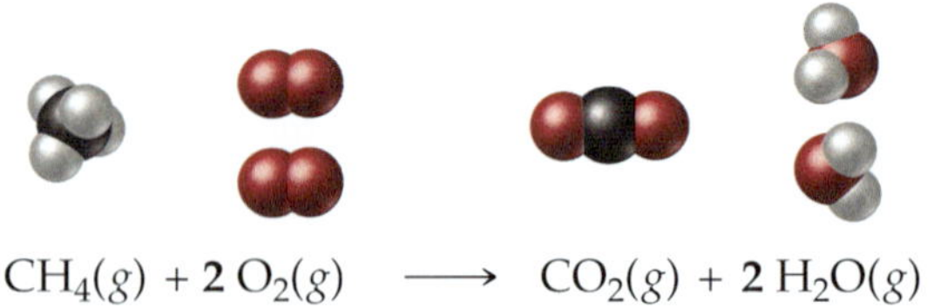

The equation is now balanced because the numbers of atoms of each element on both sides of the equation are equal. We can verify this by summing the number of atoms of each element.

We determine the number of a particular type of atom within a chemical formula in an equation by multiplying the subscript for the atom by the coefficient for the chemical formula. If there is no coefficient or subscript, a 1 is implied. So the balanced equation for the combustion of natural gas is:

$$CH_4(g) + 2\,O_2(g) \rightarrow CO_2(g) + 2\,H_2O(g)$$

Reactants	Products
1 C atom (1 × $\underline{C}H_4$)	1 C atom (1 × $\underline{C}O_2$)
4 H atoms (1 × $C\underline{H_4}$)	4 H atoms (2 × $\underline{H_2}O$)
4 O atoms (2 × $\underline{O_2}$)	4 O atoms (1 × $C\underline{O_2}$ + 2 × $H_2\underline{O}$)

The numbers of atoms of each element on both sides of the equation are equal—the equation is balanced.

▶ A balanced chemical equation represents a chemical reaction. In this equation, methane molecules combine with oxygen to form carbon dioxide and water. The photo shows a stove burning methane and the balanced equation for the reaction above it. © Sami Sarkis/PhotoDisc/Getty Images.

CONCEPTUAL CHECKPOINT 8.2

In photosynthesis, plants make the sugar glucose, $C_6H_{12}O_6$, from carbon dioxide and water. The equation for the reaction is:

$$6\,CO_2 + 6\,H_2O \rightarrow C_6H_{12}O_6 + x\,O_2$$

In order for this equation to be balanced, the coefficient x must be

(a) 3 **(b)** 6 **(c)** 9 **(d)** 12

8.4 Conservation of Matter: Why Chemical Equations must be Balanced

LO: Apply the law of conservation of mass.

This law is a slight oversimplification. In nuclear reactions, (which are beyond the scope of this book) significant changes in mass can occur. In chemical reactions, however, the changes are so minute that they can be ignored.

As we have seen, our planet, our air, and even our own bodies are composed of matter. Physical and chemical changes do not destroy matter, nor do they create new matter. This is articulated in the law of conservation of mass, which states:

Matter is neither created nor destroyed in a chemical reaction.

During physical and chemical changes, the total amount of matter remains constant even though it may not initially appear that it has. When we burn butane in a lighter, for example, the butane slowly disappears. Where does it go? It combines with oxygen to form carbon dioxide and water that travel into the surrounding air. The mass of the carbon dioxide and water that forms, however, exactly equals the mass of the butane and oxygen that combined.

We will examine the quantitative relationships in chemical reactions in Module 10.

Suppose that we burn 58 g of butane in a lighter. It will react with 208 g of oxygen to form 176 g of carbon dioxide and 90 g of water.

$$\underbrace{\underset{58\text{ g} + 208\text{ g}}{\text{Butane} + \text{Oxygen}}}_{266\text{ g}} \rightarrow \underbrace{\underset{176\text{ g} + 90\text{ g}}{\text{Carbon Dioxide} + \text{Water}}}_{266\text{ g}}$$

The sum of the masses of the butane and oxygen, 266 g, is equal to the sum of the masses of the carbon dioxide and water, which is also 266 g. In this chemical reaction, as in all chemical reactions, matter is conserved.

EXAMPLE 8.2 CONSERVATION OF MASS

A chemist forms 16.6 g of potassium iodide by combining 3.9 g of potassium with 12.7 g of iodine. Show that these results are consistent with the law of conservation of mass.

SOLUTION

The sum of the masses of the potassium and iodine is:

$$3.9 \text{ g} + 12.7 \text{ g} = 16.6 \text{ g}$$

The sum of the masses of potassium and iodine equals the mass of the product, potassium iodide. The results are consistent with the law of conservation of mass.

▶SKILLBUILDER 8.2 | Conservation of Mass

Suppose 12 g of natural gas combines with 48 g of oxygen in a flame. The chemical change produces 33 g of carbon dioxide. How many grams of water form?

▶FOR MORE PRACTICE Example 8.26; Problems 48, 49, 50, 51.

CONCEPTUAL CHECKPOINT 8.3

Consider a drop of water that is put into a flask, sealed with a cap, and heated until the droplet vaporizes. Is the mass of the container and water different after heating?

8.5 How to Write Balanced Chemical Equations

LO: Write balanced chemical equations.

The following procedure box details the steps for writing balanced chemical equations. As in other procedures in the book, we show the steps in the left column and examples of applying each step in the center and right columns. Remember, change only the *coefficients* to balance a chemical equation; *never change the subscripts because changing the subscripts changes the kinds of molecules, not the number of molecules.*

WRITING BALANCED CHEMICAL EQUATIONS	EXAMPLE 8.3	EXAMPLE 8.4
	Write a balanced equation for the reaction between solid silicon dioxide and solid carbon to produce solid silicon carbide and carbon monoxide gas.	Write a balanced equation for the combustion reaction between liquid octane (C_8H_{18}), a component of gasoline, and gaseous oxygen to form gaseous carbon dioxide and gaseous water.
1. Write the unbalanced equation by writing chemical formulas for each of the reactants and products. Review Module 3 for nomenclature rules. (If the unbalanced equation is provided in the problem, skip this step and go to Step 2.)	**SOLUTION** $SiO_2(s) + C(s) \rightarrow SiC(s) + CO(g)$	**SOLUTION** $C_8H_{18}(l) + O_2(g) \rightarrow CO_2(g) + H_2O(g)$
2. If an element occurs in only one compound on both sides of the equation, balance it first. If there is more than one such element, balance metals before nonmetals.	**BEGIN WITH Si** $SiO_2(s) + C(s) \rightarrow SiC(s) + CO(g)$ **1 Si atom → 1 Si atom** Si is already balanced. **BALANCE O NEXT** $SiO_2(s) + C(s) \rightarrow SiC(s) + CO(g)$ **2 O atoms → 1 O atom** To balance O, put a 2 before $CO(g)$. $SiO_2(s) + C(s) \rightarrow SiC(s) + 2\,CO(g)$ **2 O atoms → 2 O atoms**	**BEGIN WITH C** $C_8H_{18}(l) + O_2(g) \rightarrow CO_2(g) + H_2O(g)$ **8 C atoms → 1 C atom** To balance C, put an 8 before $CO_2(g)$. $C_8H_{18}(l) + O_2(g) \rightarrow 8\,CO_2(g) + H_2O(g)$ **8 C atoms → 8 C atoms** **BALANCE H NEXT** $C_8H_{18}(l) + O_2(g) \rightarrow 8\,CO_2(g) + H_2O(g)$ **18 H atoms → 2 H atoms** To balance H, put a 9 before $H_2O(g)$. $C_8H_{18}(l) + O_2(g) \rightarrow 8\,CO_2(g) + 9\,H_2O(g)$ **18 H atoms → 18 H atoms**
3. If an element occurs as a free element on either side of the chemical equation, balance it last. Always balance free elements by adjusting the coefficient *on the free element*.	**BALANCE C** $SiO_2(s) + C(s) \rightarrow SiC(s) + 2\,CO(g)$ **1 C atom → 1 C + 2 C = 3 C atoms** To balance C, put a 3 before $C(s)$. $SiO_2(s) + 3\,C(s) \rightarrow SiC(s) + 2\,CO(g)$ **3 C atoms → 1 C + 2 C = 3 C atoms**	**BALANCE O** $C_8H_{18}(l) + O_2(g) \rightarrow 8\,CO_2(g) + 9\,H_2O(g)$ **2 O atoms** → 16 O + 9 O = **25 O atoms** To balance O, put a $\frac{25}{2}$ before $O_2(g)$. $C_8H_{18}(l) + \frac{25}{2}\,O_2(g) \rightarrow 8\,CO_2(g) + 9\,H_2O(g)$ **25 O atoms** → 16 O + 9 O = **25 O atoms**
4. If the balanced equation contains coefficient fractions, change these into whole numbers by multiplying the entire equation by the appropriate factor.	This step is not necessary in this example. Proceed to Step 5.	$[C_8H_{18}(l) + \frac{25}{2}\,O_2(g) \rightarrow 8\,CO_2(g) + 9\,H_2O(g)] \times 2$ $2\,C_8H_{18}(l) + 25\,O_2(g) \rightarrow 16\,CO_2(g) + 18\,H_2O(g)$

5. Check to make certain the equation is balanced by summing the total number of atoms of every element on each side of the equation.

$SiO_2(s) + 3\ C(s) \rightarrow SiC(s) + 2\ CO(g)$

Reactants		Products
1 Si atom	→	1 Si atom
2 O atoms	→	2 O atoms
3 C atoms	→	3 C atoms

The equation is balanced.

▶**SKILLBUILDER 8.3** | Write a balanced equation for the reaction between solid chromium(III) oxide and solid carbon to produce solid chromium and carbon dioxide gas.

$2\ C_8H_{18}(l) + 25\ O_2(g) \rightarrow 16\ CO_2(g) + 18\ H_2O(g)$

Reactants		Products
16 C atoms	→	16 C atoms
36 H atoms	→	36 H atoms
50 O atoms	→	50 O atoms

The equation is balanced.

▶**SKILLBUILDER 8.4** | Write a balanced equation for the combustion reaction of gaseous C_4H_{10} and gaseous oxygen to form gaseous carbon dioxide and gaseous water.

▶**FOR MORE PRACTICE** Example 8.27; Problems 54, 55, 58, 59, 60, 61.

EXAMPLE 8.5 BALANCING CHEMICAL EQUATIONS

Write a balanced equation for the reaction of solid aluminum with aqueous sulfuric acid to form aqueous aluminum sulfate and hydrogen gas.

	SOLUTION
Use your knowledge of chemical nomenclature from Module 3 to write a skeletal equation containing formulas for each of the reactants and products. The formulas for each compound MUST BE CORRECT before you begin to balance the equation.	$Al(s) + H_2SO_4(aq) \rightarrow Al_2(SO_4)_3(aq) + H_2(g)$
Since both aluminum and hydrogen occur as pure elements, balance those last. Sulfur and oxygen occur in only one compound on each side of the equation, so balance these first. Sulfur and oxygen are also part of a polyatomic ion that stays intact on both sides of the equation. *Balance polyatomic ions such as these as a unit*. There are 3 SO_4^{2-} ions on the right side of the equation, so put a 3 in front of H_2SO_4.	$Al(s) + 3\ H_2SO_4(aq) \rightarrow Al_2(SO_4)_3(aq) + H_2(g)$
Balance Al next. Since there are 2 Al atoms on the right side of the equation, place a 2 in front of Al on the left side of the equation.	$2\ Al(s) + 3\ H_2SO_4(aq) \rightarrow Al_2(SO_4)_3(aq) + H_2(g)$
Balance H next. Since there are 6 H atoms on the left side, place a 3 in front of $H_2(g)$ on the right side.	$2\ Al(s) + 3\ H_2SO_4(aq) \rightarrow Al_2(SO_4)_3(aq) + 3\ H_2(g)$
Finally, sum the number of atoms on each side to make sure that the equation is balanced.	$2\ Al(s) + 3\ H_2SO_4(aq) \rightarrow Al_2(SO_4)_3(aq) + 3\ H_2(g)$

Reactants		Products
2 Al atoms	→	2 Al atoms
6 H atoms	→	6 H atoms
3 S atoms	→	3 S atoms
12 O atoms	→	12 O atoms

▶**SKILLBUILDER 8.5** | **Balancing Chemical Equations**

Write a balanced equation for the reaction of aqueous lead(II) acetate with aqueous potassium iodide to form solid lead(II) iodide and aqueous potassium acetate.

▶**FOR MORE PRACTICE** Problems 62, 63, 64, 65.

EXAMPLE 8.6 BALANCING CHEMICAL EQUATIONS

Balance the chemical equation.

$Fe(s) + HCl(aq) \rightarrow FeCl_3(aq) + H_2(g)$

	SOLUTION
Since Cl occurs in only one compound on each side of the equation, balance it first. There are 1 Cl atom on the left side of the equation and 3 Cl atoms on the right side. To balance Cl, place a 3 in front of HCl.	$Fe(s) + 3\,HCl(aq) \rightarrow FeCl_3(aq) + H_2(g)$
Since H and Fe occur as free elements, balance them last. There is 1 Fe atom on the left side of the equation and 1 Fe atom on the right, so Fe is balanced. There are 3 H atoms on the left and 2 H atoms on the right. Balance H by placing a $\frac{3}{2}$ in front of H_2. (That way you don't alter other elements that are already balanced.)	$Fe(s) + 3\,HCl(aq) \rightarrow FeCl_3(aq) + \frac{3}{2}\,H_2(g)$
The equation now contains a coefficient fraction; clear it by multiplying the entire equation (both sides) by 2.	$[Fe(s) + 3\,HCl(aq) \rightarrow FeCl_3(aq) + \frac{3}{2}\,H_2(g)] \times 2$ $2\,Fe(s) + 6\,HCl(aq) \rightarrow 2\,FeCl_3(aq) + 3\,H_2(g)$
Finally, sum the number of atoms on each side to check that the equation is balanced.	$2\,Fe(s) + 6\,HCl(aq) \rightarrow 2\,FeCl_3(aq) + 3\,H_2(g)$

Reactants		Products
2 Fe atoms	→	2 Fe atoms
6 Cl atoms	→	6 Cl atoms
6 H atoms	→	6 H atoms

▶**SKILLBUILDER 8.6 | Balancing Chemical Equations**

Balance the chemical equation.

$HCl(g) + O_2(g) \rightarrow H_2O(l) + Cl_2(g)$

▶**FOR MORE PRACTICE** Problems 68, 69, 70, 71.

CONCEPTUAL CHECKPOINT 8.4

Which quantity must always be the same on both sides of a balanced chemical equation?

(a) the number of atoms of each element

(b) the number of each type of molecule

(c) the sum of all of the coefficients

8.6 Classifying Chemical Reactions

LO: Classify chemical reactions.

No single classification scheme is perfect because all chemical reactions are unique in some sense. However, there are two classification schemes—one that focuses on the type of chemistry occurring and the other that focuses on what atoms or groups of atoms are doing—are helpful because they help us see differences and similarities among chemical reactions.

Classifying Chemical Reactions by What Atoms Do

A flowchart for this classification scheme of chemical reactions is as follows:

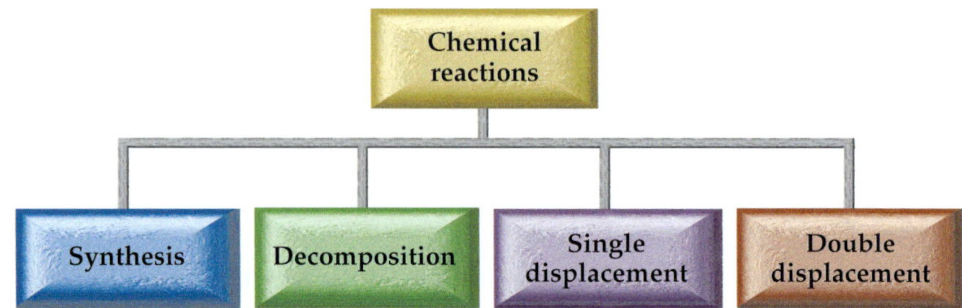

In this classification scheme, the letters (A, B, C, D) represent atoms or groups of atoms.

Type of Reaction	Generic Equation
synthesis or combination	$A + B \rightarrow AB$
decomposition	$AB \rightarrow A + B$
single-displacement	$A + BC \rightarrow AC + B$
double-displacement	$AB + CD \rightarrow AD + CB$

SYNTHESIS OR COMBINATION REACTIONS. In a **synthesis** or **combination reaction**, simpler substances combine to form more complex substances. The simpler substances may be elements, such as sodium and chlorine combining to form sodium chloride.

$$2\,Na(s) + Cl_2(g) \rightarrow 2\,NaCl(s)$$

The simpler substances may also be compounds, such as calcium oxide and carbon dioxide combining to form calcium carbonate.

$$CaO(s) + CO_2(g) \rightarrow CaCO_3(s)$$

In either case, a synthesis reaction follows the general equation:

$$A + B \rightarrow AB$$

Other examples of synthesis reactions include:

$$2\,H_2(g) + O_2(g) \rightarrow 2\,H_2O(l)$$

$$2\,Mg(s) + O_2(g) \rightarrow 2\,MgO(s)$$

$$SO_3(g) + H_2O(l) \rightarrow H_2SO_4(aq)$$

▲ In a synthesis reaction, two simpler substances combine to make a more complex substance. In this series of photographs we see sodium metal and chlorine gas. When they combine, a chemical reaction occurs that forms sodium chloride. © Richard Megna/ Fundamental Photographs.

▲ When electrical current is passed through water, the water undergoes a decomposition reaction to form hydrogen gas and oxygen gas. © Richard Megna/ Fundamental Photographs.

DECOMPOSITION REACTIONS. In a **decomposition reaction**, a complex substance decomposes to form simpler substances. The simpler substances may be elements, such as the hydrogen and oxygen gases that form upon the decomposition of water when electrical current passes through it.

$$2\,H_2O(l) \xrightarrow[\text{electrical current}]{} 2\,H_2(g) + O_2(g)$$

The simpler substances may also be compounds, such as the calcium oxide and carbon dioxide that form upon heating calcium carbonate.

$$CaCO_3(s) \xrightarrow[\text{heat}]{} CaO(s) + CO_2(g)$$

In either case, a decomposition reaction follows the general equation:

$$AB \rightarrow A + B$$

Other examples of decomposition reactions include:

$$2\,HgO(s) \xrightarrow[\text{heat}]{} 2\,Hg(l) + O_2(g)$$
$$2\,KClO_3(s) \xrightarrow[\text{heat}]{} 2\,KCl(s) + 3\,O_2(g)$$
$$2\,HI(g) \xrightarrow[\text{light}]{} H_2(g) + I_2(g)$$

Notice that these decomposition reactions require energy in the form of heat, electrical current, or light to make them happen. This is because compounds are normally stable and energy is required to decompose them. A number of decomposition reactions require *ultraviolet* or *UV light,* which is light in the ultraviolet region of the spectrum. UV light carries more energy than visible light and can therefore initiate the decomposition of many compounds.

DISPLACEMENT REACTIONS. In a **displacement** or **single-displacement reaction**, one element displaces another in a compound. For example, when we add metallic zinc to a solution of copper(II) chloride, the zinc replaces the copper.

$$Zn(s) + CuCl_2(aq) \rightarrow ZnCl_2(aq) + Cu(s)$$

A displacement reaction follows the general equation:

$$A + BC \rightarrow AC + B$$

Other examples of displacement reactions include:

$$Mg(s) + 2\ HCl(aq) \rightarrow MgCl_2(aq) + H_2(g)$$

$$2\ Na(s) + 2\ H_2O(l) \rightarrow 2\ NaOH(aq) + H_2(g)$$

The last reaction can be identified more easily as a displacement reaction if we write water as HOH(*l*).

$$2\ Na(s) + 2\ HOH(l) \rightarrow 2\ NaOH(aq) + H_2(g)$$

DOUBLE-DISPLACEMENT REACTIONS. In a **double-displacement reaction**, two elements or groups of elements in two different compounds exchange places to form two new compounds. For example, in aqueous solution, the silver in silver nitrate changes places with the sodium in sodium chloride and solid silver chloride and aqueous sodium nitrate form.

$$AgNO_3(aq) + NaCl(aq) \rightarrow AgCl(s) + NaNO_3(aq)$$

A double-displacement reaction follows the general form:

$$AB + CD \rightarrow AD + CB$$

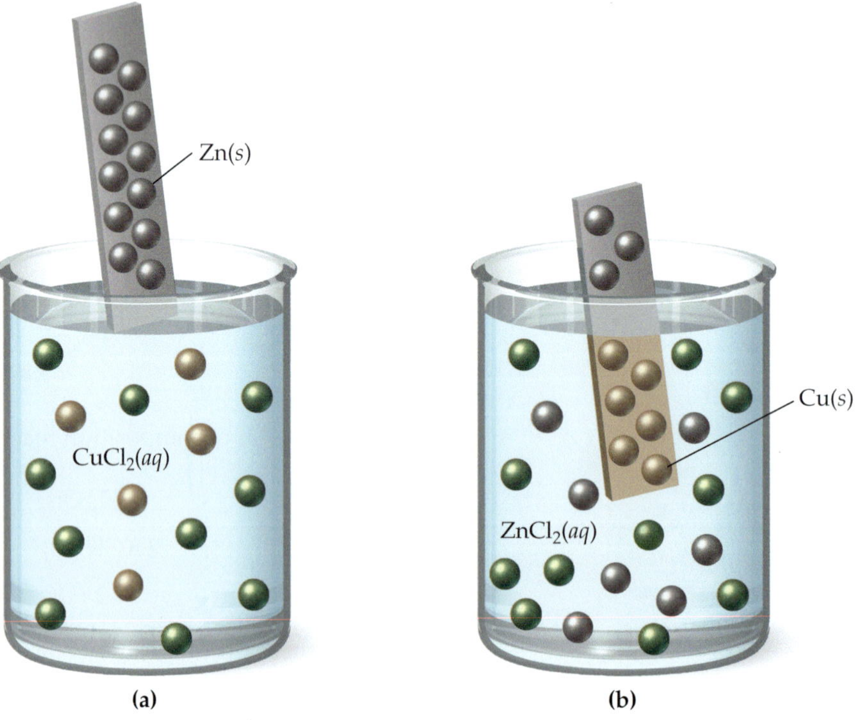

▶ In a single-displacement reaction, one element displaces another in a compound. When zinc metal is immersed in a copper(II) chloride solution, the zinc atoms displace the copper ions in solution.

Other examples of double-displacement reactions include:

$$HCl(aq) + NaOH(aq) \rightarrow H_2O(l) + NaCl(aq)$$

$$2\ HCl(aq) + Na_2CO_3(aq) \rightarrow H_2CO_3(aq) + 2\ NaCl(aq)$$

Since $H_2CO_3(aq)$ is not stable, it decomposes to form $H_2O(l)$ and $CO_2(g)$, so the overall equation is:

$$2\ HCl(aq) + Na_2CO_3(aq) \rightarrow H_2O(l) + CO_2(g) + 2\ NaCl(aq)$$

EXAMPLE 8.7 **CLASSIFYING CHEMICAL REACTIONS ACCORDING TO WHAT ATOMS DO**

Classify each reaction as a synthesis, decomposition, single-displacement, or double-displacement reaction.

(a) $Na_2O(s) + H_2O(l) \rightarrow 2\ NaOH(aq)$
(b) $Ba(NO_3)_2(aq) + K_2SO_4(aq) \rightarrow BaSO_4(s) + 2\ KNO_3(aq)$
(c) $2\ Al(s) + Fe_2O_3(s) \rightarrow Al_2O_3(s) + 2\ Fe(l)$
(d) $2\ H_2O_2(aq) \rightarrow 2\ H_2O(l) + O_2(g)$
(e) $Ca(s) + Cl_2(g) \rightarrow CaCl_2(s)$

SOLUTION

(a) Synthesis; a more complex substance forms from two simpler ones.
(b) Double-displacment; Ba and K switch places to form two new compounds.
(c) Single-displacement; Al displaces Fe in Fe_2O_3.
(d) Decomposition; a complex substance decomposes into simpler ones.
(e) Synthesis; a more complex substance forms from two simpler ones.

▶SKILLBUILDER 8.7 | Classifying Chemical Reactions According to What Atoms Do

Classify each reaction as a synthesis, decomposition, single-displacement, or double-displacement reaction.

(a) $2\ Al(s) + 2\ H_3PO_4(aq) \rightarrow 2\ AlPO_4(aq) + 3\ H_2(g)$
(b) $CuSO_4(aq) + 2\ KOH(aq) \rightarrow Cu(OH)_2(s) + K_2SO_4(aq)$
(c) $2\ K(s) + Br_2(l) \rightarrow 2\ KBr(s)$
(d) $CuCl_2(aq) \xrightarrow[\text{electrical current}]{} Cu(s) + Cl_2(g)$

▶FOR MORE PRACTICE Example 8.28; Problems 80, 81, 82, 83.

Classifying Chemical Reactions by the Chemistry Occurring during the Reaction

Several different types of chemical reactions will be discussed in this and later modules. It can be helpful to organize the different types of chemical reactions according to the chemistry or phenomenon that is occurring during the reaction. This flowchart shows classifications of chemical reactions in this way.

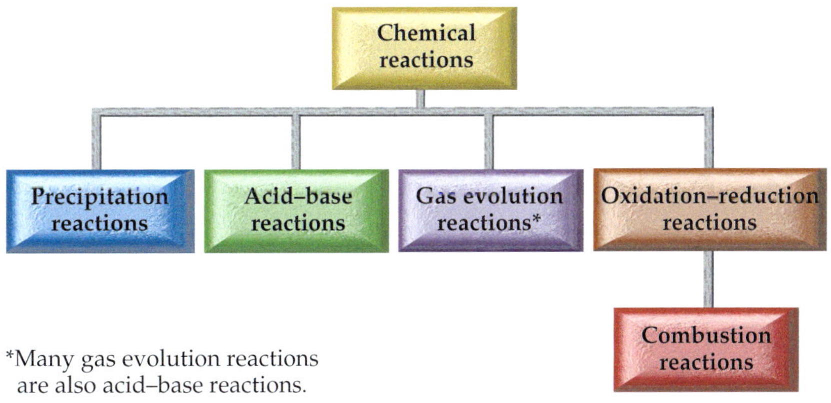

CONCEPTUAL CHECKPOINT 8.5

Precipitation reactions and acid–base reactions can both also be classified as:

(a) synthesis reactions

(b) decomposition reactions

(c) single-displacement reactions

(d) double-displacement reactions

CHEMISTRY IN THE ENVIRONMENT

▶ The Reactions Involved in Ozone Depletion

Chemistry in the Environment: Chlorine in Chlorofluorocarbons in Module 5 explained that chlorine atoms from chlorofluorocarbons deplete the ozone layer, which protects life on Earth from harmful ultraviolet light. Through research, chemists have discovered the reactions by which this depletion occurs.

Ozone normally forms in the upper atmosphere according to this reaction:

(a) $O_2(g) + O(g) \longrightarrow O_3(g)$

When chlorofluorocarbons drift to the upper atmosphere, they are exposed to ultraviolet light and undergo this reaction:

(b) $CF_2Cl_2(g) \xrightarrow[\text{UV light}]{} CF_2Cl(g) + Cl(g)$

UV light

Atomic chlorine then reacts with and depletes ozone according to this cycle of reactions:

(c) $Cl(g) + O_3(g) \longrightarrow ClO(g) + O_2(g)$

(d) $O_3(g) \xrightarrow[\text{UV light}]{} O_2(g) + O(g)$

UV light

(e) $O(g) + ClO(g) \longrightarrow O_2(g) + Cl(g)$

Notice that in the final reaction, atomic chlorine is regenerated and can go through the cycle again to deplete more ozone. Through this cycle of reactions, a single chlorofluorocarbon molecule can deplete thousands of ozone molecules.

B8.1 CAN YOU ANSWER THIS? *Classify each of the reactions involved in ozone depletion (a–e) as a synthesis, decomposition, single-displacement, or double-displacement reaction.*

8.7 Precipitation Reactions

LO: Determine whether a compound is soluble.

LO: Predict and write equations for precipitation reactions.

LO: Write molecular, complete ionic, and net ionic equations.

We now turn to investigating several types of reactions. Since many of these occur in water, we must first understand *aqueous solutions*. Reactions occurring in aqueous solution are among the most common and important. An **aqueous solution** is a homogeneous mixture of a substance with water. For example, a sodium chloride (NaCl) solution (also called a saline solution) is composed of sodium chloride dissolved in water. Sodium chloride solutions are common both in the oceans and in living cells. You can form a sodium chloride solution yourself by adding table salt to water. As you stir the salt into the water, it seems to disappear. However, you know the salt is still there because if you taste the water, it has a salty flavor. How does sodium chloride dissolve in water?

Solubility

A compound is **soluble** in a particular liquid if it dissolves in that liquid. NaCl, for example, is soluble in water. If we mix solid sodium chloride into water, it dissolves and forms a strong electrolyte solution. AgCl, on the other hand, is not very soluble in water. If we mix solid silver chloride into water, most of it remains as a solid within the liquid water.

There is no easy way to predict whether a particular compound will be highly soluble or sparingly soluble in water. For ionic compounds, however, empirical rules have been deduced from observations of many compounds. These **solubility rules** are summarized in Table 8.2 and ▼ Figure 8.6. For example, the solubility rules indicate that compounds containing the lithium ion are *soluble*. That means that compounds such as LiBr, $LiNO_3$, Li_2SO_4, LiOH, and Li_2CO_3 all dissolve in water to form strong electrolyte solutions. If a compound contains Li^+, it is soluble. Similarly, the solubility rules state that compounds containing the NO_3^- ion are soluble. Compounds such as $AgNO_3$, $Pb(NO_3)_2$, $NaNO_3$, $Ca(NO_3)_2$, and $Sr(NO_3)_2$ all dissolve in water to form strong electrolyte solutions.

The solubility rules apply only to the solubility of the compounds in water.

The solubility rules also state that, with some exceptions, compounds containing the CO_3^{2-} ion are sparingly soluble; some textbooks incorrectly refer to them as *insoluble*. Compounds such as $CuCO_3$, $CaCO_3$, $SrCO_3$, and $FeCO_3$ do not dissolve very well in water. Note that the solubility rules contain many exceptions.

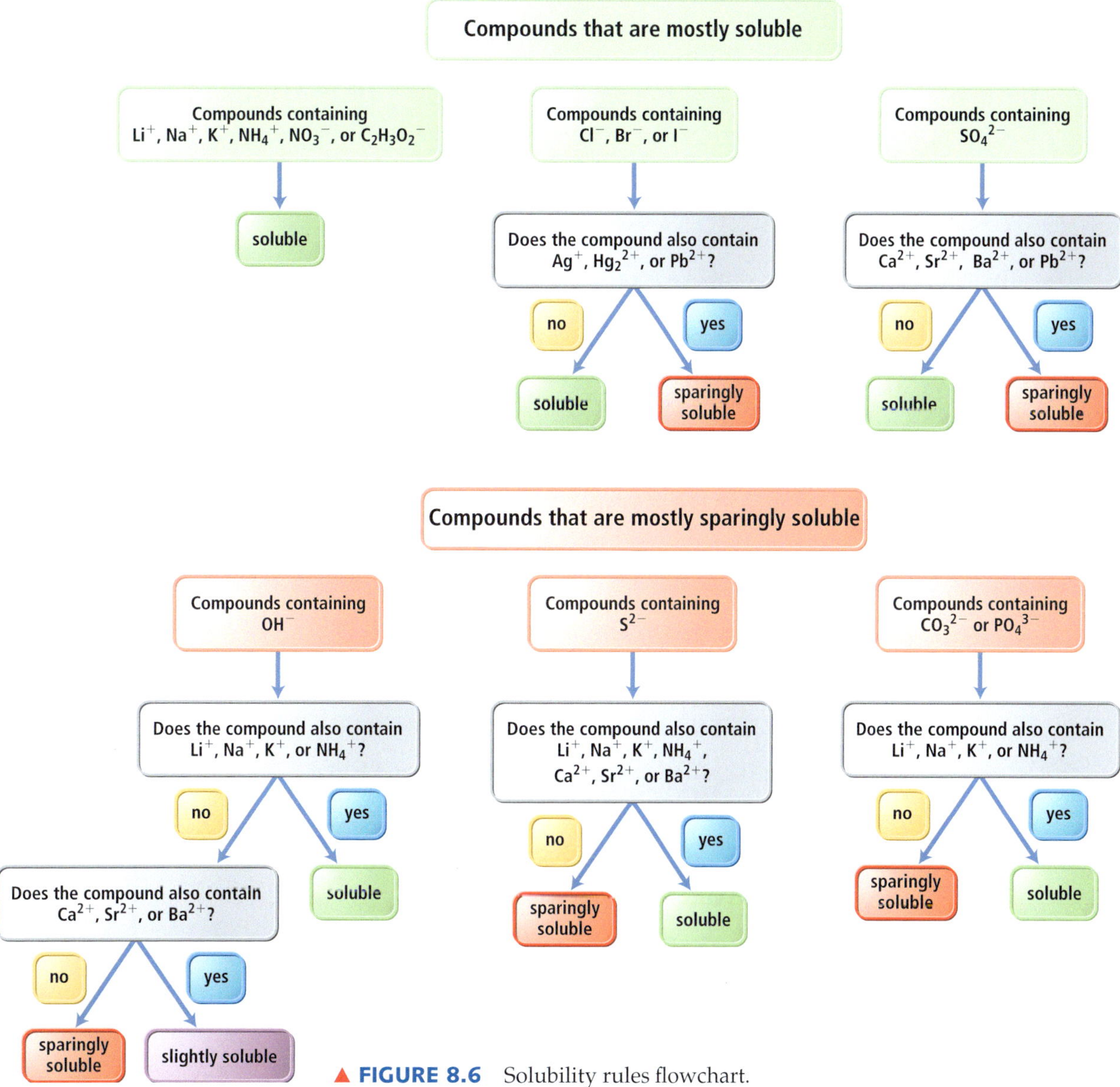

▲ **FIGURE 8.6** Solubility rules flowchart.

For example, compounds containing CO_3^{2-} are *soluble when paired with* Li^+, Na^+, K^+, or NH_4^+. Thus Li_2CO_3, Na_2CO_3, K_2CO_3, and $(NH_4)_2CO_3$ are all soluble.

TABLE 8.2 Solubility Rules

Compounds Containing the Following Ions Are Mostly Soluble	Exceptions
Li^+, Na^+, K^+, NH_4^+	None
NO_3^-, $C_2H_3O_2^-$	None
Cl^-, Br^-, I^-	When any of these ions pairs with Ag^+, Hg_2^{2+}, or Pb^{2+}, the compound is sparingly soluble.
SO_4^{2-}	When SO_4^{2-} pairs with Ca^{2+}, Sr^{2+}, Ba^{2+}, or Pb^{2+}, the compound is sparingly soluble.
Compounds Containing the Following Ions Are Mostly Sparingly Soluble	**Exceptions**
OH^-, S^{2-}	When either of these ions pairs with Li^+, Na^+, K^+, or NH_4^+, the compound is soluble. When S^{2-} pairs with Ca^{2+}, Sr^{2+}, or Ba^{2+}, the compound is soluble. When OH^- pairs with Ca^{2+}, Sr^{2+}, or Ba^{2+}, the compound is slightly soluble.
CO_3^{2-}, PO_4^{3-}	When either of these ions pairs with Li^+, Na^+, K^+, or NH_4^+, the compound is soluble.

EXAMPLE 8.8 DETERMINING WHETHER A COMPOUND IS SOLUBLE OR SPARINGLY SOLUBLE

Is each compound soluble or sparingly soluble?

(a) AgBr **(b)** $CaCl_2$ **(c)** $Pb(NO_3)_2$ **(d)** $PbSO_4$

SOLUTION

(a) Sparingly soluble; compounds containing Br^- are normally soluble, but Ag^+ is an exception.
(b) Soluble; compounds containing Cl^- are normally soluble, and Ca^{2+} is not an exception.
(c) Soluble; compounds containing NO_3^- are always soluble.
(d) Sparingly soluble; compounds containing SO_4^{2-} are normally soluble, but Pb^{2+} is an exception.

▶SKILLBUILDER 8.8 | Determining Whether a Compound Is Soluble or Sparingly Soluble

Is each compound soluble or sparingly soluble?

(a) CuS **(b)** $FeSO_4$ **(c)** $PbCO_3$ **(d)** NH_4Cl

▶FOR MORE PRACTICE Example 8.29; Problems 84, 85, 86, 89.

CONCEPTUAL CHECKPOINT 8.6

Which image best depicts a mixture of $BaCl_2$ and water?

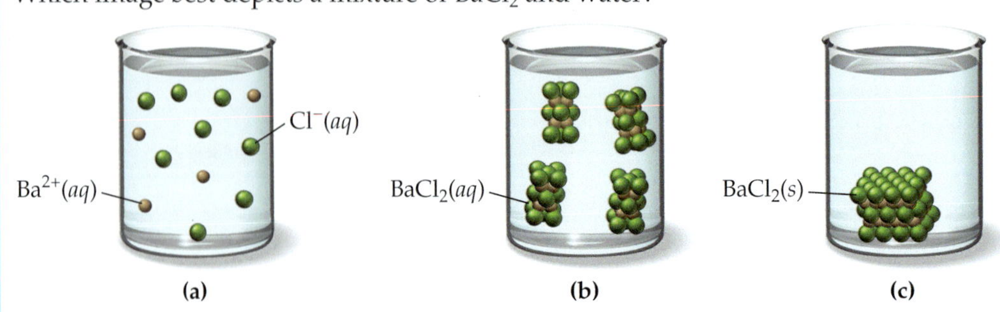

Precipitation Reactions: Reactions in Aqueous Solution That Form a Solid

Recall from Section 8.1 that sodium carbonate in laundry detergent reacts with dissolved Mg^{2+} and Ca^{2+} ions to form solids that precipitate (come out of) solution. These reactions are examples of **precipitation reactions**—reactions that form a solid, called a **precipitate**, upon mixing two aqueous solutions.

Precipitation reactions are common in chemistry. Potassium iodide and lead nitrate, for example, both form colorless, strong electrolyte solutions when dissolved in water (see the solubility rules). When the two solutions are combined, however, a brilliant yellow precipitate forms (▼ Figure 8.7). We describe this precipitation reaction with the chemical equation:

$$2\,KI(aq) + Pb(NO_3)_2(aq) \rightarrow PbI_2(s) + 2\,KNO_3(aq)$$

Precipitation reactions do not always occur when mixing two aqueous solutions. For example, when we combine solutions of KI(*aq*) and NaCl(*aq*), nothing happens (▼ Figure 8.8).

$$KI(aq) + NaCl(aq) \rightarrow \text{NO REACTION}$$

$$2\,KI(aq) + Pb(NO_3)_2(aq) \longrightarrow PbI_2(s) + 2\,KNO_3(aq)$$

▲ **FIGURE 8.7 Precipitation** When we mix a potassium iodide solution with a lead(II) nitrate solution, a brilliant yellow precipitate of $PbI_2(s)$ forms. © Richard Megna/Fundamental Photographs.

▲ **FIGURE 8.8 No reaction** When we mix a potassium iodide solution with a sodium chloride solution, no reaction occurs. © Fundamental Photographs.

Predicting Precipitation Reactions

The key to predicting precipitation reactions is understanding that *only sparingly soluble compounds form precipitates*. In a precipitation reaction, two solutions containing soluble compounds combine and a sparingly soluble compound precipitates. Consider the precipitation reaction from Figure 8.7 shown here:

$$\underset{\text{soluble}}{2\,KI(aq)} + \underset{\text{soluble}}{Pb(NO_3)_2(aq)} \rightarrow \underset{\text{sparingly soluble}}{PbI_2(s)} + \underset{\text{soluble}}{2\,KNO_3(aq)}$$

KI and $Pb(NO_3)_2$ are both soluble, but the precipitate, PbI_2, is *sparingly soluble*. Before mixing, KI(*aq*) and $Pb(NO_3)_2(aq)$ are both completely dissociated in their respective solutions.

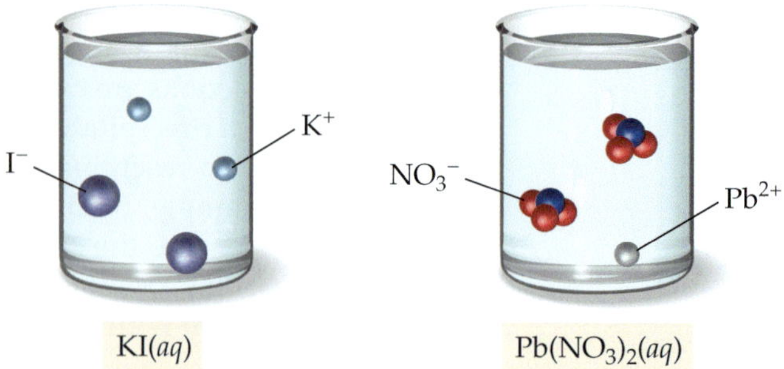

The instant that the solutions are mixed, all four ions are present.

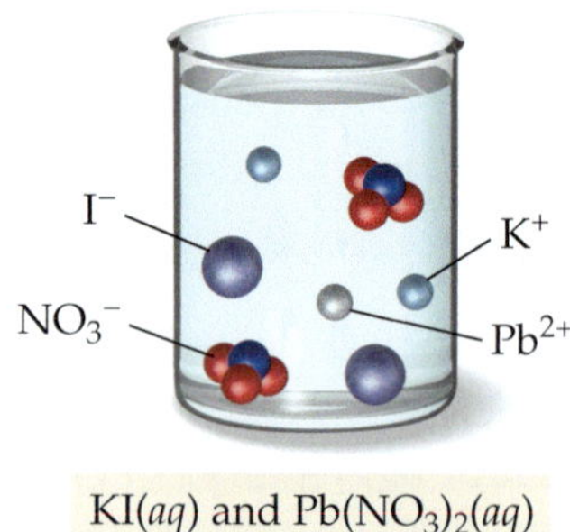

However, new compounds—potentially sparingly soluble ones—are now possible. Specifically, the cation from one compound can pair with the anion from the other compound to form new (and potentially sparingly soluble) products.

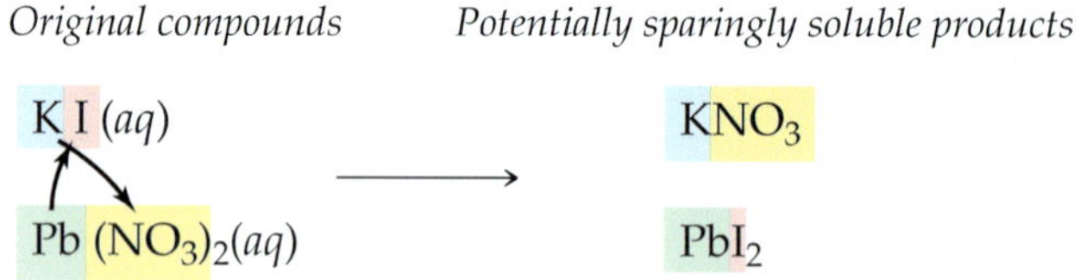

If the *potentially sparingly soluble* products are both *soluble*, no reaction occurs. If, on the other hand, one or both of the potentially sparingly soluble products are *indeed sparingly soluble*, a precipitation reaction occurs. In this case, KNO_3 is soluble, but PbI_2 is sparingly soluble. Consequently, PbI_2 precipitates.

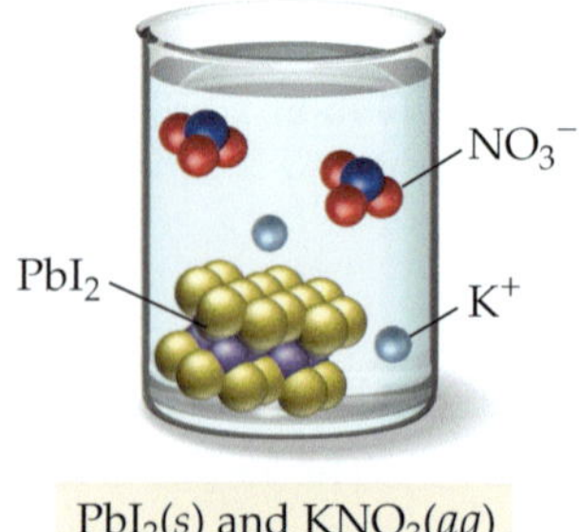

To predict whether a precipitation reaction will occur when two solutions are mixed and to write an equation for the reaction, we follow the steps in the procedure box that accompanies Examples 8.9 and 8.10. As usual, the steps are shown in the left column, and two examples of applying the procedure are shown in the center and right columns.

WRITING EQUATIONS FOR PRECIPITATION REACTIONS	EXAMPLE 8.9	EXAMPLE 8.10
	Write an equation for the precipitation reaction that occurs (if any) when solutions of sodium carbonate and copper(II) chloride are mixed.	Write an equation for the precipitation reaction that occurs (if any) when solutions of lithium nitrate and sodium sulfate are mixed.
1. Write the formulas of the two compounds being mixed as reactants in a chemical equation.	**SOLUTION** $Na_2CO_3(aq) + CuCl_2(aq) \rightarrow$	**SOLUTION** $LiNO_3(aq) + Na_2SO_4(aq) \rightarrow$
2. Below the equation, write the formulas of the products that could form from the reactants. Obtain these by combining the cation from one reactant with the anion from the other. Make sure to write correct (charge-neutral) formulas for these ionic compounds as described in Section 3.5.	$Na_2CO_3(aq) + CuCl_2(aq) \longrightarrow$ **Products** NaCl $CuCO_3$	$LiNO_3(aq) + Na_2SO_4(aq) \longrightarrow$ **Products** $NaNO_3$ Li_2SO_4
3. Use the solubility rules to determine whether any of the products are indeed sparingly soluble.	NaCl is *soluble* (compounds containing Cl^- are usually soluble, and Na^+ is not an exception). $CuCO_3$ is *sparingly soluble* (compounds containing CO_3^{2-} are usually sparingly soluble, and Cu^{2+} is not an exception).	$NaNO_3$ is *soluble* (compounds containing NO_3^- are soluble, and Na^+ is not an exception). Li_2SO_4 is *soluble* (compounds containing SO_4^{2-} are soluble and Li^+ is not an exception).
4. If all of the products are soluble, there will be no precipitate. Write NO REACTION next to the arrow.	Because this example has an sparingly soluble product, you proceed to the next step.	$LiNO_3(aq) + Na_2SO_4(aq) \rightarrow$ NO REACTION
5. If one or both of the products are sparingly soluble, write their formula(s) as the product(s) of the reaction, using (*s*) to indicate solid. Follow any soluble products with (*aq*) to indicate aqueous.	$Na_2CO_3(aq) + CuCl_2(aq) \rightarrow CuCO_3(s) + NaCl(aq)$	
6. Balance the equation. Remember to adjust only coefficients, not subscripts.	$Na_2CO_3(aq) + CuCl_2(aq) \rightarrow CuCO_3(s) + 2\ NaCl(aq)$	

▶**SKILLBUILDER 8.9** | Write an equation for the precipitation reaction that occurs (if any) when solutions of potassium hydroxide and nickel(II) bromide are mixed.

▶**SKILLBUILDER 8.10** | Write an equation for the precipitation reaction that occurs (if any) when solutions of ammonium chloride and iron(III) nitrate are mixed.

▶**FOR MORE PRACTICE** Example 8.30; Problems 90, 91, 92, 93.

EXAMPLE 8.11 **PREDICTING AND WRITING EQUATIONS FOR PRECIPITATION REACTIONS**

Write an equation for the precipitation reaction (if any) that occurs when solutions of lead(II) acetate and sodium sulfate are mixed. If no reaction occurs, write *NO REACTION*.

1. Write the formulas of the two compounds being mixed as reactants in a chemical equation.	**SOLUTION** $Pb(C_2H_3O_2)_2(aq) + Na_2SO_4(aq) \rightarrow$
2. Below the equation, write the formulas of the products that could form from the reactants. Determine these by combining the cation from one reactant with the anion from the other. Make sure to adjust the subscripts so that all formulas are charge-neutral.	$Pb(C_2H_3O_2)_2(aq) + Na_2SO_4(aq) \longrightarrow$ **Products** $NaC_2H_3O_2$ $PbSO_4$
3. Use the solubility rules to determine whether any of the products are indeed sparingly soluble.	$NaC_2H_3O_2$ is *soluble* (compounds containing Na^+ are always soluble). $PbSO_4$ is *sparingly soluble* (compounds containing SO_4^{2-} are normally soluble, but Pb^{2+} is an exception).
4. If all of the products are soluble, there will be no precipitate. Write *NO REACTION* next to the arrow.	This reaction has a sparingly soluble product so you proceed to the next step.
5. If one or both of the products are indeed sparingly soluble, write their formula(s) as the product(s) of the reaction, using (*s*) to indicate solid. Follow any soluble products with (*aq*) to indicate aqueous.	$Pb(C_2H_3O_2)_2(aq) + Na_2SO_4(aq) \rightarrow PbSO_4(s) + NaC_2H_3O_2(aq)$
6. Balance the equation.	$Pb(C_2H_3O_2)_2(aq) + Na_2SO_4(aq) \rightarrow PbSO_4(s) + 2\ NaC_2H_3O_2(aq)$

▶SKILLBUILDER 8.11 | Predicting and Writing Equations for Precipitation Reactions

Write an equation for the precipitation reaction (if any) that occurs when solutions of potassium sulfate and strontium nitrate are mixed. If no reaction occurs, write *NO REACTION*.

▶FOR MORE PRACTICE Problems 94, 95.

CONCEPTUAL CHECKPOINT 8.7

Which of these reactions results in the formation of a precipitate?

(a) $NaNO_3(aq) + CaS(aq)$

(b) $MgSO_4(aq) + CaS(aq)$

(c) $NaNO_3(aq) + MgSO_4(aq)$

Writing Chemical Equations for Reactions in Solution: Molecular, Complete Ionic, and Net Ionic Equations

Consider the following equation for a precipitation reaction:

$$AgNO_3(aq) + NaCl(aq) \rightarrow AgCl(s) + NaNO_3(aq)$$

This equation is written as a **molecular equation**, an equation showing the complete neutral formulas for every compound in the reaction. We can also write equations for reactions occurring in aqueous solution to show that aqueous ionic compounds normally dissociate in solution. For example, we can write the previous equation as:

$$Ag^+(aq) + NO_3^-(aq) + Na^+(aq) + Cl^-(aq) \rightarrow AgCl(s) + Na^+(aq) + NO_3^-(aq)$$

When writing complete ionic equations, separate only aqueous ionic compounds into their constituent ions. Do NOT separate solid, liquid, or gaseous compounds.

Equations such as this one, which show the reactants and products as they are actually present in solution, are called **complete ionic equations**.

Notice that in the complete ionic equation, some of the ions in solution appear unchanged on both sides of the equation. These ions are called **spectator ions** because they do not participate in the reaction.

$$Ag^+(aq) + NO_3^-(aq) + Na^+(aq) + Cl^-(aq) \longrightarrow AgCl(s) + Na^+(aq) + NO_3^-(aq)$$

Spectator ions

To simplify the equation, and to more clearly show what is happening, spectator ions can be omitted.

$$Ag^+(aq) + Cl^-(aq) \rightarrow AgCl(s)$$

Species refers to a kind or sort of thing. In this case, the species are the different molecules and ions that are present during the reaction.

Equations such as this one, which show only the *species* that actually participate in the reaction, are called **net ionic equations**.

As another example, consider the reaction between HCl(*aq*) and NaOH(*aq*).

$$HCl(aq) + NaOH(aq) \rightarrow H_2O(l) + NaCl(aq)$$

HCl, NaOH, and NaCl exist in solution as independent ions. The *complete ionic equation* for this reaction is:

$$H^+(aq) + Cl^-(aq) + Na^+(aq) + OH^-(aq) \rightarrow H_2O(l) + Na^+(aq) + Cl\ (aq)$$

To write the *net ionic equation*, we remove the spectator ions, those that are unchanged on both sides of the equation.

$$H^+(aq) + Cl^-(aq) + Na^+(aq) + OH^-(aq) \longrightarrow H_2O(l) + Na^+(aq) + Cl^-(aq)$$

Spectator ions

The net ionic equation is $H^+(aq) + OH^-(aq) \rightarrow H_2O(l)$.

To summarize:

- A molecular equation is a chemical equation showing the complete, neutral formulas for every compound in a reaction.
- A complete ionic equation is a chemical equation showing all of the species as they are actually present in solution.
- A net ionic equation is an equation showing only the species that actually participate in the reaction.

EXAMPLE 8.12 WRITING COMPLETE IONIC AND NET IONIC EQUATIONS

Consider this precipitation reaction occurring in aqueous solution.

$$Pb(NO_3)_2(aq) + 2\ LiCl(aq) \rightarrow PbCl_2(s) + 2\ LiNO_3(aq)$$

Write a complete ionic equation and a net ionic equation for the reaction.

Write the complete ionic equation by separating aqueous ionic compounds into their constituent ions. The $PbCl_2(s)$ remains as one unit.

SOLUTION

Complete ionic equation

$$Pb^{2+}(aq) + 2\ NO_3^-(aq) + 2\ Li^+(aq) + 2\ Cl^-(aq) \rightarrow PbCl_2(s) + 2\ Li^+(aq) + 2\ NO_3^-(aq)$$

Write the net ionic equation by eliminating the spectator ions, those that do not change during the reaction.

Net ionic equation

$$Pb^{2+}(aq) + 2\ Cl^-(aq) \rightarrow PbCl_2(s)$$

▶SKILLBUILDER 8.12 | Writing Complete Ionic and Net Ionic Equations

Consider this reaction occurring in aqueous solution.

$$2\ HBr(aq) + Ca(OH)_2(aq) \rightarrow 2\ H_2O(l) + CaBr_2(aq)$$

Write a complete ionic equation and a net ionic equation for the reaction.

▶FOR MORE PRACTICE Example 8.31; Problems 96, 97, 98, 99.

8.8 Acid–Base Reactions

LO: Identify common acids and describe their key characteristics.
LO: Identify common bases and describe their key characteristics.
LO: Identify Arrhenius acids and bases.
LO: Identify Brønsted–Lowry acids and bases and their conjugates.
LO: Write equations for neutralization reactions.
LO: Write equations for the reactions of acids with metals and with metal oxides.
LO: Identify strong and weak acids and strong and weak bases.
LO: Determine $[H_3O^+]$ in acid solutions.
LO: Determine $[OH^-]$ in base solutions.

Acids have the following properties:

- Acids have a sour taste.
- Acids dissolve many metals.
- Acids turn blue litmus paper red.

We just discussed examples of the sour taste of acids and their ability to dissolve metals. Acids also turn blue litmus paper red. Litmus paper contains a dye that turns red in acidic solutions (◀ Figure 8.9). In the laboratory, litmus paper is used routinely to test the acidity of solutions.

Table 8.3 lists some common acids. Hydrochloric acid is found in most chemistry laboratories. It is used in industry to clean metals, to prepare and process foods, and to refine metal ores.

Hydrochloric acid

NEVER taste or touch laboratory chemicals.

For a review of naming acids, see Section 3.9.

Hydrochloric acid is also the main component of stomach acid. In the stomach, hydrochloric acid breaks down food and kills harmful bacteria that might enter the body through food. The sour taste sometimes associated with indigestion

▲ **FIGURE 8.9 Acids turn blue litmus paper red.**

TABLE 8.3 Some Common Acids

Name	Uses
hydrochloric acid (HCl)	metal cleaning; food preparation; ore refining; main component of stomach acid
sulfuric acid (H_2SO_4)	fertilizer and explosive manufacturing; dye and glue production; automobile batteries
nitric acid (HNO_3)	fertilizer and explosive manufacturing; dye and glue production
acetic acid ($HC_2H_3O_2$)	plastic and rubber manufacturing; food preservation; component of vinegar
carbonic acid (H_2CO_3)	component of carbonated beverages due to the reaction of carbon dioxide with water
hydrofluoric acid (HF)	metal cleaning; glass frosting and etching

is caused by the stomach's hydrochloric acid refluxing up into the esophagus (the tube that joins the stomach and the mouth) and throat.

Annual U.S. production of sulfuric acid exceeds 40 million tons.

Sulfuric acid—the most widely produced chemical in the United States—and nitric acid are commonly used in the laboratory. In addition, they are used in the manufacture of fertilizers, explosives, dyes, and glue. Sulfuric acid is contained in most automobile batteries.

H_2SO_4 O=S(=O)(–O–H)–O–H

Sulfuric acid

HNO_3 O–N(=O)–O–H

Nitric acid

Acetic acid is present in vinegar and is also produced in improperly stored wines. The word *vinegar* originates from the French *vin aigre*, which means "sour wine." The presence of vinegar in wines is considered a serious fault, making the wine taste like salad dressing.

▲ Vinegar is a solution of acetic acid and water. © Ivaylo Ivanov/Shutterstock.

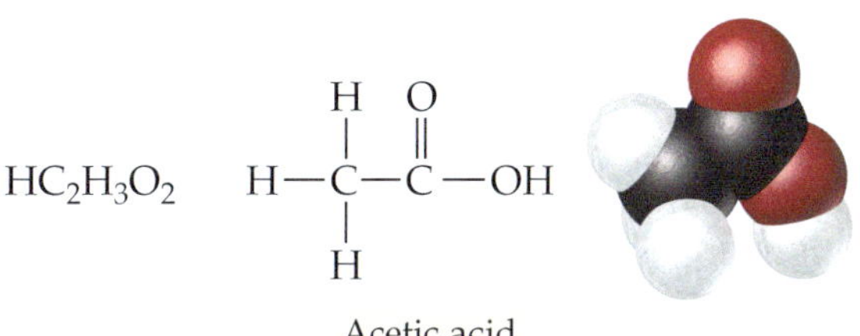

Acetic acid

Acetic acid is an example of a **carboxylic acid**, an acid containing the grouping of atoms known as the carboxylic acid group.

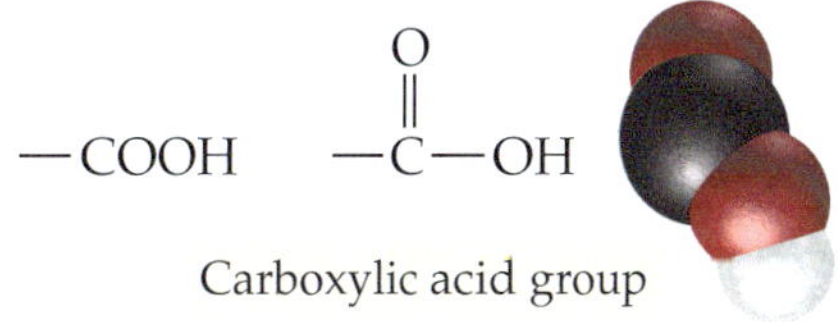

Carboxylic acid group

We often find carboxylic acids in substances derived from living organisms. Other carboxylic acids include citric acid, the main acid in lemons and limes, and malic acid, an acid found in apples, grapes, and wine.

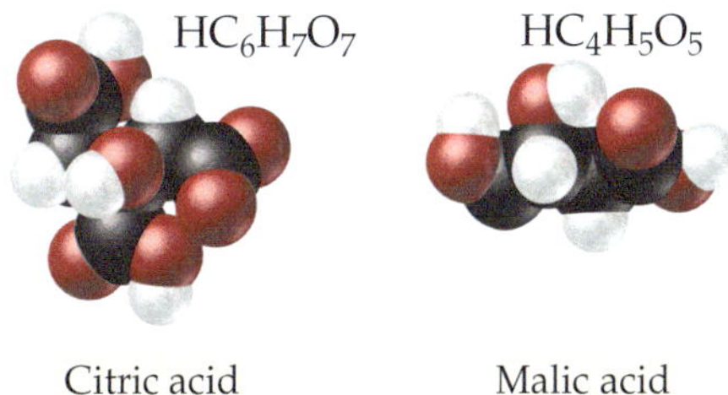

Citric acid Malic acid

Bases have the following properties:

NEVER taste or touch laboratory chemicals.

- Bases have a bitter taste.
- Bases have a slippery feel.
- Bases turn red litmus paper blue.

Coffee is acidic overall, but bases present in coffee—such as caffeine—impart a bitter flavor.

Bases are less common in foods than acids because of their bitter taste. A Sour Patch Kid coated with a base would never sell. Our aversion to the taste of bases is probably an adaptation to protect us against **alkaloids**, organic bases found in plants. Alkaloids are often poisonous—the toxic component of hemlock, for example, is the alkaloid coniine—and their bitter taste warns us against eating them. Nonetheless, some foods, such as coffee, contain small amounts of base (caffeine is a base). Many people enjoy the bitterness, but only after acquiring the taste over time.

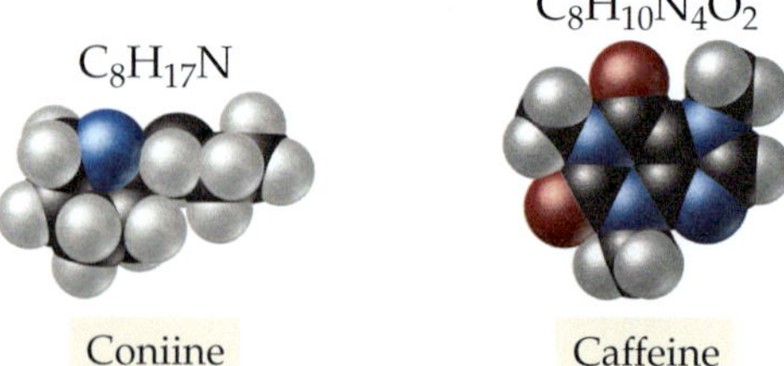

▲ All these consumer products contain bases. © Maxwell Art And Photo/Pearson.

Bases feel slippery because they react with oils on our skin to form soaplike substances. Soap itself is basic, and its slippery feel is characteristic of bases. Some household cleaning solutions, such as ammonia, are also basic and have the typical slippery feel of a base. Bases turn red litmus paper blue (◀ Figure 8.10). In the laboratory, litmus paper is routinely used to test the basicity of solutions.

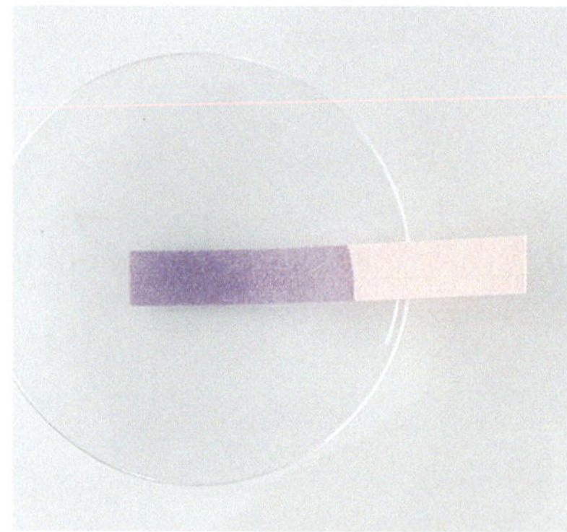

▲ **FIGURE 8.10 Bases turn red litmus paper blue.** © Richard Megna/Fundamental Photographs.

Table 8.4 lists some common bases. Sodium hydroxide and potassium hydroxide are found in most chemistry laboratories. They are also used in processing petroleum and cotton and in soap and plastic manufacturing. Sodium hydroxide is the active ingredient in products such as Drano that work to unclog drains. Sodium hydrogencarbonate (also known as sodium bicarbonate) can be found in most homes as baking soda and is also an active ingredient in many antacids. When taken as an antacid, sodium bicarbonate/hydrogencarbonate neutralizes stomach acid relieving heartburn and sour stomach.

TABLE 8.4 Some Common Bases

Name	Uses
sodium hydroxide ($NaOH$)	petroleum processing; soap and plastic manufacturing
potassium hydroxide (KOH)	cotton processing; electroplating; soap production
sodium hydrogencarbonate ($NaHCO_3$)	antacid; ingredient of baking soda; source of CO_2
ammonia (NH_3)	detergent; fertilizer and explosive manufacturing; synthetic fiber production

* Sodium hydrogencarbonate is also known as sodium bicarbonate.

CONCEPTUAL CHECKPOINT 8.8

Which substance is most likely to have a sour taste?

(a) $HCl(aq)$ **(b)** $NH_3(aq)$ **(c)** $KOH(aq)$

Molecular Definitions of Acids and Bases

We have just seen some of the properties of acids and bases. In this section we examine two different models that explain the molecular basis for acid and base behavior: the Arrhenius model and the Brønsted–Lowry model. The Arrhenius model, which was developed earlier, is more limited in its scope. The Brønsted–Lowry model was developed later and is more broadly applicable.

The Arrhenius Definition

In the 1880s, the Swedish chemist Svante Arrhenius proposed the following molecular definitions of acids and bases:

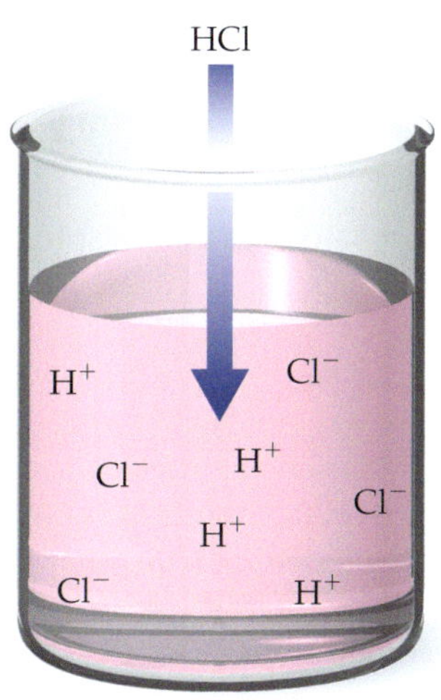

$HCl(aq) \longrightarrow H^+(aq) + Cl^-(aq)$

▲ **FIGURE 8.11 Arrhenius definition of an acid** The Arrhenius definition states that an acid is a substance that produces H^+ ions in solution. These H^+ ions associate with H_2O to form H_3O^+ ions.

ARRHENIUS DEFINITION

Acid—An acid produces $\mathbf{H^+}$ ions in aqueous solution.
Base—A base produces $\mathbf{OH^-}$ ions in aqueous solution.

For example, according to the **Arrhenius definition**, HCl is an **Arrhenius acid** because it produces H^+ ions in solution (◀ Figure 8.11).

$$HCl(aq) \rightarrow H^+(aq) + Cl^-(aq)$$

HCl is a covalent compound and does not contain ions. However, in water it **ionizes** to form $H^+(aq)$ ions and $Cl^-(aq)$ ions. The H^+ ions are highly reactive. In aqueous solution, they bond to water molecules according to this reaction:

$$H^+ + :\ddot{\underset{\cdot\cdot}{O}}:H \longrightarrow \left[H:\ddot{\underset{\cdot\cdot}{O}}:H\right]^+$$

The H_3O^+ ion is the **hydronium ion**. In water, H^+ ions *always* associate with H_2O molecules. Chemists often use $H^+(aq)$ and $H_3O^+(aq)$ interchangeably, however, to refer to the same thing—a hydronium ion.

In the molecular formula for an acid, we often write the ionizable hydrogen first. For example, we write the formula for formic acid as follows:

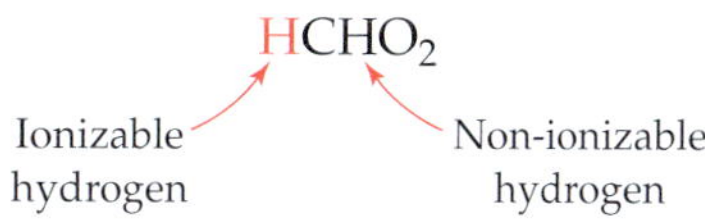

The structure of formic acid, however, is not indicated by the molecular formula in the preceding figure. We represent the *structure* of formic acid with its structural formula:

O
||
HC—OH
Ionizable hydrogen

Notice that the structural formula indicates how the atoms are bonded together; the molecular formula, by contrast, indicates only the number of atoms of each element.

NaOH is an **Arrhenius base** because it produces OH^- ions in solution (◀ Figure 8.12).

$$NaOH(aq) \rightarrow Na^+(aq) + OH^-(aq)$$

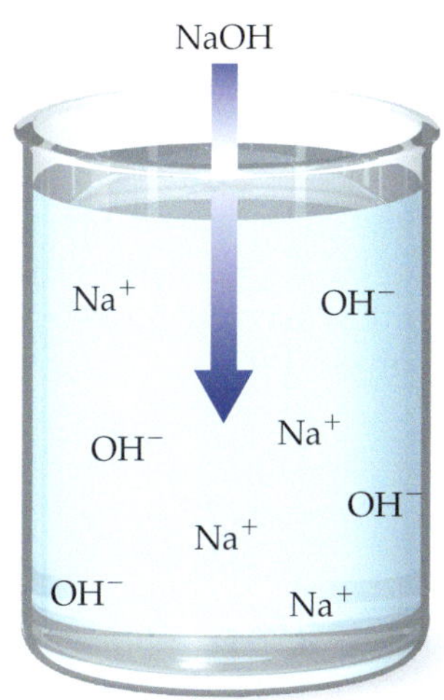

$NaOH(aq) \longrightarrow Na^+(aq) + OH^-(aq)$

▲ **FIGURE 8.12 Arrhenius definition of a base** The Arrhenius definition states that a base is a substance that produces OH^- ions in solution.

Ionic compounds such as NaOH are composed of positive and negative ions. In solution, soluble ionic compounds dissociate into their component ions. Molecular compounds containing an OH group, such as methanol CH_3OH, do not dissociate and therefore do not act as bases.

NaOH is an ionic compound and therefore contains Na^+ and OH^- ions. When NaOH is added to water, it **dissociates**, or breaks apart into its component ions.

Under the Arrhenius definition, acids and bases naturally combine to form water, neutralizing each other in the process.

$$H^+(aq) + OH^-(aq) \rightarrow H_2O(l)$$

The Brønsted–Lowry Definition

Although the Arrhenius definition of acids and bases works in many cases, it cannot easily explain why some substances act as bases even though they do not contain OH^-. The Arrhenius definition also does not apply to nonaqueous solvents. A second definition of acids and bases, called the **Brønsted–Lowry definition**, introduced in 1923, applies to a wider range of acid–base phenomena. This definition focuses on the *transfer* of H^+ ions in an acid–base reaction. Since an H^+ ion is a proton—a hydrogen atom with its electron taken away—this definition focuses on the idea of a proton donor and a proton acceptor.

BRØNSTED–LOWRY DEFINITION

Acid—An acid is a proton (**H^+ ion**) *donor*.
Base—A base is a proton (**H^+ ion**) *acceptor*.

According to this definition, HCl is a **Brønsted–Lowry acid** because, in solution, it donates a proton to water.

Johannes Brønsted, working in Denmark, and Thomas Lowry, working in England, developed the concept of proton transfer in acid–base behavior independently and simultaneously.

$$HCl(aq) + H_2O(l) \rightarrow H_3O^+(aq) + Cl^-(aq)$$

This definition more clearly accounts for what happens to the H^+ ion from an acid: It associates with a water molecule to form H_3O^+ (a hydronium ion). The Brønsted–Lowry definition also works well with bases (such as NH_3) that do not inherently contain OH^- ions but that still produce OH^- ions in solution. NH_3 is a **Brønsted–Lowry base** because it accepts a proton from water.

The double arrows in this equation indicate that the reaction does not go to completion. We discuss this concept in more detail later in this Module.

$$NH_3(aq) + H_2O(l) \rightleftharpoons NH_4^+(aq) + OH^-(aq)$$

According to the Brønsted–Lowry definition, acids (proton donors) and bases (proton acceptors) always occur together. In the reaction between HCl and H_2O, HCl is the proton donor (acid), and H_2O is the proton acceptor (base).

$$\underset{\substack{\text{Acid} \\ \text{(Proton donor)}}}{HCl(aq)} + \underset{\substack{\text{Base} \\ \text{(Proton acceptor)}}}{H_2O(l)} \rightarrow H_3O^+(aq) + Cl^-(aq)$$

In the reaction between NH_3 and H_2O, H_2O is the proton donor (acid) and NH_3 is the proton acceptor (base).

$$\underset{\substack{\text{Base} \\ \text{(Proton acceptor)}}}{NH_3(aq)} + \underset{\substack{\text{Acid} \\ \text{(Proton donor)}}}{H_2O(l)} \rightleftharpoons NH_4^+(aq) + OH^-(aq)$$

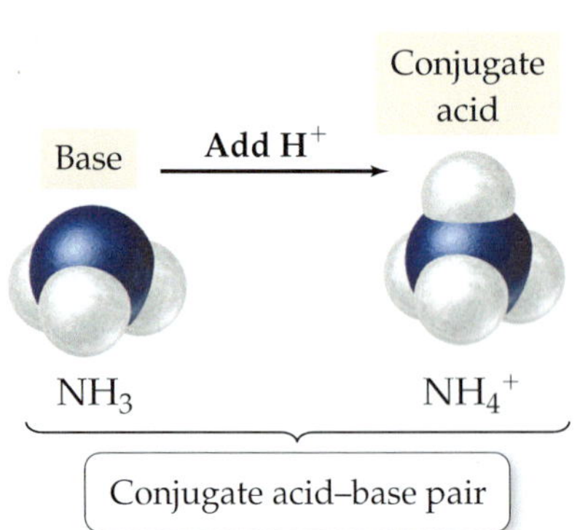

Notice that under the Brønsted–Lowry definition, some substances—such as water in the previous two equations—can act as acids *or* bases. Substances that can act as acids or bases are **amphoteric**. Notice also what happens when an equation representing Brønsted–Lowry acid–base behavior is reversed.

$$\underset{\substack{\text{Acid} \\ \text{(Proton donor)}}}{NH_4^+(aq)} + \underset{\substack{\text{Base} \\ \text{(Proton acceptor)}}}{OH^-(aq)} \rightleftharpoons NH_3(aq) + H_2O(l)$$

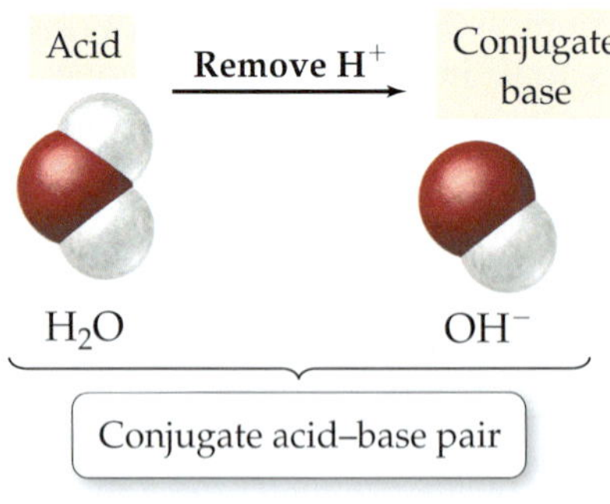

▲ **FIGURE 8.13 A conjugate acid–base pair** Any two substances related to each other by the transfer of a proton can be considered a conjugate acid–base pair.

In this reaction, NH_4^+ is the proton donor (acid) and OH^- is the proton acceptor (base). The substance that was the base (NH_3) becomes the acid (NH_4^+), and vice versa. NH_4^+ and NH_3 are often referred to as a **conjugate acid–base pair**, two substances related to each other by the transfer of a proton (◀ Figure 8.13). Going back to the original forward reaction, we can identify the conjugate acid–base pairs as follows:

$$\underset{\text{Base}}{NH_3(aq)} + \underset{\text{Acid}}{H_2O(l)} \rightleftharpoons \underset{\text{Conjugate acid}}{NH_4^+(aq)} + \underset{\text{Conjugate base}}{OH^-(aq)}$$

In an acid–base reaction, a base accepts a proton and becomes a conjugate acid. An acid donates a proton and becomes a conjugate base.

EXAMPLE 8.13 IDENTIFYING BRØNSTED–LOWRY ACIDS AND BASES AND THEIR CONJUGATES

In each reaction, identify the Brønsted–Lowry acid, the Brønsted–Lowry base, the conjugate acid, and the conjugate base.

(a) $H_2SO_4(aq) + H_2O(l) \rightarrow H_3O^+(aq) + HSO_4^-(aq)$
(b) $HCO_3^-(aq) + H_2O(l) \rightleftharpoons H_2CO_3(aq) + OH^-(aq)$

SOLUTION

(a) Since H_2SO_4 donates a proton to H_2O in this reaction, it is the acid (the proton donor). After H_2SO_4 donates the proton, it becomes HSO_4^-, the conjugate base. Since H_2O accepts a proton, it is the base (the proton acceptor). After H_2O accepts the proton, it becomes H_3O^+, the conjugate acid.

$$\underset{\text{Acid}}{H_2SO_4(aq)} + \underset{\text{Base}}{H_2O(l)} \longrightarrow \underset{\text{Conjugate base}}{HSO_4^-(aq)} + \underset{\text{Conjugate acid}}{H_3O^+(aq)}$$

(b) Since H_2O donates a proton to HCO_3^- in this reaction, it is the acid (the proton donor). After H_2O donates the proton, it becomes OH^-, the conjugate base. Since HCO_3^- accepts a proton, it is the base (the proton acceptor). After HCO_3^- accepts the proton, it becomes H_2CO_3, the conjugate acid.

$$\underset{\text{Base}}{HCO_3^-(aq)} + \underset{\text{Acid}}{H_2O(l)} \rightleftharpoons \underset{\text{Conjugate acid}}{H_2CO_3(aq)} + \underset{\text{Conjugate base}}{OH^-(aq)}$$

▶**SKILLBUILDER 8.13 | Identifying Brønsted–Lowry Acids and Bases and Their Conjugates**

In each reaction, identify the Brønsted–Lowry acid, the Brønsted–Lowry base, the conjugate acid, and the conjugate base.

(a) $C_5H_5N(aq) + H_2O(l) \rightleftharpoons C_5H_5NH^+(aq) + OH^-(aq)$
(b) $HNO_3(aq) + H_2O(l) \rightarrow NO_3^-(aq) + H_3O^+(aq)$

▶**FOR MORE PRACTICE** Example 8.32; Problems 102, 104, 106, 108, 110.

CONCEPTUAL CHECKPOINT 8.9

Which species is the conjugate base of H_2SO_3?

(a) $H_3SO_3^+$ **(b)** HSO_3^- **(c)** SO_3^{2-}

Reactions of Acids and Bases

Acids and bases are typically reactive substances. In this section, we examine how they react with one another, as well as with other substances.

Neutralization Reactions

One of the most important reactions of acids and bases is **neutralization**. When we mix an acid and a base, the $H^+(aq)$ from the acid combines with the $OH^-(aq)$ from the base to form $H_2O(l)$. For example, consider the reaction between hydrochloric acid and potassium hydroxide:

The reaction between HCl and KOH is also a double-displacement reaction (see Section 8.6).

$$\underset{\text{Acid}}{HCl(aq)} + \underset{\text{Base}}{KOH(aq)} \rightarrow \underset{\text{Water}}{H_2O(l)} + \underset{\text{Salt}}{KCl(aq)}$$

Acid–base reactions generally form water and a **salt**—an ionic compound—that usually remains dissolved in the solution. The salt contains the cation from the base and the anion from the acid.

$$\text{Acid} + \text{Base} \longrightarrow \text{Water} + \text{Salt}$$

Ionic compound that contains the cation from the base and the anion from the acid

The net ionic equation for many neutralization reactions is:

Net ionic equations are explained in Section 8.7.

$$H^+(aq) + OH^-(aq) \rightarrow H_2O(l)$$

A slightly different but common type of neutralization reaction involves an acid reacting with carbonates or hydrogencarbonates, also known as bicarbonates; these compounds contain CO_3^{2-} or HCO_3^-. This type of neutralization reaction produces water, gaseous carbon dioxide, and a salt. As an example, consider the reaction of hydrochloric acid and sodium hydrogencarbonate (also known as sodium bicarbonate).

$$HCl(aq) + NaHCO_3(aq) \rightarrow H_2O(l) + CO_2(g) + NaCl(aq)$$

Since this reaction produces gaseous CO_2, it is also called a *gas evolution reaction.*

$$HCl(aq) + NaHCO_3(aq) \longrightarrow H_2O(l) + CO_2(g) + NaCl(aq)$$

▲ The reaction of carbonates or hydrogencarbonates (also known as bicarbonates) with acids produces water, gaseous carbon dioxide, and a salt. © Paul Silverman/Fundamental Photographs.

EXAMPLE 8.14 WRITING EQUATIONS FOR NEUTRALIZATION REACTIONS

Write a molecular equation for the reaction between aqueous HCl and aqueous $Ca(OH)_2$.

SOLUTION First identify the acid and the base and write the skeletal reaction showing the production of water and the salt. The formulas for the ionic compounds in the equation must be charge neutral (see Section 3.5).	$HCl(aq) + Ca(OH)_2(aq) \rightarrow H_2O(l) + CaCl_2(aq)$
Balance the equation. Notice that $Ca(OH)_2$ contains 2 mol of OH^- for every 1 mol of $Ca(OH)_2$ and therefore requires 2 mol of H^+ to neutralize it.	$2\,HCl(aq) + Ca(OH)_2(aq) \rightarrow 2\,H_2O(l) + CaCl_2(aq)$

▶SKILLBUILDER 8.14 | Writing Equations for Neutralization Reactions

Write a molecular equation for the reaction that occurs between aqueous H_3PO_4 and aqueous NaOH. *Hint:* H_3PO_4 is a triprotic acid, meaning that 1 mol of H_3PO_4 requires 3 mol of OH^- to completely react with it.

▶FOR MORE PRACTICE Example 8.33; Problems 112, 113.

$$2\,HCl(aq) + Mg(s) \longrightarrow H_2(g) + MgCl_2(aq)$$

▲ The reaction between an acid and a metal usually produces hydrogen gas and a dissolved salt containing the metal ion. © Richard Megna/Fundamental Photographs.

Acid Reactions

Acids dissolve metals, or more precisely, that acids react with metals in a way that causes metals to go into solution. The reaction between an acid and a metal usually produces hydrogen gas and a dissolved salt containing the metal ion as the cation. For example, hydrochloric acid reacts with magnesium metal to form hydrogen gas and magnesium chloride.

$$\underset{\text{Acid}}{2\,HCl(aq)} + \underset{\text{Metal}}{Mg(s)} \rightarrow \underset{\text{Hydrogen gas}}{H_2(g)} + \underset{\text{Salt}}{MgCl_2(aq)}$$

Similarly, sulfuric acid reacts with zinc to form hydrogen gas and zinc sulfate.

$$\underset{\text{Acid}}{H_2SO_4(aq)} + \underset{\text{Metal}}{Zn(s)} \rightarrow \underset{\text{Hydrogen gas}}{H_2(g)} + \underset{\text{Salt}}{ZnSO_4(aq)}$$

Hydrochloric acid reacts with iron to form hydrogen gas and iron(II) chloride.

$$\underset{\text{Acid}}{2\,HCl(aq)} + \underset{\text{Metal}}{Fe(s)} \rightarrow \underset{\text{Hydrogen gas}}{H_2(g)} + \underset{\text{Salt}}{FeCl_2(aq)}$$

Acids also react with metal oxides to produce water and a dissolved salt. For example, hydrochloric acid reacts with potassium oxide to form water and potassium chloride.

$$\underset{\text{Acid}}{2\,HCl(aq)} + \underset{\text{Metal oxide}}{K_2O(s)} \rightarrow \underset{\text{Water}}{H_2O(l)} + \underset{\text{Salt}}{2\,KCl(aq)}$$

Similarly, hydrobromic acid reacts with magnesium oxide to form water and magnesium bromide.

$$\underset{\text{Acid}}{2\,HBr(aq)} + \underset{\text{Metal oxide}}{MgO(s)} \rightarrow \underset{\text{Water}}{H_2O(l)} + \underset{\text{Salt}}{MgBr_2(aq)}$$

EXAMPLE 8.15 **WRITING EQUATIONS FOR ACID REACTIONS**

Write an equation for each reaction.

(a) The reaction of hydroiodic acid with potassium metal

(b) The reaction of hydrobromic acid with sodium oxide

SOLUTION

(a) The reaction of hydroiodic acid with potassium metal forms hydrogen gas and a salt. The salt contains the ionized form of the metal (K^+) as the cation and the anion of the acid (I^-). Write the skeletal equation and then balance it.	$HI(aq) + K(s) \rightarrow H_2(g) + KI(aq)$ $2\,HI(aq) + 2\,K(s) \rightarrow H_2(g) + 2\,KI(aq)$
(b) The reaction of hydrobromic acid with sodium oxide forms water and a salt. The salt contains the cation from the metal oxide (Na^+) and the anion of the acid (Br^-). Write the skeletal equation and then balance it.	$HBr(aq) + Na_2O(s) \rightarrow H_2O(l) + NaBr(aq)$ $2\,HBr(aq) + Na_2O(s) \rightarrow H_2O(l) + 2\,NaBr(aq)$

▶**SKILLBUILDER 8.15** | **Writing Equations for Acid Reactions**

Write an equation for each reaction.

(a) The reaction of hydrochloric acid with strontium metal

(b) The reaction of hydroiodic acid with barium oxide

▶**FOR MORE PRACTICE** Example 8.34; Problems 114, 115, 116, 117.

Base Reactions

The most important base reactions are those in which a base neutralizes an acid (see the beginning of this section). The only other kind of base reaction that we cover in this book is the reaction of sodium hydroxide with aluminum and water.

$$2\,NaOH(aq) + 2\,Al(s) + 6\,H_2O(l) \rightarrow 2\,NaAl(OH)_4(aq) + 3\,H_2(g)$$

EVERYDAY CHEMISTRY

▶ What Is in My Antacid?

Heartburn, a burning sensation in the lower throat and above the stomach, is caused by the reflux or backflow of stomach acid into the esophagus (the tube that joins the stomach to the throat). In most individuals, this occurs only occasionally, typically after large meals. Physical activity—such as bending, stooping, or lifting—after meals also aggravates heartburn. In some people, the flap between the esophagus and the stomach that normally prevents acid reflux becomes damaged, in which case heartburn becomes a regular occurrence.

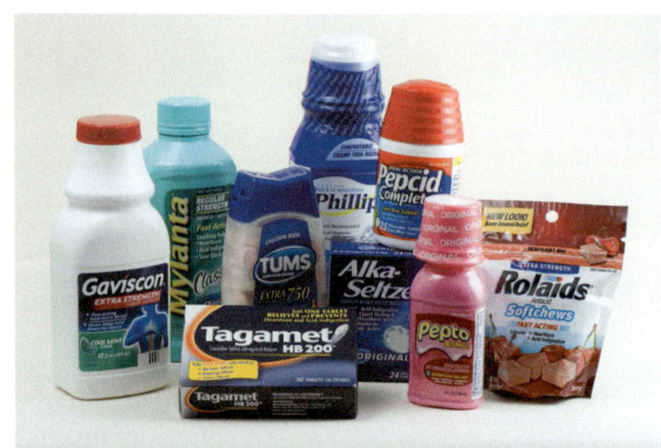

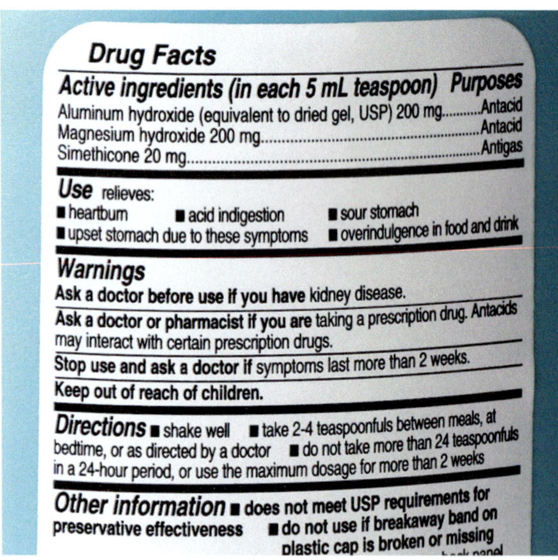

Drugstores carry many products that either reduce the secretion of stomach acid or neutralize the acid that is produced. Antacids such as Mylanta or Phillips' milk of magnesia contain bases that neutralize the refluxed stomach acid, alleviating heartburn.

B8.2 CAN YOU ANSWER THIS? *Look at the label of Mylanta shown in the photograph. Can you identify the bases responsible for the antacid action? Write chemical equations showing the reactions of these bases with stomach acid (HCl).* © (top right) Maxwell Art And Photo/Pearson. (top left): Richard Megna/Fundamental Photographs.

Aluminum is one of the few metals that dissolves in a base. Consequently, it is safe to use NaOH (the main ingredient in many drain-opening products) to unclog your drain as long as your pipes are not made of aluminum, which is generally the case as the use of aluminum pipe is forbidden by most building codes.

Strong and Weak Acids and Bases

Acids and bases can each be categorized as strong or weak, depending on how much they ionize or dissociate in aqueous solution. In this section, we first look at strong and weak acids, and then turn to strong and weak bases.

Strong Acids

Hydrochloric acid (HCl) and hydrofluoric acid (HF) appear to be similar, but there is an important difference between these two acids. HCl is an example of a **strong acid**, one that completely ionizes in solution.

Single arrow indicates complete ionization

$$HCl(aq) + H_2O(l) \longrightarrow H_3O^+(aq) + Cl^-(aq)$$

[X] means "molar concentration of X."

We show the *complete* ionization of HCl with a single arrow pointing to the right in the equation. An HCl solution contains almost no intact HCl; virtually all the HCl has reacted with water to form $H_3O^+(aq)$ and $Cl^-(aq)$ (◀ Figure 8.14). A 1.0 mol/L HCl solution therefore has an H_3O^+ concentration of 1.0 mol/L. We often abbreviate the concentration of H_3O^+ as $[H_3O^+]$. Using this notation, a 1.0 mol/L HCl solution has $[H_3O^+] = 1.0$ mol/L.

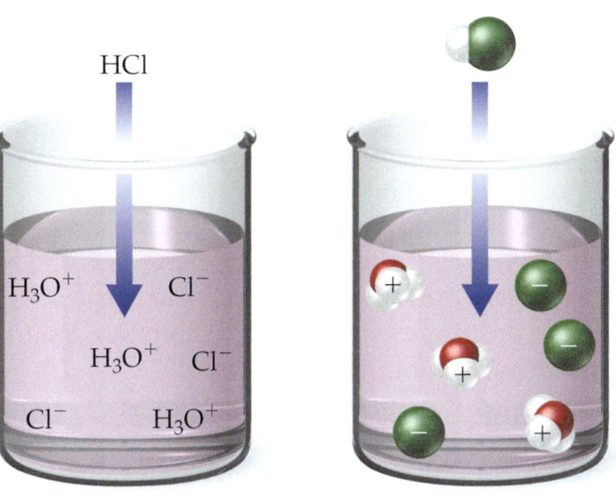

▲ **FIGURE 8.14 A strong acid** When HCl dissolves in water, it completely ionizes into H_3O^+ and Cl^- ions. The solution contains no intact HCl.

An ionizable proton is one that becomes an H^+ ion in solution.

A strong acid is also a **strong electrolyte** (first defined in Section 7.5), a substance whose aqueous solutions are good conductors of electricity (▼ Figure 8.15). Aqueous solutions require the presence of charged particles to conduct electricity. Strong acid solutions are also strong electrolyte solutions because each acid molecule ionizes into positive and negative ions. These mobile ions are good conductors of electricity. Pure water is not a good conductor of electricity because it has relatively few charged particles. The danger of using electrical devices—such as a hair dryer—while sitting in the bathtub is that water is seldom pure and often contains dissolved ions. If the device were to come in contact with the water, dangerously high levels of electricity could flow through the water and through your body.

Table 8.5 lists the six strong acids. The first five acids in the table are **monoprotic acids**, acids containing only one ionizable proton. Sulfuric acid is an example of a **diprotic acid**, an acid that contains two ionizable protons.

TABLE 8.5 Strong Acids

hydrochloric acid (HCl)	nitric acid (HNO_3)
hydrobromic acid (HBr)	perchloric acid ($HClO_4$)
hydroiodic acid (HI)	sulfuric acid (H_2SO_4) (*diprotic*)

(a) Pure water

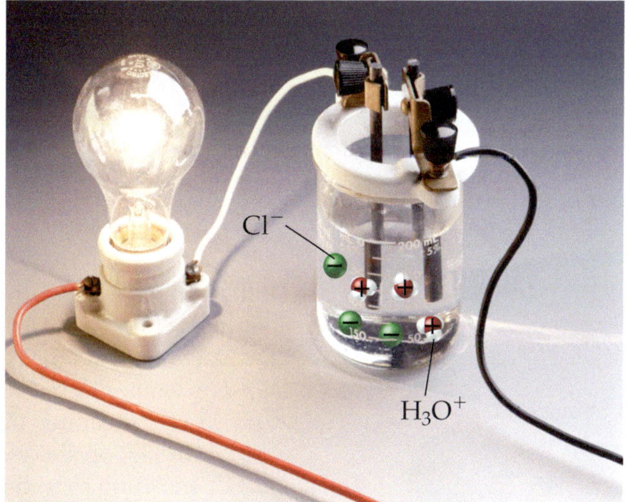

(b) HCl solution

▲ **FIGURE 8.15 Conductivity of a strong electrolyte solution** **(a)** Pure water does not conduct electricity. **(b)** The presence of ions in an HCl solution results in the conduction of electricity, causing the lightbulb to light. Solutions such as these are strong electrolyte solutions. © Richard Megna/Fundamental Photographs.

Weak Acids

It is a common mistake to confuse the terms *strong* and *weak acids* with the terms *concentrated* and *dilute acids*. Can you state the difference between these terms?

In contrast to HCl, HF is a **weak acid**, one that does not completely ionize in solution.

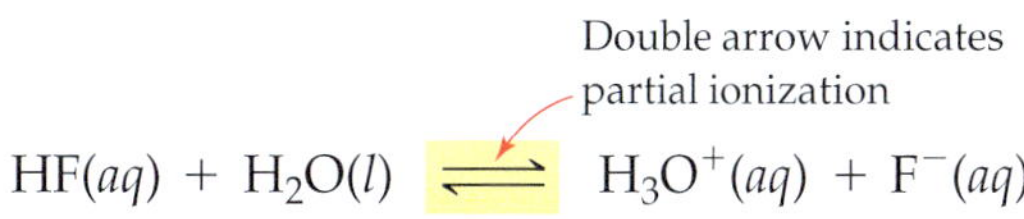

$$HF(aq) + H_2O(l) \rightleftharpoons H_3O^+(aq) + F^-(aq)$$

Calculating exact $[H_3O^+]$ for weak acids is beyond the scope of this text.

▲ **FIGURE 8.16 A weak acid** When HF dissolves in water, only a fraction of the dissolved molecules ionize into H_3O^+ and F^- ions. The solution contains many intact HF molecules.

To show that HF does not completely ionize in solution, the equation for its ionization has two opposing arrows, indicating that the reverse reaction occurs to some degree. An HF solution contains a lot of intact HF; it also contains some $H_3O^+(aq)$ and $F^-(aq)$ (◄ Figure 8.16). In other words, a 1.0 mol/L HF solution has $[H_3O^+] < 1.0$ mol/L because only some of the HF molecules ionize to form H_3O^+.

A weak acid is also a **weak electrolyte**, a substance whose aqueous solutions are poor conductors of electricity (▼ Figure 8.17). Weak acid solutions contain few charged particles because only a small fraction of the acid molecules ionize into positive and negative ions.

The degree to which an acid is strong or weak depends in part on the attraction between the anion of the acid (the conjugate base) and the hydrogen ion. Suppose *HA* is a generic formula for an acid. Then, the degree to which the following reaction proceeds in the forward direction depends in part on the strength of the attraction between H^+ and A^-.

$$\underset{\text{Acid}}{HA(aq)} + H_2O(l) \rightarrow H_3O^+(aq) + \underset{\text{Conjugate base}}{A^-(aq)}$$

(a) Pure water

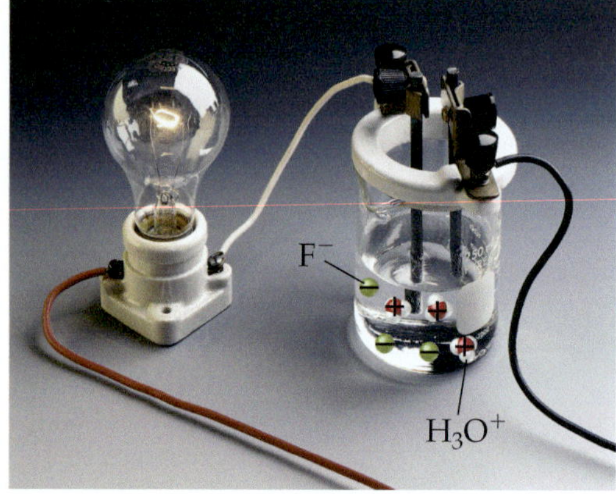

(b) HF solution

▲ **FIGURE 8.17 Conductivity of a weak electrolyte solution** **(a)** Pure water does not conduct electricity. **(b)** An HF solution contains some ions, but most of the HF is intact. The light glows only dimly. Solutions such as these are weak electrolyte solutions. © Richard Megna/Fundamental Photographs.

If the attraction between H^+ and A^- is *weak*, the reaction favors the forward direction and the acid is *strong* (◄ Figure 8.18a). If the attraction between H^+ and A^- is *strong*, the reaction favors the reverse direction and the acid is *weak* (◄ Figure 8.18b).

For example, in HCl, the conjugate base (Cl^-) has a relatively weak attraction to H^+, meaning that the reverse reaction does not occur to any significant extent. In HF, on the other hand, the conjugate base (F^-) has a greater attraction to H^+, meaning that the reverse reaction occurs to a significant degree. *In general, the stronger the acid, the weaker the conjugate base and vice versa.* This means that if the forward reaction (that of the acid) has a high tendency to occur, then the reverse reaction (that of the conjugate base) has a low tendency to occur. Table 8.6 lists some common weak acids.

(a) Strong acid

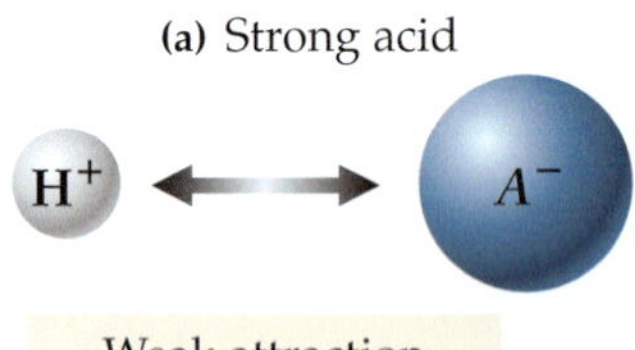

Weak attraction
Complete ionization

(b) Weak acid

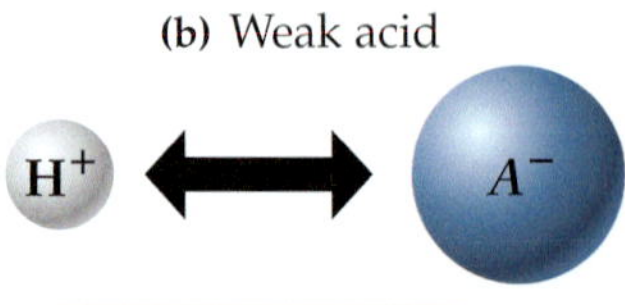

Strong attraction
Partial ionization

◄ **FIGURE 8.18 Strong and weak acids** **(a)** In a strong acid, the attraction between H^+ and A^- is low, resulting in complete ionization. **(b)** In a weak acid, the attraction between H^+ and A^- is high, resulting in partial ionization.

TABLE 8.6 Weak Acids

hydrofluoric acid (HF)	sulfurous acid (H_2SO_3) *(diprotic)*
acetic acid ($HC_2H_3O_2$)	carbonic acid (H_2CO_3) *(diprotic)*
formic acid ($HCHO_2$)	phosphoric acid (H_3PO_4) *(triprotic)*

Notice that two of the weak acids in Table 8.6 are diprotic (they have two ionizable protons) and one is triprotic (it has three ionizable protons). Let us return to sulfuric acid (Table 8.5) for a moment. Sulfuric acid is a diprotic acid that is strong in its first ionizable proton:

$$H_2SO_4(aq) + H_2O(l) \rightarrow H_3O^+(aq) + HSO_4^-(aq)$$

but weak in its second ionizable proton:

$$HSO_4^-(aq) + H_2O(l) \rightleftharpoons H_3O^+(aq) + SO_4^{2-}(aq)$$

Sulfurous acid and carbonic acid are weak in both of their ionizable protons, and phosphoric acid is weak in all three of its ionizable protons.

EXAMPLE 8.16 DETERMINING [H_3O^+] IN ACID SOLUTIONS

Determine the H_3O^+ molar concentration in each solution.

(a) 1.5 mol/L HCl **(b)** 3.0 mol/L $HC_2H_3O_2$ **(c)** 2.5 mol/L HNO_3

SOLUTION

(a) Since HCl is a strong acid, it completely ionizes. The molar concentration of H_3O^+ will be 1.5 mol/L.

$$[H_3O^+] = 1.5 \text{ mol/L}$$

(b) Since $HC_2H_3O_2$ is a weak acid, it partially ionizes. The calculation of the exact molar concentration of H_3O^+ is beyond the scope of this text, but we know that it will be less than 3.0 mol/L.

$$[H_3O^+] < 3.0 \text{ mol/L}$$

(c) Since HNO_3 is a strong acid, it completely ionizes. The molar concentration of H_3O^+ will be 2.5 mol/L.

$$[H_3O^+] = 2.5 \text{ mol/L}$$

▶SKILLBUILDER 8.16 | Determining [H_3O^+] in Acid Solutions

Determine the H_3O^+ molar concentration in each solution.

(a) 0.50 mol/L $HCHO_2$ **(b)** 1.25 mol/L HI **(c)** 0.75 mol/L HF

▶FOR MORE PRACTICE Example 8.35; Problems 122, 123.

CONCEPTUAL CHECKPOINT 8.10

Examine these molecular views of three different acid solutions. Which of these acids is a weak acid?

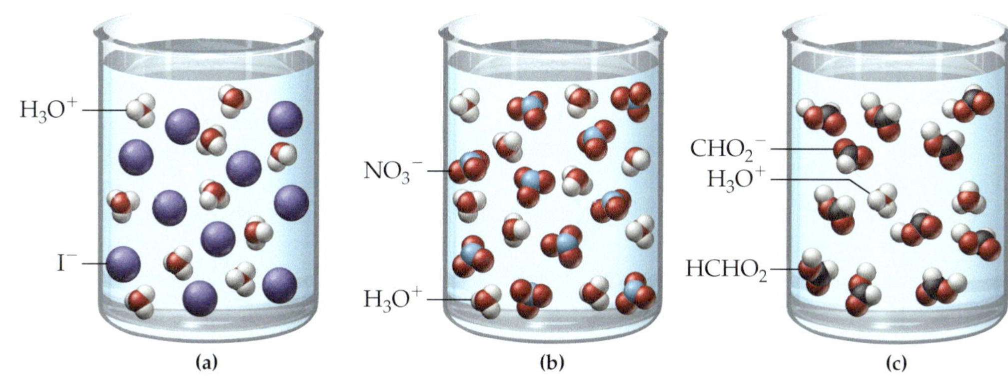

Strong Bases

By analogy to the definition of a strong acid, a **strong base** is one that completely dissociates in solution. NaOH, for example, is a strong base.

$$NaOH(aq) \rightarrow Na^+(aq) + OH^-(aq)$$

An NaOH solution contains no intact NaOH—it has all dissociated to form $Na^+(aq)$ and $OH^-(aq)$ (▼ Figure 8.19), and a 1.0 mol/L NaOH solution has $[OH^-] =$ 1.0 mol/L and $[Na^+] = 1.0$ mol/L. Table 8.7 lists some common strong bases.

▶ **FIGURE 8.19 A strong base** When NaOH dissolves in water, it completely dissociates into Na^+ and OH^-. **Question: The solution contains no intact NaOH. Is NaOH a strong or weak electrolyte?**

TABLE 8.7 Strong Bases

lithium hydroxide (LiOH)	strontium hydroxide ($Sr(OH)_2$)
sodium hydroxide (NaOH)	calcium hydroxide ($Ca(OH)_2$)
potassium hydroxide (KOH)	barium hydroxide ($Ba(OH)_2$)

Unlike diprotic acids, which ionize in two steps, bases containing 2 OH^- ions dissociate in one step.

Some strong bases, such as $Sr(OH)_2$, contain two OH^- ions. These bases completely dissociate, producing two moles of OH^- per mole of base. For example, $Sr(OH)_2$ dissociates as follows:

$$Sr(OH)_2(aq) \rightarrow Sr^{2+}(aq) + 2\,OH^-(aq)$$

Weak Bases

A **weak base** is analogous to a weak acid. In contrast to strong bases that contain OH^- and dissociate in water, the most common weak bases produce OH^- by accepting a proton from water, ionizing water to form OH^-.

Calculating exact $[OH^-]$ for weak bases is beyond the scope of this text.

$$B(aq) + H_2O(l) \rightleftharpoons BH^+(aq) + OH^-(aq)$$

In this equation, B is generic for a weak base. Ammonia, for example, ionizes water according to the reaction:

$$NH_3(aq) + H_2O(l) \rightleftharpoons NH_4^+(aq) + OH^-(aq)$$

The double arrow indicates that the ionization is not complete. An NH_3 solution contains NH_3, NH_4^+, and OH^- (◀ Figure 8.20). A 1.0 mol/L NH_3 solution has $[OH^-] < 1.0$ mol/L. Table 8.8 lists some common weak bases.

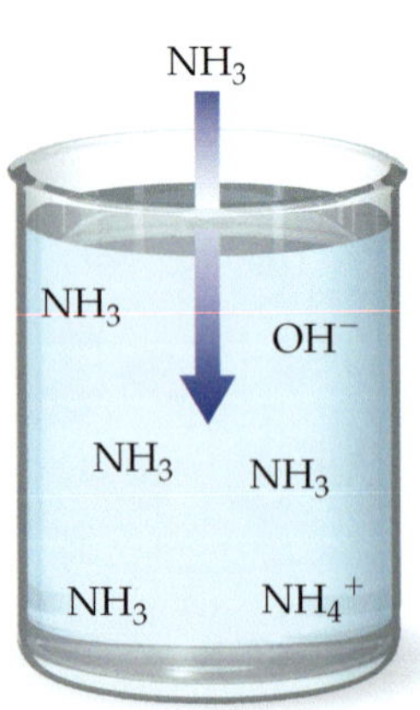

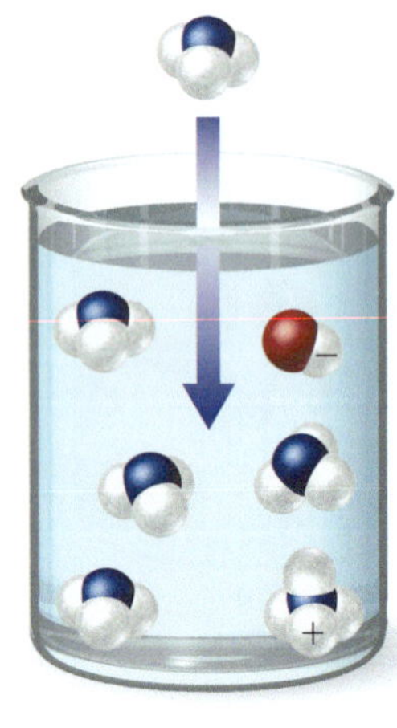

◀ **FIGURE 8.20 A weak base** When NH_3 dissolves in water, it partially ionizes to form NH_4^+ and OH^- However, only a fraction of the molecules ionize. Most NH_3 molecules remain as NH_3.
Question: Is NH_3 a strong or weak electrolyte?

TABLE 8.8 Some Weak Bases

Base	Ionization Reaction
ammonia (NH_3)	$NH_3(aq) + H_2O(l) \rightleftharpoons NH_4^+(aq) + OH^-(aq)$
pyridine (C_5H_5N)	$C_5H_5N(aq) + H_2O(l) \rightleftharpoons C_5H_5NH^+(aq) + OH^-(aq)$
methylamine (CH_3NH_2)	$CH_3NH_2(aq) + H_2O(l) \rightleftharpoons CH_3NH_3^+(aq) + OH^-(aq)$
ethylamine ($C_2H_5NH_2$)	$C_2H_5NH_2(aq) + H_2O(l) \rightleftharpoons C_2H_5NH_3^+(aq) + OH^-(aq)$
hydrogencarbonate or bicarbonate ion (HCO_3^-)*	$HCO_3^-(aq) + H_2O(l) \rightleftharpoons H_2CO_3(aq) + OH^-(aq)$

*The hydrogencarbonate or bicarbonate ion must occur with a positively charged ion such as Na^+ that serves to balance the charge but does not have any part in the ionization reaction. It is the hydrogencarbonate/bicarbonate ion that makes sodium hydrogencarbonate/bicarbonate ($NaHCO_3$) basic.

EXAMPLE 8.17 DETERMINING [OH⁻] IN BASE SOLUTIONS

Determine the OH^- molar concentration in each solution.
(a) 2.25 mol/L KOH **(b)** 0.35 mol/L CH_3NH_2 **(c)** 0.025 mol/L $Sr(OH)_2$

SOLUTION

(a) Since KOH is a strong base, it completely dissociates into K^+ and OH^- in solution. The molar concentration of OH^- is 2.25 mol/L.

$$[OH^-] = 2.25 \text{ mol/L}$$

(b) Since CH_3NH_2 is a weak base, it only partially ionizes water. We cannot calculate the exact molar concentration of OH^-, but we know it is less than 0.35 mol/L.

$$[OH^-] < 0.35 \text{ mol/L}$$

(c) Since $Sr(OH)_2$ is a strong base, it completely dissociates into $Sr^{2+}(aq)$ and 2 $OH^-(aq)$. $Sr(OH)_2$ forms 2 mol of OH^- for every 1 mol of $Sr(OH)_2$. Consequently, the molar concentration of OH^- is twice the molar concentration of $Sr(OH)_2$.

$$[OH^-] = 2(0.025 \text{ mol/L}) = 0.050 \text{ mol/L}$$

▶SKILLBUILDER 8.17 | Determining [OH⁻] in Base Solutions

Determine the OH^- molar concentration in each solution.
(a) 0.055 mol/L $Ba(OH)_2$ **(b)** 1.05 mol/L C_5H_5N **(c)** 0.45 mol/L NaOH

▶FOR MORE PRACTICE Example 8.36; Problems 126, 127.

8.9 Oxidation and Reduction

▲ **FIGURE 8.21 A fuel-cell car** The Honda FCX Clarity is a hydrogen-powered, fuel-cell automobile. Its only emission is water. © YOSHIKAZU TSUNO/AFP/Getty Images/Newscom.

It is possible, even likely, that you will see the end of the internal combustion engine within your lifetime. Although it has served us well—powering our airplanes, automobiles, and trains—its time is running out. What will replace it? If our cars don't run on gasoline, what will fuel them? The answers to these questions are not completely settled, but new and better technologies are on the horizon. The most promising of these is the use of **fuel cells** to power electric vehicles. Such whisper-quiet, environmentally friendly supercars are currently available only on a limited basis (◀ Figure 8.21), but they should become more widely available in the years to come.

In 2007, Honda's next-generation FCX Clarity fuel-cell vehicle made its European debut at the Geneva Motorshow. The four-passenger car has a top speed of 100 mph and a range of 270 miles on one tank of fuel. The electric motor is powered by hydrogen, stored as a compressed gas, and its only emission is water—which is so clean you can drink it. The FCX Clarity is available for lease on a limited basis, but it is not yet available for purchase because of the limited availability of hydrogen fuel. Other automakers have similar prototype models in development. In addition, several cities around the world—including Palm Springs, California; Washington, D.C.; Vancouver, British Columbia; Whistler, British Columbia, and São Paulo, Brazil—currently run buses powered by fuel cells.

Fuel-cell technology is possible because of the tendency of some elements to gain electrons from other elements. The most common type of fuel cell—called the *hydrogen–oxygen fuel cell*—is based on the reaction between hydrogen and oxygen.

$$2\ H_2(g) + O_2(g) \rightarrow 2\ H_2O(g)$$

In this reaction, hydrogen and oxygen form covalent bonds with one another. Recall from Section 3.12 that a single covalent bond is a shared electron pair. However, since oxygen is more electronegative than hydrogen (see Section 3.18), the electron pair in a hydrogen–oxygen bond is *unequally* shared, with oxygen getting the larger portion. In effect, oxygen has more electrons in H_2O than in elemental O_2—it has *gained electrons* in the reaction.

In a typical reaction between hydrogen and oxygen, oxygen atoms gain the electrons directly from hydrogen atoms as the reaction proceeds. In a hydrogen–oxygen fuel cell, the same reaction occurs, but the hydrogen and oxygen are separated, forcing the electrons to move through an external wire to get from hydrogen to oxygen. These moving electrons constitute an electrical current, which is used to power the electric motor of a fuel-cell vehicle. In effect, fuel cells use the electron-gaining tendency of oxygen and the electron-losing tendency of hydrogen to force electrons to move through a wire, creating the electricity that powers the car.

Reactions involving the transfer of electrons are **oxidation–reduction reactions** or **redox reactions**. In addition to their application to fuel-cell vehicles, redox reactions are prevalent in nature, in industry, and in many everyday processes. For example, the rusting of iron, the bleaching of hair, and the reactions occurring in batteries all involve redox reactions. Redox reactions are also responsible for providing the energy our bodies need to move, think, and stay alive.

LO: Define and identify oxidation and reduction.
LO: Identify oxidizing agents and reducing agents.
LO: Assign oxidation states.
LO: Use oxidation states to identify oxidation and reduction.
LO: Balance redox reactions.

Oxidation and Reduction: Some Definitions

Consider the following redox reactions:

$$2\ H_2(g) + O_2(g) \rightarrow 2\ H_2O(g) \text{ (hydrogen–oxygen fuel-cell reaction)}$$

$$4\ Fe(s) + 3\ O_2(g) \rightarrow 2\ Fe_2O_3(s) \text{ (rusting of iron)}$$

$$CH_4(g) + 2\ O_2(g) \rightarrow CO_2(g) + 2\ H_2O(g) \text{ (combustion of methane)}$$

What do they all have in common? Each of these reactions involves one or more elements gaining oxygen. In the hydrogen–oxygen fuel-cell reaction, *hydrogen* gains oxygen as it turns into water. In the rusting of iron, *iron* gains oxygen as it turns into iron oxide, the familiar orange substance we call rust (▼ Figure 8.22). In the combustion of methane, *carbon* gains oxygen to form carbon dioxide, producing the brilliant blue flame we see on gas stoves (▼ Figure 8.23). In each case,

▲ **FIGURE 8.22 Slow oxidation** Rust is produced by the oxidation of iron to form iron(III) oxide. © Ron Zmiri/Shutterstock.

▲ **FIGURE 8.23 Rapid oxidation** The flame on a gas stove results from the oxidation of carbon in natural gas. © Ryan McVay/Photodisc/Getty Images.

the substance that gains oxygen is oxidized in the reaction. One definition of *oxidation*—though not the most fundamental one—is the *gaining of oxygen.*

These definitions of oxidation and reduction are useful because they show the origin of the term *oxidation*, and they allow us to quickly identify reactions involving elemental oxygen as oxidation and reduction reactions. However, as you will see, these definitions are *not* the most fundamental.

Now consider these same three reactions in reverse:

$$2\,H_2O(g) \rightarrow 2\,H_2(g) + O_2(g)$$

$$2\,Fe_2O_3(s) \rightarrow 4\,Fe(s) + 3\,O_2(g)$$

$$CO_2(g) + 2\,H_2O(g) \rightarrow CH_4(g) + 2\,O_2(g)$$

Each of these reactions involves loss of oxygen. In the first reaction, hydrogen loses oxygen; in the second reaction, iron loses oxygen; and in the third reaction, carbon loses oxygen. In each case, the substance that loses oxygen is reduced in the reaction. One definition of *reduction* is the *loss of oxygen.*

Redox reactions need not involve oxygen, however. Consider, for example, the similarities between the following two reactions:

In redox reactions between a metal and a nonmetal, the metal is oxidized and the nonmetal is reduced.

$$4\,Li(s) + O_2(g) \rightarrow 2\,Li_2O(s)$$

$$2\,Li(s) + Cl_2(g) \rightarrow 2\,LiCl(s)$$

In both cases, lithium (a metal with a strong tendency to lose electrons) reacts with a nonmetal (which has a tendency to gain electrons). In both cases, lithium atoms lose electrons to become positive ions—lithium is oxidized.

$$Li \rightarrow Li^+ + e^-$$

The electrons lost by lithium are gained by the nonmetals, which become negative ions—the nonmetals are reduced.

$$O_2 + 4e^- \rightarrow 2\,O^{2-}$$

$$Cl_2 + 2e^- \rightarrow 2\,Cl^-$$

Helpful mnemonics: OIL RIG—**O**xidation **I**s **L**oss (of electrons); **R**eduction **I**s **G**ain (of electrons). LEO the lion says GER—**L**ose **E**lectrons **O**xidation; **G**ain **E**lectrons **R**eduction.

A more fundamental definition of **oxidation**, then, is the *loss of electrons*, and a more fundamental definition of **reduction** is the *gain of electrons.*

Notice that *oxidation and reduction must occur together*. If one substance loses electrons (oxidation), then another substance must gain electrons (reduction) (◀ Figure 8.24). The substance that is oxidized is the **reducing agent** because it causes the reduction of the other substance. Similarly, the substance that is reduced is the **oxidizing agent** because it causes the oxidation of the other substance. For example, consider our hydrogen–oxygen fuel-cell reaction:

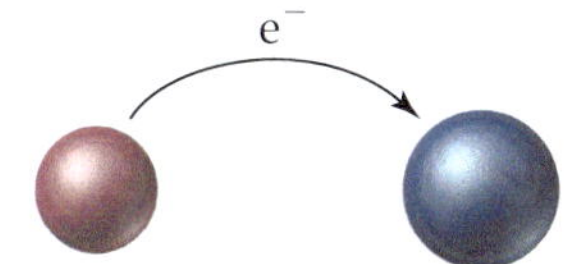

▲ **FIGURE 8.24 Oxidation and reduction** In a redox reaction, one substance loses electrons and another substance gains electrons.

$$\underset{\text{Reducing agent}}{2\,H_2(g)} + \underset{\text{Oxidizing agent}}{O_2(g)} \rightarrow 2\,H_2O(g)$$

In this reaction, hydrogen is oxidized, making it the reducing agent. Oxygen is reduced, making it the oxidizing agent. Substances such as oxygen, which have a strong tendency to attract electrons, are good oxidizing agents—they tend to cause the oxidation of other substances. Substances such as hydrogen, which have a strong tendency to give up electrons, are good reducing agents—they tend to cause the reduction of other substances.

To summarize:

The oxidizing agent may be an element that is itself reduced, or a compound or ion containing an element that is reduced. The reducing agent may be an element that is itself oxidized, or a compound or ion containing an element that is oxidized.

- Oxidation—the loss of electrons
- Reduction—the gain of electrons
- Oxidizing agent—the substance being reduced
- Reducing agent—the substance being oxidized

EXAMPLE 8.18 IDENTIFYING OXIDATION AND REDUCTION

Identify the substance being oxidized and the substance being reduced in each reaction.

(a) $2\,Mg(s) + O_2(g) \rightarrow 2\,MgO(s)$
(b) $Fe(s) + Cl_2(g) \rightarrow FeCl_2(s)$
(c) $Zn(s) + Fe^{2+}(aq) \rightarrow Zn^{2+}(aq) + Fe(s)$

SOLUTION

(a) In this reaction, magnesium is gaining oxygen and losing electrons to oxygen. Mg is therefore oxidized, and O_2 is reduced.
(b) In this reaction, a metal (Fe) is reacting with a nonmetal (Cl_2). Fe loses electrons and is therefore oxidized, while Cl_2 gains electrons and is therefore reduced.
(c) In this reaction, electrons are transferred from the Zn to the Fe^{2+}. Zn loses electrons and is oxidized. Fe^{2+} gains electrons and is reduced.

▶SKILLBUILDER 8.18 | Identifying Oxidation and Reduction

Identify the substance being oxidized and the substance being reduced in each reaction.
(a) $2\,K(s) + Cl_2(g) \rightarrow 2\,KCl(s)$
(b) $2\,Al(s) + 3\,Sn^{2+}(aq) \rightarrow 2\,Al^{3+}(aq) + 3\,Sn(s)$
(c) $C(s) + O_2(g) \rightarrow CO_2(g)$

▶FOR MORE PRACTICE Example 8.37; Problems 128, 129, 130, 131.

EXAMPLE 8.19 IDENTIFYING OXIDIZING AND REDUCING AGENTS

Identify the oxidizing agent and the reducing agent in each reaction.
(a) $2\,Mg(s) + O_2(g) \rightarrow 2\,MgO(s)$
(b) $Fe(s) + Cl_2(g) \rightarrow FeCl_2(s)$
(c) $Zn(s) + Fe^{2+}(aq) \rightarrow Zn^{2+}(aq) + Fe(s)$

SOLUTION

In the previous example, we identified the substance being oxidized and reduced for these reactions. Recall that the substance being oxidized is the reducing agent, and the substance being reduced is the oxidizing agent.

(a) Mg is oxidized and is therefore the reducing agent; O_2 is reduced and is therefore the oxidizing agent.
(b) Fe is oxidized and is therefore the reducing agent; Cl_2 is reduced and is therefore the oxidizing agent.
(c) Zn is oxidized and is therefore the reducing agent; Fe^{2+} is reduced and is therefore the oxidizing agent.

▶SKILLBUILDER 8.19 | Identifying Oxidizing and Reducing Agents

Identify the oxidizing agent and the reducing agent in each reaction.

(a) $2\,K(s) + Cl_2(g) \rightarrow 2\,KCl(s)$
(b) $2\,Al(s) + 3\,Sn^{2+}(aq) \rightarrow 2\,Al^{3+}(aq) + 3\,Sn(s)$
(c) $C(s) + O_2(g) \rightarrow CO_2(g)$

▶FOR MORE PRACTICE Example 8.38; Problems 132, 133.

CONCEPTUAL CHECKPOINT 8.11

An oxidizing agent:

(a) is always oxidized.

(b) is always reduced.

(c) can either be oxidized or reduced, depending on the reaction.

Oxidation States: Electron Bookkeeping

For many redox reactions, such as those involving oxygen or other highly electronegative elements, we can readily identify the substances being oxidized and reduced by inspection. For other redox reactions, identification is more difficult. For example, consider the redox reaction between carbon and sulfur:

$$C + O_2 \rightarrow CO_2$$

What is oxidized here? What is reduced? In order to readily identify oxidation and reduction, chemists have devised a scheme to track electrons and where they go in chemical reactions. In this scheme—which is like bookkeeping for electrons—we assign all shared electrons to the most electronegative element. Then we calculate a number—called the **oxidation state** or **oxidation number**—for each element based on the number of electrons assigned to it.

Do not confuse oxidation state with ionic charge. A substance need not be ionic to have an assigned oxidation state.

The procedure just described is a bit cumbersome in practice. However, we can summarize its main results in a series of rules. The easiest way to assign oxidation states is to follow these rules.

Rules for Assigning Oxidation States	Examples
1. The oxidation state of an atom in a free element is 0.	Cu: 0 ox state; Cl_2: 0 ox state
2. The oxidation state of a monoatomic ion is equal to its charge.	Ca^{2+}: +2 ox state; Cl^-: −1 ox state
3. The sum of the oxidation states of all atoms in: (i) a neutral molecule or formula unit is 0. (ii) an ion is equal to the charge of the ion.	H_2O 2(H ox state) + 1(O ox state) = 0 NO_3^- 1(N ox state) + 3(O ox state) = −1
4. In their compounds, (i) Group I metals have an oxidation state of +1. (ii) Group II metals have an oxidation state of +2.	NaCl +1 ox state CaF_2 +2 ox state
(iii) Hydrogen has an oxidation state of +1 (exceptions include −1 in metal hydride).*	in NH_3 : H has +1 ox state, N has −3 ox state (Rule 3i) in NaH : Na has +1 ox state (Rule 4i) , H has −1 ox state (Rule 3i)
(iv) Oxygen has an oxidation state of −2 (exceptions include −1 in peroxide and positive oxidation states in compounds containing F).*	in CO_2 : O has −2 ox state, C has +4 ox state (Rule 3i) in Na_2O_2 (sodium peroxide): Na has +1 ox state (Rule 4i), O has −1 ox state (Rule 3i) in OF_2 (oxygen difluoride): F has −1 ox state (Rule 5), O has +2 ox state (Rule 3i)
5. In compounds, the element with the higher electronegativity is assigned an oxidation state equal to its charge as an anion.	in CCl_4 : Cl has −1 ox state, C has +4 ox state (Rule 3i)

* Fundamentals of Chemistry does not cover any chemical species involving exceptions.

EXAMPLE 8.20 ASSIGNING OXIDATION STATES

Assign an oxidation state to each atom in each species.

(a) Br_2 **(b)** K^+ **(c)** LiF **(d)** CO_2 **(e)** SO_4^{2-} **(f)** Na_2O_2

SOLUTION

Since Br_2 is a free element, the oxidation state of both Br atoms is 0 (Rule 1).

(a) Br_2

Br Br
0 0

Since K^+ is a monoatomic ion, the oxidation state of the K^+ ion is +1 (Rule 2).

(b) K^+

K^+
+1

The oxidation state of Li is +1 (Rule 4i). The oxidation state of F is –1 (Rule 5). Since this is a neutral compound, the sum of the oxidation states is 0 (Rule 3i).

(c) LiF

Li F
+1 –1

sum: $+1 - 1 = 0$

The oxidation state of oxygen is –2 (Rule 4iv or Rule 5). You deduce the oxidation state of carbon from Rule 3i, which states that the sum of the oxidation states of all the atoms must be 0. Since there are two oxygen atoms, you multiply the oxidation state of O by 2 when calculating the sum.

(d) CO_2

(C ox state) + 2(O ox state) = 0
(C ox state) + 2(–2) = 0
(C ox state) – 4 = 0
C ox state = + 4

C O_2
+4 –2

sum: $+4 + 2(-2) = 0$

The oxidation state of oxygen is –2 (Rule 4iv or Rule 5). You calculate the oxidation state of sulfur from Rule 3ii by setting the sum of all of the oxidation states equal to –2 (the charge of the ion).

(e) SO_4^{2-}

(S ox state) + 4(O ox state) = –2
(S ox state) + 4(–2) = –2
(S ox state) – 8 = –2
S ox state = –2 + 8
S ox state = +6

S O_4^{2-}
+6 –2

sum: $+6 + 4(-2) = -2$

The oxidation state of sodium is +1 (Rule 4i). The oxidation state of O is expected to be –2 (Rule 4iv). However, if that were the case, the sum of the oxidation states would not equal zero. Therefore, you deduce the oxidation state of O by setting the sum of all of the oxidation states equal to 0 (Rule 3i).

(f) Na_2O_2 (sodium peroxide)

2(Na ox state) + 2(O ox state) = 0
2(+1) + 2(O ox state) = 0
(Sodium in compounds always has the ox state +1.)
+2 + 2(O ox state) = 0
O ox state = –1 (As stated in Rule 4iv, oxygen has an oxidation state of -1 in peroxide.)

Na_2 O_2
+1 –1

sum: $2(+1) + 2(-1) = 0$

▶SKILLBUILDER 8.20 | Assigning Oxidation States

Assign an oxidation state to each atom in each substance.

(a) Zn **(b)** Cu^{2+} **(c)** $CaCl_2$ **(d)** CF_4 **(e)** NO_2^- **(f)** SO_3

▶FOR MORE PRACTICE Example 8.39; Problems 140, 142, 144, 146, 148.

EVERYDAY CHEMISTRY

▶ The Bleaching of Hair

College students, both male and female, with bleached hair are a common sight on most campuses. Many students bleach their own hair with home-bleaching kits available at drugstores and supermarkets. These kits normally contain hydrogen peroxide (H_2O_2), an excellent oxidizing agent. When applied to hair, hydrogen peroxide oxidizes melanin, the dark pigment that gives hair color. Once melanin is oxidized, it no longer imparts a dark color to hair, leaving the hair with the familiar bleached look.

Hydrogen peroxide also oxidizes other components of hair. For example, the protein molecules in hair contain SH groups called *thiols*. Thiols are normally slippery (they slide across each other). Hydrogen peroxide oxidizes these thiol groups to sulfonic acid groups ($—SO_3H$). Sulfonic acid groups are stickier, causing hair to tangle more easily. Consequently, people with heavily bleached hair often use conditioners. Conditioners contain compounds that form thin, lubricating coatings on individual hair shafts. These coatings prevent tangling and make hair softer and more manageable.

B8.3 CAN YOU ANSWER THIS? *Assign oxidation states to the atoms of H_2O_2. Which atoms in H_2O_2 do you think change oxidation state when H_2O_2 oxidizes hair?*

▲ Hydrogen peroxide, a good oxidizing agent, bleaches hair. © Sean De Burca/Photodisc/Getty Images.

Now let's return to our original question. What is being oxidized and what is being reduced in the following reaction?

$$C + O_2 \rightarrow CO_2$$

We use the oxidation state rules to assign oxidation states to all elements on both sides of the equation:

$$\underset{0}{C} + \underset{0}{O_2} \rightarrow \underset{+4\ -2}{CO_2}$$

The oxidation state of carbon changed from 0 to +4. In terms of our electron bookkeeping scheme (the assigned oxidation state), carbon *lost electrons* and was *oxidized*. The oxidation state of oxygen changed from 0 to −2. In terms of our electron bookkeeping scheme, oxygen *gained electrons* and was *reduced*.

$$\underset{0}{C} + \underset{0}{O_2} \longrightarrow \underset{+4\ -2}{CO_2}$$

In terms of oxidation states, we define oxidation and reduction as follows:

Oxidation—an increase in oxidation state
Reduction—a decrease in oxidation state

EXAMPLE 8.21 **USING OXIDATION STATES TO IDENTIFY OXIDATION AND REDUCTION**

Use oxidation states to identify the element that is being oxidized and the element that is being reduced in the redox reaction.

$$Ca(s) + 2\ H_2O(l) \rightarrow Ca(OH)_2(aq) + H_2(g)$$

SOLUTION

Assign an oxidation state to each atom in the reaction. Ca increased in oxidation state; it was oxidized. H decreased in oxidation state; it was reduced. (Note that oxygen has the same oxidation state on both sides of the equation and was therefore neither oxidized nor reduced.)

$$Ca(s) + 2\ H_2O(l) \longrightarrow Ca(OH)_2(aq) + H_2(g)$$

Oxidation states: Ca: 0; H: +1, O: −2 → Ca: +2, O: −2, H: +1; H: 0

▶SKILLBUILDER 8.21 | Using Oxidation States to Identify Oxidation and Reduction

Use oxidation states to identify the element that is being oxidized and the element that is being reduced in the redox reaction.

$$Sn(s) + 4\ HNO_3(aq) \rightarrow SnO_2(s) + 4\ NO_2(g) + 2\ H_2O(g)$$

▶FOR MORE PRACTICE Problems 152, 153, 154, 155.

CONCEPTUAL CHECKPOINT 8.12

In which substance does nitrogen have the *lowest* oxidation state?

(a) N_2

(b) NO

(c) NO_2

(d) NH_3

Balancing Redox Equations

Earlier in this module, we learned how to balance chemical equations by inspection. We can balance some redox reactions in this way. However, redox reactions occurring in aqueous solutions are usually difficult to balance by inspection and require a special procedure called the *half-reaction method of balancing*. In this procedure, we break down the overall equation into two **half-reactions**: one for oxidation and one for reduction. We balance the half-reactions individually and then add them together. For example, consider the redox reaction:

$$Al(s) + Ag^+(aq) \rightarrow Al^{3+}(aq) + Ag(s)$$

We assign oxidation numbers to all atoms to determine what is being oxidized and what is being reduced.

$$Al(s) + Ag^+(aq) \longrightarrow Al^{3+}(aq) + Ag(s)$$

Oxidation states: Al: 0; Ag: +1 → Al: +3; Ag: 0

We then divide the reaction into two half-reactions, one for oxidation and one for reduction.

Oxidation: $Al(s) \rightarrow Al^{3+}(aq)$

Reduction: $Ag^+(aq) \rightarrow Ag(s)$

Next we balance the two half-reactions individually. In this case, the half-reactions are already balanced with respect to mass—the number of each type of atom on both sides of each half-reaction is the same. However, the equations are not balanced with respect to charge—in the oxidation half-reaction, the left side of the equation has 0 charge while the right side has +3 charge, and in the reduction half-reaction, the left side has +1 charge and the right side has 0 charge. We balance the charge of each half-reaction individually by adding the appropriate number of electrons to make the charges on both sides equal.

$$Al(s) \rightarrow Al^{3+}(aq) + 3e^- \qquad \text{(zero charge on both sides)}$$

$$1e^- + Ag^+(aq) \rightarrow Ag(s) \qquad \text{(zero charge on both sides)}$$

Since these half-reactions must occur together, the number of electrons lost in the oxidation half-reaction must equal the number gained in the reduction half-reaction. We equalize these by multiplying one or both half-reactions by appropriate whole numbers to balance the electrons lost and gained. In this case, we multiply the reduction half-reaction by 3:

$$Al(s) \rightarrow Al^{3+}(aq) + 3e^-$$

$$3 \times [1e^- + Ag^+(aq) \rightarrow Ag(s)]$$

We then add the half-reactions together, canceling electrons and other species as necessary.

$$Al(s) \rightarrow Al^{3+}(aq) + 3e^-$$

$$3e^- + 3Ag^+(aq) \rightarrow 3\,Ag(s)$$

$$\overline{Al(s) + 3\,Ag^+(aq) \rightarrow Al^{3+}(aq) + 3\,Ag(s)}$$

Reactants	Products
1 Al	1 Al
3 Ag	3 Ag
+3 charge	+3 charge

Lastly, we verify that the equation is balanced, with respect to both mass and charge as shown in the margin. Notice that the charge need not be zero on both sides of the equation—it just has to be *equal* on both sides. The equation is balanced.

A general procedure for balancing redox reactions is given in the following procedure box. Because aqueous solutions are often acidic or basic, the procedure must account for the presence of H^+ ions or OH^- ions. We only cover acidic solutions in Examples 8.22 through 8.24.

BALANCING REDOX EQUATIONS USING THE HALF-REACTION METHOD	**EXAMPLE 8.22** Balance the redox reaction.	**EXAMPLE 8.23** Balance the redox reaction.
1. Assign oxidation states to all atoms and identify the substances being oxidized and reduced.	**SOLUTION** $Al(s) + Cu^{2+}(aq) \longrightarrow Al^{3+}(aq) + Cu(s)$ Al: 0 → +3 (Oxidation); Cu: +2 → 0 (Reduction)	**SOLUTION** $Fe^{2+}(aq) + MnO_4^-(aq) \longrightarrow Fe^{3+}(aq) + Mn^{2+}(aq)$ Fe: +2 → +3 (Oxidation); Mn: +7 → +2 (Reduction); O: −2
2. Separate the overall reaction into two half-reactions, one for oxidation and one for reduction.	**OXIDATION** $Al(s) \rightarrow Al^{3+}(aq)$ **REDUCTION** $Cu^{2+}(aq) \rightarrow Cu(s)$	**OXIDATION** $Fe^{2+}(aq) \rightarrow Fe^{3+}(aq)$ **REDUCTION** $MnO_4^-(aq) \rightarrow Mn^{2+}(aq)$

3. Balance each half-reaction with respect to mass in the following order:

- Balance all elements other than H and O.

All elements other than hydrogen and oxygen are balanced, so you can proceed to the next step.

All elements other than hydrogen and oxygen are balanced, so you can proceed to the next step.

- Balance O by adding H_2O.

No oxygen; proceed to the next step.

$$Fe^{2+}(aq) \rightarrow Fe^{3+}(aq)$$
$$MnO_4^-(aq) \rightarrow Mn^{2+}(aq) + 4\,H_2O(l)$$

- Balance H by adding H^+.

No hydrogen; proceed to the next step.

$$Fe^{2+}(aq) \rightarrow Fe^{3+}(aq)$$
$$8\,H^+(aq) + MnO_4^-(aq) \rightarrow Mn^{2+}(aq) + 4\,H_2O(l)$$

4. Balance each half-reaction with respect to charge by adding electrons to the right side of the oxidation half-reaction and the left side of the reduction half-reaction. (The sum of the charges on both sides of each equation should be equal.)

$$Al(s) \rightarrow Al^{3+}(aq) + 3e^-$$
$$2e^- + Cu^{2+}(aq) \rightarrow Cu(s)$$

$$Fe^{2+}(aq) \rightarrow Fe^{3+}(aq) + 1e^-$$
$$5e^- + 8\,H^+(aq) + MnO_4^-(aq) \rightarrow Mn^{2+}(aq) + 4\,H_2O(l)$$

5. Make the number of electrons in both half-reactions equal by multiplying one or both half-reactions by a small whole number.

$$2 \times [Al(s) \rightarrow Al^{3+}(aq) + 3e^-]$$
$$3 \times [2e^- + Cu^{2+}(aq) \rightarrow Cu(s)]$$

$$5 \times [Fe^{2+}(aq) \rightarrow Fe^{3+}(aq) + 1e^-]$$
$$5e^- + 8\,H^+(aq) + MnO_4^-(aq) \rightarrow Mn^{2+}(aq) + 4\,H_2O(l)$$

6. Add the two half-reactions together, canceling electrons and other species as necessary.

$$2\,Al(s) \rightarrow 2\,Al^{3+}(aq) + 6e^-$$
$$6e^- + 3\,Cu^{2+}(aq) \rightarrow 3\,Cu(s)$$
$$2\,Al(s) + 3\,Cu^{2+}(aq) \rightarrow 2\,Al^{3+}(aq) + 3\,Cu(s)$$

$$5\,Fe^{2+}(aq) \rightarrow 5\,Fe^{3+}(aq) + 5e^-$$
$$5e^- + 8\,H^+(aq) + MnO_4^-(aq) \rightarrow Mn^{2+}(aq) + 4\,H_2O(l)$$
$$5\,Fe^{2+}(aq) + 8\,H^+(aq) + MnO_4^-(aq) \rightarrow 5\,Fe^{3+}(aq) + Mn^{2+}(aq) + 4\,H_2O(l)$$

7. Verify that the reaction is balanced with respect to both mass and charge.

Reactants	Products
2 Al	2 Al
3 Cu	3 Cu
+6 charge	+6 charge

Reactants	Products
5 Fe	5 Fe
8 H	8 H
1 Mn	1 Mn
4 O	4 O
+17 charge	+17 charge

▶**SKILLBUILDER 8.22** | Balance the redox reaction occurring in acidic solution.

$$H^+(aq) + Cr(s) \rightarrow H_2(g) + Cr^{3+}(aq)$$

▶**SKILLBUILDER 8.23** | Balance the redox reaction occurring in acidic solution.

$$Cu(s) + NO_3^-(aq) \rightarrow Cu^{2+}(aq) + NO_2(g)$$

▶**FOR MORE PRACTICE** Example 8.40; Problems 156, 157, 158, 159.

EXAMPLE 8.24 BALANCING REDOX REACTIONS

Balance the following redox reaction occurring in acidic solution.

$$I^-(aq) + Cr_2O_7^{2-}(aq) \rightarrow Cr^{3+}(aq) + I_2(s)$$

1. Follow the half-reaction method for balancing redox reactions. Begin by assigning oxidation states.	**SOLUTION** $I^-(aq) + Cr_2O_7^{2-}(aq) \longrightarrow Cr^{3+}(aq) + I_2(s)$ with oxidation states −1; +6 −2; +3; 0. Reduction (Cr: +6 → +3); Oxidation (I: −1 → 0)
2. Separate the overall reaction into two half-reactions.	**OXIDATION** $I^-(aq) \rightarrow I_2(s)$ **REDUCTION** $Cr_2O_7^{2-}(aq) \rightarrow Cr^{3+}(aq)$
3. Balance each half-reaction with respect to mass.	
• Balance all elements other than H and O.	$2\,I^-(aq) \rightarrow I_2(aq)$ $Cr_2O_7^{2-}(aq) \rightarrow 2\,Cr^{3+}(s)$
• Balance O by adding H_2O.	$2\,I^-(aq) \rightarrow I_2(s)$ $Cr_2O_7^{2-}(aq) \rightarrow 2\,Cr^{3+}(aq) + 7\,H_2O(l)$
• Balance H by adding H^+.	$2\,I^-(aq) \rightarrow I_2(s)$ $14\,H^+(aq) + Cr_2O_7^{2-}(aq) \rightarrow 2\,Cr^{3+}(aq) + 7\,H_2O(l)$
4. Balance each half-reaction with respect to charge.	$2\,I^-(aq) \rightarrow I_2(s) + 2e^-$ $6e^- + 14\,H^+(aq) + Cr_2O_7^{2-}(aq) \rightarrow 2\,Cr^{3+}(aq) + 7\,H_2O(l)$
5. Make the number of electrons in both half-reactions equal.	$3 \times [2\,I^-(aq) \rightarrow I_2(s) + 2e^-]$ $6e^- + 14\,H^+(aq) + Cr_2O_7^{2-}(aq) \rightarrow 2\,Cr^{3+}(aq) + 7\,H_2O(l)$
6. Add the half-reactions together.	$6\,I^-(aq) \rightarrow 3\,I_2(s) + \cancel{6e^-}$ $\cancel{6e^-} + 14\,H^+(aq) + Cr_2O_7^{2-}(aq) \rightarrow 2\,Cr^{3+}(aq) + 7\,H_2O(l)$ $6\,I^-(aq) + 14\,H^+(aq) + Cr_2O_7^{2-}(aq) \rightarrow 3\,I_2(s) + 2\,Cr^{3+}(aq) + 7\,H_2O(l)$
7. Verify that the reaction is balanced.	

Reactants	Products
6 I	6 I
14 H	14 H
2 Cr	2 Cr
7 O	7 O
+6 charge	+6 charge

▶SKILLBUILDER 8.24 | Balancing Redox Reactions

Balance the redox reaction occurring in acidic solution.

$$Sn(s) + MnO_4^-(aq) \rightarrow Sn^{2+}(aq) + Mn^{2+}(aq)$$

▶FOR MORE PRACTICE Problems 160, 161, 162, 163.

CHEMISTRY **IN THE ENVIRONMENT**

▶ Photosynthesis and Respiration: Energy for Life

All living things require energy, and most of that energy comes from the sun. Solar energy reaches Earth in the form of electromagnetic radiation. This radiation keeps our planet at a temperature that allows life as we know it to flourish. But the wavelengths that make up visible light have an additional and very crucial role to play in the maintenance of life. Plants capture this light and use it to make energy-rich organic molecules such as carbohydrates. Animals get their energy by eating plants or by eating other animals that have eaten plants. So ultimately, virtually all of the energy for life comes from sunlight.

But in chemical terms, how is this energy captured, transferred from organism to organism, and used? *The key reactions in these processes all involve oxidation and reduction.*

Most living things use chemical energy through a process known as *respiration*. In respiration, energy-rich molecules, typified by the sugar glucose, are "burned" in a reaction that we summarize as follows:

$$\underset{\text{Glucose}}{C_6H_{12}O_6} + \underset{\text{Oxygen}}{6\,O_2} \rightarrow \underset{\text{Water}}{6\,H_2O} + \underset{\text{Carbon dioxide}}{6\,CO_2} + \text{energy}$$

We can readily see that respiration is a redox reaction. On the simplest level, it is clear that some of the atoms in glucose are gaining oxygen. More precisely, we can use the rules for assigning oxidation states to show that carbon is oxidized from an oxidation number of 0 in glucose to +4 in carbon dioxide.

Respiration is also an exothermic reaction—it releases energy. If we burn glucose in a test tube, the energy is lost as heat. Living things, however, have devised ways to capture the energy released and use it to power their life processes, such as movement, growth, and the synthesis of other life-sustaining molecules.

Respiration is one half of a larger cycle; the other half is *photosynthesis*. Photosynthesis is the series of reactions by which green plants capture the energy of sunlight and store it as chemical energy in compounds such as glucose. We summarize photosynthesis as follows:

$$\underset{\text{Carbon dioxide}}{6\,CO_2} + \underset{\text{Water}}{6\,H_2O} + \text{energy (sunlight)} \rightarrow \underset{\text{Glucose}}{C_6H_{12}O_6} + \underset{\text{Oxygen}}{6\,O_2}$$

This reaction—the exact reverse of respiration—is the ultimate source of the molecules that are oxidized in respiration. And just as the key process in respiration is the oxidation of carbon, the key process in photosynthesis is the reduction of carbon. This reduction is driven by solar energy—energy that is then stored in the resulting glucose molecule. Living things harvest that energy when they "burn" glucose in the respiration half of the cycle.

Thus, oxidation and reduction reactions are at the very center of all life on Earth.

B8.4 CAN YOU ANSWER THIS? *What is the oxidation state of the oxygen atoms in CO_2, H_2O, and O_2? What does this information tell you about the reactions of photosynthesis and respiration?*

▲ Sunlight, captured by plants in photosynthesis, is the ultimate source of the chemical energy for nearly all living things on Earth. © Yellowj/Shutterstock.

MODULE IN REVIEW

Self-Assessment Quiz

Q1. Which process is a chemical reaction?

(a) Gasoline evaporating from a gasoline tank
(b) Iron rusting when left outdoors
(c) Dew condensing on grass during the night
(d) Water boiling on a stove top

Q2. A 35 g sample of potassium completely reacts with chlorine to form 67 g of potassium chloride. How many grams of chlorine must have reacted?

(a) 67 g
(b) 35 g
(c) 32 g
(d) 12 g

Q3. How many oxygen atoms are on the reactant side of this chemical equation?

$$K_2CO_3(aq) + Pb(NO_3)_2(aq) \rightarrow 2\ KNO_3(aq) + PbCO_3(s)$$

(a) 3
(b) 6
(c) 9
(d) 12

Q4. What is the coefficient for hydrogen in the balanced equation for the reaction of solid iron(III) oxide with gaseous hydrogen to form solid iron and liquid water?

(a) 2
(b) 3
(c) 4
(d) 6

Q5. Determine the correct set of coefficients to balance the chemical equation.

$$\_C_6H_6(l) + \_O_2(g) \rightarrow \_CO_2(g) + \_H_2O(g)$$

(a) 2, 15, 12, 6
(b) 1, 15, 6, 3
(c) 2, 15, 12, 12
(d) 1, 7, 6, 3

Q6. Which compound is soluble in water?

(a) $Fe(OH)_2$
(b) CuS
(c) AgCl
(d) $CuCl_2$

Q7. Name the precipitate that forms (if any) when aqueous solutions of barium nitrate and potassium sulfate are mixed.

(a) BaK(*s*)
(b) $NO_3SO_4(s)$
(c) $BaSO_4(s)$
(d) $KNO_3(s)$

Q8. Which set of reactants forms a solid precipitate when mixed?

(a) $NaNO_3(aq)$ and KCl(*aq*)
(b) KOH(*aq*) and $Na_2CO_3(aq)$
(c) $CuCl_2(aq)$ and $NaC_2H_3O_2(aq)$
(d) $Na_2CO_3(aq)$ and $CaCl_2(aq)$

Q9. What is the net ionic equation for the reaction between $Pb(C_2H_3O_2)_2(aq)$ and KBr(*aq*)?

(a) $Pb(C_2H_3O_2)_2(aq) + 2\ KBr(aq) \rightarrow PbBr_2(s) + 2\ KC_2H_3O_2(aq)$
(b) $Pb^{2+}(aq) + 2\ Br^-(aq) \rightarrow PbBr_2(s)$
(c) $K^+(aq) + C_2H_3O_2^-(aq) \rightarrow KC_2H_3O_2(s)$
(d) $Pb(C_2H_3O_2)_2(aq) + 2\ KBr(aq) \rightarrow 2\ KC_2H_3O_2(s) + PbBr_2\ (aq)$

Q10. Which substance is most likely to have a bitter taste?

(a) HCl(*aq*)
(b) NaCl(*aq*)
(c) NaOH(*aq*)
(d) None of the above

Q11. Identify the Brønsted–Lowry base in the reaction.

$$HClO_2(aq) + H_2O(aq) \rightleftharpoons H_3O^+(aq) + ClO_2^-(aq)$$

(a) $HClO_2$
(b) H_2O
(c) H_3O^+
(d) ClO_2^-

Q12. What is the conjugate base of the acid $HClO_4$?

(a) HCl
(b) HO_3^+
(c) ClO_3^-
(d) ClO_4^-

Q13. What are the products of the reaction between HBr(*aq*) and KOH(*aq*)?

(a) HK(*aq*) and BrOH(*aq*)
(b) $H_2(g)$ and KBr(*aq*)
(c) $H_2O(l)$ and KBr(*aq*)
(d) $H_2O(l)$ and BrOH(*aq*)

Q14. What are the products of the reaction between HCl and Sn?

(a) $SnCl_2(aq)$ and $H_2(g)$
(b) SnH(*aq*) and $Cl_2(g)$
(c) $H_2O(l)$ and SnCl(*aq*)
(d) $H_2(g)$, $Cl_2(g)$ and Sn(*s*)

Q15. In which solution is $[H_3O^+]$ less than 0.100 mol/L?

(a) 0.100 mol/L $H_2SO_4(aq)$

(b) 0.100 mol/L $HCHO_2(aq)$

(c) 0.100 mol/L $HClO_4(aq)$

(d) None of the above (all have $[H_3O^+]$ less than 0.100 mol/L.)

Q16. In which solution is $[OH^-]$ equal to 0.100 mol/L?

(a) 0.100 mol/L $NH_3(aq)$

(b) 0.100 mol/L $Ba(OH)_2(aq)$

(c) 0.100 mol/L $NaOH(aq)$

(d) All of the above (all have $[OH^-]$ equal to 0.100 mol/L).

Q17. Which substance is being oxidized in the reaction?

$$2\ Na(s) + Br_2(g) \rightarrow 2\ NaBr(s)$$

(a) Na

(b) Br_2

(c) NaBr

(d) None of the above

Q18. What always happens to an oxidizing agent during a redox reaction?

(a) It is oxidized.

(b) It is reduced.

(c) It becomes a liquid.

(d) It becomes a gas.

Q19. What is the oxidation state of carbon in CO_3^{2-}?

(a) –3

(b) –2

(c) +3

(d) +4

Q20. In which compound does phosphorus have the lowest oxidation state?

(a) P_2S_5

(b) P_2O_3

(c) K_3PO_4

(d) None of the above (the oxidation state of phosphorus is the same in all three compounds)

Q21. Sodium reacts with water according to the reaction:

$$2\ Na(s) + 2\ H_2O(l) \rightarrow 2\ NaOH(aq) + H_2(g)$$

Identify the element that is reduced.

(a) Na

(b) O

(c) H

(d) None of the above

Q22. How many electrons are exchanged when this reaction is balanced?

$$Mn(s) + Cr^{3+}(aq) \rightarrow Mn^{2+}(aq) + Cr(s)$$

(a) 2

(b) 3

(c) 4

(d) 6

Q23. Balance the redox reaction equation (occurring in acidic solution) and choose the correct coefficients for each reactant and product.

$$\_VO_2^+(aq) + \_Sn(s) + \_H^+(aq) \rightarrow \_VO^{2+}(aq) + \_Sn^{2+}(aq) + \_H_2O(l)$$

(a) 1,1,2 → 1,1,1

(b) 2,1,2 → 2,1,1

(c) 2,1,4 → 2,1,2

(d) 2,1,2 → 2,1,2

Answers: 1:b, 2:c, 3:c, 4:b, 5:a, 6:d, 7:c, 8:d, 9:b, 10:c, 11:b, 12:d, 13:c, 14:a, 15:b, 16:c, 17:a, 18:b, 19:d, 20:b, 21:c, 22:d, 23:c

Chemical Principles

Chemical Reactions: In a chemical reaction, one or more substances—either elements or compounds—change into a different substance.

Evidence of a Chemical Reaction: The only absolute evidence for a chemical reaction is chemical analysis showing that one or more substances have changed into another substance. However, one or more of the following often accompany a chemical reaction: a color change; the formation of a solid or precipitate; the formation of a gas; the emission of light; and the emission or absorption of heat.

Conservation of Mass: Whether the changes in matter are chemical or physical, matter is always conserved. In a chemical change, the masses of the matter undergoing the chemical change must equal the sum of the masses of matter resulting from the chemical change.

Relevance

Chemical Reactions: Chemical reactions are central to many processes, including transportation, energy generation, manufacture of household products, vision, and life itself.

Evidence of a Chemical Reaction: We can often perceive the changes that accompany chemical reactions. In fact, we often employ chemical reactions for the changes they produce. For example, we use the heat emitted by the combustion of fossil fuels to heat our homes, drive our cars, and generate electricity.

Conservation of Mass: The conservation of matter is relevant to, for example, pollution. We often think that humans create pollution, but, actually, we are powerless to create anything. Matter cannot be created. So, pollution is simply misplaced matter—matter that we have put into places where it does not belong.

Chemical Equations: Chemical equations represent chemical reactions. They include formulas for the reactants (the substances present before the reaction) and for the products (the new substances formed by the reaction). Chemical equations must be balanced to reflect the conservation of matter in nature; atoms do not spontaneously appear or disappear.

Chemical Equations: Chemical equations allow us to represent and understand chemical reactions. For example, the equations for the combustion reactions of fossil fuels let us see that carbon dioxide, a gas that contributes to global warming, is one of the products of these reactions.

Aqueous Solutions and Solubility: Aqueous solutions are mixtures of a substance dissolved in water. If a substance dissolves in water it is soluble; when it remains mostly as solid, it is sparingly soluble.

Aqueous Solutions and Solubility: Aqueous solutions are common. Oceans, lakes, and most of the fluids in our bodies are aqueous solutions.

Classifying Chemical Reactions: We can classify many chemical reactions into one of the following four categories according to what atoms or groups of atoms do:

- synthesis: $(A + B \rightarrow AB)$
- decomposition: $(AB \rightarrow A + B)$
- single-displacement: $(A + BC \rightarrow AC + B)$
- double-displacement: $(AB + CD \rightarrow AD + CB)$

Classifying Chemical Reactions: We classify chemical reactions to better understand them and to recognize similarities and differences among reactions.

Molecular Definitions of Acids and Bases:
Arrhenius definition

Acid—substance that produces H^+ ions in solution

Base—substance that produces OH^- ions in solution

Brønsted–Lowry definition

Acid—proton donor

Base—proton acceptor

Molecular Definitions of Acids and Bases: The Arrhenius definition is simpler and easier to use. It also shows how an acid and a base neutralize each other to form water ($H^+ + OH^- \rightarrow H_2O$). The more generally applicable Brønsted–Lowry definition helps us see that, in water, H^+ ions usually associate with water molecules to form H_3O^+. It also shows how bases that do not contain OH^- ions can still act as bases by accepting a proton from water.

Reactions of Acids and Bases:
Neutralization Reactions An acid and a base react to form water and a salt.

$$\underset{\text{Acid}}{HCl(aq)} + \underset{\text{Base}}{KOH(aq)} \rightarrow \underset{\text{Water}}{H_2O(l)} + \underset{\text{Salt}}{KCl(aq)}$$

Acid–Metal Reactions Acids react with many metals to form hydrogen gas and a salt.

$$\underset{\text{Acid}}{2\,HCl(aq)} + \underset{\text{Metal}}{Mg(s)} \rightarrow \underset{\text{Hydrogen gas}}{H_2(g)} + \underset{\text{Salt}}{MgCl_2(aq)}$$

Acid–Metal Oxide Reactions Acids react with many metal oxides to form water and a salt.

$$\underset{\text{Acid}}{2\,HCl(aq)} + \underset{\text{Metal oxide}}{K_2O(s)} \rightarrow \underset{\text{Water}}{H_2O(l)} + \underset{\text{Salt}}{2\,KCl(aq)}$$

Reactions of Acids and Bases: Neutralization reactions are common in our everyday lives. Antacids, for example, are bases that react with acids from the stomach to alleviate heartburn and sour stomach.

Acid–metal and acid–metal oxide reactions show the corrosive nature of acids. In both of these reactions, the acid dissolves the metal or the metal oxide. Some of the effects of these kinds of reactions can be seen in the damage to building materials caused by acid rain. Acids dissolve metals and metal oxides, so building materials composed of these substances are susceptible to acid rain.

Strong and Weak Acids and Bases: Strong acids completely ionize, and strong bases completely dissociate in aqueous solutions. For example:

$$HCl(aq) + H_2O(l) \rightarrow H_3O^+(aq) + Cl^-(aq)$$
$$NaOH(aq) \rightarrow Na^+(aq) + OH^-(aq)$$

A 1 mol/L HCl solution has $[H_3O^+] = 1$ mol/L and a 1 mol/L NaOH solution has $[OH^-] = 1$ mol/L.

Weak acids only partially ionize in solution. Most weak bases partially ionize water in solution. For example:

$$HF(aq) + H_2O(l) \rightleftharpoons H_3O^+(aq) + F^-(aq)$$
$$NH_3(aq) + H_2O(l) \rightleftharpoons NH_4^+(aq) + OH^-(aq)$$

A 1 mol/L HF solution has $[H_3O^+] < 1$ mol/L, and a 1 mol/L NH_3 solution has $[OH^-] < 1$ mol/L.

Strong and Weak Acids and Bases: Whether an acid is strong or weak depends on the conjugate base: The stronger the conjugate base, the weaker the acid. Since the acidity or basicity of a solution depends on $[H_3O^+]$ and $[OH^-]$, we must know whether an acid is strong or weak to know the degree of acidity or basicity.

Oxidation and Reduction:
Oxidation is:
- the loss of electrons.
- an increase in oxidation state.

Reduction is:
- the gain of electrons.
- a decrease in oxidation state.

Oxidation and reduction reactions always occur together and are sometimes called redox reactions. The substance that is oxidized is the reducing agent, and the substance that is reduced is the oxidizing agent.

Oxidation and Reduction: Redox reactions are common in nature, in industry, and in many everyday processes. Batteries use redox reactions to generate electrical current. Our bodies use redox reactions to obtain energy from glucose. In addition, the bleaching of hair, the rusting of iron, and the electroplating of metals all involve redox reactions.

Good oxidizing agents, such as oxygen, hydrogen peroxide, and chlorine, have a strong tendency to gain electrons. Good reducing agents, such as sodium and hydrogen, have a strong tendency to lose electrons.

Oxidation States: The oxidation state is a fictitious charge assigned to each atom in a compound. It is calculated by assigning all bonding electrons in a compound to the most electronegative element.

Oxidation States: Oxidation states help us more easily identify substances being oxidized and reduced in a chemical reaction.

Chemical Skills

LO: Identify a chemical reaction (Section 8.2).

To identify a chemical reaction, determine whether one or more of the initial substances changed into a different substance. If so, a chemical reaction occurred. One or more of the following often accompanies a chemical reaction: a color change; the formation of a solid or precipitate; the formation of a gas; the emission of light; and the emission or absorption of heat.

LO: Apply the law of conservation of mass (Section 8.4).

The sum of the masses of the substances involved in a chemical change must be the same before and after the change.

Examples

EXAMPLE 8.25 IDENTIFYING A CHEMICAL REACTION

Which of these are chemical reactions?

(a) Copper turns green on exposure to air.
(b) When sodium hydrogencarbonate (also known as sodium bicarbonate) is combined with hydrochloric acid, bubbling is observed.
(c) Liquid water freezes to form solid ice.
(d) A pure copper penny forms bubbles of a dark brown gas when dropped into nitric acid. The nitric acid solution turns blue.

SOLUTION

(a) Chemical reaction, as evidenced by the color change.
(b) Chemical reaction, as evidenced by the evolution of a gas.
(c) Not a chemical reaction; solid ice is still water.
(d) Chemical reaction, as evidenced by the evolution of a gas and by a color change.

EXAMPLE 8.26 APPLYING THE LAW OF CONSERVATION OF MASS

An automobile runs for 10 minutes and burns 47 g of gasoline. The gasoline combines with oxygen from air and forms 132 g of carbon dioxide and 34 g of water. How much oxygen is consumed in the process?

SOLUTION

The total mass after the chemical change is:

$$132 \text{ g} + 34 \text{ g} = 166 \text{ g}$$

The total mass before the change must also be 166 g.

$$47 \text{ g} + \text{oxygen} = 166 \text{ g}$$

So, the mass of oxygen consumed is the total mass (166 g) minus the mass of gasoline (47 g).

$$\text{grams of oxygen} = 166 \text{ g} - 47 \text{ g} = 119 \text{ g}$$

LO: Write balanced chemical equations (Section 8.5).

To write balanced chemical equations, follow these steps.

1. Write a skeletal equation by writing chemical formulas for each of the reactants and products. (If a skeletal equation is provided, proceed to Step 2.)
2. If an element occurs in only one compound on both sides of the equation, balance that element first. If there is more than one such element, and the equation contains both metals and nonmetals, balance metals before nonmetals.
3. If an element occurs as a free element on either side of the chemical equation, balance that element last.
4. If the balanced equation contains coefficient fractions, clear these by multiplying the entire equation by the appropriate factor.
5. Check to make certain the equation is balanced by summing the total number of atoms of every element on each side of the equation.

Reminders

- Change only the *coefficients* to balance a chemical equation, *never the subscripts*. Changing the subscripts would change the compound itself.
- If the equation contains polyatomic ions that stay intact on both sides of the equation, balance the polyatomic ions as a group.

EXAMPLE 8.27 WRITING BALANCED CHEMICAL EQUATIONS

Write a balanced chemical equation for the reaction of solid vanadium(V) oxide with hydrogen gas to form solid vanadium(III) oxide and liquid water.

$$V_2O_5(s) + H_2(g) \rightarrow V_2O_3(s) + H_2O(l)$$

SOLUTION

A skeletal equation is given. Proceed to Step 2.

Vanadium occurs in only one compound on both sides of the equation. However, it is balanced, so you can proceed and balance oxygen by placing a 2 in front of H_2O on the right side.

$$V_2O_5(s) + H_2(g) \rightarrow V_2O_3(s) + 2\ H_2O(l)$$

Hydrogen occurs as a free element; balance it last by placing a 2 in front of H_2 on the left side.

$$V_2O_5(s) + 2\ H_2(g) \rightarrow V_2O_3(s) + 2\ H_2O(l)$$

Equation does not contain coefficient fractions. Proceed to Step 5.

Check the equation.

$$V_2O_5(s) + 2\ H_2(g) \rightarrow V_2O_3(s) + 2\ H_2O(l)$$

Reactants		Products
2 V atoms	→	2 V atoms
5 O atoms	→	5 O atoms
4 H atoms	→	4 H atoms

LO: Classify chemical reactions (Section 8.6).

You can can classify chemical reactions by inspection. The four major categories are:

Synthesis or combination

$$A + B \rightarrow AB$$

Decomposition

$$AB \rightarrow A + B$$

Single-displacement

$$A + BC \rightarrow AC + B$$

Double-displacement

$$AB + CD \rightarrow AD + CB$$

EXAMPLE 8.28 CLASSIFYING CHEMICAL REACTIONS

Classify each chemical reaction as a synthesis, decomposition, single-displacement, or double-displacement reaction.

(a) $2\ K(s) + Br_2(g) \rightarrow 2\ KBr(s)$
(b) $Fe(s) + 2\ AgNO_3(aq) \rightarrow Fe(NO_3)_2(aq) + 2\ Ag(s)$
(c) $CaSO_3(s) \rightarrow CaO(s) + SO_2(g)$
(d) $CaCl_2(aq) + Li_2SO_4(aq) \rightarrow CaSO_4(s) + 2\ LiCl(aq)$

SOLUTION

(a) Synthesis; KBr, a more complex substance, is formed from simpler substances.
(b) Single-displacement; Fe displaces Ag in $AgNO_3$.
(c) Decomposition; $CaSO_3$ decomposes into simpler substances.
(d) Double-displacement; Ca and Li switch places to form new compounds.

LO: Determine whether a compound is soluble (Section 8.7).

To determine whether a compound is soluble or sparingly soluble, refer to the solubility rules in Table 8.2. It is simplest to begin by looking for those ions that always form soluble compounds (Li^+, Na^+, K^+, NH_4^+, NO_3^-, and $C_2H_3O_2^-$). If a compound contains one of those, it is soluble. If it does not, determine if the anion is mostly soluble (Cl^-, Br^-, I^-, or SO_4^{2-}) or mostly sparingly soluble (OH^-, S^{2-}, CO_3^{2-}, or PO_4^{3-}). Look at the cation as well to determine whether it is one of the exceptions.

EXAMPLE 8.29 DETERMINING WHETHER A COMPOUND IS SOLUBLE

Is each compound soluble or sparingly soluble?

(a) $CuCO_3$
(b) $BaSO_4$
(c) $Fe(NO_3)_3$

SOLUTION

(a) Sparingly soluble; compounds containing CO_3^{2-} are sparingly soluble, and Cu^{2+} is not an exception.
(b) Sparingly soluble; compounds containing SO_4^{2-} are usually soluble, but Ba^{2+} is an exception.
(c) Soluble; all compounds containing NO_3^- are soluble.

LO: Predict and write equations for precipitation reactions (Section 8.7).

To predict whether a precipitation reaction occurs when two solutions are mixed and to write an equation for the reaction, follow these steps.

1. Write the formulas of the two compounds being mixed as reactants in a chemical equation.
2. Below the equation, write the formulas of the products that could form from the reactants. These are obtained by combining the cation from one reactant with the anion from the other. Make sure to adjust the subscripts so that all formulas are charge-neutral.
3. Use the solubility rules to determine whether any of the products are indeed sparingly soluble.
4. If all of the products are soluble, there will be no precipitate. Write *NO REACTION* next to the arrow.
5. If one or both of the products are sparingly soluble, write their formula(s) as the product(s) of the reaction using (*s*) to indicate *solid*. Write any soluble products with (*aq*) to indicate *aqueous*.
6. Balance the equation.

EXAMPLE 8.30 PREDICTING PRECIPITATION REACTIONS

Write an equation for the precipitation reaction that occurs, if any, when solutions of sodium phosphate and cobalt(II) chloride are mixed.

SOLUTION

$$Na_3PO_4(aq) + CoCl_2(aq) \rightarrow$$

Products:

$$NaCl \qquad Co_3(PO_4)_2$$

NaCl is soluble.
$Co_3(PO_4)_2$ is sparingly soluble.

Reaction contains a sparingly soluble product; proceed to Step 5.

$$Na_3PO_4(aq) + CoCl_2(aq) \rightarrow Co_3(PO_4)_2(s) + NaCl(aq)$$

$$2\ Na_3PO_4(aq) + 3\ CoCl_2(aq) \rightarrow Co_3(PO_4)_2(s) + 6\ NaCl(aq)$$

LO: Write molecular, complete ionic, and net ionic equations (Section 8.7).

To write a molecular equation, include the complete, neutral formulas for every compound in the reaction.

To write a complete ionic equation from a molecular equation, separate all aqueous ionic compounds into independent ions. Do not separate solid, liquid, or gaseous compounds.

To write a net ionic equation from a complete ionic equation, eliminate all species that do not change (spectator ions) in the course of the reaction.

EXAMPLE 8.31 WRITING COMPLETE IONIC AND NET IONIC EQUATIONS

Write a complete ionic and a net ionic equation for the reaction.

$$2\ NH_4Cl(aq) + Hg_2(NO_3)_2(aq) \rightarrow Hg_2Cl_2(s) + 2\ NH_4NO_3(aq)$$

SOLUTION

Complete ionic equation:

$$2\ NH_4^+(aq) + 2\ Cl^-(aq) + Hg_2^{2+}(aq) + 2\ NO_3^-(aq) \rightarrow Hg_2Cl_2(s) + 2\ NH_4^+(aq) + 2\ NO_3^-(aq)$$

Net ionic equation:

$$2\ Cl^-(aq) + Hg_2^{2+}(aq) \rightarrow Hg_2Cl_2(s)$$

LO: Identify Brønsted–Lowry acids and bases and their conjugates (Section 8.8).

The substance that donates the proton is the acid (proton donor) and becomes the conjugate base (as a product). The substance that accepts the proton (proton acceptor) is the base and becomes the conjugate acid (as a product).

EXAMPLE 8.32 IDENTIFYING BRØNSTED–LOWRY ACIDS AND BASES AND THEIR CONJUGATES

Identify the Brønsted–Lowry acid, the Brønsted–Lowry base, the conjugate acid, and the conjugate base in this reaction:

$$HNO_3(aq) + H_2O(l) \rightarrow H_3O^+(aq) + NO_3^-(aq)$$

SOLUTION

$$\underset{\text{Acid}}{HNO_3(aq)} + \underset{\text{Base}}{H_2O(l)} \rightarrow \underset{\text{Conjugate acid}}{H_3O^+(aq)} + \underset{\text{Conjugate base}}{NO_3^-(aq)}$$

LO: Write equations for neutralization reactions (Section 8.8).

In a neutralization reaction, an acid and a base usually react to form water and a salt (ionic compound).

$$\text{Acid} + \text{Base} \rightarrow \text{Water} + \text{Salt}$$

Write the skeletal equation first, making sure to write the formula of the salt so that it is charge-neutral. Then balance the equation.

EXAMPLE 8.33 WRITING EQUATIONS FOR NEUTRALIZATION REACTIONS

Write a molecular equation for the reaction between aqueous HBr and aqueous $Ca(OH)_2$.

SOLUTION

Skeletal equation:

$$HBr(aq) + Ca(OH)_2(aq) \rightarrow H_2O(l) + CaBr_2(aq)$$

Balanced equation:

$$2\ HBr(aq) + Ca(OH)_2(aq) \rightarrow 2\ H_2O(l) + CaBr_2(aq)$$

LO: Write equations for the reactions of acids with metals and with metal oxides (Section 8.8).

Acids react with many metals to form hydrogen gas and a salt.

$$\text{Acid} + \text{Metal} \rightarrow \text{Hydrogen gas} + \text{Salt}$$

Write the skeletal equation first, making sure to write the formula of the salt so that it is charge-neutral. Then balance the equation.

Acids react with many metal oxides to form water and a salt.

$$\text{Acid} + \text{Metal oxide} \rightarrow \text{Water} + \text{Salt}$$

Write the skeletal equation first, making sure to write the formula of the salt so that it is charge-neutral. Then balance the equation.

EXAMPLE 8.34 WRITING EQUATIONS FOR THE REACTIONS OF ACIDS WITH METALS AND WITH METAL OXIDES

Write equations for the reaction of hydrobromic acid with calcium metal and for the reaction of hydrobromic acid with calcium oxide.

SOLUTION

Skeletal equation:

$$HBr(aq) + Ca(s) \rightarrow H_2(g) + CaBr_2(aq)$$

Balanced equation:

$$2\ HBr(aq) + Ca(s) \rightarrow H_2(g) + CaBr_2(aq)$$

Skeletal equation:

$$HBr(aq) + CaO(s) \rightarrow H_2O(l) + CaBr_2(aq)$$

Balanced equation:

$$2\ HBr(aq) + CaO(s) \rightarrow H_2O(l) + CaBr_2(aq)$$

LO: Determine $[H_3O^+]$ in acid solutions (Section 8.8).

In a strong acid, $[H_3O^+]$ is equal to the molar concentration of the acid times the number of hydrogen ions in the acid. In a weak acid, $[H_3O^+]$ is less than the molar concentration of the acid times the number of hydrogen ions in the acid.

EXAMPLE 8.35 DETERMINING $[H_3O^+]$ IN ACID SOLUTIONS

What is the $[H_3O^+]$ concentration in a 0.25 mol/L HCl solution and in a 0.25 mol/L HF solution?

SOLUTION

In the 0.25 mol/L HCl solution (strong acid), $[H_3O^+] = 0.25$ mol/L. In the 0.25 mol/L HF solution (weak acid), $[H_3O^+] < 0.25$ mol/L.

LO: Determine $[OH^-]$ in base solutions (Section 8.8).

In a strong base, $[OH^-]$ is equal to the molar concentration of the base times the number of hydroxide ions in the base. In a weak base, $[OH^-]$ is less than the molar concentration of the base times the number of hydroxide ions in the base.

EXAMPLE 8.36 DETERMINING $[OH^-]$ IN BASE SOLUTIONS

What is the OH^- molar concentration in a 0.25 mol/L NaOH solution, in a 0.25 mol/L $Sr(OH)_2$ solution, and in a 0.25 mol/L NH_3 solution?

SOLUTION

In the 0.25 mol/L NaOH solution (strong base), $[OH^-] = 0.25$ mol/L. In the 0.25 mol/L $Sr(OH)_2$ solution (strong base), $[OH^-] = 0.50$ mol/L. In the 0.25 mol/L NH_3 solution (weak base), $[OH^-] < 0.25$ mol/L.

LO: Define and identify oxidation and reduction (Section 8.9).

Oxidation can be identified as the gain of oxygen, the loss of electrons, or an increase in oxidation state. Reduction can be identified as the loss of oxygen, the gain of electrons, or a decrease in oxidation state.

When a substance gains oxygen, the substance is oxidized and the oxygen is reduced.

When a metal reacts with a nonmetal, the metal is oxidized and the nonmetal is reduced.

When a metal transfers electrons to a metal ion, the metal is oxidized and the metal ion is reduced.

EXAMPLE 8.37 IDENTIFYING OXIDATION AND REDUCTION

Determine the substance being oxidized and the substance being reduced in each redox reaction.

(a) $Sn(s) + O_2(g) \rightarrow SnO_2(s)$
(b) $2\ Na(s) + F_2(g) \rightarrow 2\ NaF(s)$
(c) $Mg(s) + Cu^{2+}(aq) \rightarrow Mg^{2+}(aq) + Cu(s)$

SOLUTION

(a) Sn oxidized; O_2 reduced
(b) Na oxidized; F_2 reduced
(c) Mg oxidized; Cu^{2+} reduced

LO: Identify oxidizing agents and reducing agents (Section 8.9).

The reducing agent is the substance that is oxidized. The oxidizing agent is the substance that is reduced.

EXAMPLE 8.38 IDENTIFYING OXIDIZING AGENTS AND REDUCING AGENTS

Identify the oxidizing and reducing agents in each redox reaction.

(a) $Sn(s) + O_2(g) \rightarrow SnO_2(s)$
(b) $2\ Na(s) + F_2(g) \rightarrow 2\ NaF(s)$
(c) $Mg(s) + Cu^{2+}(aq) \rightarrow Mg^{2+}(aq) + Cu(s)$

SOLUTION

(a) Sn is the reducing agent; O_2 is the oxidizing agent.
(b) Na is the reducing agent; F_2 is the oxidizing agent.
(c) Mg is the reducing agent; Cu^{2+} is the oxidizing agent.

LO: Assign oxidation states (Section 8.9).

Rules for Assigning Oxidation States

1. The oxidation state of an atom in a free element is 0.
2. The oxidation state of a monoatomic ion is equal to its charge.
3. The sum of the oxidation states of all atoms in:
   (i) a neutral molecule or formula unit is 0.
   (ii) an ion is equal to the charge of the ion.
4. In their compounds,
   (i) Group I metals have an oxidation state of +1.
   (ii) Group II metals have an oxidation state of +2.
   (iii) Hydrogen has an oxidation state of +1 (exceptions include −1 in metal hydride).
   (iv) Oxygen has an oxidation state of −2 (exceptions include −1 in peroxide and positive oxidation states in compounds containing F).
5. In compounds, the element with the higher electronegativity is assigned an oxidation state equal to its charge as an anion.

EXAMPLE 8.39 ASSIGNING OXIDATION STATES

Assign an oxidation state to each atom in each species.

(a) Al
(b) Al^{3+}
(c) N_2O
(d) CO_3^{2-}

SOLUTION

(a) $\underset{0}{Al}$ (Rule 1)

(b) $\underset{+3}{Al^{3+}}$ (Rule 2)

(c) $\underset{+1\ -2}{N_2O}$ For O, (Rule 4iv or Rule 5, O has higher electronegativity than N; then Rule 3i for N)

(d) $\underset{+4\ -2}{CO_3^{2-}}$ (Rule 4iv or Rule 5 for O; then Rule 3ii for C)

LO: Balance redox reactions (Section 8.9).

To balance redox reactions in aqueous acidic solutions, follow this procedure (brief version).

EXAMPLE 8.40 BALANCING REDOX REACTIONS

Balance the reaction occurring in acidic solution.

$$IO_3^-(aq) + Fe^{2+}(aq) \rightarrow I_2(s) + Fe^{3+}(aq)$$

SOLUTION

1. Assign oxidation states.

$$\underset{+5\ -2}{IO_3^-}(aq) + \underset{+2}{Fe^{2+}}(aq) \longrightarrow \underset{0}{I_2}(s) + \underset{+3}{Fe^{3+}}(aq)$$

Reduction (I: +5 → 0); Oxidation (Fe: +2 → +3)

2. Separate the overall reaction into two half-reactions.

OXIDATION $Fe^{2+}(aq) \rightarrow Fe^{3+}(aq)$

REDUCTION $IO_3^-(aq) \rightarrow I_2(s)$

3. Balance each half-reaction with respect to mass.

- Balance all elements other than H and O.

$Fe^{2+}(aq) \rightarrow Fe^{3+}(aq)$

$\mathbf{2}\ IO_3^-(aq) \rightarrow I_2(s)$

- Balance O by adding H_2O.

$Fe^{2+}(aq) \rightarrow Fe^{3+}(aq)$

$2\ IO_3^-(aq) \rightarrow I_2(s) + \mathbf{6\ H_2O(l)}$

- Balance H by adding H^+.

$Fe^{2+}(aq) \rightarrow Fe^{3+}(aq)$

$\mathbf{12\ H^+}(aq) + 2\ IO_3^-(aq) \rightarrow I_2(s) + 6\ H_2O(l)$

4. Balance each half-reaction with respect to charge by adding electrons.

$Fe^{2+}(aq) \rightarrow Fe^{3+}(aq) + \mathbf{e^-}$

$\mathbf{10\ e^-} + 12\ H^+(aq) + 2\ IO_3^-(aq) \rightarrow I_2(s) + 6\ H_2O(l)$

5. Make the number of electrons in both half-reactions equal.

$\mathbf{10} \times [Fe^{2+}(aq) \rightarrow Fe^{3+}(aq) + e^-]$

$10\ e^- + 12\ H^+(aq) + 2\ IO_3^-(aq) \rightarrow I_2(s) + 6\ H_2O$

6. Add the two half-reactions together.

$$10\ Fe^{2+}(aq) \rightarrow 10\ Fe^{3+}(aq) + 10e^-$$

$$10\ e^- + 12\ H^+(aq) + 2\ IO_3^-(aq) \rightarrow I_2(s) + 6\ H_2O$$

$$10\ Fe^{2+}(aq) + 12\ H^+(aq) + 2\ IO_3^-(aq) \rightarrow 10\ Fe^{3+}(aq) + I_2(s) + 6\ H_2O$$

7. Verify that the reaction is balanced.

Reactants	Products
10 Fe	10 Fe
12 H	12 H
2 I	2 I
6 O	6 O
+30 charge	+30 charge

KEY TERMS

aqueous solution **[8.7]**
acid **[8.8]**
alkaloid **[8.8]**
amphoteric **[8.8]**
Arrhenius acid **[8.8]**
Arrhenius base **[8.8]**
Arrhenius definition **[8.8]**
balanced equation **[8.3]**
base **[8.8]**
Brønsted–Lowry acid **[8.8]**
Brønsted–Lowry base **[8.8]**
Brønsted–Lowry definition **[8.8]**
chemical reaction **[8.3]**
carboxylic acid **[8.8]**
combination reaction **[8.6]**
complete ionic equation **[8.7]**
conjugate acid–base pair **[8.8]**
decomposition reaction **[8.6]**
diprotic acid **[8.8]**
dissociation **[8.8]**
displacement reaction **[8.6]**
double-displacement reaction **[8.6]**
fuel cell **[8.9]**
half-reaction **[8.9]**
hydronium ion **[8.8]**
ionize **[8.8]**
molecular equation **[8.7]**
monoprotic acid **[8.8]**
net ionic equation **[8.7]**
neutralization **[8.8]**
oxidation **[8.9]**
oxidation state (oxidation number) **[8.9]**
oxidizing agent **[8.9]**
precipitate **[8.7]**
precipitation reaction **[8.7]**
product **[8.3]**
reactant **[8.3]**
redox (oxidation–reduction) reaction **[8.9]**
reducing agent **[8.9]**
reduction **[8.9]**
salt **[8.8]**
single-displacement reaction **[8.6]**
solubility rules **[8.7]**
soluble **[8.7]**
sparingly soluble **[8.7]**
spectator ion **[8.7]**
strong acid **[8.8]**
strong base **[8.8]**
strong electrolyte **[8.7, 8.8]**
strong electrolyte solution **[8.7, 8.8]**
synthesis reaction **[8.6]**
weak acid **[8.8]**
weak base **[8.8]**
weak electrolyte **[8.8]**

EXERCISES

QUESTIONS

1. What is a chemical reaction? List some examples.

2. If you could observe atoms and molecules with the naked eye, what would you look for as conclusive evidence of a chemical reaction?

3. What are the main indications that a chemical reaction has occurred?

4. What is a chemical equation? Provide an example and identify the reactants and products.

5. What does each abbreviation, often used in chemical equations, represent?
(a) (g) (b) (l) (c) (s) (d) (aq)

6. What is the law of conservation of mass?

7. To balance a chemical equation, adjust the ______ as necessary to make the numbers of each type of atom on both sides of the equation equal. Never adjust the ______ to balance a chemical equation.

8. Is the chemical equation balanced? Why or why not?

$$2\ Ag_2O(s) + C(s) \rightarrow CO_2(g) + 4\ Ag(s)$$

9. Explain the difference between a synthesis reaction and a decomposition reaction and provide an example of each.

10. Explain the difference between a single-displacement reaction and a double-displacement reaction and provide an example of each.

11. What does it mean if a compound is referred to as soluble? Sparingly soluble?

12. What are the solubility rules, and how are they useful?

13. What is a precipitation reaction? Provide an example and identify the precipitate.

14. Is the precipitate in a precipitation reaction always a compound that is soluble or sparingly soluble? Explain.

15. Describe the differences between a molecular equation, a complete ionic equation, and net ionic equation. Give an example of each to illustrate the differences.

16. What are the properties of bases? Provide some examples of common substances that contain bases.

17. Restate the Arrhenius definition of an acid and demonstrate the definition with a chemical equation.

18. Restate the Arrhenius definition of a base and demonstrate the definition with a chemical equation.

19. Restate the Brønsted–Lowry definitions of acids and bases and demonstrate the definitions with a chemical equation.

20. According to the Brønsted–Lowry definition of acids and bases, what is a conjugate acid–base pair? Provide an example.

21. What is an acid–base neutralization reaction? Provide an example.

22. Provide an example of a reaction between an acid and a metal.

23. List an example of a reaction between an acid and a metal oxide.

24. Name a metal that a base can dissolves and write an equation for the reaction.

25. What is the difference between a strong acid and a weak acid?

26. How is the strength of an acid related to the strength of its conjugate base?

27. What are monoprotic and diprotic acids?

28. What is the difference between a strong base and a weak base?

29. What is an oxidation–reduction or redox reaction?

30. Define oxidation and reduction with respect to:
- **(a)** oxygen
- **(b)** electrons
- **(c)** oxidation state

31. What is an oxidizing agent? What is a reducing agent?

32. Good oxidizing agents have a strong tendency to _____ electrons in reactions.

33. Good reducing agents have a strong tendency to _____ electrons in reactions.

34. What is the oxidation state of a free element? Of a monoatomic ion?

35. For a neutral molecule, the oxidation states of the individual atoms must add up to _____.

36. For an ion, the oxidation states of the individual atoms must add up to _____.

37. In their compounds, elements have oxidation states equal to _____. Are there exceptions to this rule? Explain.

38. In a redox reaction, an atom that undergoes an increase in oxidation state is _____. An atom that undergoes a decrease in oxidation state is _____.

39. When balancing redox equations, the number of electrons lost in the oxidation half-reaction must _____ the number of electrons gained in the reduction half-reaction.

40. When balancing aqueous redox reactions, oxygen is balanced using _____, and hydrogen is balanced using _____.

41. When balancing aqueous redox reactions, charge is balanced using _____.

PROBLEMS

EVIDENCE OF CHEMICAL REACTIONS

42. Which observation is consistent with a chemical reaction occurring? Why?
- **(a)** Solid copper deposits on a piece of aluminum foil when the foil is placed in a blue copper nitrate solution. The blue color of the solution fades.
- **(b)** Liquid ethyl alcohol turns into a solid when placed in a low-temperature freezer.
- **(c)** A white precipitate forms when solutions of barium nitrate and sodium sulfate are mixed.
- **(d)** A mixture of sugar and water bubbles when yeasts are added. After several days, the sugar is gone and ethyl alcohol is found in the water.

43. Which observation is consistent with a chemical reaction occurring? Why?
- **(a)** Propane forms a flame and emits heat as it burns.
- **(b)** Acetone feels cold as it evaporates from the skin.
- **(c)** Bubbling occurs when potassium carbonate and hydrochloric acid solutions are mixed.
- **(d)** Heat is felt when a warm object is placed in your hand.

44. Vinegar forms bubbles when it is poured onto the calcium deposits on a faucet, and some of the calcium dissolves. Has a chemical reaction occurred? Explain your answer.

45. When a chemical drain opener is added to a clogged sink, bubbles form and the water in the sink gets warmer. Has a chemical reaction occurred? Explain your answer.

46. When a commercial hair bleaching mixture is applied to brown hair, the hair turns blond. Has a chemical reaction occurred? Explain your answer.

47. When water is boiled in a pot, it bubbles. Has a chemical reaction occurred? Explain your answer.

THE CONSERVATION OF MASS

48. An automobile gasoline tank holds 42 kg of gasoline. When the gasoline burns, 168 kg of oxygen are consumed and carbon dioxide and water are produced. What total combined mass of carbon dioxide and water is produced?

49. In the explosion of a hydrogen-filled balloon, 0.50 g of hydrogen reacts with 4.0 g of oxygen. How many grams of water vapor are formed? (Water vapor is the only product.)

50. Are these data sets on chemical changes consistent with the law of conservation of mass?

(a) A 7.5 g sample of hydrogen gas completely reacts with 60.0 g of oxygen gas to form 67.5 g of water.

(b) A 60.5 g sample of gasoline completely reacts with 243 g of oxygen to form 206 g of carbon dioxide and 88 g of water.

51. Are these data sets on chemical changes consistent with the law of conservation of mass?

(a) A 12.8 g sample of sodium completely reacts with 19.6 g of chlorine to form 32.4 g of sodium chloride.

(b) An 8 g sample of natural gas completely reacts with 32 g of oxygen gas to form 17 g of carbon dioxide and 16 g of water.

52. In a butane lighter, 9.7 g of butane combine with 34.7 g of oxygen to form 29.3 g carbon dioxide and how many grams of water?

53. A 56 g sample of iron reacts with 24 g of oxygen to form how many grams of iron oxide?

WRITING AND BALANCING CHEMICAL EQUATIONS

54. For each chemical equation (which may or may not be balanced), list the number of atoms of each element on each side of the equation, and determine if the equation is balanced.

(a) $Pb(NO_3)_2(aq) + 2\ NaCl(aq) \rightarrow PbCl_2(s) + 2\ NaNO_3(aq)$

(b) $C_3H_8(g) + O_2(g) \rightarrow 3\ CO_2(g) + 4\ H_2O(g)$

55. For each chemical equation (which may or may not be balanced), list the number of atoms of each element on each side of the equation, and determine if the equation is balanced.

(a) $MgS(aq) + 2\ CuCl_2(aq) \rightarrow 2\ CuS(s) + MgCl_2(aq)$

(b) $2\ C_6H_{14}(l) + 19\ O_2(g) \rightarrow 12\ CO_2(g) + 14\ H_2O(g)$

56. Consider the unbalanced chemical equation.

$$H_2O(l) \xrightarrow[\text{electrical current}]{} H_2(g) + O_2(g)$$

A chemistry student tries to balance the equation by placing the subscript 2 after the oxygen atom in H_2O. Explain why this is not correct. What is the correct balanced equation?

57. Consider the unbalanced chemical equation.

$$Al(s) + Cl_2(g) \rightarrow AlCl_3(s)$$

A student tries to balance the equation by changing the subscript 2 on Cl to a 3. Explain why this is not correct. What is the correct balanced equation?

58. Write a balanced chemical equation for each chemical reaction.

(a) Solid lead(II) sulfide reacts with aqueous hydrochloric acid to form solid lead(II) chloride and dihydrogen sulfide gas.

(b) Gaseous carbon monoxide reacts with hydrogen gas to form gaseous methane (CH_4) and liquid water.

(c) Solid iron(III) oxide reacts with hydrogen gas to form solid iron and liquid water.

(d) Gaseous ammonia (NH_3) reacts with gaseous oxygen to form gaseous nitrogen monoxide and gaseous water.

59. Write a balanced chemical equation for each chemical reaction.

(a) Solid copper reacts with solid sulfur to form solid copper(I) sulfide.

(b) Sulfur dioxide gas reacts with oxygen gas to form sulfur trioxide gas.

(c) Aqueous hydrochloric acid reacts with solid manganese(IV) oxide to form aqueous manganese(II) chloride, liquid water, and chlorine gas.

(d) Liquid benzene (C_6H_6) reacts with gaseous oxygen to form carbon dioxide and liquid water.

60. Write a balanced chemical equation for each chemical reaction.

(a) Solid magnesium reacts with aqueous copper(I) nitrate to form aqueous magnesium nitrate and solid copper.

(b) Gaseous dinitrogen pentoxide decomposes to form nitrogen dioxide and oxygen gas.

(c) Solid calcium reacts with aqueous nitric acid to form aqueous calcium nitrate and hydrogen gas.

(d) Liquid methanol (CH_3OH) reacts with oxygen gas to form gaseous carbon dioxide and gaseous water.

61. Write a balanced chemical equation for each chemical reaction.

(a) Gaseous acetylene (C_2H_2) reacts with oxygen gas to form gaseous carbon dioxide and gaseous water.

(b) Chlorine gas reacts with aqueous potassium iodide to form solid iodine and aqueous potassium chloride.

(c) Solid lithium oxide reacts with liquid water to form aqueous lithium hydroxide.

(d) Gaseous carbon monoxide reacts with oxygen gas to form carbon dioxide gas.

62. When solid sodium is added to liquid water, it reacts with the water to produce hydrogen gas and aqueous sodium hydroxide. Write a balanced chemical equation for this reaction.

63. When iron rusts, solid iron reacts with gaseous oxygen to form solid iron(III) oxide. Write a balanced chemical equation for this reaction.

64. Sulfuric acid in acid rain forms when gaseous sulfur dioxide pollutant reacts with gaseous oxygen and liquid water to form aqueous sulfuric acid. Write a balanced chemical equation for this reaction.

65. Nitric acid in acid rain forms when gaseous nitrogen dioxide pollutant reacts with gaseous oxygen and liquid water to form aqueous nitric acid. Write a balanced chemical equation for this reaction.

66. Write a balanced chemical equation for the reaction of solid vanadium(V) oxide with hydrogen gas to form solid vanadium(III) oxide and liquid water.

67. Write a balanced chemical equation for the reaction of gaseous nitrogen dioxide with hydrogen gas to form gaseous ammonia and liquid water.

68. Write a balanced chemical equation for the fermentation of sugar ($C_{12}H_{22}O_{11}$) by yeasts in which the aqueous sugar reacts with water to form aqueous ethyl alcohol (C_2H_5OH) and carbon dioxide gas.

69. Write a balanced chemical equation for the photosynthesis reaction in which gaseous carbon dioxide and liquid water react in the presence of chlorophyll to produce aqueous glucose ($C_6H_{12}O_6$) and oxygen gas.

70. Balance each chemical equation.

(a) $Na_2S(aq) + Cu(NO_3)_2(aq) \rightarrow NaNO_3(aq) + CuS(s)$

(b) $HCl(aq) + O_2(g) \rightarrow H_2O(l) + Cl_2(g)$

(c) $H_2(g) + O_2(g) \rightarrow H_2O(l)$

(d) $FeS(s) + HCl(aq) \rightarrow FeCl_2(aq) + H_2S(g)$

71. Balance each chemical equation.

(a) $N_2H_4(l) \rightarrow NH_3(g) + N_2(g)$

(b) $H_2(g) + N_2(g) \rightarrow NH_3(g)$

(c) $Cu_2O(s) + C(s) \rightarrow Cu(s) + CO(g)$

(d) $H_2(g) + Cl_2(g) \rightarrow HCl(g)$

72. Balance each chemical equation.

(a) $BaO_2(s) + H_2SO_4(aq) \rightarrow BaSO_4(s) + H_2O_2(aq)$

(b) $Co(NO_3)_3(aq) + (NH_4)_2S(aq) \rightarrow Co_2S_3(s) + NH_4NO_3(aq)$

(c) $Li_2O(s) + H_2O(l) \rightarrow LiOH(aq)$

(d) $Hg_2(C_2H_3O_2)_2(aq) + KCl(aq) \rightarrow Hg_2Cl_2(s) + KC_2H_3O_2(aq)$

73. Balance each chemical equation.

(a) $MnO_2(s) + HCl(aq) \rightarrow Cl_2(g) + MnCl_2(aq) + H_2O(l)$

(b) $CO_2(g) + CaSiO_3(s) + H_2O(l) \rightarrow SiO_2(s) + Ca(HCO_3)_2(aq)$

(c) $Fe(s) + S(l) \rightarrow Fe_2S_3(s)$

(d) $NO_2(g) + H_2O(l) \rightarrow HNO_3(aq) + NO(g)$

74. Is each chemical equation correctly balanced? If not, correct it.

(a) $Rb(s) + H_2O(l) \rightarrow RbOH(aq) + H_2(g)$

(b) $2\ N_2H_4(g) + N_2O_4(g) \rightarrow 3\ N_2(g) + 4\ H_2O(g)$

(c) $NiS(s) + O_2(g) \rightarrow NiO(s) + SO_2(g)$

(d) $PbO(s) + 2\ NH_3(g) \rightarrow Pb(s) + N_2(g) + H_2O(l)$

75. Is each chemical equation correctly balanced? If not, correct it.

(a) $SiO_2(s) + 4\ HF(aq) \rightarrow SiF_4(g) + 2\ H_2O(l)$

(b) $2\ Cr(s) + 3\ O_2(g) \rightarrow Cr_2O_3(s)$

(c) $Al_2S_3(s) + H_2O(l) \rightarrow 2\ Al(OH)_3(s) + 3\ H_2S(g)$

(d) $Fe_2O_3(s) + CO(g) \rightarrow 2\ Fe(s) + CO_2(g)$

76. Human cells obtain energy from a reaction called cellular respiration. Balance the skeletal equation for cellular respiration.

$$C_6H_{12}O_6(aq) + O_2(g) \rightarrow CO_2(g) + H_2O(l)$$

77. Propane camping stoves produce heat by the combustion of gaseous propane (C_3H_8). Balance the skeletal equation for the combustion of propane.

$$C_3H_8(g) + O_2(g) \rightarrow CO_2(g) + H_2O(g)$$

78. Catalytic converters work to remove nitrogen oxides and carbon monoxide from exhaust. Balance the skeletal equation for one of the reactions that occurs in a catalytic converter.

$$NO(g) + CO(g) \rightarrow N_2(g) + CO_2(g)$$

79. Billions of pounds of urea are produced annually for use as fertilizer. Balance the skeletal equation for the synthesis of urea.

$$NH_3(g) + CO_2(g) \rightarrow CO(NH_2)_2(s) + H_2O(l)$$

CLASSIFYING CHEMICAL REACTIONS BY WHAT ATOMS DO

80. Classify each chemical reaction as a synthesis, decomposition, single-displacement, or double-displacement reaction.

(a) $K_2S(aq) + Co(NO_3)_2(aq) \rightarrow 2\ KNO_3(aq) + CoS(s)$

(b) $3\ H_2(g) + N_2(g) \rightarrow 2\ NH_3(g)$

(c) $Zn(s) + CoCl_2(aq) \rightarrow ZnCl_2(aq) + Co(s)$

(d) $CH_3Br(g) \xrightarrow[\text{UV light}]{} CH_3(g) + Br(g)$

81. Classify each chemical reaction as a synthesis, decomposition, single-displacement, or double-displacement reaction.

(a) $CaSO_4(g) \xrightarrow[\text{heat}]{} CaO(s) + SO_3(g)$

(b) $2\ Na(s) + O_2(g) \rightarrow Na_2O_2(s)$

(c) $Pb(s) + 2\ AgNO_3(aq) \rightarrow Pb(NO_3)_2(aq) + 2\ Ag(s)$

(d) $HI(aq) + NaOH(aq) \rightarrow H_2O(l) + NaI(aq)$

82. NO is a pollutant emitted by motor vehicles. It is formed by the reaction:

(a) $N_2(g) + O_2(g) \rightarrow 2\ NO(g)$

Once in the atmosphere, NO (through a series of reactions) adds one oxygen atom to form NO_2. NO_2 then interacts with UV light according to the reaction:

(b) $NO_2(g) \xrightarrow[\text{UV light}]{} NO(g) + O(g)$

These freshly formed oxygen atoms then react with O_2 in the air to form ozone (O_3), a main component of smog:

(c) $O(g) + O_2(g) \rightarrow O_3(g)$

Classify each of the preceding reactions (a, b, c) as a synthesis, decomposition, single-displacement, or double-displacement reaction.

83. A main source of sulfur oxide pollutants are smelters where sulfide ores are converted into metals. The first step in this process is the reaction of the sulfide ore with oxygen in reactions such as:

(a) $2\ PbS(s) + 3\ O_2(g) \rightarrow 2\ PbO(s) + 2\ SO_2(g)$

Sulfur dioxide can then react with oxygen in air to form sulfur trioxide:

(b) $2\ SO_2(g) + O_2(g) \rightarrow 2\ SO_3(g)$

Sulfur trioxide can then react with water from rain to form sulfuric acid that falls as acid rain:

(c) $SO_3(g) + H_2O(l) \rightarrow H_2SO_4(aq)$

Classify each of the preceding reactions (a, b, c) as a synthesis, decomposition, single-displacement, or double-displacement reaction.

SOLUBILITY

84. Is each compound soluble or sparingly soluble? For the soluble compounds, identify the ions present in solution.

(a) $NaC_2H_3O_2$

(b) $Sn(NO_3)_2$

(c) AgI

(d) Na_3PO_4

85. Is each compound soluble or sparingly soluble? For the soluble compounds, identify the ions present in solution.

(a) $(NH_4)_2S$

(b) $CuCO_3$

(c) ZnS

(d) $Pb(C_2H_3O_2)_2$

86. Pair each cation on the left with an anion on the right that will form a *sparingly soluble* compound with it and write a formula for the sparingly soluble compound. Use each anion only once.

Ag^+ SO_4^{2-}
Ba^{2+} Cl^-
Cu^{2+} CO_3^{2-}
Fe^{3+} S^{2-}

87. Pair each cation on the left with an anion on the right that will form a *soluble* compound with it and write a formula for the soluble compound. Use each anion only once.

Na^+ NO_3^-
Sr^{2+} SO_4^{2-}
Co^{2+} S^{2-}
Pb^{2+} CO_3^{2-}

88. Move any misplaced compounds to the correct column.

Soluble	sparingly soluble
K_2S	K_2SO_4
$PbSO_4$	Hg_2I_2
BaS	$Cu_3(PO_4)_2$
$PbCl_2$	MgS
Hg_2Cl_2	$CaSO_4$
NH_4Cl	SrS
Na_2CO_3	Li_2S

89. Move any misplaced compounds to the correct column.

Soluble	sparingly soluble
$LiOH$	$CaCl_2$
Na_2CO_3	$Cu(OH)_2$
$AgCl$	$Ca(C_2H_3O_2)_2$
K_3PO_4	$SrSO_4$
CuI_2	Hg_2Br_2
$Pb(NO_3)_2$	$PbBr_2$
$CoCO_3$	PbI_2

PRECIPITATION REACTIONS

90. Complete and balance each equation. If no reaction occurs, write *NO REACTION*.
(a) $KI(aq) + BaS(aq) \rightarrow$
(b) $K_2SO_4(aq) + BaBr_2(aq) \rightarrow$
(c) $NaCl(aq) + Hg_2(C_2H_3O_2)_2(aq) \rightarrow$
(d) $NaC_2H_3O_2(aq) + Pb(NO_3)_2(aq) \rightarrow$

91. Complete and balance each equation. If no reaction occurs, write *NO REACTION*.
(a) $NaOH(aq) + FeBr_3(aq) \rightarrow$
(b) $BaCl_2(aq) + AgNO_3(aq) \rightarrow$
(c) $Na_2CO_3(aq) + CoCl_2(aq) \rightarrow$
(d) $K_2S(aq) + BaCl_2(aq) \rightarrow$

92. Write a molecular equation for the precipitation reaction that occurs (if any) when each pair of solutions is mixed. If no reaction occurs, write *NO REACTION*.
(a) sodium carbonate and lead(II) nitrate
(b) potassium sulfate and lead(II) acetate
(c) copper(II) nitrate and barium sulfide
(d) calcium nitrate and sodium iodide

93. Write a molecular equation for the precipitation reaction that occurs (if any) when each pair of solutions is mixed. If no reaction occurs, write *NO REACTION*.
(a) potassium chloride and lead(II) acetate
(b) lithium sulfate and strontium chloride
(c) potassium bromide and calcium sulfide
(d) chromium(III) nitrate and potassium phosphate

94. Correct any incorrect equations. If no reaction occurs, write *NO REACTION*.
(a) $Ba(NO_3)_2(aq) + (NH_4)_2SO_4(aq) \rightarrow BaSO_4(s) + 2\ NH_4NO_3(aq)$
(b) $BaS(aq) + 2\ KCl(aq) \rightarrow BaCl_2(s) + K_2S(aq)$
(c) $2\ KI(aq) + Pb(NO_3)_2(aq) \rightarrow PbI_2(s) + 2\ KNO_3(aq)$
(d) $Pb(NO_3)_2(aq) + 2\ LiCl(aq) \rightarrow 2\ LiNO_3(s) + PbCl_2(aq)$

95. Correct any incorrect equations. If no reaction occurs, write *NO REACTION*.
(a) $AgNO_3(aq) + NaCl(aq) \rightarrow NaCl(s) + AgNO_3(aq)$
(b) $K_2SO_4(aq) + Co(NO_3)_2(aq) \rightarrow CoSO_4(s) + 2\ KNO_3(aq)$
(c) $Cu(NO_3)_2(aq) + (NH_4)_2S(aq) \rightarrow CuS(s) + 2\ NH_4NO_3(aq)$
(d) $Hg_2(NO_3)_2(aq) + 2\ LiCl(aq) \rightarrow Hg_2Cl_2(s) + 2\ LiNO_3(aq)$

IONIC AND NET IONIC EQUATIONS

96. Identify the spectator ions in the complete ionic equation.

$$2\ K^+(aq) + S^{2-}(aq) + Pb^{2+}(aq) + 2\ NO_3^-(aq) \rightarrow PbS(s) + 2\ K^+(aq) + 2\ NO_3^-(aq)$$

97. Identify the spectator ions in the complete ionic equation.

$$Ba^{2+}(aq) + 2\ I^-(aq) + 2\ Na^+(aq) + SO_4^{2-}(aq) \rightarrow BaSO_4(s) + 2\ I^-(aq) + 2\ Na^+(aq)$$

98. Write balanced complete ionic and net ionic equations for each reaction.
(a) $AgNO_3(aq) + KCl(aq) \rightarrow AgCl(s) + KNO_3(aq)$
(b) $CaS(aq) + CuCl_2(aq) \rightarrow CuS(s) + CaCl_2(aq)$
(c) $NaOH(aq) + HNO_3(aq) \rightarrow H_2O(l) + NaNO_3(aq)$
(d) $2\ K_3PO_4(aq) + 3\ NiCl_2(aq) \rightarrow Ni_3(PO_4)_2(s) + 6\ KCl(aq)$

99. Write balanced complete ionic and net ionic equations for each reaction.
(a) $HI(aq) + KOH(aq) \rightarrow H_2O(l) + KI(aq)$
(b) $Na_2SO_4(aq) + CaI_2(aq) \rightarrow CaSO_4(s) + 2\ NaI(aq)$
(c) $2\ HC_2H_3O_2(aq) + Na_2CO_3(aq) \rightarrow H_2O(l) + CO_2(g) + 2\ NaC_2H_3O_2(aq)$
(d) $NH_4Cl(aq) + NaOH(aq) \rightarrow H_2O(l) + NH_3(g) + NaCl(aq)$

100. Write complete ionic and net ionic equations for each of the reactions in Problem 92.

101. Write complete ionic and net ionic equations for each of the reactions in Problem 93.

ACID AND BASE DEFINITIONS

102. Identify each substance as an acid or a base and write a chemical equation showing how it is an acid or a base according to the Arrhenius definition.
(a) $H_2SO_4(aq)$
(b) $Sr(OH)_2(aq)$
(c) $HBr(aq)$
(d) $NaOH(aq)$

103. Identify each substance as an acid or a base and write a chemical equation showing how it is an acid or a base according to the Arrhenius definition.
(a) $Ca(OH)_2(aq)$
(b) $HC_2H_3O_2(aq)$
(c) $KOH(aq)$
(d) $HNO_3(aq)$

104. For each reaction, identify the Brønsted–Lowry acid, the Brønsted–Lowry base, the conjugate acid, and the conjugate base.

(a) $HBr(aq) + H_2O(l) \rightarrow H_3O^+(aq) + Br^-(aq)$
(b) $NH_3(aq) + H_2O(l) \rightleftharpoons NH_4^+(aq) + OH^-(aq)$
(c) $HNO_3(aq) + H_2O(l) \rightarrow H_3O^+(aq) + NO_3^-(aq)$
(d) $C_2H_5N(aq) + H_2O(l) \rightleftharpoons C_2H_5NH^+(aq) + OH^-(aq)$

105. For each reaction, identify the Brønsted–Lowry acid, the Brønsted–Lowry base, the conjugate acid, and the conjugate base.

(a) $HI(aq) + H_2O(l) \rightarrow H_3O^+(aq) + I^-(aq)$
(b) $CH_3NH_2(aq) + H_2O(l) \rightleftharpoons CH_3NH_3^+(aq) + OH^-(aq)$
(c) $CO_3^{2-}(aq) + H_2O(l) \rightleftharpoons HCO_3^-(aq) + OH^-(aq)$
(d) $H_2CO_3(aq) + H_2O(l) \rightleftharpoons H_3O^+(aq) + HCO_3^-(aq)$

106. Determine whether each pair is a conjugate acid–base pair.

(a) NH_3, NH_4^+
(b) HCl, HBr
(c) $C_2H_3O_2^-$, $HC_2H_3O_2$
(d) HCO_3^-, NO_3^-

107. Determine whether each pair is a conjugate acid–base pair.

(a) HI, I^-
(b) $HCHO_2$, SO_4^{2-}
(c) PO_4^{3-}, HPO_4^{2-}
(d) CO_3^{2-}, HCl

108. Write the formula for the conjugate base of each acid.

(a) HCl
(b) H_2SO_3
(c) $HCHO_2$
(d) HF

109. Write the formula for the conjugate base of each acid.

(a) HBr
(b) H_2CO_3
(c) $HClO_4$
(d) $HC_2H_3O_2$

110. Write the formula for the conjugate acid of each base.

(a) NH_3
(b) ClO_4^-
(c) HSO_4^-
(d) CO_3^{2-}

111. Write the formula for the conjugate acid of each base.

(a) CH_3NH_2
(b) C_5H_5N
(c) Cl^-
(d) F^-

ACID–BASE REACTIONS

112. Write a neutralization reaction for each acid and base pair.

(a) $HI(aq)$ and $NaOH(aq)$
(b) $HBr(aq)$ and $KOH(aq)$
(c) $HNO_3(aq)$ and $Ba(OH)_2(aq)$
(d) $HClO_4(aq)$ and $Sr(OH)_2(aq)$

113. Write a neutralization reaction for each acid and base pair.

(a) $HF(aq)$ and $Ba(OH)_2(aq)$
(b) $HClO_4(aq)$ and $NaOH(aq)$
(c) $HBr(aq)$ and $Ca(OH)_2(aq)$
(d) $HCl(aq)$ and $KOH(aq)$

114. Write a balanced chemical equation showing how each metal reacts with HBr.

(a) Rb
(b) Mg
(c) Ba
(d) Al

115. Write a balanced chemical equation showing how each metal reacts with HCl.

(a) K
(b) Ca
(c) Na
(d) Sr

116. Write a balanced chemical equation showing how each metal oxide reacts with HI.

(a) MgO
(b) K_2O
(c) Rb_2O
(d) CaO

117. Write a balanced chemical equation showing how each metal oxide reacts with HCl.

(a) SrO
(b) Na_2O
(c) Li_2O
(d) BaO

118. Predict the products of each reaction.

(a) $HClO_4(aq) + Fe_2O_3(s) \rightarrow$
(b) $H_2SO_4(aq) + Sr(s) \rightarrow$
(c) $H_3PO_4(aq) + KOH(aq) \rightarrow$

119. Predict the products of each reaction.

(a) $HI(aq) + Al(s) \rightarrow$
(b) $H_2SO_4(aq) + TiO_2(s) \rightarrow$
(c) $H_2CO_3(aq) + LiOH(aq) \rightarrow$

STRONG AND WEAK ACIDS AND BASES

120. Classify each acid as strong or weak.
- **(a)** HCl
- **(b)** HF
- **(c)** HBr
- **(d)** H_2SO_3

121. Classify each acid as strong or weak.
- **(a)** $HCHO_2$
- **(b)** H_2SO_4
- **(c)** HNO_3
- **(d)** H_2CO_3

122. Determine $[H_3O^+]$ in each acid solution. If the acid is weak, indicate the value that $[H_3O^+]$ is less than.
- **(a)** 1.7 mol/L HBr
- **(b)** 1.5 mol/L HNO_3
- **(c)** 0.38 mol/L H_2CO_3
- **(d)** 1.75 mol/L $HCHO_2$

123. Determine $[H_3O^+]$ in each acid solution. If the acid is weak, indicate the value that $[H_3O^+]$ is less than.
- **(a)** 0.125 mol/L $HClO_2$
- **(b)** 1.25 mol/L H_3PO_4
- **(c)** 2.77 mol/L HCl
- **(d)** 0.95 mol/L H_2SO_3

124. Classify each base as strong or weak.
- **(a)** LiOH
- **(b)** NH_4OH
- **(c)** $Ca(OH)_2$
- **(d)** NH_3

125. Classify each base as strong or weak.
- **(a)** C_5H_5N
- **(b)** NaOH
- **(c)** $Ba(OH)_2$
- **(d)** KOH

126. Determine $[OH^-]$ in each base solution. If the base is weak, indicate the value that $[OH^-]$ is less than.
- **(a)** 0.25 mol/L NaOH
- **(b)** 0.25 mol/L NH_3
- **(c)** 0.25 mol/L $Sr(OH)_2$
- **(d)** 1.25 mol/L KOH

127. Determine $[OH^-]$ in each base solution. If the base is weak, indicate the value that $[OH^-]$ is less than.
- **(a)** 2.5 mol/L KOH
- **(b)** 1.95 mol/L NH_3
- **(c)** 0.225 mol/L $Ba(OH)_2$
- **(d)** 1.8 mol/L C_5H_5N

OXIDATION AND REDUCTION

128. Which substance is oxidized in each reaction?
- **(a)** $2\,H_2(g) + O_2(g) \rightarrow 2\,H_2O(l)$
- **(b)** $4\,Al(s) + 3\,O_2(g) \rightarrow 2\,Al_2O_3(s)$
- **(c)** $2\,Al(s) + 3\,Cl_2(g) \rightarrow 2\,AlCl_3(s)$

129. Which substance is oxidized in each reaction?
- **(a)** $2\,Zn(s) + O_2(g) \rightarrow 2\,ZnO(s)$
- **(b)** $CH_4(g) + 2\,O_2(g) \rightarrow CO_2(g) + 2\,H_2O(g)$
- **(c)** $Sr(s) + F_2(g) \rightarrow SrF_2(s)$

130. For each reaction, identify the substance being oxidized and the substance being reduced.
- **(a)** $2\,Sr(s) + O_2(g) \rightarrow 2\,SrO(s)$
- **(b)** $Ca(s) + Cl_2(g) \rightarrow CaCl_2(s)$
- **(c)** $Ni^{2+}(aq) + Mg(s) \rightarrow Mg^{2+}(aq) + Ni(s)$

131. For each reaction, identify the substance being oxidized and the substance being reduced.
- **(a)** $Mg(s) + Br_2(g) \rightarrow MgBr_2(s)$
- **(b)** $2\,Cr^{3+}(aq) + 3\,Mn(s) \rightarrow 2\,Cr(s) + 3\,Mn^{2+}(aq)$
- **(c)** $2\,H^+(aq) + Ni(s) \rightarrow H_2(g) + Ni^{2+}(aq)$

132. For each of the reactions in Problem 130, identify the oxidizing agent and the reducing agent.

133. For each of the reactions in Problem 131, identify the oxidizing agent and the reducing agent.

134. Based on periodic trends, which elements would you expect to be good oxidizing agents?
- **(a)** potassium
- **(b)** fluorine
- **(c)** iron
- **(d)** chlorine

135. Based on periodic trends, which elements would you expect to be good oxidizing agents?
- **(a)** oxygen
- **(b)** bromine
- **(c)** lithium
- **(d)** sodium

136. Based on periodic trends, which elements in Problem 134 (in their elemental form) would you expect to be good reducing agents?

137. Based on periodic trends, which elements in Problem 135 (in their elemental form) would you expect to be good reducing agents?

138. For each redox reaction, identify the substance being oxidized, the substance being reduced, the oxidizing agent, and the reducing agent.
(a) $N_2(g) + O_2(g) \rightarrow 2\,NO(g)$
(b) $2\,CO(g) + O_2(g) \rightarrow 2\,CO_2(g)$
(c) $SbCl_3(g) + Cl_2(g) \rightarrow SbCl_5(g)$

139. For each redox reaction, identify the substance being oxidized, the substance being reduced, the oxidizing agent, and the reducing agent.
(a) $H_2(g) + I_2(g) \rightarrow 2\,HI(g)$
(b) $CO(g) + H_2(g) \rightarrow C(s) + H_2O(g)$
(c) $2\,Al(s) + 6\,H^+(aq) \rightarrow 2\,Al^{3+}(aq) + 3\,H_2(g)$

OXIDATION STATES

140. Assign an oxidation state to each element or ion.
(a) V
(b) Mg^{2+}
(c) Cr^{3+}
(d) O_2

141. Assign an oxidation state to each element or ion.
(a) Ne
(b) Br_2
(c) Cu^+
(d) Fe^{3+}

142. Assign an oxidation state to each atom in each compound.
(a) NaCl
(b) CaF_2
(c) SO_2
(d) H_2S

143. Assign an oxidation state to each atom in each compound.
(a) CH_4
(b) CH_2Cl_2
(c) $CuCl_2$
(d) HI

144. What is the oxidation state of nitrogen in each compound?
(a) NO
(b) NO_2
(c) N_2O

145. What is the oxidation state of Cr in each compound?
(a) CrO
(b) CrO_3
(c) Cr_2O_3

146. Assign an oxidation state to each atom in each polyatomic ion.
(a) CO_3^{2-}
(b) OH^-
(c) NO_3^-
(d) NO_2^-

147. Assign an oxidation state to each atom in each polyatomic ion.
(a) CrO_4^{2-}
(b) $Cr_2O_7^{2-}$
(c) PO_4^{3-}
(d) MnO_4^-

148. What is the oxidation state of Cl in each ion?
(a) ClO^-
(b) ClO_2^-
(c) ClO_3^-
(d) ClO_4^-

149. What is the oxidation state of S in each ion?
(a) SO_4^{2-}
(b) SO_3^{2-}
(c) HSO_3^-
(d) HSO_4^-

150. Assign an oxidation state to each element in each compound.
(a) $Cu(NO_3)_2$
(b) $Sr(OH)_2$
(c) $K_2Cr_2O_7$
(d) $NaHCO_3$

151. Assign an oxidation state to each element in each compound.
(a) Na_3PO_4
(b) Hg_2S
(c) $Fe(CN)_3$
(d) NH_4Cl

152. Assign an oxidation state to each element in each reaction and use the change in oxidation state to determine which element is being oxidized and which element is being reduced.
(a) $SbCl_5(g) \rightarrow SbCl_3(g) + Cl_2(g)$
(b) $CO(g) + Cl_2(g) \rightarrow COCl_2(g)$
(c) $H_2(g) + CO_2(g) \rightarrow H_2O(g) + CO(g)$

153. Assign an oxidation state to each element in each reaction and use the change in oxidation state to determine which element is being oxidized and which element is being reduced.
(a) $2\,H_2S(g) \rightarrow 2\,H_2(g) + S_2(g)$
(b) $C_6H_{12}O_6(s) + 6\,O_2(g) \rightarrow 6\,CO_2(g) + 6\,H_2O(g)$
(c) $C_2H_4(g) + Cl_2(g) \rightarrow C_2H_4Cl_2\,(g)$

154. Use oxidation states to identify the oxidizing agent and the reducing agent in the redox reaction.

$$2\,Na(s) + 2\,H_2O(l) \rightarrow 2\,NaOH(aq) + H_2(g)$$

155. Use oxidation states to identify the oxidizing agent and the reducing agent in the redox reaction.

$$N_2(g) + 3\,H_2(g) \rightarrow 2\,NH_3(g)$$

BALANCING REDOX REACTIONS

156. Balance each redox reaction using the half-reaction method.
(a) $K(s) + Cr^{3+}(aq) \rightarrow Cr(s) + K^+(aq)$
(b) $Mg(s) + Ag^+(aq) \rightarrow Mg^{2+}(aq) + Ag(s)$
(c) $Al(s) + Fe^{2+}(aq) \rightarrow Al^{3+}(aq) + Fe(s)$

157. Balance each redox reaction using the half-reaction method.
(a) $Zn(s) + Sn^{2+}(aq) \rightarrow Zn^{2+}(aq) + Sn(s)$
(b) $Mg(s) + Cr^{3+}(aq) \rightarrow Mg^{2+}(aq) + Cr(s)$
(c) $Al(s) + Ag^+(aq) \rightarrow Al^{3+}(aq) + Ag(s)$

158. Classify each half-reaction occurring in acidic aqueous solution as an oxidation or a reduction and balance the half-reaction.
(a) $MnO_4^-(aq) \rightarrow Mn^{2+}(aq)$
(b) $Pb^{2+}(aq) \rightarrow PbO_2(s)$
(c) $IO_3^-(aq) \rightarrow I_2(s)$
(d) $SO_2(g) \rightarrow SO_4^{2-}(aq)$

159. Classify each half-reaction occurring in acidic aqueous solution as an oxidation or a reduction and balance the half-reaction.
(a) $S(s) \rightarrow H_2S(g)$
(b) $S_2O_8^{2-}(aq) \rightarrow 2\ SO_4^{2-}(aq)$
(c) $Cr_2O_7^{2-}(aq) \rightarrow Cr^{3+}(aq)$
(d) $NO(g) \rightarrow NO_3^-(aq)$

160. Use the half-reaction method to balance each redox reaction occurring in acidic aqueous solution.
(a) $PbO_2(s) + I^-(aq) \rightarrow Pb^{2+}(aq) + I_2(s)$
(b) $SO_3^{2-}(aq) + MnO_4^-(aq) \rightarrow SO_4^{2-}(aq) + Mn^{2+}(aq)$
(c) $S_2O_3^{2-}(aq) + Cl_2(g) \rightarrow SO_4^{2-}(aq) + Cl^-(aq)$

161. Use the half-reaction method to balance each redox reaction occurring in acidic aqueous solution.
(a) $I^-(aq) + NO_2^-(aq) \rightarrow I_2(s) + NO(g)$
(b) $BrO_3^-(aq) + N_2H_4(g) \rightarrow Br^-(aq) + N_2(g)$
(c) $NO_3^-(aq) + Sn^{2+}(aq) \rightarrow Sn^{4+}(aq) + NO(g)$

162. Use the half-reaction method to balance each redox reaction occurring in acidic aqueous solution.
(a) $ClO_4^-(aq) + Cl^-(aq) \rightarrow ClO_3^-(aq) + Cl_2(g)$
(b) $MnO_4^-(aq) + Al(s) \rightarrow Mn^{2+}(aq) + Al^{3+}(aq)$
(c) $Br_2(aq) + Sn(s) \rightarrow Sn^{2+}(aq) + Br^-(aq)$

163. Use the half-reaction method to balance each redox reaction occurring in acidic aqueous solution.
(a) $IO_3^-(aq) + SO_2(g) \rightarrow I_2(s) + SO_4^{2-}(aq)$
(b) $Sn^{4+}(aq) + H_2(g) \rightarrow Sn^{2+}(aq) + H^+(aq)$
(c) $Cr_2O_7^{2-}(aq) + Br^-(aq) \rightarrow Cr^{3+}(aq) + Br_2(aq)$

CUMULATIVE PROBLEMS

164. What solution can you add to each cation mixture to precipitate one cation while keeping the other cation in solution? Write a net ionic equation for the precipitation reaction that occurs.
(a) $Fe^{2+}(aq)$ and $Pb^{2+}(aq)$
(b) $K^+(aq)$ and $Ca^{2+}(aq)$
(c) $Ag^+(aq)$ and $Ba^{2+}(aq)$
(d) $Cu^{2+}(aq)$ and $Hg_2^{2+}(aq)$

165. What solution can you add to each cation mixture to precipitate one cation while keeping the other cation in solution? Write a net ionic equation for the precipitation reaction that occurs.
(a) $Sr^{2+}(aq)$ and $Hg_2^{2+}(aq)$
(b) $NH_4^+(aq)$ and $Ca^{2+}(aq)$
(c) $Ba^{2+}(aq)$ and $Mg^{2+}(aq)$
(d) $Ag^+(aq)$ and $Zn^{2+}(aq)$

166. A solution contains one or more of the following ions: Ag^+, Ca^{2+}, and Cu^{2+}. When sodium chloride is added to the solution, no precipitate occurs. When sodium sulfate is added to the solution, a white precipitate occurs. The precipitate is filtered off and sodium carbonate is added to the remaining solution, producing a precipitate. Which ions were present in the original solution? Write net ionic equations for the formation of each of the precipitates observed.

167. A solution contains one or more of the following ions: Hg_2^{2+}, Ba^{2+}, and Fe^{2+}. When potassium chloride is added to the solution, a precipitate forms. The precipitate is filtered off and potassium sulfate is added to the remaining solution, producing no precipitate. When potassium carbonate is added to the remaining solution, a precipitate occurs. Which ions were present in the original solution? Write net ionic equations for the formation of each of the precipitates observed.

168. Determine whether each reaction is a redox reaction. For those reactions that are redox reactions, identify the substance being oxidized and the substance being reduced.
(a) $Zn(s) + CoCl_2(aq) \rightarrow ZnCl_2(aq) + Co(s)$
(b) $HI(aq) + NaOH(aq) \rightarrow H_2O(l) + NaI(aq)$
(c) $AgNO_3(aq) + NaCl(aq) \rightarrow AgCl(s) + NaNO_3(aq)$
(d) $2\ K(s) + Br_2(l) \rightarrow 2\ KBr(s)$

169. Determine whether each reaction is a redox reaction. For those reactions that are redox reactions, identify the substance being oxidized and the substance being reduced.
(a) $Pb(NO_3)_2(aq) + 2\ LiCl(aq) \rightarrow PbCl_2(s) + 2\ LiNO_3(aq)$
(b) $2\ HBr(aq) + Ca(OH)_2(aq) \rightarrow 2\ H_2O(l) + CaBr_2(aq)$
(c) $2\ Al(s) + Fe_2O_3(s) \rightarrow Al_2O_3(s) + 2\ Fe(l)$
(d) $Na_2O(s) + H_2O(l) \rightarrow 2\ NaOH(aq)$

HIGHLIGHT PROBLEMS

170. Shown here are molecular views of two different possible mechanisms by which an automobile airbag might function. One of these mechanisms involves a chemical reaction and the other does not. By looking at the molecular views, can you tell which mechanism operates via a chemical reaction?

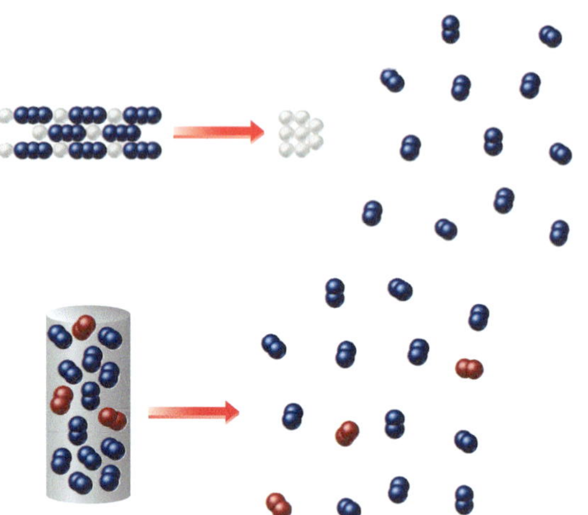

▲ When an airbag is detonated, the bag inflates. These figures show two possible ways in which the inflation may happen.

171. Precipitation reactions often produce brilliant colors. Look at the photographs of each precipitation reaction and write molecular, complete ionic, and net ionic equations for each one.

▲ **(a)** The precipitation reaction that occurs when aqueous iron(III) nitrate is added to aqueous sodium hydroxide.
(b) The precipitation reaction that occurs when aqueous cobalt(II) chloride is added to aqueous potassium hydroxide.
(c) The precipitation reaction that occurs when aqueous $AgNO_3$ is added to aqueous sodium iodide.

172. Based on the molecular view of each acid solution, determine whether the acid is weak or strong.

(a)

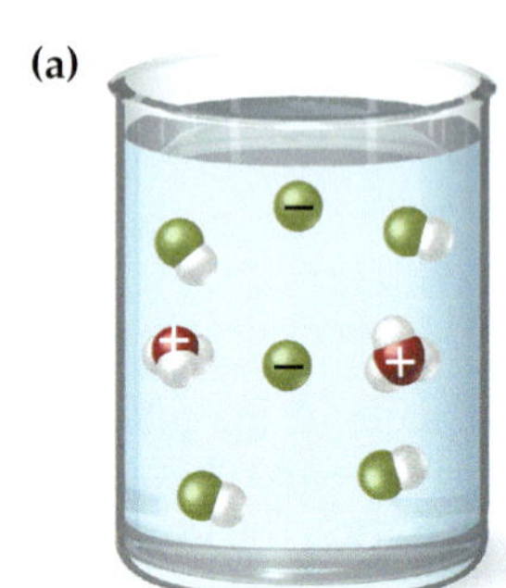

(b)

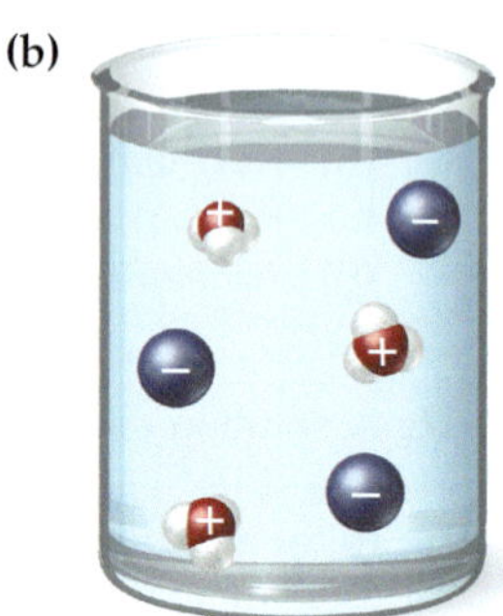

(c)

(d)

Answers to Skillbuilder Exercises

Skillbuilder 8.1

(a) Chemical reaction; heat and light are emitted.
(b) Not a chemical reaction; gaseous and liquid butane are both butane.
(c) Chemical reaction; heat and light are emitted.
(d) Not a chemical reaction; solid dry ice is made of carbon dioxide, which sublimes (evaporates) as carbon dioxide gas.

Skillbuilder 8.2 27 g

Skillbuilder 8.3

$2\,Cr_2O_3(s) + 3\,C(s) \rightarrow 4\,Cr(s) + 3\,CO_2(g)$

Skillbuilder 8.4

$2\,C_4H_{10}(g) + 13\,O_2(g) \rightarrow 8\,CO_2(g) + 10\,H_2O(g)$

Skillbuilder 8.5

$Pb(C_2H_3O_2)_2(aq) + 2\,KI(aq) \rightarrow PbI_2(s) + 2\,KC_2H_3O_2(aq)$

Skillbuilder 8.6

$4\,HCl(g) + O_2(g) \rightarrow 2\,H_2O(l) + 2\,Cl_2(g)$

Skillbuilder 8.7

(a) single-displacement **(b)** double-displacement **(c)** synthesis **(d)** decomposition

Skillbuilder 8.8

(a) sparingly soluble
(b) soluble
(c) sparingly soluble
(d) soluble

Skillbuilder 8.9

$2\,KOH(aq) + NiBr_2(aq) \rightarrow Ni(OH)_2(s) + 2\,KBr(aq)$

Skillbuilder 8.10

$NH_4Cl(aq) + Fe(NO_3)_3(aq) \rightarrow$ NO REACTION

Skillbuilder 8.11

$K_2SO_4(aq) + Sr(NO_3)_2(aq) \rightarrow SrSO_4(s) + 2\,KNO_3(aq)$

Skillbuilder 8.12

Complete ionic equation:

$2\,H^+(aq) + 2\,Br^-(aq) + Ca^{2+}(aq) + 2\,OH^-(aq) \rightarrow 2\,H_2O(l) + Ca^{2+}(aq) + 2\,Br^-(aq)$

Net ionic equation:

$2\,H^+(aq) + 2\,OH^-(aq) \rightarrow 2\,H_2O(l)$, or simply

$H^+(aq) + OH^-(aq) \rightarrow H_2O(l)$

Skillbuilder 8.13

(a) $\underset{\text{Base}}{C_5H_5N(aq)} + \underset{\text{Acid}}{H_2O(l)} \rightleftharpoons \underset{\text{Conjugate acid}}{C_5H_5NH^+(aq)} + \underset{\text{Conjugate base}}{OH^-(aq)}$

(b) $\underset{\text{Acid}}{HNO_3(aq)} + \underset{\text{Base}}{H_2O(l)} \rightarrow \underset{\text{Conjugate acid}}{H_3O^+(aq)} + \underset{\text{Conjugate base}}{NO_3^-(aq)}$

Skillbuilder 8.14

$H_3PO_4(aq) + 3\,NaOH(aq) \rightarrow 3\,H_2O(l) + Na_3PO_4(aq)$

Skillbuilder 8.15

(a) $2\,HCl(aq) + Sr(s) \rightarrow H_2(g) + SrCl_2(aq)$
(b) $2\,HI(aq) + BaO(s) \rightarrow H_2O(l) + BaI_2(aq)$

Skillbuilder 8.16

(a) $[H_3O^+] < 0.50$ mol/L
(b) $[H_3O^+] = 1.25$ mol/L
(c) $[H_3O^+] < 0.75$ mol/L

Skillbuilder 8.17

(a) $[OH^-] = 0.11$ mol/L
(b) $[OH^-] < 1.05$ mol/L
(c) $[OH^-] = 0.45$ mol/L

Skillbuilder 8.18

(a) K is oxidized; Cl_2 is reduced.
(b) Al is oxidized; Sn^{2+} is reduced.
(c) C is oxidized; O_2 is reduced.

Skillbuilder 8.19

(a) K is the reducing agent; Cl_2 is the oxidizing agent.
(b) Al is the reducing agent; Sn^{2+} is the oxidizing agent.
(c) C is the reducing agent; O_2 is the oxidizing agent.

Skillbuilder 8.20

(a) $\underset{0}{Zn}$
(b) $\underset{+2}{Cu^{2+}}$
(c) $\underset{+2}{Ca}\underset{-1}{Cl_2}$
(d) $\underset{+4}{C}\underset{-1}{F_4}$
(e) $\underset{+3}{N}\underset{-2}{O_2^-}$
(f) $\underset{+6}{S}\underset{-2}{O_3}$

Skillbuilder 8.21

Sn oxidized ($0 \rightarrow +4$); N reduced ($+5 \rightarrow +4$)

Skillbuilder 8.22

$6\,H^+(aq) + 2\,Cr(s) \rightarrow 3\,H_2(g) + 2\,Cr^{3+}(aq)$

Skillbuilder 8.23

$Cu(s) + 4\,H^+(aq) + 2\,NO_3^-(aq) \rightarrow Cu^{2+}(aq) + 2\,NO_2(g) + 2\,H_2O(l)$

Skillbuilder 8.24

$5\,Sn(s) + 16\,H^+(aq) + 2\,MnO_4^-(aq) \rightarrow 5\,Sn^{2+}(aq) + 2\,Mn^{2+}(aq) + 8\,H_2O(l)$

Answers to Conceptual Checkpoints

8.1 (a) No reaction occurred. The molecules are the same before and after the change.

(b) A reaction occurred; the molecules have changed.

(c) A reaction occurred; the molecules have changed.

8.2 (b) There are 18 oxygen atoms on the left side of the equation, so the same number is needed on the right: $6 + 6(2) = 18$.

8.3 No In the vaporization, the liquid water becomes gaseous, but its mass does not change. Like chemical changes, physical changes also follow the law of conservation of mass.

8.4 (a) (a) The number of atom must of each element be the same on both sides of a balanced chemical equation. Since molecules change during a chemical reaction, their number is not the same on both sides (b), nor is the sum of all of the coefficients the same (c).

8.5 (d) In a precipitation reaction, cations and anions "exchange partners" to produce at least one product. In an acid–base reaction, H^+ and OH^- combine to form water, and their partners pair off to a salt.

8.6 (a) Since chlorides are usually soluble and Ba^{2+} is not an exception, $BaCl_2$ is soluble and will dissolve in water. When it dissolves, it dissociates into its component ions, as shown in (a).

8.7 (b) Both of the possible products, MgS and $CaSO_4$, are sparingly soluble. The possible products of the other reactions—Na_2S, $Ca(NO_3)_2$, Na_2SO_4, and $Mg(NO_3)_2$—are all soluble.

8.8 (a) HCl is an acid. Acids have a sour taste. The other two compounds are bases.

8.9 (b) The conjugate base of an acid always has one fewer proton and is one charge unit lower (more negative) than the acid.

8.10 (c) Both **(a)** and **(b)** show complete ionization and are therefore strong acids. Only the acid depicted in **(c)** undergoes partial ionization and is therefore a weak acid.

8.11 (b) The oxidizing agent oxidizes another species and is itself always reduced.

8.12 (d) From Rule 1, you know that the oxidation state of nitrogen in N_2 is 0. According to Rule 3, the sum of the oxidation states of all atoms in a compound = 0. Therefore, by applying Rule 5, you can determine that the oxidation state of nitrogen in NO is +2; in NO_2 it is +4; and in NH_3 it is –3.

▲ Dynamic equilibrium involves two opposing processes that occur at the same rate. This image draws an analogy between a chemical equilibrium ($N_2O_4 \rightleftharpoons 2NO_2$), in which the two opposing reactions occur at the same rate, and a freeway with traffic moving in opposing directions at the same rate.

Chemical Reactions 2

9

MODULE OUTLINE

9.1 Energy

LO: Recognize the different forms of energy.

LO: Identify energy units.

Matter is one of the two major components of our universe. The other major component is **energy**, *the capacity to do work*. **Work** is defined as the result of a force acting on a distance. For example, if you push this book across your desk, you have done work. You may at first think that in chemistry we are only concerned with matter, but the behavior of matter is driven in large part by energy, so understanding energy is critical to understanding chemistry. Like matter, energy is conserved. The **law of conservation of energy** states that *energy is neither created nor destroyed*. The total amount of energy is constant; energy can be changed from one form to another or transferred from one object to another, but it cannot be created out of nothing, and it does not vanish into nothing.

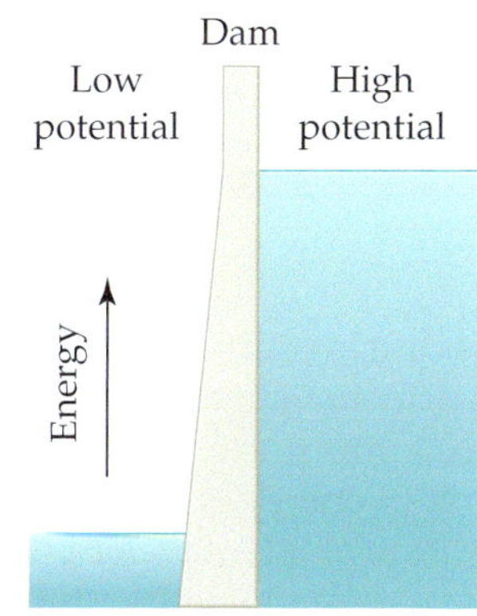

▲ Water behind a dam contains potential energy. © Jose Luis Gutierrez/iStockphoto/Getty Images.

Virtually all samples of matter have energy. The total energy of a sample of matter is the sum of its **kinetic energy**, the energy associated with its motion, and its **potential energy**, the energy associated with its position or composition. For example, a moving billiard ball contains *kinetic energy*, because it is *moving* at some speed across the billiard table. Water behind a dam contains *potential energy*, because it is held at a high *position* in the Earth's gravitational field by the dam. When the water flows through the dam from a higher position to a lower position, it can turn a turbine and produce electrical energy. **Electrical energy** is the energy associated with the flow of electrical charge. **Thermal energy** is the energy associated with the random motions of atoms and molecules in matter. The hotter an object, the more thermal energy it contains.

CHEMISTRY IN THE ENVIRONMENT

▶ Getting Energy out of Nothing?

The law of conservation of energy has significant implications for energy use. The best we can do with energy is break even (and even that is not really possible); we can't continually draw energy from a device without putting energy into it. A device that supposedly produces energy without the need for energy input is sometimes called a *perpetual motion machine* (▶ Figure 9.1). According to the law of conservation of energy, such a machine cannot exist. Occasionally, the media report or speculate on the discovery of a system that appears to produce more energy than it consumes. For example, I once heard a radio talk show on the subject of energy and gasoline costs. The reporter suggested that we simply design an electric car that recharges itself while being driven. The battery in the electric car would charge during operation in the same way that the battery in a conventional car recharges, except the electric car would run with energy from the battery. Although people have dreamed of machines such as this for decades, such ideas violate the law of conservation of energy because they produce energy without any energy input. In the case of the perpetually moving electric car, the fault lies in the idea that driving the electric car can recharge the battery—it can't.

The battery in a conventional car recharges because energy from gasoline combustion is converted into electrical energy that then charges the battery. The electric car needs energy to move forward, and the battery eventually discharges as it provides that energy. Hybrid cars (electric and gasoline-powered) such as the Toyota Prius can capture some limited energy from braking and use that energy to recharge the battery. However, they could never run indefinitely without the addition of fuel. Our society has a continual need for energy, and as our current energy resources dwindle, new energy sources are required. Unfortunately, those sources must also follow the law of conservation of energy—energy must be conserved.

◀ **FIGURE 9.1 A proposed perpetual motion machine** The rolling balls supposedly keep the wheel perpetually spinning. **Question: Can you explain why this would not work?**

B9.1 CAN YOU ANSWER THIS? *A friend asks you to invest in a new flashlight he invented that never needs batteries. What questions should you ask before writing a check?*

Chemical systems contain **chemical energy**, a form of potential energy associated with the positions of the particles that compose the chemical system. For example, the molecules that compose gasoline contain a substantial amount of chemical energy. They are a bit like the water behind a dam. Burning the gasoline is analogous to releasing the water from the dam. The chemical energy present in the gasoline is released upon burning. When we drive a car, we use that chemical energy to move the car forward. When we heat a home, we use chemical energy stored in natural gas to produce heat and warm the air in the house.

Units of Energy

Several different energy units are in common use. The SI unit of energy is the joule (J), named after the English scientist James Joule (1818–1889), who demonstrated that energy could be converted from one type to another as long as the total energy was conserved. A second unit of energy is the **calorie (cal)**, the amount of energy required to raise the temperature of 1 g of water by 1 °C. A calorie is a larger unit than a joule: 1 cal = 4.184 J. A related energy unit is the nutritional or *capital C* **Calorie (Cal)**, equivalent to 1000 *little c* calories. Electricity bills usually come in yet another energy unit, the **kilowatt-hour (kWh).** Table 9.1 lists various energy units and their conversion factors. Table 9.2 shows the amount of energy required for various processes in each of these units.

TABLE 9.1 Energy Conversion Factors

1 calorie (cal)	=	4.184 joules (J)
1 Calorie (Cal)	=	1000 calories (cal)
1 kilowatt-hour (kWh)	=	3.60×10^6 joules (J)

TABLE 9.2 Energy Use in Various Units

Unit	Energy Required to Raise Temperature of 1 g of Water by 1 °C	Energy Required to Light 100-W Bulb for 1 Hour
joule (J)	4.18	3.6×10^5
calorie (cal)	1.00	8.60×10^4
Calorie (Cal)	0.00100	86.0
kilowatt-hour (kWh)	1.16×10^{-6}	0.100

CONCEPTUAL CHECKPOINT 9.1

Suppose a salesperson wants to make an appliance seem as efficient as possible. In which units does the yearly energy consumption of the appliance have the lowest numerical value and therefore seem most efficient?

(a) J **(b)** cal

(c) Cal **(d)** kWh

9.2 Energy and Chemical and Physical Change

LO: Distinguish between exothermic and endothermic reactions.

When discussing energy transfer, we often define the object of our study (such as a flask in which a chemical reaction is occurring) as the *system*. The system then exchanges energy with its *surroundings*. In other words, we view energy changes as an exchange of energy between the system and the surroundings.

The physical and chemical changes that we discussed in Module 1 are usually accompanied by energy changes. For example, when water evaporates from skin (a physical change), the water molecules absorb energy, cooling the skin. When we burn natural gas on a stove (a chemical change), energy is released, heating the food we are cooking.

The release of energy during a chemical reaction is analogous to the release of energy that occurs when a weight falls to the ground. When you lift a weight, you raise its potential energy; when you drop it, you release the potential energy (◀ Figure 9.2). *Systems with high potential energy—like the raised weight—have a tendency to change in a way that lowers their potential energy.* For this reason, objects or systems with high potential energy tend to be *unstable*. A weight lifted several metres from the ground is unstable because it contains a significant amount of localized potential energy. Unless restrained, the weight will fall, lowering its potential energy.

▲ **FIGURE 9.2 Potential energy of raised weight** A weight lifted off the ground has a high potential and will tend to fall toward the ground to lower its potential energy.

Some chemical substances are like a raised weight. For example, the molecules that compose TNT (trinitrotoluene) have a relatively high potential energy—energy is concentrated in them just as energy is concentrated in the raised weight. TNT molecules therefore tend to undergo rapid chemical changes that lower their potential energy, which is why TNT is explosive. Chemical reactions that *release* energy, like the explosion of TNT, are **exothermic**.

Some chemical reactions behave in just the opposite way—they *absorb* energy from their surroundings as they occur. Such reactions are **endothermic**. The reaction that occurs in a chemical cold pack is a good example of an endothermic reaction. When you break the barrier separating the reactants in the chemical cold pack, the substances mix, react, and absorb heat from the surroundings. The surroundings—possibly including your bruised ankle—get colder.

We can represent the energy changes that occur during a chemical reaction with an energy diagram, as shown in ▶ Figure 9.3. In an exothermic reaction (▶ Figure 9.3(a), the reactants have greater energy than the products, and energy is released as the reaction occurs. In an endothermic reaction (▶ Figure 9.3(b), the products have more energy than the reactants, and energy is absorbed as the reaction occurs.

If a particular reaction or process is exothermic, then the reverse process must be endothermic. For example, the evaporation of water from skin is endothermic (and therefore cools you off), but the condensation of water onto skin is exothermic (which is why 9.3(a), steam burns can be so painful).

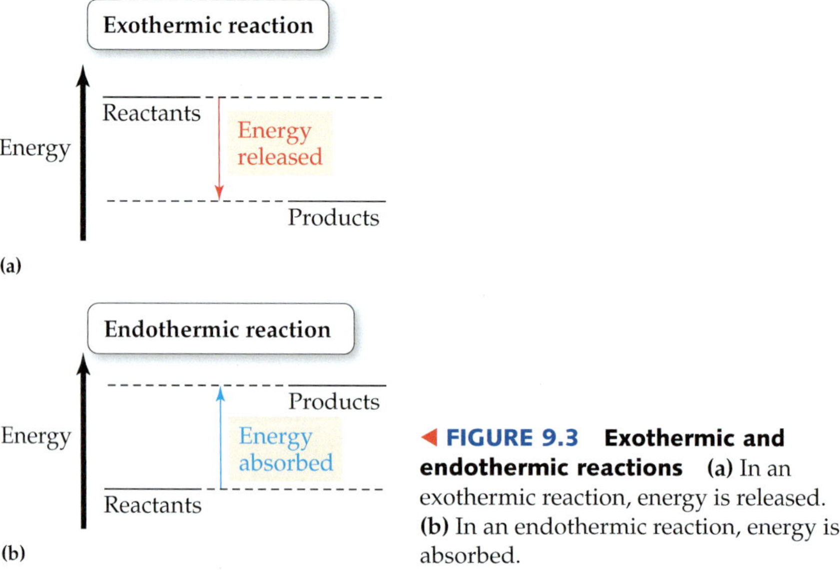

◀ **FIGURE 9.3 Exothermic and endothermic reactions** **(a)** In an exothermic reaction, energy is released. **(b)** In an endothermic reaction, energy is absorbed.

EXAMPLE 9.1 EXOTHERMIC AND ENDOTHERMIC PROCESSES

Classify each change as exothermic or endothermic.

(a) wood burning in a fire
(b) ice melting

SOLUTION

(a) When wood burns, it emits heat into the surroundings. Therefore, the process is exothermic.
(b) When ice melts, it absorbs heat from the surroundings. For example, when ice melts in a glass of water, it cools the water as the melting ice absorbs heat from the water. Therefore, the process is endothermic.

▶SKILLBUILDER 9.1 | Exothermic and Endothermic Processes

Classify each change as exothermic or endothermic.

(a) water freezing into ice
(b) natural gas burning

▶FOR MORE PRACTICE Problems 41, 42, 43, 44.

9.3 Enthalpy

LO: Describe enthalpy and interpret the sign of enthalpy changes.

The previous sections in this module describe how chemical reactions can be *exothermic* (in which case they *emit* thermal energy when they occur) or *endothermic* (in which case they *absorb* thermal energy when they occur). The *amount* of thermal energy emitted or absorbed by a chemical reaction, under conditions of constant pressure (which are common for most everyday reactions), can be quantified with a function called **enthalpy**. Specifically, we define a quantity called the **enthalpy of reaction** (ΔH_{rxn}) as the amount of thermal energy (or heat) that flows when a reaction occurs at constant pressure.

Sign of ΔH_{rxn}

The *sign* of ΔH_{rxn} (positive or negative) depends on the *direction* in which thermal energy flows when the reaction occurs. If thermal energy flows out of the reaction and into the surroundings (as in an exothermic reaction), then ΔH_{rxn} is negative. For example, we can specify the enthalpy of reaction for the combustion of CH_4, the main component in natural gas, as:

$$CH_4(g) + 2\,O_2(g) \rightarrow CO_2(g) + 2\,H_2O(g) \qquad \Delta H_{rxn} = -802.3\text{ kJ}$$

This reaction is exothermic and therefore has a negative enthalpy of reaction. The magnitude of ΔH_{rxn} tells us that 802.3 kJ of heat are emitted when 1 mol of CH_4 reacts with 2 mol of O_2.

If, by contrast, thermal energy flows into the reaction and out of the surroundings (as in an endothermic reaction), then ΔH_{rxn} is positive. For example, we specify the enthalpy of reaction for the reaction between nitrogen and oxygen gas to form nitrogen monoxide as:

$$N_2(g) + O_2(g) \rightarrow 2\,NO(g) \qquad \Delta H_{rxn} = +182.6\text{ kJ}$$

This reaction is endothermic and therefore has a positive enthalpy of reaction. When 1 mol of N_2 reacts with 1 mol of O_2, 182.6 kJ of heat are absorbed from the surroundings.

We can think of the energy of a chemical system in the same way that you think about the balance in a checking account. Energy flowing *out* of the chemical system is like a withdrawal and carries a negative sign as shown in ▲Figure 9.4(a). Energy flowing *into* the system is like a deposit and carries a positive sign as shown in ▲ Figure 9.4(b).

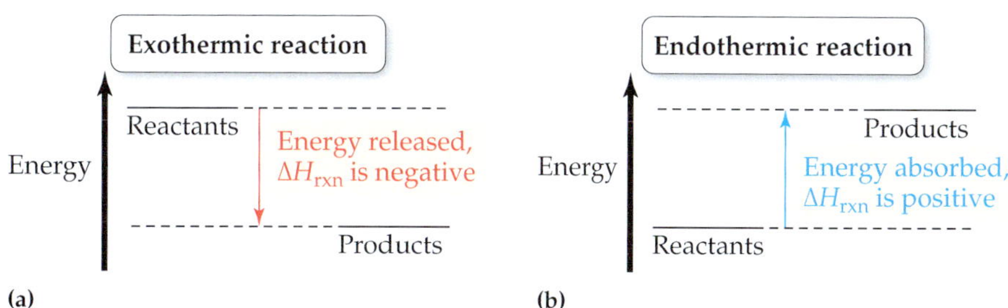

▶ **FIGURE 9.4 Exothermic and endothermic reactions** **(a)** In an exothermic reaction, energy is released into the surroundings. **(b)** In an endothermic reaction, energy is absorbed from the surroundings.

9.4 Life: Controlled Disequilibrium

Have you ever tried to define life? If you have, you know that it is not easily defined. How are living things different from nonliving things? You may try to define living things as those things that can move. But of course many living things do not move—many plants, for example, do not move very much—and some nonliving things, such as glaciers and Earth itself, do move. So motion is neither unique to nor definitive of life. You may try to define living things as those things that can reproduce. But again, many living things, such as mules or sterile humans, cannot reproduce; yet they are alive. In addition, some nonliving things—such as crystals—reproduce (in some sense). So what is unique about living things?

The concept of equilibrium underlies one definition of life. We define *chemical* equilibrium more carefully soon; for now, we can think generally of equilibrium as *sameness and constancy*. When an object is in equilibrium with its surroundings, some property of the object has reached sameness with the surroundings and is no longer changing. For example, a cup of hot water is not in equilibrium with its surroundings with respect to temperature. If left undisturbed, the cup of hot water will slowly cool until it reaches equilibrium with its surroundings. At that point, the temperature of the water is the *same as* that of the surroundings (sameness) and *no longer changes* (constancy).

So equilibrium involves sameness and constancy. Part of a definition for living things, then, is that living things *are not* in equilibrium with their surroundings. Our body temperature, for example, is not the same as the temperature of our surroundings. When we jump into a swimming pool, the pH of our blood does not become the same as the pH of the surrounding water. Living things, even the simplest ones, maintain some measure of *disequilibrium* with their environment.

We must add one more concept, however, to complete our definition of life with respect to equilibrium. A cup of hot water is in disequilibrium with its environment, yet it is not alive. However, the cup of hot water has no control over its disequilibrium and will slowly come to equilibrium with its environment. In contrast, living things—as long as they are alive—maintain and *control* their disequilibrium. Your body temperature, for example, is not only in disequilibrium with your surroundings—it is in controlled disequilibrium. Your body maintains your temperature within a specific range that is not in equilibrium with the surrounding temperature.

So one definition of life is that living things are in *controlled disequilibrium* with their environment. A living thing comes into equilibrium with its surroundings only after it dies. In this module, we examine the concept of equilibrium, especially chemical equilibrium—the state that involves sameness and constancy.

9.5 The Rate of a Chemical Reaction

LO: Identify and understand the relationship between concentration and temperature and the rate of a chemical reaction.

Reaction rates are related to chemical equilibrium because, as we will see in Section 9.6, a chemical system is at equilibrium when the rate of the forward reaction equals the rate of the reverse reaction.

A reaction rate can also be defined as the amount of a product that forms in a given period of time.

Before we probe more deeply into the concept of chemical equilibrium, we must first understand something about the rates of chemical reactions. The **rate of a chemical reaction**—a measure of how fast the reaction proceeds—is defined as the amount of reactant that changes to product in a given period of time. A reaction with a fast rate proceeds quickly; a large amount of reactant is converted to product in a certain period of time (▶ Figure 9.5(a). A reaction with a slow rate proceeds slowly; only a small amount of reactant is converted to product in the same period of time (▶ Figure 9.5(b).

Chemists seek to control reaction rates for many chemical reactions. For example, rockets can be propelled by the reaction of hydrogen and oxygen to form water. If the reaction proceeds too slowly, the rocket will not lift off the ground. If, however, the reaction proceeds too quickly, the rocket can explode. Reaction rates can be controlled if we understand the factors that influence them.

Collision Theory

According to **collision theory**, chemical reactions occur through collisions between molecules or atoms. For example, consider the gas-phase chemical reaction between $H_2(g)$ and $I_2(g)$ to form $HI(g)$.

$$H_2(g) + I_2(g) \rightarrow 2\ HI(g)$$

The gas-phase reaction between hydrogen and iodine can proceed by other mechanisms, but the mechanism here is valid for the low-temperature thermal reaction.

Whether a collision leads to a reaction also depends on the *orientation* of the colliding molecules, but this topic is beyond the scope of this text.

The reaction begins when an H_2 molecule collides with an I_2 molecule. If the collision occurs with enough energy—that is, if the colliding molecules are moving fast enough—the reaction can proceed to form the products. If the collision occurs with insufficient energy, the reactant molecules (H_2 and I_2) simply bounce off of one another. Since gas-phase molecules have a wide distribution of velocities, collisions occur with a wide distribution of energies. High-energy collisions lead to products, and low-energy collisions do not.

Higher-energy collisions are more likely to lead to products because most chemical reactions have an *activation energy* (or an activation barrier). We discuss the activation energy for chemical reactions in more detail in Section 9.13. For now, we can think of the activation energy as an energy barrier that must be overcome for the reaction to proceed. For example, in the case of H_2 reacting with I_2 to form HI, the product (HI) can begin to form only after the H—H bond and the I—I bond each begin to break. The activation energy is the energy required to begin to break these bonds.

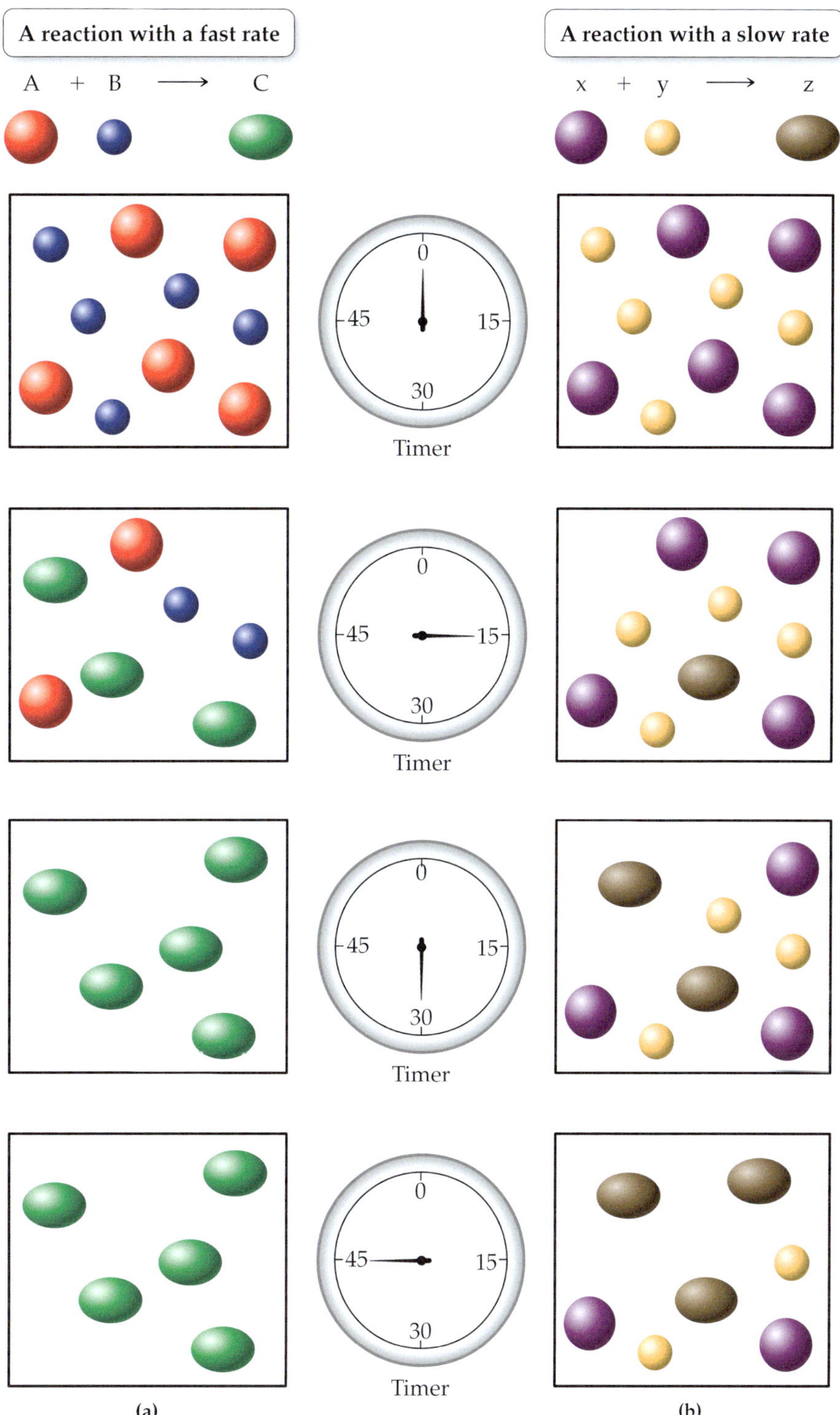

▲ **FIGURE 9.5 Reaction rates** **(a)** In a reaction with a fast rate, the reactants react to form products in a short period of time. **(b)** In a reaction with a slow rate, the reactants react to form products over a long period of time.

If molecules react via high-energy collisions, then the factors that influence the rate of a reaction must be the same factors that affect the number of high-energy collisions that occur per unit time. Here, we focus on the two most important factors that affect collisions: the *concentration* of the reacting molecules and the *temperature* of the reaction mixture.

How Concentration Affects the Rate of a Reaction

▼ Figure 9.6 shows various mixtures of H_2 and I_2 at the same temperature but different concentrations. If H_2 and I_2 react via collisions to form HI, which mixture do you think has the highest reaction rate? Since ▼ Figure 9.6(c) has the highest concentration of H_2 and I_2, it has the most collisions per unit time and therefore the fastest reaction rate. This idea holds true for most chemical reactions.

> The rate of a chemical reaction generally increases with increasing concentration of the reactants.

The exact relationship between increases in concentration and increases in reaction rate varies for different reactions and is beyond the scope of this text. For our purposes, we just need to know that for most reactions the reaction rate increases with increasing reactant concentration.

Armed with this knowledge, what can we say about the rate of a reaction as the reaction proceeds? Since reactants turn into products in the course of a reaction, their concentration decreases. Consequently, the reaction rate decreases as well. In other words, as a reaction proceeds, there are fewer reactant molecules (because they have turned into products), and the reaction slows down.

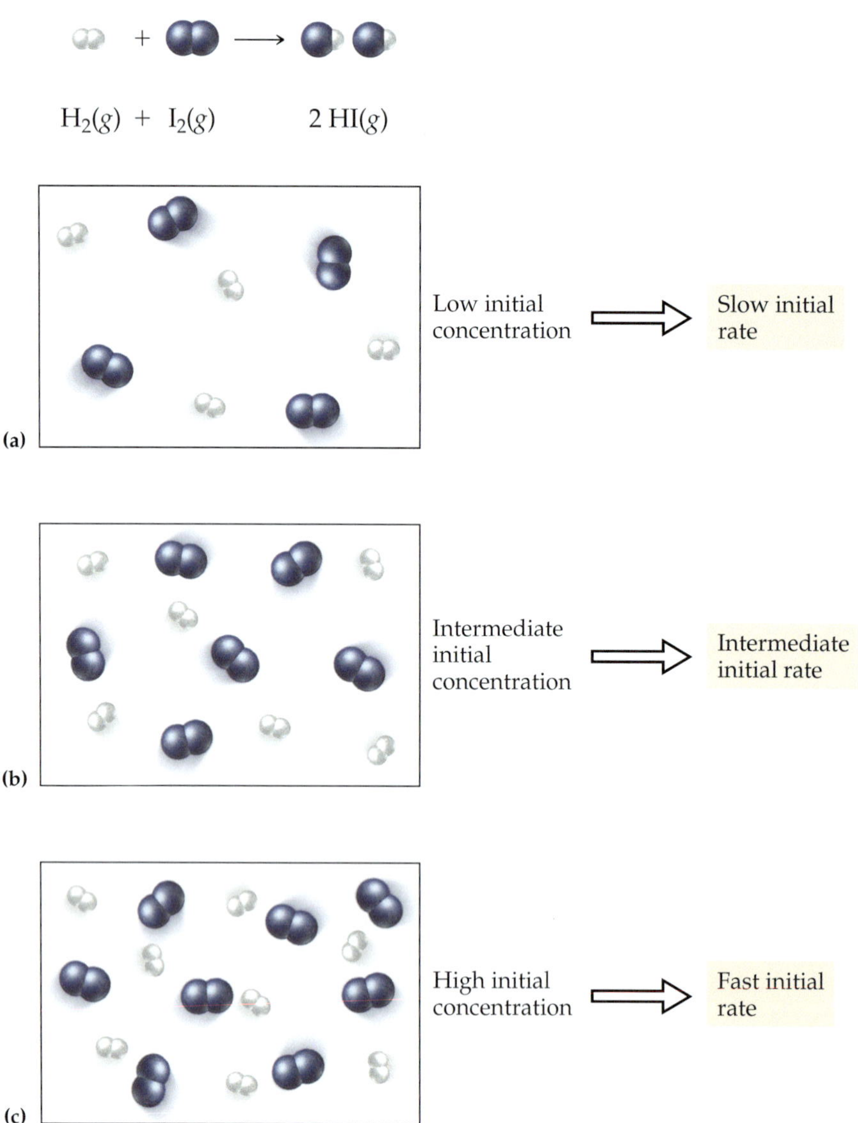

▲ **FIGURE 9.6 Effect of concentration on reaction rate** **Question: Which reaction mixture will have the fastest initial rate? The mixture in (c) is fastest because it has the highest concentration of reactants and therefore the highest rate of collisions.**

How Temperature Affects the Rate of a Reaction

Reaction rates also depend on temperature. ▼ Figure 9.7 shows various mixtures of H_2 and I_2 at the same concentration, but different temperatures. Which will have the fastest rate? Recall that raising the temperature makes the molecules move faster (Section 6.4). They therefore experience more collisions per unit time, resulting in a faster reaction rate. In addition, a higher temperature results in more collisions that are (on average) of higher energy. Because the high-energy collisions are the ones that result in products, this also produces a faster rate. Consequently, ▼ Figure 9.7(c) (which has the highest temperature) has the fastest reaction rate. This relationship holds true for most chemical reactions.

> The rate of a chemical reaction generally increases with increasing temperature of the reaction mixture.

The temperature dependence of reaction rates is the reason that cold-blooded animals become more sluggish at lower temperatures. The reactions required for them to think and move become slower, resulting in the sluggish behavior.

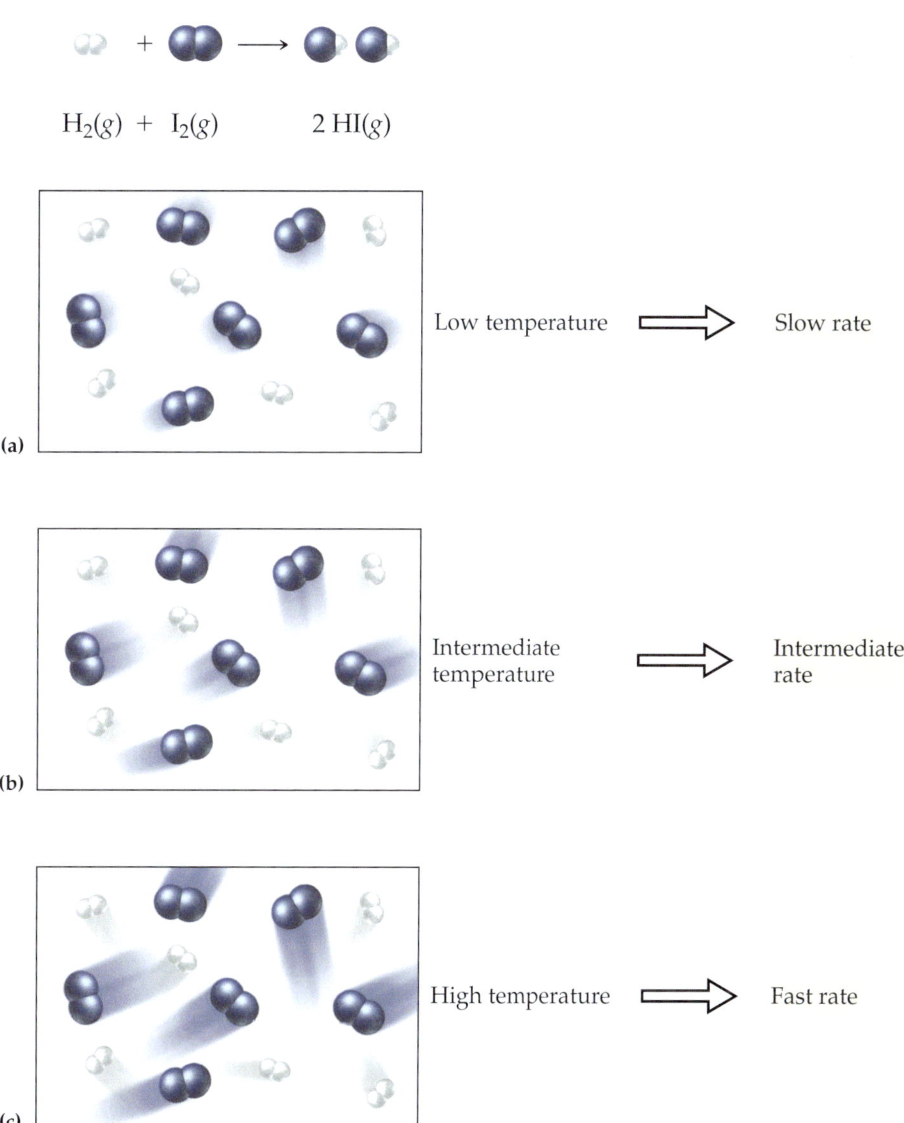

▲ **FIGURE 9.7 Effect of temperature on reaction rate** **Question: Which reaction mixture will have the fastest initial rate? The mixture in (c) is fastest because it has the highest temperature.**

To summarize:

- Reaction rates generally increase with increasing reactant concentration.
- Reaction rates generally increase with increasing temperature.
- Reaction rates generally decrease as a reaction proceeds.

CONCEPTUAL CHECKPOINT 9.2

In a chemical reaction between two gases, what would you expect to be the result of increasing the pressure of the gases?

(a) An increase in the reaction rate.

(b) A decrease in the reaction rate.

(c) No affect on the reaction rate.

9.6 The Idea of Dynamic Chemical Equilibrium

LO: Define dynamic equilibrium.

What would happen if our reaction between H_2 and I_2 to form HI were able to proceed in both the forward and reverse directions?

$$H_2(g) + I_2(g) \rightleftharpoons 2\ HI\ (g)$$

In this case, H_2 and I_2 would collide and react to form 2 HI molecules, but the 2 HI molecules also collide and react to reform H_2 and I_2. A reaction that can proceed in both the forward and reverse directions is a **reversible reaction**.

Suppose we begin with only H_2 and I_2 in a container ▶ Figure 9.8(a). What happens initially? The H_2 and I_2 molecules begin to react to form HI ▶ Figure 9.8(b). However, as H_2 and I_2 react, their concentration decreases, which in turn decreases the rate of the forward reaction. At the same time, HI begins to form. As the concentration of HI increases, the reverse reaction begins to occur at an increasingly faster rate because there are more HI collisions with other HI molecules. Eventually, the rate of the reverse reaction (which is increasing) equals the rate of the forward reaction (which is decreasing). At that point, **dynamic equilibrium** is reached ▶ Figure 9.8(c) and ▶ Figure 9.8(d).

> **Dynamic equilibrium**—In a chemical reaction, the condition in which the rate of the forward reaction equals the rate of the reverse reaction.

This condition is not static—it is dynamic because the forward and reverse reactions are still occurring but at the same constant rate. When dynamic equilibrium is reached, the concentrations of H_2, I_2, and HI no longer change. They remain the same because the reactants and products are being depleted at the same rate at which they are being formed.

Notice that dynamic equilibrium includes the concepts of sameness and constancy that we discussed earlier in this module. When dynamic equilibrium is reached, the forward reaction rate is the same as the reverse reaction rate (sameness). Because the reaction rates are the same, the concentrations of the reactants and products no longer change (constancy). However, just because the concentrations of reactants and products no longer change at equilibrium does *not* imply that the concentrations of reactants and products are *equal* to one another at equilibrium. Some reactions reach equilibrium only after most of the reactants have formed products. Others reach equilibrium when only a small

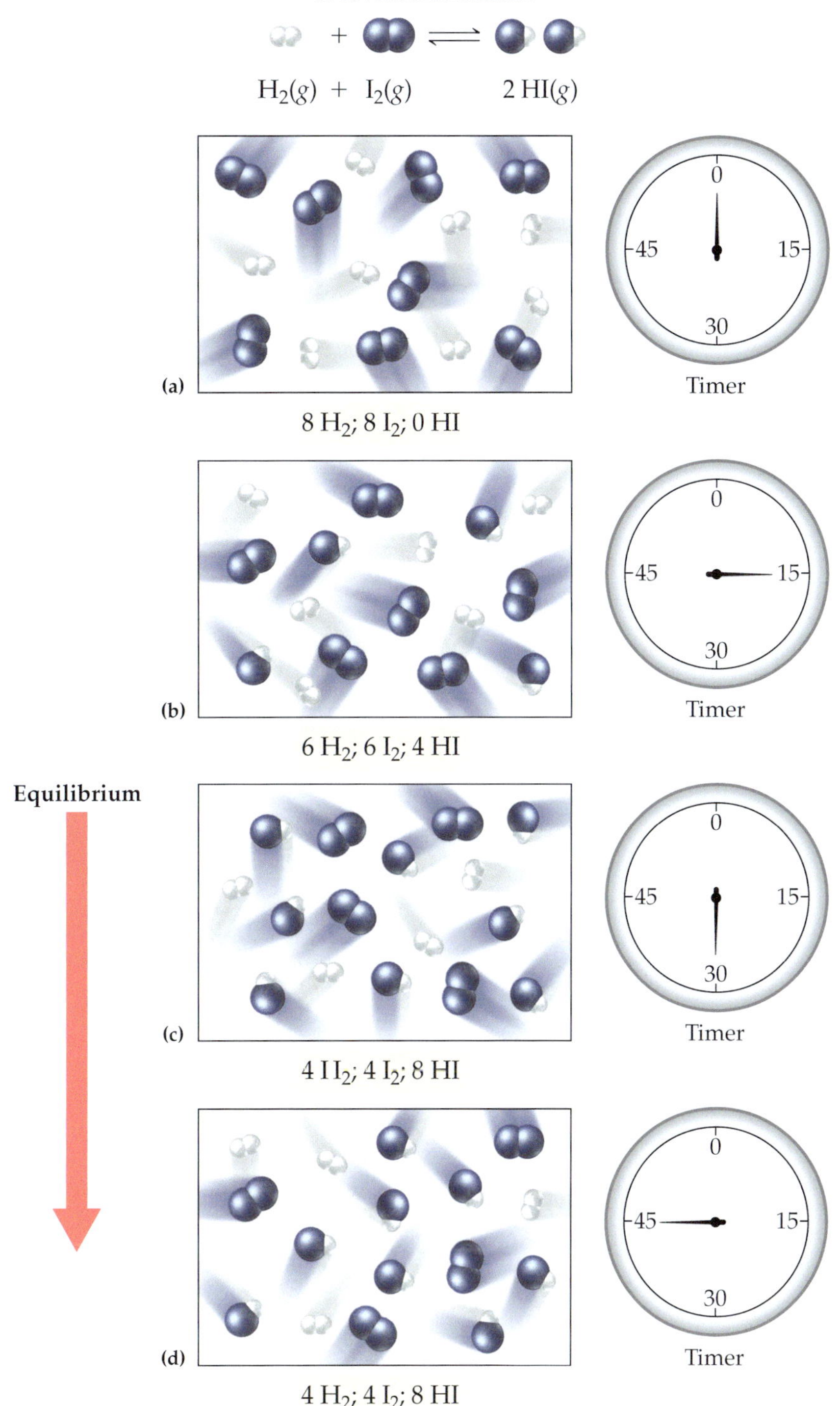

▶ **FIGURE 9.8 Equilibrium**
When the concentrations of the reactants and products no longer change, equilibrium has been reached.

fraction of the reactants have formed products (recall our discussion of weak acids in Module 8.) It depends on the reaction.

We can better understand dynamic equilibrium with a simple analogy. Imagine that Narnia and Middle Earth are two neighboring kingdoms (▶ Figure 9.9). Narnia is overpopulated, and Middle Earth is underpopulated. One day, however, the border between the two kingdoms opens, and people immediately begin to leave Narnia for Middle Earth (call this the forward reaction).

$$\text{Narnia} \rightarrow \text{Middle Earth} \quad \text{(forward reaction)}$$

Narnia is the fictitious world featured in C.S. Lewis's *The Chronicles of Narnia*, and Middle Earth is the fictitious world featured in J.R.R. Tolkien's *The Lord of the Rings*.

FIGURE 9.9 Population analogy for a chemical reaction proceeding to equilibrium

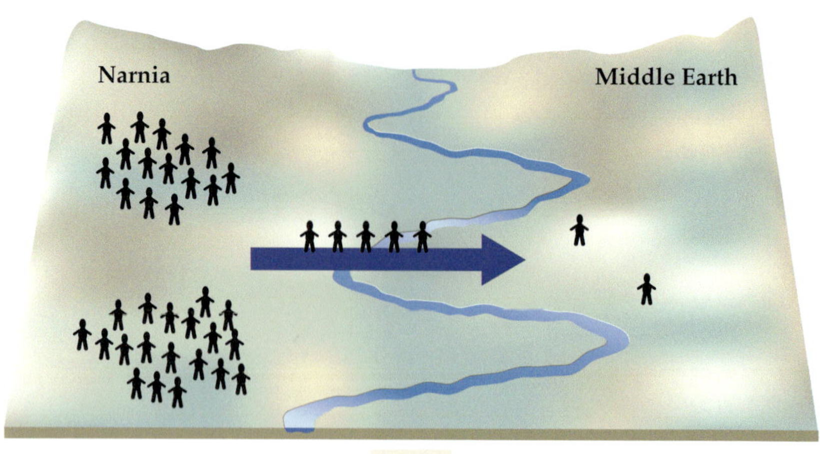

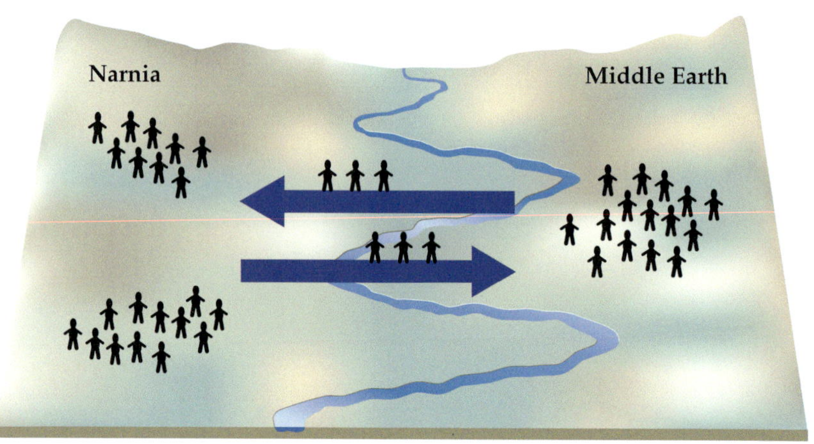

The population of Narnia decreases as the population of Middle Earth increases. As people leave Narnia, however, the *rate* at which they leave begins to slow down (because Narnia becomes less crowded). On the other hand, as people move into Middle Earth, some decide it was not for them and begin to move back (call this the reverse reaction).

$$\text{Narnia} \leftarrow \text{Middle Earth} \quad \text{(reverse reaction)}$$

As Middle Earth fills, the rate of people moving back to Narnia accelerates. Eventually, the *rate* of people moving out of Narnia (which has been slowing down as people leave) equals the *rate* of people moving back to Narnia (which has been increasing as Middle Earth gets more crowded). Dynamic equilibrium has been reached.

$$\text{Narnia} \rightleftharpoons \text{Middle Earth}$$

Notice that when the two kingdoms reach dynamic equilibrium, their populations no longer change because the number of people moving out equals the number of people moving in. However, one kingdom—because of its charm, or the character of its leader, or a lower tax rate, or whatever other reason—may have a higher population than the other kingdom, even when dynamic equilibrium is reached.

Similarly, when a chemical reaction reaches dynamic equilibrium, the rate of the forward reaction (analogous to people moving out of Narnia) equals the rate of the reverse reaction (analogous to people moving back into Narnia), and the relative concentrations of reactants and products (analogous to the relative populations of the two kingdoms) become constant. Also, like our two kingdoms, the concentrations of reactants and products are not necessarily equal at equilibrium, just as the populations of the two kingdoms are not equal at equilibrium.

9.7 The Equilibrium Constant: A Measure of How Far a Reaction Goes

LO: Write equilibrium constant expressions for chemical reactions.

We have just learned that the *concentrations* of reactants and products are not equal at equilibrium; rather, it is the *rates* of the forward and reverse reactions that are equal. But what about the concentrations? What can we know about them? The equilibrium constant (K_{eq}) is a way to quantify the relative concentrations of the reactants and products at equilibrium. Consider the generic chemical reaction:

$$aA + bB \rightleftharpoons cC + dD$$

where A and B are reactants, C and D are products, and a, b, c, and d are the respective stoichiometric coefficients in the chemical equation. The **equilibrium constant (K_{eq})** for the reaction is defined as the ratio—at equilibrium—of the concentrations of the products raised to their stoichiometric coefficients divided by the concentrations of the reactants raised to their stoichiometric coefficients.

$$K_{eq} = \frac{[C]^c\,[D]^d}{[A]^a\,[B]^b}$$

Products

Reactants

Notice that the equilibrium constant is a measure of the relative concentrations of reactants and products at equilibrium; the larger the equilibrium constant, the greater the concentration of products relative to reactants at equilibrium.

Writing Equilibrium Constant Expressions for Chemical Reactions

To write an equilibrium constant expression for a chemical reaction, we examine the chemical equation and follow the definition for the equilibrium constant. For example, suppose we want to write an equilibrium expression for this reaction:

$$2\,N_2O_5(g) \rightleftharpoons 4\,NO_2(g) + O_2(g)$$

The equilibrium constant is $[NO_2]$ raised to the fourth power multiplied by $[O_2]$ raised to the first power divided by $[N_2O_5]$ raised to the second power.

$$K_{eq} = \frac{[NO_2]^4[O_2]}{[N_2O_5]^2}$$

Notice that the *coefficients* in the chemical equation become the *exponents* in the equilibrium expression.

$$2\,N_2O_5(g) \rightleftharpoons 4\,NO_2(g) + O_2(g)$$

Implied 1

$$K_{eq} = \frac{[NO_2]^4\,[O_2]}{[N_2O_5]^2}$$

The Significance of the Equilibrium Constant

The symbol >> means *much greater than.*

What does an equilibrium constant tell us? For example, what does a large equilibrium constant ($K_{eq} >> 1$) imply about a reaction? It indicates that the forward reaction is largely favored and that there will be more products than reactants when equilibrium is reached. For example, consider the reaction:

$$H_2(g) + Br_2(g) \rightleftharpoons 2\,HBr(g) \quad K_{eq} = 1.9 \times 10^{19} \text{ at } 25\ ^\circ C$$

The equilibrium constant is large, meaning that at equilibrium the reaction lies far to the right—high concentrations of products, tiny concentrations of reactants (▼ Figure 9.10).

The symbol << means *much less than.*

Conversely, what does a *small* equilibrium constant ($K_{eq} << 1$) mean? It indicates that the reverse reaction is favored and that there will be more reactants than products when equilibrium is reached. For example, consider the reaction:

$$N_2(g) + O_2(g) \rightleftharpoons 2\,NO(g) \quad K_{eq} = 4.1 \times 10^{-31} \text{ at } 25\ ^\circ C$$

The equilibrium constant is very small, meaning that at equilibrium the reaction lies far to the left—high concentrations of reactants, low concentrations of products

$$H_2(g) + Br_2(g) \rightleftharpoons 2\,HBr(g)$$

$$K_{eq} = \frac{[HBr]^2}{[H_2][Br_2]} = \textbf{Large Number}$$

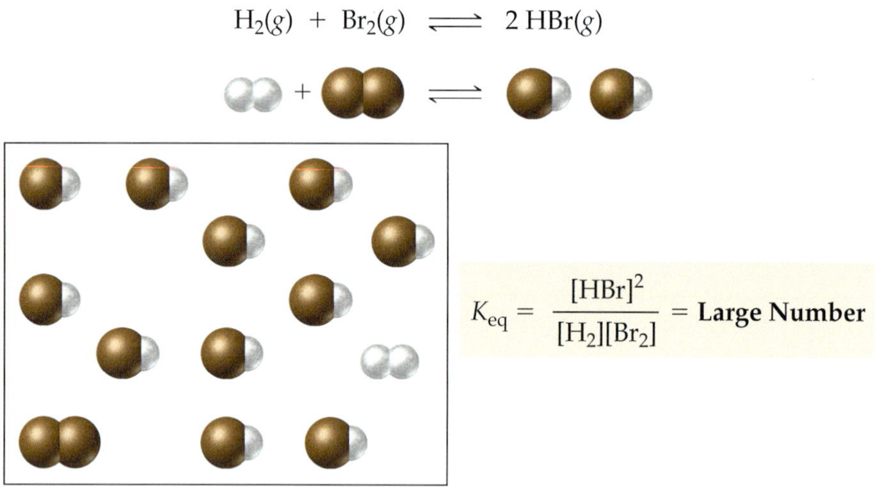

▶ **FIGURE 9.10 The meaning of a large equilibrium constant** A large equilibrium constant means that there is a high concentration of products and a low concentration of reactants at equilibrium.

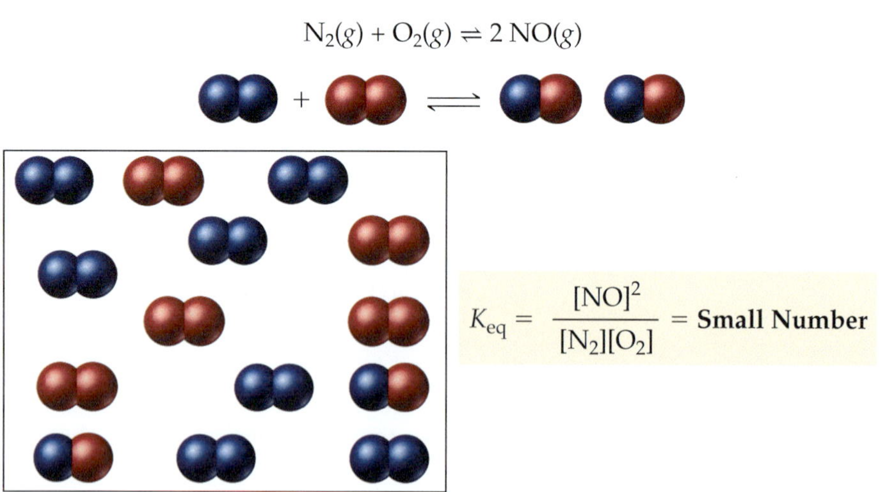

▲ **FIGURE 9.11 The meaning of a small equilibrium constant** A small equilibrium constant means that there is a high concentration of reactants and a low concentration of products at equilibrium.

(◀ Figure 9.11). This is fortunate because N_2 and O_2 are the main components of air. If this equilibrium constant were large, much of the N_2 and O_2 in air would react to form NO, a toxic gas.

To summarize:

- $K_{eq} << 1$ Reverse reaction is favored; forward reaction does not proceed very far.
- $K_{eq} \approx 1$ Neither direction is favored; forward reaction proceeds about halfway (significant amounts of both reactants and products are present at equilibrium).
- $K_{eq} >> 1$ Forward reaction is favored; forward reaction proceeds virtually to completion.

The symbol ≈ means "approximately equal to."

EXAMPLE 9.2 WRITING EQUILIBRIUM CONSTANT EXPRESSIONS FOR CHEMICAL REACTIONS

Write an equilibrium expression for the chemical equation.

$$CO(g) + 2\,H_2(g) \rightleftharpoons CH_3OH(g)$$

SOLUTION

The equilibrium expression is the concentration of the products raised to their stoichiometric coefficients divided by the concentration of the reactants raised to their stoichiometric coefficients. Notice that the expression is a ratio of products over reactants. Notice also that the coefficients in the chemical equation are the exponents in the equilibrium expression.

$$K_{eq} = \frac{[CH_3OH]}{[CO][H_2]^2}$$

(Product: $[CH_3OH]$; Reactants: $[CO][H_2]^2$)

▶SKILLBUILDER 9.2 | Writing Equilibrium Expressions for Chemical Reactions

Write an equilibrium expression for the chemical equation.

$$H_2(g) + F_2(g) \rightleftharpoons 2\,HF(g)$$

▶FOR MORE PRACTICE Example 9.7; Problems 53, 54, 57, 59.

CONCEPTUAL CHECKPOINT 9.3

Consider a generic chemical reaction in which the reactant, A, turns directly into the product, B.

$$A(g) \rightleftharpoons B(g)$$

The reaction is allowed to come to equilibrium at three different temperatures as represented in the three circles. At which temperature is the equilibrium constant the largest?

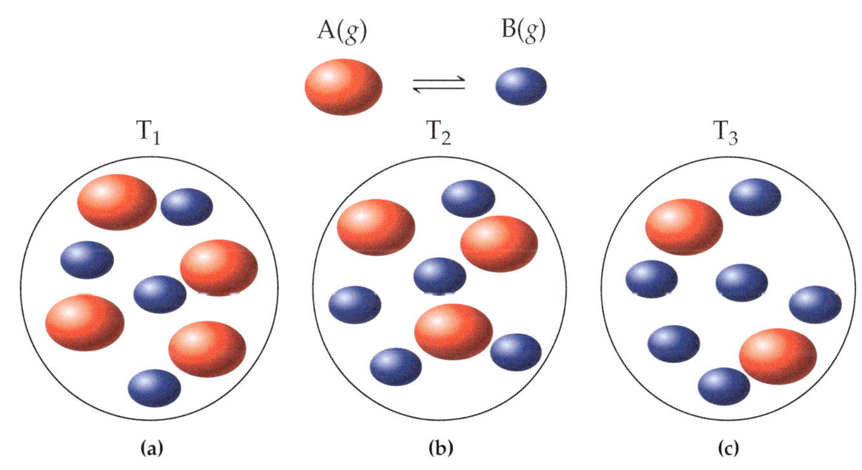

9.8 Heterogeneous Equilibria: The Equilibrium Expression for Reactions Involving a Solid or a Liquid

LO: **Write equilibrium expressions for chemical reactions involving a solid or a liquid.**

Consider the chemical reaction shown here:

$$2\,CO(g) \rightleftharpoons CO_2(g) + C(s)$$

We might expect the expression for the equilibrium constant to be:

$$K_{eq} = \frac{[CO_2][C]}{[CO]^2} \quad \text{(incorrect)}$$

However, since carbon is a solid, its concentration is constant—it does not change. Adding more or less carbon to the reaction mixture does not change the concentration of carbon. The concentration of a solid does not change because a solid does not expand to fill its container. The concentration of a solid, therefore, depends only on its density, which (except for slight variations due to temperature) is constant as long as *some solid is present*. Consequently, pure solids—those reactants or products labeled in the chemical equation with an (*s*)—are not included in the equilibrium expression. The correct equilibrium expression is:

We exclude the concentrations of pure solids and pure liquids from equilibrium expressions because they are constant.

$$K_{eq} = \frac{[CO_2]}{[CO]^2} \quad \text{(correct)}$$

Similarly, the concentration of a pure liquid does not change. Consequently, pure liquids—those reactants or products labeled in the chemical equation with an (*l*)—are also excluded from the equilibrium expression. For example, what is the equilibrium expression for the following reaction?

$$CO_2(g) + H_2O(l) \rightleftharpoons H^+(aq) + HCO_3^-(aq)$$

Because $H_2O(l)$ is pure liquid, it is omitted from the equilibrium expression.

$$K_{eq} = \frac{[H^+][HCO_3^-]}{[CO_2]}$$

EXAMPLE 9.3 WRITING EQUILIBRIUM EXPRESSIONS FOR REACTIONS INVOLVING A SOLID OR A LIQUID

Write an equilibrium expression for the chemical equation.

$$CaCO_3(s) \rightleftharpoons CaO(s) + CO_2(g)$$

SOLUTION

Since $CaCO_3(s)$ and $CaO(s)$ are both solids, they are omitted from the equilibrium expression.

$$K_{eq} = [CO_2]$$

▶SKILLBUILDER 9.3 | Writing Equilibrium Expressions for Reactions Involving a Solid or a Liquid

Write an equilibrium expression for the chemical equation.

$$4\,HCl(g) + O_2(g) \rightleftharpoons 2\,H_2O(l) + 2\,Cl_2(g)$$

▶FOR MORE PRACTICE Problems 55, 56.

9.9 Disturbing a Reaction at Equilibrium: Le Châtelier's Principle

LO: Restate Le Châtelier's principle.

We have seen that a chemical system not in equilibrium tends to go toward equilibrium and that the concentrations of the reactants and products at equilibrium correspond to the equilibrium constant, K_{eq}. What happens, however, when a chemical system already at equilibrium is disturbed? **Le Châtelier's principle** states that the chemical system will respond to minimize the disturbance.

Pronounced "le-sha-te-lyay."

> **Le Châtelier's principle**—When a chemical system at equilibrium is disturbed, the system shifts in a direction that minimizes the disturbance.

In other words, a system at equilibrium tries to maintain that equilibrium—it fights back when disturbed.

We can understand Le Châtelier's principle by returning to our Narnia and Middle Earth analogy. Suppose the populations of Narnia and Middle Earth are at equilibrium. This means that the rate of people moving out of Narnia (and into Middle Earth) is equal to the rate of people moving into Narnia (and out of Middle Earth). It also means that the populations of the two kingdoms are stable. Now imagine disturbing that balance (▼ Figure 9.12). Suppose we add extra

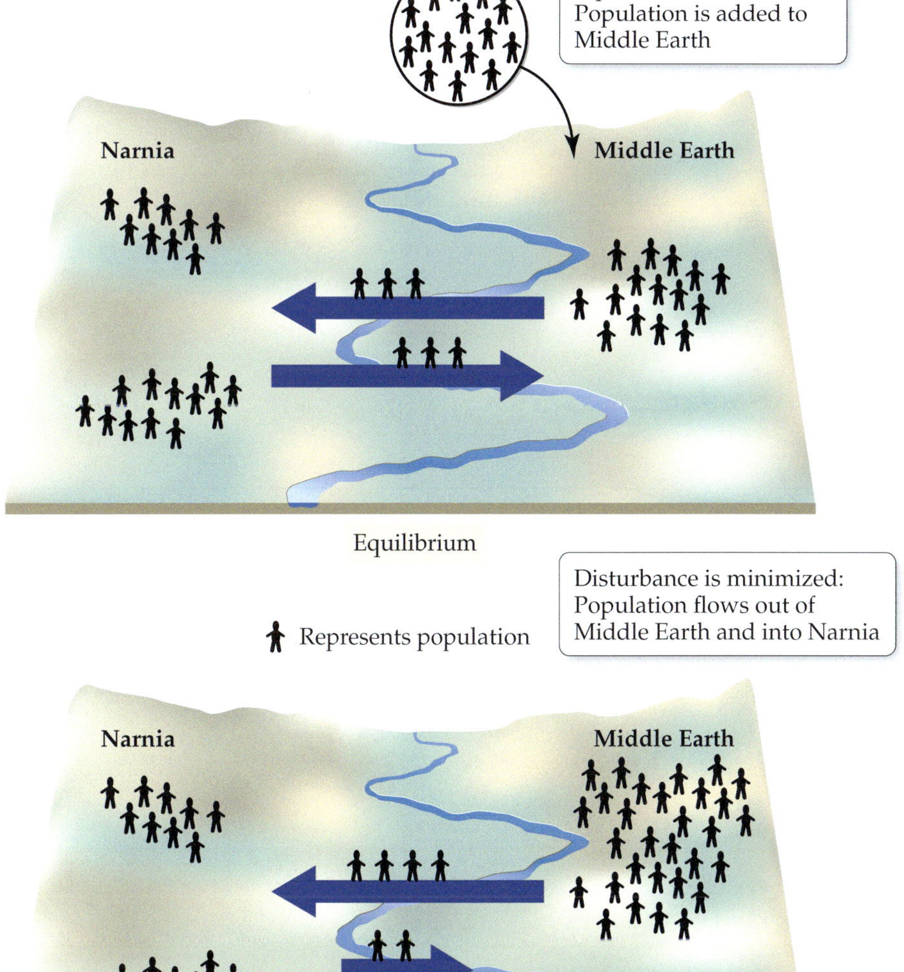

▶ **FIGURE 9.12 Population analogy for Le Châtelier's principle** When a system at equilibrium is disturbed, it shifts to minimize the disturbance. In this case, adding population to Middle Earth (the disturbance) causes population to move out of Middle Earth (minimizing the disturbance). **Question: What would happen if you disturbed the equilibrium by taking population out of Middle Earth? In which direction would the population move to minimize the disturbance?**

people to Middle Earth. What happens? Since Middle Earth suddenly becomes more crowded, the rate of people leaving Middle Earth increases. The net flow of people is *out of Middle Earth and into Narnia.* Notice what happened. We disturbed the equilibrium by adding more people to Middle Earth. The system responded by moving people out of Middle Earth—it shifted in the direction that minimized the disturbance.

On the other hand, what happens if we add extra people to Narnia? Since Narnia suddenly gets more crowded, the rate of people leaving Narnia goes up. The net flow of people is out of Narnia and into Middle Earth. We added people to Narnia, and the system responded by moving people out of Narnia. When systems at equilibrium are disturbed, they react to counter the disturbance. Chemical systems behave similarly. There are several ways to disturb a system in chemical equilibrium. We consider each of these separately in the next three sections of the module.

9.10 The Effect of a Concentration Change on Equilibrium

LO: Apply Le Châtelier's principle in the case of a change in concentration.

Consider the following reaction at chemical equilibrium:

$$N_2O_4(g) \rightleftharpoons 2\,NO_2(g)$$

Suppose we disturb the equilibrium by adding NO_2 to the equilibrium mixture (▼ Figure 9.13). In other words, we increase the concentration of NO_2. What happens? According to Le Châtelier's principle, the system shifts in a direction to minimize the disturbance. The shift is caused by the increased concentration of NO_2, which in turn increases the rate of the reverse reaction because reaction rates generally increase with increasing concentration (as we discussed earlier in this module).

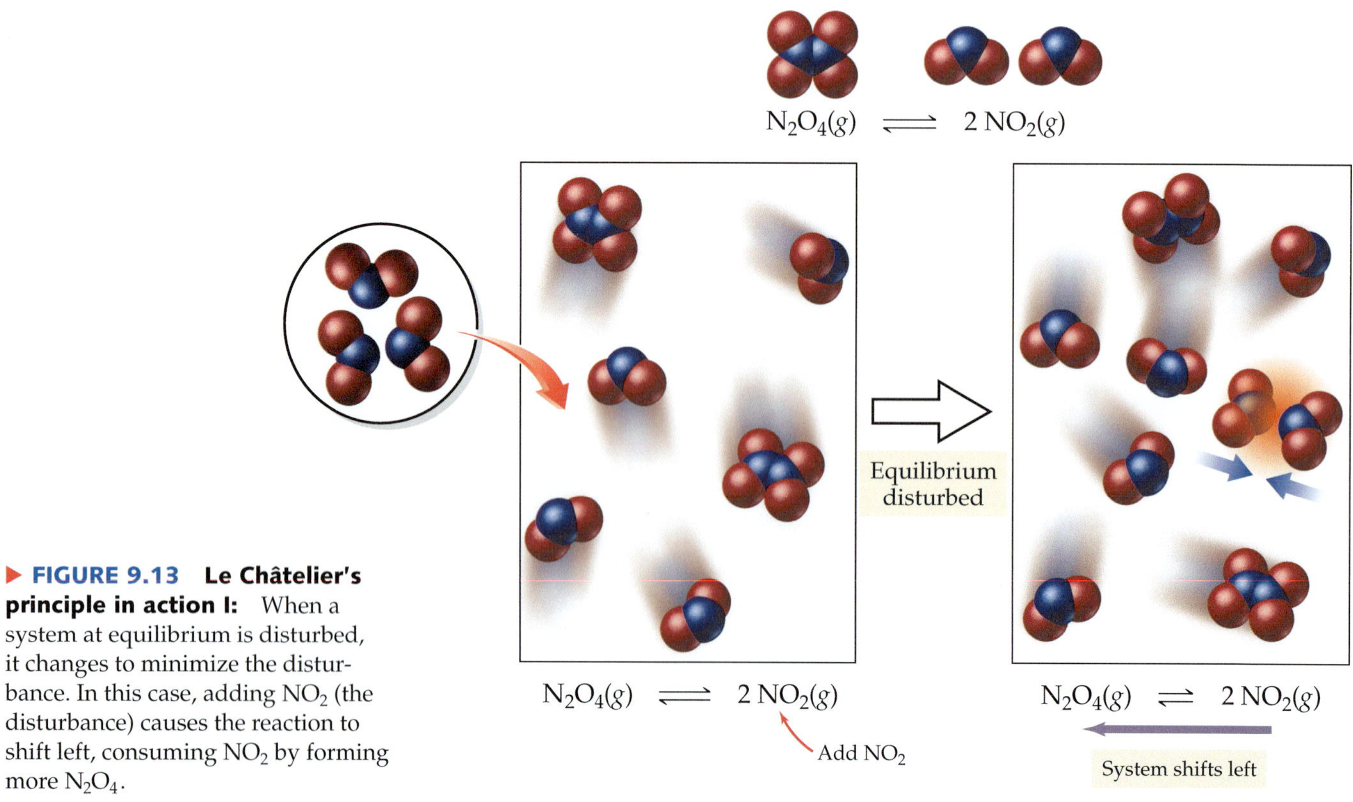

▶ **FIGURE 9.13 Le Châtelier's principle in action I:** When a system at equilibrium is disturbed, it changes to minimize the disturbance. In this case, adding NO_2 (the disturbance) causes the reaction to shift left, consuming NO_2 by forming more N_2O_4.

When we say that a reaction *shifts to the left*, we mean that it proceeds in the reverse direction, consuming products and forming reactants.

The reaction shifts to the left (it proceeds in the reverse direction), consuming some of the added NO_2 and bringing its concentration back down.

$$N_2O_4(g) \rightleftharpoons 2\,NO_2(g)$$

Reaction shifts left   Add NO_2

When we say that a reaction *shifts to the right*, we mean that it proceeds in the forward direction, consuming reactants and forming products.

In contrast, what happens if we add extra N_2O_4, increasing its concentration? In this case, the rate of the forward reaction increases and the reaction shifts to the right, consuming some of the added N_2O_4 and bringing *its* concentration back down (▼ Figure 9.14).

$$N_2O_4(g) \rightleftharpoons 2\,NO_2(g)$$

Add N_2O_4 Reaction shifts right

In each case, the system shifts in a direction that minimizes the disturbance.

To summarize, if a chemical system is at equilibrium:

- Increasing the concentration of one or more of the reactants causes the reaction to shift to the right (in the direction of the products).
- Increasing the concentration of one or more of the products causes the reaction to shift to the left (in the direction of the reactants).

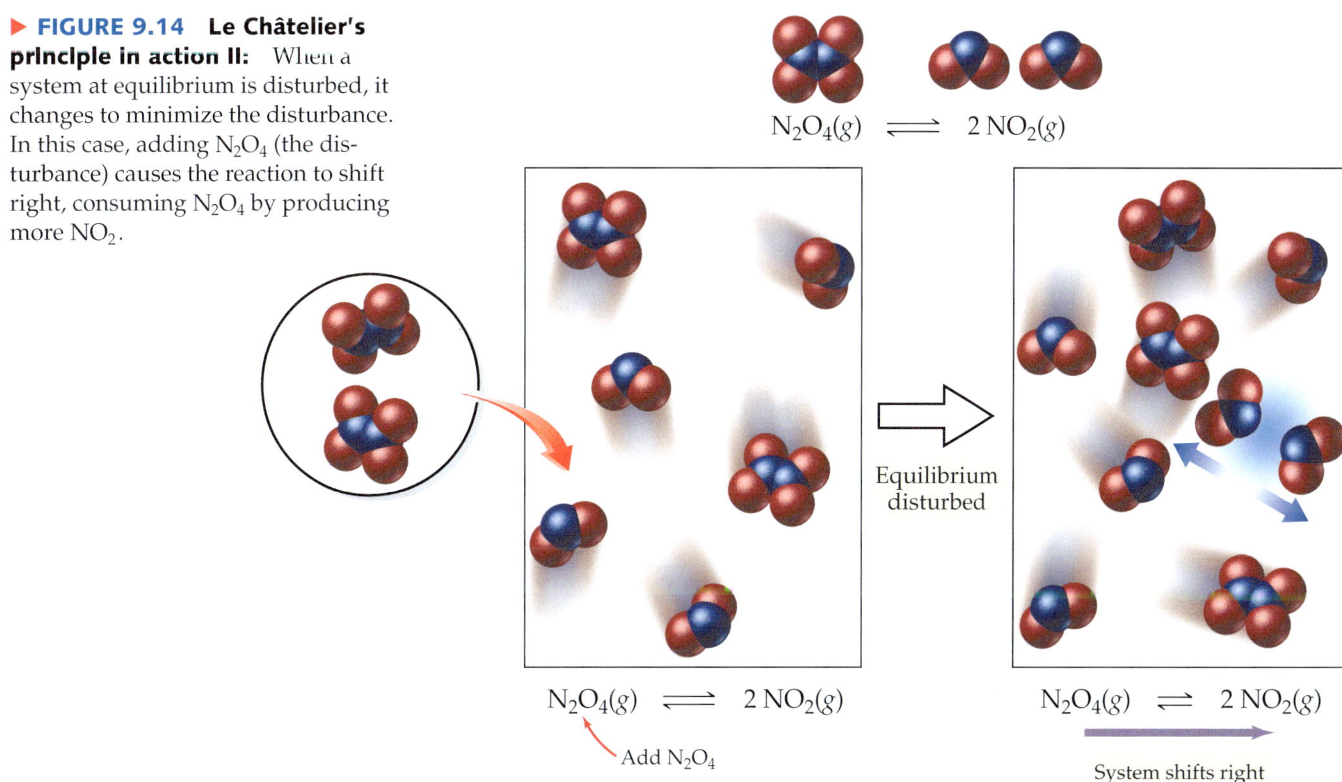

▶ **FIGURE 9.14 Le Châtelier's principle in action II:** When a system at equilibrium is disturbed, it changes to minimize the disturbance. In this case, adding N_2O_4 (the disturbance) causes the reaction to shift right, consuming N_2O_4 by producing more NO_2.

EXAMPLE 9.4 THE EFFECT OF A CONCENTRATION CHANGE ON EQUILIBRIUM

Consider the following reaction at equilibrium:

$$CaCO_3(s) \rightleftharpoons CaO(s) + CO_2(g)$$

What is the effect of adding more CO_2 to the reaction mixture? What is the effect of adding more $CaCO_3$?

SOLUTION

Adding more CO_2 increases the concentration of CO_2 and causes the reaction to shift to the left. Adding more $CaCO_3$ does not increase the concentration of $CaCO_3$ because $CaCO_3$ is a solid and thus has a constant concentration. It is therefore not included in the equilibrium expression and has no effect on the position of the equilibrium.

▶SKILLBUILDER 9.4 | The Effect of a Concentration Change on Equilibrium

Consider the following reaction in chemical equilibrium:

$$2\ BrNO(g) \rightleftharpoons 2\ NO(g) + Br_2(g)$$

What is the effect of adding more Br_2 to the reaction mixture? What is the effect of adding more BrNO?

▶SKILLBUILDER PLUS

What is the effect of removing some Br_2 from the preceding reaction mixture?

▶FOR MORE PRACTICE Example 9.8a, b; Problems 61, 62, 63, 64.

CONCEPTUAL CHECKPOINT 9.4

Consider the equilibrium reaction between carbon monoxide and hydrogen gas to form methanol.

$$CO(g) + 2\ H_2(g) \rightleftharpoons CH_3OH(g)$$

Suppose you have a reaction mixture of these three substances at equilibrium. Which causes a greater shift toward products: doubling the carbon monoxide concentration or doubling the hydrogen gas concentration?

9.11 The Effect of a Volume Change on Equilibrium

LO: Apply Le Châtelier's principle in the case of a change in volume.

See Section 6.3 for a complete description of Boyle's law.

How does a system in chemical equilibrium respond to a volume change? Recall from Module 6 that changing the volume of a gas (or a gas mixture) results in a change in pressure. Remember also that pressure and volume are inversely related: A *decrease* in volume causes an *increase* in pressure, and an *increase* in volume causes a *decrease* in pressure. So, if the volume of a gaseous reaction mixture at chemical equilibrium changes, the pressure changes and the system shifts in a direction to minimize that change. For example, consider the following reaction at equilibrium in a cylinder equipped with a moveable piston:

$$N_2(g) + 3\ H_2(g) \rightleftharpoons 2\ NH_3(g)$$

What happens if we push down on the piston, lowering the volume and raising the pressure (▶ Figure 9.15)? How can the chemical system bring the pressure back down? Look carefully at the reaction coefficients in the balanced equation. If the reaction shifts to the right, 4 mol of gas particles (1 mol of N_2 and 3 mol of H_2) are converted to 2 mol of gas particles (2 mol of NH_3). Therefore, as the reaction shifts toward products, the pressure is lowered (because the reaction mixture contains fewer gas particles). So the system then shifts to the right, bringing the pressure back down and minimizing the disturbance.

From the ideal gas law ($PV = nRT$), we can see that lowering the number of moles of a gas (n) results in a lower pressure (P) at constant temperature and volume.

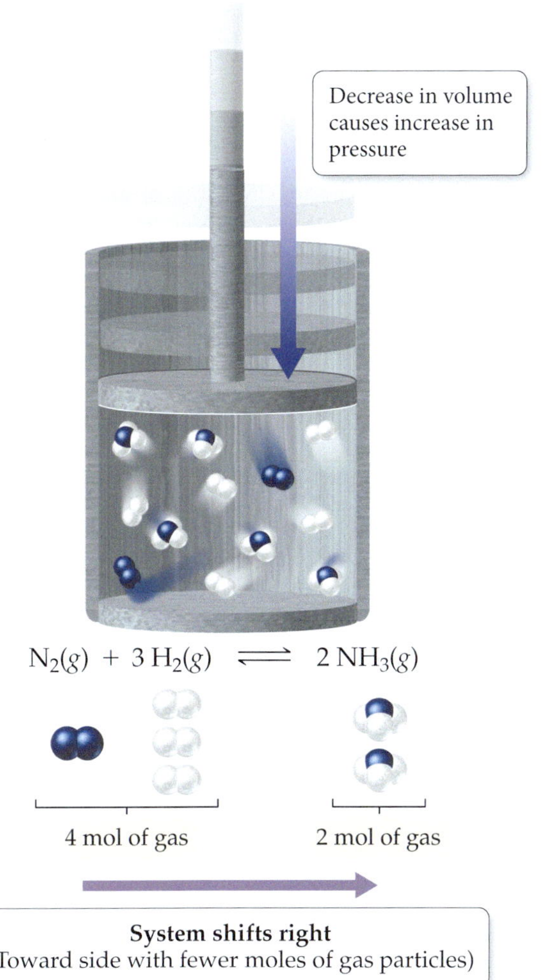

▲ **FIGURE 9.15 Effect of volume decrease on equilibrium** When the volume of an equilibrium mixture decreases, the pressure increases. The system responds (to bring the pressure back down) by shifting to the right, the side of the reaction with fewer moles of gas particles.

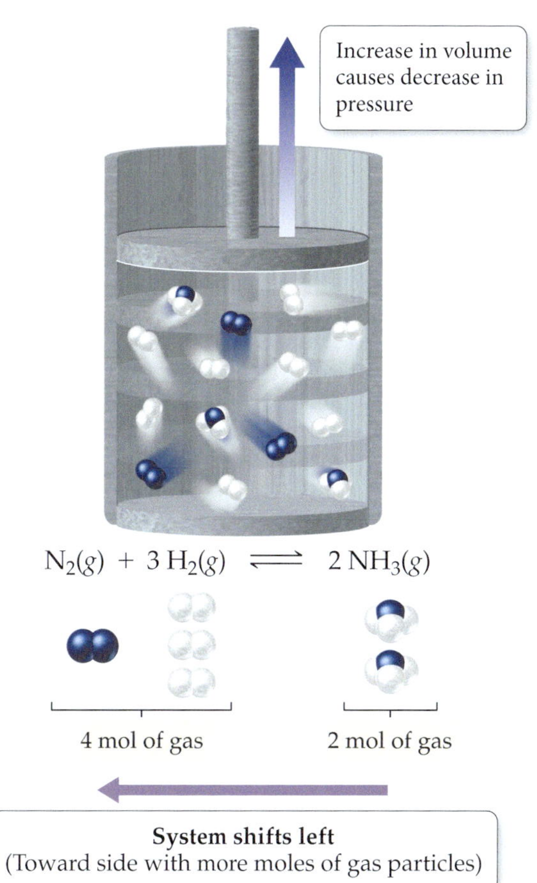

▲ **FIGURE 9.16 Effect of volume increase on equilibrium** When the volume of an equilibrium mixture increases, the pressure decreases. The system responds (to raise the pressure) by shifting to the left, the side of the reaction with more moles of gas particles.

Consider the same reaction mixture at equilibrium again. What happens if, this time, we pull *up* on the piston, *increasing* the volume (▲ Figure 9.16)? The higher volume results in a lower pressure, and the system responds by trying to bring the pressure back up. It can do this by shifting to the left, converting 2 mol of gas particles into 4 mol of gas particles. As the reaction shifts toward reactants, the pressure increases again (because the reaction mixture contains more gas particles), minimizing the disturbance.

To summarize, if a chemical system is at equilibrium:

- Decreasing the volume causes the reaction to shift in the direction that has fewer moles of gas particles.
- Increasing the volume causes the reaction to shift in the direction that has more moles of gas particles.

Notice that if a chemical reaction has an equal number of moles of gas particles on both sides of the chemical equation, a change in volume has no effect. For example, consider the following reaction:

$$H_2(g) + I_2(g) \rightleftharpoons 2\,HI(g)$$

Both the left and the right sides of the equation contain 2 mol of gas particles, so a change in volume has no effect on this reaction. In addition, a change in volume has no effect on a reaction that has no gaseous reactants or products.

CHEMISTRY **AND HEALTH**

▶ How a Developing Fetus Gets Oxygen from Its Mother

Have you ever wondered how a fetus in the womb gets oxygen? Unlike you and me, a fetus cannot breathe. Yet like you and me, a fetus needs oxygen. Where does that oxygen come from? In adults, oxygen is absorbed in the lungs and carried in the blood by a protein molecule called hemoglobin, which is abundantly present in red blood cells. Hemoglobin (Hb) reacts with oxygen according to the equilibrium equation:

$$Hb + O_2 \rightleftharpoons HbO_2$$

The equilibrium constant for this reaction is neither large nor small, but intermediate. Consequently, the reaction shifts toward the right or the left, depending on the concentration of oxygen. As blood flows through the lungs, where oxygen concentrations are high, the equilibrium shifts to the right—hemoglobin loads oxygen.

Lung $[O_2]$ high

$$Hb + O_2 \rightleftharpoons HbO_2$$

Reaction shifts right

As blood flows through muscles and organs that are using oxygen (where oxygen concentrations have been depleted), the equilibrium shifts to the left—hemoglobin unloads oxygen.

Muscle $[O_2]$ low

$$Hb + O_2 \rightleftharpoons HbO_2$$

Reaction shifts left

A fetus has its own blood circulatory system. The mother's blood never flows into the fetus's body, and the fetus cannot get any air in the womb. So how does the fetus get oxygen?

The answer lies in fetal hemoglobin (HbF), which is slightly different from adult hemoglobin. Like adult hemoglobin, fetal hemoglobin is in equilibrium with oxygen.

$$HbF + O_2 \rightleftharpoons HbFO_2$$

However, the equilibrium constant for fetal hemoglobin is larger than the equilibrium constant for adult hemoglobin. In other words, fetal hemoglobin loads oxygen at a lower oxygen concentration than adult hemoglobin. So, when the mother's hemoglobin flows through the placenta, it unloads oxygen into the placenta. The baby's blood also flows into the placenta, and even though the baby's blood never mixes with the mother's blood, the fetal hemoglobin within the baby's blood loads the oxygen (that the mother's hemoglobin unloaded) and carries it to the baby. Nature has thus engineered a chemical system where the mother's hemoglobin can in effect *hand off* oxygen to the baby's hemoglobin.

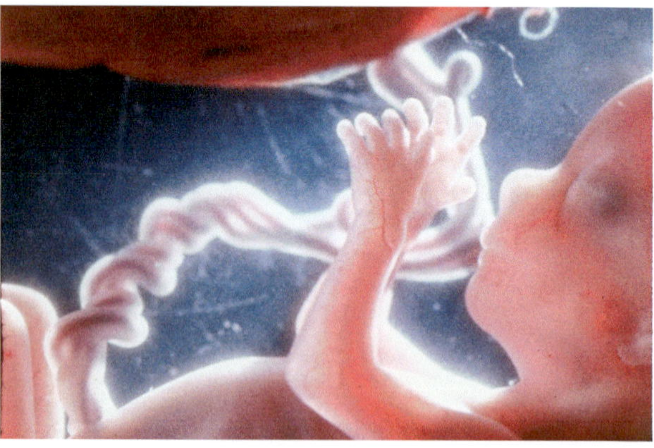

▲ A human fetus. **Question: How does the fetus get oxygen?**
© Claude Edelmann/Science Source.

B9.2 CAN YOU ANSWER THIS? *What would happen if fetal hemoglobin had the same equilibrium constant for the reaction with oxygen as adult hemoglobin?*

EXAMPLE 9.5 THE EFFECT OF A VOLUME CHANGE ON EQUILIBRIUM

Consider the reaction at chemical equilibrium:

$$2\ KClO_3(s) \rightleftharpoons 2\ KCl(s) + 3\ O_2(g)$$

What is the effect of decreasing the volume of the reaction mixture? Increasing the volume of the reaction mixture?

SOLUTION

The chemical equation has 3 mol of gas on the right and 0 mol of gas on the left. Decreasing the volume of the reaction mixture increases the pressure and causes the reaction to shift to the left (toward the side with fewer moles of gas particles). Increasing the volume of the reaction mixture decreases the pressure and causes the reaction to shift to the right (toward the side with more moles of gas particles).

▶SKILLBUILDER 9.5 | The Effect of a Volume Change on Equilibrium

Consider the reaction at chemical equilibrium:

$$2\ SO_2(g) + O_2(g) \rightleftharpoons 2\ SO_3(g)$$

What is the effect of decreasing the volume of the reaction mixture? Increasing the volume of the reaction mixture?

▶FOR MORE PRACTICE Example 9.8c; Problems 65, 66, 67, 68.

CONCEPTUAL CHECKPOINT 9.5

Consider the reaction:

$$H_2(g) + I_2(g) \rightleftharpoons 2\ HI(g)$$

Which change would cause the reaction to shift to the right (toward products)?

(a) Decreasing the volume

(b) Increasing the volume

(c) Increasing the concentration of hydrogen gas

(d) Decreasing the concentration of hydrogen gas

9.12 The Effect of a Temperature Change on Equilibrium

LO: Apply Le Châtelier's principle in the case of a change in temperature.

According to Le Châtelier's principle, if the temperature of a system at equilibrium is changed, the system should shift in a direction to counter that change. So if the temperature is increased, the reaction should shift in the direction that attempts to lower the temperature and vice versa. Recall from earlier in this module that energy changes are often associated with chemical reactions. If we want to predict the direction in which a reaction will shift upon a temperature change, we must understand how a shift in the reaction affects the temperature.

In Section 9.2, we classify chemical reactions according to whether they absorb or emit heat energy in the course of the reaction. Recall that an *exothermic reaction* (one with a negative ΔH_{rxn}) emits heat.

$$\textbf{Exothermic reaction:}\quad A + B \rightleftharpoons C + D + \text{Heat}$$

In an exothermic reaction, we can think of heat as a product. Consequently, raising the temperature of an exothermic reaction—think of this as adding heat—causes the reaction to shift left. For example, the reaction of nitrogen with hydrogen to form ammonia is exothermic.

$$N_2(g) + 3\,H_2(g) \rightleftharpoons 2\,NH_3(g) + \text{Heat}$$

2) Reaction shifts left 1) Add heat

Raising the temperature of an equilibrium mixture of these three gases causes the reaction to shift left, absorbing some of the added heat. Conversely, lowering the temperature of an equilibrium mixture of these three gases causes the reaction to shift right, releasing heat.

$$N_2(g) + 3\,H_2(g) \rightleftharpoons 2\,NH_3(g) + \text{Heat}$$

2) Reaction shifts right 1) Remove heat

In contrast, an *endothermic reaction* (one with a positive ΔH_{rxn}) absorbs heat.

Endothermic reaction: $A + B + \text{Heat} \rightleftharpoons C + D$

In an endothermic reaction, we can think of heat as a reactant. Consequently, raising the temperature (or adding heat) causes an endothermic reaction to shift right. For example, the following reaction is endothermic:

Colorless Brown

$$N_2O_4(g) + \text{Heat} \rightleftharpoons 2\,NO_2(g)$$

1) Add heat 2) Reaction shifts right

Raising the temperature of an equilibrium mixture of these two gases causes the reaction to shift right, absorbing some of the added heat. Since N_2O_4 is colorless and NO_2 is brown, we can see the effects of changing the temperature of this reaction (▼ Figure 9.17). On the other hand, lowering the temperature of a reaction mixture of these two gases causes the reaction to shift left, releasing heat.

Colorless Brown

$$N_2O_4(g) + \text{Heat} \rightleftharpoons 2\,NO_2(g)$$

1) Remove heat 2) Reaction shifts left

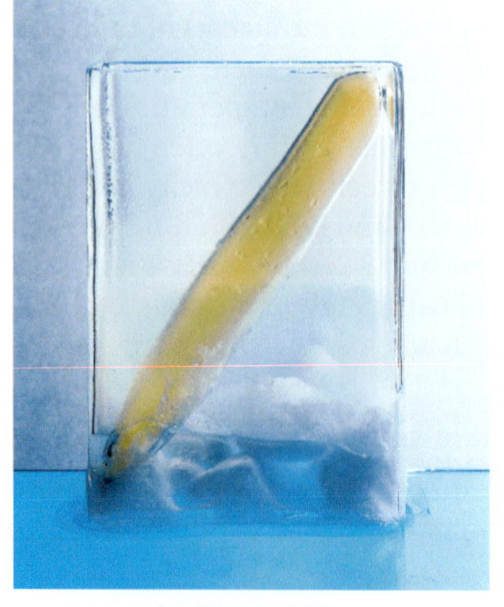

(a) Warm: NO_2

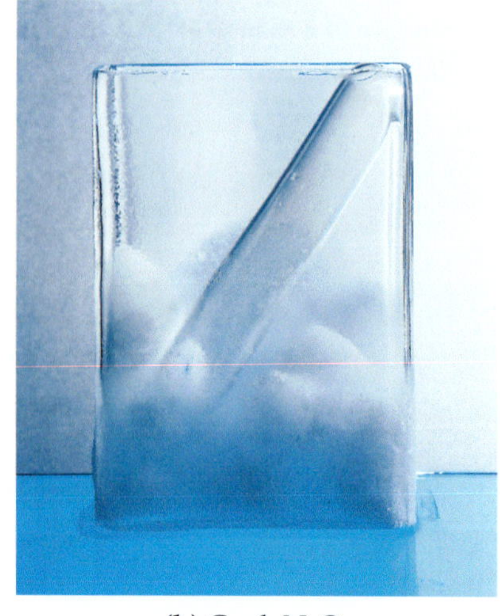

(b) Cool: N_2O_4

▶ **FIGURE 9.17 Equilibrium as a function of temperature** Since the reaction $N_2O_4(g) \rightleftharpoons 2\,NO_2(g)$ is endothermic, warm temperatures **(a)** cause a shift to the right, toward the production of brown NO_2. Cool temperatures **(b)** cause a shift to the left, to colorless N_2O_4. © Richard Megna/Fundamental Photographs.

To summarize:

In an exothermic chemical reaction, heat is a product and:

- Increasing the temperature causes the reaction to shift left (in the direction of the reactants).
- Decreasing the temperature causes the reaction to shift right (in the direction of the products).

In an endothermic chemical reaction, heat is a reactant and:

- Increasing the temperature causes the reaction to shift right (in the direction of the products).
- Decreasing the temperature causes the reaction to shift left (in the direction of the reactants).

EXAMPLE 9.6 **THE EFFECT OF A TEMPERATURE CHANGE ON EQUILIBRIUM**

The following reaction is endothermic:

$$CaCO_3(s) \rightleftharpoons CaO(s) + CO_2(g)$$

What is the effect of increasing the temperature of the reaction mixture? Decreasing the temperature?

SOLUTION

Because the reaction is endothermic, we can think of heat as a reactant.

$$\text{Heat} + CaCO_3(s) \rightleftharpoons CaO(s) + CO_2(g)$$

Raising the temperature is adding heat, causing the reaction to shift to the right. Lowering the temperature is removing heat, causing the reaction to shift to the left.

▶SKILLBUILDER 9.6 | The Effect of a Temperature Change on Equilibrium

The following reaction is exothermic.

$$2\,SO_2(g) + O_2(g) \rightleftharpoons 2\,SO_3(g)$$

What is the effect of increasing the temperature of the reaction mixture? Decreasing the temperature?

▶FOR MORE PRACTICE Example 9.8d; Problems 69, 70, 71, 72.

CONCEPTUAL CHECKPOINT 9.6

Consider the endothermic reaction.

$$Cl_2(g) \rightleftharpoons 2\,Cl(g)$$

If the reaction mixture is at equilibrium, which disturbances increase the amount of product the most?

(a) Increasing the temperature and increasing the volume

(b) Increasing the temperature and decreasing the volume

(c) Decreasing the temperature and increasing the volume

(d) Decreasing the temperature and decreasing the volume

9.13 The Path of a Reaction and the Effect of a Catalyst

LO: Describe the relationship between activation energy and reaction rates and the role catalysts play in reactions.

Warning: Hydrogen gas is explosive and should never be handled without proper training.

In this module, we have learned that the equilibrium constant describes the ultimate fate of a chemical reaction. Large equilibrium constants indicate that the reaction favors the products. Small equilibrium constants indicate that the reaction favors the reactants. But the equilibrium constant by itself does not tell the whole story. For example, consider the reaction between hydrogen gas and oxygen gas to form water:

$$2\,H_2(g) + O_2(g) \rightleftharpoons 2\,H_2O(g) \quad K_{eq} = 3.2 \times 10^{81} \text{ at } 25\ ^\circ\text{C}$$

The equilibrium constant is huge, meaning that the forward reaction is heavily favored. Yet we can mix hydrogen and oxygen in a balloon at room temperature, and no reaction occurs. Hydrogen and oxygen peacefully coexist together inside of the balloon and form virtually no water. Why?

The equilibrium constant describes *how far* a chemical reaction will go. The reaction rate describes *how fast* it will get there.

To answer this question, we revisit a topic from the beginning of this module—*the reaction rate.* At 25 °C, the reaction rate between hydrogen gas and oxygen gas is virtually zero. Even though the equilibrium constant is large, the reaction rate is small and no reaction occurs. The reaction rate between hydrogen and oxygen is slow because the reaction has a large *activation energy.* The **activation energy** (or activation barrier) for a reaction is the energy barrier that must be overcome in order for the reactants to be converted into products. Activation energies exist for most chemical reactions because the original bonds must begin to break before new bonds begin to form, and this requires energy. For example, for H_2 and O_2 to react to form H_2O, the H — H and O — O bonds must begin to break before the new bonds can form. The initial weakening of H_2 and O_2 bonds takes energy—this is the activation energy of the reaction.

The activation energy is sometimes called the *activation barrier*.

How Activation Energies Affect Reaction Rates

We can illustrate how activation energies affect reaction rates with a graph showing the energy progress of a reaction (▼ Figure 9.18). In the figure the products have less energy than the reactants, so we know the reaction is exothermic (it releases energy when it occurs). However, before the reaction can take place, some energy must first be *added*—the energy of the reactants must be raised by an amount that we call the activation energy. The activation energy is a kind of "energy hump" that normally exists between the reactants and products.

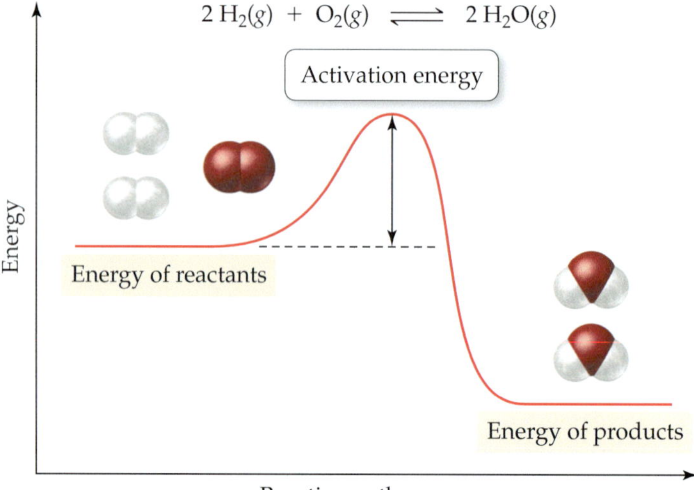

▶ **FIGURE 9.18 Activation energy** This plot represents the energy of the reactants and products along the reaction pathway (as the reaction occurs). Notice that the energy of the products is lower than the energy of the reactants, so this is an exothermic reaction. However, notice that the reactants must get over an energy hump—called the *activation energy*—to proceed from reactants to products.

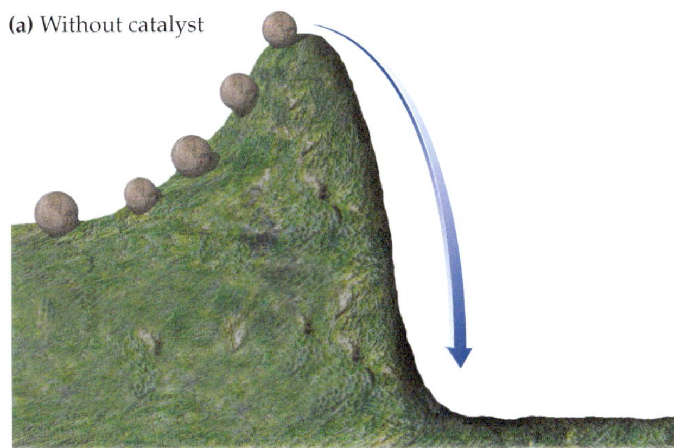

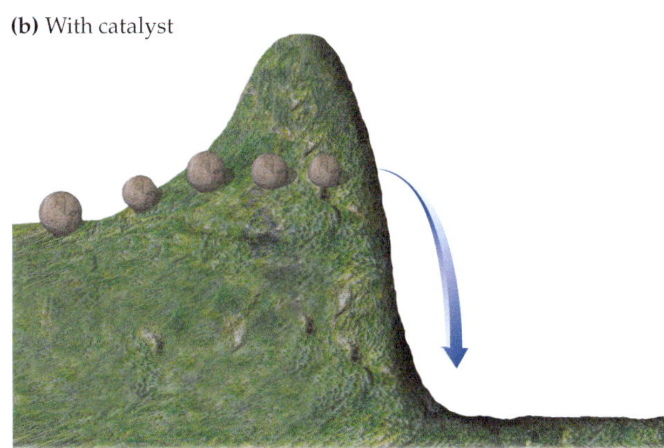

▲ **FIGURE 9.19 Hill analogy for activation energy** There are several ways to get these boulders over the hill as fast as possible. **(a)** One way is simply to push them harder—this is analogous to an increase in temperature for a chemical reaction. **(b)** Another way is to find a path that goes *around* the hill—this is analogous to the role of a catalyst for a chemical reaction.

We can understand this concept better by means of a simple analogy—getting a chemical reaction to occur is much like trying to push a bunch of boulders over a hill (▲ Figure 9.19(a). We can think of each collision that occurs between reactant molecules as an attempt to roll a boulder over the hill. A successful collision between two molecules (one that leads to product) is like a successful attempt to roll a boulder over the hill and down the other side.

The higher the hill is, the harder it is to get the boulders over the hill, and the fewer the number of boulders that make it over the hill in a given period of time. Similarly, for chemical reactions, the higher the activation energy, the fewer the number of reactant molecules that make it over the barrier, and the slower the reaction rate. In general:

> At a given temperature, the higher the activation energy for a chemical reaction, the slower the reaction rate.

Are there any ways to speed up a slow reaction (one with a high activation barrier)? In Section 9.5 we discussed two ways to increase reaction rates. The first way is to increase the concentrations of the reactants, which results in more collisions per unit time. This is analogous to pushing more boulders toward the hill in a given period of time. The second way is to increase the temperature. This results in more collisions per unit time and in higher-energy collisions. Higher-energy collisions are analogous to pushing the boulders harder (with more force), which results in more boulders making it over the hill per unit time—a faster reaction rate. There is, however, a third way to speed up a slow chemical reaction: by using a *catalyst*.

CONCEPTUAL CHECKPOINT 9.7

Reaction A has an activation barrier of 35 kJ/mol, and reaction B has an activation barrier of 55 kJ/mol. Which of the two reactions is likely to have the faster rate at room temperature?

A catalyst does not change the *position* of equilibrium, only *how fast* equilibrium is reached.

Upper-atmospheric ozone forms a shield against harmful ultraviolet light that would otherwise enter Earth's atmosphere. See the *Chemistry in the Environment* box in Module 5.

Catalysts Lower the Activation Energy

A **catalyst** is a substance that increases the rate of a chemical reaction but is not consumed by the reaction. A catalyst works by lowering the activation energy for the reaction, making it easier for reactants to get over the energy barrier (▼ Figure 9.20). In our boulder analogy, a catalyst creates another path for the boulders to travel—a path with a smaller hill (▼ see Figure 9.19(b). For example, consider the noncatalytic destruction of ozone in the upper atmosphere.

$$O_3 + O \rightarrow 2\,O_2$$

We have a protective ozone layer because this reaction has a fairly high activation barrier and therefore proceeds at a fairly slow rate. The ozone layer does not rapidly decompose into O_2.

However, the addition of Cl (from synthetic chlorofluorocarbons) to the upper atmosphere has resulted in another pathway by which O_3 can be destroyed. The first step in this pathway—called the *catalytic* destruction of ozone—is the reaction of Cl with O_3 to form ClO and O_2.

$$Cl + O_3 \rightarrow ClO + O_2$$

This is followed by a second step in which ClO reacts with O, regenerating Cl.

$$ClO + O \rightarrow Cl + O_2$$

Notice that, if we add the two reactions, the overall reaction is identical to the noncatalytic reaction.

$$\begin{array}{lcl} \cancel{Cl} + O_3 & \rightarrow & \cancel{ClO} + O_2 \\ \cancel{ClO} + O & \rightarrow & \cancel{Cl} + O_2 \\ \hline O_3 + O & \rightarrow & 2\,O_2 \end{array}$$

However, the activation energies for the two reactions in this pathway are much smaller than for the first, uncatalyzed pathway, and therefore the reaction occurs at a much faster rate. The Cl is not consumed in the overall reaction; this is characteristic of a catalyst.

In the case of the catalytic destruction of ozone, the catalyst speeds up a reaction that we do *not* want to happen. Most of the time, however, we use catalysts to speed up reactions that we *do* want to happen. For example, your car most likely has a catalytic converter in its exhaust system. The catalytic converter contains a catalyst that converts exhaust pollutants (such as carbon monoxide) into less

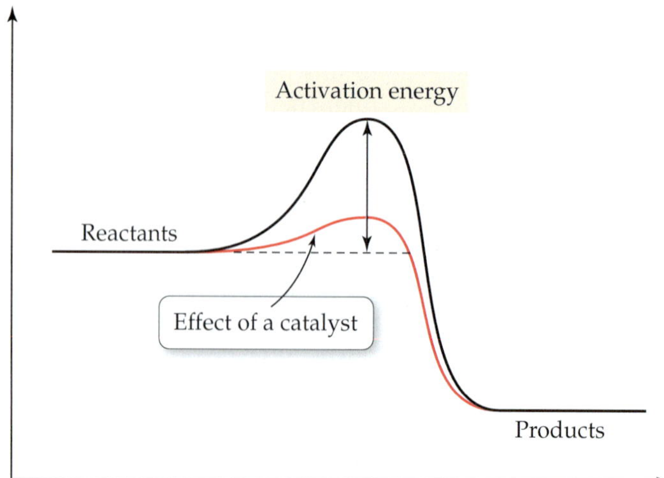

▲ **FIGURE 9.20 Function of a catalyst** A catalyst provides an alternate pathway with a lower activation energy barrier for the reaction.

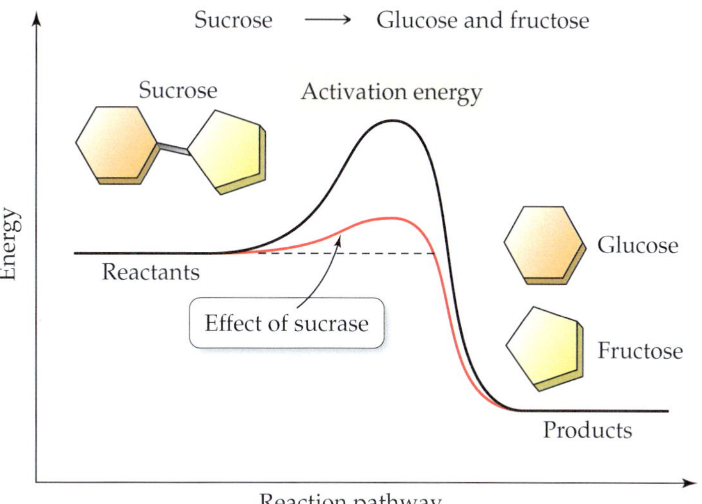

▶ **FIGURE 9.21 An enzyme catalyst** The enzyme sucrase creates a pathway with a lower activation energy for the conversion of sucrose to glucose and fructose.

harmful substances (such as carbon dioxide). These reactions occur only with the help of a catalyst because they are too slow to occur otherwise.

A catalyst cannot change the value of K_{eq} for a reaction—it affects only the *rate* of the reaction.

The role of catalysis in chemistry cannot be overstated. Without catalysts, chemistry would be a different field. For many reactions, increasing the reaction rate in another way—such as raising the temperature—is simply not feasible. Many reactants are thermally sensitive—increasing the temperature can destroy them. The only way to carry out many reactions is to use catalysts.

Enzymes: Biological Catalysts

Perhaps the best example of chemical catalysis is found in living organisms. Most of the thousands of reactions that must occur for a living organism to survive would be too slow at normal temperatures. So living organisms use **enzymes**—biological catalysts that increase the rates of biochemical reactions. For example, when we eat sucrose (table sugar), our bodies must break it into two smaller molecules: glucose and fructose. The equilibrium constant for this reaction is large, favoring the products. However, at room temperature, or even at body temperature, the sucrose does not break down into glucose and fructose because the activation energy is high, resulting in a slow reaction rate. In other words, table sugar remains table sugar at room temperature, even though the equilibrium constant for its reaction to glucose and fructose is relatively large (▲ Figure 9.21). In the body, however, an enzyme called *sucrase* catalyzes the conversion of sucrose to glucose and fructose. Sucrase has a pocket—called the active site—into which sucrose snugly fits (like a key into a lock). When sucrose is in the active site, the bond between the glucose and fructose units weakens, lowering the activation energy for the reaction and increasing the reaction rate (◀ Figure 9.22). The reaction can then proceed toward equilibrium—which favors the products—at a much lower temperature.

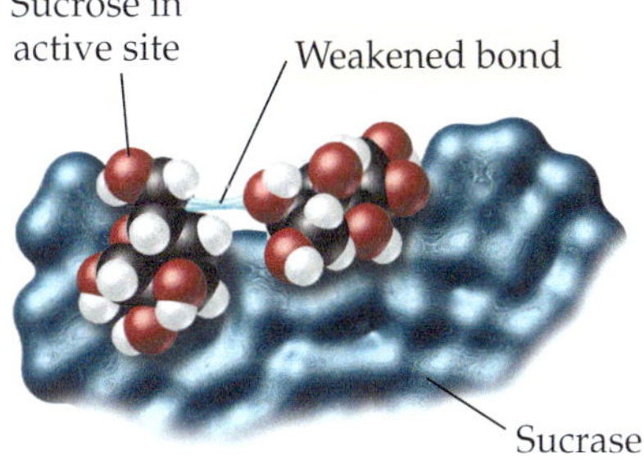

▲ **FIGURE 9.22 How an enzyme works** Sucrase has a pocket called the active site where sucrose binds. When a molecule of sucrose enters the active site, the bond between glucose and fructose is weakened, lowering the activation energy for the reaction.

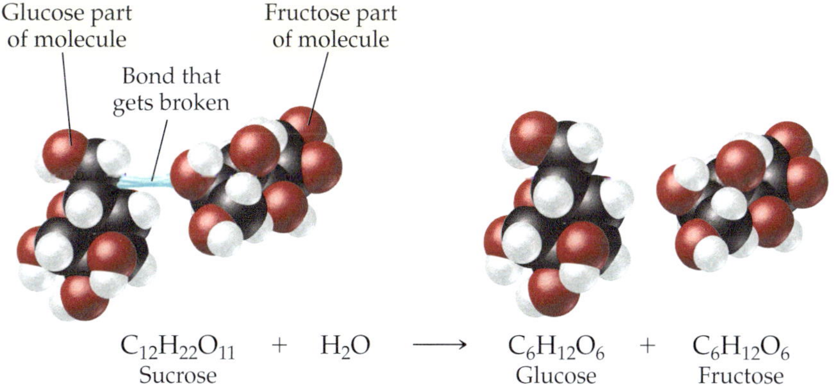

MODULE IN REVIEW

Self-Assessment Quiz

Q1. Which statement best describes an exothermic reaction?
(a) An exothermic reaction gives off heat.
(b) An exothermic reaction absorbs heat.
(c) An exothermic reaction produces only small amounts of products.
(d) None of the above.

Q2. Which change is likely to increase the *rate* of reaction in a reaction mixture?
(a) Decreasing the temperature
(b) Increasing the concentration of the reactants
(c) Increasing the volume of the reaction vessel
(d) None of the above

Q3. The equilibrium constants at a fixed temperature for several reactions are listed here. Which equilibrium constant indicates a reaction that is favored in the forward direction?
(a) $K_{eq} = 1.0 \times 10^5$ **(b)** $K_{eq} = 1.0 \times 10^{-5}$
(c) $K_{eq} = 1.0$ **(d)** None of the above

Q4. What is the correct expression for the equilibrium constant (Keq) for the reaction between carbon and hydrogen gas to form methane?

$$C(s) + 2\,H_2(g) \rightleftharpoons CH_4(g)$$

(a) $K_{eq} = \dfrac{[CH_4]}{[H_2]}$ **(b)** $K_{eq} = \dfrac{[CH_4]}{[C][H_2]}$

(c) $K_{eq} = \dfrac{[CH_4]}{[C][H_2]^2}$ **(d)** $K_{eq} = \dfrac{[CH_4]}{[H_2]^2}$

Q5. What is the effect of adding chlorine gas (at constant volume and temperature) to an equilibrium mixture of this reaction?

$$CO(g) + Cl_2(g) \rightleftharpoons COCl_2(g)$$

(a) The reaction shifts toward the products.
(b) The reaction shifts toward the reactants.
(c) The reaction does not shift in either direction.
(d) The reaction slows down.

Q6. The decomposition of NH_4HS is endothermic.

$$NH_4HS(s) \rightleftharpoons NH_3(g) + H_2S(g)$$

Which change to an equilibrium mixture of this reaction results in the formation of more H_2S?
(a) A decrease in the volume of the reaction vessel (at constant temperature)
(b) An increase in the amount of NH_3 in the reaction vessel
(c) An increase in temperature
(d) All of the above

Answers: 1:a, 2:b, 3:a, 4:d, 5:a, 6:c

Chemical Principles

Energy: Besides matter, energy is the other major component of our universe. Like matter, energy is conserved—it can be neither created nor destroyed. Energy exists in various different types, and these can be converted from one to another. Some common units of energy are the joule (J), the calorie (cal), the nutritional Calorie (Cal), and the kilowatt-hour (kWh). Chemical reactions that emit energy are exothermic; those that absorb energy are endothermic.

Enthalpy of Reaction: The amount of heat released or absorbed by a chemical reaction under conditions of constant pressure is the enthalpy of reaction (ΔH_{rxn}).

Relevance

Energy: Our society's energy sources will not last forever because as we burn fossil fuels—our primary energy source—we convert chemical energy, stored in molecules, to kinetic and thermal energy. The kinetic and thermal energy is not readily available to be used again. Consequently, our energy resources are dwindling, and the conservation of energy implies that we will not be able simply to create new energy—it must come from somewhere. All of the chemical reactions that we use for energy are exothermic.

Enthalpy of Reaction: The enthalpy of reaction describes the relationship between the amount of reactant that undergoes reaction and the amount of thermal energy produced. This is important, for example, in determining quantities such as the amount of fuel needed to produce a given amount of energy.

The Concept of Equilibrium: Equilibrium involves the ideas of sameness and constancy. When a system is in equilibrium, some property of the system remains the same and does not change.

The Concept of Equilibrium: The equilibrium concept explains many phenomena such as the human body's oxygen delivery system. Life itself can be defined as controlled disequilibrium with the environment.

Rates of Chemical Reactions: The rate of a chemical reaction is the amount of reactant(s) that goes to product(s) in a given period of time. In general, reaction rates increase with increasing reactant concentration and increasing temperature. Reaction rates depend on the concentration of reactants, and the concentration of reactants decreases as a reaction proceeds, so reaction rates usually slow down as a reaction proceeds.

Rates of Chemical Reactions: The rate of a chemical reaction determines how fast a reaction will reach its equilibrium. Chemists want to understand the factors that influence reaction rates so they can control them.

Dynamic Chemical Equilibrium: Dynamic chemical equilibrium occurs when the rate of the forward reaction equals the rate of the reverse reaction.

Dynamic Chemical Equilibrium: When dynamic chemical equilibrium is reached, the concentrations of the reactants and products become constant.

The Equilibrium Constant: For the generic reaction

$$aA + bB \rightleftharpoons cC + dD$$

we define the equilibrium constant (K_{eq}) as:

$$K_{eq} = \frac{[C]^c[D]^d}{[A]^a[B]^b}$$

Only the concentrations of gaseous or aqueous reactants and products are included in the equilibrium constant—the concentrations of solid or liquid reactants or products are omitted.

The Equilibrium Constant: The equilibrium constant is a measure of how far a reaction will proceed. A large K_{eq} indicates that the forward reaction is favored (lots of products at equilibrium). A small K_{eq} indicates that the reverse reaction is favored (lots of reactants at equilibrium). An intermediate K_{eq} indicates that there will be significant amounts of both reactants and products at equilibrium.

Le Châtelier's Principle: Le Châtelier's principle states that when a chemical system at equilibrium is disturbed, the system shifts in a direction that minimizes the disturbance.

Le Châtelier's Principle: Le Châtelier's principle helps us predict what happens to a chemical system at equilibrium when conditions change. This allows chemists to modify the conditions of a chemical reaction to obtain a desired result.

Effect of a Concentration Change on Equilibrium:

- Increasing the concentration of one or more of the *reactants* causes the reaction to shift to the *right*.
- Increasing the concentration of one or more of the *products* causes the reaction to shift to the *left*.

Effect of a Concentration Change on Equilibrium: In many cases, a chemist may want to drive a reaction in one direction or another. For example, suppose a chemist is carrying out a reaction to make a desired compound. The reaction can be pushed to the right by continuously removing the product from the reaction mixture as it forms, maximizing the amount of product that can be made.

Effect of a Volume Change on Equilibrium:

- Decreasing the volume causes the reaction to shift in the direction that has *fewer* moles of gas particles.
- Increasing the volume causes the reaction to shift in the direction that has *more* moles of gas particles.

Effect of a Volume Change on Equilibrium: Like the effect of concentration, the effect of pressure on equilibrium allows a chemist to choose the best conditions under which to carry out a chemical reaction. Some reactions are favored in the forward direction by high pressure (those with fewer moles of gas particles in the products), and others (those with fewer moles of gas particles in the reactants) are favored in the forward direction by low pressure.

Effect of a Temperature Change on Equilibrium:

Exothermic chemical reaction (heat is a product):

- *Increasing* the temperature causes the reaction to shift *left*.
- *Decreasing* the temperature causes the reaction to shift *right*.

Endothermic chemical reaction (heat is a reactant):

- *Increasing* the temperature causes the reaction to shift *right*.
- *Decreasing* the temperature causes the reaction to shift *left*.

Effect of a Temperature Change on Equilibrium: Again, the effect of temperature on a reaction allows chemists to choose conditions that favor desired reactions. Higher temperatures favor endothermic reactions, while lower temperatures favor exothermic reactions. Most reactions occur faster at higher temperature, so the effect of temperature on the rate, not just on the equilibrium constant, must be considered.

Reaction Paths and Catalysts: Most chemical reactions must overcome an energy hump, called the *activation energy*, as they proceed from reactants to products. In general, increasing the temperature of a reaction mixture increases the fraction of reactant molecules that make it over the energy hump, therefore increasing the rate. A catalyst—a substance that increases the rate of the reaction but is not consumed by it—lowers the activation energy so that it is easier to get over the energy hump without increasing the temperature.

Reaction Paths and Catalysts: Catalysts are used in many chemical reactions to increase the rates. Without catalysts, many reactions occur too slowly to be of any value. The thousands of reactions that occur in living organisms are controlled by biological catalysts called *enzymes*.

Chemical Skills

LO: Write equilibrium expressions for chemical reactions (Section 9.7).

To write the equilibrium expression for a reaction, write the concentrations of the products raised to their stoichiometric coefficients divided by the concentrations of the reactants raised to their stoichiometric coefficients. Remember that reactants or products that are liquids or solids are omitted from the equilibrium expression.

LO: Apply Le Châtelier's principle (Sections 9.10, 9.11, 9.12).

To apply Le Châtelier's principle, review the effects of concentration, volume, and temperature in the end-of-chapter Le Chatelier's principle section. For each disturbance, predict how the reaction changes to counter the disturbance.

Examples

EXAMPLE 9.7 WRITING EQUILIBRIUM EXPRESSIONS FOR CHEMICAL REACTIONS

Write an equilibrium expression for the chemical equation:

$$2\ NO(g) + Br_2(g) \rightleftharpoons 2\ NOBr(g)$$

SOLUTION

$$K_{eq} = \frac{[NOBr]^2}{[NO]^2[Br_2]}$$

EXAMPLE 9.8 USING LE CHÂTELIER'S PRINCIPLE

Consider the *endothermic* chemical reaction:

$$C(s) + H_2O(g) \rightleftharpoons CO(g) + H_2(g)$$

Predict the effect of:

(a) increasing [CO]
(b) increasing $[H_2O]$
(c) increasing the reaction volume
(d) increasing the temperature

SOLUTION

(a) Shift left
(b) Shift right
(c) Shift right (more moles of gas on right)
(d) Shift right (heat is a reactant)

KEY TERMS

activation energy **[9.5, 9.13]**
calorie (cal) **[9.1]**
Calorie (Cal) **[9.1]**
catalyst **[9.13]**
chemical energy **[9.1]**
collision theory **[9.5]**
dynamic equilibrium **[9.6]**
electrical energy **[9.1]**
endothermic **[9.2, 9.3, 9.12]**
enthalpy **[9.3]**
enthalpy of reaction (ΔH_{rxn}) **[9.3]**
energy **[9.1]**
enzyme **[9.13]**
equilibrium constant (K_{eq}) **[9.7]**
exothermic **[9.2, 9.3, 9.12]**
kilowatt-hour (kWh) **[9.1]**
kinetic energy **[9.1]**
law of conservation of energy **[9.1]**
Le Châtelier's principle **[9.9]**
potential energy **[9.1]**
rate of a chemical reaction(reaction rate) **[9.5]**
reversible reaction **[9.6]**
thermal energy **[9.1]**
work **[9.1]**

EXERCISES

QUESTIONS

1. What is the definition of energy?

2. What is the law of conservation of energy?

3. Explain the difference between kinetic energy and potential energy.

4. What is chemical energy? List some examples of common substances that contain chemical energy.

5. List three common units for energy.

6. What is an exothermic reaction? Which has greater energy in an exothermic reaction, the reactants or the products?

7. What is an endothermic reaction? Which has greater energy in an endothermic reaction, the reactants or the products?

8. What is the enthalpy of reaction (ΔH_{rxn})? Why is this quantity important?

9. Explain the relationship between the sign of ΔH_{rxn} and whether a reaction is exothermic or endothermic.

10. What are the two *general* concepts involved in equilibrium?

11. What is the rate of a chemical reaction? What is the difference between a chemical reaction with a fast rate and one with a slow rate?

12. Why do chemists seek to control reaction rates?

13. How do most chemical reactions occur?

14. What factors influence reaction rates? How?

15. What normally happens to the rate of the forward reaction as a reaction proceeds?

16. What is dynamic chemical equilibrium?

17. Explain how dynamic chemical equilibrium involves the concepts of sameness and constancy.

18. Explain why the concentrations of reactants and products are not necessarily the same at equilibrium.

19. Devise your own analogy—like the Narnia and Middle Earth analogy in the chapter—to explain chemical equilibrium.

20. What is the equilibrium constant? Why is it significant?

21. Write the expression for the equilibrium constant for the following generic chemical equation.

$$aA + bB \rightleftharpoons cC + dD$$

22. What does a small equilibrium constant tell you about a reaction? A large equilibrium constant?

23. Why are solids and liquids omitted from the equilibrium expression?

24. What is Le Châtelier's principle?

25. Apply Le Châtelier's principle to your analogy from Question 21.

26. What is the effect of *increasing* the concentration of a reactant in a reaction mixture at equilibrium?

27. What is the effect of *decreasing* the concentration of a reactant in a reaction mixture at equilibrium?

28. What is the effect of *increasing* the concentration of a product in a reaction mixture at equilibrium?

29. What is the effect of *decreasing* the concentration of a product in a reaction mixture at equilibrium?

30. What is the effect of *increasing* the pressure of a reaction mixture at equilibrium if the reactant side has fewer moles of gas particles than the product side?

31. What is the effect of *increasing* the pressure of a reaction mixture at equilibrium if the product side has fewer moles of gas particles than the reactant side?

32. What is the effect of *decreasing* the pressure of a reaction mixture at equilibrium if the reactant side has fewer moles of gas particles than the product side?

33. What is the effect of *decreasing* the pressure of a reaction mixture at equilibrium if the product side has fewer moles of gas particles than the reactant side?

34. What is the effect of *increasing* the temperature of an endothermic reaction mixture at equilibrium? Of *decreasing* the temperature?

35. What is the effect of *increasing* the temperature of an exothermic reaction mixture at equilibrium? Of *decreasing* the temperature?

36. What is activation energy for a chemical reaction?

37. Explain why two reactants with a large K_{eq} for a particular reaction might not react immediately when combined.

38. What is the effect of a catalyst on a reaction? Why are catalysts so important to chemistry?

39. Does a catalyst affect the value of the equilibrium constant?

40. What are enzymes

PROBLEMS

ENERGY AND CHEMICAL AND PHYSICAL CHANGE

41. A common type of handwarmer contains iron powder that reacts with oxygen to form an oxide of iron. As soon as the handwarmer is exposed to air, the reaction begins and heat is emitted. Is the reaction between the iron and oxygen exothermic or endothermic? Draw an energy diagram showing the relative energies of the reactants and products in the reaction.

42. In a chemical cold pack, two substances are kept separate by a divider. When the divider is broken, the substances mix and absorb heat from the surroundings. The chemical cold pack feels cold. Is the reaction exothermic or endothermic? Draw an energy diagram showing the relative energies of the reactants and products in the reaction.

43. Classify each process as exothermic or endothermic.
- **(a)** gasoline burning in a car
- **(b)** isopropyl alcohol evaporating from skin
- **(c)** water condensing as dew during the night

44. Classify each process as exothermic or endothermic.
- **(a)** dry ice subliming (changing from a solid directly to a gas)
- **(b)** the wax in a candle burning
- **(c)** a match burning

ENTHALPY

45. Classify each process as exothermic or endothermic and indicate the sign of ΔH_{rxn}.
- **(a)** butane gas burning in a lighter
- **(b)** the reaction that occurs in the chemical cold packs used to ice athletic injuries
- **(c)** the burning of wax in a candle

46. Classify each process as exothermic or endothermic and indicate the sign of ΔH_{rxn}.
- **(a)** ice melting
- **(b)** a sparkler burning
- **(c)** acetone evaporating from skin

THE RATE OF REACTION

47. Two gaseous reactants are allowed to react in a 1 L flask, and the reaction rate is measured. The experiment is repeated with the same amount of each reactant and at the same temperature in a 2 L flask (so the concentration of each reactant is less). What is likely to happen to the measured reaction rate in the second experiment compared to the first?

48. The rate of phosphorus pentachloride decomposition is measured at a PCl_5 pressure of 0.015 atm and then again at a PCl_5 pressure of 0.30 atm. The temperature is identical in both measurements. Which rate is likely to be faster?

49. The body temperature of cold-blooded animals varies with the ambient temperature. From the point of view of reaction rates, explain why cold-blooded animals are more sluggish at cold temperatures.

50. The rate of a particular reaction doubles when the temperature increases from 25 °C to 35 °C. Explain why this happens.

51. The initial rate of a chemical reaction was measured, and one of the reactants was found to be reacting at a rate of 0.0011 mol/L s. The reaction was allowed to proceed for 15 minutes, and the rate was measured again. What would you predict about the second measured rate relative to the first?

52. When vinegar is added to a solution of sodium bicarbonate, the mixture immediately begins to bubble furiously. As time passes, however, there is less and less bubbling. Explain why this happens.

THE EQUILIBRIUM CONSTANT

53. Write an equilibrium expression for each chemical equation.
- **(a)** $2\,NO_2(g) \rightleftharpoons N_2O_4(g)$
- **(b)** $2\,BrNO(g) \rightleftharpoons 2\,NO(g) + Br_2(g)$
- **(c)** $H_2O(g) + CO(g) \rightleftharpoons H_2(g) + CO_2(g)$
- **(d)** $CH_4(g) + 2\,H_2S(g) \rightleftharpoons CS_2(g) + 4\,H_2(g)$

54. Write an equilibrium expression for each chemical equation.
- **(a)** $2\,CO(g) + O_2(g) \rightleftharpoons 2\,CO_2(g)$
- **(b)** $N_2(g) + O_2(g) \rightleftharpoons 2\,NO(g)$
- **(c)** $SbCl_5(g) \rightleftharpoons SbCl_3(g) + Cl_2(g)$
- **(d)** $CO(g) + Cl_2(g) \rightleftharpoons COCl_2(g)$

55. Write an equilibrium expression for each chemical equation involving one or more solid or liquid reactants or products.
(a) $PCl_5(g) \rightleftharpoons PCl_3(l) + Cl_2(g)$
(b) $2\ KClO_3(s) \rightleftharpoons 2\ KCl(s) + 3\ O_2(g)$
(c) $HF(aq) + H_2O(l) \rightleftharpoons H_3O^+(aq) + F^-(aq)$
(d) $NH_3(aq) + H_2O(l) \rightleftharpoons NH_4^+(aq) + OH^-(aq)$

56. Write an equilibrium expression for each chemical equation involving one or more solid or liquid reactants or products.
(a) $HCHO_2(aq) + H_2O(l) \rightleftharpoons H_3O^+(aq) + CHO_2^-(aq)$
(b) $CO_3^{2-}(aq) + H_2O(l) \rightleftharpoons HCO_3^-(aq) + OH^-(aq)$
(c) $2\ C(s) + O_2(g) \rightleftharpoons 2\ CO(g)$
(d) $C(s) + CO_2(g) \rightleftharpoons 2\ CO(g)$

57. Consider the reaction.

$$2\ H_2S(g) \rightleftharpoons 2\ H_2(g) + S_2(g)$$

Find the mistakes in the equilibrium expression and fix them.

$$K_{eq} = \frac{[H_2][S_2]}{[H_2S]}$$

58. Consider the reaction.

$$CO(g) + Cl_2(g) \rightleftharpoons COCl_2(g)$$

Find the mistake in the equilibrium expression and fix it.

$$K_{eq} = \frac{[CO][Cl_2]}{[COCl_2]}$$

59. For each equilibrium constant, indicate if you would expect an equilibrium reaction mixture to be dominated by reactants or by products, or to contain significant amounts of both.
(a) $K_{eq} = 5.2 \times 10^{17}$
(b) $K_{eq} = 1.24$
(c) $K_{eq} = 3.22 \times 10^{-21}$
(d) $K_{eq} = 0.47$

60. For each equilibrium constant, indicate if you would expect an equilibrium reaction mixture to be dominated by reactants or by products, or to contain significant amounts of both.
(a) $K_{eq} = 0.75$
(b) $K_{eq} = 8.5 \times 10^{-7}$
(c) $K_{eq} = 1.4 \times 10^{19}$
(d) $K_{eq} = 4.7 \times 10^{-9}$

LE CHÂTELIER'S PRINCIPLE

61. Consider this reaction at equilibrium.

$$CO(g) + Cl_2(g) \rightleftharpoons COCl_2(g)$$

Predict the effect (shift right, shift left, or no effect) of these changes.
(a) adding Cl_2 to the reaction mixture
(b) adding $COCl_2$ to the reaction mixture
(c) adding CO to the reaction mixture

62. Consider this reaction at equilibrium.

$$2\ BrNO(g) \rightleftharpoons 2\ NO(g) + Br_2(g)$$

Predict the effect (shift right, shift left, or no effect) of these changes.
(a) adding BrNO to the reaction mixture
(b) adding NO to the reaction mixture
(c) adding Br_2 to the reaction mixture

63. Consider this reaction at equilibrium.

$$C(s) + H_2O(g) \rightleftharpoons CO(g) + H_2(g)$$

Predict the effect (shift right, shift left, or no effect) of these changes.
(a) adding C to the reaction mixture
(b) condensing H_2O and removing it from the reaction mixture
(c) adding CO to the reaction mixture
(d) removing H_2 from the reaction mixture

64. Consider this reaction at equilibrium.

$$2\ KClO_3(s) \rightleftharpoons 2\ KCl(s) + 3\ O_2(g)$$

Predict the effect (shift right, shift left, or no effect) of these changes.
(a) adding KCl to the reaction mixture
(b) adding $KClO_3$ to the reaction mixture
(c) adding O_2 to the reaction mixture
(d) removing O_2 from the reaction mixture

65. Consider the effect of a volume change on this reaction at equilibrium.

$$I_2(g) \rightleftharpoons 2\ I(g)$$

Predict the effect (shift right, shift left, or no effect) of these changes.
(a) increasing the reaction volume
(b) decreasing the reaction volume

66. Consider the effect of a volume change on this reaction at equilibrium.

$$2\ H_2S(g) \rightleftharpoons 2\ H_2(g) + S_2(g)$$

Predict the effect (shift right, shift left, or no effect) of these changes.
(a) increasing the reaction volume
(b) decreasing the reaction volume

67. Consider the effect of a volume change on this reaction at equilibrium.

$$I_2(g) + Cl_2(g) \rightleftharpoons 2\ ICl(g)$$

Predict the effect (shift right, shift left, or no effect) of these changes.
(a) increasing the reaction volume
(b) decreasing the reaction volume

68. Consider the effect of a volume change on this reaction at equilibrium.

$$CO(g) + H_2O(g) \rightleftharpoons CO_2(g) + H_2(g)$$

Predict the effect (shift right, shift left, or no effect) of these changes.
(a) increasing the reaction volume
(b) decreasing the reaction volume

69. This reaction is endothermic.

$$C(s) + CO_2(g) \rightleftharpoons 2\ CO(g)$$

Predict the effect (shift right, shift left, or no effect) of these changes.
(a) increasing the reaction temperature
(b) decreasing the reaction temperature

70. This reaction is endothermic.

$$I_2(g) \rightleftharpoons 2\ I(g)$$

Predict the effect (shift right, shift left, or no effect) of these changes.
(a) increasing the reaction temperature
(b) decreasing the reaction temperature

71. This reaction is exothermic.

$$C_6H_{12}O_6(s) + 6\ O_2(g) \rightleftharpoons 6\ CO_2(g) + 6\ H_2O(g)$$

Predict the effect (shift right, shift left, or no effect) of these changes.
(a) increasing the reaction temperature
(b) decreasing the reaction temperature

72. The following reaction is exothermic.

$$C_2H_4(g) + Br_2(g) \rightleftharpoons C_2H_4Br_2(g)$$

Predict the effect (shift right, shift left, or no effect) of these changes.
(a) increasing the reaction temperature
(b) decreasing the reaction temperature

73. Coal, which is primarily carbon, can be converted to natural gas, primarily CH_4, by this exothermic reaction.

$$C(s) + 2\ H_2(g) \rightleftharpoons CH_4(g)$$

If this reaction mixture is at equilibrium, predict the effect (shift right, shift left, or no effect) of these changes.
(a) adding more C to the reaction mixture
(b) adding more H_2 to the reaction mixture
(c) raising the temperature of the reaction mixture
(d) lowering the volume of the reaction mixture
(e) adding a catalyst to the reaction mixture

74. Coal can be used to generate hydrogen gas (a potential fuel) by this endothermic reaction.

$$C(s) + H_2O(g) \rightleftharpoons CO(g) + H_2(g)$$

If this reaction mixture is at equilibrium, predict the effect (shift right, shift left, or no effect) of these changes.
(a) adding more C to the reaction mixture
(b) adding more $H_2O(g)$ to the reaction mixture
(c) raising the temperature of the reaction mixture
(d) increasing the volume of the reaction mixture
(e) adding a catalyst to the reaction mixture

CUMULATIVE PROBLEMS

75. This reaction is exothermic.

$$C_2H_4(g) + Cl_2(g) \rightleftharpoons C_2H_4Cl_2(g)$$

If you were a chemist trying to maximize the amount of $C_2H_4Cl_2$ produced, which of the following might you try? Assume that the reaction mixture reaches equilibrium.
(a) increasing the reaction volume
(b) removing $C_2H_4Cl_2$ from the reaction mixture as it forms
(c) lowering the reaction temperature
(d) adding Cl_2

76. This reaction is endothermic.

$$C_2H_4(g) + I_2(g) \rightleftharpoons C_2H_4I_2(g)$$

If you were a chemist trying to maximize the amount of $C_2H_4I_2$ produced, which of the following might you try? Assume that the reaction mixture reaches equilibrium.
(a) decreasing the reaction volume
(b) removing I_2 from the reaction mixture
(c) raising the reaction temperature
(d) adding C_2H_4 to the reaction mixture

HIGHLIGHT PROBLEMS

77. H_2 and I_2 are combined in a flask and allowed to react according to the reaction:

$$H_2(g) + I_2(g) \rightleftharpoons 2\,HI(g)$$

Examine the figures (sequential in time) and determine which figure represents the point where equilibrium is reached.

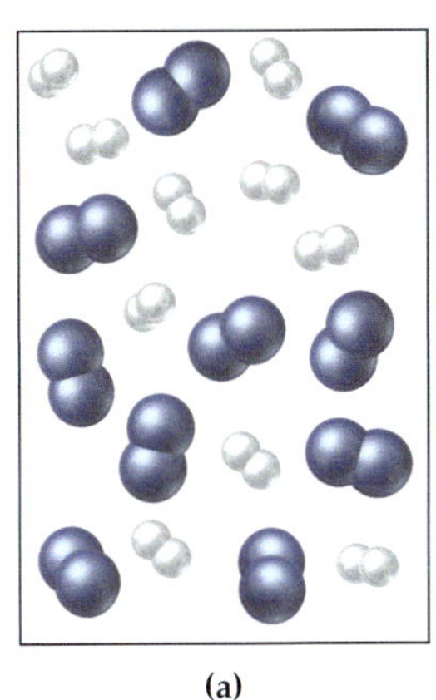

(a)

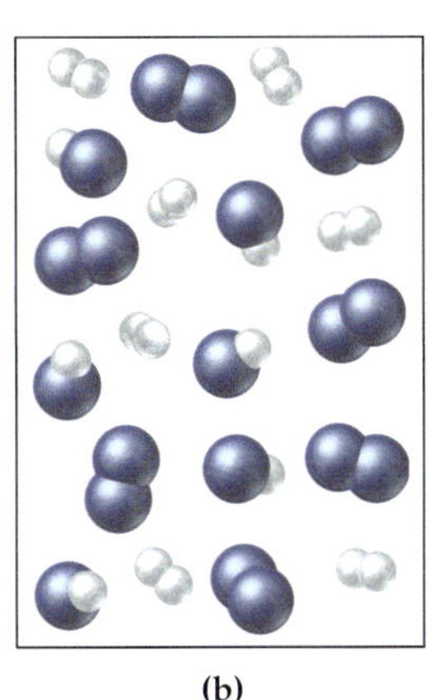

(b)

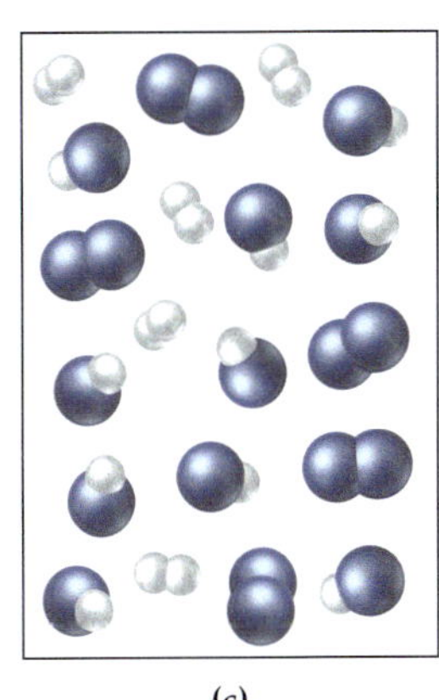

(c)

(d)

(e)

(f)

78. Ethene (C_2H_4) can be halogenated by the reaction:

$$C_2H_4(g) + X_2(g) \rightleftharpoons C_2H_4X_2(g)$$

where X_2 can be Cl_2, Br_2, or I_2. Examine the figures representing equilibrium concentrations of this reaction at the same temperature for the three different halogens. Rank the equilibrium constants for these three reactions from largest to smallest.

$C_2H_4 + Cl_2 \rightleftharpoons C_2H_4Cl_2$

(a)

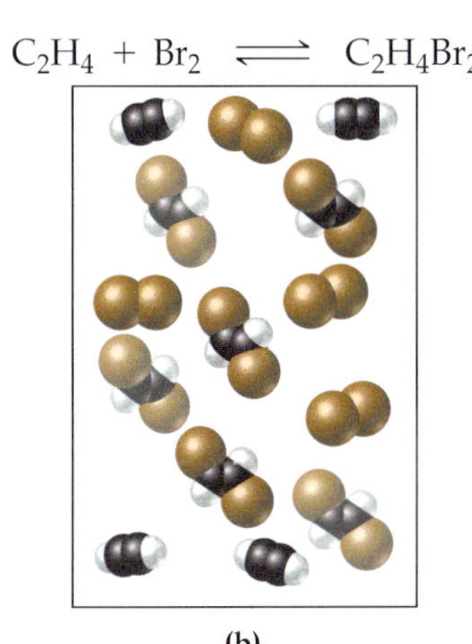

(b)

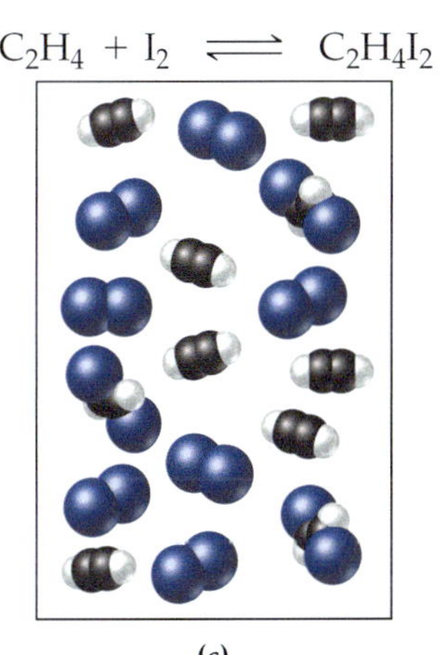

(c)

Answers to Skillbuilder Exercises

Skillbuilder 9.1
(a) exothermic
(b) exothermic

Skillbuilder 9.2 $K_{eq} = \dfrac{[HF]^2}{[H_2][F_2]}$

Skillbuilder 9.3 $K_{eq} = \dfrac{[Cl_2]^2}{[HCl]^4[O_2]}$

Skillbuilder 9.4 Adding Br_2 causes a shift to the left; adding BrNO causes a shift to the right.

Skillbuilder Plus, p. 400 Removing Br_2 causes a shift to the right.

Skillbuilder 9.5 Decreasing volume causes a shift to the right; increasing volume causes a shift to the left.

Skillbuilder 9.6 Increasing the temperature shifts the reaction to the left; decreasing the temperature shifts the reaction to the right.

Answers to Conceptual Checkpoints

9.1 (d) kWh is the largest of the four units listed, so the numerical value of the yearly energy consumption is lowest if expressed in kWh.

9.2 (a) In accordance with the gas laws (Module 6), increasing the pressure would increase the temperature, decrease the volume, or both. Increasing the temperature would increase the reaction rate. Decreasing the volume would increase the concentration of the reactants, which would also increase the reaction rate. Therefore, we would expect that increasing the pressure would speed up the reaction.

9.3 (c) Since the image in (c) has the greatest amount of product, the equilibrium constant must be largest at T_3.

9.4 Doubling the hydrogen concentration leads to a larger shift toward products because the concentration of hydrogen in the denominator is squared.

9.5 (c) Increasing the concentration of a reactant causes the reaction to shift right. For this reaction, changing the volume has no effect because the number of gas particles on both sides of the equation is equal.

9.6 (a) Since the reaction is endothermic, increasing the temperature drives it to the right (toward the product). Since the reaction has 1 mol particles on the left and 2 mol particles on the right, increasing the volume drives it to the right. Therefore, increasing the temperature and increasing the volume will create the greatest amount of product.

9.7 Reaction A is likely to have the faster rate. In general, the lower the activation barrier, the faster the rate at a given temperature.

▲ The combustion of fossil fuels such as octane (shown here) produces water and carbon dioxide as products. Carbon dioxide is a greenhouse gas that is believed to be responsible for climate change.

Chemical Quantities 1 10

"Man masters nature not by force but by understanding. That is why science has succeeded where magic failed: because it has looked for no spell to cast." —

Jacob Bronowski (1908–1974)

MODULE OUTLINE

10.1 Climate Change: Too Much Carbon Dioxide

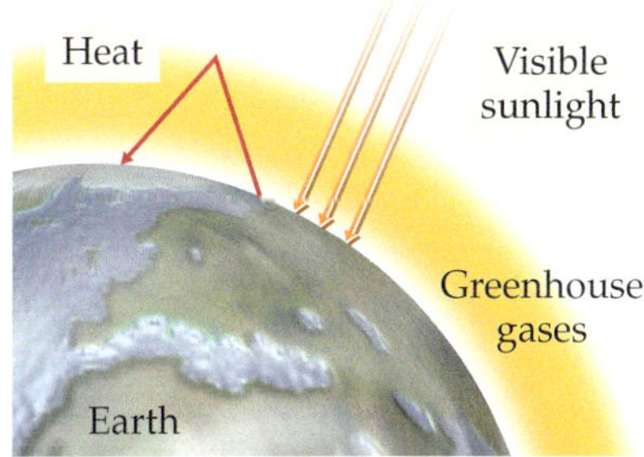

▲ **FIGURE 10.1 The greenhouse effect** Greenhouse gases act like glass in a greenhouse, allowing visible-light energy to enter the atmosphere but preventing heat energy from escaping.

Average global temperatures depend on the balance between incoming sunlight, which warms Earth, and outgoing heat lost to space, which cools it. Certain gases in Earth's atmosphere, called **greenhouse gases**, affect that balance by acting like glass in a greenhouse. They allow sunlight into the atmosphere to warm Earth but prevent heat from escaping (◀ Figure 10.1). Without greenhouse gases, more heat would escape, and Earth's average temperature would be about 16 °C colder. Caribbean tourists would freeze at an icy –6 °C, instead of baking at a tropical 27 °C. On the other hand, if the concentration of greenhouse gases in the atmosphere were to increase, Earth's average temperature would rise.

In recent decades, scientists have become concerned because the atmospheric concentration of carbon dioxide (CO_2)—Earth's most significant greenhouse gas in terms of its contribution to climate—is rising. This rise in CO_2 concentration enhances the atmosphere's ability to hold heat and may therefore lead to **climate change**, seen most clearly as an increase in Earth's average temperature. Since 1860, atmospheric CO_2 levels have risen by 25 %, and Earth's average temperature has increased by 0.7 °C (▶ Figure 10.2).

The primary cause of rising atmospheric CO_2 concentration is the burning of fossil fuels. Fossil fuels—natural gas, petroleum, and coal—provide approximately 90 % of our society's energy. Combustion of fossil fuels, however, produces CO_2. As an example, consider the combustion of octane (C_8H_{18}), a component of gasoline.

$$2\,C_8H_{18}(l) + 25\,O_2(g) \rightarrow 16\,CO_2(g) + 18\,H_2O(g)$$

The balanced chemical equation shows that 16 mol of CO_2 are produced for every 2 mol of octane burned. Because we know the world's annual fossil fuel consumption, we can estimate the world's annual CO_2 production. A simple calculation shows that the world's annual CO_2 production—from fossil fuel combustion—matches the measured annual atmospheric CO_2 increase. This implies that fossil fuel combustion is indeed responsible for increased atmospheric CO_2 levels.

The numerical relationship between chemical quantities in a balanced chemical equation is called reaction **stoichiometry**. Stoichiometry allows us to predict

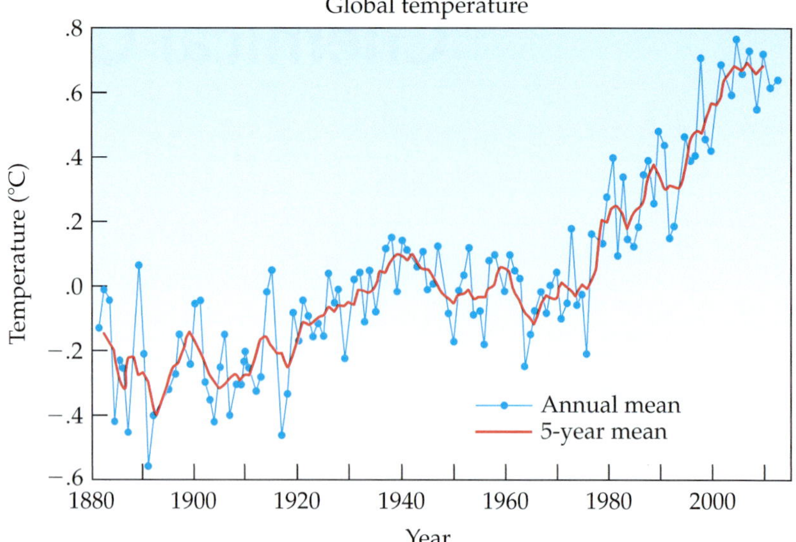

▶ **FIGURE 10.2 Climate change** Yearly temperature differences from the 120-year average temperature. Earth's average temperature has increased by about 0.7 °C since 1880. (*Source*: NASA GISS Surface Temperature Analysis)

the amounts of products that form in a chemical reaction based on the amounts of reactants. Stoichiometry also allows us to predict how much of the reactants is necessary to form a given amount of product, or how much of one reactant is required to completely react with another reactant. These calculations are central to chemistry, allowing chemists to plan and carry out chemical reactions to obtain products in desired quantities.

10.2 Making Pancakes: Relationships between Ingredients

LO: Recognize the numerical relationship between chemical quantities in a balanced chemical equation.

The concepts of stoichiometry are similar to the concepts we use in following a cooking recipe. Calculating the amount of carbon dioxide produced by the combustion of a given amount of a fossil fuel is similar to calculating the number of pancakes that can be made from a given number of eggs. For example, suppose we use the following pancake recipe:

For the sake of simplicity, this recipe omits liquid ingredients.

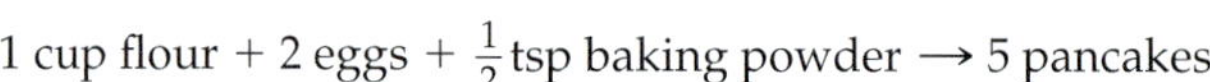

$$1 \text{ cup flour} + 2 \text{ eggs} + \tfrac{1}{2} \text{tsp baking powder} \longrightarrow 5 \text{ pancakes}$$

▲ A recipe gives numerical relationships between the ingredients and the number of pancakes. © Maxwell Art And Photo/Pearson.

The recipe shows the numerical relationships between the pancake ingredients. It says that if we have 2 eggs—and enough of everything else—we can make 5 pancakes. We can write this relationship as a ratio:

2 eggs : 5 pancakes

What if we have 8 eggs? Assuming that we have enough of everything else, how many pancakes can we make? If the preceding ratio holds, we can determine that 8 eggs are sufficient to make 20 pancakes.

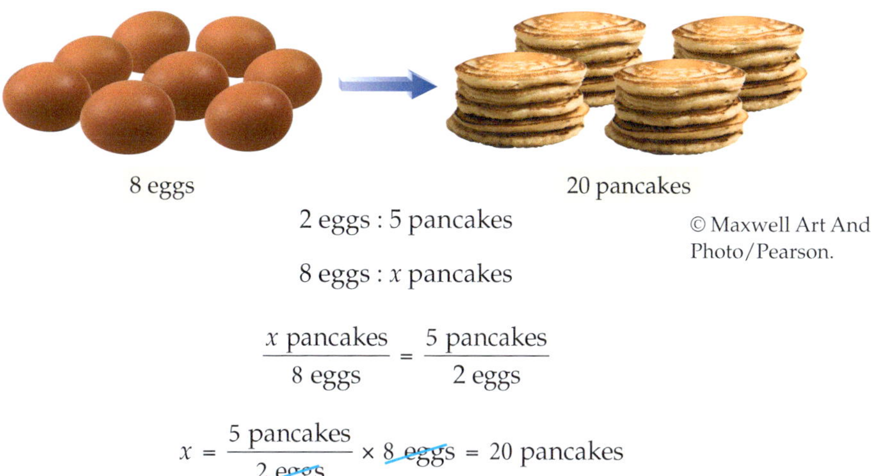

2 eggs : 5 pancakes

8 eggs : x pancakes

$$\frac{x \text{ pancakes}}{8 \text{ eggs}} = \frac{5 \text{ pancakes}}{2 \text{ eggs}}$$

$$x = \frac{5 \text{ pancakes}}{2 \text{ eggs}} \times 8 \text{ eggs} = 20 \text{ pancakes}$$

The pancake recipe contains numerical ratios between the pancake ingredients and the number of pancakes. Other ratios from this recipe include:

1 cup flour : 5 pancakes

$\frac{1}{2}$ tsp baking powder : 5 pancakes

The recipe also gives us relationships among the ingredients themselves. For example, how much baking powder is required to go with 3 cups of flour? From the recipe:

1 cup flour : $\frac{1}{2}$ tsp baking powder

With this ratio, we can calculate the appropriate amount of baking powder.

1 cup flour : $\frac{1}{2}$ tsp baking powder

3 cup flour : x tsp baking powder

$$\frac{x \text{ tsp of baking powder}}{3 \text{ cups flour}} = \frac{\frac{1}{2} \text{ tsp of baking powder}}{1 \text{ cup flour}}$$

$$x \text{ tsp baking powder} = \frac{\frac{1}{2} \text{ tsp baking powder}}{1 \text{ cup flour}} \times 3 \text{ cups flour}$$

$$= \frac{3}{2} \text{ tsp baking powder}$$

10.3 Mole to Mole Calculations

LO: Carry out mole-to-mole calculations between reactants and products based on the numerical relationship between chemical quantities in a balanced chemical equation.

In a balanced chemical equation, we have a "recipe" for how reactants combine to form products. For example, the following equation shows how hydrogen and nitrogen combine to form ammonia (NH_3):

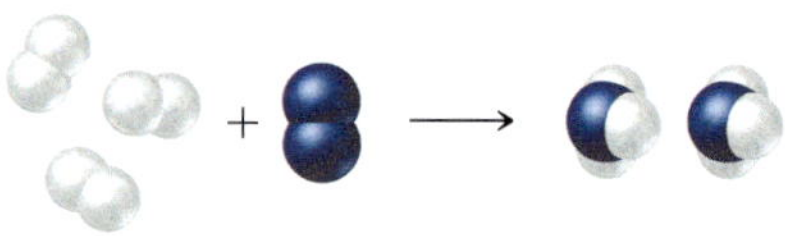

$$3\,H_2(g) + N_2(g) \rightarrow 2\,NH_3(g)$$

The balanced equation shows that 3 H_2 molecules react with 1 N_2 molecule to form 2 NH_3 molecules. We can express these relationships as the following ratios:

$$3\ H_2 \text{ molecules} : 1\ N_2 \text{ molecule} : 2\ NH_3 \text{ molecules}$$

Since we do not ordinarily deal with individual molecules, we can express the same ratios in moles.

$$3 \text{ mol } H_2 : 1 \text{ mol } N_2 : 2 \text{ mol } NH_3$$

If we have 3 mol of N_2, and more than enough H_2, how much NH_3 can we make? We first sort the information in the problem.

GIVEN: 3 mol N_2

FIND: mol NH_3

SOLUTION MAP

We then strategize by drawing a solution map that begins with mol N_2 and ends with mol NH_3. The ratio comes from the balanced chemical equation.

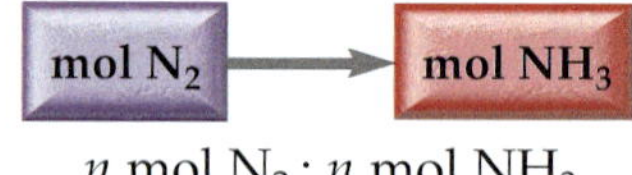

$$n \text{ mol } N_2 : n \text{ mol } NH_3$$

RELATIONSHIPS USED

1 mol N_2 : 2 mol NH_3 (from balanced equation)

SOLUTION

We can then do the calculation.

$$1 \text{ mol } N_2 : 2 \text{ mol } NH_3$$

$$3 \text{ mol } N_2 : x \text{ mol } NH_3$$

$$\text{where } x \text{ mol } NH_3 = n(NH_3)$$

$$\frac{n(NH_3)}{3 \text{ mol } N_2} = \frac{2 \text{ mol } NH_3}{1 \text{ mol } N_2}$$

$$n(NH_3) = \frac{2 \text{ mol } NH_3}{1 \cancel{\text{ mol } N_2}} \times 3 \cancel{\text{ mol } N_2}$$

$$= 6 \text{ mol } NH_3$$

We have enough N_2 to make 6 mol of NH_3. This example is relatively straight forward. What do we do when the ratio is more complicated looking or when the moles of the known chemical isn't a nice round number like in the above example?

As the information in a balanced chemical reaction equation represent the ratios of the chemicals involved in the reaction, they can be treated mathematically precisely the same way as the ratios in Module 4 when converting between different units.

EXAMPLE 10.1 MOLE-TO-MOLE CALCULATIONS

Sodium chloride, NaCl, forms by this reaction between sodium and chlorine.

$$2\ \text{Na}(s) + \text{Cl}_2(g) \rightarrow 2\ \text{NaCl}(s)$$

How many moles of NaCl result from the complete reaction of 3.4 mol of Cl_2? Assume that there is more than enough Na.

SORT

You are given the number of moles of a reactant (Cl_2) and asked to find the number of moles of product (NaCl) that will form if the reactant completely reacts.

GIVEN: 3.4 mol Cl_2

FIND: mol NaCl

STRATEGIZE

Draw the solution map beginning with moles of chlorine and using the stoichiometric ratio to calculate moles of sodium chloride. The ratio comes from the coefficients in balanced chemical equation.

SOLUTION MAP

n mol Cl_2 : n mol NaCl

RELATIONSHIPS USED

1 mol Cl_2 : 2 mol NaCl (from balanced chemical equation)

SOLVE

Follow the solution map to solve the problem.

SOLUTION

1 mol Cl_2 : 2 mol NaCl

3.4 mol Cl_2 : x mol NaCl

Where x mol NaCl = n(NaCl)

$$\frac{n(\text{NaCl})}{3.4\ \text{mol Cl}_2} = \frac{2\ \text{mol NaCl}}{1\ \text{mol Cl}_2}$$

$$n(\text{NaCl}) = \frac{2\ \text{mol NaCl}}{1\ \cancel{\text{mol Cl}_2}} \times 3.4\ \cancel{\text{mol Cl}_2}$$

$$= 6.8\ \text{mol NaCl}$$

There is enough Cl_2 to produce 6.8 mol of NaCl.

CHECK

Check your answer. Are the units correct? Does the answer make physical sense?

The answer has the correct units, mol. The answer is reasonable because each mole of Cl_2 makes two moles of NaCl.

▶SKILLBUILDER 10.1 | Mole-to-Mole Calculations

Water forms when hydrogen gas reacts explosively with oxygen gas according to the balanced equation:

$$\text{O}_2(g) + 2\ \text{H}_2(g) \rightarrow 2\ \text{H}_2\text{O}(g)$$

How many moles of H_2O result from the complete reaction of 24.6 mol of O_2? Assume that there is more than enough H_2.

▶FOR MORE PRACTICE Example 10.7; Problems 13, 14, 15, 16.

CONCEPTUAL CHECKPOINT 10.1

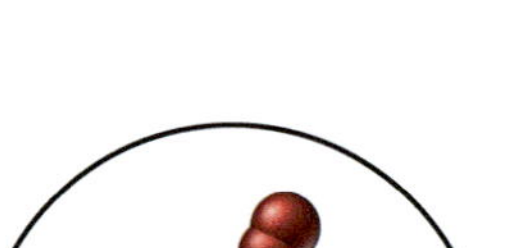

Methane (CH_4) undergoes combustion according to this reaction.

$$CH_4(g) + 2\,O_2(g) \rightarrow CO_2(g) + 2\,H_2O(g)$$

If the figure shown in the left margin represents the amount of oxygen available to react, which of the following figures best represents the amount of CH_4 required to completely react with all of the oxygen?

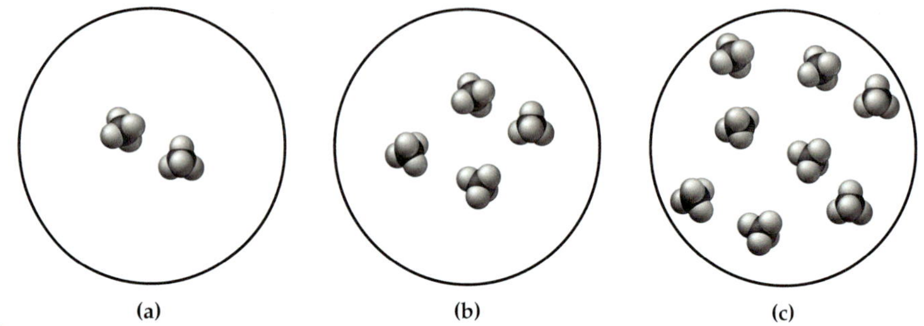

10.4 Mass to Mass Calculations

LO: Carry out mass-to-mass calculations between reactants and products based on the numerical relationship between chemical quantities in a balanced chemical equation and molar masses.

In Module 5, we learned how a chemical *formula* contains the ratio of moles of a compound to moles of its constituent elements. In this module, we have seen how a chemical *equation* contains ratios between moles of reactants and moles of products. However, we are often interested in relationships between *mass* of reactants and *mass* of products. For example, we might want to know the mass of carbon dioxide emitted by an automobile per kilogram of gasoline used. Or we might want to know the mass of each reactant required to obtain a certain mass of a product in a synthesis reaction. These calculations are similar to calculations covered in Section 5.4, where we converted between mass of a compound and mass of a constituent element. The general outline for these types of calculations is:

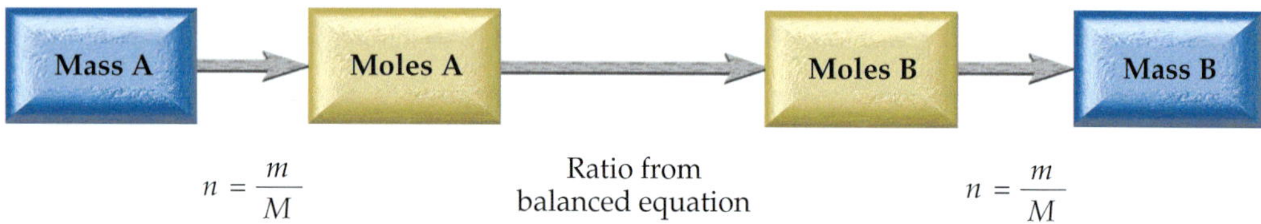

where A and B are two different substances involved in the reaction. We use the molar mass of A to calculate moles of A from the mass of A. We use the ratio from the balanced equation to calculate moles of B from A, and we use the molar mass of B to calculate mass of B from the moles of B. For example, suppose we want to calculate the mass of CO_2 emitted upon the combustion of 5.0×10^2 g of pure octane. The balanced chemical equation for octane combustion is:

$$2\,C_8H_{18}(l) + 25\,O_2(g) \rightarrow 16\,CO_2(g) + 18\,H_2O(g)$$

CHEMISTRY IN THE MEDIA

▶ The Controversy over Oxygenated Fuels

We have seen that the balanced chemical equation for the combustion of octane, a component of gasoline, is:

$$2\ C_8H_{18}(l) + 25\ O_2(g) \rightarrow 16\ CO_2(g) + 18\ H_2O(g)$$

We have also learned how balanced chemical equations give numerical relationships between reactants. This equation shows that 25 mol of O_2 are required to completely react with 2 mol of C_8H_{18}. What if there is not enough O_2 in the cylinders of an automobile engine to fully react with the amount of octane flowing into them? For many reactions, a shortage of one reactant simply means that less product forms, something we will learn more about later in this module. However, for some reactions, a shortage of one reactant causes other reactions—called *side reactions*—to occur along with the desired reaction. In the case of octane and the other major components of gasoline, those side reactions result in pollutants such as carbon monoxide (CO) and ozone (O_3).

In 1990, the U.S. Congress, in efforts to lower air pollution levels, passed amendments to the Clean Air Act requiring oil companies to add substances to gasoline that prevent these side reactions. Because these additives increase the amount of oxygen during combustion, the resulting gasoline is called oxygenated fuel. The additive of choice among oil companies used to be a compound called MTBE (methyl tertiary butyl ether). The immediate results were positive: carbon monoxide and ozone levels in many major cities decreased significantly.

▲ The 1990 amendments to the Clean Air Act required oil companies to put additives in gasoline that increased its oxygen content. © Maxwell Art And Photo/Pearson.

▲ At one time, MTBE was the gasoline additive of choice.

Over time, however, MTBE—a compound that does not readily biodegrade—began to appear in drinking-water supplies across the nation. MTBE made its way into drinking water through gasoline spills at gas stations, from boat motors, and from leaking underground storage tanks. The consequences have been significant. MTBE, even at low levels, imparts a turpentine-like odor and foul taste to drinking water. It is also a suspected carcinogen.

Public response was swift and dramatic. Several multimillion dollar class-action lawsuits were filed and settled against the manufacturers of MTBE, against gas stations suspected of leaking it, and against the oil companies that put the additive into gasoline. Most states have completely banned MTBE from gasoline. Ethanol, made from the fermentation of grains, has been used as a substitute for MTBE because it has many of the same pollution-reducing effects without the associated health hazards. Oil companies did not use ethanol originally because it was more expensive than MTBE, but now ethanol has become the additive of choice.

B10.1 CAN YOU ANSWER THIS? *How many moles of oxygen (O_2) are required to completely react with 425 mol of octane (approximate capacity of a 57 L automobile fuel tank)?*

We begin by sorting the information in the problem.

GIVEN: 5.0×10^2 g C_8H_{18}

FIND: g CO_2

Notice that we are given g C_8H_{18} and asked to find g CO_2. The balanced chemical equation, however, gives us a relationship between moles of C_8H_{18} and moles of CO_2. Consequently, before using that relationship, we must calculate moles.

The solution map uses the general outline:

$$\text{Mass A} \rightarrow \text{Moles A} \rightarrow \text{Moles B} \rightarrow \text{Mass B}$$

where A is octane and B is carbon dioxide.

SOLUTION MAP

We strategize by drawing the solution map, which begins with mass of octane and ends with mass of carbon dioxide.

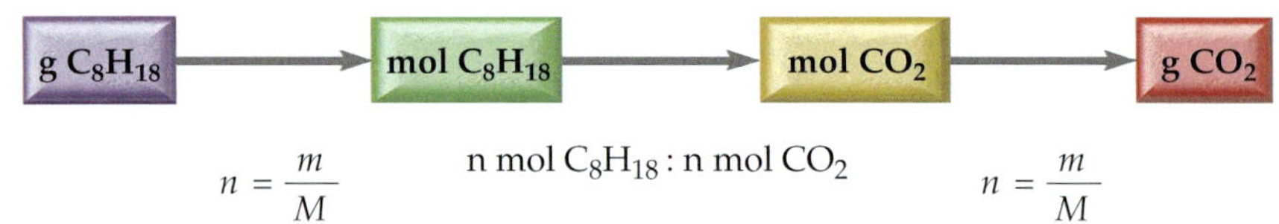

$n = \frac{m}{M}$ n mol C_8H_{18} : n mol CO_2 $n = \frac{m}{M}$

RELATIONSHIPS USED

2 mol C_8H_{18} : 16 mol CO_2 (from chemical equation)

Molar mass C_8H_{18} = 114.3 g/mol

Molar mass CO_2 = 44.01 g/mol

SOLUTION

We then follow the solution map to solve the problem, beginning with g C_8H_{18} and canceling units to arrive at g CO_2.

Step 1 Calculate moles C_8H_{18}

$$n = \frac{m}{M} = \frac{5.0 \times 10^2 \cancel{g}}{114.3 \cancel{g}/\text{mol}} = 4.4 \text{ mol } C_8H_{18}$$

Step 2 Calculate moles CO_2

2 mol C_8H_{18} : 16 mol CO_2

4.374 mol C_8H_{18} : n CO_2

$$n(CO_2) = \frac{16 \text{ mol } CO_2}{2 \cancel{\text{mol}}\, C_8H_{18}} \times 4.4 \cancel{\text{mol}}\, C_8H_{18}$$
$$= 35 \text{ mol } CO_2$$

Step 3 Calculate mass CO_2

$$n = \frac{m}{M}$$

$$m = n \times M = 35 \cancel{\text{mol}} \times 44.01 \text{ g}/\cancel{\text{mol}}$$
$$= 1.5 \times 10^3 \text{ g } CO_2$$

Upon combustion, 5.0×10^2 g of octane produces 1.5×10^3 g of carbon dioxide. With practice you will be able to omit the first part of step 2 when working through these problems. The worked examples that follow only show the second part of step 2. You can do the first part of step 2 in your working if you need to see the ratio clearly.

EXAMPLE 10.2 MASS-TO-MASS CALCULATIONS

In photosynthesis, plants convert carbon dioxide and water into glucose ($C_6H_{12}O_6$) according to the reaction:

$$6\ CO_2(g) + 6\ H_2O(l) \xrightarrow[\text{sunlight}]{} 6\ O_2(g) + C_6H_{12}O_6(aq)$$

How many grams of glucose can be synthesized from 58.5 g of CO_2? Assume that there is more than enough water present to react with all of the CO_2.

SORT

You are given the mass of carbon dioxide and asked to find the mass of glucose that can form if the carbon dioxide completely reacts.

GIVEN: 58.5 g CO_2

FIND: g $C_6H_{12}O_6$

STRATEGIZE

The solution map uses the general outline:

Mass A → Moles A → Moles B → Mass B

where A is carbon dioxide and B is glucose.

The stoichiometric relationship between moles of carbon dioxide and moles of glucose comes from the balanced equation. The molar masses of carbon dioxide and glucose are also needed.

SOLUTION MAP

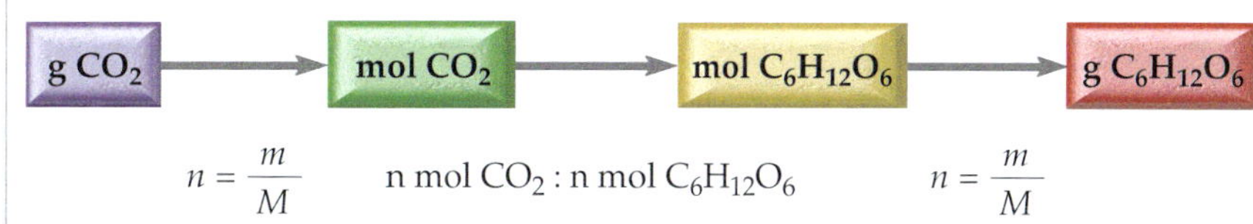

RELATIONSHIPS USED

6 mol CO_2 : 1 mol $C_6H_{12}O_6$ (from balanced chemical equation)

Molar mass CO_2 = 44.01 g/mol

Molar mass $C_6H_{12}O_6$ = 180.2 g/mol

SOLVE

Follow the solution map to solve the problem. Begin with grams of carbon dioxide and using the appropriate equations and ratios to arrive at grams of glucose.

SOLUTION

Step 1 moles CO_2

$$n = \frac{m}{M} = \frac{58.5\ \cancel{g}}{44.01\ \cancel{g}/\text{mol}} = 1.33\ \text{mol } CO_2$$

Step 2 moles $C_6H_{12}O_6$

$$n(C_6H_{12}O_6) = \frac{1\ \text{mol } C_6H_{12}O_6}{6\ \cancel{\text{mol } CO_2}} \times 1.33\ \cancel{\text{mol } CO_2}$$
$$= 0.222\ \text{mol } C_6H_{12}O_6$$

Step 3 mass $C_6H_{12}O_6$

$$n = \frac{m}{M}$$

$$m(C_6H_{12}O_6) = n \times M = 0.222\ \cancel{\text{mol}} \times 180.2\ \text{g}/\cancel{\text{mol}}$$
$$= 39.9\ \text{g } C_6H_{12}O_6$$

CHECK

Are the units correct? Does the answer make physical sense?

The units, g $C_6H_{12}O_6$, are correct. The magnitude of the answer seems reasonable because it is of the same order of magnitude as the given mass of carbon dioxide. An answer that is orders of magnitude different would immediately be suspect.

▶SKILLBUILDER 10.2 | Mass-to-Mass Calculations

Magnesium hydroxide, the active ingredient in milk of magnesia, neutralizes stomach acid, primarily HCl, according to the reaction:

$$Mg(OH)_2(aq) + 2\ HCl(aq) \rightarrow 2\ H_2O(l) + MgCl_2(aq)$$

How much HCl in grams can be neutralized by 5.50 g of $Mg(OH)_2$?

▶FOR MORE PRACTICE Example 10.8; Problems 29, 30, 31, 32.

EXAMPLE 10.3 MASS-TO-MASS CALCULATIONS

One of the components of acid rain is nitric acid, which forms when NO_2, a pollutant, reacts with oxygen and rainwater according to the following simplified reaction:

$$4\ NO_2(g) + O_2(g) + 2\ H_2O(l) \rightarrow 4\ HNO_3(aq)$$

Assuming that there is more than enough O_2 and H_2O, how much HNO_3 in kilograms forms from 1.5×10^3 kg of NO_2 pollutant?

SORT

You are given the mass of nitrogen dioxide (a reactant) and asked to find the mass of nitric acid that can form if the nitrogen dioxide completely reacts.

GIVEN: 1.5×10^3 kg NO_2

FIND: kg HNO_3

STRATEGIZE

The solution map follows the general format of:

Mass → Moles → Moles → Mass

However, because the original quantity of NO_2 is given in kilograms, you must first convert to grams. The final quantity is requested in kilograms, so you must convert back to kilograms at the end. The stoichiometric relationship between moles of nitrogen dioxide and moles of nitric acid comes from the balanced equation. The molar masses of nitrogen dioxide and nitric acid and the relationship between kilograms and grams are also required.

SOLUTION MAP

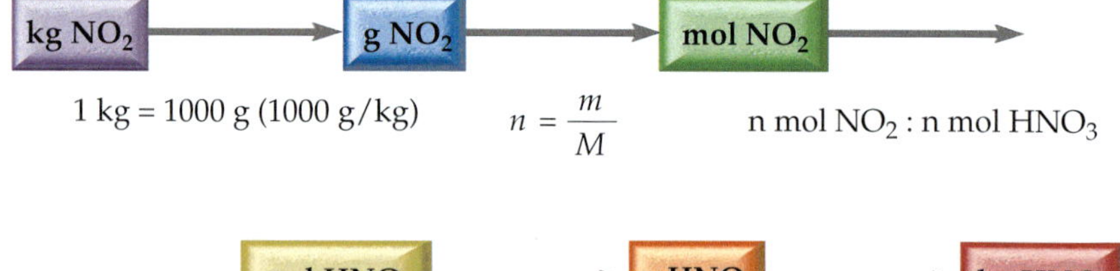

$n = \frac{m}{M}$ 1 g = 10^{-3} kg (1000 g/kg)

RELATIONSHIPS USED

4 mol NO_2 : 4 mol HNO_3 (from balanced chemical equation)

Molar mass NO_2 = 46.01 g/mol

Molar mass HNO_3 = 63.02 g/mol

1 kg = 1000 g

SOLVE
Follow the solution map to solve the problem. Begin with kilograms of nitrogen dioxide and use the appropriate equations to and ratios arrive at kilograms of nitric acid.

SOLUTION
Convert kg NO_2 to g NO_2

$$m(NO_2) = 1000\text{ g}/\cancel{\text{kg}} \times 1.5 \times 10^3\ \cancel{\text{kg}})$$
$$= 1.5 \times 10^6\text{ g } NO_2$$

Step 1 moles NO_2

$$n(NO_2) = \frac{m}{M} = \frac{1.5 \times 10^6\ \cancel{\text{g}}}{46.01\ \cancel{\text{g}}/\text{mol}} = 3.3 \times 10^4\text{ mol } NO_2$$

Step 2 moles HNO_3

$$n(HNO_3) = \frac{4\text{ mol } HNO_3}{4\ \cancel{\text{mol } NO_2}} \times 3.3 \times 10^4\ \cancel{\text{mol } NO_2}$$
$$= 3.3 \times 10^4\text{ mol } HNO_3$$

Step 3 mass HNO_3

$$n = \frac{m}{M}$$

$$m(HNO_3) = n \times M = 3.3 \times 10^4\ \cancel{\text{mol}}\ HNO_3 \times 63.02\text{ g}/\cancel{\text{mol}}\ HNO_3$$
$$= 2.1 \times 10^6\text{ g } HNO_3$$

Convert g HNO_3 to kg HNO_3

$$m(HNO_3) = \frac{2.1 \times 10^6\ \cancel{\text{g}}\ HNO_3}{1000\ \cancel{\text{g}}\ /\ \text{kg}} = 2.1 \times 10^3\text{ kg } HNO_3$$

CHECK
Are the units correct? Does the answer make physical sense?

The units, kg HNO_3 are correct. The magnitude of the answer seems reasonable because it is of the same order of magnitude as the given mass of nitrogen dioxide. An answer that is orders of magnitude different would immediately be suspect.

▶SKILLBUILDER 10.3 | Mass-to-Mass Calculations

Another component of acid rain is sulfuric acid, which forms when SO_2, also a pollutant, reacts with oxygen and rainwater according to the following reaction:

$$2\ SO_2(g) + O_2(g) + 2\ H_2O(l) \rightarrow 2\ H_2SO_4(aq)$$

Assuming that there is more than enough O_2 and H_2O, how much H_2SO_4 in kilograms forms from 2.6×10^3 kg of SO_2?

▶FOR MORE PRACTICE Problems 33, 34, 35, 36.

10.5 Limiting Reactant, Theoretical Yield, and Percent Yield

LO: Calculate limiting reactant, theoretical yield, and percent yield for a given amount of reactants in a balanced chemical equation.

Let's return to our pancake analogy to understand two more concepts important in reaction stoichiometry: limiting reactant and percent yield. Recall our pancake recipe:

$$1\text{ cup flour} + 2\text{ eggs} + \tfrac{1}{2}\text{ tsp baking powder} \rightarrow 5\text{ pancakes}$$

Suppose we have 3 cups flour, 10 eggs, and 4 tsp baking powder. How many pancakes can we make?

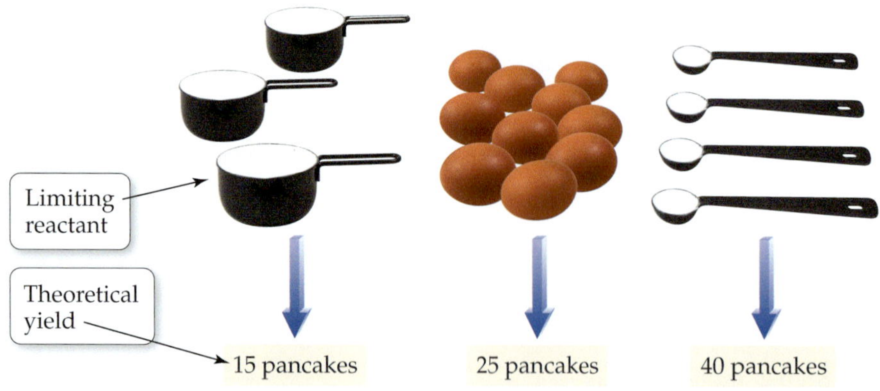

▲ If this were a chemical reaction, the flour would be the limiting reactant and 15 pancakes would be the theoretical yield. © Maxwell Art And Photo/Pearson.

We have enough flour for 15 pancakes, enough eggs for 25 pancakes, and enough baking powder for 40 pancakes. Consequently, unless we get more ingredients, *we can make only 15 pancakes*. The amount of flour we have *limits* the number of pancakes we can make. If this were a chemical reaction, the flour would be the *limiting reactant*, the reactant that limits the amount of product in a chemical reaction. Notice that the **limiting reactant** is simply the reactant that makes *the least amount of product*. If this were a chemical reaction, 15 pancakes would be the **theoretical yield**, the amount of product that can be made in a chemical reaction based on the amount of limiting reactant.

The term *limiting reagent* is sometimes used in place of limiting reactant.

Let us carry this analogy one step further. Suppose we go on to make our pancakes. We accidentally burn three of them and one falls on the floor. So even though we had enough flour for 15 pancakes, we finished with only 11 pancakes. If this were a chemical reaction, the 11 pancakes would be our **actual yield**, the amount of product actually produced by a chemical reaction. Finally, our **percent yield**, the percentage of the theoretical yield that was actually attained, would be:

$$\text{Percent yield} = \frac{11 \text{ pancakes}}{15 \text{ pancakes}} \times 100\,\% = 73\,\%$$

The actual yield of a chemical reaction, which must be determined experimentally, often depends in various ways on the reaction conditions.

Since four of the pancakes were ruined, we ended up with only 73 % of our theoretical yield. In a chemical reaction, the actual yield is almost always less than 100 % because at least some of the product does not form or is lost in the process of recovering it (in analogy to some of the pancakes being burned).

To summarize:

- **Limiting reactant (or limiting reagent)**—the reactant that is completely consumed in a chemical reaction.
- **Theoretical yield**—the amount of product that can be made in a chemical reaction based on the amount of limiting reactant.
- **Actual yield**—the amount of product actually produced by a chemical reaction.
- **Percent yield** $= \dfrac{\textbf{Actual yield}}{\textbf{Theoretical yield}} \times \mathbf{100\,\%}$

Consider this reaction:

$$Ti(s) + 2\,Cl_2(g) \rightarrow TiCl_4(s)$$

If we begin with 1.8 mol of titanium and 3.2 mol of chlorine, what is the limiting reactant and theoretical yield of $TiCl_4$ in moles? We begin by sorting the information in the problem according to our standard problem-solving procedure.

GIVEN: 1.8 mol Ti
3.2 mol Cl_2

FIND: limiting reactant
theoretical yield

SOLUTION MAP

As in our pancake analogy, we determine the limiting reactant by calculating how much product can be made from each reactant. The reactant that makes the *least amount of product* is the limiting reactant.

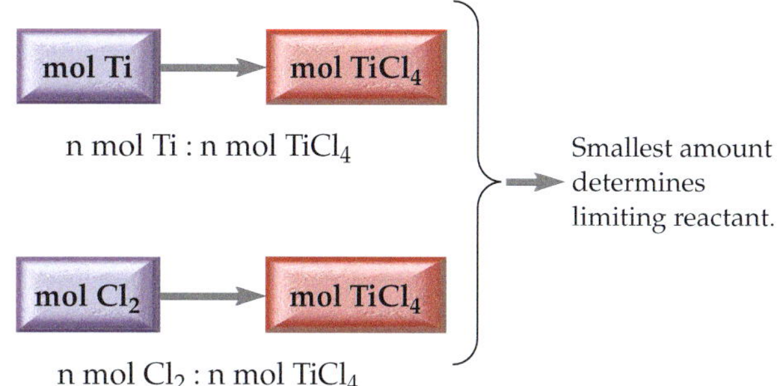

RELATIONSHIPS USED

From the balanced chemical equation we have the relationships between moles of each of the reactants and moles of product.

$$1 \text{ mol Ti} : 1 \text{ mol TiCl}_4$$

$$2 \text{ mol Cl}_2 : 1 \text{ mol TiCl}_4$$

SOLUTION

$TiCl_4$ that can form from Ti available:

$$n(\text{TiCl}_4) = \frac{1 \text{ mol TiCl}_4}{1 \text{ mol Ti}} \times 1.8 \text{ mol Ti} = 1.8 \text{ mol TiCl}_4$$

$TiCl_4$ that can form from Cl_2 available:

$$n(\text{TiCl}_4) = \frac{1 \text{ mol TiCl}_4}{2 \text{ mol Cl}_2} \times 3.2 \text{ mol Cl}_2 = 1.6 \text{ mol TiCl}_4$$

Limiting reactant

Least amount of product

In many industrial applications, the more costly reactant or the reactant that is most difficult to remove from the product mixture is chosen to be the limiting reactant.

Since the 3.2 mol of Cl_2 make the least amount of $TiCl_4$, Cl_2 is the limiting reactant. Notice that we began with more moles of Cl_2 than Ti, but because the reaction requires 2 Cl_2 for each Ti, Cl_2 is still the limiting reactant. The theoretical yield is 1.6 mol of $TiCl_4$.

EXAMPLE 10.4 LIMITING REACTANT AND THEORETICAL YIELD FROM INITIAL MOLES OF REACTANTS

Consider this reaction:

$$2 \text{ Al}(s) + 3 \text{ Cl}_2(g) \rightarrow 2 \text{ AlCl}_3(s)$$

If you begin with 0.552 mol of aluminum and 0.887 mol of chlorine, what is the limiting reactant and theoretical yield of $AlCl_3$ in moles?

SORT You are given the number of moles of aluminum and chlorine and asked to find the limiting reactant and theoretical yield of aluminum chloride.	**GIVEN:** 0.552 mol Al 0.887 mol Cl_2 **FIND:** limiting reactant theoretical yield of $AlCl_3$

STRATEGIZE

Draw a solution map that shows how to get from moles of each reactant to moles of $AlCl_3$.

The reactant that makes the *least amount of $AlCl_3$* is the limiting reactant. The stoichiometric relationships are based on the balanced equation.

SOLUTION MAP

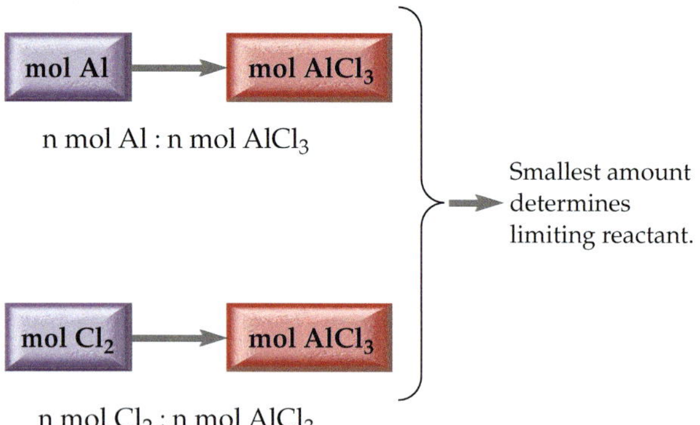

RELATIONSHIPS USED

2 mol Al:2 mol $AlCl_3$ (from balanced equation)

3 mol Cl_2:2 mol $AlCl_3$ (from balanced equation)

SOLVE

Follow the solution map to solve the problem.

SOLUTION

$$n(AlCl_3) = \frac{2 \text{ mol } AlCl_3}{2 \text{ mol Al}} \times 0.552 \text{ mol Al} = 0.552 \text{ mol } AlCl_3$$

Limiting reactant (0.552 mol Al); Least amount of product (0.552 mol $AlCl_3$)

$$n(AlCl_3) = \frac{2 \text{ mol } AlCl_3}{3 \text{ mol } Cl_2} \times 0.887 \text{ mol } Cl_2 = 0.591 \text{ mol } AlCl_3$$

Because the 0.552 mol of Al makes the least amount of $AlCl_3$, Al is the limiting reactant. The theoretical yield is 0.552 mol of $AlCl_3$.

$n(AlCl_3)$ from Al < $n(AlCl_3)$ from Cl_2

Therefore Al is limiting reactant.

Theoretical yield is 0.552 mol $AlCl_3$

CHECK

Are the units correct? Does the answer make physical sense?

The units, mol $AlCl_3$, are correct. The magnitude of the answer seems reasonable because it is of the same order of magnitude as the given number of moles of Al and Cl_2. An answer that is orders of magnitude different would immediately be suspect.

▶SKILLBUILDER 10.4 | Limiting Reactant and Theoretical Yield from Initial Moles of Reactants

Consider the reaction:

$$2\,Na(s) + F_2(g) \rightarrow 2\,NaF(s)$$

If you begin with 4.8 mol of sodium and 2.6 mol of fluorine, what is the limiting reactant and theoretical yield of NaF in moles?

▶FOR MORE PRACTICE Problems 41, 42, 43, 44, 45, 46, 47, 48.

CONCEPTUAL CHECKPOINT 10.2

Consider the reaction:

$$N_2(g) + 3\,H_2(g) \rightarrow 2\,NH_3(g)$$

If the flask in the left margin represents the mixture before the reaction, which flask represents the products after the limiting reactant has completely reacted?

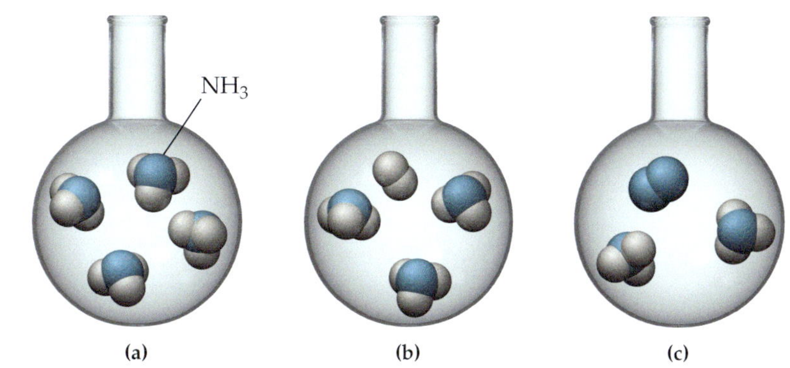

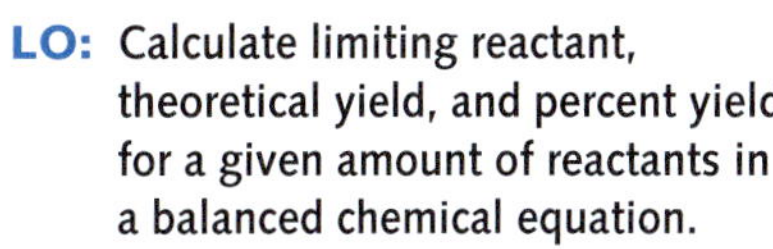

LO: Calculate limiting reactant, theoretical yield, and percent yield for a given amount of reactants in a balanced chemical equation.

When working in the laboratory, we normally measure the initial amounts of reactants in grams. To find limiting reactants and theoretical yields from initial masses, we must add two steps to our calculations. Consider, for example, the synthesis reaction:

$$2\,Na(s) + Cl_2(g) \rightarrow 2\,NaCl(s)$$

If we have 53.2 g of Na and 65.8 g of Cl_2, what is the limiting reactant and theoretical yield? We begin by sorting the information in the problem.

GIVEN: 53.2 g Na
65.8 g Cl_2

FIND: limiting reactant
theoretical yield

SOLUTION MAP

Again, we find the limiting reactant by calculating how much product can be made from each reactant. Since we are given the initial amounts in grams, we must first calculate moles. After we calculate moles of product, we calculate grams of product. The reactant that makes the *least amount of product* is the limiting reactant.

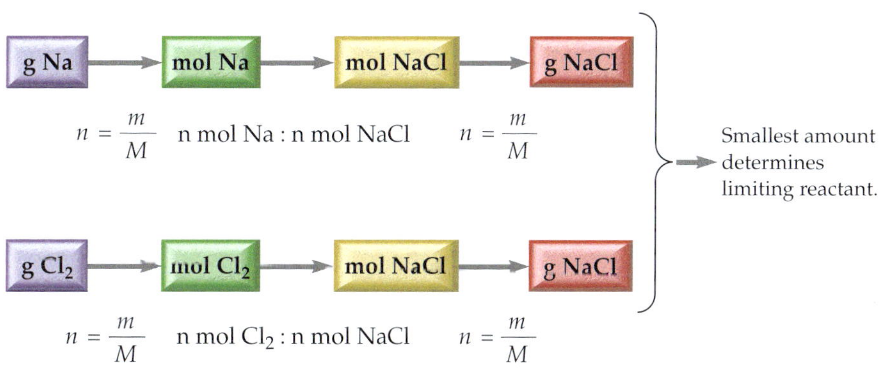

RELATIONSHIPS USED

From the balanced chemical equation, we know:

$$2 \text{ mol Na} : 2 \text{ mol NaCl}$$
$$1 \text{ mol Cl}_2 : 2 \text{ mol NaCl}$$

We also use these molar masses:

$$\text{Molar mass Na} = 22.99 \text{ g/mol}$$

$$\text{Molar mass Cl}_2 = 70.90 \text{ g/mol}$$

$$\text{Molar mass NaCl} = 58.44 \text{ g/mol}$$

SOLUTION

Beginning with the actual amounts of each reactant, we follow the solution map to calculate how much product can be made from each.

Possible yield of NaCl from Na:

$$n(\text{Na}) = \frac{m}{M} = \frac{53.2 \cancel{\text{g}} \text{Na}}{22.99 \cancel{\text{g}}/\text{mol Na}} = 2.31 \text{ mol Na}$$

We can also find the limiting reactant by calculating the number of moles of NaCl (rather than grams) that can be made from each reactant. However, since theoretical yields are normally calculated in grams, we take the calculation all the way to grams to determine limiting reactant.

$$n(\text{NaCl}) = \frac{2 \text{ mol NaCl}}{2 \cancel{\text{mol Na}}} \times 2.31 \cancel{\text{mol Na}} = 2.31 \text{ mol NaCl}$$

$$m(\text{NaCl}) = n \times M = 2.31 \cancel{\text{mol}} \text{ NaCl} \times 58.44 \text{ g}/\cancel{\text{mol}} \text{ NaCl} = 135 \text{ g NaCl}$$

Possible yield of NaCl from Cl_2:

Limiting reactant

$$n(\text{Cl}_2) = \frac{m}{M} = \frac{65.8 \cancel{\text{g}} \text{ Cl}_2}{70.90 \cancel{\text{g}}/\text{mol Cl}_2} = 0.928 \text{ mol Cl}_2$$

$$n(\text{NaCl}) = \frac{2 \text{ mol NaCl}}{1 \cancel{\text{mol Cl}_2}} \times 0.928 \cancel{\text{mol Cl}_2} = 1.86 \text{ mol NaCl}$$

$$m(\text{NaCl}) = n \times M = 1.86 \cancel{\text{mol NaCl}} \times 58.44 \text{ g}/\cancel{\text{mol NaCl}} = 108 \text{ g NaCl}$$

Least amount of product

The limiting reactant is not necessarily the reactant with the least mass.

Since Cl_2 makes the least amount of product, it is the limiting reactant. Notice that the limiting reactant is not necessarily the reactant with the least mass. In this case, we had fewer grams of Na than Cl_2, yet Cl_2 was the limiting reactant because it made less NaCl. The theoretical yield is therefore 108 g of NaCl, the amount of product possible based on the limiting reactant.

Now suppose that when the synthesis is carried out, the actual yield of NaCl is 86.4 g. What is the percent yield? The percent yield is:

The actual yield is always less than the theoretical yield because at least a small amount of product is usually lost or does not form during a reaction.

$$\text{Percent yield} = \frac{\text{Actual yield}}{\text{Theoretical yield}} \times 100\,\% = \frac{86.4\text{ g}}{108\text{ g}} \times 100\,\% = 79.7\,\%$$

EXAMPLE 10.5 FINDING LIMITING REACTANT AND THEORETICAL YIELD

Ammonia, NH_3, can be synthesized by this reaction:

$$2\,NO(g) + 5\,H_2(g) \rightarrow 2\,NH_3(g) + 2\,H_2O(g)$$

What maximum amount of ammonia in grams can be synthesized from 45.8 g of NO and 12.4 g of H_2?

SORT

You are given the masses of two reactants and asked to find the maximum mass of ammonia that forms. Although this problem does not specifically ask for the limiting reactant, it must be found to determine the theoretical yield, which is the maximum amount of ammonia that can be synthesized.

GIVEN: 45.8 g NO, 12.4 g H_2

FIND: maximum amount of NH_3 in g (this is the theoretical yield)

STRATEGIZE

Identify the limiting reactant by calculating how much product can be made from each reactant. The reactant that makes the *least amount of product* is the limiting reactant. The mass of ammonia formed by the limiting reactant is the maximum amount of ammonia that can be synthesized.

SOLUTION MAP

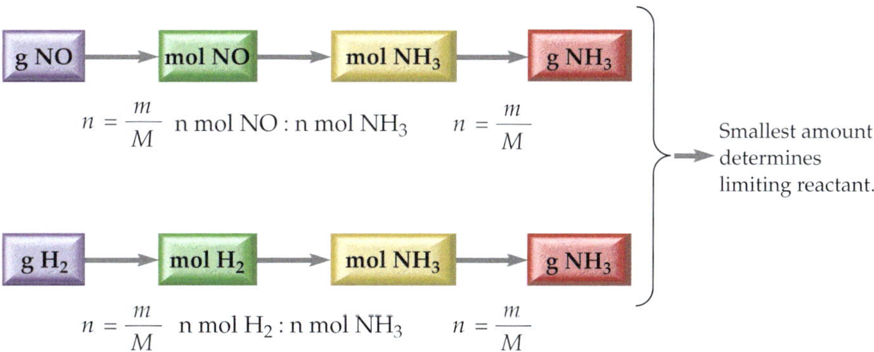

The stoichiometric relationship between moles of each reactant and moles of ammonia is from the balanced chemical equation. The molar masses of nitrogen monoxide, hydrogen gas, and ammonia are also required.

RELATIONSHIPS USED

2 mol NO : 2 mol NH_3

5 mol H_2 : 2 mol NH_3

Molar mass NO = 30.01 g/mol

Molar mass H_2 = 2.02 g/mol

Molar mass NH_3 = 17.04 g/mol

SOLVE

Follow the solution map, beginning with the actual amount of each reactant given, to calculate the amount of product that can be made from each reactant.

SOLUTION

Possible yield of NH_3 from NO:

Limiting reactant

$$n(\text{NO}) = \frac{m}{M} = \frac{45.8\text{ g NO}}{30.01\text{ g/mol NO}} = 1.53\text{ mol NO}$$

$$n(\text{NH}_3) = \frac{2\text{ mol NH}_3}{2\text{ mol NO}} \times 1.53\text{ mol NO} = 1.53\text{ mol NH}_3$$

$$m(\text{NH}_3) = n \times M = 1.53\text{ mol NH}_3 \times 17.04\text{ g/mol NH}_3 = 26.0\text{ g NH}_3$$

Least amount of product

Possible yield of NH_3 from H_2

$$n(\text{H}_2) = \frac{m}{M} = \frac{12.4\text{ g H}_2}{2.02\text{ g/mol H}_2} = 6.14\text{ mol H}_2$$

$$n(\text{NH}_3) = \frac{2\text{ mol NH}_3}{5\text{ mol H}_2} \times 6.14\text{ mol H}_2 = 2.46\text{ mol NH}_3$$

$$m(\text{NH}_3) = n \times M = 2.46\text{ mol NH}_3 \times 17.04\text{ g/mol NH}_3 = 41.8\text{ g NH}_3$$

There is enough NO to make 26.0 g of NH_3 and enough H_2 to make 41.8 g of NH_3. Therefore, NO is the limiting reactant, and the maximum amount of ammonia that can possibly be made is 26.0 g, which is the theoretical yield.

$m(NH_3)$ from NO < $m(NH_3)$ from H_2

NO is limiting reactant

Theoretical yield = 26.0 g NH_3

CHECK

Are the units correct? Does the answer make physical sense?

The units of the answer, g NH_3, are correct. The magnitude of the answer seems reasonable because it is of the same order of magnitude as the given masses of NO and H_2. An answer that is orders of magnitude different would immediately be suspect.

▶SKILLBUILDER 10.5 | Finding Limiting Reactant and Theoretical Yield

Ammonia can also be synthesized by this reaction:

$$3\,\text{H}_2(g) + \text{N}_2(g) \rightarrow 2\,\text{NH}_3(g)$$

What maximum amount of ammonia in grams can be synthesized from 25.2 g of N_2 and 8.42 g of H_2?

▶SKILLBUILDER PLUS What maximum amount of ammonia in kilograms can be synthesized from 5.22 kg of H_2 and 31.5 kg of N_2?

▶FOR MORE PRACTICE Problems 53, 54, 55, 56.

EXAMPLE 10.6 FINDING LIMITING REACTANT, THEORETICAL YIELD, AND PERCENT YIELD

Consider this reaction:

$$Cu_2O(s) + C(s) \rightarrow 2\,Cu(s) + CO(g)$$

When 11.5 g of C are allowed to react with 114.5 g of Cu_2O, 87.4 g of Cu are obtained. Determine the limiting reactant, theoretical yield, and percent yield.

SORT

You are given the mass of the reactants, carbon and copper(I) oxide, as well as the mass of copper formed by the reaction. You are asked to find the limiting reactant, theoretical yield, and percent yield.

GIVEN: 11.5 g C
114.5 g Cu_2O
87.4 g Cu produced

FIND: limiting reactant
theoretical yield
percent yield

STRATEGIZE

The solution map shows how to find the mass of Cu formed by the initial masses of Cu_2O and C. The reactant that makes the *least amount of product* is the limiting reactant and determines the theoretical yield.

SOLUTION MAP

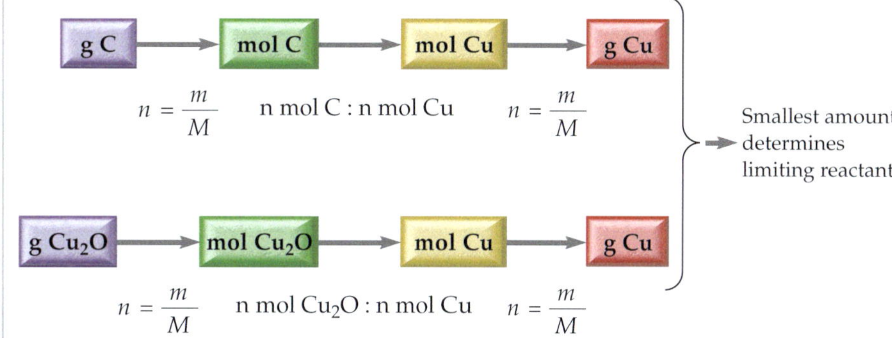

The stoichiometric relationships between moles of each reactant and moles of copper are found in the balanced chemical equation. The molar masses of copper(I) oxide, carbon, and copper are also necessary.

RELATIONSHIPS USED

1 mol Cu_2O:2 mol Cu
1 mol C:2 mol Cu
Molar mass Cu_2O = 143.10 g/mol
Molar mass C = 12.01 g/mol
Molar mass Cu = 63.55 g/mol

SOLVE Follow the solution map, beginning with the actual amount of each reactant given, to calculate the amount of product that can be made from each reactant.	**SOLUTION** Possible yield of Cu from C: $n(\text{C}) = \frac{m}{M} = \frac{11.5\,\cancel{\text{g}}\,\text{C}}{12.01\,\cancel{\text{g}}/\text{mol C}} = 0.958\,\text{mol C}$ $n(\text{Cu}) = \frac{2\,\text{mol Cu}}{1\,\cancel{\text{mol C}}} \times 0.958\,\cancel{\text{mol C}} = 1.92\,\text{mol Cu}$ $m(\text{Cu}) = n \times M = 1.92\,\cancel{\text{mol}}\,\text{Cu} \times 63.55\,\text{g}/\cancel{\text{mol}}\,\text{Cu} = 122\,\text{g Cu}$ Possible yield of Cu from CuO_2: **Limiting reactant** $n(\text{Cu}_2\text{O}) = \frac{m}{M} = \frac{114.5\,\cancel{\text{g}}\,\text{Cu}_2\text{O}}{143.10\,\cancel{\text{g}}/\text{mol Cu}_2\text{O}} = 0.8001\,\text{mol Cu}_2\text{O}$ $n(\text{Cu}) = \frac{2\,\text{mol Cu}}{1\,\cancel{\text{mol Cu}_2\text{O}}} \times 0.8001\,\cancel{\text{mol Cu}_2\text{O}} = 1.600\,\text{mol Cu}$ $m(\text{Cu}) = n \times M = 1.600\,\cancel{\text{mol Cu}} \times 63.55\,\text{g}/\cancel{\text{mol Cu}} = 101.7\,\text{g Cu}$ **Least amount of product** $m(\text{Cu})$ from CuO_2 < $m(\text{Cu})$ from C
Since Cu_2O makes the least amount of product, Cu_2O is the limiting reactant. The theoretical yield is then the amount of product made by the limiting reactant. The percent yield is the actual yield (87.4 g Cu) divided by the theoretical yield (101.7 g Cu) multiplied by 100 %.	Cu_2O is limiting reactant $\text{Theoretical yield} = 101.7\,\text{g Cu}$ $\text{Percent yield} = \frac{\text{Actual yield}}{\text{Theoretical yield}} \times 100\,\%$ $= \frac{87.4\,\text{g}}{101.7\,\text{g}} \times 100\,\% = 85.9\,\%$
CHECK Are the units correct? Does the answer make physical sense?	The theoretical yield has the right units (g Cu). The magnitude of the theoretical yield seems reasonable because it is of the same order of magnitude as the given masses of C and Cu_2O. The percent yield is reasonable because it is less than 100 %. Any calculated percent yield above 100 % is incorrect.

▶**SKILLBUILDER 10.6 | Finding Limiting Reactant, Theoretical Yield, and Percent Yield**

This reaction is used to obtain iron from iron ore:

$$Fe_2O_3(s) + 3\,CO(g) \rightarrow 2\,Fe(s) + 3\,CO_2(g)$$

The reaction of 185 g of Fe_2O_3 with 95.3 g of CO produces 87.4 g of Fe. Determine the limiting reactant, theoretical yield, and percent yield.

▶**FOR MORE PRACTICE** Example 10.9; Problems 59, 60, 61 62, 63, 64.

CONCEPTUAL CHECKPOINT 10.3

Limiting Reactant and Theoretical Yield

Ammonia can be synthesized by the reaction of nitrogen monoxide and hydrogen gas.

$$2\ NO(g) + 5\ H_2(g) \rightarrow 2\ NH_3(g) + 2\ H_2O(g)$$

A reaction vessel initially contains 4.0 mol of NO and 15.0 mol of H_2. What is in the reaction vessel once the reaction has occurred to the fullest extent possible?

(a) 2 mol NO; 5 mol H_2; 2 mol NH_3; and 2 mol H_2O

(b) 0 mol NO; 0 mol H_2; 6 mol NH_3; and 6 mol H_2O

(c) 2 mol NO; 0 mol H_2; 4 mol NH_3; and 2 mol H_2O

(d) 0 mol NO; 5 mol H_2; 4 mol NH_3; and 4 mol H_2O

EVERYDAY CHEMISTRY

▶ Bunsen Burners

In the laboratory, we often use Bunsen burners as heat sources. These burners are normally fueled by methane. The balanced equation for methane (CH_4) combustion is:

$$CH_4(g) + 2\ O_2(g) \rightarrow CO_2(g) + 2\ H_2O(g)$$

Most Bunsen burners have a mechanism to adjust the amount of air (and therefore of oxygen) that is mixed with the methane. If you light the burner with the air completely closed off, you get a yellow, smoky flame that is not very hot. As you increase the amount of air going into the burner, the flame becomes bluer, less smoky, and hotter. When you reach the optimum adjustment, the flame has a sharp, inner blue triangle, no smoke, and is hot enough to melt glass easily. Continuing to increase the air beyond this point causes the flame to become cooler again and may actually extinguish it.

B10.2 CAN YOU ANSWER THIS? *Can you use the concepts from this module to explain the changes in the Bunsen burner as the air intake is adjusted?*

(a) No air

(b) Small amount of air

(c) Optimum

(d) Too much air

▲ Bunsen burner at various stages of air-intake adjustment.

MODULE IN REVIEW

Self-Assessment Quiz

Q1. Sulfur and fluorine react to form sulfur hexafluoride according to the reaction shown here. How many moles of F_2 are required to react completely with 2.55 mol of S?

$$S(s) + 3\,F_2(g) \rightarrow SF_6(g)$$

(a) 0.85 mol F_2 **(b)** 2.55 mol F_2

(c) 7.65 mol F_2 **(d)** 15.3 mol F_2

Q2. Hydrogen chloride gas and oxygen gas react to form gaseous water and chlorine gas according to the reaction shown here.

$$4\,HCl(g) + O_2(g) \rightarrow 2\,H_2O(g) + 2\,Cl_2(g)$$

If the first image below represents the amount of HCl available for the reaction, which image represents the amount of oxygen required to react completely with the amount of available HCl?

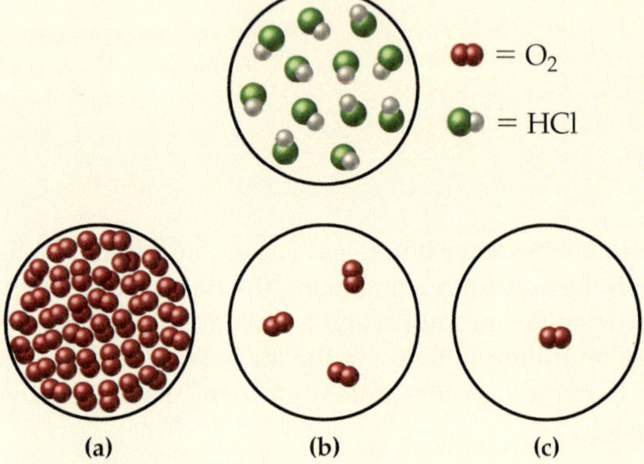

Q3. Sodium reacts with fluorine to form sodium fluoride. What mass of sodium fluoride forms from the complete reaction of 12.5 g of fluorine with enough sodium to completely react with it?

$$2\,Na(s) + F_2(g) \rightarrow 2\,NaF(s)$$

(a) 0.658 g NaF **(b)** 13.8 g NaF

(c) 6.91 g NaF **(d)** 27.6 g NaF

Q4. Consider the hypothetical reaction shown here. If 11 moles of A are combined with 16 moles of B and the reaction occurs to the greatest extent possible, how many moles of C form?

$$3\,A + 4\,B \rightarrow 2\,C$$

(a) 7.3 mol C **(b)** 8.0 mol C

(c) 17 mol C **(d)** 32 mol C

Q5. Consider the generic reaction:

$$2\,A + 3\,B + C \rightarrow 2\,D$$

A reaction mixture contains 6 mol A; 8 mol B; and 10 mol C. What is the limiting reactant?

(a) A **(b)** B

(c) C **(d)** D

Q6. Methanol (CH_3OH) reacts with oxygen to form carbon dioxide and water according to the reaction shown here.

$$2\,CH_3OH(g) + 3\,O_2(g) \rightarrow 2\,CO_2(g) + 4\,H_2O(g)$$

If the first image below represents a reaction mixture of methanol and oxygen, which image represents the reaction mixture after the reaction has occurred to the maximum extent possible?

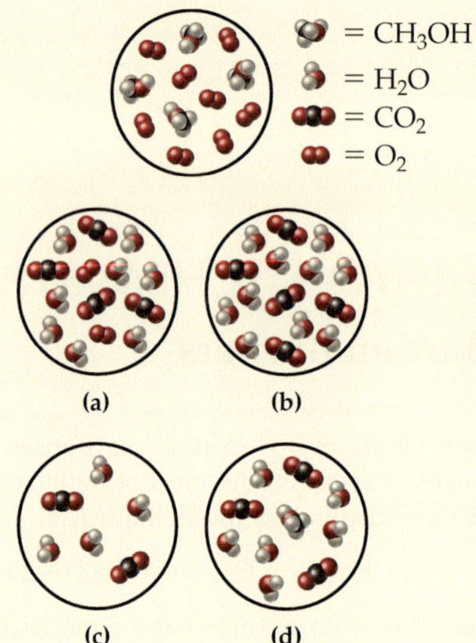

Q7. Sodium and chlorine react to form sodium chloride.

$$2\,Na(s) + Cl_2(g) \rightarrow 2\,NaCl(s)$$

What is the theoretical yield of sodium chloride for the reaction of 55.0 g Na with 67.2 g Cl_2?

(a) 1.40×10^2 g NaCl

(b) 111 g NaCl

(c) 55.4 g NaCl

(d) 222 g NaCl

Q8. A reaction has a theoretical yield of 22.8 g. When the reaction is carried out, 15.1 g of the product is obtained. What is the percent yield?

(a) 151 % **(b)** 66.2 % **(c)** 344 % **(d)** 88.2 %

Q9. Titanium can be obtained from its oxide by the reaction shown here. When 42.0 g of TiO_2 react with 11.5 g C, 18.7 g Ti are obtained. What is the percent yield for the reaction?

$$TiO_2(s) + 2\,C(s) \rightarrow Ti(s) + CO(g)$$

(a) 74.2 % **(b)** 81.6 %

(c) 122 % **(d)** 41 %

Answers: 1:c; 2:b; 3:d; 4:a; 5:b; 6:a; 7:b; 8:b; 9:b

Chemical Principles

Stoichiometry: A balanced chemical equation indicates quantitative relationships between the amounts of reactants and products. For example, the reaction $2\ H_2 + O_2 \rightarrow 2\ H_2O$ tells us that 2 mol of H_2 reacts with 1 mol of O_2 to form 2 mol of H_2O. We can use these relationships to calculate quantities such as the amount of product possible with a certain amount of reactant, or the amount of one reactant required to completely react with a certain amount of another reactant. The quantitative relationship between reactants and products in a chemical reaction is reaction stoichiometry.

Relevance

Stoichiometry: Reaction stoichiometry is important because we often want to know the numerical relationship between the reactants and products in a chemical reaction. For example, we might want to know how much carbon dioxide, a greenhouse gas, is formed when a certain amount of a particular fossil fuel burns.

Limiting Reactant, Theoretical Yield, and Percent Yield: The limiting reactant in a chemical reaction is the reactant that limits the amount of product that can be made. The theoretical yield in a chemical reaction is the amount of product that can be made based on the amount of the limiting reactant. The actual yield in a chemical reaction is the amount of product actually produced. The percent yield in a chemical reaction is the actual yield divided by theoretical yield times 100 %.

Limiting Reactant, Theoretical Yield, and Percent Yield: Calculations of limiting reactant, theoretical yield, and percent yield are central to chemistry because they allow for quantitative understanding of chemical reactions. Just as we need to know relationships between ingredients to follow a recipe, so we must know relationships between reactants and products to carry out a chemical reaction. The percent yield in a chemical reaction is often used as a measure of the success of the reaction. Imagine following a recipe and making only 1 % of the final product—your cooking would be a failure. Similarly, low percent yields in chemical reactions are usually considered poor, and high percent yields are considered good.

Chemical Skills

LO: Carry out mole-to-mole calculations between reactants and products based on the numerical relationship between chemical quantities in a balanced chemical equation (Section 10.3).

Examples

EXAMPLE 10.7 MOLE-TO-MOLE CALCULATIONS

How many moles of sodium oxide can be synthesized from 4.8 mol of sodium? Assume that more than enough oxygen is present. The balanced equation is:

$$4\ Na(s) + O_2(g) \rightarrow 2\ Na_2O(s)$$

SORT

You are given the number of moles of sodium and asked to find the number of moles of sodium oxide formed by the reaction.

GIVEN: 4.8 mol Na

FIND: mol Na_2O

STRATEGIZE

Draw a solution map beginning with the number of moles of the given substance and then use relationships in the balanced chemical equation to determine the number of moles of the substance you are trying to find.

SOLUTION MAP

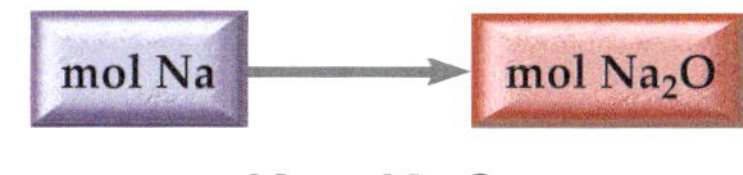

n Na : n Na_2O

RELATIONSHIPS USED

4 mol Na : 2 mol Na_2O

SOLVE

Follow the solution map to get to the number of moles of the substance you are trying to find.

SOLUTION

$$n(Na_2O) = \frac{2\ \text{mol}\ Na_2O}{4\ \text{mol Na}} \times 4.8\ \text{mol Na} = 2.4\ \text{mol}\ Na_2O$$

CHECK

Are the units correct? Does the answer make physical sense?

The units of the answer, mol Na_2O, are correct. The magnitude of the answer seems reasonable because it is of the same order of magnitude as the given number of moles of Na.

LO: Carry out mass-to-mass calculations between reactants and products based on the numerical relationship between chemical quantities in a balanced chemical equation and molar masses (Section 10.4).

EXAMPLE 10.8 MASS-TO-MASS CALCULATIONS

How many grams of sodium oxide can be synthesized from 17.4 g of sodium? Assume that more than enough oxygen is present. The balanced equation is:

$$4\,Na(s) + O_2(g) \rightarrow 2\,Na_2O(s)$$

SORT
You are given the mass of sodium and asked to find the mass of sodium oxide that forms upon reaction.

GIVEN: 17.4 g Na

FIND: g Na_2O

STRATEGIZE
Draw the solution map by beginning with the mass of the given substance. Calculate moles using the molar mass and then calculate moles of the substance you are trying to find, using the stoichiometric relationship in the balanced chemical equation.

Finally, stoichiometric relationship in mass of the substance you are trying to find, using its molar mass.

SOLUTION MAP

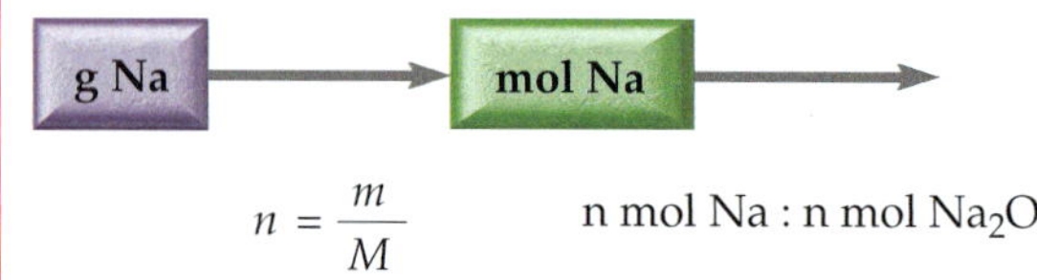

$n = \frac{m}{M}$ n mol Na : n mol Na_2O

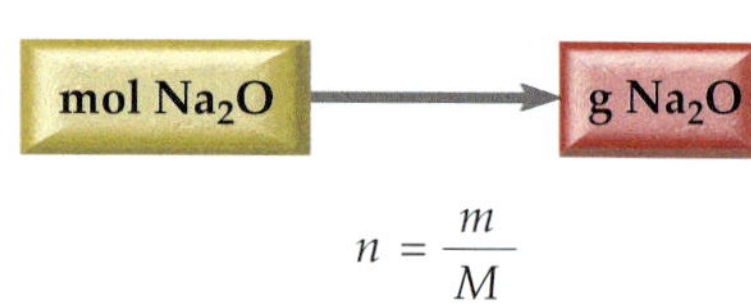

$n = \frac{m}{M}$

RELATIONSHIPS USED

4 mol Na:2 mol Na_2O (from balanced equation)

Molar mass Na = 22.99 g/mol

Molar mass Na_2O = 61.98 g/mol

SOLVE
Follow the solution map and calculate the answer by beginning with the mass of the given substance and using the appropriate ratios and equations to determine the mass of the substance you are trying to find.

SOLUTION

$$n(Na) = \frac{m}{M} = \frac{17.4\,\cancel{g}\,Na}{22.99\,\cancel{g}/mol\,Na} = 0.757\text{ mol Na}$$

$$n(Na_2O) = \frac{2\,mol\,Na_2O}{4\,\cancel{mol\,Na}} \times 0.757\,\cancel{mol\,Na}$$

$$= 0.378\,mol\,Na_2O$$

$$m(Na_2O) = n \times M = 0.378\,\cancel{mol}\,Na_2O \times 61.98\,g\,\cancel{mol}\,Na_2O$$

$$= 23.5\,g\,Na_2O$$

CHECK
Are the units correct? Does the answer make physical sense?

The units of the answer, g Na_2O, are correct. The magnitude of the answer seems reasonable because it is of the same order of magnitude as the given mass of Na.

LO: Calculate limiting reactant, theoretical yield, and percent yield for a given amount of reactants in a balanced chemical equation (Sections 10.5).

EXAMPLE 10.9 LIMITING REACTANT, THEORETICAL YIELD, AND PERCENT YIELD

10.4 g of As reacts with 11.8 g of S to produce 14.2 g of As_2S_3. Find the limiting reactant, theoretical yield, and percent yield for this reaction. The balanced chemical equation is:

$$2\,As(s) + 3\,S(l) \rightarrow As_2S_3(s)$$

SORT
You are given the masses of arsenic and sulfur as well as the mass of arsenic sulfide formed by the reaction. You are asked to find the limiting reactant, theoretical yield, and percent yield.

GIVEN: 10.4 g As
11.8 g S
14.2 g As_2S_3

FIND: limiting reactant
theoretical yield
percent yield

STRATEGIZE

The solution map for limiting-reactant problems shows how to calculate the mass of the product for each reactant. These are mass-to-mass calculations with the basic outline of

Mass → Moles → Moles → Mass

The reactant that forms the least amount of product is the limiting reactant.

The stoichiometric relationships between each of the reactants and the product are sourced from the balanced chemical reaction equation. You also need the molar masses of each reactant and product.

SOLUTION MAP

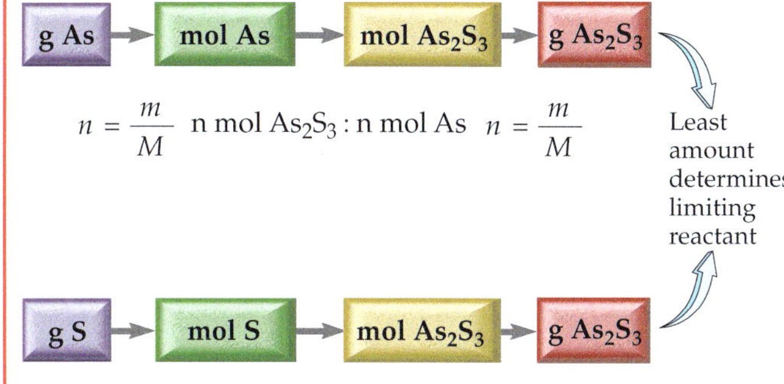

$n = \frac{m}{M}$ n mol As_2S_3 : n mol S $n = \frac{m}{M}$

RELATIONSHIPS USED

2 mol As:1 mol As_2S_3

3 mol S:1 mol As_2S_3

Molar mass As = 74.92 g/mol

Molar mass S = 32.07 g/mol

Molar mass As_2S_3 = 246.05 g/mol

SOLVE

To calculate the amount of product formed by each reactant, begin with the given amount (mass) of each reactant and calculate the number of moles of each reactant. Use the stoichiometric relationship between each reactant and the product to calculate the theoretical yield, in moles, of the product from each reactant. Finally, calculate the mass of product from each reactant using the appropriate formula. The reactant that forms the least amount of product is the limiting reactant.

SOLUTION

Possible yield of As_2S_3 from As:

$$n(\text{As}) = \frac{m}{M} = \frac{10.4\ \text{g As}}{74.92\ \text{g/mol As}} = 0.139\ \text{mol As}$$

$$n(\text{As}_2\text{S}_3) = \frac{1\ \text{mol As}_2\text{S}_3}{2\ \text{mol As}} \times 0.139\ \text{mol As} = 0.0694\ \text{mol As}_2\text{S}_3$$

$$m(\text{As}_2\text{S}_3) = n \times M = 0.0694\ \text{mol As}_2\text{S}_3 \times 246.06\ \text{g/mol As}_2\text{S}_3 = 17.1\ \text{g As}_2\text{S}_3$$

Possible yield of As_2S_3 from S

$$n(\text{S}) = \frac{m}{M} = \frac{11.8\ \text{g S}}{32.07\ \text{g/mol S}} = 0.368\ \text{mol S}$$

$$n(\text{As}_2\text{S}_3) = \frac{1\ \text{mol As}_2\text{S}_3}{3\ \text{mol S}} \times 0.368\ \text{mol S} = 0.123\ \text{mol As}_2\text{S}_3$$

$$m(\text{As}_2\text{S}_3) = n \times M = 0.123\ \text{mol As}_2\text{S}_3 \times 246.05\ \text{g/mol As}_2\text{S}_3 = 30.2\text{g As}_2\text{S}_3$$

The theoretical yield is the amount of product formed by the limiting reactant.

m (As_2S_3) from As < m (As_2S_3) from S
Therefore the limiting reactant is As.
The theoretical yield is 17.1 g of As_2S_3.

The percent yield is the actual yield divided by the theoretical yield times 100 %.

$$\text{Percent yield} = \frac{\text{Actual yield}}{\text{Theoretical yield}} \times 100\ \%$$

$$= \frac{14.2\ \text{g}}{17.1\ \text{g}} \times 100\ \% = 83.0\ \%$$

The percent yield is 83.0 %.

CHECK

Check your answer. Are the units correct? Does the answer make physical sense?

The theoretical yield has the right units (g As_2S_3). The magnitude of the theoretical yield seems reasonable because it is of the same order of magnitude as the given masses of As and S. The percent yield is reasonable because it is less than 100 %. Any calculated percent yield above 100 % would be suspect.

KEY TERMS

actual yield **[10.5]**
climate change **[10.1]**
greenhouse gases **[10.1]**
limiting reactant **[10.5]**
percent yield **[10.5]**
stoichiometry **[10.1]**
theoretical yield **[10.5]**

EXERCISES

QUESTIONS

1. Why is reaction stoichiometry important? Cite some examples in your answer.

2. Nitrogen and hydrogen can react to form ammonia:

$$N_2(g) + 3\,H_2(g) \rightarrow 2\,NH_3(g)$$

(a) Write ratios showing the relationships between moles of each of the reactants and products in the reaction.

(b) How many molecules of H_2 are required to completely react with two molecules of N_2?

(c) How many moles of H_2 are required to completely react with 2 mol of N_2?

3. Write how you would calculate the moles of NaCl from the moles of Cl_2 in the reaction:

$$2\,Na(s) + Cl_2(g) \rightarrow 2\,NaCl(s)$$

4. What is wrong with this statement in reference to the reaction in the previous problem? "2 g of Na react with 1 g of Cl_2 to form 2 g of NaCl." Correct the statement to make it true.

5. What is the general form of the solution map for problems in which you are given the mass of a reactant in a chemical reaction and asked to find the mass of the product that can be made from the given amount of reactant?

6. Consider the recipe for making tomato and garlic pasta.

2 cups noodles + 12 tomatoes + 3 cloves garlic
→ 4 servings pasta

If you have 7 cups of noodles, 27 tomatoes, and 9 cloves of garlic, how many servings of pasta can you make? Which ingredient limits the amount of pasta that it is possible to make?

7. In a chemical reaction, what is the limiting reactant?

8. In a chemical reaction, what is the theoretical yield?

9. In a chemical reaction, what are the actual yield and percent yield?

10. If you are given a chemical equation and specific amounts for each reactant in grams, how do you determine the maximum amount of product that can be made?

11. Consider the generic chemical reaction:

$$A + 2\,B \rightarrow C + D$$

Suppose you have 12 g of A and 24 g of B. Which statement is true?

(a) A will definitely be the limiting reactant.

(b) B will definitely be the limiting reactant.

(c) A will be the limiting reactant if its molar mass is less than B.

(d) A will be the limiting reactant if its molar mass is greater than B.

12. Consider the generic chemical equation:

$$A + B \rightarrow C$$

Suppose 25 g of A were allowed to react with 8 g of B. Analysis of the final mixture showed that A was completely used up and 4 g of B remained. What was the limiting reactant?

PROBLEMS

MOLE-TO-MOLE CALCULATIONS

13. Consider the generic chemical reaction:

$$A + 2\,B \rightarrow C$$

How many moles of C are formed upon complete reaction of:

(a) 2 mol of A
(b) 2 mol of B
(c) 3 mol of A
(d) 3 mol of B

14. Consider the generic chemical reaction:

$$2\,A + 3\,B \rightarrow 3\,C$$

How many moles of B are required to completely react with:

(a) 6 mol of A
(b) 2 mol of A
(c) 7 mol of A
(d) 11 mol of A

15. For the reaction shown, calculate how many moles of NO_2 form when each amount of reactant completely reacts.

$$2\,N_2O_5(g) \rightarrow 4\,NO_2(g) + O_2(g)$$

(a) 1.3 mol N_2O_5
(b) 5.8 mol N_2O_5
(c) 4.45×10^3 mol N_2O_5
(d) 1.006×10^{-3} mol N_2O_5

16. For the reaction shown, calculate how many moles of NH_3 form when each amount of reactant completely reacts.

$$3\,N_2H_4(l) \rightarrow 4\,NH_3(g) + N_2(g)$$

(a) 5.3 mol N_2H_4
(b) 2.28 mol N_2H_4
(c) 5.8×10^{-2} mol N_2H_4
(d) 9.76×10^7 mol N_2H_4

17. Dihydrogen monosulfide reacts with sulfur dioxide according to the balanced equation:

$$2\,H_2S(g) + SO_2(g) \rightarrow 3\,S(s) + 2\,H_2O(g)$$

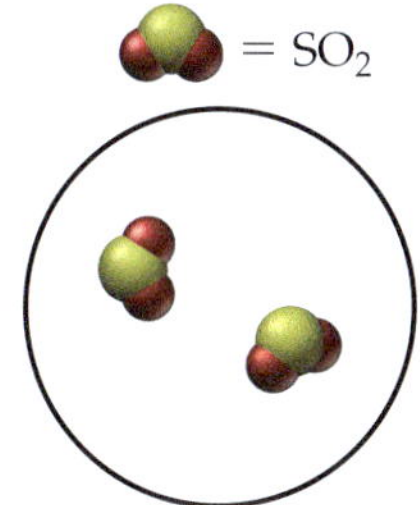

If the first figure represents the amount of SO_2 available to react, which figure best represents the amount of H_2S required to completely react with all of the SO_2?

= H_2S

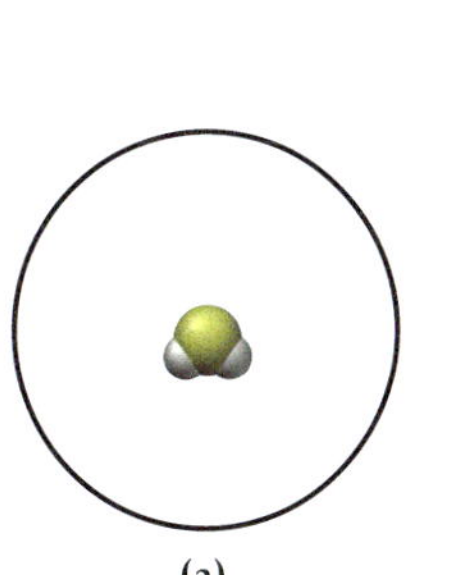
(a)

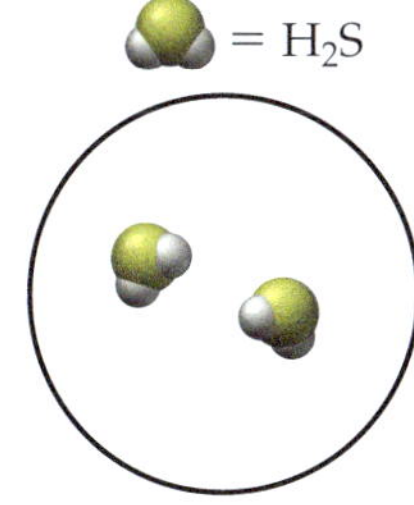
(b)

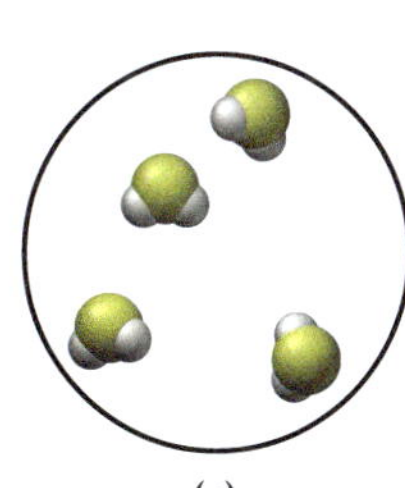
(c)

18. Chlorine gas reacts with fluorine gas according to the balanced equation:

$$Cl_2(g) + 3\,F_2(g) \rightarrow 2\,ClF_3(g)$$

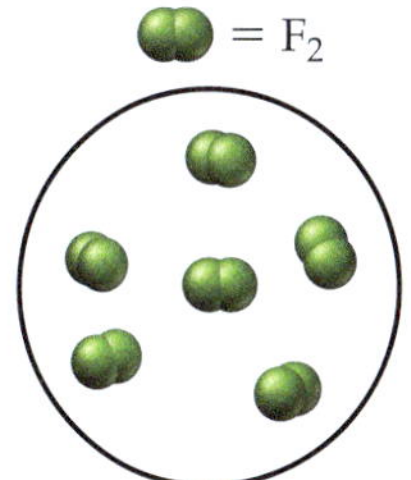

If the first figure represents the amount of fluorine available to react, and assuming that there is more than enough chlorine, which figure best represents the amount of chlorine trifluoride that would form upon complete reaction of all of the fluorine?

= ClF_3

(a)

(b)

(c)

19. For each reaction, calculate how many moles of product form when 1.75 mol of the reactant in color completely reacts. Assume there is more than enough of the other reactant.

(a) $H_2(g) + Cl_2(g) \rightarrow 2\ HCl(g)$

(b) $2\ H_2(g) + O_2(g) \rightarrow 2\ H_2O(l$

(c) $2\ Na(s) + O_2(g) \rightarrow Na_2O_2(s)$

(d) $2\ S(s) + 3\ O_2(g) \rightarrow 2\ SO_3(g)$

20. For each reaction, calculate how many moles of the product form when 0.112 mol of the reactant in color completely reacts. Assume there is more than enough of the other reactant.

(a) $2\ Ca(s) + O_2(g) \rightarrow 2\ CaO(s)$

(b) $4\ Fe(s) + 3\ O_2(g) \rightarrow 2\ Fe_2O_3(s)$

(c) $4\ K(s) + O_2(g) \rightarrow 2\ K_2O(s)$

(d) $4\ Al(s) + 3\ O_2(g) \rightarrow 2\ Al_2O_3(s)$

21. For the reaction shown, calculate how many moles of each product form when the given amount of each reactant completely reacts. Assume there is more than enough of the other reactant.

$$2\ PbS(s) + 3\ O_2(g) \rightarrow 2\ PbO(s) + 2\ SO_2(g)$$

(a) 2.4 mol PbS

(b) 2.4 mol O_2

(c) 5.3 mol PbS

(d) 5.3 mol O_2

22. For the reaction shown, calculate how many moles of each product form when the given amount of each reactant completely reacts. Assume there is more than enough of the other reactant.

$$C_3H_8(g) + 5\ O_2(g) \rightarrow 3\ CO_2(g) + 4\ H_2O(g)$$

(a) 4.6 mol C_3H_8

(b) 4.6 mol O_2

(c) 0.0558 mol C_3H_8

(d) 0.0558 mol O_2

23. Consider the balanced equation:

$$2\ N_2H_4(g) + N_2O_4(g) \rightarrow 3\ N_2(g) + 4\ H_2O(g)$$

Complete the table with the appropriate number of moles of reactants and products. If the number of moles of a reactant is provided, fill in the required amount of the other reactant, as well as the moles of each product formed. If the number of moles of a product is provided, fill in the required amount of each reactant to make that amount of product, as well as the amount of the other product that is made.

mol N_2H_4	mol N_2O_4	mol N_2	mol H_2O
______	2	______	______
6	______	______	______
______	______	______	8
______	5.5	______	______
3	______	______	______
______	______	12.4	______

24. Consider the balanced equation:

$$SiO_2(s) + 3\ C(s) \rightarrow SiC(s) + 2\ CO(g)$$

Complete the table with the appropriate number of moles of reactants and products. If the number of moles of a reactant is provided, fill in the required amount of the other reactant, as well as the moles of each product formed. If the number of moles of a product is provided, fill in the required amount of each reactant to make that amount of product, as well as the amount of the other product that is made.

mol SiO_2	mol C	mol SiC	mol CO
______	6	______	______
3	______	______	______
______	______	______	10
______	9.5	______	______
3.2	______	______	______

25. Consider the unbalanced equation for the combustion of butane:

$$C_4H_{10}(g) + O_2(g) \rightarrow CO_2(g) + H_2O(g)$$

Balance the equation and determine how many moles of O_2 are required to react completely with 4.9 mol of C_4H_{10}.

26. Consider the unbalanced equation for the neutralization of acetic acid:

$$HC_2H_3O_2(aq) + Ca(OH)_2(aq) \rightarrow H_2O(l) + Ca(C_2H_3O_2)_2(aq)$$

Balance the equation and determine how many moles of $Ca(OH)_2$ are required to completely neutralize 1.07 mol of $HC_2H_3O_2$.

27. Consider the unbalanced equation for the reaction of solid lead with silver nitrate:

$$Pb(s) + AgNO_3(aq) \rightarrow Pb(NO_3)_2(aq) + Ag(s)$$

(a) Balance the equation.
(b) How many moles of silver nitrate are required to completely react with 9.3 mol of lead?
(c) How many moles of Ag are formed by the complete reaction of 28.4 mol of Pb?

28. Consider the unbalanced equation for the reaction of aluminum with sulfuric acid:

$$Al(s) + H_2SO_4(aq) \rightarrow Al_2(SO_4)_3(aq) + H_2(g)$$

(a) Balance the equation.
(b) How many moles of H_2SO_4 are required to completely react with 8.3 mol of Al?
(c) How many moles of H_2 are formed by the complete reaction of 0.341 mol of Al?

MASS-TO-MASS CALCULATIONS

29. For the reaction shown, calculate how many grams of oxygen form when each quantity of reactant completely reacts.

$$2\ HgO(s) \rightarrow 2\ Hg(l) + O_2(g)$$

(a) 2.13 g HgO
(b) 6.77 g HgO
(c) 1.55 kg HgO
(d) 3.87 mg HgO

30. For the reaction shown, calculate how many grams of oxygen form when each quantity of reactant completely reacts.

$$2\ KClO_3(s) \rightarrow 2\ KCl(s) + 3\ O_2(g)$$

(a) 2.72 g $KClO_3$
(b) 0.361g $KClO_3$
(c) 83.6 kg $KClO_3$
(d) 22.4 mg $KClO_3$

31. For each of the reactions, calculate how many grams of the product form when 2.4 g of the reactant in color completely reacts. Assume there is more than enough of the other reactant.
(a) $2\ Na(s) + Cl_2(g) \rightarrow 2\ NaCl(s)$
(b) $CaO(s) + CO_2(g) \rightarrow CaCO_3(s)$
(c) $2\ Mg(s) + O_2(g) \rightarrow 2\ MgO(s)$
(d) $Na_2O(s) + H_2O(l) \rightarrow 2\ NaOH(aq)$

32. For each of the reactions, calculate how many grams of the product form when 17.8 g of the reactant in color completely reacts. Assume there is more than enough of the other reactant.
(a) $Ca(s) + Cl_2(g) \rightarrow CaCl_2(s)$
(b) $2\ K(s) + Br_2(l) \rightarrow 2\ KBr(s)$
(c) $4\ Cr(s) + 3\ O_2(g) \rightarrow 2\ Cr_2O_3(s)$
(d) $2\ Sr(s) + O_2(g) \rightarrow 2\ SrO(s)$

33. For the reaction shown, calculate how many grams of each product form when the given amount of each reactant completely reacts to form products. Assume there is more than enough of the other reactant.

$$2\ Al(s) + Fe_2O_3(s) \rightarrow Al_2O_3(s) + 2\ Fe(l)$$

(a) 4.7 g Al
(b) 4.7 g Fe_2O_3

34. For the reaction shown, calculate how many grams of each product form when the given amount of each reactant completely reacts to form products. Assume there is more than enough of the other reactant.

$$2\ HCl(aq) + Na_2CO_3(aq) \rightarrow 2\ NaCl(aq) + H_2O(l) + CO_2(g)$$

(a) 10.8 g HCl
(b) 10.8 g Na_2CO_3

35. Consider the balanced equation for the combustion of methane, a component of natural gas:

$$CH_4(g) + 2\ O_2(g) \rightarrow CO_2(g) + 2\ H_2O(g)$$

Complete the table with the appropriate masses of reactants and products. If the mass of a reactant is provided, fill in the mass of other reactants required to completely react with the given mass, as well as the mass of each product formed. If the mass of a product is provided, fill in the required masses of each reactant to make that amount of product, as well as the mass of the other product that forms.

Mass CH_4	Mass O_2	Mass CO_2	Mass H_2O
______	2.57 g	______	______
22.32 g	______	______	______
______	______	______	11.32 g
______	______	2.94 g	______
3.18 kg	______	______	______
______	______	2.35×10^3 kg	______

36. Consider the balanced equation for the combustion of butane, a fuel often used in lighters:

$$2\ C_4H_{10}(g) + 13\ O_2(g) \rightarrow 8\ CO_2(g) + 10\ H_2O(g)$$

Complete the table showing the appropriate masses of reactants and products. If the mass of a reactant is provided, fill in the mass of other reactants required to completely react with the given mass, as well as the mass of each product formed. If the mass of a product is provided, fill in the required masses of each reactant to make that amount of product, as well as the mass of the other product that forms.

Mass C_4H_{10}	Mass O_2	Mass CO_2	Mass H_2O
______	1.11 g	______	______
5.22 g	______	______	______
______	______	10.12 g	______
______	______	______	9.04 g
232 mg	______	______	______
______	______	118 mg	______

37. For each acid–base reaction, calculate how many grams of acid are necessary to completely react with and neutralize 2.5 g of the base.

(a) $HCl(aq) + NaOH(aq) \rightarrow H_2O(l) + NaCl(aq)$
(b) $2\ HNO_3(aq) + Ca(OH)_2(aq) \rightarrow 2\ H_2O(l) + Ca(NO_3)_2(aq)$
(c) $H_2SO_4(aq) + 2\ KOH(aq) \rightarrow 2\ H_2O(l) + K_2SO_4(aq)$

38. For each precipitation reaction, calculate how many grams of the first reactant are necessary to completely react with 17.3 g of the second reactant.

(a) $2\ KI(aq) + Pb(NO_3)_2(aq) \rightarrow PbI_2(s) + 2\ KNO_3(aq)$
(b) $Na_2CO_3(aq) + CuCl_2(aq) \rightarrow CuCO_3(s) + 2\ NaCl(aq)$
(c) $K_2SO_4(aq) + Sr(NO_3)_2(aq) \rightarrow SrSO_4(s) + 2\ KNO_3(aq)$

39. Sulfuric acid can dissolve aluminum metal according to the reaction:

$$2\ Al(s) + 3\ H_2SO_4(aq) \rightarrow Al_2(SO_4)_3(aq) + 3\ H_2(g)$$

Suppose you wanted to dissolve an aluminum block with a mass of 22.5 g. What minimum amount of H_2SO_4 in grams do you need? How many grams of H_2 gas will be produced by the complete reaction of the aluminum block?

40. Hydrochloric acid can dissolve solid iron according to the reaction:

$$Fe(s) + 2\ HCl(aq) \rightarrow FeCl_2(aq) + H_2(g)$$

What minimum mass of HCl in grams dissolves a 2.8-g iron bar on a padlock? How much H_2 is produced by the complete reaction of the iron bar?

LIMITING REACTANT, THEORETICAL YIELD, AND PERCENT YIELD

41. Consider the generic chemical equation:

$$2\ A + 4\ B \rightarrow 3\ C$$

What is the limiting reactant when each of the initial quantities of A and B is allowed to react?

(a) 2 mol A; 5 mol B
(b) 1.8 mol A; 4 mol B
(c) 3 mol A; 4 mol B
(d) 22 mol A; 40 mol B

42. Consider the generic chemical equation:

$$A + 3\ B \rightarrow C$$

What is the limiting reactant when each of the initial quantities of A and B is allowed to react?

(a) 1 mol A; 4 mol B
(b) 2 mol A; 3 mol B
(c) 0.5 mol A; 1.6 mol B
(d) 24 mol A; 75 mol B

43. Determine the theoretical yield of C when each of the initial quantities of A and B is allowed to react in the generic reaction:

$$A + 2\ B \rightarrow 3\ C$$

(a) 1 mol A; 1 mol B
(b) 2 mol A; 2 mol B
(c) 1 mol A; 3 mol B
(d) 32 mol A; 68 mol B

44. Determine the theoretical yield of C when each of the initial quantities of A and B is allowed to react in the generic reaction:

$$2\ A + 3\ B \rightarrow 2\ C$$

(a) 2 mol A; 4 mol B
(b) 3 mol A; 3 mol B
(c) 5 mol A; 6 mol B
(d) 4 mol A; 5 mol B

45. For the reaction shown, find the limiting reactant for each of the initial quantities of reactants.

$$2\ K(s) + Cl_2(g) \rightarrow 2\ KCl(s)$$

(a) 1 mol K; 1 mol Cl_2
(b) 1.8 mol K; 1 mol Cl_2
(c) 2.2 mol K; 1 mol Cl_2
(d) 14.6 mol K; 7.8 mol Cl_2

46. For the reaction shown, find the limiting reactant for each of the initial quantities of reactants.

$$4\ Cr(s) + 3\ O_2(g) \rightarrow 2\ Cr_2O_3(s)$$

(a) 1 mol Cr; 1 mol O_2
(b) 4 mol Cr; 2.5 mol O_2
(c) 12 mol Cr; 10 mol O_2
(d) 14.8 mol Cr; 10.3 mol O_2

47. For the reaction shown, calculate the theoretical yield of product in moles for each of the initial quantities of reactants.

$$2\ Mn(s) + 3\ O_2(g) \rightarrow 2\ MnO_3(s)$$

(a) 2 mol Mn; 2 mol O_2
(b) 4.8 mol Mn; 8.5 mol O_2
(c) 0.114 mol Mn; 0.161 mol O_2
(d) 27.5 mol Mn; 43.8 mol O_2

48. For the reaction shown, calculate the theoretical yield of the product in moles for each of the initial quantities of reactants.

$$Ti(s) + 2\ Cl_2(g) \rightarrow TiCl_4(s)$$

(a) 2 mol Ti; 2 mol Cl_2
(b) 5 mol Ti; 9 mol Cl_2
(c) 0.483 mol Ti; 0.911 mol Cl_2
(d) 12.4 mol Ti; 15.8 mol Cl_2

49. Consider the generic reaction between reactants A and B:

$$3\,A + 4\,B \rightarrow 2\,C$$

If a reaction vessel initially contains 9 mol A and 8 mol B, how many moles of A, B, and C will be in the reaction vessel after the reactants have reacted as much as possible? (Assume 100 % yield.)

50. Consider the reaction between reactants S and O_2:

$$2\,S(s) + 3\,O_2(g) \rightarrow 2\,SO_3(g)$$

If a reaction vessel initially contains 5 mol S and 9 mol O_2, how many moles of S, O_2, and SO_3 will be in the reaction vessel after the reactants have reacted as much as possible? (Assume 100 % yield.)

51. Consider the reaction:

$$4\,HCl(g) + O_2(g) \rightarrow 2\,H_2O(g) + 2\,Cl_2(g)$$

Each molecular diagram represents an initial mixture of the reactants. How many molecules of Cl_2 are formed by complete reaction in each case? (Assume 100 % yield.)

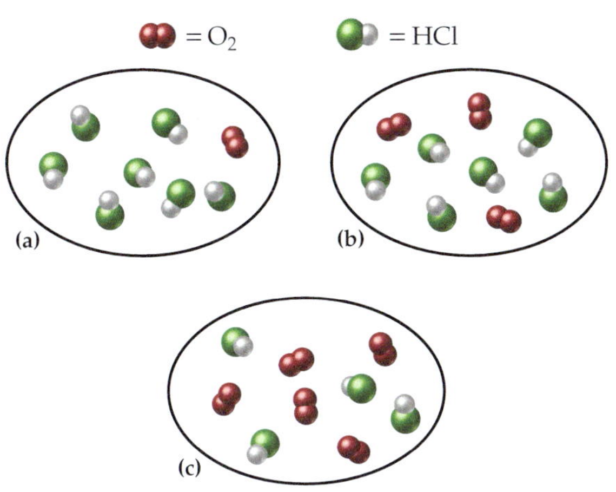

52. Consider the reaction:

$$2\,CH_3OH(g) + 3\,O_2(g) \rightarrow 2\,CO_2(g) + 4\,H_2O(g)$$

Each molecular diagram represents an initial mixture of the reactants. How many CO_2 molecules are formed by complete reaction in each case? (Assume 100 % yield.)

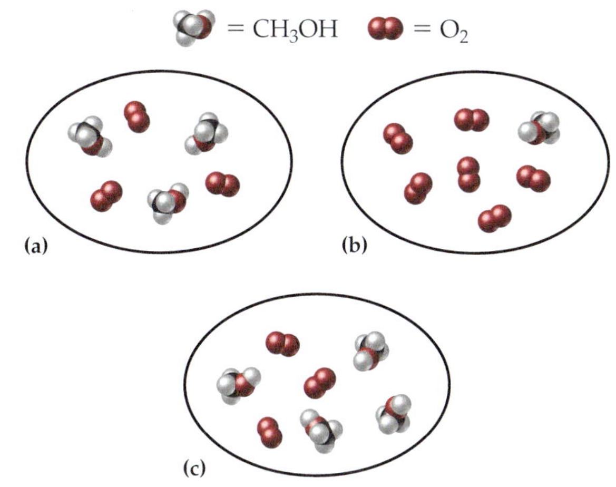

53. For the reaction shown, find the limiting reactant for each of the initial quantities of reactants.

$$2\,Li(s) + F_2(g) \rightarrow 2\,LiF(s)$$

(a) 1.0 g Li; 1.0 g F_2
(b) 10.5 g Li; 37.2 g F_2
(c) 2.85×10^3 g Li; 6.79×10^3 g F_2

54. For the reaction shown, find the limiting reactant for each of the initial quantities of reactants.

$$4\,Al(s) + 3\,O_2(g) \rightarrow 2\,Al_2O_3(s)$$

(a) 1.0 g Al; 1.0 g O_2
(b) 2.2 g Al; 1.8 g O_2
(c) 0.353 g Al; 0.482 g O_2

55. For the reaction shown, calculate the theoretical yield of the product in grams for each of the initial quantities of reactants.

$$2\,Al(s) + 3\,Cl_2(g) \rightarrow 2\,AlCl_3(s)$$

(a) 1.0 g Al; 1.0 g Cl_2
(b) 5.5 g Al; 19.8 g Cl_2
(c) 0.439 g Al; 2.29 g Cl_2

56. For the reaction shown, calculate the theoretical yield of the product in grams for each of the initial quantities of reactants.

$$Ti(s) + 2\,F_2(g) \rightarrow TiF_4(s)$$

(a) 1.0 g Ti; 1.0 g F_2
(b) 4.8 g Ti; 3.2 g F_2
(c) 0.388 g Ti; 0.341 g F_2

57. If the theoretical yield of a reaction is 24.8 g and the actual yield is 18.5 g, what is the percent yield?

58. If the theoretical yield of a reaction is 0.118 g and the actual yield is 0.104 g, what is the percent yield?

59. Consider the reaction between calcium oxide and carbon dioxide:

$$CaO(s) + CO_2(g) \rightarrow CaCO_3(s)$$

A chemist allows 14.4 g of CaO and 13.8 g of CO_2 to react. When the reaction is finished, the chemist collects 19.4 g of $CaCO_3$. Determine the limiting reactant, theoretical yield, and percent yield for the reaction.

60. Consider the reaction between sulfur trioxide and water:

$$SO_3(g) + H_2O(l) \rightarrow H_2SO_4(aq)$$

A chemist allows 61.5 g of SO_3 and 11.2 g of H_2O to react. When the reaction is finished, the chemist collects 54.9 g of H_2SO_4. Determine the limiting reactant, theoretical yield, and percent yield for the reaction.

61. Consider the reaction between NiS_2 and O_2:

$$2\ NiS_2(s) + 5\ O_2(g) \rightarrow 2\ NiO(s) + 4\ SO_2(g)$$

When 11.2 g of NiS_2 react with 5.43 g of O_2, 4.86 g of NiO are obtained. Determine the limiting reactant, theoretical yield of NiO, and percent yield for the reaction.

62. Consider the reaction between HCl and O_2:

$$4\ HCl(g) + O_2(g) \rightarrow 2\ H_2O(l) + 2\ Cl_2(g)$$

When 63.1 g of HCl react with 17.2 g of O_2, 49.3 g of Cl_2 are collected. Determine the limiting reactant, theoretical yield of Cl_2, and percent yield for the reaction.

63. Lead ions can be precipitated from solution with NaCl according to the reaction:

$$Pb^{2+}(aq) + 2\ NaCl(aq) \rightarrow PbCl_2(s) + 2\ Na^+(aq)$$

When 135.8 g of NaCl are added to a solution containing 195.7 g of Pb^{2+}, a $PbCl_2$ precipitate forms. The precipitate is filtered and dried and found to have a mass of 252.4 g. Determine the limiting reactant, theoretical yield of $PbCl_2$, and percent yield for the reaction.

64. Magnesium oxide can be produced by heating magnesium metal in the presence of oxygen. The balanced equation for the reaction is:

$$2\ Mg(s) + O_2(g) \rightarrow 2\ MgO(s)$$

When 10.1 g of Mg react with 10.5 g of O_2, 11.9 g of MgO are collected. Determine the limiting reactant, theoretical yield, and percent yield for the reaction.

65. Consider the reaction between TiO_2 and C:

$$TiO_2(s) + 2\ C(s) \rightarrow Ti(s) + 2\ CO(g)$$

A reaction vessel initially contains 10.0 g of each of the reactants. Calculate the masses of TiO_2, C, Ti, and CO that will be in the reaction vessel after the reactants have reacted as much as possible. (Assume 100 % yield.)

66. Consider the reaction between N_2H_4 and N_2O_4:

$$2\ N_2H_4(g) + N_2O_4(g) \rightarrow 3\ N_2(g) + 4\ H_2O(g)$$

A reaction vessel initially contains 27.5 g N_2H_4 and 74.9 g of N_2O_4. Calculate the masses of N_2H_4, N_2O_4, N_2, and H_2O that will be in the reaction vessel after the reactants have reacted as much as possible. (Assume 100 % yield.)

CUMULATIVE PROBLEMS

67. Consider the reaction:

$$2\ N_2(g) + 5\ O_2(g) + 2\ H_2O(g) \rightarrow 4\ HNO_3(g)$$

If a reaction mixture contains 28 g of N_2, 150 g of O_2, and 36 g of H_2O, what is the limiting reactant? (Try to do this problem in your head without any written calculations.)

68. Consider the reaction:

$$2\ CO(g) + O_2(g) \rightarrow 2\ CO_2(g)$$

If a reaction mixture contains 28 g of CO and 32 g of O_2, what is the limiting reactant? (Try to do this problem in your head without any written calculations.)

69. A solution contains an unknown mass of dissolved barium ions. When sodium sulfate is added to the solution, a white precipitate forms. The precipitate is filtered and dried and found to have a mass of 258 mg. What mass of barium was in the original solution? (Assume that all of the barium was precipitated out of solution by the reaction.)

70. A solution contains an unknown mass of dissolved silver ions. When potassium chloride is added to the solution, a white precipitate forms. The precipitate is filtered and dried and found to have a mass of 212 mg. What mass of silver was in the original solution? (Assume that all of the silver was precipitated out of solution by the reaction.)

71. Sodium hydrogencarbonate is often used as an antacid to neutralize excess hydrochloric acid in an upset stomach. How much hydrochloric acid in grams can be neutralized by 3.5 g of sodium hydrogencarbonate? (*Hint:* Begin by writing a balanced equation for the reaction between aqueous sodium bicarbonate and aqueous hydrochloric acid.)

72. Toilet bowl cleaners often contain hydrochloric acid to dissolve the calcium carbonate deposits that accumulate within a toilet bowl. How much calcium carbonate in grams can be dissolved by 5.8 g of HCl? (*Hint:* Begin by writing a balanced equation for the reaction between hydrochloric acid and calcium carbonate.)

73. The combustion of gasoline produces carbon dioxide and water. Assume gasoline to be pure octane (C_8H_{18}) and calculate how many kilograms of carbon dioxide are added to the atmosphere per 1.0 kg of octane burned. (*Hint:* Begin by writing a balanced equation for the combustion reaction.)

74. Many home barbecues are fueled with propane gas (C_3H_8). How much carbon dioxide in kilograms is produced upon the complete combustion of 18.9 L of propane (approximate contents of one 19 L tank)? Assume that the density of the liquid propane in the tank is 0.621 g/mL. (*Hint:* Begin by writing a balanced equation for the combustion reaction.)

75. A hard water solution contains 4.8 g of calcium chloride. How much sodium phosphate in grams should be added to the solution to completely precipitate all of the calcium?

76. Magnesium ions can be precipitated from seawater by the addition of sodium hydroxide. How much sodium hydroxide in grams must be added to a sample of seawater to completely precipitate the 88.4 mg of magnesium present?

77. Hydrogen gas can be prepared in the laboratory by a single-displacement reaction in which solid zinc reacts with hydrochloric acid. How much zinc in grams is required to make 14.5 g of hydrogen gas through this reaction?

78. Sodium peroxide (Na_2O_2) reacts with water to form sodium hydroxide and oxygen gas. Write a balanced equation for the reaction and determine how much oxygen in grams is formed by the complete reaction of 35.23 g of Na_2O_2.

79. Ammonium nitrate reacts explosively upon heating to form nitrogen gas, oxygen gas, and gaseous water. Write a balanced equation for this reaction and determine how much oxygen in grams is produced by the complete reaction of 1.00 kg of ammonium nitrate.

80. Pure oxygen gas can be prepared in the laboratory by the decomposition of solid potassium chlorate to form solid potassium chloride and oxygen gas. How much oxygen gas in grams can be prepared from 45.8 g of potassium chlorate?

81. Urea (CH_4N_2O), a common fertilizer, can be synthesized by the reaction of ammonia (NH_3) with carbon dioxide:

$$2\ NH_3(aq) + CO_2(aq) \rightarrow CH_4N_2O(aq) + H_2O(l)$$

An industrial synthesis of urea produces 87.5 kg of urea upon reaction of 68.2 kg of ammonia with 105 kg of carbon dioxide. Determine the limiting reactant, theoretical yield of urea, and percent yield for the reaction.

82. Silicon, which occurs in nature as SiO_2, is the material from which most computer chips are made. If SiO_2 is heated until it melts into a liquid, it reacts with solid carbon to form liquid silicon and carbon monoxide gas. In an industrial preparation of silicon, 52.8 kg of SiO_2 reacts with 25.8 kg of carbon to produce 22.4 kg of silicon. Determine the limiting reactant, theoretical yield, and percent yield for the reaction.

83. An emergency breathing apparatus placed in mines or caves works via the chemical reaction:

$$4\ KO_2(s) + 2\ CO_2(g) \rightarrow 2\ K_2CO_3(s) + 3\ O_2(g)$$

If the oxygen supply becomes limited or if the air becomes poisoned, a worker can use the apparatus to breathe while exiting the mine. Notice that the reaction produces O_2, which can be breathed, and absorbs CO_2, a product of respiration. What minimum amount of KO_2 is required for the apparatus to produce enough oxygen to allow the user 15 minutes to exit the mine in an emergency? Assume that an adult consumes approximately 4.4 g of oxygen in 15 minutes of normal breathing.

HIGHLIGHT PROBLEMS

84. As we have seen, scientists have grown progressively more worried about the potential for climate change caused by increasing atmospheric carbon dioxide levels. The world burns the fossil fuel equivalent of approximately 9.0×10^{12} kg of petroleum per year. Assume that all of this petroleum is in the form of octane (C_8H_{18}) and calculate how much CO_2 in kilograms is produced by world fossil fuel combustion per year. (*Hint:* Begin by writing a balanced equation for the combustion of octane.) If the atmosphere currently contains approximately 3.0×10^{15} kg of CO_2, how long will it take for the world's fossil fuel combustion to double the amount of atmospheric carbon dioxide?

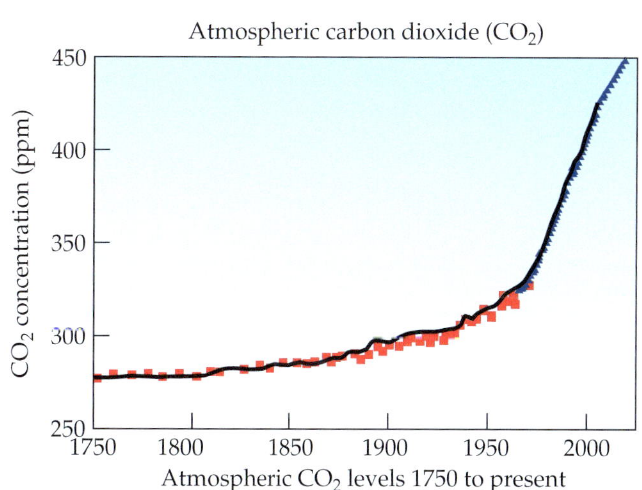

Atmospheric CO_2 levels 1750 to present

Answers to Skillbuilder Exercises

Skillbuilder 10.1 49.2 mol H_2O

Skillbuilder 10.2 6.88 g HCl

Skillbuilder 10.3 4.0×10^3 kg H_2SO4

Skillbuilder 10.4 Limiting reactant is Na; theoretical yield is 4.8 mol of NaF

Skillbuilder 10.5 30.7 g NH_3

Skillbuilder Plus, p. 470 29.4 kg NH_3

Skillbuilder 10.6 Limiting reactant is CO; theoretical yield = 127 g Fe; percent yield = 68.8 %

Answers to Conceptual Checkpoints

10.1 (a) Because the reaction requires 2 O_2 molecules to react with 1 CH_4 molecule, and there are 4 O_2 molecules available to react, 2 CH_4 molecules are required for complete reaction.

10.2 (c) Hydrogen is the limiting reactant. The reaction mixture contains 3 H_2 molecules; therefore 2 NH_3 molecules will form when the reactants have reacted as completely as possible. Nitrogen is in excess, and there is one leftover nitrogen molecule.

10.3 (d) NO is the limiting reagent. The reaction mixture initially contains 4 mol NO; therefore 10 moles of H_2O will be consumed, leaving 5 mol H_2 unreacted. The products will be 4 mol NH_3 and 4 mol H_2O.

▲ Acids have been used in spy movies and other thrillers to dissolve the metal bars of a prison cell.

Chemical Quantities 2 11

"The differences between the various acid–base concepts are not concerned with which is 'right,' but which is most convenient to use in a particular situation." —James E. Huheey

MODULE OUTLINE

11.1 Gases in Chemical Reactions

LO: Apply the principles of stoichiometry to chemical reactions involving gases.

Module 10 describes how the coefficients in chemical equations give the ratios between moles of reactants and moles of products in a chemical reaction. We can use these ratios to determine, for example, the amount of product obtained in a chemical reaction based on a given amount of reactant or the amount of one reactant we need to completely react with a given amount of another reactant. The general solution map for these kinds of calculations is:

$$\text{Moles A} \rightarrow \text{Moles B}$$

where A and B are two different substances involved in the reaction and the ratio between them comes from the stoichiometric coefficients in the balanced chemical equation.

In reactions involving gaseous reactants or products, the amount of gas is often specified in terms of its volume at a given temperature and pressure. In cases like this, we can use the ideal gas law to calculate moles from pressure, volume, and temperature.

$$n = \frac{PV}{RT}$$

We can then use the stoichiometric coefficients to calculate to other quantities in the reaction. For example, consider the reaction for the synthesis of ammonia.

$$3\,H_2(g) + N_2(g) \rightarrow 2\,NH_3(g)$$

How many moles of NH_3 are formed by the complete reaction of 2.5 L of hydrogen at 381 K and 1.32 atm? Assume that there is more than enough N_2.

We begin by sorting the information in the problem statement.

GIVEN: $V = 2.5\text{ L}$

$T = 381\text{ K}$

$P = 1.32\text{ atm (of }H_2)$

FIND: mol NH_3

SOLUTION MAP

We strategize by drawing a solution map. The solution map for this problem is similar to the solution maps for other stoichiometric problems (see Sections 10.3 and 10.4). However, we first use the ideal gas law to find mol H_2 from P, V, and T. Then we use the stoichiometric coefficients from the equation to calculate mol NH_3 from mol H_2.

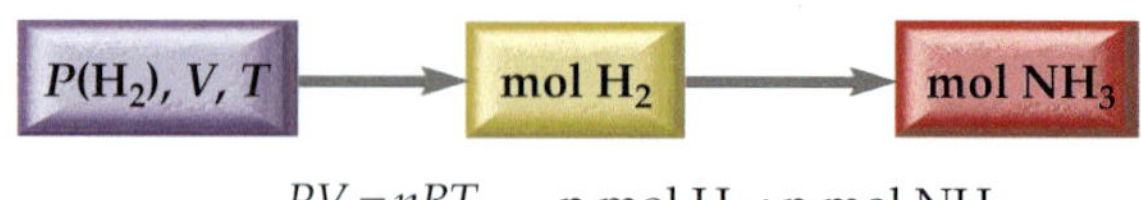

$PV = nRT$ n mol H_2 : n mol NH_3

RELATIONSHIPS USED

$PV = nRT$ (ideal gas law)

3 mol H_2 : 2 mol NH_3 (from balanced equation given in problem)

SOLUTION

We first solve the ideal gas equation for n.

$$PV = nRT$$

$$n = \frac{PV}{RT}$$

Then we substitute in the appropriate values.

$$n = \frac{1.32 \text{ atm} \times 2.5 \text{ L}}{0.0821 \frac{\text{L} \cdot \text{atm}}{\text{mol} \cdot \text{K}} \times 381 \text{ K}}$$

$$= 0.11 \text{ mol } H_2$$

Next, we calculate mol NH_3 from mol H_2.

$$n(NH_3) = \frac{2 \text{ mol } NH_3}{3 \text{ mol } H_2} \times 0.11 \text{ mol } H_2 = 0.070 \text{ mol } NH_3$$

There is enough H_2 to form 0.070 mol NH_3.

EXAMPLE 11.1 GASES IN CHEMICAL REACTIONS

How many litres of oxygen gas form when 294 g of $KClO_3$ completely react in this reaction (which is used in the ignition of fireworks)?

$$2\ KClO_3(s) \rightarrow 2\ KCl(s) + 3\ O_2(g)$$

Assume that the oxygen gas is collected at $P = 755$ mm Hg and $T = 305$ K.

SORT You are given the mass of a reactant in a chemical reaction. You are asked to find the volume of a gaseous product at a given pressure and temperature.	**GIVEN:** 294 g $KClO_3$ $P = 755$ mm Hg (of oxygen gas) $T = 305$ K **FIND:** Volume of O_2 in litres

STRATEGIZE

The solution map has three parts. In the first part, calculate mol $KClO_3$ from mass of $KClO_3$. In the second part, mol O_2 is calculated from mol $KClO_3$. Finally, the volume of O_2 is calculated using the ideal gas law.

SOLUTION MAP

g $KClO_3$ → mol $KClO_3$

$$n = \frac{m}{M}$$

mol $KClO_3$ → mol O_2

n mol $KClO_3$: n mol O_2

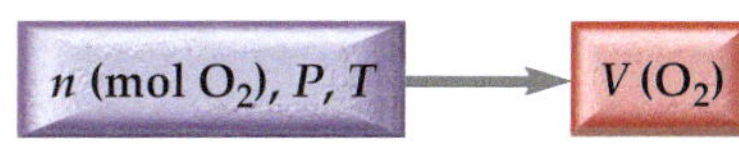

$$PV = nRT$$

You need the molar mass of $KClO_3$ and the stoichiometric relationship between $KClO_3$ and O_2 (from the balanced chemical equation). You also use the ideal gas law.

RELATIONSHIPS USED

$M(KClO_3)$ = 122.55 g/mol

2 mol $KClO_3$: 3 mol O_2 (from balanced equation given in problem)

$PV = nRT$ (ideal gas law, Section 6.8)

SOLVE

Begin by calculating mol $KClO_3$.

SOLUTION

Calculate moles $KClO_3$

$$n = \frac{m}{M} = \frac{294\ \text{g}}{122.55\ \text{g/mol}} = 2.40\ \text{mol KClO}_3$$

Then calculate moles of O_2.

Calculate moles O_2

$$n(\text{O}_2) = \frac{3\ \text{mol O}_2}{2\ \text{mol KClO}_3} \times 2.40\ \text{mol KClO}_3 = 3.60\ \text{mol O}_2$$

Then solve the ideal gas equation for V. Before substituting the values into this equation, convert the pressure to atm.

Calculate volume O_2

$$P = 755\ \text{mm Hg} \times \frac{1\ \text{atm}}{760\ \text{mm Hg}} = 0.993\ \text{atm}$$

$$PV = nRT$$

$$V = \frac{nRT}{P}$$

$$V = \frac{3.60\ \text{mol} \times 0.0821\ \frac{\text{L}\cdot\text{atm}}{\text{mol}\cdot\text{K}} \times 305\ \text{K}}{0.993\ \text{atm}}$$

$$= 90.7\ \text{L}$$

CHECK

Check your answer. Are the units correct? Does the answer make physical sense?

The answer has the correct units, litres. The *value* of the answer is a bit more difficult to judge. Again, it is helpful to know that at standard temperature and pressure (t = 0 °C or 273.15 K and P = 1 atm), 1 mol of gas occupies 22.4 L. A 90.7 L sample of gas at STP would contain about 4 mol; since we started with a little more than 2 mol $KClO_3$, and since 2 mol $KClO_3$ forms 3 mol O_2, an answer that corresponds to about 4 mol O_2 is reasonable.

▶SKILLBUILDER 11.1 | Gases in Chemical Reactions

In this reaction, 4.58 L of O_2 were formed at 745 mm Hg and 308 K. How many grams of Ag_2O decomposed?

$$2\ Ag_2O(s) \rightarrow 4\ Ag(s) + O_2(g)$$

▶FOR MORE PRACTICE Problems 11, 12, 13, 14, 15, 16.

11.2 Solution Stoichiometry

LO: Use volume and concentration to calculate the number of moles of reactants or products and then use stoichiometric coefficients to calculate chemical quantities in a reaction.

As we discussed in Module 8, many chemical reactions take place in aqueous solutions. Precipitation reactions, neutralization reactions, and gas evolution reactions, for example, all occur in aqueous solutions.

As already shown in Section 11.1, the amount of a chemical involved in a reaction doesn't have to be given as a mass. If you have a gas, you can express the amount with a volume at a known temperature and pressure. In reactions involving aqueous reactant and products, it is often convenient to specify the amount of reactants or products in terms of their volume and concentration. We can use the volume and concentration to calculate the number of moles of reactants or products, and then use the stoichiometric coefficients to calculate the other quantities in the reaction. The general solution map for these kinds of calculations is:

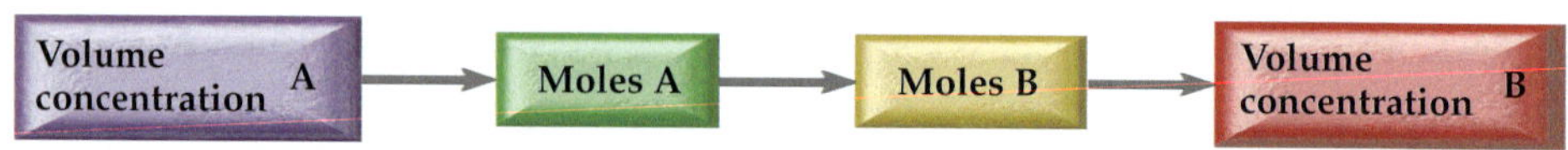

where the calculations of moles are achieved using the concentrations and volumes of the solutions. For example, consider the reaction for the neutralization of sulfuric acid.

$$H_2SO_4(aq) + 2\ NaOH(aq) \rightarrow Na_2SO_4(aq) + 2\ H_2O(l)$$

How much 0.125 mol/L NaOH solution do we need to completely neutralize 0.225 L of 0.175 mol/L H_2SO_4 solution? We begin by sorting the information in the problem statement.

GIVEN: $V(H_2SO_4) = 0.225$ L
$c(H_2SO_4) = 0.175$ mol/L
$c(NaOH) = 0.125$ mol/L

FIND: V NaOH solution

SOLUTION MAP

We strategize by drawing a solution map, which is similar to those for other stoichiometric problems. We first use the volume and concentration of H_2SO_4 solution to get mol H_2SO_4. Then we use the stoichiometric coefficients from the equation to calculate mol NaOH. Finally, we use the concentration of NaOH to get to L NaOH solution.

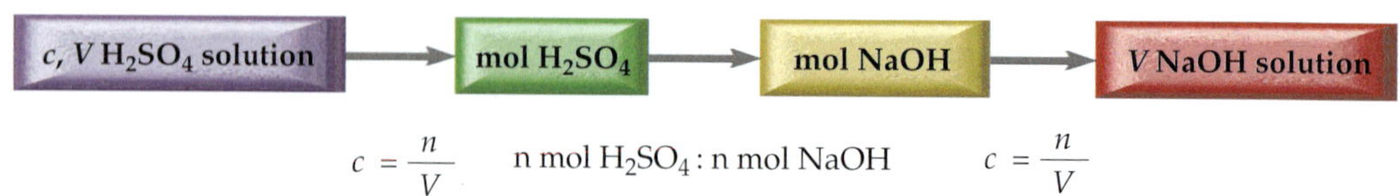

RELATIONSHIPS USED

1 mol H_2SO_4:2 mol NaOH (stoichiometric relationship between H_2SO_4 and NaOH, from balanced chemical equation)

SOLUTION

To solve the problem, we follow the solution map and calculate the answer.

Calculate moles H_2SO_4

$$c = \frac{n}{V}$$

$$n = cV = 0.175\,\text{mol}/\cancel{\text{L}}\ H_2SO_4 \times 0.225\,\cancel{\text{L}}\ H_2SO_4$$
$$= 0.0394\ \text{mol}$$

Calculate moles NaOH

$$n(\text{NaOH}) = \frac{2\ \text{mol NaOH}}{1\ \cancel{\text{mol}\ H_2SO_4}} \times 0.0394\ \cancel{\text{mol}\ H_2SO_4}$$
$$= 0.0788\ \text{mol NaOH}$$

Calculate volume NaOH solution

$$c = \frac{n}{V}$$

$$V(\text{NaOH}) = \frac{n}{c} = \frac{0.0788\ \cancel{\text{mol}}\ \text{NaOH}}{0.125\ \cancel{\text{mol}}/\text{L NaOH}}$$
$$= 0.630\text{L NaOH}$$

It will take 0.630 L of the NaOH solution to completely neutralize the H_2SO_4.

EXAMPLE 11.2 SOLUTION STOICHIOMETRY

Consider the precipitation reaction.

$$2\ \text{KI}(aq) + \text{Pb(NO}_3)_2(aq) \rightarrow \text{PbI}_2(s) + 2\ \text{KNO}_3(aq)$$

What volume 0.115 mol/L KI solution in litres will completely precipitate the Pb^{2+} in 0.104 L of 0.225 mol/L $Pb(NO_3)_2$ solution?

SORT

You are given the concentration of a reactant, KI, in a chemical reaction. You are also given the volume and concentration of a second reactant, $Pb(NO_3)_2$. You are asked to find the volume of the first reactant that completely reacts with the given amount of the second.

GIVEN: 0.115 mol/L KI
0.104 L $Pb(NO_3)_2$ solution
0.225 mol/L $Pb(NO_3)_2$

FIND: V KI solution

STRATEGIZE

The solution map for this problem is similar to the solution maps for other stoichiometric problems. First use the volume and concentration of $Pb(NO_3)_2$ solution to determine mol $Pb(NO_3)_2$. Then use the stoichiometric coefficients from the equation to calculate mol KI from mol $Pb(NO_3)_2$. Finally, use mol KI to find the volume of the KI solution.

SOLUTION MAP

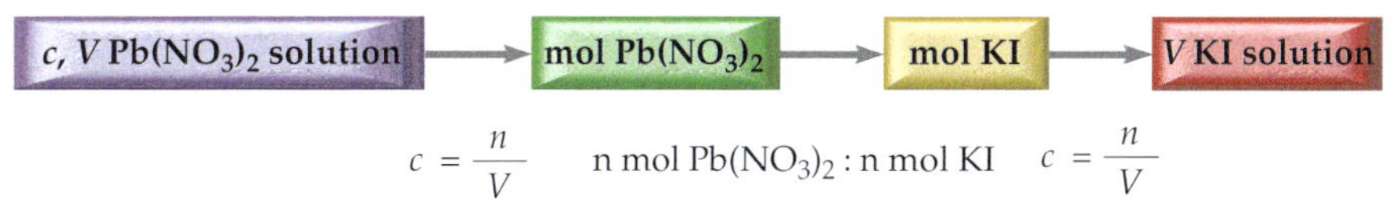

$c = \frac{n}{V}$ n mol $Pb(NO_3)_2$: n mol KI $c = \frac{n}{V}$

RELATIONSHIP USED

2 mol KI : 1 mol $Pb(NO_3)_2$ (stoichiometric relationship between KI and $Pb(NO_3)_2$, from balanced chemical equation)

SOLVE
Follow the solution map to solve the problem beginning with volume of $Pb(NO_3)_2$ solution.

SOLUTION

Calculate moles $Pb(NO_3)_2$

$$c = \frac{n}{V}$$

$$n(Pb(NO_3)_2) = c \times V = 0.225\ \text{mol}/\cancel{\text{L}}\ Pb(NO_3)_2 \times 0.104\ \cancel{\text{L}}\ Pb(NO_3)$$
$$= 0.0234\ \text{mol}\ Pb(NO_3)_2$$

Calculate moles KI

$$n(\text{KI}) = \frac{2\ \text{mol KI}}{1\ \cancel{\text{mol}\ Pb(NO_3)_2}} \times 0.0234\ \cancel{\text{mol}\ Pb(NO_3)_2}$$
$$= 0.0468\ \text{mol KI}$$

Calculate volume KI solution

$$c = \frac{n}{V}$$

$$V(\text{KI}) = \frac{n}{c} = \frac{0.0468\ \cancel{\text{mol}}\ \text{KI}}{0.115\ \cancel{\text{mol}}/\text{L KI}} = 0.407\ \text{L KI}$$

CHECK
Check your answer. Are the units correct? Does the answer make physical sense?

The units (L KI solution) are correct. The magnitude of the answer makes sense because the lead nitrate solution is about twice as concentrated as the potassium iodide solution and 2 mol of potassium iodide are required to react with 1 mol of lead(II) nitrate. Therefore you would expect the volume of the potassium solution required to completely react with a given volume of the $Pb(NO_3)_2$ solution to be about four times as much.

▶**SKILLBUILDER 11.2 | Solution Stoichiometry**

How many millilitres of 0.112 mol/L Na_2CO_3 will completely react with 27.2 mL of 0.135 mol/L HNO_3 according to the reaction?

$$2\ HNO_3(aq) + Na_2CO_3(aq) \rightarrow H_2O(l) + CO_2(g) + 2\ NaNO_3(aq)$$

▶**SKILLBUILDER PLUS** A 25.0 mL sample of HNO_3 solution requires 35.7 mL of 0.108 mol/L Na_2CO_3 to completely react with all of the HNO_3 in the solution. What is the concentration of the HNO_3 solution?

▶**FOR MORE PRACTICE** Example 11.8; Problems 21, 22, 23, 24.

CONCEPTUAL CHECKPOINT 11.1

Consider the following reaction occurring in aqueous solution.

$$A(aq) + 2\ B(aq) \rightarrow \text{Products}$$

What volume of a 0.100 mol/L solution of B is required to completely react with 50.0 mL of a 0.200 mol/L solution of A?

(a) 25.0 mL
(b) 50.0 mL
(c) 100.0 mL
(d) 200.0 mL

11.3 Acid–Base Titration: A Way to Quantify the Amount of Acid or Base in a Solution

LO: Use acid–base titration to determine the concentration of an unknown solution.

We can apply the principles we learned in the previous section on solution stoichiometry to a common laboratory procedure called a titration. In a **titration**, we react a substance in a solution of known concentration with another substance in a solution of unknown concentration. For example, consider the acid–base reaction between hydrochloric acid and sodium hydroxide:

$$HCl(aq) + NaOH(aq) \rightarrow H_2O(l) + NaCl(aq)$$

The net ionic equation for this reaction is:

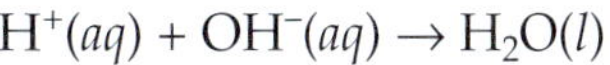

$$H^+(aq) + OH^-(aq) \rightarrow H_2O(l)$$

H^+

Suppose we have an HCl solution represented by the molecular diagram at left. (The Cl^- ions and the H_2O molecules not involved in the reaction have been omitted from this representation for clarity.)

In titrating this sample, we slowly add a solution of known OH^- concentration. This process is represented by the following molecular diagrams:

The OH^- solution also contains Na^+ cations that we do not show in this figure for clarity.

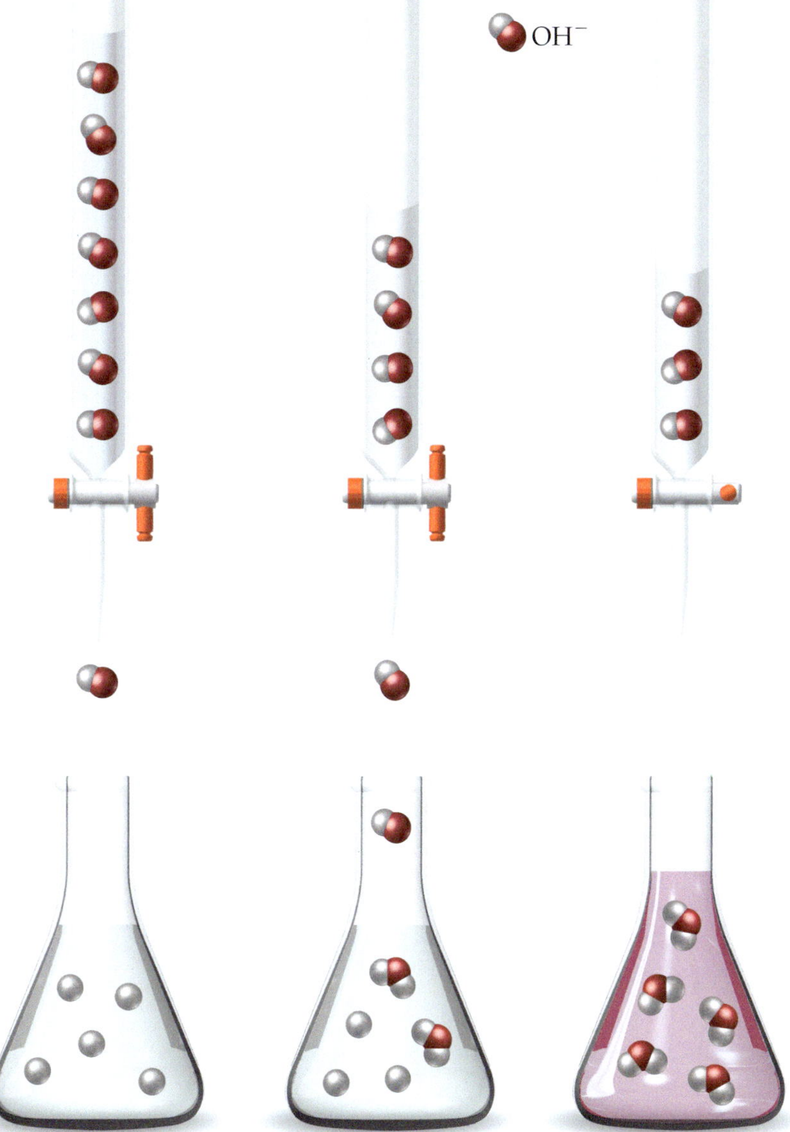

Beginning of titration

Equivalence point

As we add the OH^-, it reacts with and neutralizes the H^+, forming water. At the **equivalence point**—*the point in the titration when the number of moles of OH^- added equals the number of moles of H^+ originally in solution*—the titration is complete. The equivalence point is usually signaled by an **indicator**, a dye whose color depends on the acidity of the solution (▶ Figure 11.1). In most laboratory titrations, the concentration of one of the reactant solutions is unknown, and the concentration of the other is precisely known. By carefully measuring the volume of each solution required to reach the equivalence point, we can determine the concentration of the unknown solution, as demonstrated in Example 11.3.

At the equivalence point, neither reactant is present in excess, and both are limiting. The number of moles of the reactants are related by the reaction stoichiometry (see Module 10).

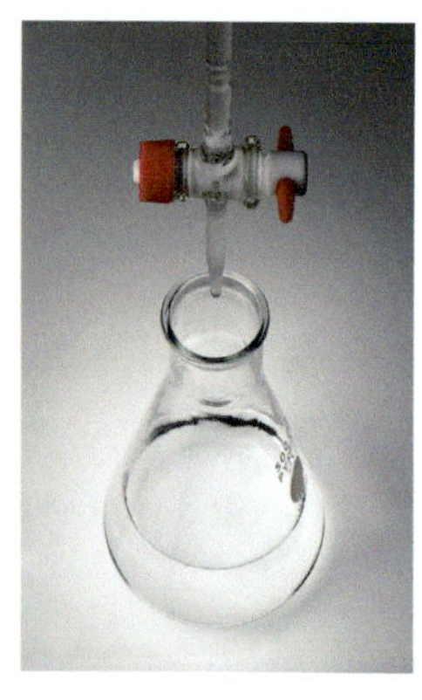
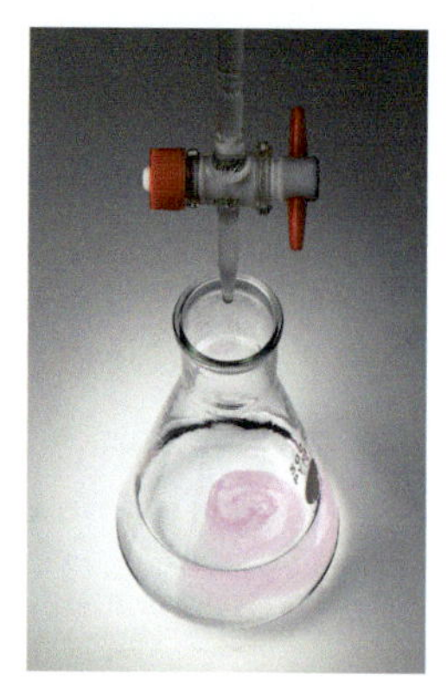

▶ **FIGURE 11.1 Acid–base titration** In this titration, NaOH is added to an HCl solution. When the NaOH and HCl reach stoichiometric proportions (1 mol of OH^- for every 1 mol of H^+), the indicator (phenolphthalein) changes to pink, signaling the equivalence point of the titration. (Phenolphthalein is an indicator that is colorless in acidic solution and pink in basic solution.) © Richard Megna/Fundamental Photographs.

EXAMPLE 11.3 ACID–BASE TITRATION

The titration of 10.00 mL of an HCl solution of unknown concentration requires 12.54 mL of a 0.100 mol/L NaOH solution to reach the equivalence point. What is the concentration of the unknown HCl solution?

SORT

You are given the volume of an unknown HCl solution and the volume of a known NaOH solution required to titrate the unknown solution. You are asked to find the concentration of the unknown solution.

GIVEN: 10.00 mL HCl solution
12.54 mL of a 0.100 mol/L NaOH solution

FIND: concentration of HCl solution (mol/L)

STRATEGIZE

First write the balanced chemical equation for the reaction between the acid and the base.

The solution map has two parts. In the first part, the volume of NaOH is used to calculate the number of moles of HCl in the original solution. The stoichimetric coefficients in the balanced neutralization equation can be used to calculate the moles of HCl from moles of NaOH.

In the second part, use the number of moles of HCl and the volume of HCl solution to determine the concentration of the HCl solution.

SOLUTION MAP

$$HCl(aq) + NaOH(aq) \rightarrow H_2O(l) + NaCl(aq)$$

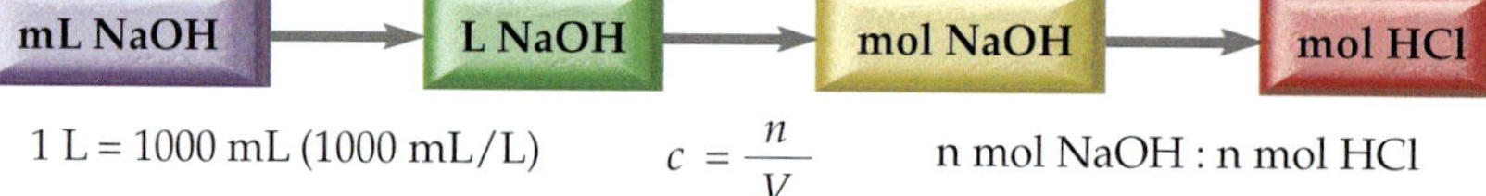

1 L = 1000 mL (1000 mL/L) $c = \frac{n}{V}$ n mol NaOH : n mol HCl

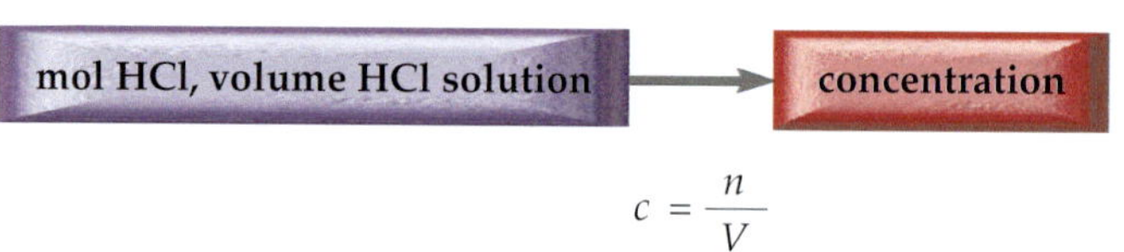

$c = \frac{n}{V}$

RELATIONSHIPS USED

1 mol HCl : 1 mol NaOH (from balanced chemical equation)

SOLVE

Calculate the moles of HCl in the unknown solution by following the first part of the solution map.

Note that volumes must be in L rather than mL.

SOLUTION

Convert volumes to L

$$V(\text{NaOH}) = \frac{12.54\,\cancel{\text{mL}}}{1000\,\cancel{\text{mL}}\,/\,\text{L}} = 0.0125\,\text{L}$$

$$\text{V(HCl)} = \frac{10.00\,\cancel{\text{ml}}}{1000\,\cancel{\text{ml}}\,/\,\text{l}} = 0.0100\text{L}$$

Calculate moles NaOH

$$c = \frac{n}{V}$$

$$n(\text{NaOH}) = c \times V = 0.100\,\text{mol}\,/\,\cancel{\text{L}}\,\text{NaOH} \times 0.0125\,\cancel{\text{L}}\,\text{NaOH}$$
$$= 0.00125\,\text{mol NaOH}$$

Calculate moles HCl

$$n(\text{HCl}) = \frac{1 \text{ mol HCl}}{1 \text{ mol NaOH}} \times 0.00125 \text{ mol NaOH}$$
$$= 0.00125 \text{ mol HCl}$$

The unknown HCl solution therefore has a concentration of 0.125 M.

Calculate concentration HCl solution

$$c(\text{HCl}) = \frac{n}{V} = \frac{0.00125 \text{ mol HCl}}{0.0100 \text{ L HCl}}$$
$$= 0.125 \text{ mol/L HCl}$$

CHECK

Check your answer. Are the units correct? Does the answer make physical sense?

The units (mol/L) are correct. The magnitude of the answer makes sense because the reaction has a one-to-one stoichiometry and the volumes of the two solutions are similar; therefore, their concentrations should also be similar.

▶SKILLBUILDER 11.3 | Acid–Base Titration

The titration of a 20.0 mL sample of an H_2SO_4 solution of unknown concentration requires 22.87 mL of a 0.158 mol/L KOH solution to reach the equivalence point. What is the concentration of the unknown H_2SO_4 solution?

▶FOR MORE PRACTICE Example 11.9; Problems 29, 30, 31, 32, 33, 34.

CONCEPTUAL CHECKPOINT 11.2

The flask shown here represents a sample of acid to be titrated.

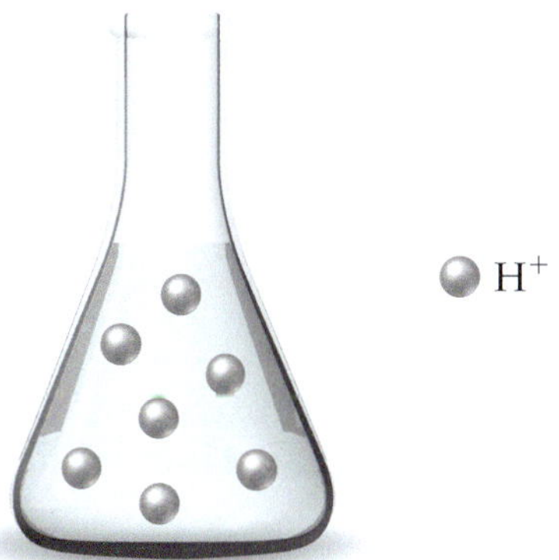

Which of the beakers contains the amount of OH^- required to reach the equivalence point in the titration?

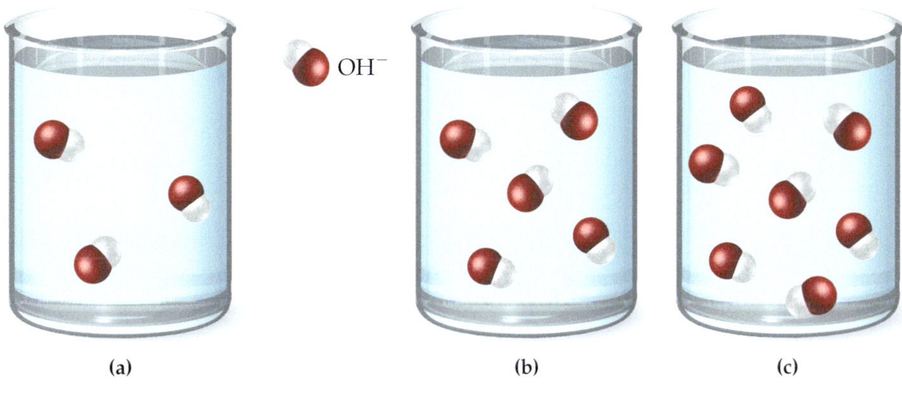

11.4 Water: Acid and Base in One

Recall that water acts as a base when it reacts with HCl and as an acid when it reacts with NH_3.

Water acting as a base

$$\underset{\substack{\text{Acid} \\ \text{(Proton donor)}}}{HCl(aq)} + \underset{\substack{\text{Base} \\ \text{(Proton acceptor)}}}{H_2O(l)} \longrightarrow H_3O^+(aq) + Cl^-(aq)$$

Water acting as an acid

$$\underset{\substack{\text{Base} \\ \text{(Proton acceptor)}}}{NH_3(aq)} + \underset{\substack{\text{Acid} \\ \text{(Proton donor)}}}{H_2O(l)} \rightleftharpoons NH_4^+(aq) + OH^-(aq)$$

Water is *amphoteric;* it can act as either an acid or a base. Even in pure water, water acts as an acid and a base with itself, a process called self-ionization.

Water acting as both an acid and a base

$$\underset{\substack{\text{Acid} \\ \text{(Proton donor)}}}{H_2O(l)} + \underset{\substack{\text{Base} \\ \text{(Proton acceptor)}}}{H_2O(l)} \rightleftharpoons H_3O^+(aq) + OH^-(aq)$$

In pure water, at 25 °C, the preceding reaction occurs only to a very small extent, resulting in equal and small concentrations of H_3O^+ and OH^-.

$$[H_3O^+] = [OH^-] = 1.0 \times 10^{-7} \text{ mol/L} \qquad \text{(in pure water at 25 °C)}$$

where $[H_3O^+]$ = the concentration of H_3O^+ in mol/L

and $[OH^-]$ = the concentration of OH^- in mol/L

So all samples of water contain some hydronium ions (H_3O^+) and some hydroxide ions. The *product* of the concentration of these two ions in aqueous solutions is called the **ion product constant for water (K_w)**.

$$K_w = [H_3O^+][OH^-]$$

The units of K_w are normally dropped.

We can find the value of K_w at 25 °C by multiplying the hydronium and hydroxide concentrations for pure water listed earlier.

$$\begin{aligned} K_w &= [H_3O^+][OH^-] \\ &= (1.0 \times 10^{-7})(1.0 \times 10^{-7}) \\ &= (1.0 \times 10^{-7})^2 \\ &= 1.0 \times 10^{-14} \end{aligned}$$

In a neutral solution, $[H_3O^+] = [OH^-]$.

The preceding equation holds true for all aqueous solutions at 25 °C. The concentration of H_3O^+ times the concentration of OH^- is 1.0×10^{-14}. In pure water, since H_2O is the only source of these ions, there is one H_3O^+ ion for every OH^- ion. Consequently, the concentrations of H_3O^+ and OH^- are equal. Such a solution is a **neutral solution**.

$$[H_3O^+] = [OH^-] = \sqrt{K_w} = 1.0 \times 10^{-7} \text{ mol/L (in pure water)}$$

An **acidic solution** contains an acid that creates additional H_3O^+ ions, causing H_3O^+ to increase. However, the *ion product constant still applies.*

$$[H_3O^+][OH^-] = K_w = 1.0 \times 10^{-14}$$

In an acidic solution, $[H_3O^+] > [OH^-]$.

If $[H_3O^+]$ increases, then $[OH^-]$ must decrease for the ion product to remain 1.0×10^{-14}. For example, suppose $[H_3O^+] = 1.0 \times 10^{-3}$ mol/L; then $[OH^-]$ can be found by solving the ion product expression for $[OH^-]$.

$$(1.0 \times 10^{-3})[OH^-] = 1.0 \times 10^{-14}$$

$$[OH^-] = \frac{1.0 \times 10^{-14}}{1.0 \times 10^{-3}} = 1.0 \times 10^{-11}\ \text{mol/L}$$

In an acidic solution, $[H_3O^+]$ is greater than 1.0×10^{-7} mol/L, and $[OH^-]$ is less than 1.0×10^{-7} mol/L.

In a basic solution, $[H_3O^+] < [OH^-]$.

A **basic solution** contains a base that creates additional OH^- ions, causing the $[OH^-]$ to increase and the $[H_3O^+]$ to decrease. For example, suppose $[OH^-] = 1.0 \times 10^{-2}$ mol/L; $[H_3O^+]$ can be found by solving the ion product expression for $[H_3O^+]$.

$$[H_3O^+](1.0 \times 10^{-2}) = 1.0 \times 10^{-14}$$

$$[H_3O^+] = \frac{1.0 \times 10^{-14}}{1.0 \times 10^{-2}} = 1.0 \times 10^{-12}\ \text{mol/L}$$

In a basic solution, $[OH^-]$ is greater than 1.0×10^{-7} mol/L and $[H_3O^+]$ is less than 1.0×10^{-7} mol/L.

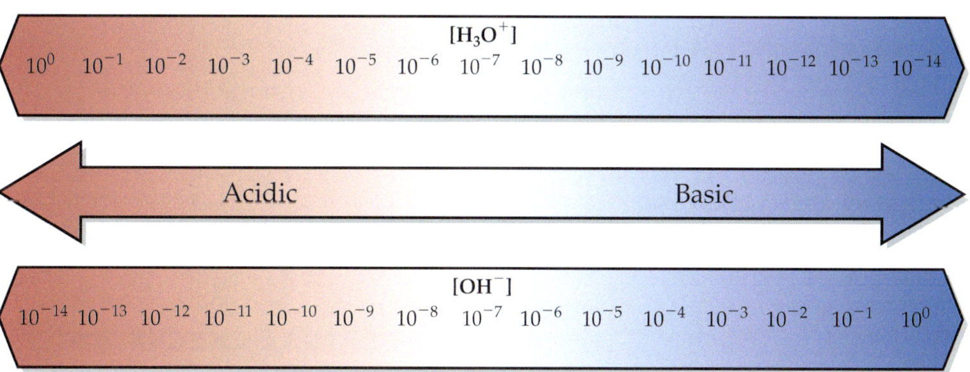

▶ **FIGURE 11.2 Acidic and basic solutions**

To summarize, at 25 °C (▲ Figure 11.2):

- In a neutral solution, $[H_3O^+] = [OH^-] = 1.0 \times 10^{-7}$ mol/L
- In an acidic solution, $[H_3O^+] > 1.0 \times 10^{-7}$ mol/L $[OH^-] < 1.0 \times 10^{-7}$ mol/L
- In a basic solution, $[H_3O^+] < 1.0 \times 10^{-7}$ mol/L $[OH^-] > 1.0 \times 10^{-7}$ mol/L
- In all aqueous solutions, $[H_3O^+][OH^-] = K_w = 1.0 \times 10^{-14}$

EXAMPLE 11.4 USING K_W IN CALCULATIONS

Calculate $[OH^-]$ for each solution and determine whether the solution is acidic, basic, or neutral.

(a) $[H_3O^+] = 7.5 \times 10^{-5}$ mol/L
(b) $[H_3O^+] = 1.5 \times 10^{-9}$ mol/L
(c) $[H_3O^+] = 1.0 \times 10^{-7}$ mol/L

To find $[OH^-]$ use the ion product constant, K_W. Substitute the given value for $[H_3O^+]$ and solve the equation for $[OH^-]$. Since $[H_3O^+] > 1.0 \times 10^{-7}$ mol/L and $[OH^-] < 1.0 \times 10^{-7}$ mol/L, the solution is acidic.

SOLUTION

(a) $[H_3O^+][OH^-] = K_W = 1.0 \times 10^{-14}$

$$[7.5 \times 10^{-5}][OH^-] = K_W = 1.0 \times 10^{-14}$$

$$[OH^-] = \frac{1.0 \times 10^{-14}}{7.5 \times 10^{-5}} = 1.3 \times 10^{-10}\ \text{mol/L}$$

acidic solution

Substitute the given value for $[H_3O^+]$ into the ion product constant equation and solve the equation for $[OH^-]$. Since $[H_3O^+] < 1.0 \times 10^{-7}$ mol/L and $[OH^-] > 1.0 \times 10^{-7}$ mol/L, the solution is basic.

(b) $[H_3O^+][OH^-] = K_W = 1.0 \times 10^{-14}$

$$[1.5 \times 10^{-9}][OH^-] = 1.0 \times 10^{-14}$$

$$[OH^-] = \frac{1.0 \times 10^{-14}}{1.5 \times 10^{-9}} = 6.7 \times 10^{-6}\ \text{mol/L}$$

basic solution

Substitute the given value for $[H_3O^+]$ into the ion product constant equation and solve the equation for $[OH^-]$. Since $[H_3O^+] = 1.0 \times 10^{-7}$ mol/L and $[OH^-] = 1.0 \times 10^{-7}$ mol/L, the solution is neutral.

(c) $[H_3O^+][OH^-] = K_W = 1.0 \times 10^{-14}$

$$[1.0 \times 10^{-7}][OH^-] = 1.0 \times 10^{-14}$$

$$[OH^-] = \frac{1.0 \times 10^{-14}}{1.0 \times 10^{-7}} = 1.0 \times 10^{-7}\ \text{mol/L}$$

neutral solution

▶SKILLBUILDER 11.4 | Using K_W in Calculations

Calculate $[H_3O^+]$ in each solution and determine whether the solution is acidic, basic, or neutral.

(a) $[OH^-] = 1.5 \times 10^{-2}$ mol/L
(b) $[OH^-] = 1.0 \times 10^{-7}$ mol/L
(c) $[OH^-] = 8.2 \times 10^{-10}$ mol/L

▶FOR MORE PRACTICE Example 11.10; Problems 37, 38, 39, 40.

CONCEPTUAL CHECKPOINT 11.3

Which substance is least likely to act as a base?

(a) H_2O
(b) OH^-
(c) NH_3
(d) NH_4^+

11.5 The pH and pOH Scales: Ways to Express Acidity and Basicity

LO: Calculate pH from $[H_3O^+]$.
LO: Calculate $[H_3O^+]$ from pH.
LO: Calculate $[OH^-]$ from pOH.
LO: Compare and contrast the pOH scale and the pH scale.

Notice that an increase of 1 in pH corresponds to a tenfold *decrease* in $[H_3O^+]$.

Chemists have devised a scale, called the **pH** scale, based on hydrogen ion (or hydronium ions) concentration to express the acidity or basicity of solutions. At 25 °C according to the pH scale, a solution has these general characteristics:

- pH < 7 ***acidic*** solution
- pH > 7 ***basic*** solution
- pH = 7 ***neutral*** solution

TABLE 11.1 The pH of Some Common Substances

Substance	pH
gastric (human stomach) acid	1.0–3.0
limes	1.8–2.0
lemons	2.2–2.4
soft drinks	2.0–4.0
plums	2.8–3.0
wine	2.8–3.8
apples	2.9–3.3
peaches	3.4–3.6
cherries	3.2–4.0
beer	4.0–5.0
rainwater (unpolluted)	5.6
human blood	7.3–7.4
egg whites	7.6–8.0
milk of magnesia	10.5
household ammonia	10.5–11.5
4 % NaOH solution	14

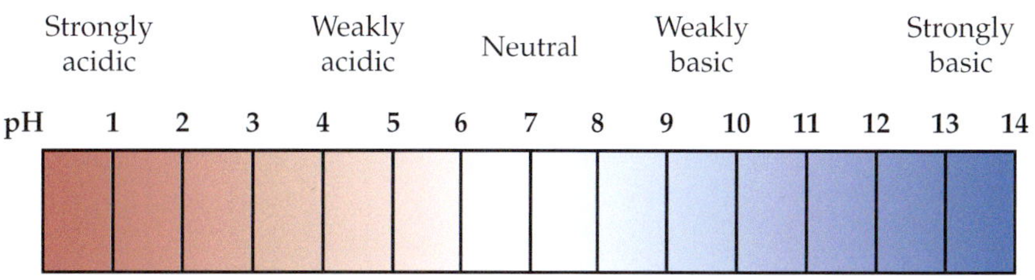

Strongly acidic — Weakly acidic — Neutral — Weakly basic — Strongly basic

pH 1 2 3 4 5 6 7 8 9 10 11 12 13 14

Table 11.1 lists the pH of some common substances. Notice that many foods, especially fruits, are acidic and therefore have low pH values. The foods with the lowest pH values are limes and lemons, and they are among the sourest. Relatively few foods, however, are basic.

The pH scale is a **logarithmic scale**; a change of 1 pH unit corresponds to a tenfold change in H_3O^+ concentration. For example, a lime with a pH of 2.0 is 10 times more acidic than a plum with a pH of 3.0 and 100 times more acidic than a cherry with a pH of 4.0. Each change of 1 in pH scale corresponds to a change of 10 in $[H_3O^+]$ (▼ Figure 11.3).

▶ **FIGURE 11.3 The pH scale is a logarithmic scale** A *decrease* of 1 unit on the pH scale corresponds to an *increase* in $[H_3O^+]$ concentration by a factor of 10. Each circle stands for 10^{-4} mol H^+/L, or 6.022×10^{19} H^+ ions per litre.
Question: How much of an increase in H_3O^+ concentration corresponds to a decrease of 2 pH units?

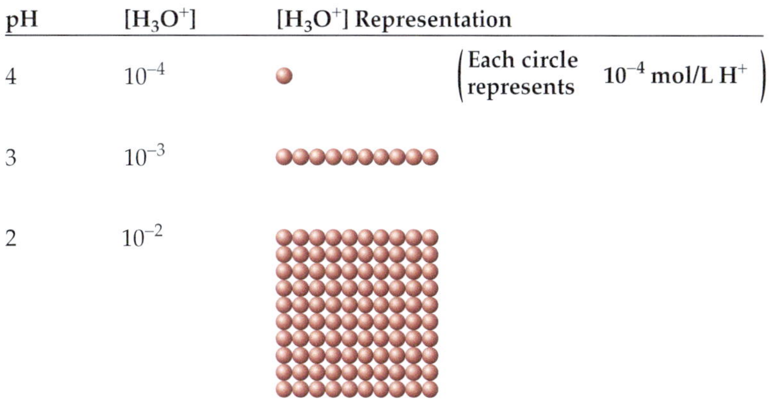

pH	$[H_3O^+]$	$[H_3O^+]$ Representation
4	10^{-4}	(Each circle represents 10^{-4} mol/L H^+)
3	10^{-3}	
2	10^{-2}	

Calculating pH from $[H_3O^+]$

We define the pH of a solution as the negative of the log of the hydronium ion concentration:

$$pH = -\log[H_3O^+]$$

Note that pH is defined using the log function (base ten), which is different from the natural log (abbreviated ln).

To calculate pH, we must use logarithms. Recall that the log of a number is the exponent to which 10 must be raised to obtain that number, as shown in these examples:

$$\log 10^1 = 1; \log 10^2 = 2; \log 10^3 = 3$$

$$\log 10^{-1} = -1; \log 10^{-2} = -2; \log 10^{-3} = -3$$

In the next example, we calculate the log of 1.5×10^{-7}. A solution having an $[H_3O^+] = 1.5 \times 10^{-7}$ mol/L (acidic) has a pH of:

$$\begin{aligned} pH &= -\log[H_3O^+] \\ &= -\log(1.5 \times 10^{-7}) \\ &= -(-6.82) \\ &= 6.82 \end{aligned}$$

Notice that the pH is reported to two decimal places here. This is because only the numbers to the right of the decimal place are significant in a log. Because our original value for the concentration had two significant figures, the log of that number has two decimal places:

$$\log(1.0 \times 10^{-3}) = 3.00$$

2 decimal places

When we take the log of a quantity, the result should have the same number of decimal places as the number of significant figures in the original quantity.

If the original number has three significant figures, we report the log to three decimal places:

$$-\log(1.00 \times 10^{-3}) = 3.000$$

3 decimal places

A solution having $[H_3O^+] = 1.0 \times 10^{-7}$ mol/L (neutral) has a pH of:

$$\begin{aligned} pH &= -\log[H_3O^+] \\ &= -\log(1.0 \times 10^{-7}) \\ &= -(-7.00) \\ &= 7.00 \end{aligned}$$

EXAMPLE 11.5 CALCULATING pH FROM $[H_3O^+]$

Calculate the pH of each solution and indicate whether the solution is acidic or basic.

(a) $[H_3O^+] = 1.8 \times 10^{-4}$ mol/L
(b) $[H_3O^+] = 7.2 \times 10^{-9}$ mol/L

SOLUTION

To calculate pH, substitute the given $[H_3O^+]$ into the pH equation. Since the pH < 7, this solution is acidic.	**(a)** $pH = -\log[H_3O^+] = -\log(1.8 \times 10^{-4}) = -(-3.74) = 3.74$
Again, substitute the given $[H_3O^+]$ into the pH equation. Since the pH > 7, this solution is basic.	**(b)** $pH = -\log[H_3O^+] = -\log(7.2 \times 10^{-9}) = -(-8.14) = 8.14$

▶SKILLBUILDER 11.5 | Calculating pH from $[H_3O^+]$

Calculate the pH of each solution and indicate whether the solution is acidic or basic.

(a) $[H_3O^+] = 9.5 \times 10^{-9}$ mol/L
(b) $[H_3O^+] = 6.1 \times 10^{-3}$ mol/L

▶SKILLBUILDER PLUS

Calculate the pH of a solution with $[OH^-] = 1.3 \times 10^{-2}$ mol/L and indicate whether the solution is acidic or basic. *Hint:* Begin by using K_w to find $[H_3O^+]$.

▶FOR MORE PRACTICE Example 11.11; Problems 43, 44.

Calculating $[H_3O^+]$ from pH

Ten raised to the log of a number is equal to that number: $10^{\log x} = x$.

To calculate $[H_3O^+]$ from a pH value, we *undo* the log. The log can be undone using the inverse log function (*Method 1*) on most calculators or using the 10^x key (*Method 2*). Both methods do the same thing; the one you use depends on your calculator.

The invlog function "undoes" log: $\text{invlog}(\log x) = x$.

Method 1: Inverse Log Function	Method 2: 10^x Function
$pH = -\log[H_3O^+]$	$pH = -\log[H_3O^+]$
$-pH = \log[H_3O^+]$	$-pH = \log[H_3O^+]$
$\text{invlog}(-pH) = \text{invlog}(\log[H_3O^+])$	$10^{-pH} = 10^{\log[H_3O^+]}$
$\text{invlog}(-pH) = [H_3O^+]$	$10^{-pH} = [H_3O^+]$

The inverse log is sometimes called the antilog.

So, to calculate $[H_3O^+]$ from a pH value, we take the inverse log of the negative of the pH value (*Method 1*) or raise 10 to the negative of the pH value (*Method 2*).

CALCULATING [H3O+] FROM PH

Calculate the H_3O^+ concentration for a solution with a pH of 4.80.

SOLUTION

To find the $[H_3O^+]$ from pH, undo the log function. Use either Method 1 or Method 2.

The number of significant figures in the inverse log of a number is equal to the number of decimal places in the number.

Method 1: Inverse Log Function	Method 2: 10^x Function
$pH = -\log[H_3O^+]$	$pH = -\log[H_3O^+]$
$4.80 = -\log[H_3O^+]$	$4.80 = -\log[H_3O^+]$
$-4.80 = \log[H_3O^+]$	$-4.80 = \log[H_3O^+]$
$\text{invlog}(-4.80) = \text{invlog}(\log[H_3O^+])$	$10^{-4.80} = 10^{\log[H_3O^+]}$
$\text{invlog}(-4.80) = [H_3O^+]$	$10^{-4.80} = [H_3O^+]$
$[H_3O^+] = 1.6 \times 10^{-5}$ mol/L	$[H_3O^+] = 1.6 \times 10^{-5}$ mol/L

▶SKILLBUILDER 11.6 | Calculating $[H_3O^+]$ from pH

Calculate the $[H_3O^+]$ concentration for a solution with a pH of 8.37.

▶SKILLBUILDER PLUS

Calculate the OH^- concentration for a solution with a pH of 3.66.

▶FOR MORE PRACTICE Example 11.12; Problems 45, 46.

CONCEPTUAL CHECKPOINT 11.4

Solution A has a pH of 13. Solution B has a pH of 10. The concentration of H_3O^+ in solution B is ______ times that in solution A.

(a) 0.001

(b) $\frac{1}{3}$

(c) 3

(d) 1000

The pOH Scale

The **pOH** scale is analogous to the pH scale but is defined with respect to $[OH^-]$ instead of $[H_3O^+]$.

Notice that *p* is the mathematical function; –log; thus, $pX = -\log X$.

$$pOH = -\log[OH^-]$$

A solution with an $[OH^-]$ of 1.0×10^{-3} mol/L (basic) has a pOH of 3.00. On the pOH scale, a pOH less than 7 is basic and a pOH greater than 7 is acidic. A pOH of 7 is neutral. We can find the $[OH^-]$ concentration from the pOH just as the $[H_3O^+]$ concentration is found from the pH, as shown in Example 11.7.

EXAMPLE 11.7 CALCULATING [OH⁻] FROM pOH

Calculate the $[OH^-]$ concentration for a solution with a pOH of 8.55.

SOLUTION

To find the $[OH^-]$ from pOH, undo the log function. Use either Method 1 or Method 2.

Method 1: Inverse Log Function	Method 2: 10^x Function
$pOH = -\log[OH^-]$	$pOH = -\log[OH^-]$
$8.55 = -\log[OH^-]$	$8.55 = -\log[OH^-]$
$-8.55 = \log[OH^-]$	$-8.55 = \log[OH^-]$
$\text{invlog}(-8.55) = \text{invlog}(\log[OH^-])$	$10^{-8.55} = 10^{\log[OH^-]}$
$\text{invlog}(-8.55) = [OH^-]$	$10^{-8.55} = [OH^-]$
$[OH^-] = 2.8 \times 10^{-9}$ mol/L	$[OH^-] = 2.8 \times 10^{-9}$ mol/L

▶SKILLBUILDER 11.7 | Calculating [OH⁻] from pOH

Calculate the OH^- concentration for a solution with a pOH of 4.25.

▶SKILLBUILDER PLUS

Calculate the H_3O^+ concentration for a solution with a pOH of 5.68.

▶FOR MORE PRACTICE Problems 53, 54, 55, 56.

We can derive a relationship between pH and pOH at 25 °C from the expression for K_w.

$$[H_3O^+][OH^-] = 1.0 \times 10^{-14}$$

log (AB) = log A + log B

Taking the log of both sides, we get:

$$\log\{[H_3O^+][OH^-]\} = \log(1.0 \times 10^{-14})$$

$$\log[H_3O^+] + \log[OH^-] = -14.00$$

$$-\log[H_3O^+] - \log[OH^-] = 14.00$$

$$pH + pOH = 14.00$$

The sum of pH and pOH is always equal to 14.00 at 25 °C. Therefore, a solution with a pH of 3 has a pOH of 11.

CONCEPTUAL CHECKPOINT 11.5

A solution has a pH of 5. What is the pOH of the solution?

(a) 5

(b) 10

(c) 14

(d) 9

11.6 Buffers: Solutions That Resist pH Change

LO: Describe how buffers resist pH change.

Most solutions rapidly become more acidic (lower pH) upon addition of an acid or more basic (higher pH) upon addition of a base. A **buffer**, however, resists pH change by neutralizing added acid or added base. Human blood, for example, is a buffer. Acid or base that is added to blood gets neutralized by components within blood, resulting in a nearly constant pH. In healthy individuals, blood pH is between 7.36 and 7.40. If blood pH drops below 7.0 or rises above 7.8, death results.

How does blood maintain such a narrow pH range? Like all buffers, blood contains *significant* amounts of *both a weak acid and its conjugate base*. When additional base is added to blood, the weak acid reacts with the base, neutralizing it. When additional acid is added to blood, the conjugate base reacts with the acid, neutralizing it. In this way, blood maintains a constant pH.

Buffers can also be composed of a weak base and its conjugate acid.

CHEMISTRY AND HEALTH

▶ Alkaloids

Alkaloids are organic bases that occur naturally in many plants that often have medicinal qualities. Morphine, for example, is a powerful alkaloid drug that occurs in the opium poppy (▶ Figure 11.4) and is used to relieve severe pain. Morphine is an example of a *narcotic*, a drug that dulls the senses and induces sleep. It produces relief from and indifference to pain. Morphine can also produce feelings of euphoria and contentment, which leads to its abuse. Morphine is highly addictive, both psychologically and physically. A person who abuses morphine over long periods of time becomes physically dependent on the drug and suffers severe withdrawal symptoms upon termination of use.

▲ **FIGURE 11.4 Opium poppy** The opium poppy contains the alkaloids morphine and codeine.© Konstantin Christian/Shutterstock.

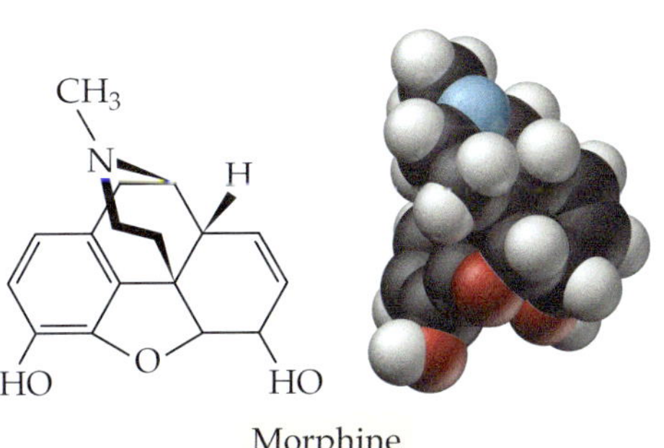

Morphine

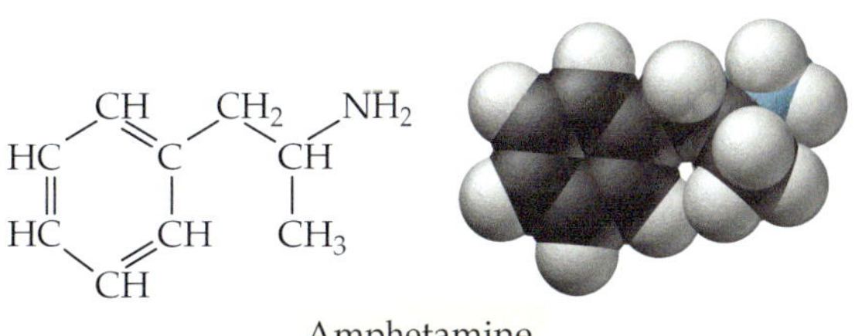

Amphetamine

Amphetamine is another powerful drug related to the alkaloid ephedrine. Whereas morphine slows down nerve signal transmissions, amphetamine enhances them. Amphetamine is an example of a *stimulant*, a drug that increases alertness and wakefulness. Amphetamine is widely used to treat Attention-Deficit Hyperactivity Disorder (ADHD) and is prescribed under the trade name Adderall. Patients suffering from ADHD find that amphetamine helps them to focus and concentrate more effectively. However, because amphetamine produces alertness and increased stamina, it, too, is often abused.

Other common alkaloids include caffeine and nicotine, both of which are stimulants. Caffeine is found in the coffee bean, and nicotine is found in tobacco. Although both have some addictive qualities, nicotine is by far the most addictive. A nicotine addiction is among the most difficult to break, as any smoker can attest.

B11.1 CAN YOU ANSWER THIS? *What part of the amphetamine and morphine molecules makes them bases?*

We can make a simple buffer by mixing acetic acid ($HC_2H_3O_2$) and its conjugate base, sodium acetate ($NaC_2H_3O_2$), in water (▶ Figure 11.5) (The sodium in sodium acetate is just a spectator ion and does not contribute to buffering action). Because $HC_2H_3O_2$ is a weak acid and $C_2H_3O_2^-$ is its conjugate base, a solution containing both of these is a buffer. Note that a weak acid by itself, even though it partially ionizes to form some of its conjugate base, does not contain sufficient base to be a buffer. A buffer must contain *significant* amounts of *both* a weak acid and its conjugate base. If we add more base, in the form of NaOH, to the buffer solution containing acetic acid and sodium acetate, the acetic acid neutralizes the base according to the reaction:

$$\underset{\text{Base}}{OH^-(aq)} + \underset{\text{Acid}}{HC_2H_3O_2(aq)} \rightarrow C_2H_3O_2^-(aq) + H_2O(l)$$

As long as the amount of NaOH that we add is less than the amount of $HC_2H_3O_2$ in solution, the solution neutralizes the NaOH, and the resulting pH change is small. Suppose, on the other hand, that we add more acid, in the form of HCl, to the solution. In this case, the conjugate base, $NaC_2H_3O_2$, neutralizes the added HCl according to the reaction:

$$\underset{\text{Acid}}{H^+(aq)} + \underset{\text{Base}}{C_2H_3O_2^-(aq)} \rightarrow HC_2H_3O_2(aq)$$

As long as the amount of HCl that we add is less than the amount of $NaC_2H_3O_2$ in solution, the solution neutralizes the HCl and the resulting pH change is small.

To summarize:

- Buffers resist pH change.
- Buffers contain significant amounts of both a weak acid and its conjugate base.
- The weak acid in a buffer neutralizes added base.
- The conjugate base in a buffer neutralizes added acid.

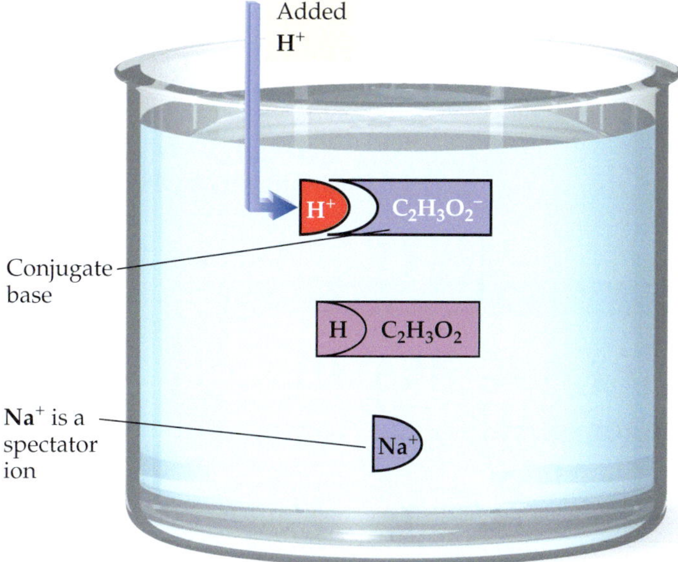

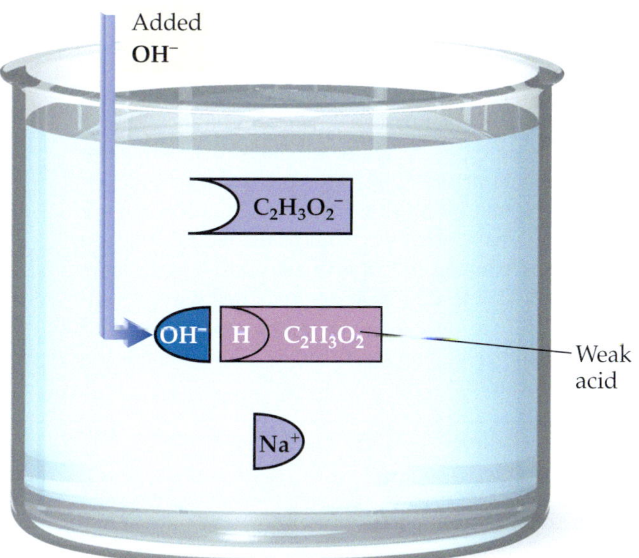

▲ **FIGURE 11.5** **How buffers resist pH change** A buffer contains significant amounts of a weak acid and its conjugate base. The acid consumes any added base, and the base consumes any added acid. In this way, a buffer resists pH change.

CONCEPTUAL CHECKPOINT 11.6

Which of the following is a buffer solution?

(a) $H_2SO_4(aq)$ and $H_2SO_3(aq)$

(b) $HF(aq)$ and $NaF(aq)$

(c) $HCl(aq)$ and $NaCl(aq)$

(d) $NaCl(aq)$ and $NaOH(aq)$

CHEMISTRY AND HEALTH

▶ The Danger of Antifreeze

Most types of antifreeze used in cars are solutions of ethylene glycol. Every year, thousands of dogs and cats die from ethylene glycol poisoning because they consume improperly stored antifreeze or antifreeze that has leaked out of a radiator. The antifreeze has a somewhat sweet taste, which attracts a curious dog or cat. Young children are also at risk for ethylene glycol poisoning.

The first stage of ethylene glycol poisoning is a drunken state. Ethylene glycol is an alcohol, and it affects the brain of a dog or cat much as an alcoholic beverage would. Once ethylene glycol begins to metabolize, however, the second and more deadly stage begins. Ethylene glycol is metabolized in the liver into glycolic acid ($HC_2H_3O_3$), which enters the bloodstream. If the original quantities of consumed antifreeze are significant, the glycolic acid overwhelms the blood's natural buffering system, causing blood pH to drop to dangerously low levels. At this point, the cat or dog may begin hyperventilating in an effort to overcome the acidic blood's reduced ability to carry oxygen. If no treatment is administered, the animal will eventually go into a coma and die.

One treatment for ethylene glycol poisoning is the administration of ethyl alcohol (the alcohol found in alcoholic beverages). The liver enzyme that metabolizes ethylene glycol is the same one that metabolizes ethyl alcohol, but it has a higher affinity for ethyl alcohol than for ethylene glycol. Consequently, the enzyme preferentially metabolizes ethyl alcohol, allowing the unmetabolized ethylene glycol to escape through the urine. If administered early, this treatment can save the life of a dog or cat that has consumed ethylene glycol.

B11.2 CAN YOU ANSWER THIS? *One of the main buffering systems found in blood consists of carbonic acid (H_2CO_3) and bicarbonate ion (HCO_3^-). Write an equation showing how this buffering system could neutralize glycolic acid ($HC_2H_3O_3$) that might enter the blood from ethylene glycol poisoning. Suppose a cat has 0.15 mol of HCO_3^- and 0.15 mol of H_2CO_3 in its bloodstream. How many grams of $HC_2H_3O_3$ could be neutralized before the buffering system in the cat's blood is overwhelmed?*

MODULE IN REVIEW

Self-Assessment Quiz

Q1. Aluminum reacts with chlorine gas to form aluminium chloride.

$$2\,Al(s) + 3\,Cl_2(g) \rightarrow 2\,AlCl_3(s)$$

What minimum volume of chlorine gas (at 298 K and 225 mm Hg) is required to completely react with 7.85 g of aluminum?

(a) 36.0 L
(c) 0.0474 L
(b) 24.0 L
(d) 16.0 L

Q2. Potassium iodide reacts with lead(II) nitrate in the following precipitation reaction:

$$2\,KI(aq) + Pb(NO_3)_2(aq) \rightarrow 2\,KNO_3(aq) + PbI_2(s)$$

What minimum volume of 0.200 mol/L potassium iodide solution is required to completely precipitate all of the lead in 155.0 mL of a 0.112 mol/L lead(II) nitrate solution?

(a) 348 mL
(b) 86.8 mL
(c) 43.4 mL
(d) 174 mL

Q3. A 25.00 mL sample of an HNO_3 solution is titrated with 0.102 mol/L NaOH. The titration requires 28.52 mL to reach the equivalence point. What is the concentration of the HNO_3 solution?

(a) 0.116 mol/L
(b) 8.89 mol/L
(c) 0.0894 mol/L
(d) 0.102 mol/L

Q4. What is $[H_3O^+]$ in a solution with $[OH^-] = 2.5 \times 10^{-4}$ mol/L?

(a) 2.5×10^{10} mol/L
(b) 4.0×10^{-11} mol/L
(c) 1.0×10^{-7} mol/L
(d) 1.0×10^{-14} mol/L

Q5. What is the pH of a solution with $[H_3O^+] = 2.8 \times 10^{-5}$ mol/L?

(a) –4.55
(b) 4.55
(c) 10.48
(d) 1.00

Q6. What is $[OH^-]$ in a solution with a pH of 9.55?

(a) 2.82×10^{-10} mol/L
(b) 4.45 mol/L
(c) 2.82×10^{4} mol/L
(d) 3.55×10^{-5} mol/L

Q7. A buffer contains $HCHO_2(aq)$ and $KCHO_2(aq)$. Which statement correctly summarizes the action of this buffer?

(a) $HCHO_2(aq)$ neutralizes added acid, and $KCHO_2(aq)$ neutralizes added base.
(b) Both $HCHO_2(aq)$ and $KCHO_2(aq)$ neutralize added acid.
(c) Both $HCHO_2(aq)$ and $KCHO_2(aq)$ neutralize added base.
(d) $HCHO_2(aq)$ neutralizes added base, and $KCHO_2(aq)$ neutralizes added acid.

Answers: 1:a; 2:d; 3:a; 4:b; 5:b; 6:d; 7:d

Chemical Principles

Gases in Chemical Reactions: Stoichiometric calculations involving gases are similar to those that do not involve gases in that the coefficients in a balanced chemical equation provide ratios among moles of reactants and products in the reaction. For gases, the amount of a reactant or product is often specified by the volume of reactant or product at a given temperature and pressure. The ideal gas law is then used to convert from these quantities to moles of reactant or product.

Relevance

Gases in Chemical Reactions: Reactions involving gases are common in chemistry. For example, many atmospheric reactions—some of which are important to the environment—occur as gaseous reactions.

Acid–Base Titration: In an acid–base titration, we add an acid (or base) of known concentration to a base (or acid) of unknown concentration. We combine the two reactants until they are in exact stoichiometric proportions (moles of H^+ = moles of OH^-), which marks the equivalence point of the titration. In titration, since we know the moles of H^+ (or OH^-) that we added, we can determine the moles of OH^- (or H^+) in the unknown solution.

Acid–Base Titration: An acid–base titration is a laboratory procedure often used to determine the unknown concentration of an acid or a base.

Self-Ionization of Water: Water can act as both an acid and a base with itself.

$$\underset{\text{Acid}}{H_2O(l)} + \underset{\text{Base}}{H_2O(l)} \rightleftharpoons H_3O^+(aq) + OH^-(aq)$$

The product of $[H_3O^+]$ and $[OH^-]$ in aqueous solutions is always equal to the ion product constant, $K_w(1.0 \times 10^{-14})$.

$$[H_3O^+][OH^-] = K_w = 1.0 \times 10^{-14}$$

Self-Ionization of Water: The self-ionization of water occurs because aqueous solutions always contain some H_3O^+ and some OH^-. In a neutral solution, the concentrations of these are equal (1.0×10^{-7} mol/L). When an acid is added to water, $[H_3O^+]$ increases and $[OH^-]$ decreases. When a base is added to water, the opposite happens. The ion product constant, however, still equals 1.0×10^{-14}, allowing us to calculate $[H_3O^+]$ given $[OH^-]$ and vice versa.

pH and pOH Scales:

$$pH = -\log[H_3O^+]$$
$$pH > 7 \text{ (basic)}$$
$$pH < 7 \text{ (acidic)}$$
$$pH = 7 \text{ (neutral)}$$
$$pOH = -\log[OH^-]$$
$$pH + pOH = pKw = 14$$

pH and pOH Scales: pH is a convenient way to specify acidity or basicity. Since the pH scale is logarithmic, a change of one on the pH scale corresponds to a tenfold change in the $[H_3O^+]$. The pOH scale, defined with respect to $[OH^-]$ instead of $[H_3O^+]$, is less commonly used.

Buffers: Buffers are solutions containing significant amounts of both a weak acid and its conjugate base. Buffers resist pH change by neutralizing added acid or base.

Buffers: Buffers are important in blood chemistry because blood must stay within a narrow pH range in order to carry oxygen.

Chemical Skills

LO: Use volume and concentration to calculate the number of moles of reactants or products and then use stoichiometric coefficients to calculate other quantities in a reaction (Section 11.2).

Examples

EXAMPLE 11.8 SOLUTION STOICHIOMETRY

Consider the reaction:

$$HCl(aq) + NaOH(aq) \rightarrow NaCl(aq) + H_2O(l)$$

How much 0.113 mol/L NaOH solution will completely neutralize 1.25 L of 0.228 mol/L HCl solution?

SORT

You are given the volume and concentration of a hydrochloric acid solution as well as the concentration of a sodium hydroxide solution with which it reacts. You are asked to find the volume of the sodium hydroxide solution that will completely react with the hydrochloric acid.

GIVEN: 1.25 L HCl solution

0.228 mol/L HCl

0.113 mol/L NaOH

FIND: V NaOH solution

STRATEGIZE
Draw a solution map. Use the volume and concentration of HCl to get to mol HCl. Then use the stoichiometric coefficients to calculate mol NaOH. Finally, calculate volume of NaOH using the concentration of NaOH.

SOLUTION MAP

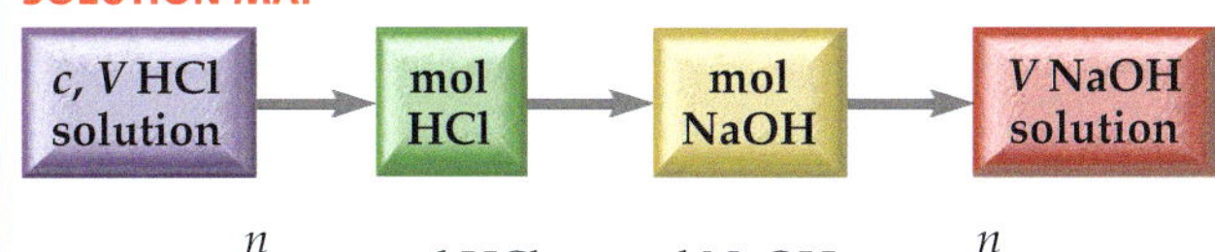

$c = \frac{n}{V}$ n mol HCl : n mol NaOH $c = \frac{n}{V}$

RELATIONSHIPS USED

1 mol HCl : 1 mol NaOH (from balanced equation)

SOLVE
Follow the solution map to calculate the answer.

SOLUTION

moles HCl

$$c = \frac{n}{V}$$

$$n(\text{HCl}) = c \times V = 0.228\,\text{mol/L HCl} \times 1.25\,\text{L HCl} = 0.285\text{ mol HCl}$$

moles NaOH

$$n(\text{NaOH}) = \frac{1\,\text{mol NaOH}}{1\,\text{mol HCl}} \times 0.285\,\text{mol HCl} = 0.285\,\text{mol NaOH}$$

Volume NaOH solution

$$c = \frac{n}{V}$$

$$V(\text{NaOH}) = \frac{n}{c} = \frac{0.285\,\text{mol NaOH}}{0.113\,\text{mol/L NaOH}} = 2.52\,\text{L NaOH}$$

CHECK
Check your answer. Are the units correct? Does the answer make physical sense?

The units (L NaOH solution) are correct. The magnitude of the answer makes sense because the sodium hydroxide solution is about half as concentrated as the hydrochloric acid. Since the reaction stoichiometry is 1:1, the volume of the sodium hydroxide solution should be about twice the volume of the hydrochloric acid.

LO: Use acid–base titration to determine the concentration of an unknown solution (Section 11.3).

EXAMPLE 11.9 ACID–BASE TITRATIONS

A 15.00 mL sample of a NaOH solution of unknown concentration requires 17.88 mL of a 0.1053 mol/L H_2SO_4 solution to reach the equivalence point in a titration. What is the concentration of the NaOH solution?

SORT
You are given the volume of a sodium hydroxide solution and the volume and concentration of the sulfuric acid solution required for its titration. You are asked to find the concentration of the sodium hydroxide solution.

GIVEN: 15.00 mL NaOH
17.88 mL of a 0.1053 mol/L H_2SO_4 solution

FIND: concentration of NaOH solution (mol/L)

STRATEGIZE
Begin by writing the balanced equation for the neutralization reaction.

Next draw a solution map. Use the volume and concentration of the known reactant to determine moles of the known reactant. (You have to convert from millilitres to litres first.) Then use the stoichiometric ratio from the balanced equation to get moles of the unknown reactant.

Then add a second part to the solution map indicating how to use moles and volume to determine concentration.

SOLUTION MAP

$$H_2SO_4(aq) + 2NaOH(aq) \rightarrow 2\,H_2O(l) + Na_2SO_4(aq)$$

1 L = 1000 mL (1000 mL/L) $c = \frac{n}{V}$ n mol H_2SO_4 : n mol NaOH

$c = \frac{n}{V}$

RELATIONSHIPS USED

2 mol NaOH : 1 mol H_2SO_4 (from balanced equation)

SOLVE

Follow the solution map to solve the problem. The first part of the solution gives you moles of the unknown reactant. In the second part of the solution, divide the moles from the first part by the volume to obtain concentration.

SOLUTION

Volume H_2SO_4 (L)

$$V(H_2SO_4) = \frac{17.88\ \text{mL}}{1000\ \text{mL/L}} = 0.01788\ \text{L}$$

moles H_2SO_4

$$c = \frac{n}{V}$$

$$n(H_2SO_4) = c \times V$$
$$= 0.1053\ \text{mol/L}\ H_2SO_4 \times 0.01788\ \text{L}\ H_2SO_4$$
$$= 0.001883\ \text{mol}\ H_2SO_4$$

moles NaOH

$$n(\text{NaOH}) = \frac{2\ \text{mol NaOH}}{1\ \text{mol}\ H_2SO_4} \times 0.001883\ \text{mol}\ H_2SO_4$$
$$= 0.00377\ \text{mol NaOH}$$

Volume NaOH

$$V(\text{NaOH}) = \frac{15.00\ \text{mL}}{1000\ \text{mL/L}} = 0.01500\ \text{L}$$

Concentration NaOH solution

$$c(\text{NaOH}) = \frac{n}{V} = \frac{0.00377\ \text{mol NaOH}}{0.01500\ \text{L NaOH}}$$
$$= 0.2510\ \text{mol/L NaOH}$$

The unknown NaOH solution has a concentration of 0.2510 mol/L.

CHECK

Check your answer. Are the units correct? Does the answer make physical sense?

The units (mol/L) are correct. The magnitude of the answer makes sense because the reaction has a two-to-one stoichiometry and the volumes of the two solutions are similar; therefore, the concentration of the NaOH solution must be approximately twice the concentration of the H_2SO_4 solution.

LO: Calculate [H_3O^+] or [OH^-] from K_W (Section 11.4).

EXAMPLE 11.10 FINDING [H_3O^+] OR [OH^-] FROM K_W

Calculate [OH^-] in a solution with [H_3O^+] = 1.5×10^{-4} mol/L.

SOLUTION

To find [H_3O^+] or [OH^-], use the ion product constant expression.

$$[H_3O^+][OH^-] = 1.0 \times 10^{-14}$$

Substitute the known quantity into the equation ([H_3O^+] or [OH^-]) and solve for the unknown quantity.

$$[H_3O^+][OH^-] = 1.0 \times 10^{-14}$$

$$[1.5 \times 10^{-4}][OH^-] = 1.0 \times 10^{-14}$$

$$[OH^-] = \frac{1.0 \times 10^{-14}}{1.5 \times 10^{-4}} = 6.7 \times 10^{-11}\ \text{mol/L}$$

LO: Calculate pH from $[H_3O^+]$ (Section 11.5).

To calculate the pH of a solution from $[H_3O^+]$, take the negative log of $[H_3O^+]$.

$$pH = -\log[H_3O^+]$$

EXAMPLE 11.11 CALCULATING pH FROM $[H_3O^+]$

Calculate the pH of a solution with $[H_3O^+] = 2.4 \times 10^{-5}$ mol/L.

SOLUTION

$$\begin{aligned} pH &= -\log[H_3O^+] \\ &= -\log(2.4 \times 10^{-5}) \\ &= -(-4.62) \\ &= 4.62 \end{aligned}$$

LO: Calculate $[H_3O^+]$ from pH (Section 11.5).

Calculate $[H_3O^+]$ from pH by taking the inverse log of the negative of the pH value (Method 1):

$$[H_3O^+] = \text{invlog}(-pH)$$

You can also calculate $[H_3O^+]$ from pH by raising 10 to the negative of the pH (Method 2):

$$[H_3O^+] = 10^{-pH}$$

EXAMPLE 11.12 CALCULATING $[H_3O^+]$ FROM pH

Calculate $[H_3O^+]$ for a solution with a pH of 6.22.

SOLUTION

Method 1: Inverse Log Function

$$\begin{aligned} [H_3O^+] &= \text{invlog}(-pH) \\ &= \text{invlog}(-6.22) \\ &= 6.0 \times 10^{-7} \text{ mol/L} \end{aligned}$$

Method 2: 10^x Function

$$\begin{aligned} [H_3O^+] &= 10^{-pH} \\ &= 10^{-6.22} \\ &= 6.0 \times 10^{-7} \text{ mol/L} \end{aligned}$$

KEY TERMS

acidic solution **[11.4]**
basic solution **[11.4]**
equivalence point **[11.3]**
indicator **[11.3]**
ion product constant for water (K_w) **[11.5]**
logarithmic scale **[11.5]**
neutral solution **[11.4]**
pH **[11.5]**
pOH **[11.5]**
titration **[11.3]**

EXERCISES

QUESTIONS

1. What is a titration? What is the equivalence point?

2. If a solution contains 0.85 mol of OH^-, how many moles of H^+ are required to reach the equivalence point in a titration?

3. Does pure water contain any H_3O^+ ions? Explain your answer.

4. What happens to $[OH^-]$ in an aqueous solution when $[H_3O^+]$ increases?

5. Give a possible value of $[OH^-]$ and $[H_3O^+]$ in a solution at 25 °C that is:
(a) acidic **(b)** basic **(c)** neutral

6. How is pH defined? A change of 1.0 pH unit corresponds to how much of a change in $[H_3O^+]$?

7. How is pOH defined? A change of 2.0 pOH units corresponds to how much of a change in $[OH^-]$?

8. In any aqueous solution at 25 °C, the sum of pH and pOH is 14.00. Explain why this is so.

9. What is a buffer?

10. What are the main components in a buffer?

PROBLEMS

GASES IN CHEMICAL REACTIONS

11. Consider the chemical reaction:

$$C(s) + H_2O(g) \rightarrow CO(g) + H_2(g)$$

How many litres of hydrogen gas are formed from the complete reaction of 1.07 mol of C? Assume that the hydrogen gas is collected at 1.0 atm and 315 K.

12. Consider the chemical reaction:

$$2\,H_2O(l) \rightarrow 2\,H_2(g) + O_2(g)$$

How many moles of H_2O are required to form 1.3 L of O_2 at 325 K and 0.988 atm?

13. CH_3OH can be synthesized by the reaction:

$$CO(g) + 2\ H_2(g) \rightarrow CH_3OH(g)$$

How many litres of H_2 gas, measured at 748 mm Hg and 86 °C, are required to synthesize 0.55 mol of CH_3OH? How many litres of CO gas, measured under the same conditions, are required?

14. Oxygen gas reacts with powdered aluminum according to the reaction:

$$4\ Al(s) + 3\ O_2(g) \rightarrow 2\ Al_2O_3(s)$$

How many litres of O_2 gas, measured at 782 mm Hg and 25 °C, are required to completely react with 2.4 mol of Al?

15. Nitrogen reacts with powdered aluminum according to the reaction:

$$2\ Al(s) + N_2(g) \rightarrow 2\ AlN(s)$$

How many litres of N_2 gas, measured at 1.17 atm and 95 °C, are required to completely react with 18.5 g of Al?

16. Sodium reacts with chlorine gas according to the reaction:

$$2\ Na(s) + Cl_2(g) \rightarrow 2\ NaCl(s)$$

What volume of Cl_2 gas, measured at 0.904 atm and 35 °C, is required to form 28 g of NaCl?

17. How many grams of NH_3 form when 24.8 L of $H_2(g)$ (measured at STP) reacts with N_2 to form NH_3 according to this reaction?

$$N_2(g) + 3\ H_2(g) \rightarrow 2\ NH_3(g)$$

18. Lithium reacts with nitrogen gas according to the reaction:

$$6\ Li(s) + N_2(g) \rightarrow 2\ Li_3N(s)$$

How many grams of lithium are required to completely react with 58.5 mL of N_2 gas measured at STP?

19. How many grams of calcium are consumed when 156.8 mL of oxygen gas, measured at STP, reacts with calcium according to this reaction?

$$2\ Ca(s) + O_2(g) \rightarrow 2\ CaO(s)$$

20. How many grams of magnesium oxide form when 14.8 L of oxygen gas, measured at STP, completely reacts with magnesium metal according to this reaction?

$$2\ Mg(s) + O_2(g) \rightarrow 2\ MgO(s)$$

SOLUTION STOICHIOMETRY

21. Determine the volume of 0.150 mol/L NaOH solution required to neutralize each sample of hydrochloric acid. The neutralization reaction is:

$$NaOH(aq) + HCl(aq) \rightarrow H_2O(l) + NaCl(aq)$$

(a) 25 mL of a 0.150 mol/L HCl solution
(b) 55 mL of a 0.055 mol/L HCl solution
(c) 175 mL of a 0.885 mol/L HCl solution

22. Determine the volume of 0.225 mol/L KOH solution required to neutralize each sample of sulfuric acid. The neutralization reaction is:

$$H_2SO_4(aq) + 2\ KOH(aq) \rightarrow K_2SO_4(aq) + 2\ H_2O(l)$$

(a) 45 mL of 0.225 mol/L H_2SO_4
(b) 185 mL of 0.125 mol/L H_2SO_4
(c) 75 mL of 0.100 mol/L H_2SO_4

23. Consider the reaction:

$$2\ K_3PO_4(aq) + 3\ NiCl_2(aq) \rightarrow Ni_3(PO_4)_2(s) + 6\ KCl(aq)$$

What volume of 0.225 mol/L K_3PO_4 solution is necessary to completely react with 134 mL of 0.0112 mol/L $NiCl_2$?

24. Consider the reaction:

$$K_2S(aq) + Co(NO_3)_2(aq) \rightarrow 2\ KNO_3(aq) + CoS(s)$$

What volume of 0.225 mol/L K_2S solution is required to completely react with 175 mL of 0.115 mol/L $Co(NO_3)_2$?

25. A 10.0 mL sample of an unknown H_3PO_4 solution requires 112 mL of 0.100 mol/L KOH to completely react with the H_3PO_4. What was the concentration of the unknown H_3PO_4 solution?

$$H_3PO_4(aq) + 3\ KOH(aq) \rightarrow 3\ H_2O(l) + K_3PO_4(aq)$$

26. A 25.0 mL sample of an unknown $HClO_4$ solution requires 45.3 mL of 0.101 mol/L NaOH for complete neutralization. What was the concentration of the unknown $HClO_4$ solution? The neutralization reaction is:

$$HClO_4(aq) + NaOH(aq) \rightarrow H_2O(l) + NaClO_4(aq)$$

27. What is the minimum amount of 6.0 mol/L H_2SO_4 necessary to produce 15.0 g of $H_2(g)$ according to the reaction:

$$2\ Al(s) + 3\ H_2SO_4(aq) \rightarrow Al_2(SO_4)_3(aq) + 3\ H_2(g)$$

28. What is the concentration of $ZnCl_2(aq)$ that forms when 15.0 g of zinc completely reacts with $CuCl_2(aq)$ according to the following reaction? (Assume a final volume of 175 mL.)

$$Zn(s) + CuCl_2(aq) \rightarrow ZnCl_2 + Cu(s)$$

ACID–BASE TITRATIONS

29. Four solutions of unknown HCl concentration are titrated with solutions of NaOH. The following table lists the volume of each unknown HCl solution, the volume of NaOH solution required to reach the equivalence point, and the concentration of each NaOH solution. Calculate the concentration (in mol/L) of the unknown HCl solution in each case.

HCl Volume (mL)	NaOH Volume (mL)	[NaOH] (mol/L)
(a) 25.00 mL	28.44 mL	0.1231 mol/L
(b) 15.00 mL	21.22 mL	0.0972 mol/L
(c) 20.00 mL	14.88 mL	0.1178 mol/L
(d) 5.00 mL	6.88 mL	0.1325 mol/L

30. Four solutions of unknown NaOH concentration are titrated with solutions of HCl. The following table lists the volume of each unknown NaOH solution, the volume of HCl solution required to reach the equivalence point, and the concentration of each HCl solution. Calculate the concentration (in mol/L) of the unknown NaOH solution in each case.

NaOH Volume (mL)	HCl Volume (mL)	[HCl] (mol/L)
(a) 5.00 mL	9.77 mL	0.1599 mol/L
(b) 15.00 mL	11.34 mL	0.1311 mol/L
(c) 10.00 mL	10.55 mL	0.0889 mol/L
(d) 30.00 mL	36.18 mL	0.1021 mol/L

31. A 25.00 mL sample of an H_2SO_4 solution of unknown concentration is titrated with a 0.1322 mol/L KOH solution. A volume of 41.22 mL of KOH is required to reach the equivalence point. What is the concentration of the unknown H_2SO_4 solution?

32. A 5.00 mL sample of an H_3PO_4 solution of unknown concentration is titrated with a 0.1090 mol/L NaOH solution. A volume of 7.12 mL of the NaOH solution is required to reach the equivalence point. What is the concentration of the unknown H_3PO_4 solution?

33. What volume in millilitres of a 0.121 mol/L sodium hydroxide solution is required to reach the equivalence point in the complete titration of a 10.0 mL sample of 0.102 mol/L sulfuric acid?

34. What volume in millilitres of 0.0985 mol/L sodium hydroxide solution is required to reach the equivalence point in the complete titration of a 15.0 mL sample of 0.124 mol/L phosphoric acid?

ACIDITY, BASICITY, AND K_W

35. Determine if each solution is acidic, basic, or neutral.
(a) $[H_3O^+] = 1 \times 10^{-5}$ mol/L; $[OH^-] = 1 \times 10^{-9}$ mol/L
(b) $[H_3O^+] = 1 \times 10^{-6}$ mol/L; $[OH^-] = 1 \times 10^{-8}$ mol/L
(c) $[H_3O^+] = 1 \times 10^{-7}$ mol/L; $[OH^-] = 1 \times 10^{-7}$ mol/L
(d) $[H_3O^+] = 1 \times 10^{-8}$ mol/L; $[OH^-] = 1 \times 10^{-6}$ mol/L

36. Determine if each solution is acidic, basic, or neutral.
(a) $[H_3O^+] = 1 \times 10^{-9}$ mol/L; $[OH^-] = 1 \times 10^{-5}$ mol/L
(b) $[H_3O^+] = 1 \times 10^{-10}$ mol/L; $[OH^-] = 1 \times 10^{-4}$ mol/L
(c) $[H_3O^+] = 1 \times 10^{-2}$ mol/L; $[OH^-] = 1 \times 10^{-12}$ mol/L
(d) $[H_3O^+] = 1 \times 10^{-13}$ mol/L; $[OH^-] = 1 \times 10^{-1}$ mol/L

37. Calculate $[OH^-]$ given $[H_3O^+]$ in each aqueous solution and classify the solution as acidic or basic.
(a) $[H_3O^+] = 1.5 \times 10^{-9}$ mol/L
(b) $[H_3O^+] = 9.3 \times 10^{-9}$ mol/L
(c) $[H_3O^+] = 2.2 \times 10^{-6}$ mol/L
(d) $[H_3O^+] = 7.4 \times 10^{-4}$ mol/L

38. Calculate $[OH^-]$ given $[H_3O^+]$ in each aqueous solution and classify the solution as acidic or basic.
(a) $[H_3O^+] = 1.3 \times 10^{-3}$ mol/L
(b) $[H_3O^+] = 9.1 \times 10^{-12}$ mol/L
(c) $[H_3O^+] = 5.2 \times 10^{-4}$ mol/L
(d) $[H_3O^+] = 6.1 \times 10^{-9}$ mol/L

39. Calculate $[H_3O^+]$ given $[OH^-]$ in each aqueous solution and classify each solution as acidic or basic.
(a) $[OH^-] = 2.7 \times 10^{-12}$ mol/L
(b) $[OH^-] = 2.5 \times 10^{-2}$ mol/L
(c) $[OH^-] = 1.1 \times 10^{-10}$ mol/L
(d) $[OH^-] = 3.3 \times 10^{-4}$ mol/L

40. Calculate $[H_3O^+]$ given $[OH^-]$ in each aqueous solution and classify each solution as acidic or basic.
(a) $[OH^-] = 2.1 \times 10^{-11}$ mol/L
(b) $[OH^-] = 7.5 \times 10^{-9}$ mol/L
(c) $[OH^-] = 2.1 \times 10^{-4}$ mol/L
(d) $[OH^-] = 1.0 \times 10^{-2}$ mol/L

pH

41. Classify each solution as acidic, basic, or neutral according to its pH value.
(a) pH = 8.0
(b) pH = 7.0
(c) pH = 3.5
(d) pH = 6.1

42. Classify each solution as acidic, basic, or neutral according to its pH value.
(a) pH = 4.0
(b) pH = 3.5
(c) pH = 13.0
(d) pH = 0.85

43. Calculate the pH of each solution.
(a) $[H_3O^+] = 1.7 \times 10^{-8}$ mol/L
(b) $[H_3O^+] = 1.0 \times 10^{-7}$ mol/L
(c) $[H_3O^+] = 2.2 \times 10^{-6}$ mol/L
(d) $[H_3O^+] = 7.4 \times 10^{-4}$ mol/L

44. Calculate the pH of each solution.
(a) $[H_3O^+] = 2.4 \times 10^{-10}$ mol/L
(b) $[H_3O^+] = 7.6 \times 10^{-2}$ mol/L
(c) $[H_3O^+] = 9.2 \times 10^{-13}$ mol/L
(d) $[H_3O^+] = 3.4 \times 10^{-5}$ mol/L

45. Calculate $[H_3O^+]$ for each solution.
(a) pH = 8.55
(b) pH = 11.23
(c) pH = 2.87
(d) pH = 1.22

46. Calculate $[H_3O^+]$ for each solution.
(a) pH = 1.76
(b) pH = 3.88
(c) pH = 8.43
(d) pH = 12.32

47. Calculate the pH of each solution.
(a) $[OH^-] = 1.9 \times 10^{-7}$ mol/L
(b) $[OH^-] = 2.6 \times 10^{-8}$ mol/L
(c) $[OH^-] = 7.2 \times 10^{-11}$ mol/L
(d) $[OH^-] = 9.5 \times 10^{-2}$ mol/L

48. Calculate the pH of each solution.
(a) $[OH^-] = 2.8 \times 10^{-11}$ mol/L
(b) $[OH^-] = 9.6 \times 10^{-3}$ mol/L
(c) $[OH^-] = 3.8 \times 10^{-12}$ mol/L
(d) $[OH^-] = 6.4 \times 10^{-4}$ mol/L

49. Calculate $[OH^-]$ for each solution.
(a) pH = 4.25
(b) pH = 12.53
(c) pH = 1.50
(d) pH = 8.25

50. Calculate $[OH^-]$ for each solution.
(a) pH = 1.82
(b) pH = 13.28
(c) pH = 8.29
(d) pH = 2.32

51. Calculate the pH of each solution:
(a) 0.0155 mol/L HBr
(b) 1.28×10^{-3} mol/L KOH
(c) 1.89×10-$^{-3}$ mol/L HNO_3
(d) 1.54×10^{-4} mol/L $Sr(OH)_2$

52. Calculate the pH of each solution:
(a) 1.34×10^{-3} mol/L $HClO_4$
(b) 0.0211 mol/L NaOH
(c) 0.0109 mol/L HBr
(d) 7.02×10^{-5} mol/L $Ba(OH)_2$

pOH

53. Detemine the pOH of each solution and classify it as acidic, basic, or neutral.
(a) $[OH^-] = 1.5 \times 10^{-9}$ mol/L
(b) $[OH^-] = 7.0 \times 10^{-5}$ mol/L
(c) $[OH^-] = 1.0 \times 10^{-7}$ mol/L
(d) $[OH^-] = 8.8 \times 10^{-3}$ mol/L

54. Detemine the pOH of each solution and classify it as acidic, basic, or neutral.
(a) $[OH^-] = 4.5 \times 10^{-2}$ mol/L
(b) $[OH^-] = 3.1 \times 10^{-12}$ mol/L
(c) $[OH^-] = 5.4 \times 10^{-5}$ mol/L
(d) $[OH^-] = 1.2 \times 10^{-2}$ mol/L

55. Determine the pOH of each solution.
(a) $[H_3O^+] = 1.2 \times 10^{-8}$ mol/L
(b) $[H_3O^+] = 5.5 \times 10^{-2}$ mol/L
(c) $[H_3O^+] = 3.9 \times 10^{-9}$ mol/L
(d) $[H_3O^+] = 1.88 \times 10^{-13}$ mol/L

56. Determine the pOH of each solution.
(a) $[H_3O^+] = 8.3 \times 10^{-10}$ mol/L
(b) $[H_3O^+] = 1.6 \times 10^{-7}$ mol/L
(c) $[H_3O^+] = 7.3 \times 10^{-2}$ mol/L
(d) $[OH^-] = 4.32 \times 10^{-4}$ mol/L

57. Determine the pH of each solution and classify it as acidic, basic, or neutral.
(a) pOH = 8.5
(b) pOH = 4.2
(c) pOH = 1.7
(d) pOH = 7.0

58. Determine the pH of each solution and classify it as acidic, basic, or neutral.
(a) pOH = 12.5
(b) pOH = 5.5
(c) pOH = 0.55
(d) pOH = 7.98

BUFFERS

59. Determine whether or not each mixture is a buffer.
(a) HCl and HF
(b) NaOH and NH_3
(c) HF and NaF
(d) $HC_2H_3O_2$ and $KC_2H_3O_2$

60. Determine whether or not each mixture is a buffer.
(a) HBr and NaCl
(b) $HCHO_2$ and $NaCHO_2$
(c) HCl and HBr
(d) KOH and NH_3

61. Write reactions showing how each of the buffers in Problem 59 would neutralize added HCl.

62. Write reactions showing how each of the buffers in Problem 60 would neutralize added NaOH.

63. What substance could you add to each solution to make it a buffer solution?
(a) 0.100 mol/L $NaC_2H_3O_2$
(b) 0.500 mol/L H_3PO_4
(c) 0.200 mol/L $HCHO_2$

64. What substance could you add to each solution to make it a buffer solution?
(a) 0.050 mol/L $NaHSO_3$
(b) 0.150 mol/L HF
(c) 0.200 mol/L $KCHO_2$

CUMULATIVE PROBLEMS

65. When hydrochloric acid is poured over a sample of sodium hydrogencarbonate, 28.2 mL of carbon dioxide gas is produced at a pressure of 0.954 atm and a temperature of 22.7 °C. Write an equation for the gas evolution reaction and determine how much sodium hydrogencarbonate (in grams) reacted.

66. When hydrochloric acid is poured over potassium sulfide, 42.9 mL of hydrogen sulfide gas is produced at a pressure of 0.989 atm and a temperature of 25.8 °C. Write an equation for the gas evolution reaction and determine how much potassium sulfide (in grams) reacted.

67. Consider the reaction:

$$2\ SO_2(g) + O_2(g) \rightarrow 2\ SO_3(s)$$

(a) If 285.5 mL of SO_2 is allowed to react with 158.9 mL of O_2 (both measured at STP), what is the limiting reactant and the theoretical yield of SO_3?
(b) If 187.2 mL of SO_3 is collected (measured at STP), what is the percent yield for the reaction?

68. Consider the reaction:

$$P_4(s) + 6\ H_2(g) \rightarrow 4\ PH_3(g)$$

(a) If 88.6 L of $H_2(g)$, measured at STP, is allowed to react with 158.3 g of P_4, what is the limiting reactant?
(b) If 48.3 L of PH_3, measured at STP, forms, what is the percent yield?

69. Consider the reaction for the synthesis of nitric acid:

$$3\ NO_2(g) + H_2O(l) \rightarrow 2\ HNO_3(aq) + NO(g)$$

(a) If 12.8 L of $NO_2(g)$, measured at STP, is allowed to react with 14.9 g of water, find the limiting reagent and the theoretical yield of HNO_3 in grams.
(b) If 14.8 g of HNO_3 forms, what is the percent yield?

70. Consider the reaction for the production of NO_2 from NO:

$$2\ NO(g) + O_2(g) \rightarrow 2\ NO_2(g)$$

(a) If 84.8 L of $O_2(g)$, measured at 35 °C and 632 mm Hg, is allowed to react with 158.2 g of NO, find the limiting reagent.
(b) If 97.3 L of NO_2 forms, measured at 35 °C and 632 mm Hg, what is the percent yield?

71. Ammonium carbonate decomposes upon heating according to the balanced equation:

$$(NH_4)_2CO_3(s) \rightarrow 2\ NH_3(g) + CO_2(g) + H_2O(g)$$

Calculate the total volume of gas produced at 22 °C and 1.02 atm by the complete decomposition of 11.83 g of ammonium carbonate.

72. Ammonium nitrate decomposes explosively upon heating according to the balanced equation:

$$2\ NH_4NO_3(s) \rightarrow 2\ N_2(g) + O_2(g) + 4\ H_2O(g)$$

Calculate the total volume of gas (at 25 °C and 748 mm Hg) produced by the complete decomposition of 1.55 kg of ammonium nitrate.

73. Consider the reaction:

$$2\ Al(s) + 3\ H_2SO_4(aq) \rightarrow Al_2(SO_4)_3(aq) + 3\ H_2(g)$$

What minimum volume of 4.0 mol/L H_2SO_4 is required to produce 15.0 L of H_2 at STP?

74. Consider the reaction:

$$Mg(s) + 2\ HCl(aq) \rightarrow MgCl_2(aq) + H_2(g)$$

What minimum amount of 1.85 mol/L HCl is necessary to produce 28.5 L of H_2 at STP?

75. How much of a 1.25 mol/L sodium chloride solution in millilitres is required to completely precipitate all of the silver in 25.0 mL of a 0.45 mol/L silver nitrate solution?

76. How much of a 1.50 mol/L sodium sulfate solution in millilitres is required to completely precipitate all of the barium in 150.0 mL of a 0.250 mol/L barium nitrate solution?

77. For each $[H_3O^+]$, determine the pH and state whether the solution is acidic or basic.
(a) $[H_3O^+] = 0.0025$ mol/L
(b) $[H_3O^+] = 1.8 \times 10^{-12}$ mol/L
(c) $[H_3O^+] = 9.6 \times 10^{-9}$ mol/L
(d) $[H_3O^+] = 0.0195$ mol/L

78. For each $[OH^-]$, determine the pH and state whether the solution is acidic or basic.
(a) $[OH^-] = 1.8 \times 10^{-5}$ mol/L
(b) $[OH^-] = 8.9 \times 10^{-12}$ mol/L
(c) $[OH^-] = 3.1 \times 10^{-2}$ mol/L
(d) $[OH^-] = 1.96 \times 10^{-9}$ mol/L

79. Complete the table. (The first row is completed for you.)

$[H_3O^+]$	$[OH^-]$	pOH	pH	Acidic or Basic
1.0×10^{-4}	1.0×10^{-10}	10.00	4.00	acidic
5.5×10^{-3}	____	____	____	____
____	3.2×10^{-6}	____	____	____
4.8×10^{-9}	____	____	____	____
____	____	____	7.55	____

80. Complete the table. (The first row is completed for you.)

$[H_3O^+]$	$[OH^-]$	pOH	pH	Acidic or Basic
1.0×10^{-8}	1.0×10^{-6}	6.00	8.00	basic
____	____	____	3.55	____
1.7×10^{-9}	____	____	____	____
____	____	____	13.5	____
____	8.6×10^{-11}	____	____	____

81. For each strong acid solution, determine $[H_3O^+]$, $[OH^-]$, and pH.
(a) 0.0088 mol/L $HClO_4$
(b) 1.5×10^{-3} mol/L HBr
(c) 9.77×10^{-4} mol/L HI
(d) 0.0878 mol/L HNO_3

82. For each strong acid solution, determine $[H_3O^+]$, $[OH^-]$, and pH.
(a) 0.0150 mol/L HCl
(b) 1.9×10^{-4} mol/L HI
(c) 0.0226 mol/L HBr
(d) 1.7×10^{-3} mol/L HNO_3

83. For each strong base solution, determine $[OH^-]$, $[H_3O^+]$, pH, and pOH.
(a) 0.15 mol/L NaOH
(b) 1.5×10^{-3} mol/L $Ca(OH)_2$
(c) 4.8×10^{-4} mol/L $Sr(OH)_2$
(d) 8.7×10^{-5} mol/L KOH

84. For each strong base solution, determine $[OH^-]$, $[H_3O^+]$, pH, and pOH.
(a) 8.77×10^{-3} mol/L LiOH
(b) 0.0112 mol/L $Ba(OH)_2$
(c) 1.9×10^{-4} mol/L KOH
(d) 5.0×10^{-4} mol/L $Ca(OH)_2$

85. How many H^+ (or H_3O^+) ions are present in one drop (0.050 mL) of pure water at 25 °C?

86. Calculate the number of H^+ (or H_3O^+) ions and OH^- ions in 1.0 mL of 0.100 mol/L HCl.

HIGHLIGHT PROBLEMS

87. Automobile air bags inflate following a serious impact. The impact triggers the chemical reaction:

$$2\ NaN_3(s) \rightarrow 2\ Na(s) + 3\ N_2(g)$$

If an automobile air bag has a volume of 11.8 L, how much NaN_3 in grams is required to fully inflate the air bag upon impact? Assume STP conditions.

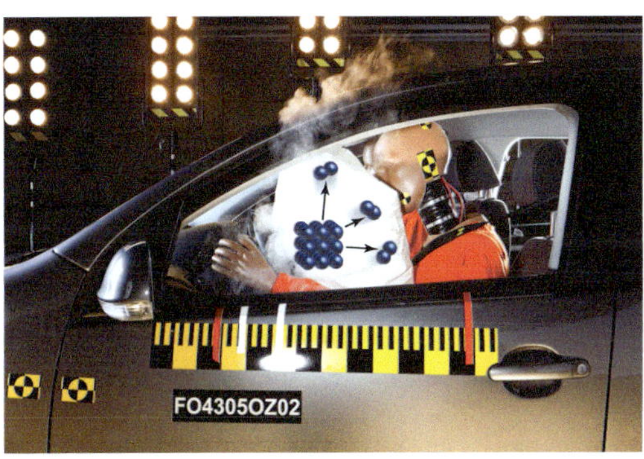

▸Answers to Skillbuilder Exercises

Skillbuilder 11.1 82.3 g
Skillbuilder 11.2 16.4 mL
Skillbuilder Plus, p. 490 0.308 mol/L
Skillbuilder 11.3 9.03×10^{-2} mol/L H_2SO_4
Skillbuilder 11.4
(a) $[H_3O^+] = 6.7 \times 10^{-13}$ mol/L; basic
(b) $[H_3O^+] = 1.0 \times 10^{-7}$ mol/L; neutral
(c) $[H_3O^+] = 1.2 \times 10^{-5}$ mol/L; acidic

Skillbuilder 11.5
(a) pH = 8.02; basic
(b) pH = 2.21; acidic
Skillbuilder Plus, p. 498 pH = 12.11; basic
Skillbuilder 11.6 4.3×10^{-9} mol/L
Skillbuilder Plus, p. 499 4.6×10^{-11} mol/L
Skillbuilder 11.7 5.6×10^{-5} mol/L
Skillbuilder Plus, p. 500 4.8×10^{-9} mol/L

▸Answers to Conceptual Checkpoints

11.1 (d) You need twice as many moles of B as you have moles of A. But since the solution of A is twice as concentrated, you need four times the volume of B as you have of A.

11.2 (c) The acid solution contains 7 H^+ ions; therefore, 7 OH^- ions are required to reach the equivalence point.

11.3 (d) Each of the others can accept a proton and thus acts as a base. NH_4^+, however, is the conjugate acid of NH_3 and therefore acts as an acid and not as a base.

11.4 (d) Because pH is the negative log of the H_3O^+ concentration, a higher pH corresponds to a lower $[H_3O^+]$, and each unit of pH represents a tenfold change in concentration.

11.5 (d) Since the pH is 5, the pOH = 14 – 5 = 9.

11.6 (b) A buffer solution consists of a weak acid and its conjugate base. Of the compounds listed, HF is the only weak acid, and F^- (from NaF in solution) is its conjugate base.

▲ The vibrant colours, flavours, distinct fragrances and waxy coating of various fruits are all due to organic compounds

Organic Compounds 12

Kate Rowen

MODULE OUTLINE

12.1 Organic Compounds and Carbon Atoms

LO: Define organic compounds and organic chemistry.

LO: Describe the bonding combinations possible for carbon atoms.

This module covers fundamentals of the diverse and fascinating world of organic compounds. Knowledge from previous modules is applied and you might find that you need to review the topics listed in Table 12.1 as you progress though Module 12. All of the topics listed relate to molecular structure and this is critically important because understanding organic compounds stems from an understanding of their molecular structure.

Table 12.1 Topics from Previous Modules that are Applied in Module 12

Module 3 Topics	Module 6 Topics
Covalent bonding	Intermolecular forces
Lewis structures	• Dispersion
Electronegativity	• Dipole-dipole
Polar bonds	• Hydrogen bonding
Valence shell electron pair repulsion (VSEPR) theory	
Molecular shapes	
Polar and non-polar molecules	

Many substances encountered in everyday life are made up of organic compounds. The protein and carbohydrate in the food we eat and the materials in our clothing are organic compounds. The caffeine present in many beverages we enjoy is an organic compound. Any over the counter drug you might take for a headache or other pain is an organic compound. There are organic compounds in sunscreen that absorb UV light and protect our skin. Indeed, the majority of known compounds are classified as organic. Despite this diversity of organic compounds, the definition of them is quite simple. **Organic compounds** are compounds that are based on a framework of carbon atoms. The nature of the carbon atom frameworks will become clear as you work through this module. **Organic chemistry** is the study of organic compounds, their structure, properties and reactions.

Despite their diversity, only a fraction of the known elements occur in organic compounds. The periodic table in ▼ Figure 12.1 shows the elements most commonly observed in organic compounds. In this module we only focus on organic compounds that contain the elements highlighted in Figure 12.1: **carbon, hydrogen, oxygen, nitrogen and halogens**. You will need to remember the number of valence electrons and hence the number of covalent bonds that atoms of these elements tend to form.

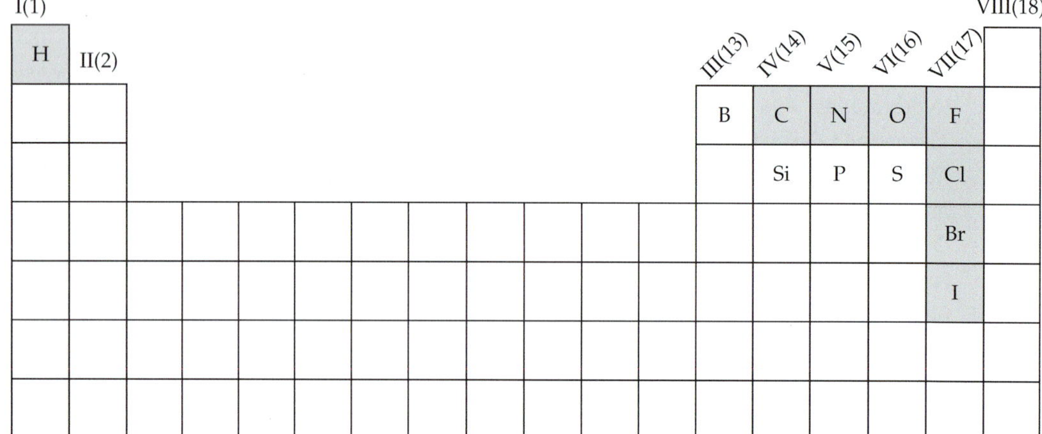

▶ **FIGURE 12.1 Periodic table outline** This shows elements most commonly observed in organic compounds.

Since organic compounds are based on a framework of carbon atoms, we'll spend a bit of time here considering their covalent bonding. Carbon is in group IV (14) of the periodic table and thus has four valence electrons. To completely fill its valence shell, carbon atoms tend to share these four valence electrons and form four covalent bonds in stable compounds. *Carbon atoms never form more than four covalent bonds.* Sometimes carbon atoms form fewer than four bonds, but these are present in highly reactive chemical species that form during reactions, and only exist as stable compounds in very rare circumstances. All the organic compounds that you encounter here will contain carbon atoms forming four covalent bonds.

The simplest carbon compound CH_4 is shown in ▼ Figure 12.2 using three types of representation. You will become accustomed to seeing organic compounds represented using all of these, and other methods will be introduced later in the module. Figure 12.2(a) is the Lewis structure for CH_4, showing the central carbon atom forming four bonds to four hydrogen atoms. Note how two of the bonds are shown using different lines, this is how a better impression of molecular shape can be included in a Lewis structure. The dashed line represents a bond projecting away from you, the bold wedge line represents a bond projecting towards you, and the lines drawn in the standard way represent bonds in the same plane as the surface they are drawn on. When a carbon atom forms four bonds to four hydrogen

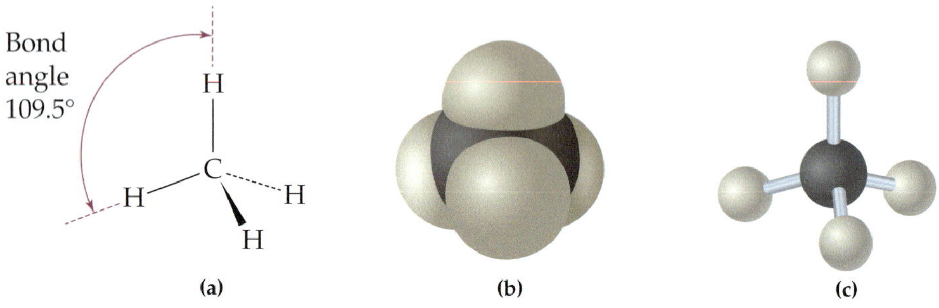

▶ **FIGURE 12.2 The simplest carbon compound CH_4** The simplest carbon compound CH_4, shown using 3 types of representation. **(a)** Lewis structure. **(b)** Space filled model. **(c)** Ball and stick model.

atoms, the electron groups in the bonds arrange themselves in a tetrahedral manner. This is because the electrons in the bonds repel each other and the tetrahedral arrangement maximises the space between them. This is the concept of valence shell electron pair repulsion (VSEPR) introduced in Module 3. Using the dashed and bold wedge lines in the Lewis structure portrays the tetrahedral geometry and 109.5° bond angles. Figure 12.2(b) is referred to as a 'spaced filled' model. These representations give an impression of the overall volume of a molecule while still showing that the carbon atom is central and forming four bonds to four hydrogen atoms. The 'ball and stick' model of CH_4 shown in Figure 12.2(c) makes use of different coloured balls to represent atoms and the sticks between them represent covalent bonds. The space filled and ball and stick models also clearly show the tetrahedral geometry of the bonds around the central carbon atom.

While carbon atoms always form four covalent bonds, there are a number of different combinations possible due to the ability of carbon atoms to form multiple (double and triple) bonds. The possibilities for carbon atom covalent bonding are shown in Table 12.2. In discussing the simplest carbon compound CH_4 above, the formation of four single bonds to a carbon atom and tetrahedral geometry of the electron groups around the central carbon atom has already been highlighted. A carbon atom may also form one double bond and two single bonds and have trigonal planar electron group geometry. If a carbon atom forms a triple bond, the fourth covalent bond must be single and the electron group geometry is linear. The same electron group geometry results when a carbon atom forms two double bonds. Note that all the possibilities shown in Table 12.2 have a total of four bonds to carbon atoms no matter what the combination of single and multiple bonds. The majority of organic compounds discussed in this module have carbon atoms forming four single bonds, or carbon atoms forming one double and two single bonds.

TABLE 12.2 Bonding Combinations for Carbon Atoms and their Associated Geometries

Bonding combination	four single bonds	one double and two single bonds	one triple bond and one single bond	two double bonds
Representation	—C— (with bonds above and below)	\C= /	≡C—	=C=
Electron group geometry	tetrahedral	trigonal planar	linear	linear

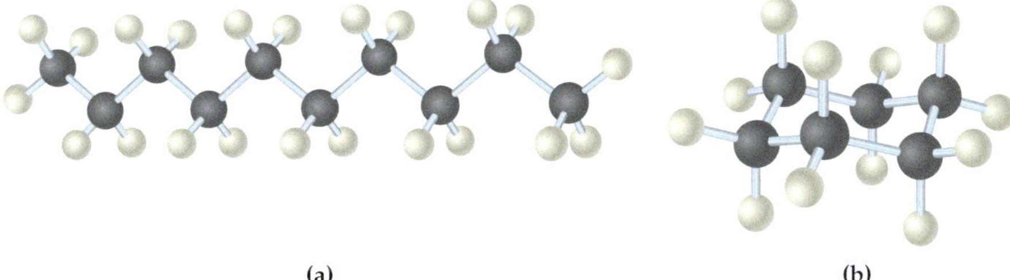

▶ FIGURE 12.3 A distinctive property of carbon atoms is the ability to form (a) chains and (b) rings The carbon atoms have tetrahedral electron group geometry whenever they form 4 single bonds.

A distinctive property of carbon atoms is the ability to form bonds to other carbon atoms. The technical name for the formation of covalent bonds between atoms of the same element to form chains and rings is **catenation**, and it is most common to carbon atoms. The ball and stick models in ◀ Figure 12.3 show examples of compounds where carbon atoms are bonded to form a chain, as in Figure 12.3(a), and a ring, as in Figure 12.3(b). Note that the tetrahedral arrangement of the single bond electron groups for the compounds shown in Figure 12.3 still applies when carbon atoms form chains and rings. This results in the zigzag arrangement of carbon atoms within organic compounds that have chains and rings of carbon atoms. Remember also that single bonds can rotate and the atoms can move in space to allow molecules to adopt different shapes. Rotation about carbon-carbon multiple bonds is not possible, and the important consequences of this will be discussed later in the module. Carbon atoms can also form covalent bonds with atoms of other elements including hydrogen, oxygen, nitrogen and the halogens. The formation of chains and rings of carbon atoms and the ability to form bonds to atoms of other elements are among the factors that result in the great diversity of organic compounds.

12.2 Representing Organic Compounds

LO: Use structural, condensed and line bond formulas to represent the molecular structures of organic compounds.

Before progressing to discover more about organic compounds, it is important to introduce the different ways their structures are represented. The molecular formula for an organic compound tells us how many and what type of atoms are present, but does not describe the actual structure of the molecule. Different organic compounds can have the same molecular formula, but they are different compounds because the atoms are connected in a different arrangement. To describe organic compounds properly, it's the molecular structure that must be shown in representations and structural formulas are used.

In a **structural formula** the atoms are represented by their elemental symbols and bonds are represented by lines. Structural formulas are analogous to Lewis structures that you are already familiar with. The examples provided show how all the atoms and bonds present in a compound are shown in a structural formula (Table 12.3). You can check for yourself that the structural formula for the straight chain form of C_6H_{14} has symbols for all six carbon atoms and all fourteen hydrogen atoms. See that the symbols for all the atoms in the branched chain form of C_7H_{16} are also shown in its structural formula.

The structures of organic compounds can be quite intricate and drawing Lewis structures to represent them can be cumbersome, taking up a lot of space on a page. A number of different ways to abbreviate, while still accurately representing a molecular structure, have developed. A **condensed formula** abbreviates the structure by omitting some or all of the bonds and indicating the number of atoms bonded to each carbon atom with subscripts. There can be several correct condensed formulas for a compound and three possibilities for the straight chain form of C_6H_{14} are shown (Table 12.3). The first one omits carbon-hydrogen bonds (C–H), but still shows carbon-carbon bonds (C–C). In the second condensed formula, the C–C bonds are not drawn, but are implied by drawing the groups of atoms next to each other. In the third condensed formula for the straight chain form of C_6H_{14} shown, the identical CH_2 groups are bracketed and a subscript is used to show that there are four of these groups linked in a chain to the CH_3 groups on each end.

TABLE 12.3 Methods of Representing Organic Compounds

Representation type	Straight chain form of C_6H_{14}	Branched chain form of C_7H_{16}
Structural formula (Lewis structure) • Atoms represented by elemental symbols • Bonds represented by lines	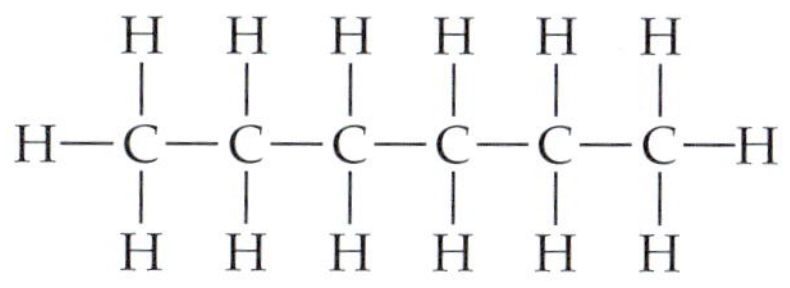 H H H H H H H—C—C—C—C—C—C—H H H H H H H	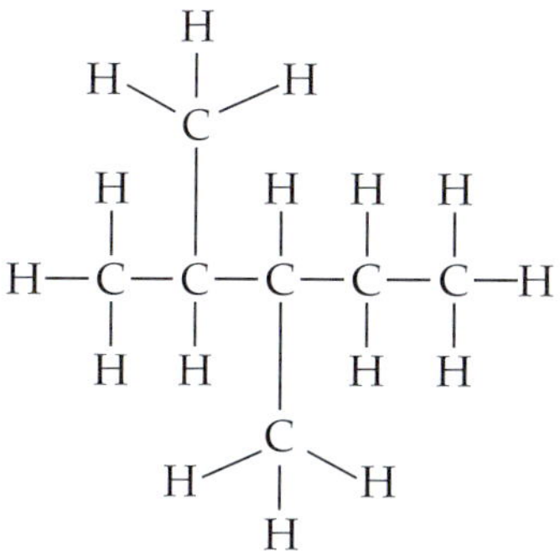 H H C H H H H H H—C—C—C—C—C—H H H H H C H H H
Condensed formula • Abbreviates the structural representation by omitting some or all of the bonds • The number of hydrogen atoms bonded to each carbon atom is indicated by subscripts • There can be several correct condensed formulas for a compound • Note how branching groups can be represented using brackets	$CH_3–CH_2–CH_2–CH_2–CH_2–CH_3$ $CH_3CH_2CH_2CH_2CH_2CH_3$ $CH_3(CH_2)_4CH_3$	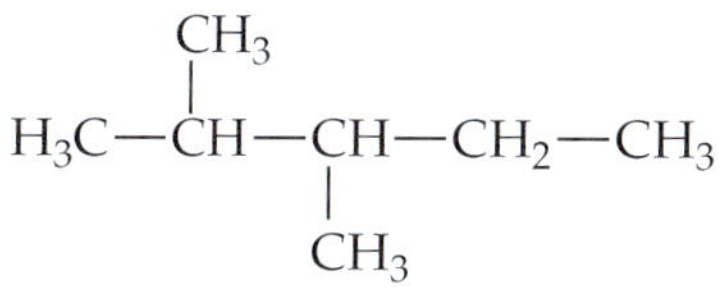 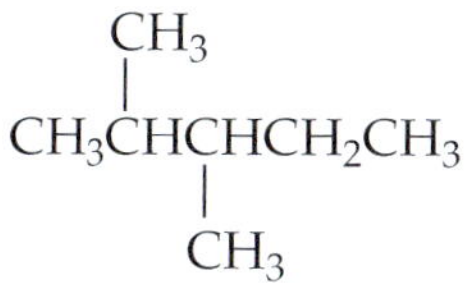 CH_3 $H_3C—CH—CH—CH_2—CH_3$ CH_3 CH_3 $CH_3CHCHCH_2CH_3$ CH_3 $CH_3CH(CH_3)CH(CH_3)CH_2CH_3$
Line bond formula • Symbols for carbon atoms are omitted • Each line represents a bond • Each bend or junction between bonds represents a carbon atom • Bonds not shown are bonds to hydrogen atoms		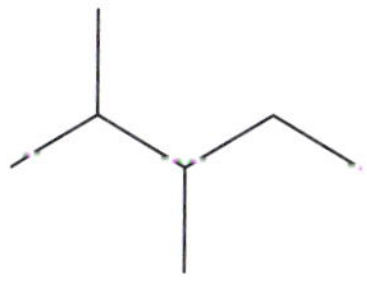

Branching molecules can also be represented using condensed formulas. Compare the structural formula to the condensed formulas shown for the branched chain form of C_7H_{16} given in Table 12.3 and see that they represent the same structure. All representations show a chain of five carbon atoms with single carbon atoms branching from the chain in specific positions. The second condensed formula shown in this group of examples omits the C-C bonds from the main chain, but shows the ones between the main chain carbon atoms and the branching carbon atoms. A condensed formula for a branching molecular structure can even be shown on one line using brackets to indicate the branching groups. Compare the final example of a condensed formula for the branched chain form of C_7H_{16} to the other representations above it and see how the branching CH_3 groups are shown in brackets.

Line bond formulas are the most convenient and simplest way to represent the structures of organic compounds. The key features of line bond formulas are highlighted in Table 12.4 with examples. Example (a) looks like a hexagon, but it represents an organic compound that has six carbon atoms connected in a ring. See how there are no symbols for carbon atoms (C) shown, each point in the hexagonal shape represents a carbon atom. Examine each carbon atom and see how only two bonds to each one are shown in the structure. Carbon atoms always form four bonds in stable compounds, so the two bonds for each carbon atom that are not shown in this structure are implied. Any bonds that are not shown in line bond formulas are bonds to hydrogen atoms. By omitting symbols for carbon and hydrogen atoms, and bonds between carbon and hydrogen atoms, the structures for organic compounds can be represented efficiently.

TABLE 12.4 The Key Features of Line Bond Formulas

Line bond formula for an organic compound that has six carbon atoms connected in a ring	(a)	• Each line represents a carbon-carbon bond • Each point in the hexagon represents a carbon atom • Only two bonds for each carbon atom are shown, the two bonds omitted are bonds to hydrogen atoms and are implied
Line bond formula for an organic compound that has four carbon atoms connected in a straight chain	(b)	• Each line represents a carbon-carbon bond • The bend or junction between each bond represents a carbon atom • Note the carbon atoms on either end of the chain have only one bond shown, the three bonds not shown are bonds to hydrogen atoms • The groups on each end are thus $-CH_3$ • The groups within the chain are $-CH_2-$
Line bond formula for an organic compound that has three carbon atoms connected in a chain and a fourth one branching from centre of the chain	(c)	• Each line represents a carbon-carbon bond • The bend or junction between each bond represents a carbon atom • Note the carbon atom in the centre of the structure, three bonds to that carbon atom are shown so the fourth bond not shown is a bond to a hydrogen atom

The structure shown in Table 12.4 (b) represents another organic compound that has four carbon atoms connected in a straight chain. Each line represents a carbon-carbon bond and the bend or junction between each line represents a carbon atom. See how the carbon atoms on each end of that structure have only one bond shown. The three bonds to those carbon atoms not shown are bonds to hydrogen atoms. The groups on each end of this organic compound are thus $-CH_3$, and terminating lines like this always represent $-CH_3$ groups. The same principles apply to example (c) in Table 12.4. This example is given to show how a carbon atom forming three bonds to other carbon atoms and one bond to a hydrogen atom is represented. The line bond formula clearly shows three bonds to the carbon atom in the centre of the structure. The fourth bond not shown is a bond to a hydrogen atom.

Examine the line bond formulas given for the straight chain form of C_6H_{14} and the branched chain form of C_7H_{16} in Table 12.3. Compare the line bond formulas to the structural formulas and see how all the atoms are accounted for. Chains of carbon atoms look like zigzag lines and each bend represents a carbon atom. Double and triple bonds between carbon atoms can also be represented using line bond formulas, with some examples shown in Table 12.5. Logically, double lines are used to show double bonds and triple lines are used to show triple bonds. The same conventions apply for the omission of bonds to hydrogen atoms in these examples. If fewer than four bonds to a carbon atom are shown, the ones not shown are bonds to hydrogen atoms.

TABLE 12.5 Line Bond Formulas for Compounds with Double and Triple Bonds

• Line bond formula for an organic compound with four carbon atoms in a chain • The carbon atom on the end to the left is connected to the next one by a double bond • Examine the end carbon atom and see that two bonds to it are shown, this carbon atom is also bonded to two hydrogen atoms • Examine the other doubly bonded carbon atom within the chain, this carbon atom shows three bonds to it and the one not shown is a bond to a hydrogen atom	• Line bond formula for an organic compound with four carbon atoms in a chain • The two carbon atoms within the chain are connected by a double bond • Examine each doubly bonded carbon atom and see that each one has three bonds to it shown • Each doubly bonded carbon atom thus has a fourth bond to a hydrogen atom that is not shown	• Line bond formula for an organic compound with four carbon atoms in a chain • The two carbon atoms within the chain are connected by a triple bond • The chain is drawn without bends because triply bonded carbon atoms have linear electron and molecular geometry

Throughout this module you will also see representations that use a combination of line bond with condensed or structural formulas. A portion of an organic compound can be represented using condensed or structural formula representations if it's important to highlight more detail of the structure. Note that it's only the symbols for carbon and hydrogen atoms that are omitted from line bond formulas. Symbols for atoms of all other elements are used in line bond formulas for organic compounds that have elements in addition to carbon and hydrogen. Shown in Table 12.6 are a few examples to demonstrate this.

TABLE 12.6 Examples of Line Bond Formulas for Compounds Containing Cl, O and N Atoms

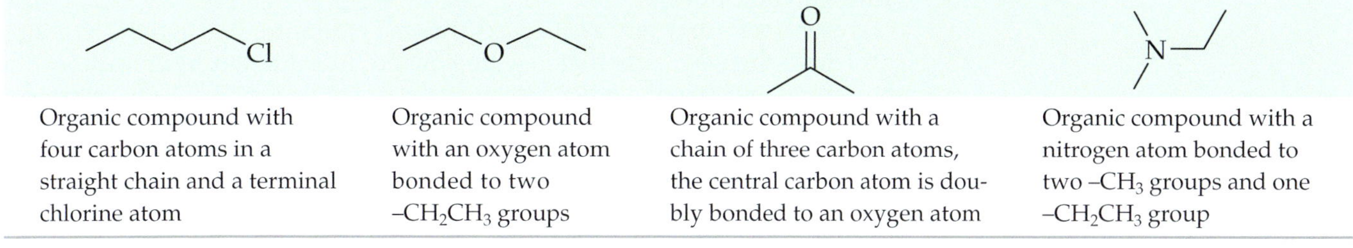

Cl	O	O	N
Organic compound with four carbon atoms in a straight chain and a terminal chlorine atom	Organic compound with an oxygen atom bonded to two $-CH_2CH_3$ groups	Organic compound with a chain of three carbon atoms, the central carbon atom is doubly bonded to an oxygen atom	Organic compound with a nitrogen atom bonded to two $-CH_3$ groups and one $-CH_2CH_3$ group

EXERCISE 12.1

The following table gives the structures for an organic compound as a condensed formula, and another as a line bond formula. Complete the table to practice drawing different structural representations.

Structural formula (Lewis structure)		
Condensed formula	$CH_3(CH_2)_2CH_3$	
Line bond formula		[line bond structure]

12.3 Hydrocarbons

LO: Identify hydrocarbons and explain the difference between saturated and unsaturated hydrocarbons.

The first specific examples of organic compounds described in this module belong to a broad grouping called the hydrocarbons. As the name suggests, **hydrocarbons** are organic compounds that contain only carbon and hydrogen atoms. You might think that compounds containing only two elements could be dull, but there is amazing diversity even within this one class of organic compounds. Abundant sources of hydrocarbons are the fossil fuels crude oil and natural gas, and these are widely used for energy production. Another source of hydrocarbons are the fragrant essential oils of plants. The main components of essential oils are terpenes, which are hydrocarbons.

Hydrocarbons may be classified as saturated or unsaturated and the terminology refers to the way carbon atoms are bonded. In **saturated hydrocarbons** the carbon atoms are linked to other carbon atoms by single bonds only. In **unsaturated hydrocarbons** there are one or more carbon atoms that are linked to others by a double or triple bond. Examples of both saturated and unsaturated hydrocarbons are shown in Table 12.7. You may have heard the terms saturated and unsaturated used to describe types of fat. The molecular structure of fats, which will be described later in this module, consists of a hydrocarbon portion. In saturated fats this hydrocarbon portion has only single bonds; in unsaturated fats the hydrocarbon portion has one or more carbon-carbon double bonds.

TABLE 12.7 Examples of Saturated and Unsaturated Hydrocarbons

Saturated hydrocarbons: The carbon atoms are linked by single bonds only.

Butane

```
    H   H   H   H
    |   |   |   |
H — C — C — C — C — H
    |   |   |   |
    H   H   H   H
```

© photohome.123rf.com

Used as a fuel in disposable lighters and barbeque gas cylinders.

Octane

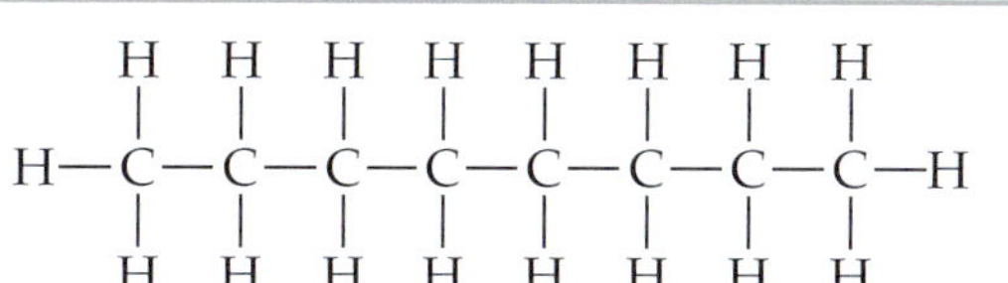

© www.BillionPhotos.com. Shutterstock

Component of petrol.

Unsaturated hydrocarbons: Carbon atoms are linked by multiple (double or triple) bonds.

Acetylene (ethyne)

$H-C\equiv C-H$

© Bogdan Vasilescu. Shutterstock

Used as a fuel for welding and cutting.

Benzene

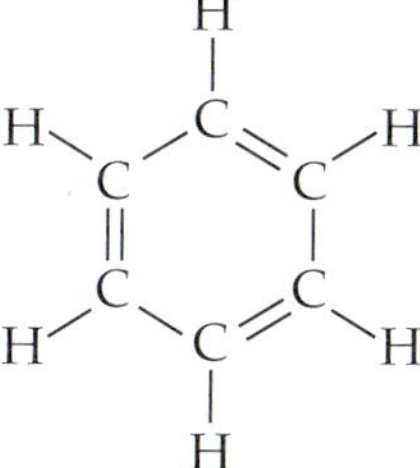

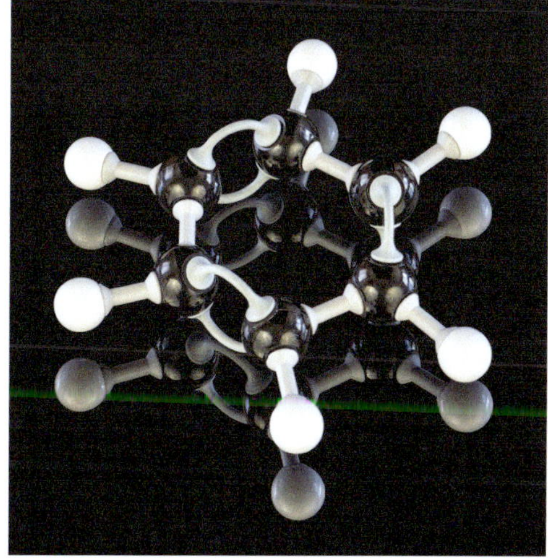

© Charles Knowles. Shutterstock

Widely used industrial chemical present in crude oil.

12.4 Alkanes and Cycloalkanes

LO: Identify and describe alkanes and cycloalkanes.
LO: Explain the physical properties of alkanes.
LO: Describe and write equations for chemical reactions of alkanes.
LO: Examine structural representations of alkanes and derive their IUPAC names.
LO: Interpret the IUPAC names of alkanes and draw representations for their molecular structures.

Alkanes

Alkanes are saturated hydrocarbons. Only carbon-carbon single bonds (C–C) are present, and the other bonds to carbon are bonds to hydrogen (H) atoms. Alkanes have a general formula C_nH_{2n+2}, for every carbon atom there are twice that number plus two hydrogen atoms. The names of compounds that are alkanes end with the suffix 'ane'. We can consider alkanes to be the 'parents' of all organic compounds, so developing a sound understanding of alkanes is a good basis for learning about organic compounds. The first four members of the alkane family are shown in Table 12.8 with their formulas, Lewis structures and names. Note that there are two alkanes with four carbon atoms. The number of possible structures for alkanes increases with the number of carbon atoms.

TABLE 12.8 The First Four Members of the Alkane Family

<table>
<tr><th>Number of C atoms (n)</th><th>Formula (C_nH_{2n+2})</th><th>Lewis structure</th><th>Name</th></tr>
<tr><td>1</td><td>CH_4</td><td>H
|
H—C—H
|
H</td><td>methane</td></tr>
<tr><td>2</td><td>C_2H_6</td><td>H H
| |
H—C—C—H
| |
H H</td><td>ethane</td></tr>
<tr><td>3</td><td>C_3H_8</td><td>H H H
| | |
H—C—C—C—H
| | |
H H H</td><td>propane</td></tr>
<tr><td rowspan="2">4</td><td rowspan="2">C_4H_{10}</td><td>H H H H
| | | |
H—C—C—C—C—H
| | | |
H H H H</td><td rowspan="2">butane*</td></tr>
<tr><td>H
H\ | /H
C
H | H
| | |
H—C—C—C—H
| | |
H H H</td></tr>
</table>

*Note that only the straight chain compound is called butane, the name of the branching compound will be explained later in the module.

The names of the first ten straight chain alkanes, given in Table 12.9, must be memorised. This task is important because these names are the basis of the names for other organic compounds. For these alkanes the suffix 'ane' identifies the compound as an alkane and the rest of the name identifies how many carbon atoms are present in the chain. The systematic naming of organic compounds is addressed throughout the module, starting later in this section for alkanes.

TABLE 12.9 The First Ten Straight Chain Alkanes

n	Formula	Name
1	CH_4	methane
2	C_2H_6	ethane
3	C_3H_8	propane
4	C_4H_{10}	butane
5	C_5H_{12}	pentane
6	C_6H_{14}	hexane
7	C_7H_{16}	heptane
8	C_8H_{18}	octane
9	C_9H_{20}	nonane
10	$C_{10}H_{22}$	decane

Cycloalkanes

It is possible for three or more carbon atoms to form a ring and thus cyclic alkanes are also possible. Cycloalkanes are named according to the same conventions as alkanes with the prefix 'cyclo'. For example, the straight chain alkane with five carbon atoms is called pentane while the cycloalkane with five carbon atoms is called cyclopentane. Cycloalkanes have a different general formula (C_nH_{2n}) to their acyclic counterparts. Note that the difference between the general formula for an alkane (C_nH_{2n+2}) and a cycloalkane (C_nH_{2n}) is that the '+2' part of the alkane general formula is missing from the cycloalkane general formula. You can see why this is by the comparison of an acyclic and a cyclic alkane that have the same number of carbon atoms in Table 12.10.

TABLE 12.10 Comparing an Acyclic and a Cyclic Alkane with the Same Number of Carbon Atoms

Acyclic alkane with 6 carbon atoms General formula C_nH_{2n+2}	Cyclic alkane with 6 carbon atoms General formula C_nH_{2n}
Number of carbon atoms (n) = 6, so the number of hydrogen atoms = 14	Number of carbon atoms (n) = 6, so the number of hydrogen atoms = 12
H—C—C—C—C—C—C—H (each C bonded to H above and H below)	Ring of six C atoms, each bonded to two H
Each carbon atom forms 4 covalent bonds. The '+2' H atoms indicated in the general formula complete the bonds for the carbon atoms on the ends of the chain.	In the cyclic alkane the carbon atoms form a ring and the '+2' H atoms are not needed to complete the bonds to carbon atoms on the ends of a chain.

EXERCISE 12.2

The table below gives the structural formula, line bond formula or name of a cycloalkane. to derive the name, find the name of the straight chain alkane with the same number of carbon atoms and add the prefix 'cyclo'. Complete the table by showing the appropriate structural representations for the first three members of the cycloalkane family.

n	Structural formula	Line bond formula	Name
3			cyclopropane
4	Four-membered ring of C atoms, each bonded to two H		
5		Pentagon	

Alkyl Groups

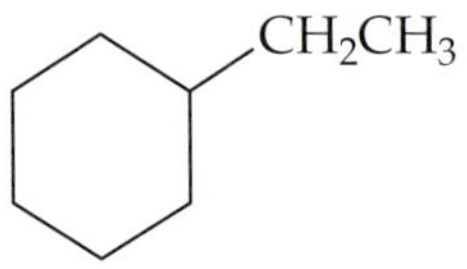

Branching saturated hydrocarbon chains that occur in the structures of organic compounds are referred to as **alkyl groups**. Alkyl groups are described in a manner that reflects the number of carbon atoms present, and the name of the alkane that contains the same number of carbon atoms is used as a basis for naming the alkyl group. This is done by working out the 'parent' alkane name based on the number of carbon atoms, dropping the suffix 'ane' from that name, and adding the suffix 'yl'. The difference between an alkyl group and its parent alkane is that one hydrogen atom of the alkane has been replaced with a carbon atom from another hydrocarbon framework. Let's examine what all this means using an example.

CH_2CH_3

Branching alkane structure

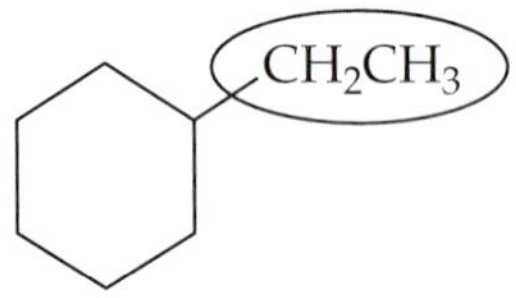

CH_2CH_3

The circled group is described using the term 'ethyl group'

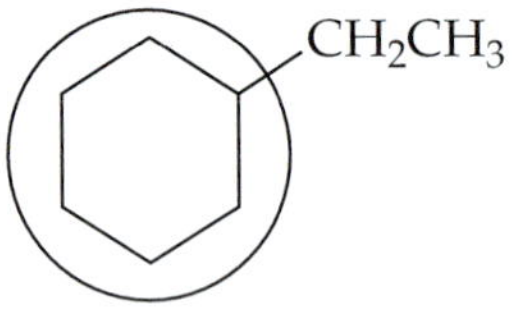

CH_2CH_3

The circled group is described using the term 'cyclohexyl group'

Consider the structure to the left that has a $-CH_2CH_3$ group branching from a cyclohexane ring. The branching $-CH_2CH_3$ group has the same number of carbon atoms as ethane, so the parent alkane name for this group is ethane. To name the branching group, the 'ane' is dropped from the parent alkane name and the suffix 'yl' is added. The branching $-CH_2CH_3$ group is therefore called an ethyl group. Note that when the formula for the ethyl group is written there must be a dash included ($-CH_2CH_3$) to represent that the group is bonded to an atom in another hydrocarbon framework. The cyclohexane ring part of the structure can also be described as an alkyl group with the specific name of the group being cyclohexyl.

Whenever you need to describe a saturated hydrocarbon group branching from another framework, consider the name of the alkane that has the same number of carbon atoms as the branching group. A $-CH_2CH_2CH_2CH_3$ branching group is called butyl, based on the name of the parent alkane with the same number of carbon atoms, butane.

EXERCISE 12.3

The table below gives the parent alkane formula/structure, parent alkane name, alkyl group formula/structure or alkyl group name. To derive the alkyl group name, drop the suffix 'ane' from the parent alkane name and add the suffix 'yl'. Complete the table by filling the blank spaces for each example.

Parent alkane formula/structure	Parent alkane name	Alkyl group formula/structure	Alkyl group name
CH_4			
			pentyl
		$-CH_2CH_2CH_3$	
	octane		

Isomerism in Alkanes

Earlier you saw that there are two structures possible for alkanes with molecular formula C_4H_{10}. This is an example of structural isomerism. **Structural isomers** are molecules that have the same molecular formula but different structures. ▼ Figure 12.4 shows that two structures for alkanes with molecular formula C_4H_{10} are possible using ball and stick models to visualise the distinctly different molecular shapes. Even though they have the same molecular formula, isomers are different compounds that have different names, physical properties and chemical reactivity. How to derive the different names for alkane structural isomers will be covered shortly, and you will see in the next section why alkane structural isomers have different physical properties.

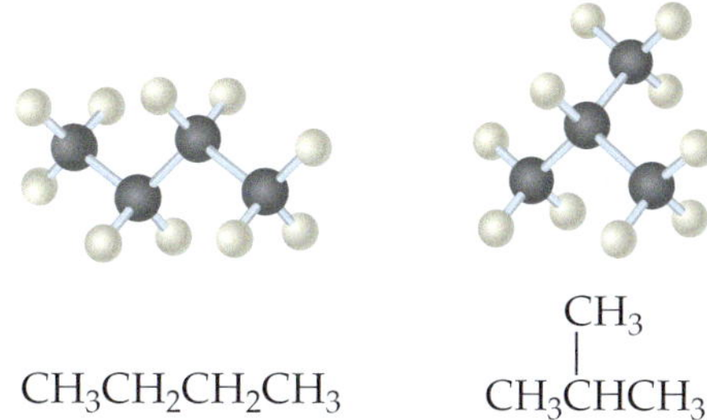

▶ **FIGURE 12.4 Ball and stick models of the two alkanes with molecular formula C_4H_{10}** Examine the structures to see that both contain 4 carbon atoms and 10 hydrogen atoms, but they are connected in a different order and have distinct shapes.

EXERCISE 12.4

Examine the images below that show ball and stick models for the three alkane structural isomers with molecular formula C_5H_{12}. Complete the table to show how the three isomers can be distinguished in their structural, condensed and line bond formulas.

C_5H_{12} isomer ball and stick model	Structural formula	Condensed formula	Line bond formula

Physical Properties of Alkanes

Throughout this module the physical properties of substances that are organic compounds will be discussed, starting in this section for alkanes. The physical properties of substances depend on the type of intermolecular forces (IMFs) that occur between the individual molecules that make up the substance. The IMFs that are possible for any given molecule depend on the molecular structure. An important factor in determining which IMFs are possible is whether a molecule is polar or non-polar. How to determine if a molecule is polar or non-polar was introduced in Module 3. To predict the physical properties of organic compounds you need to be able to examine any structural representation of a molecule and work out if it is polar or non-polar. Dispersion forces are the only possible IMF for non-polar molecules. The IMF known as dipole-dipole attraction is relevant to polar molecules, but if the molecule has highly polar O–H or N–H bonds in its structure then hydrogen bonding is possible. The type of IMFs possible for any given molecule allow for explanation and prediction of the physical properties of substances. Table 12.11 shows intermolecular forces relevant to organic compounds and some key points about them.

TABLE 12.11 Intermolecular Forces Relevant to Organic Compounds

Dispersion forces	• Possible for all organic compounds • Only possible IMF for non-polar organic compounds • Weakest IMF in relative terms, but strength increases with molecular size
Dipole-dipole attraction	• Possible for polar organic compounds • Stronger IMF relative to dispersion forces
Hydrogen bonding	• Possible for organic compounds that have highly polar O–H and N–H bonds • The strongest IMF in relative terms

Alkanes are non-polar molecules and dispersion forces are the only type of IMF possible. Dispersion forces are the weakest of the IMFs, but their strength increases with increasing molecular size. The chart in ▶ Figure 12.5 shows boiling points for straight chain alkanes with up to eight carbon atoms. Let's explore the reasons for the increase in boiling point with increasing number of carbon atoms by considering the questions in the following exercise.

EXERCISE 12.5

Cover the information in the right hand column while you consider the questions in the left column. Think about each question and write an answer before uncovering the explanations given.

Questions	Answers (don't look until you've thought about and written your own explanation)
What is it about the structure of alkane molecules that make them non-polar?	For a molecule to be polar there must be polar bonds and the shape of the molecule means that there is asymmetrical distribution of electron density throughout. Alkanes have only non-polar C–C bonds and slightly polar C–H bonds. However, the polar C–H bonds are evenly distributed throughout and alkane molecules consequently have no net dipole moment.
Why do dispersion forces increase with increasing molecular size?	Dispersion forces are the weakest of the IMFs in relative terms, but their strength increases with increasing molecular size. The reason for this is the presence of more electrons as molecular size increases. The greater electron density is more readily distorted in larger molecules and they are more likely to have temporary/instantaneous dipoles. This facilitates greater attraction between larger molecules.
Under normal conditions (25 °C and 1 atm) which straight chain alkanes with up to 8 carbon atoms would exist as gases?	The alkanes that would exist as gases at 25 °C are those that have boiling points below 25 °C. These can be read from the chart in Figure 12.5. It's alkanes with 1 – 4 carbon atoms. The small sizes of these molecules means that there is insufficient electron density to allow for dispersion forces between them. There is no attractive force to hold the molecules close to each other, the distances between them are great and thus the substances exist as gases.

Under normal conditions (25 °C and 1 atm) which straight chain alkanes with up to 8 carbon atoms would be liquids?	The alkanes that would exist as liquids at this temperature are those that have boiling points greater than 25 °C. Interpret the data in the chart (Figure 12.5) and see that alkanes with 5 – 8 carbon atoms have boiling points greater than 25 °C. Compared to the alkanes with fewer carbon atoms, these have greater electron density and the dispersion forces between them are stronger. The stronger attraction between molecules allows them to be close together and the substances thus exist as liquids.
Read the approximate boiling points of pentane (5 carbon atoms) and octane (8 carbon atoms) from the chart in Figure 12.5. Consider the reason for the difference and write a detailed explanation.	The approximate boiling points read from the chart are pentane (5 carbon atoms) 36 °C and octane (8 carbon atoms) 126 °C. The higher boiling point for octane indicates that the IMFs for that compound are stronger. More energy is required to overcome the attractive forces and allow octane molecules to leave the liquid phase and enter the gas phase. Both pentane and octane are non-polar molecules, so dispersion forces are the only IMF possible. Dispersion forces are stronger for octane because it is a larger molecule and thus has more electron density.

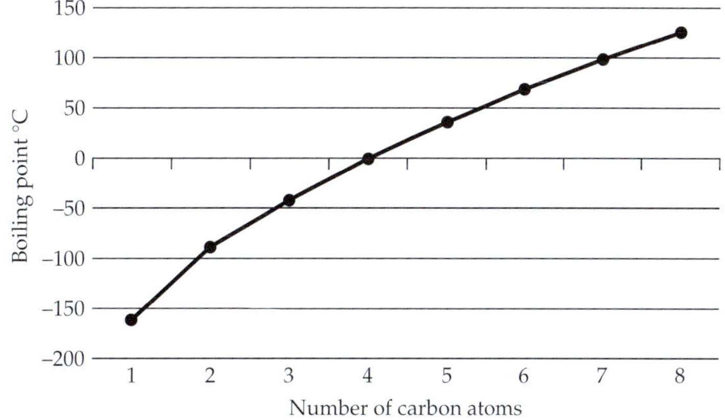

▶ **FIGURE 12.5** **Chart of boiling points for straight chain alkanes with up to eight carbon atoms**

The difference in physical properties between alkane structural isomers is very interesting because they have the same molecular formula, molar mass and therefore, the same electron density. We might expect that non-polar molecules with similar electron density experience dispersion forces to the same extent and thus have similar boiling points. However, this is not the case for alkane structural isomers. Consider the boiling points for alkane structural isomers with molecular formula C_5H_{12} in Table 12.12 . The compounds have distinct boiling points despite having similar electron density. The difference here is related to molecular shape rather than molecular size.

TABLE 12.12 Boiling Points for Alkane Structural Isomers with Molecular Formula C_5H_{12}

Isomer 1	Isomer 2	Isomer 3
$CH_3CH_2CH_2CH_2CH_3$	$H_3C-CH(CH_3)-CH_2CH_3$ (central C bearing H and CH_3)	$H_3C-C(CH_3)_2-CH_3$ (central C bearing two CH_3)
BP = 36 °C	BP = 28 °C	BP = 9 °C

One way to describe the difference in shape between the isomers is in terms of the amount of branching within the molecule. Isomer 1 is a straight chain alkane with no branching. Isomer 2 has a chain of four carbon atoms and one carbon atom branching. Isomer 3 has a chain of three carbon atoms and two carbon atoms branching. Isomer 1 with no branching has the highest boiling point, while Isomer 3 with the most branching has the lowest boiling point. For alkane structural isomers, boiling point decreases with increased branching of the carbon chain. The ball and stick models of alkane structural isomers with molecular formula C_5H_{12} shown in ▼ Figure 12.6 show more clearly the difference in shape between the molecules. Another way to describe the difference in shape between the isomers, in terms of increasingly spherical shape, is also shown in Figure 12.6. Refer back to the boiling points in Table 12.12 and check the relationship with overall shape. Isomer 1 has the highest boiling point and therefore has the strongest dispersion forces. Isomer 3 has the lowest boiling point and therefore has the weakest dispersion forces. It can be concluded that dispersion forces decrease with an increasingly spherical molecular shape. This is due to decreased surface area for spherical objects compared to more cylindrically shaped objects. There is less interaction by dispersion forces possible with the decreased surface area of the molecule.

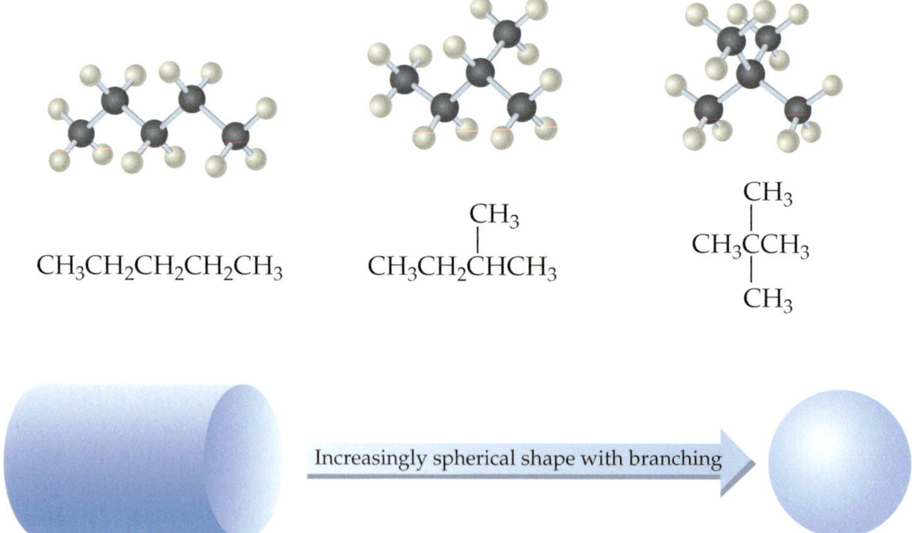

▶ **FIGURE 12.6 Ball and stick models of alkane structural isomers with molecular formula C_5H_{12}** The straight chain isomer is closer in shape to a cylinder. The more branched isomer has a more spherical shape.

Reactions of Alkanes

An introduction to the chemical reactivity of organic compounds is an important part of this module and the most common reactions of alkanes will be outlined in this section. There are some important aspects to the way equations for chemical reactions of organic compounds are written, they are not written in the ways you have seen in previous modules. The most important difference is that *equations for reactions of organic compounds are written using structural formulas rather than molecular formulas for reactants and products.* You will see very few examples of equations written using molecular formulas in this module. This is because the molecular structure of an organic compound changes during any of its reactions and the equation has to show the change very clearly. If molecular formulas are used in the equation, then the structural change that occurred during the reaction is not obvious.

In the conventional way of writing an equation for a chemical reaction, the reactants are on the left hand side of the arrow and the products are on the right. In equations for reactions of organic compounds it is permissible to show one of the reactants above the reaction arrow. Various reactions of organic compounds require the presence of another chemical to make the reaction proceed, and these are commonly referred to as reagents. The formulas for reagents are also written above the reaction arrow in these equations. Sometimes certain conditions, such as energy from heat or light, are required to start the reaction of an organic compound. These conditions can also be included in the equation by writing their symbols above the reaction arrow. The change in organic compound molecular structure that occurs in a reaction is usually the focus of attention, so equations may only show the organic compound that forms as the product and omit any by products that form. Equations that show reactions of organic compounds are not always balanced. Again, this is because the focus is on the change in molecular structure that occurs for the organic compound reacting. Table 12.13 shows a summary of the key things to note when writing or interpreting equations for reactions of organic compounds.

TABLE 12.13 Key Points for Writing and Interpreting Equations for Reactions of Organic Compounds

- Structural formulas used instead of molecular formulas
- Reactants, reagents and/or conditions required may be written above the reaction arrow
- Organic compound formed in a reaction may be the only product shown
- By products that form may be left out
- Equations for reactions of organic compounds are not always balanced

▶ **FIGURE 12.7 Natural gas** The major component of natural gas, used as a fuel for gas stoves, is methane (CH_4).

Alkanes are fairly inert and unreactive substances, but they can react with oxygen and halogens. The reaction of alkanes with oxygen is an example of a combustion reaction. The only products from complete combustion of an alkane are carbon dioxide and water. Combustion reactions of alkanes also produce energy in the form of heat and alkanes are thus used as fuels. One example of a fuel that is widely used is natural gas and the major component of this fuel is the alkane methane (CH_4) ▲ (Figure 12.7). Equations for the complete reaction of alkanes in combustion may be written using molecular formulas since the molecular structure of the reactant organic compound is completely lost in the reaction. Here is the balanced equation for the reaction of methane (CH_4) in combustion.

$$CH_4(g) + 2O_2(g) \rightarrow CO_2(g) + 2H_2O(g)$$

EXERCISE 12.6

Equations for combustion reactions of alkanes.

Write the balanced equation for the complete combustion of butane. The reactants are butane (C_4H_{10}) and oxygen gas. The only products are gaseous carbon dioxide and gaseous water. Include state symbols in the equation, refer to the chart in Figure 12.5 to determine the physical state of butane.
Write the balanced equation for the complete combustion of octane. The reactants are octane (C_8H_{18}) and oxygen gas. The only products are gaseous carbon dioxide and gaseous water. Include state symbols in the equation, refer to the chart in Figure 12.5 to determine the physical state of octane.

In the presence of ultraviolet (UV) light, alkanes can react with halogens such as chlorine (Cl_2) or bromine (Br_2) in substitution reactions. One or more of the hydrogen atoms of an alkane may be substituted with a halogen atom. The substitution of one hydrogen atom for a chlorine atom in the reaction of methane (CH_4) with Cl_2 is shown in the equation below. The *hv* above the reaction arrow indicates that light is required for the reaction to occur, it overcomes the activation energy. Note the formation of HCl in the reaction is shown and the equation is balanced.

$$\mathrm{H{-}\overset{\overset{\large H}{|}}{\underset{\underset{\large H}{|}}{C}}{-}H + Cl{-}Cl \xrightarrow{hv} H{-}\overset{\overset{\large H}{|}}{\underset{\underset{\large H}{|}}{C}}{-}Cl + H{-}Cl}$$

The reaction as shown in the equation above cannot be controlled and doesn't stop at forming CH_3Cl. Further reaction of Cl_2 with CH_3Cl formed in the initial reaction takes place, and substitution of another hydrogen atom with a chlorine atom occurs as follows.

$$\mathrm{H{-}\overset{\overset{\large H}{|}}{\underset{\underset{\large H}{|}}{C}}{-}Cl + Cl{-}Cl \xrightarrow{hv} H{-}\overset{\overset{\large H}{|}}{\underset{\underset{\large Cl}{|}}{C}}{-}Cl + H{-}Cl}$$

As long as there are hydrogen atoms present in the product that forms, further reaction and substitution of hydrogen atoms by chlorine atoms can occur. The next equation shows substitution of a third hydrogen atom from the reactant methane by a chlorine atom.

$$\mathrm{H{-}\overset{\overset{\large H}{|}}{\underset{\underset{\large Cl}{|}}{C}}{-}Cl + Cl{-}Cl \xrightarrow{hv} H{-}\overset{\overset{\large Cl}{|}}{\underset{\underset{\large Cl}{|}}{C}}{-}Cl + H{-}Cl}$$

The final product is CCl_4, as shown in the equation below.

$$\mathrm{H{-}\overset{\overset{\large Cl}{|}}{\underset{\underset{\large Cl}{|}}{C}}{-}Cl + Cl{-}Cl \xrightarrow{hv} Cl{-}\overset{\overset{\large Cl}{|}}{\underset{\underset{\large Cl}{|}}{C}}{-}Cl + H{-}Cl}$$

The actual product obtained when this reaction is carried out is a mixture of CH_3Cl, CH_2Cl_2, $CHCl_3$ and CCl_4. Examine the equations above carefully, noting the fate of the two chlorine atoms from Cl_2 in each step. One chlorine atom replaces a hydrogen atom of the alkane. The other chlorine atom from Cl_2 is in the HCl molecule that forms in each step of the reaction. The hydrogen atom in HCl is the hydrogen atom from the alkane that has undergone substitution reaction. The formation of HCl in this reaction may be observed as a white vapour.

The analogous reaction occurs with bromine (Br_2), which is a dark brown-coloured liquid, and the visible changes that occur during the reaction serve as the basis for a chemical test for alkanes. As bromine is consumed in the reaction the brown colour is lost and changes to colourless. The formation of HBr in each substitution step may be observed as the presence of a white vapour.

EXERCISE 12.7

Writing equations for reactions of alkanes with halogens.

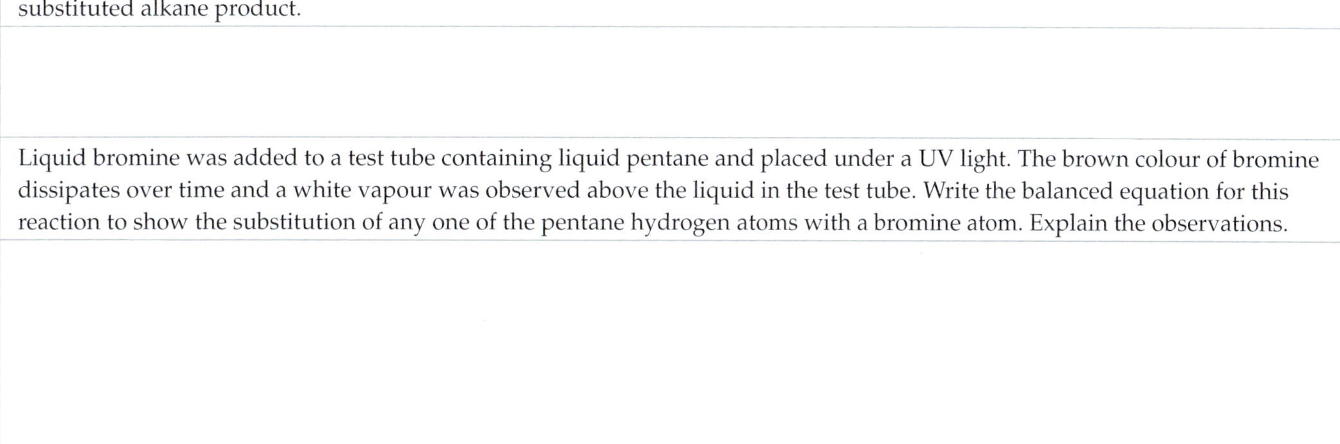

Write the equation for the first step in the reaction between ethane and chlorine in the presence of light to give only the mono-substituted alkane product.
Liquid bromine was added to a test tube containing liquid pentane and placed under a UV light. The brown colour of bromine dissipates over time and a white vapour was observed above the liquid in the test tube. Write the balanced equation for this reaction to show the substitution of any one of the pentane hydrogen atoms with a bromine atom. Explain the observations.

Naming Alkanes

Learning how to name alkanes serves as a basis for being able to name other organic compounds, so the process introduced in this section for naming alkanes will be built upon throughout the module. The system for naming organic compounds is based on guidelines developed by the IUPAC (International Union of Pure and Applied Chemistry). Table 12.14 shows that the IUPAC names of organic compounds have components (prefix-parent-suffix).

TABLE 12.14 Components of Organic Compound IUPAC Names

Prefix	• Identifies substituents, such as alkyl groups, that branch from the parent chain • The prefix can include numbers that describe the location of substituents along the parent chain
Parent	• Identifies the length of the longest continuous chain of carbon atoms in the organic compound structure
Suffix	• Identifies the type or classification of organic compound

The fact that alkane names end in the suffix 'ane' was mentioned earlier in this module. If the structure of an organic compound molecule has nothing branching from the main chain, then the name contains no prefix and the names of straight chain alkanes given earlier exemplify this. Conversely, the prefix part of an organic compound name can have several parts as it describes all components of the compound structure that branch from the parent chain. The prefix can also contain numbers to describe the location along the parent chain where substituents branch. The central part of the name is the 'parent', which describes the longest continuous chain of carbon atoms in the compound. Table 12.15 shows the parent names for chains with 1 – 10 carbon atoms. They will look familiar, as you have already seen them in the straight chain alkane names given previously.

TABLE 12.15 Parent Names for Chains with 1-10 Carbon Atoms

Parent chain length	1 C	2 C	3 C	4 C	5 C	6 C	7 C	8 C	9 C	10 C
Parent name	meth	eth	prop	but	pent	hex	hept	oct	non	dec

Naming an alkane, or any organic compound, is a step by step process. All of the steps are listed in Table 12.16 so they can be referred back to, but each step will be explained in detail using examples. The examples below are all alkanes, so the suffix (Step 0) is 'ane' in each case. Steps 1 – 4 will be demonstrated with five alkanes.

TABLE 12.16 Steps for Naming an Organic Compound

Step	Description of the step	Points to consider in applying the step
0	Examine the structural representation to identify the type or classification of the organic compound.	• This step allows the appropriate choice of suffix for the compound name.
1	Identify the parent chain, which is the longest continuous chain of carbon atoms.	• This step allows the appropriate choice of parent name. • If there are chains of equal length, the parent chain is the one with more branch points.
2	Number the carbon atoms in the parent chain.	• The correct option is the one that gives the lowest number for the first branch point. • If there is branching an equal distance from both ends, begin numbering from the end closer to the second branch point.
3	Identify the substituents and the number of the parent chain carbon atoms that they branch from.	• Substituent names are arranged in alphabetical order in the compound name, their position along the parent chain is not considered. • The presence of more than one of the same substituent is designated using 'di', 'tri', 'tetra' etc. in front of the substituent name. • If 'di', 'tri' or 'tetra' has been used, it is not considered in the alphabetical ordering of substituent names.
4	Write the whole name of the compound as one word.	• Use commas between numbers in the compound name. • Use hyphens between numbers and letters in the compound name. • Compound names are not proper nouns, so don't use upper case for the first or any letter within the name. • Only capitalise the first letter of a compound name if it is the first word in a sentence, according to the rules of syntax.

ALKANE 1

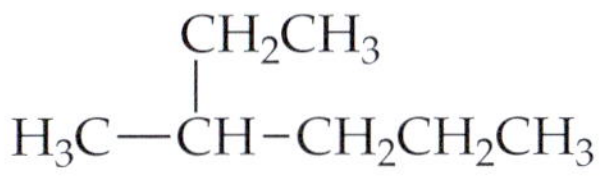

$$\begin{array}{l} \quad\quad\; CH_2CH_3 \\ \quad\quad\;\; | \\ H_3C-CH-CH_2CH_2CH_3 \end{array}$$

Step 0 The suffix part of the compound name is **ane**.

Step 1 Identify the parent chain, which is the longest continuous chain of carbon atoms. For alkane 1, there are three chains indicated below. Note how the longest chain is not necessarily drawn in a straight line. Carbon-carbon single bonds can rotate and the chains can twist, so they can be drawn in many configurations.

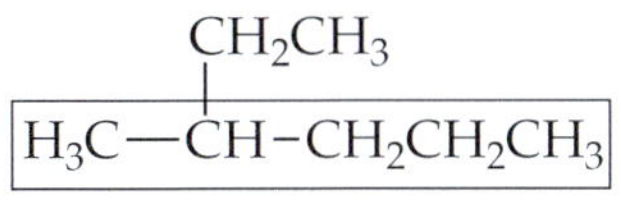

This chain has 5 carbon atoms.

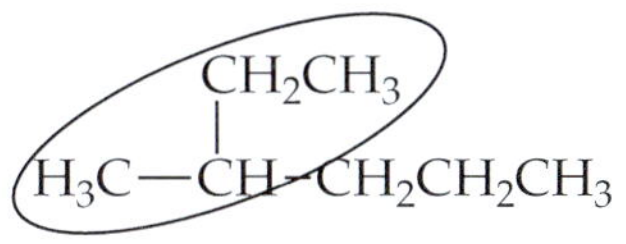

This chain has 4 carbon atoms.

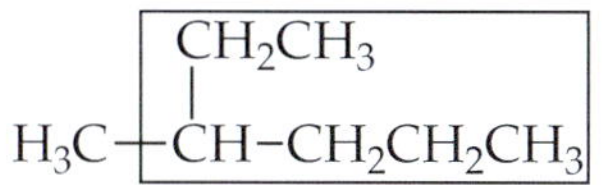

This chain has 6 carbon atoms, which is the longest. The parent part of the compound name is **hex**.

Step 2 Number the carbon atoms in the parent chain. The two options for alkane 1 are shown below. The correct one has the lowest number for the position of the first substituent branching from the main chain.

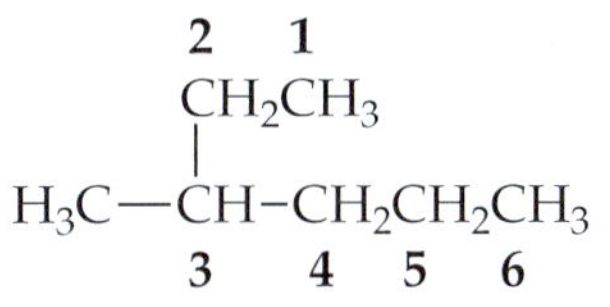

This is the correct numbering of the parent chain as it gives the lowest possible number (3) to the carbon atom with the branching $-CH_3$ group.

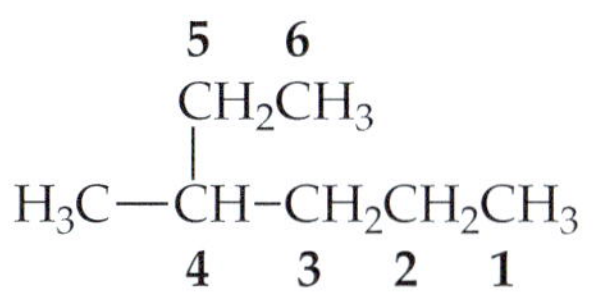

With this numbering the branching $-CH_3$ group is on carbon atom 4 of the parent chain. This is incorrect because the conventions stipulate allocation of the lowest possible number to the position of the first branch point.

Step 3 Identify the substituents and the number of the parent chain carbon atom that they branch from. Alkane 1 has a $-CH_3$ group branching from carbon atom 3 of the parent chain. This is circled on the structure of alkane 1 shown below. The name given to a $-CH_3$ group is methyl, so this will be the prefix in the compound name. The name of the compound also has to describe the location of substituents on the parent chain. In alkane 1 the methyl group is branching from carbon 3, in the name '3-methyl' is written. Hyphens are used between numbers and letters in organic compound names.

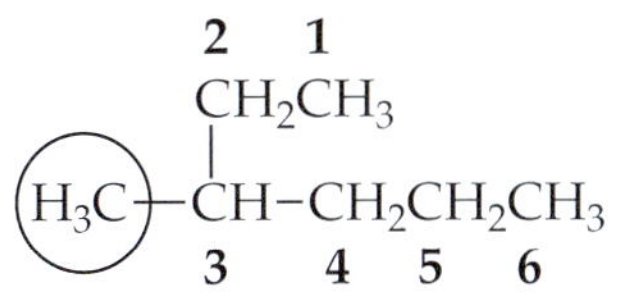

The branching $-CH_3$ group is called methyl and it is connected to carbon atom 3 of the parent chain. In the compound name this is written as **3-methyl**.

Step 4 Write the whole name of the compound as one word. At this stage all parts of the name have been identified and it's just a matter of putting the whole thing together. Note that the name is written as one word, the prefix name runs onto the parent and suffix parts.

Alkane 1	Components of the name identified	Compound name
$H_3C{-}CH(CH_2CH_3){-}CH_2CH_2CH_3$	Prefix: 3-methyl Parent: hex Suffix: ane	**3-methylhexane**

The name of the compound is a complete description of the structure. The suffix tells us it's an alkane, and the rest of the name explains that the molecule has a chain of six carbon atoms with a branching $-CH_3$ group at the third carbon atom. Note that compound names are not proper nouns, so the first letter of the name is not capitalised. Only capitalise the first letter of a compound name if it is the first word in a sentence, as per the rules for sentence construction.

ALKANE 2

$$\begin{array}{l} \quad\;\; CH_2CH_3 \\ \quad\;\; | \\ H_3C{-}CH{-}CH{-}CH_2CH_3 \\ \qquad\quad\;\; | \\ \qquad\quad\;\; CH_2CH_2CH_3 \end{array}$$

Step 0 The suffix part of the compound name is **ane**.

Step 1 Identify the parent chain, which is the longest continuous chain of carbon atoms. For alkane 2, there are several chains with the longest having seven carbon atoms. The parent part of the compound name is **hept**.

Step 2 Number the carbon atoms in the parent chain. The two options for numbering the parent chain are shown below.

$$\begin{array}{l} \quad\;\; \overset{2}{C}H_2\overset{1}{C}H_3 \\ \quad\;\; | \\ H_3C{-}\overset{3}{C}H{-}\overset{4}{C}H{-}CH_2CH_3 \\ \qquad\quad\;\; | \\ \qquad\quad\;\; \underset{5}{C}H_2\underset{6}{C}H_2\underset{7}{C}H_3 \end{array}$$

This is the correct numbering.
The first branching substituent is at carbon atom 3 and the second at carbon atom 4.

$$\begin{array}{l} \quad\;\; \overset{6}{C}H_2\overset{7}{C}H_3 \\ \quad\;\; | \\ H_3C{-}\overset{5}{C}H{-}\overset{4}{C}H{-}CH_2CH_3 \\ \qquad\quad\;\; | \\ \qquad\quad\;\; \underset{3}{C}H_2\underset{2}{C}H_2\underset{1}{C}H_3 \end{array}$$

Incorrect numbering.
The first branching substituent is at carbon atom 4 and the second at carbon atom 5.
This is not the lowest possible number for the first branch point.

Step 3 Identify the substituents and the number of the parent chain carbon atom that they branch from. Alkane 2 has two substituents, a $-CH_3$ (methyl) group branching from carbon atom 3 and a $-CH_2CH_3$ (ethyl) group branching from carbon atom 4. These are circled on the structure shown below. In the name of the compound these will appear in alphabetical order as 4-ethyl and 3-methyl, the positions for the substituents are not considered in formatting compound names. In the name of the compound these will run together with hyphens to separate the last letter of the first substituent name from the number of the second substituent name as shown below.

$$\begin{array}{l} \qquad\; \overset{2}{C}H_2\overset{1}{C}H_3 \\ \qquad\; | \\ (H_3C){-}\overset{3}{C}H{-}\overset{4}{C}H{-}(CH_2CH_3) \\ \qquad\qquad\; | \\ \qquad\qquad\; \underset{5}{C}H_2\underset{6}{C}H_2\underset{7}{C}H_3 \end{array}$$

The branching $-CH_3$ group (methyl) is connected to carbon atom 3 of the parent chain. In the compound name this is written as **3-methyl**.
The branching $-CH_2CH_3$ (ethyl) is connected to carbon atom 4 of the parent chain. In the compound name this is written as **4-ethyl**.
Different substituent names are written in alphabetical order in the compound name, for alkane 2 this will be **4-ethyl-3-methyl**.
Note how hyphens are used between letters and numbers. There are no spaces.

Step 4 Write the whole name of the compound as one word. Assemble the components of the name identified as one word.

Alkane 2	Components of the name identified	Compound name
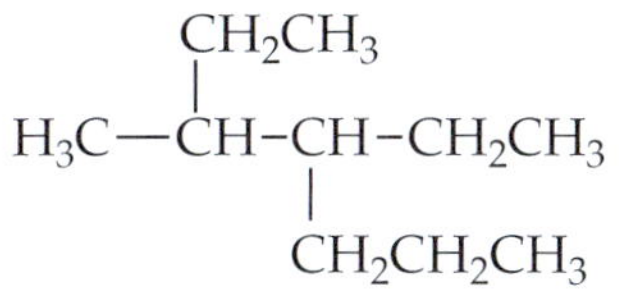 $H_3C-CH(CH_2CH_3)-CH(CH_2CH_2CH_3)-CH_2CH_3$	Prefix: 4-ethyl-3-methyl Parent: hept Suffix: ane	**4-ethyl-3-methylheptane**

Naming alkane 2 was a progression from the naming of alkane 1 as there were two substituents in the structure. Dealing with the presence of two carbon atom chains with equal length will be demonstrated with alkane 3.

ALKANE 3

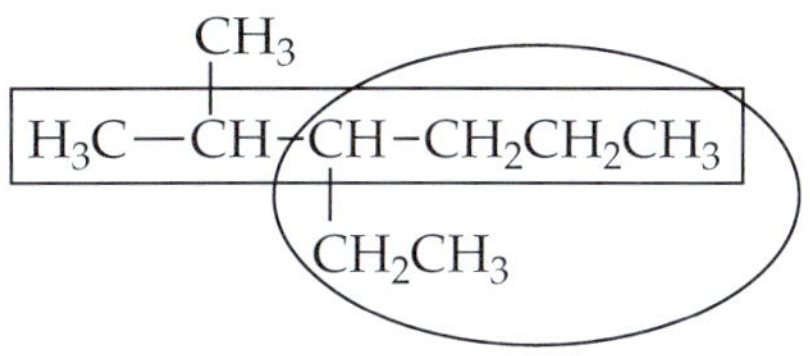

$$H_3C-CH(CH_3)-CH(CH_2CH_3)-CH_2CH_2CH_3$$

Step 0 The suffix part of the compound name is **ane**.

Step 1 Identify the parent chain, which is the longest continuous chain of carbon atoms. For alkane 3, there are two chains of six carbon atoms indicated on the structure. The one in the rectangle is the correct option as it has two branches. The alternative in the oval has only one branch. The parent part of the compound name is **hex**.

Step 2 Number the carbon atoms in the parent chain. The correct numbering is shown on the structure.

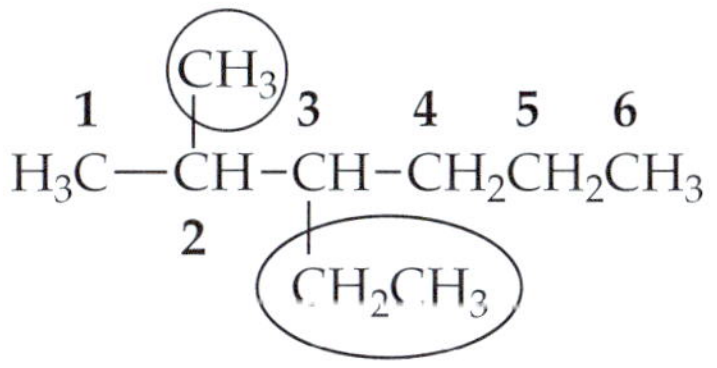

$$\overset{1}{H_3C}-\overset{2}{C}H(CH_3)-\overset{3}{C}H(CH_2CH_3)-\overset{4}{C}H_2\overset{5}{C}H_2\overset{6}{C}H_3$$

Step 3 The substituents connected to carbon atoms 2 and 3 (circled on the structure) will appear in the name as **3-ethyl-2-methyl** (alphabetical order).

Step 4 The full name of the compound is **3-ethyl-2-methylhexane**.

Dealing with the presence of more than one of the same substituent in the naming process will be shown for alkane 4. At Step 3 we find that there are two methyl groups in the structure, one at carbon atom 2 and another at carbon atom 4. Rather than write the lengthy '2-methyl-4-methyl' in the compound name, we use 2,4-dimethyl. The 'di' is used to designate the presence of two methyl groups. If there are three of the same group 'tri' is used, and 'tetra' is used if there are four. In cases where there are more than four of the same substituent, the logical prefixes are used. The numbers included to show the positions of the two methyl groups are separated by a comma, and a hyphen is used in the usual way between the number and letter in the prefix.

ALKANE 4

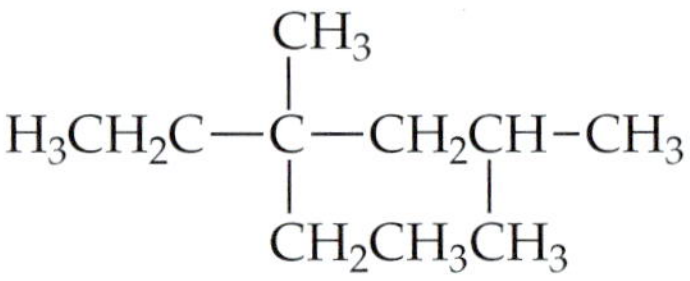

	Parent chain numbering	Substituents
$$\begin{array}{c} CH_3 \\ | \\ H_3CH_2C{-}C{-}CH_2CH{-}CH_3 \\ | \quad\quad | \\ CH_2CH_3CH_3 \end{array}$$	$$\begin{array}{c} CH_3 \\ 6\ \ 5\ \ 4|\ \ 3\ \ 2\ \ 1 \\ H_3CH_2C{-}C{-}CH_2CH{-}CH_3 \\ | \quad\quad | \\ CH_2CH_3CH_3 \end{array}$$	$$\begin{array}{c} (CH_3) \\ | \\ H_3CH_2C{-}C{-}CH_2CH{-}CH_3 \\ | \quad\quad | \\ (CH_2CH_3)(CH_3) \end{array}$$

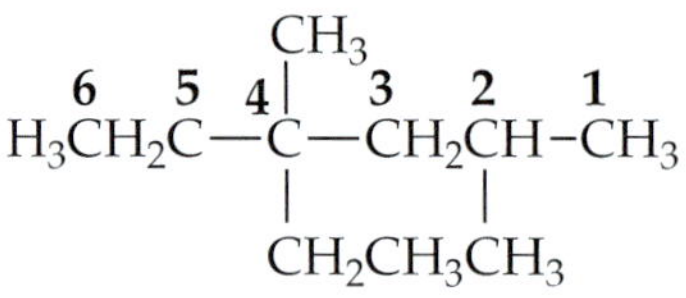

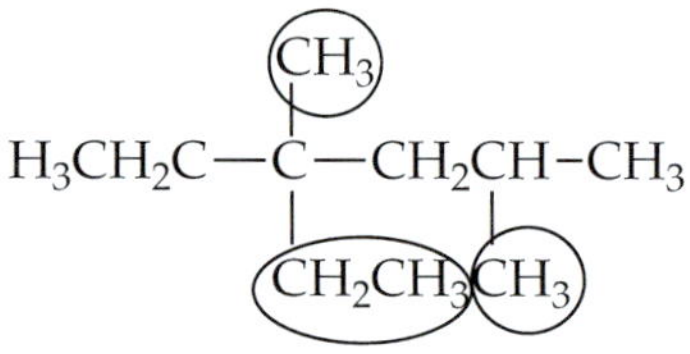

Step 0 The suffix part of the compound name is **ane**.

Step 1 Identify the parent chain. The parent part of the compound name is **hex**.

Step 2 Number the parent chain carbon atoms. The numbering shown above is correct because the first branch point is at carbon atom 2, the alternative would have the first branch point at carbon atom 3.

Step 3 Identify substituents and their position along the parent chain. These are circled in the structure above. The substituent names in alphabetical order with their position numbers are:
4-ethyl
2-methyl
4-methyl
To show there are two methyl groups we write dimethyl with the two position numbers separated by a comma as **2,4-dimethyl**. Adding the 'di' does not change the order of substituent names in the compound, ethyl still comes before methyl. The whole prefix for alkane 4 is **4-ethyl-2,4-dimethyl**.

Step 4 Write the compound name as one word.

Alkane 4	Components of the name identified	Compound name
$$\begin{array}{c} CH_3 \\ | \\ H_3CH_2C{-}C{-}CH_2CH{-}CH_3 \\ | \quad\quad | \\ CH_2CH_3CH_3 \end{array}$$	Prefix: 4-ethyl-2,4-dimethyl Parent: hex Suffix: ane	**4-ethyl-2,4-dimethylhexane**

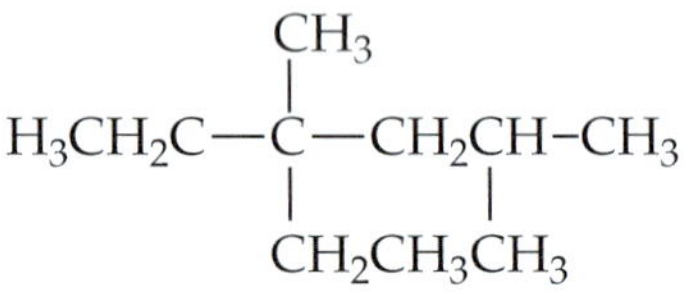

Here is another example, alkane 5, to show what to do if the first branch occurs at an equal distance from both ends of the parent chain. Logically, we continue counting along the chain and choose the option that gives the lowest number to the second branch point. To name a compound that has both the first and second branch occurring at an equal distance along the parent chain, the numbering that gives the lowest number to the third branch is used. If a difference can't be found, then the molecule must be the same from both ends and numbering from either end of the parent chain is correct.

ALKANE 5

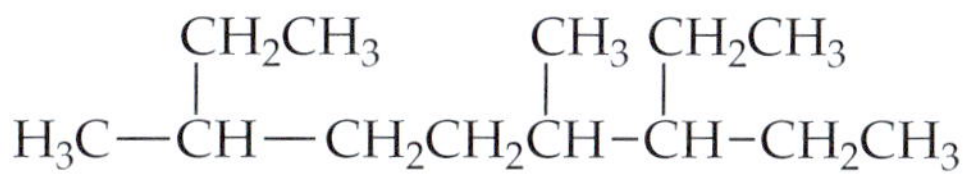

Step 0 The suffix part of the compound name is **ane**.

Step 1 The parent chain in alkane 5 has nine carbon atoms. The parent part of the compound name is **non**.

Step 2 The two options for numbering the parent chain carbon atoms are shown below.

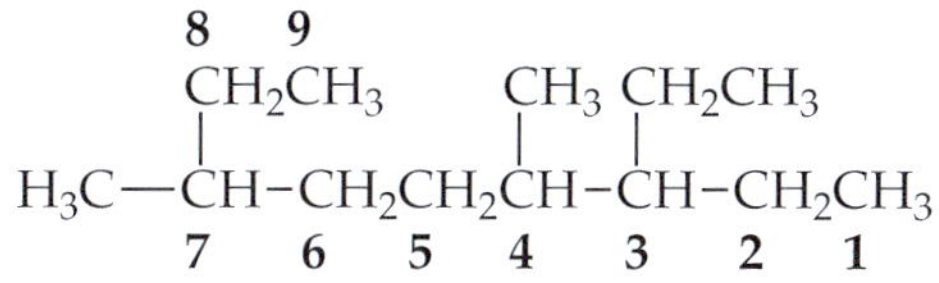

$$H_3C-CH(CH_2CH_3)-CH_2CH_2CH(CH_3)-CH(CH_2CH_3)-CH_2CH_3$$

Correct numbering (main chain 7 6 5 4 3 2 1; ethyl branch 8 9).
The first substituent is at carbon atom 3.
The second substituent is at carbon atom 4.

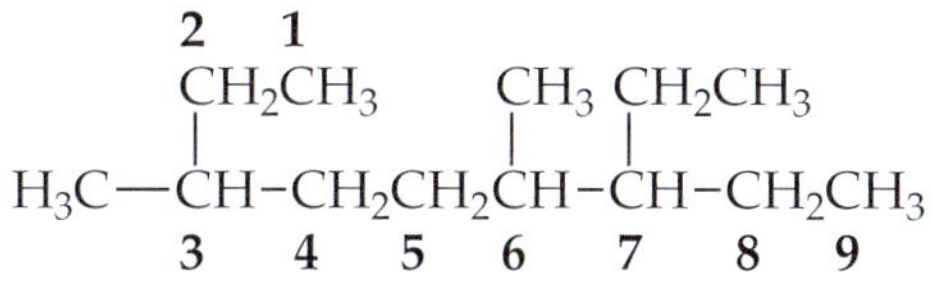

Incorrect numbering (ethyl branch 2 1; main chain 3 4 5 6 7 8 9).
The first substituent is at carbon atom 3.
The second substituent is at carbon atom 6, which is not the lowest possible number for the second substituent.

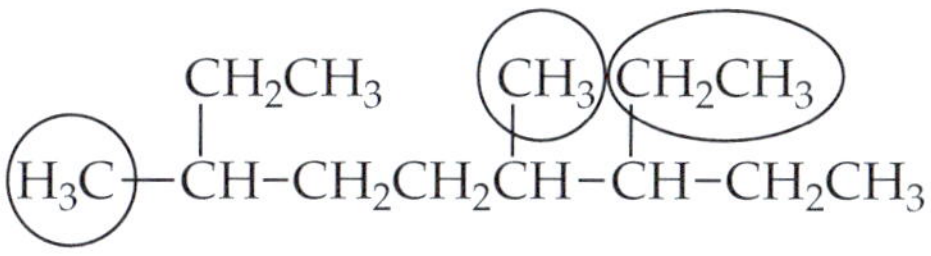

Step 3 Identify substituents and their position along the parent chain. The substituent names in alphabetical order with their position numbers are:
3-ethyl
4-methyl
7-methyl
The prefix is **3-ethyl-4,7-dimethyl**.

Step 4 Write the compound name as one word.

Alkane 5	Components of the name identified	Compound name
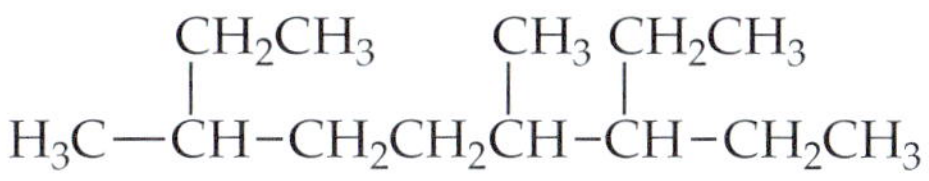 $H_3C-CH(CH_2CH_3)-CH_2CH_2CH(CH_3)-CH(CH_2CH_3)-CH_2CH_3$	Prefix: 3-ethyl-4,7-dimethyl Parent: non Suffix: ane	**3-ethyl-4,7-dimethylnonane**

EXERCISE 12.8

Name the following alkanes, which are drawn using different representations. You may find it helpful to redraw them as structural formulas.

$CH_3CH(CH_3)CH_3$	$CH_3CH(CH_3)CH(CH_3)CH_2CH_3$	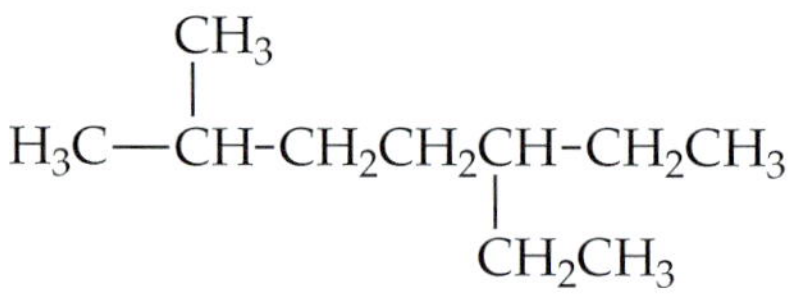 CH_3 | $H_3C—CH\text{-}CH_2CH_2CH\text{-}CH_2CH_3$ | CH_2CH_3
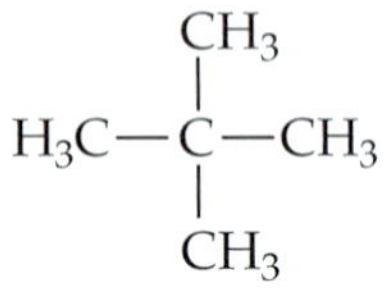 CH_3 | $H_3C—C—CH_3$ | CH_3	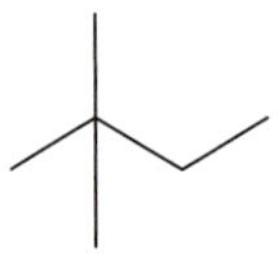	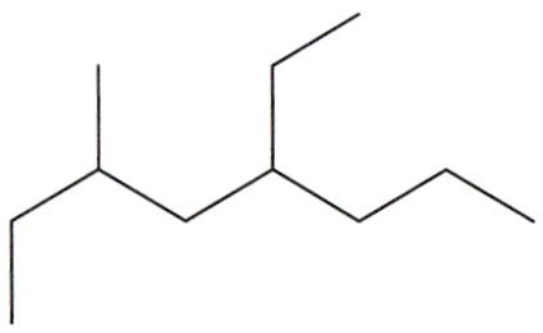

The process of naming alkanes can also be worked 'backwards' to draw a structural representation for an organic compound given its name. Here is an example to demonstrate a process for interpreting the name of an alkane and draw its structure. The steps in drawing **2-methylbutane** using both structural and line bond formulas are shown. The first step in the process is to interpret the suffix of the name and realise that you will be drawing an alkane, so the structure to be drawn will show only carbon-carbon single bonds.

Interpretation of the name	Developing the structural formula	Developing the line bond formula
Interpret the parent part of the name, this tells us how many carbon atoms in the longest continuous chain. The parent part of the name 2-methylbutane is **but**, so the parent chain has 4 carbon atoms.	C—C—C—C To start drawing the structural formula, draw a straight chain of 4 carbon atoms connected by single bonds.	 To start drawing the line bond formula, draw a zigzag line representing 4 carbon atoms connected by single bonds.
Next the prefix of the name is interpreted to determine the identity and location of the substituents. In this case we have to show a methyl ($–CH_3$) group branching from carbon atom 2.	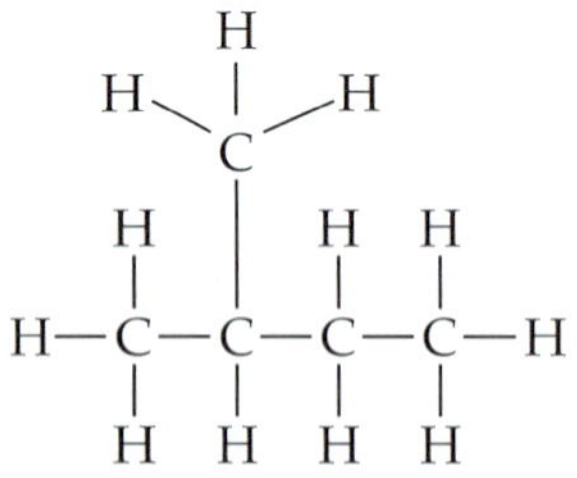 H H—C—H H | H H H—C—C—C—C—H H H H H To draw the complete structural formula a methyl group is added to carbon atom number 2. For this representation the second carbon atom from the left has been chosen. Then the bonds to hydrogen atoms have been added to complete the four bonds to each carbon atom.	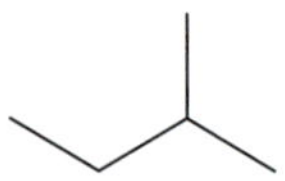 In the line bond formula the 2-methyl group is represented by drawing a single line from either of the carbon atoms within the chain. You choose which carbon atom becomes carbon atom number 2 in the representation. In this structure the second carbon atom from the right has been chosen.

Note that parts of the structural formula can be condensed as shown below on the left where just the branching 2-methyl group formula has been condensed. Other condensed formulas for 2-methylbutane are also shown. All are correct representations.

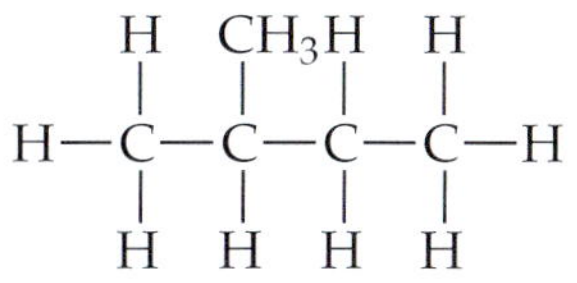

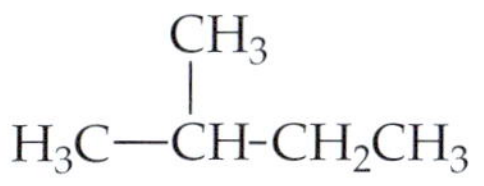

```
    H  CH3H  H
    |   |  |  |
H — C — C — C — C — H
    |   |  |  |
    H   H  H  H
```

```
        CH3
         |
H3C — CH-CH2CH3
```

$CH_3CH(CH_3)CH_2CH_3$

The following structures also correctly represent 2-methylbutane. Remember that molecules move around, so their structures can be drawn in different spatial positions. If the connectivity of the atoms is shown correctly, then the structure is correct and its positioning doesn't matter.

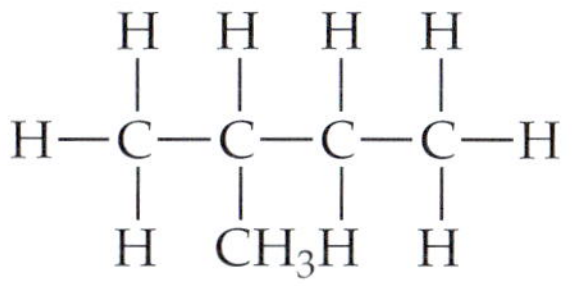

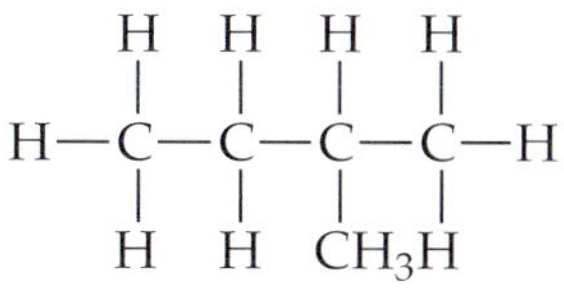

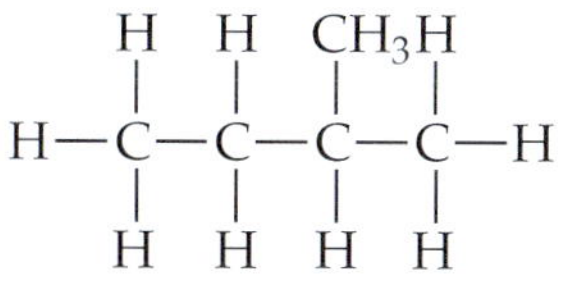

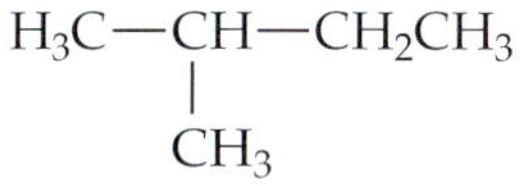

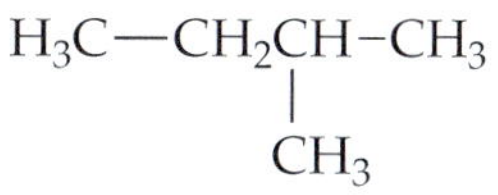

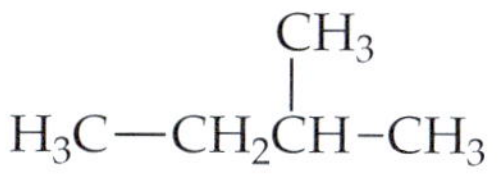

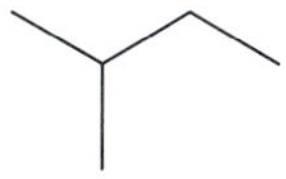

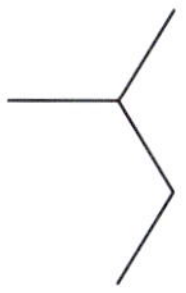

```
    H   H  H  H
    |   |  |  |
H — C — C — C — C — H
    |   |  |  |
    H  CH3H  H
```

```
    H  H  H  H
    |  |  |  |
H — C — C — C — C — H
    |  |  |  |
    H  H CH3H
```

```
    H  H CH3H
    |  |  |  |
H — C — C — C — C — H
    |  |  |  |
    H  H  H  H
```

```
H3C — CH — CH2CH3
       |
      CH3
```

```
H3C — CH2CH-CH3
          |
         CH3
```

```
           CH3
            |
H3C — CH2CH-CH3
```

$CH_3CH_2CH(CH_3)CH_3$

EXERCISE 12.9

Draw line bond formulas for the following alkanes. Also draw the structural formulas if that helps you derive the line bond formulas.

3-ethyl-2-methylhexane	2,4-dimethylpentane	2,3-dimethylbutane
2,2-dimethyl-4-propyloctane	methylcyclopropane	ethylcycloheptane

The steps demonstrated in this section for naming alkanes are the fundamental steps for naming all organic compounds. Throughout the remainder of the module you will learn how the process may be applied to name any organic compound.

12.5 Alkenes

LO: Identify and describe the molecular structures of alkenes.

LO: Examine structural representations of alkenes and derive their IUPAC names.

LO: Interpret the IUPAC names of alkenes and draw representations for their molecular structures.

LO: Describe and write equations for addition reactions of alkenes with halogens.

Alkenes are unsaturated hydrocarbons. The terms 'unsaturated' and 'hydrocarbon' have already been defined in this module and they can be interpreted to understand aspects of alkene molecular structure. Unsaturated hydrocarbons contain only carbon and hydrogen atoms and there are one or more carbon-carbon multiple (double or triple) bonds present in their molecules. For alkenes the term unsaturated refers to the presence of one or more **carbon-carbon double bonds (C=C)**. Alkene names end with the suffix 'ene'. Alkenes with only one double bond have the general formula C_nH_{2n}. The molecular formulas, names and structural representations of the first three members of the alkene family are shown in Table 12.17. Note that structural isomers are possible for alkenes with four or more carbon atoms. This is due to there being more than one possible location for the double bond. Examine all the structural representations for each alkene and see how they relate to each other.

TABLE 12.17 The First Three Members of the Alkene Family

Number of C atoms (n)	2	3	4	4
Molecular formula (C_nH_{2n})	C_2H_4	C_3H_6	C_4H_8	C_4H_8
Alkene name	ethene	propene	but-1-ene*	but-2-ene*
Structural formula	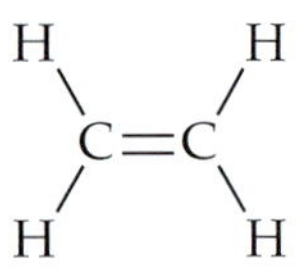 C=C	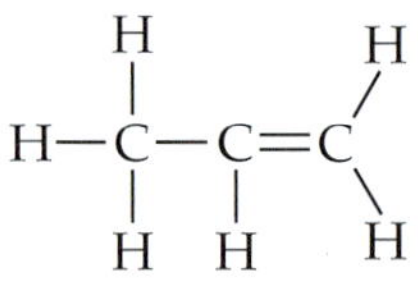 H—C—C=C	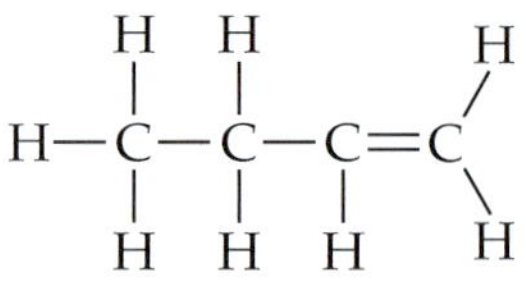 H—C—C—C=C	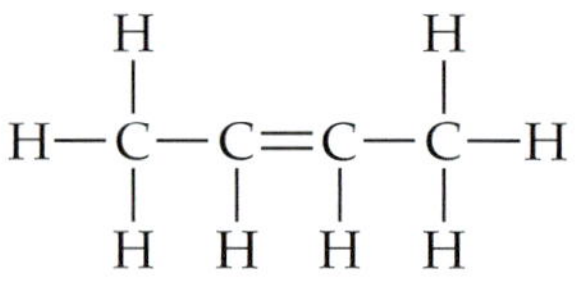 H—C—C=C—C—H
Line bond formula		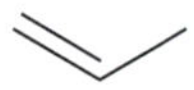		
Condensed formulas	$H_2C{=}CH_2$	$H_2C{=}CH{-}CH_3$	$H_2C{=}CH{-}CH_2CH_3$	$H_3C{-}CH{=}CH{-}CH_3$
	CH_2CH_2	CH_2CHCH_3	$CH_2CHCH_2CH_3$	$CH_3CHCHCH_3$

*The numbers in these names describe the position of the carbon-carbon double bond. The systematic naming of alkenes is fully explained in the following section.

The focus of your introduction to alkenes will be on those that only contain one carbon-carbon double bond, but it's worth noting that they can have more than one double bond. If an alkene has two double bonds it is called a diene, and an alkene with three double bonds may be described as a triene. There are even alkenes that have four or more double bonds, and the compound lycopene is an interesting example. The molecular structure of lycopene is shown in ▶ Figure 12.8(a). Have a look and count that there are 13 double bonds! The system of alternating single and double bonds in the molecular structure is responsible for the colour of the substance. Lycopene is the substance that gives tomatoes their red colour ▶ Figure 12.8(b). The double bonds absorb certain colours of the visible spectrum and transmit red light. It's the alkene double bonds of lycopene that are responsible for its colour.

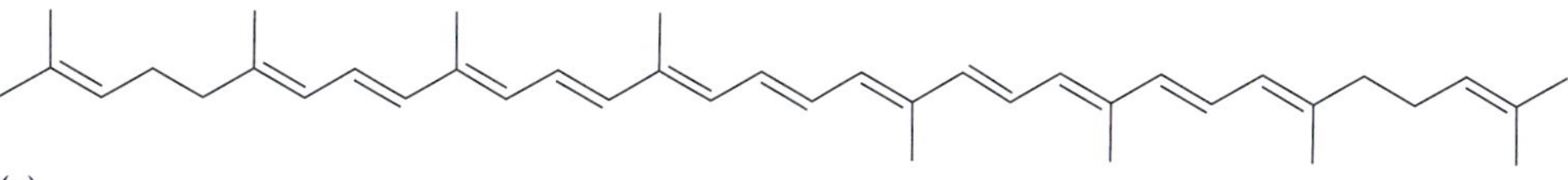

(a)

(b)

▶ **FIGURE 12.8 Lycopene** **(a)** Molecular structure of lycopene, an alkene with 13 carbon-carbon double bonds. **(b)** The molecular structure of lycopene is responsible for the red colour of tomatoes. © Kerry V. McQuaid. Shutterstock

Unlike single bonds, carbon-carbon double bonds (C=C) do not rotate, and this has interesting consequences in terms of the types of isomerism possible for alkenes. The carbon-carbon double bonds in alkenes also confer particular chemical reactivity. Isomerism and the most common reaction of alkenes will be described a little later. The following section explains how to name alkenes. You will see that you don't have to consider too much more than steps you have already learned for the naming of alkanes.

Naming Alkenes

The steps to naming organic compounds explained previously and applied to alkanes are shown in Table 12.18 with the factors to consider when applying them to naming alkenes. Applying the steps to naming alkenes will be reinforced using an example.

TABLE 12.18 The Steps to Naming Organic Compounds Applied to Alkenes

Step	Description of the step	Applying the step to naming alkenes
0	Examine the structural representation to identify the type or classification of the organic compound.	• If a hydrocarbon with a carbon-carbon double bond is represented, then the compound is an alkene and the name will have the suffix **ene.**
1	Identify the parent chain, which is the longest continuous chain of carbon atoms.	• The parent chain of an alkene contains both carbon atoms of the double bond, even if it is not the longest chain.
2	Number the carbon atoms in the parent chain.	• For alkenes the carbon atoms of the parent chain are numbered so the first carbon atom of the double bond has the lowest possible number. • The name of an alkene must describe the position of the double bond in the parent chain, this is done by inserting the number of the first carbon atom in the double bond before the suffix.
3	Identify the substituents and the number of the parent chain carbon atoms that they branch from.	• This step is applied in the same way for alkenes.
4	Write the whole name of the compound as one word.	• This step is applied in the same way for alkenes.

ALKENE 1

```
H       CH2CH3
 \     /
  C=C
 /     \
H       CH2CH2CH3
```

Step 0 The presence of the double bond means the compound is an alkene and the suffix part of its name is **ene**.

Step 1 Identify the parent chain. For alkenes, the parent chain must contain both carbon atoms of the double bond, even if it is not the longest chain. Identification of the parent chain for alkene 1 is shown below.

The longest continuous chain in alkene 1 contains 6 carbon atoms. However, it is not the parent chain because it does not include both carbon atoms of the double bond.

This is the parent chain of alkene 1 because it includes both carbon atoms of the double bond. The parent chain is 5 carbon atoms long, so the parent part of the compound name is **pent**.

Step 2 Number the carbon atoms in the parent chain. Two options for alkene 1 are shown below. The correct option has the lowest possible number for the first carbon atom in the double bond.

```
H       CH2CH3
 \     /
  C=C
 /1  2\
H       CH2CH2CH3
        3   4   5
```

This is the correct option for numbering the parent chain for alkene 1. It gives the lowest possible number to the first carbon atom of the double bond.

```
H       CH2CH3
 \     /
  C=C
 /5  4\
H       CH2CH2CH3
        3   2   1
```

In this option the first carbon atom of the double bond is carbon atom 4 and it is not the correct option for numbering the parent chain.

The name of an alkene must also describe the position of the double bond in the parent chain. This is achieved by inserting the number of the first carbon atom of the double bond before the suffix. For alkene 1 this is carbon atom 1 and the suffix is thus extended to **1-ene**.

Step 3 Identify the substituents and the number of the parent chain carbon atom that they branch from. Alkene 1 has a $-CH_2CH_3$ (ethyl) group branching from carbon atom 2 of the parent chain. The name of the compound will contain the prefix 2-ethyl.

```
H       (CH2CH3)
 \     /
  C=C
 /1  2\
H       CH2CH2CH3
        3   4   5
```

The branching $-CH_2CH_3$ group is called ethyl and it is connected to carbon atom 2 of the parent chain.
In the compound name this is written as **2-ethyl**.

Step 4 Write the whole name of the compound as one word. All parts of the name for alkene 1 have been identified and the full name can be constructed as shown below. Note the use of hyphens to separate the numbers and letters in the name, hyphens are placed on either side of numbers that appear within compound names.

Alkene 1	Components of the name identified	Compound name
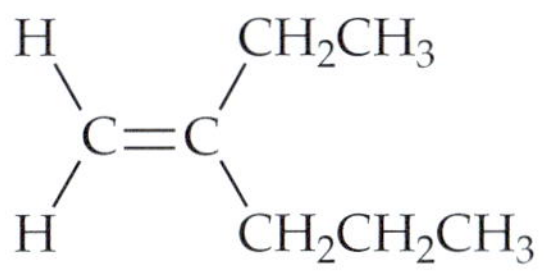 $(H)_2C=C(CH_2CH_3)(CH_2CH_2CH_3)$	Prefix: 2-ethyl Parent: pent Suffix: 1-ene	**2-ethylpent-1-ene**

EXERCISE 12.10

Derive the names of the following alkenes.

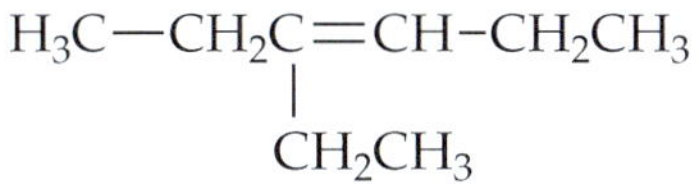 $H_3C-CH_2C(CH_2CH_3)=CH-CH_2CH_3$	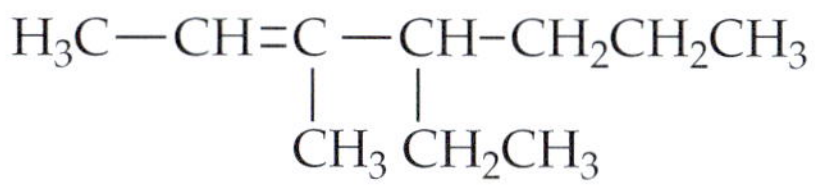 $H_3C-CH=C(CH_3)-CH(CH_2CH_3)-CH_2CH_2CH_3$
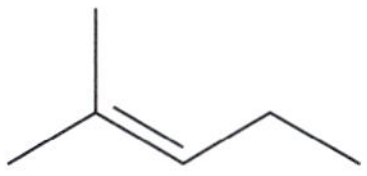	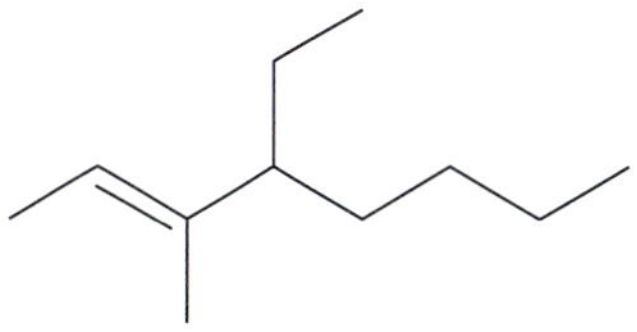

EXERCISE 12.11

Draw line bond formula structures for the following alkenes. You may find it helpful to draw a structural formula first.

2,3-dimethylpent-1-ene	3-ethyl-6-methylhept-3-ene
3,4,5-trimethyloct-1-ene	3-ethylhex-2-ene

Note that cyclic alkenes are also possible, and their names are logical based on the conventions described previously for cycloalkanes. See how the names of the cycloalkenes shown below have names based on the cycloalkane with the same number of carbon atoms. The names have the suffix ene instead of ane. Note that the number of the first carbon atom in the double bond does not have to be specified in the names of cycloalkenes that have no branching substituents. The double bond can be drawn in any position in these cases.

Isomerism in Alkenes

Structural isomers are possible for alkenes with four or more carbon atoms because the double bond can have different locations along the parent chain. Table 12.19 shows the structural isomers for alkenes with 4 and 5 carbon atoms. The compounds are differentiated by their names, which clearly describe the position of the double bond.

TABLE 12.19 Names of Alkene Structural Isomers Must Describe the Position of the Double Bond

Structural isomers of butene	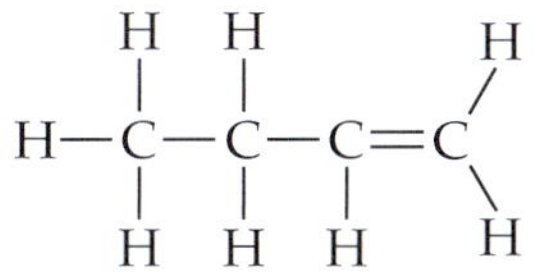 H H H H—C—C—C=C H H H H	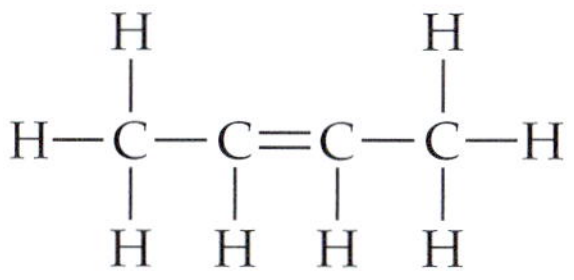 H H H—C—C=C—C—H H H H H
Compound names	but-1-ene	but-2-ene
Structural isomers of pentene	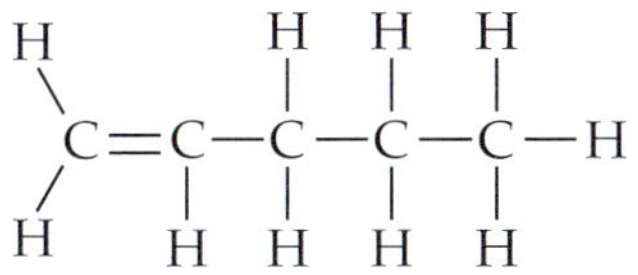 H H H H C=C—C—C—C—H H H H H H	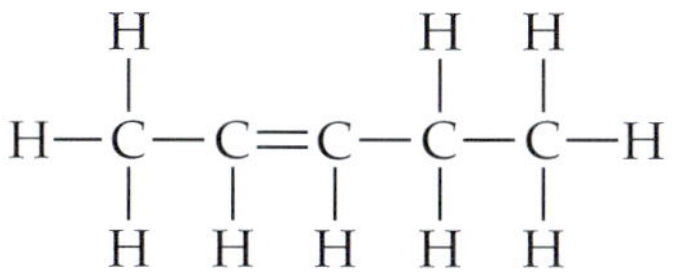 H H H H—C—C=C—C—C—H H H H H H
Compound names	pent-1-ene	pent-2-ene

Unlike single bonds, rotation of carbon-carbon double bonds is not possible. For alkenes this gives rise to another form of isomerism called geometric isomerism. Consider the three 'forms' of butane shown below. These all represent the same compound and the different structures simply result from rotation of carbon-carbon single bonds.

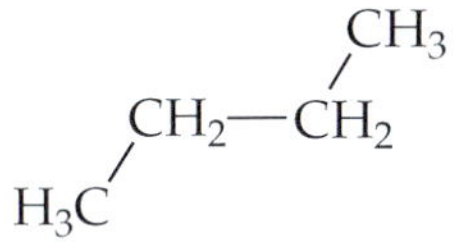

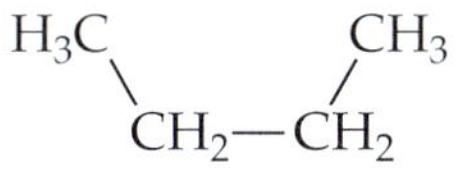

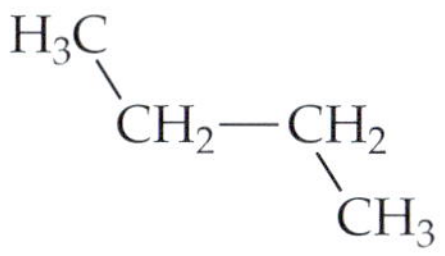

```
        CH3                H3C        CH3              H3C
       /                      \      /                    \
   CH2—CH2                    CH2—CH2                     CH2—CH2
  /                                                              \
H3C                                                              CH3
```

The two compounds shown below are both but-2-ene, but they are distinctly different compounds. It is not possible to convert from one structure to the other by bond rotation because rotation of the double bonds cannot occur. These two 'forms' of but-2-ene are different compounds called geometric isomers. Since both of the compounds shown below are correctly called but-2-ene, we need some way to differentiate them in their names. The prefixes 'cis' and 'trans' are used to differentiate alkene geometric isomers as follows. The appropriate prefix is written in italics at the front of the name, followed by a hyphen.

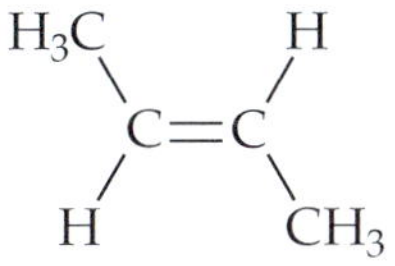

```
H3C       H                 H3C       CH3
   \     /                     \     /
    C=C                         C=C
   /     \                     /     \
  H       CH3                 H       H
```

The isomer with **identical substituents on opposite sides** of the double bond is called the **trans** isomer, the name of this compound is: ***trans*-but-2-ene**.

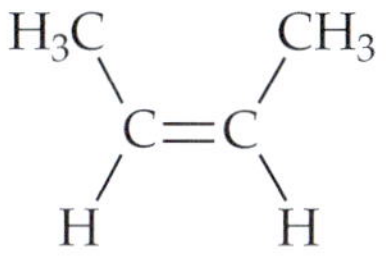

The isomer with **identical substituents on the same side** of the double bond is called the **cis** isomer, the name of this compound is: ***cis*-but-2-ene**.

In the following examples the identical substituents are hydrogen atoms bonded to each of the alkene carbon atoms. Cover the compound names given and have a go at naming these alkenes, including the term that describes the configuration of the double bond (cis or trans). Check the name only after you've tried to name it yourself.

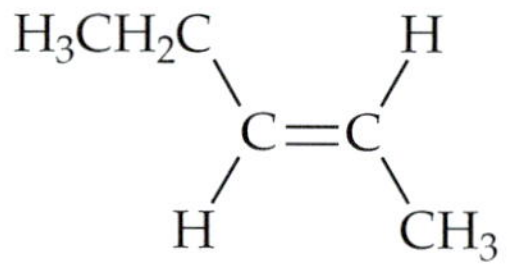

H_3CH_2C / H — C=C — H / CH_3 *trans*-pent-2-ene

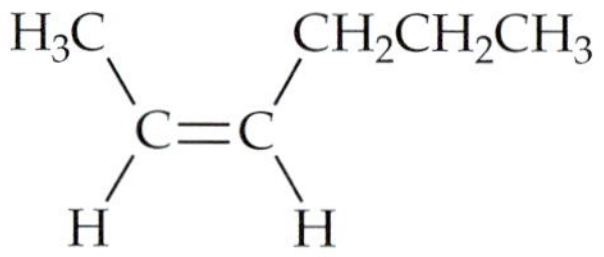

H_3C / $CH_2CH_2CH_3$ — C=C — H / H *cis*-hex-2-ene

The cis and trans isomers of an alkene can be represented using line bond formulas. It can be difficult at first to work out if an alkene is cis or trans by examining its line bond formula, but they can be distinguished using this type of representation. Consider the isomers of hept-3-ene in Table 12.20 and see that their line bond formulas look different. With experience you can differentiate geometric isomers by examining their line bond formulas, but the bonds to hydrogen atoms not shown in a line bond formula can be drawn on the structure to clearly see the relationship between them.

TABLE 12.20 The Isomers of Hept-3-ene Represented with Line Bond Formulas

Isomer name	Line bond formula	Line bond formula with bonds to hydrogen atoms shown for the alkene carbon atoms. This makes the cis or trans relationship between identical substituents obvious.	
trans-hept-3-ene	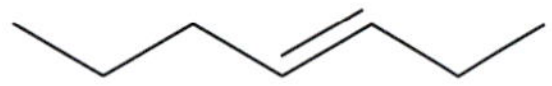	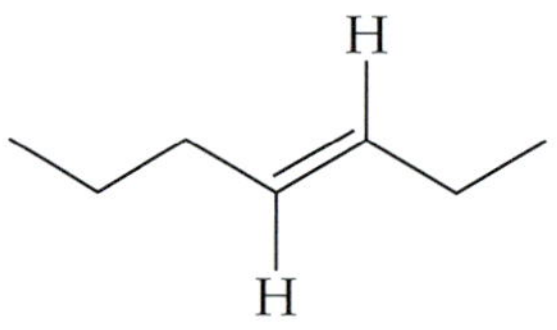 H, H	Identical substituents (H atoms) are on opposite sides of the double bond (trans.)
cis-hept-3-ene	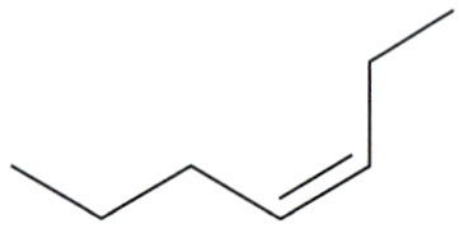	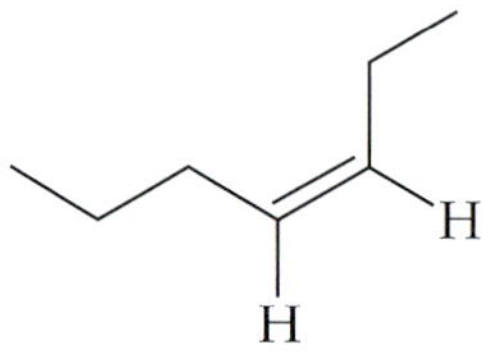 H, H	Identical substituents (H atoms) are on the same side of the double bond (cis).

The structural representation of any alkene can be examined to determine if it could have a geometric isomer. *Geometric isomerism is possible if both of the alkene carbon atoms have two different substituents. Geometric isomerism is not possible if one of the alkene carbon atoms is bonded to two identical substituents.* The examples in Table 12.21 show how these statements are true. To determine if an alkene could have a geometric isomer, it can be redrawn with the substituents on one double bond carbon atom swapped around.

TABLE 12.21 Analysis of Alkene Structures to Determine if Geometric Isomerism is Possible

Alkene	Redraw by swapping the groups of one alkene carbon atom	Redrawn alkene	Geometric isomerism possible?
(H)(Cl)C=C(H)(Cl)	(H)(Cl)C=C(H)(Cl), groups on right carbon circled	(H)(Cl)C=C(Cl)(H)	Yes. The redrawn structure is the geometric isomer (trans) of the original (cis) alkene.
(Cl)(Cl)C=C(H)(CH_3)	(Cl)(Cl)C=C(H)(CH_3), groups on right carbon circled	(Cl)(Cl)C=C(CH_3)(H)	No. The redrawn structure is identical to the original, it has simply been 'flipped'.

If the redrawn structure is identical to the original, then geometric isomerism is not possible. If the redrawn structure is an isomer, then geometric isomerism is possible. In the above examples, geometric isomerism is possible for the first alkene because both carbon atoms of the double bond have two different substituents. Geometric isomerism is not possible for the second alkene because one of the double bond carbon atoms has identical substituents.

EXERCISE 12.12

Which of the following alkenes could have a geometric isomer?

(Cl)(H)C=C(H)(CH_3) (a)	(Br)(H)C=C(CH_2CH_3)(H) (b)	(H)(H)C=C(CH_3)(Cl) (c)
(H)(Br)C=C(CH_3)(CH_3) (d)	(H)(Cl)C=C(CH_2CH_3)(CH_2CH_3) (e)	(Br)(H_3C)C=C(CH_2CH_3)(H) (f)

Reactions of Alkenes

The carbon-carbon double bond in an alkene is a site for chemical reactivity. One of the major reactions of alkenes is addition of a small molecule to the carbon atoms of the double bond. These are examples of a broad class of organic reactions called **addition reactions**. A number of small molecules can add to a double bond (eg H_2O, HBr, H_2, X_2). We will examine addition of bromine (Br_2) to alkene double bonds, but other halogens (Cl_2, I_2) react with alkenes in the same way. The equations in Table 12.22 both show the reaction of the simplest alkene (ethene) with bromine. The reactant is unsaturated and the product is saturated (it's called 1,2-dibromoethane). The net effect of the reaction is the addition of bromine across the double bond and there is only one product. The equations in Table 12.22 show two correct ways to represent this reaction.

TABLE 12.22 Equations for the Reaction of Ethene with Bromine (Br_2)

Equation	Description
$H_2C{=}CH_2 + Br_2 \longrightarrow H{-}C(H)(Br){-}C(H)(Br){-}H$	Equation for the reaction of ethene with bromine (Br_2). The bromine atoms from Br_2 add to the carbon atoms of the alkene double bond. The double bond is not present in the product.
$H_2C{=}CH_2 \xrightarrow{Br_2} H{-}C(H)(Br){-}C(H)(Br){-}H$	This is another correct way to write the equation for this reaction, with the formula for reactant Br_2 written above the reaction arrow.

▲ **FIGURE 12.9 Chemical test for alkenes**

All alkenes react with halogens in this way and it is a rapid reaction that proceeds in the absence of light, unlike the reaction of alkanes with halogens, which needs UV light to make the reaction go. The observations made during the reaction of alkenes with bromine can serve as the basis for a chemical test, as shown in ◀ Figure 12.9. The test tube on the left contains a solution of bromine, which has a distinctive colour. As bromine reacts, its colour is lost. The test tube on the right demonstrates the observation that occurs upon addition of an alkene to the bromine solution. The loss of colour occurs immediately upon addition of the alkene. This test can be used as an indicator of whether or not an unknown compound is an alkene.

Examine the following examples of equations (Table 12.23) showing reactions of alkenes with halogens and see that they all proceed in the same manner. The two atoms of the halogen molecule add one each to the two carbon atoms of the alkene double bond. The double bond is not present in the product and there is only one product, no by-product forms. Note how the equations can also be written using line bond formulas.

TABLE 12.23 Equations for Reactions of Alkenes with Halogens

Reaction	Reaction
$H_3C(H)C{=}C(H)CH_3 \xrightarrow{Br_2} H_3C{-}C(Br)(H){-}C(Br)(H){-}CH_3$	$H_3C(H)C{=}C(H)CH_3 \xrightarrow{Cl_2} H_3C{-}C(Cl)(H){-}C(H)(Cl){-}CH_3$
(line bond formula of alkene) $\xrightarrow{Cl_2}$ (line bond formula of product with Cl, Cl)	(cyclohexene) $\xrightarrow{Br_2}$ (1,2-dibromocyclohexane: Br, Br)

EXERCISE 12.13

Complete the following equations for reactions of alkenes with halogens by drawing the structure of the missing reactant, product, or halogen as appropriate.

$H_3C(H)C{=}C(H)CH_2CH_3$	$\longrightarrow$	$H_3C{-}CHI{-}CHI{-}CH_2CH_3$
$H_2C{=}C(CH_3)CH_2CH_3$	$\xrightarrow{Cl_2}$	
	$\xrightarrow{Br_2}$	1,2-dibromocyclopentane (Br, Br)

12.6 Aromatic Hydrocarbons

LO: Identify and describe aromatic hydrocarbons.

LO: Distinguish aliphatic and aromatic hydrocarbons.

LO: Derive the names of monosubstituted benzene compounds

Any organic compound that is a hydrocarbon can be further classified using the terms aliphatic and aromatic. The hydrocarbons introduced so far, alkanes and alkenes, can actually be grouped together under the term **aliphatic hydrocarbons**. These are hydrocarbons that do not contain a benzene ring within their structure. To be classified as an **aromatic hydrocarbon**, the compound must contain at least one benzene ring. While aliphatic hydrocarbons can be saturated (alkanes) or unsaturated (alkenes), all aromatic hydrocarbons are unsaturated. Aromatic hydrocarbons belong to their own classification because they exhibit special properties in terms of their chemical reactivity. The distinguishing points about aromatic and aliphatic hydrocarbons are summarised in Table 12.24.

TABLE 12.24 Distinguishing Points about Aromatic and Aliphatic Hydrocarbons

Aromatic hydrocarbons	Aliphatic hydrocarbons
• Any hydrocarbon that contains at least one benzene ring. • All aromatic hydrocarbons are unsaturated • Special properties related to chemical reactivity.	• Any hydrocarbon that does not contain a benzene ring. • May be saturated (alkanes) or unsaturated (alkenes).

To understand the difference between aliphatic and aromatic hydrocarbons, the organic compound known as benzene must be introduced. Benzene has molecular formula C_6H_6 and the structure involves the six carbon atoms connected in a ring. Each carbon atom is bonded to one hydrogen atom and there is a system of alternating double and single bonds in the ring. Structural and line bond formulas for benzene are shown in Table 12.25. Note the different resonance structures that are possible due to the alternative positioning of the double and single bonds within the ring. Examine carefully each carbon atom in the representations of benzene shown and see that they all have trigonal planar geometry. The consequence of this is that the entire benzene molecule is flat, and this can be seen more clearly in a ball and stick model of benzene ▼ Figure 12.10 (a). The representation of benzene with alternating single and double bonds is the only type used in this book. You will see another correct line bond representation of benzene ▼ Figure 12.10 (b) in other sources. This representation looks like a hexagon with a circle inside it. The science behind why this is a good representation of benzene is beyond the scope of this introduction to organic compounds. If you study chemistry further you will learn about the molecular structure of benzene and the sense of the line bond formula in Figure 12.10 (b) will become clear. The purpose of introducing it is so you know what it represents when you see it elsewhere.

TABLE 12.25 Structural and Line Bond Formulas for Benzene

Structural formulas for benzene resonance structures		
Line bond formulas for benzene resonance structures		

Benzene can be considered as the 'parent' of all aromatic compounds. If an aromatic compound only contains carbon and hydrogen, then it is an aromatic hydrocarbon. Any compound that contains at least one benzene ring and elements in addition to carbon and hydrogen may simply be classified as an aromatic compound.

When a benzene molecule has one or more hydrogen atoms replaced with another atom or group of atoms, these compounds are called substituted benzenes. Naming monosubstituted benzenes is very simple. The parent name is benzene and the substituent is identified using a prefix.

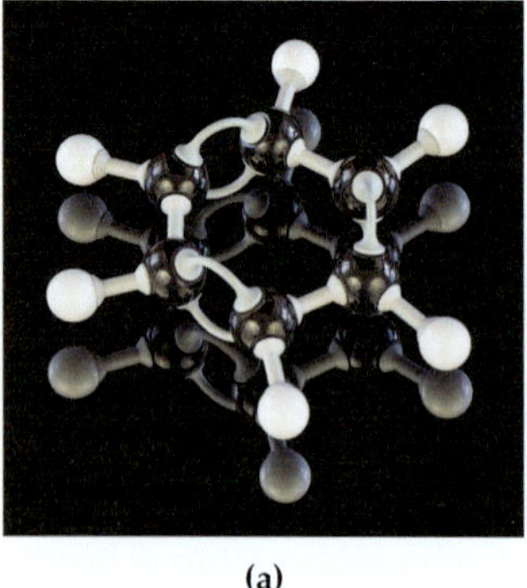

(a)

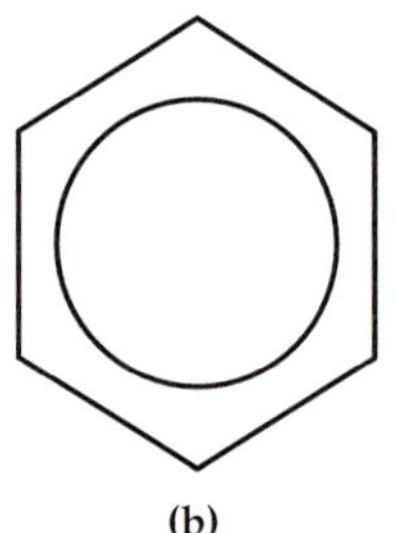

(b)

▶ **FIGURE 12.10 Benzene**
(a) Ball and stick model of benzene showing the overall planar structure resulting from the trigonal planar geometry of the individual carbon atoms. **(b)** Another correct line bond formula for benzene that you will see in other sources. This representation is not used in this book

EXERCISE 12.14

Classify the following hydrocarbons as aliphatic or aromatic.

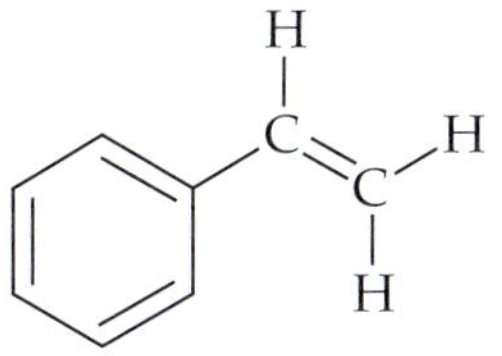 (a)	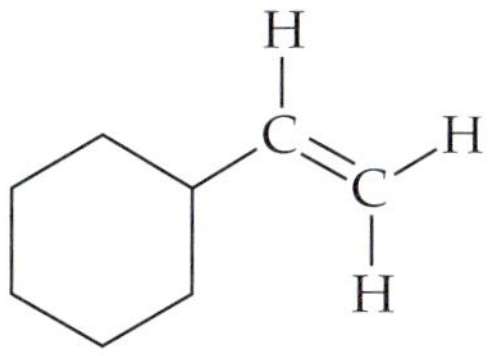 (b)	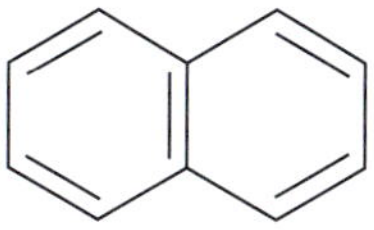 (c)
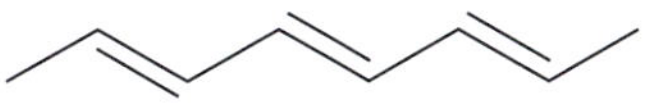 (d)	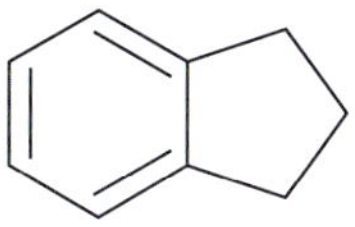 (e)	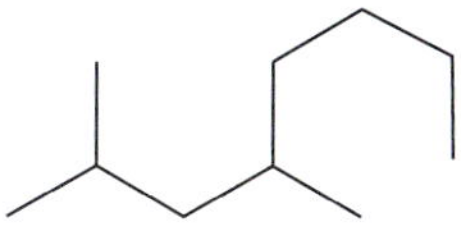 (f)
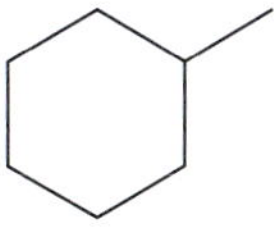 (g)	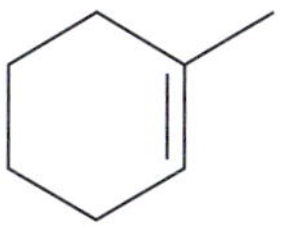 (h)	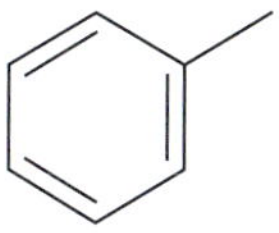 (i)

EXERCISE 12.15

Use the names of the two monosubstituted benzene compounds given as a guide to name the others.

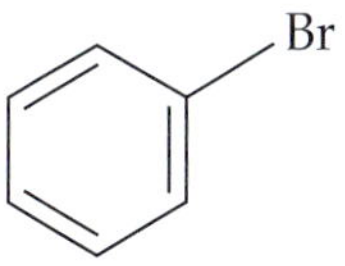

Br	bromobenzene	CH_3	methylbenzene
$CH_2CH_2CH_3$		Cl	
F			

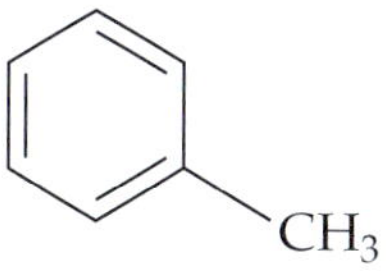

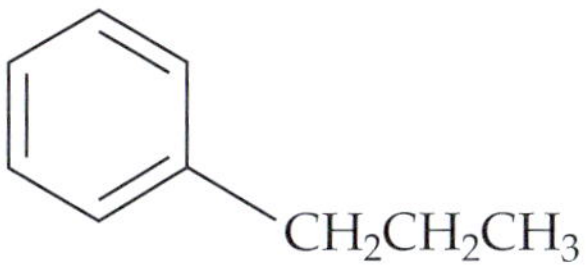

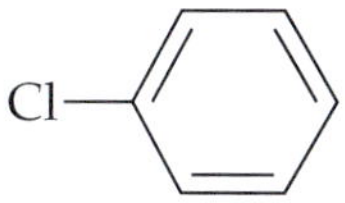

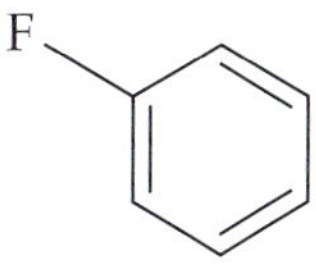

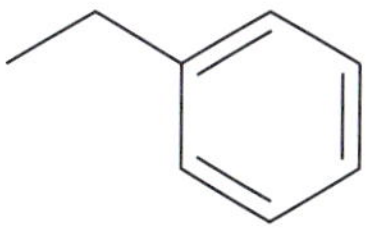

12.7 Organic Compounds with Halogen Atoms

LO: Examine structural representations of organic compounds that contain halogen atoms and derive their IUPAC names.

LO: Interpret the IUPAC names of organic compounds that contain halogen atoms and draw representations for their molecular structures.

LO: Explain the physical properties of organic compounds that contain halogen atoms.

The focus of this module so far has been on organic compounds that only contain carbon and hydrogen. It's now time to introduce organic compounds that contain additional elements. The presence of other elements has consequences on both the physical properties and chemical reactivity of organic compounds. The presence of elements in addition to carbon and hydrogen in organic compounds adds another layer of diversity in this classification. In this section, organic compounds that contain halogen atoms are introduced.

Recall that the halogens are the Group 7 (17) elements ▼ (Figure 12.11) and the letter 'X' may be used as a general representation of any halogen atom. The atoms of elements that are halogens have 7 valence electrons and they can attain a noble gas electron configuration by gaining one electron to form anions (X^-). Another way halogen atoms can complete their octet is by sharing one electron in a single covalent bond, and they can form these single covalent bonds with carbon atoms in organic compounds. When halogens are present in organic compounds they are named as a substituent using a prefix. This will be demonstrated in this section for **haloalkanes**, which are alkanes that have one or more hydrogen atoms replaced by halogen atoms. To name a haloalkane you simply add the appropriate prefix to the name of the parent alkane. You have already seen halogen prefixes in the names of compounds that consist of a benzene ring with a halogen substituent (Section 12.6). These are the prefixes used for each halogen. The prefix for astatine is not included as it is a rare, radioactive element not commonly observed in organic compounds.

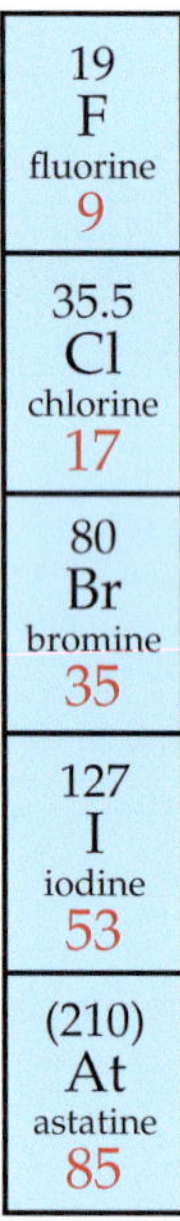

▲ **FIGURE 12.11 Halogens** The halogens are the Group 7 (17). elements.

Halogen	fluorine (F)	chlorine (Cl)	bromine (Br)	iodine (I)
Prefix	fluoro	chloro	bromo	iodo

Structures of the simplest haloalkanes are shown below, see how the names given consist of the whole parent alkane name with the appropriate halogen prefix. Using the ones given as a guide, derive the haloalkane names and condensed formulas that are not shown in the table. Note that the halogen atom can be shown in any position around a carbon atom, and for the haloalkanes with two carbon atoms the halogen can be bonded to either one.

EXERCISE 12.16

Naming haloalkanes.

Haloalkane structural formula	H | H—C—F | H	Cl | H—C—H | H	H | Br—C—H | H	H | H—C—I | H
Name	fluoromethane			
Condensed formula	CH_3F		CH_3Br	
Haloalkane structural formula	H H | | H—C—C—Br | | H H	H H | | H—C—C—H | | H F	H H | | I—C—C—H | | H H	Cl H | | H—C—C—H | | H H
Name			iodoethane	
Condensed formula	CH_3CH_2Br		CH_3CH_2I	

The names of cycloalkanes that have one halogen substituent consist of the whole parent cycloalkane name with the appropriate halogen prefix. Using the example given below as a guide, name the following halogenated cycloalkanes. Note that if there is only one halogen atom there is no need to number the carbon atoms in the cycloalkane. The naming of cycloalkanes that have halogen atoms bonded to more than one carbon atom is not covered in this module.

EXERCISE 12.17

Naming halogenated cycloalkanes.

Halogenated cycloalkane structural formula	Br (on cyclodecane ring)	F (on cyclohexane ring)	I (on cyclooctane ring)	Cl (on cyclopentane ring)
Name	bromocyclodecane			

Structural isomers are possible for haloalkanes with three or more carbon atoms because the halogen atom may be bonded to different carbon atoms within the chain. The two structural isomers of bromopropane are shown below (Table 12.26) to demonstrate how the isomers are differentiated in their structural formulas, condensed formulas and names. It's clear to see the difference between the isomers by comparing the structural formulas, one has the bromine atom bonded to a terminal carbon atom and the bromine atom is bonded to the carbon atom within the chain in the other. Another way to differentiate the isomers is in their distinct names. Information about which carbon atom in the chain the halogen atom is bonded to is built into the name. The position of the halogen atom is designated by numbering the longest carbon chain and assigning the lowest possible number to the carbon atom bonded to the halogen. This follows the naming conventions for branching alkyl groups that you have seen previously. Examine the names and structures of the bromopropane isomers and note how the isomer with the bromine atom bonded to the terminal carbon atom is called 1-bromopropane, and the one with bromine bonded to the carbon atom within the chain is called 2-bromopropane. Note that there is no isomer called 3-bromopropane because of the convention that assigns the lowest possible number to the positon of the halogen in the carbon chain.

TABLE 12.26 Names of Haloalkane Structural Isomers Must Describe the Position of the Halogen Atom

Bromopropane isomer structural formula	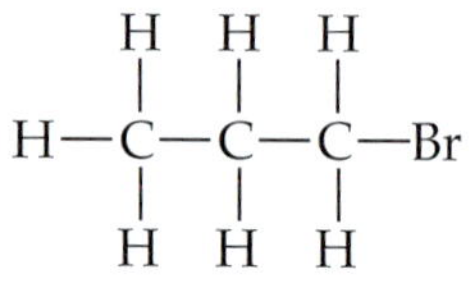 H H H / H—C—C—C—Br / H H H	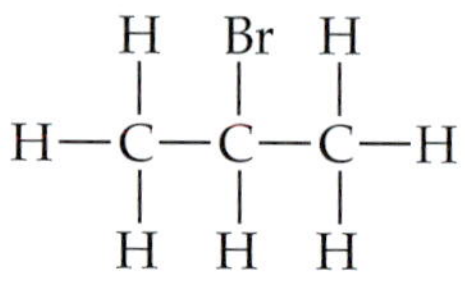 H Br H / H—C—C—C—H / H H H
Name	1-bromopropane	2-bromopropane
Condensed formula	$CH_3CH_2CH_2Br$	$CH_3CHBrCH_3$

Structural representations of haloalkanes in the following examples are drawn using line bond formulas so you can practice interpreting them. Here is an example of naming a haloalkane that has structural isomers. It's simply a matter of applying naming conventions that you have already seen for alkanes, with the halogen atom named as a substituent.

HALOALKANE 1

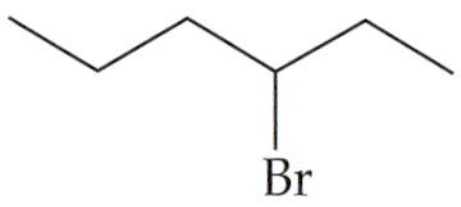

Br

Step 0 Examine the structural representation to identify the type or classification of the organic compound. For haloalkane 1, the suffix will be **ane** because the hydrocarbon framework is an alkane (saturated hydrocarbon). The halogen is named as a substituent.

Step 1 Identify the parent chain, which is the longest continuous chain of carbon atoms. Haloalkane 1 has only one carbon chain, which is six atoms long. The parent name is **hex**.

Step 2 Number the carbon atoms in the parent chain. The correct option is the one that gives the lowest possible number to the position of the halogen substituent. The two options for numbering the parent chain in haloalkane 1 are shown.

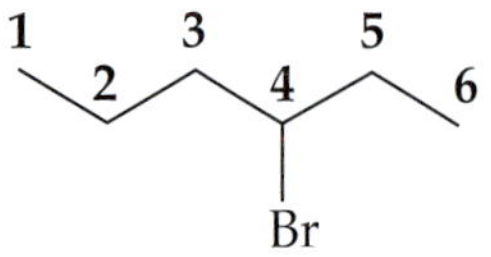

1 2 3 4 5 6, Br

This is not the correct option because it gives a higher number to the carbon atom bonded to the halogen compared to the alternative.

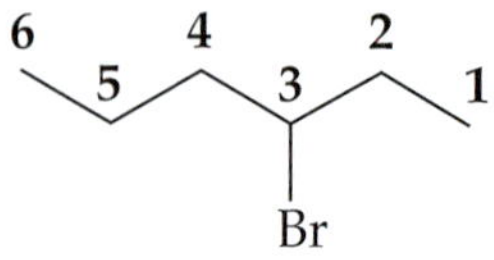

6 5 4 3 2 1, Br

This is the correct numbering of the parent chain as it gives the lowest possible number (3) to the carbon atom bonded to the halogen.

Step 3 Identify the substituents and the number of the parent chain carbon atom that they branch from. In haloalkane 1 the halogen is a bromine atom and the prefix name will be bromo. The position of the bromine atom at carbon atom 3 also has to be made clear in the name, so the complete prefix is **3-bromo**.

Step 4 Write the whole name of the compound as one word. Putting the components of the name together for haloalkane 1 gives **3-bromohexane**. The name differentiates the compound from its structural isomers that have the bromine atom bonded to carbon atom 1 (1-bromohexane) or carbon atom 2 (2-bromohexane) shown below.

1-bromohexane

Br

2-bromohexane

Br

Naming haloalkanes that have more than one of the same halogen substituent is managed in the same way as the presence of more than one of the same alkyl substituent, which was described earlier for the naming of alkanes (Section 12.4). Haloalkane 2 has two fluorine atoms and the steps to naming it are given in this example.

HALOALKANE 2

F F

Step 0 Examine the structural representation to identify the type or classification of the organic compound. For haloalkane 2, the suffix will be **ane** because the hydrocarbon framework is an alkane (saturated hydrocarbon). The two halogen atoms present are named as substituents.

Step 1 Identify the parent chain, which is the longest continuous chain of carbon atoms. Haloalkane 2 has only one carbon chain, which is seven atoms long. The parent name is **hept**.

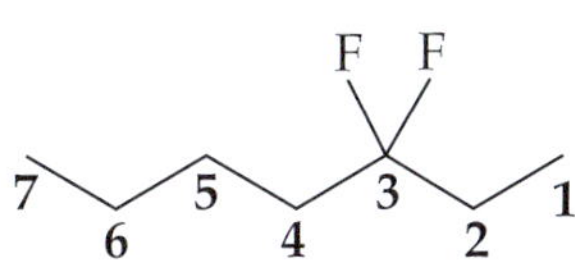

Step 2 Number the carbon atoms in the parent chain. The correct option is the one that gives the lowest possible number to the position of the halogen substituent. Only the correct numbering option is shown for haloalkane 2. Imagine the alternative numbering and see how it gives a higher number to the carbon atom bonded to the halogen atoms.

Step 3 Identify the substituents and the number of the parent chain carbon atom that they branch from. In haloalkane 2 there are two fluorine atoms bonded to carbon atom 3 of the parent chain. The prefix for substituent fluorine atoms is fluoro, and the presence of two fluorine atoms will be included in the name as **difluoro**. There are several options for where the two fluorine atoms could be bonded to the parent chain, so the name of this compound has to make it clear that they are both bonded to carbon atom 3. This is done by using the prefix **3,3-difluoro**. The 'difluoro' part of the prefix clearly states that there are two fluorine atoms in the structure, while the '3,3' part describes that both fluorine atoms are bonded to carbon atom 3 of the parent chain.

Step 4 Write the whole name of the compound as one word. Putting the components of the name together for haloalkane 2 gives **3,3-difluoroheptane**. The name differentiates the compound from its various structural isomers that could have the two fluorine atoms bonded to another carbon atom, or those with the two fluorine atoms bonded to different carbon atoms.

Below is a direct comparison of haloalkane structural isomers to highlight how they are differentiated in their names. Both compounds are called 'dichloropentane' because they have two substituent chlorine atoms connected to an alkane framework with a straight chain of five carbon atoms. The difference between the compounds is the location of the substituent chlorine atoms, and the numbers included in the prefix fully describe their locations. The numbers to include in the names to differentiate the structural isomers are derived by applying the conventions for numbering the parent carbon atom chain. The correct option gives the lowest possible number to the first chloro substituent.

2,3-dichloropentane

Cl

Cl

1,4-dichloropentane

Cl

Cl

Correct numbering – lowest possible number (2) for the first chloro substituent.

The prefix is 2,3-dichloro

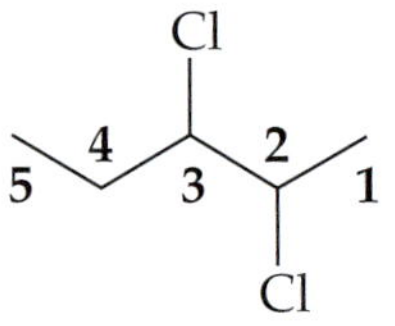

Incorrect numbering – gives a higher number for the for the first chloro substituent compared to the alternative.

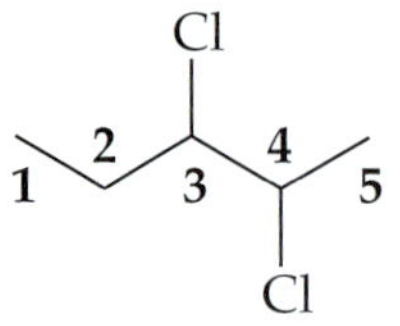

Correct numbering – lowest possible number (1) for the first chloro substituent

The prefix is 1,4-dichloro

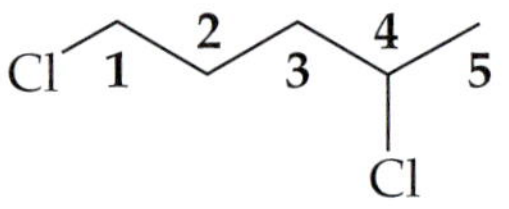

Incorrect numbering – gives a higher number for the for the first chloro substituent compared to the alternative.

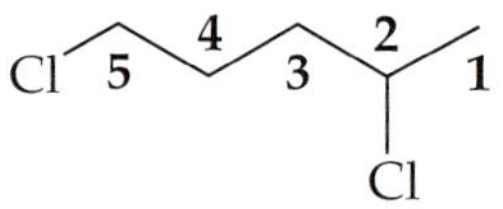

EXERCISE 12.18

Name the following haloalkanes.

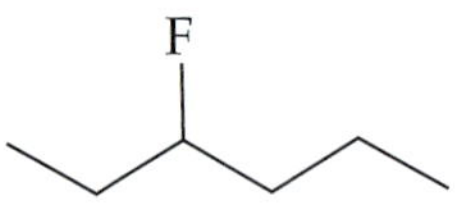	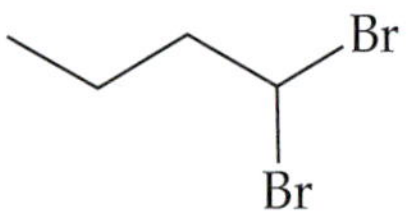
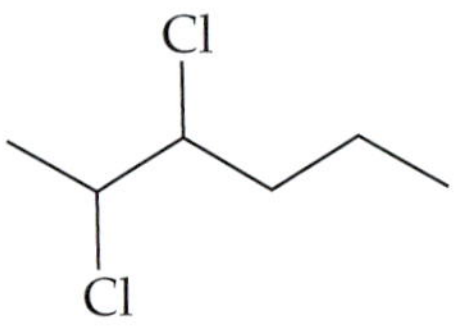	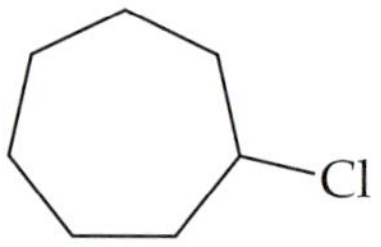

EXERCISE 12.19

Draw line bond formulas for the following haloalkanes. You may find it helpful to draw a structural formula first.

bromocyclopentane	2-chlorobutane
2,4-dibromohexane	2-fluoropentane

The presence of halogen atoms in organic compounds influences both chemical reactivity and physical properties with the focus in this module being on physical properties. The fact that physical properties of substances depend on the type of intermolecular forces (IMFs) that occur between the individual molecules that make up the substance was raised earlier in the module when the physical properties of alkanes were addressed (Section 12.4). The IMFs that are possible for any given molecule depend on the molecular structure. An important factor in determining which IMFs are possible is whether a molecule is polar or non-polar. One factor influencing the physical properties of haloalkanes is that the molecules are generally polar. Let's consider the difference between methane and chloromethane to see how the presence of a halogen can confer polarity.

For a molecule to be polar there must be polar bonds and the shape of the molecule means that there is asymmetrical distribution of electron density throughout. The difference between methane and chloromethane is just one atom, but the presence of the chlorine atom in the latter makes the molecule polar as explained below. Stronger IMFs are possible for the polar molecule.

Compound name	Structural formula	Molecular shape	Polarity and IMFs
methane	H | H—C—H | H	H | C H / ▲ \ H H	**Non-polar** due to the symmetrical distribution of the slightly polar C–H bonds throughout the tetrahedral molecule. **Dispersion** is the only IMF possible.
chloromethane	Cl | H—C—H | H	δ– Cl | C H / ▲ \ H H δ+	**Polar** due to the presence of a polar C–Cl bond (chlorine is more electronegative than carbon) and the asymmetrical distribution of electron density throughout the tetrahedral molecule. The molecule has a dipole moment as indicated by the δ– and δ+ symbols shown on the structure, and may be described as polar. **Dipole-dipole attraction**, which is relatively stronger than dispersion, is possible.

Boiling point is a common physical property to compare when discussing the differences in IMFs between organic compounds. Boiling points for an alkane and haloalkane with the same number of carbon atoms are shown in Table 12.27. The compound 1-chloropropane has a much higher boiling point than propane, so the IMFs must be stronger for 1-chloropropane. The higher boiling point is because more energy is required to overcome attractive forces between 1-chloropropane molecules and bring about the phase change from liquid to gas during boiling. There are two reasons 1-chloropropane has stronger IMFs. Firstly, it has a higher molar mass and therefore greater electron density, so dispersion forces are stronger. Secondly, dipole-dipole attraction is possible for 1-chloropropane because it is a polar molecule. This is due to the presence of the electronegative chlorine atom, which results in an overall dipole moment for the molecule. In relative terms dipole-dipole attraction is a stronger IMF than dispersion, which is the only possible IMF for non-polar propane.

TABLE 12.27 Comparison of Alkane and Haloalkane Boiling Points

Structure	$H-\underset{H}{\overset{H}{C}}-\underset{H}{\overset{H}{C}}-\underset{H}{\overset{H}{C}}-H$	$H-\underset{H}{\overset{H}{C}}-\underset{H}{\overset{H}{C}}-\underset{H}{\overset{H}{C}}-Cl$
Compound name	propane	1-chloropropane
Boiling point	–42 °C	46 °C
Molar mass	44.1 g/mol	78.5 g/mol

In summary, the presence of a halogen atom in an organic compound molecule can confer polarity. This is due to the higher electronegativity of halogen atoms compared to carbon atoms and the consequent polarity of C–X bonds. If these polar C–X bonds are present and they result in asymmetrical distribution of electron density throughout the molecule, then the molecule is polar.

12.8 Functional Groups

LO: Examine representations of organic compound molecular structures and identify functional groups.

A **functional group** is an atom or group of atoms attached to a hydrocarbon framework, and they are used as a basis for the classification of organic compounds. The physical properties and chemical reactivity of organic compounds can be predicted based on the presence of particular functional groups. The functional groups to be covered will be introduced briefly in this section with understanding about how their presence influences physical properties and chemical reactivity of organic compounds developed as you progress through the rest of this module.

R Groups

In this section, and in representations of organic compounds elsewhere, you will see the letter 'R' used in structural, condensed and line bond formulas. This is the convention used to represent any hydrocarbon framework (straight or branched chain of carbon atoms, ring of carbon atoms, benzene ring). R groups can be thought of as the 'rest' of the molecule when the focus is on a particular functional group or another part of an organic compound. In this section R groups are used in the general representations of functional groups. If more than one R group is shown in a structure, they may be the same or different.

Introduction to the Functional Groups

An **alcohol** is an organic compound with the general structure $R\diagup^{O}\diagdown H$. The oxygen and hydrogen atoms attached to a carbon atom framework in this way are called an alcohol group or simply an 'OH' (oh aitch) group. The –OH groups of alcohols are also called 'hydroxyl' groups. Alcohols can be thought of as derived from water molecules, where one hydrogen atom has been replaced by a hydrocarbon framework. Condensed formulas for alcohols are written ROH, to write the condensed formula for a specific alcohol you write the condensed formula for the R group then the –OH. Probably the best known alcohol is ethanol, which is the specific alcohol present in alcoholic beverages (▶ Figure 12.12). In ethanol the R group is CH_3CH_2–, so the condensed formula for ethanol is CH_3CH_2OH. Note that it also correct to write this formula in the opposite direction as $HOCH_2CH_3$. The names of compounds that are alcohols end in the suffix 'ol'.

▲ **FIGURE 12.12** **Ethanol** Present in alcoholic beverages, ethanol is probably the best known alcohol. The structure of ethanol is shown using structural, condensed and line bond formulas.
© Christian Draghici. Shutterstock

$$H{-}\overset{\displaystyle H}{\underset{\displaystyle H}{\overset{|}{\underset{|}{C}}}}{-}\overset{\displaystyle H}{\underset{\displaystyle H}{\overset{|}{\underset{|}{C}}}}{-}O{-}H$$

Structural formula

CH_3CH_2OH

Condensed formula

OH

Line bond formula

The general structure of an **ether** is R–O–R and we can think of these organic compounds as derivatives of water molecules where both hydrogen atoms have been replaced with hydrocarbon frameworks. If you wanted to write the condensed formula for an ether, you simply write the formulas for the R groups on either side of the symbol for an oxygen atom. A fun way to remember the general structure of ethers is to think of it as an 'O with R on ether (either) side'. Diethyl ether is a widely used laboratory solvent that was used in the past as an anaesthetic. In diethyl ether the R groups are both CH_3CH_2– and the condensed formula can thus be written as $CH_3CH_2OCH_2CH_3$. The naming conventions won't be addressed in this module, but you will be able to recognise and represent ethers.

There are a number of different functional groups that contain what is known as a **carbonyl group**, which is a carbon atom and an oxygen atom connected through a double bond. Carbonyl groups can be represented like this C=O. Since carbon atoms always form four covalent bonds, the carbon atom in a carbonyl group has to be bonded to two other groups. For functional groups that contain a carbonyl group, the identity depends on the nature of the two other groups bonded to the carbonyl group carbon atom.

An organic compound that is an **aldehyde** has the general structure $R{-}\overset{\displaystyle O}{\overset{\|}{C}}{-}H$, which contains a carbonyl group. The carbonyl group carbon atom is bonded to an R group and a hydrogen atom. Condensed formulas for aldehydes are written with the atom symbols in a particular order (RCHO) to distinguish them from alcohols. For example, the aldehyde that has CH_3CH_2– as the R group has the condensed formula CH_3CH_2CHO. The names of compounds that are aldehydes have the suffix 'al'. Many fragrant compounds contain an aldehyde functional group, with citronellal being one example. ▼ Figure 12.13 shows the line bond formula of citronellal, a component of citronella oil which is commonly used as an insect repellent.

▼ **FIGURE 12.13** **Citronellal** The structure of the aldehyde 'citronellal' is shown using a line bond formula. Citronellal is a fragrant compound present in citronella oil, which is used as an insect repellent.
© mackoflower.123rf.com

O

H

Citronellal

Examine the general structure of a **ketone** $\mathrm{R{-}C({=}O){-}R}$ and see how it is different to an aldehyde. Ketones have two R groups bonded to the carbon atom of a carbonyl group, and these R groups can be the same or different. In condensed formulas of ketones, the carbonyl group is simply represented by writing the symbols for the carbon and oxygen atoms next to each other like this CO. The R groups are written on either side of the CO to show that they are bonded to the carbonyl group carbon atom. Consider a ketone where both the R groups are CH_3CH_2–, the condensed formula is written $CH_3CH_2COCH_2CH_3$. The convention for naming ketones uses the suffix 'one', pronounced as it is in the functional group name, 'key-tone'.

The general structure of a **carboxylic acid** $\mathrm{R{-}C({=}O){-}OH}$ reveals yet another functional group that contains a carbonyl group. In a carboxylic acid the carbonyl group carbon atom is bonded to an R group and a hydroxyl (–OH) group. Condensed formulas for carboxylic acids are written like this RCOOH, so if the R group is CH_3CH_2– the condensed formula for the carboxylic acid is CH_3CH_2COOH. The names of compounds that are carboxylic acids have 'oic acid' on the end. Vinegar is an aqueous solution of a simple carboxylic acid CH_3COOH, commonly known as acetic acid (◀ Figure 12.14). How to derive the systematic name of acetic acid later is covered later in the module.

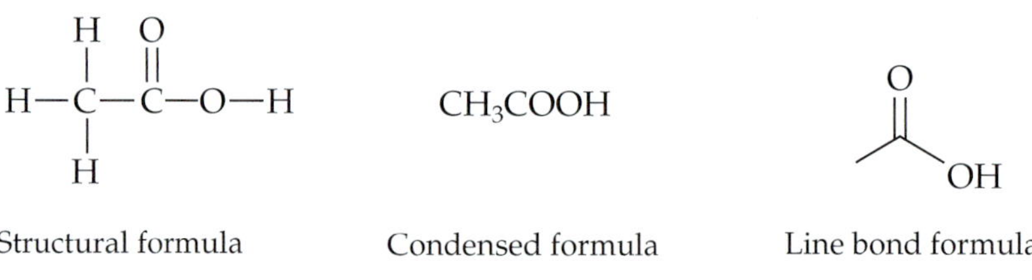

▲ **FIGURE 12.14 Vinegar** This is an aqueous solution of acetic acid, which is the common name for a particular carboxylic acid. The structure of acetic acid is shown using structural, condensed and line bond formulas.

The **ester** functional group has the general structure $\mathrm{R{-}C({=}O){-}O{-}R}$. In an ester the carbonyl group carbon atom is bonded to an R group and an –OR group. An –OR group involves an oxygen atom bonded to an R group and is known as an 'alkoxy' group. In condensed formulas the general structure for an ester is represented as RCOOR and the R groups can be the same or different. The condensed formula for an ester where the R groups are both CH_3CH_2– would be written like this $CH_3CH_2COOCH_2CH_3$. The names of esters have the suffix 'ate'. Esters can be thought of as being derived from carboxylic acids, esters are commonly made from carboxylic acids using chemical reactions and you will see this later in the module.

The **primary amine**, with general structure 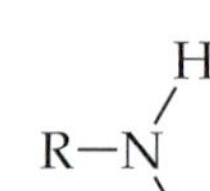, is the first functional group considered that contains nitrogen. Primary amines consist of a nitrogen atom bonded to one R group and two hydrogen atoms. There are amines that have more than one R group and these will be discussed later in the module, but you will only have to learn the naming conventions for primary amines. The amine functional group is present in many important and useful organic compounds, but one thing amines are unfortunately better known for is their generally unpleasant odours. Whenever you can smell something 'fishy' it's normally due to the presence of amines.

The **amide** functional group, which has the general structure R–C(=O)–NR–R, contains a carbonyl group and a nitrogen atom. The carbonyl group carbon atom is bonded to an R group and a nitrogen atom. The nitrogen atom has two R groups. The R groups can be the same or different. This group of atoms is very important in the structures of proteins and some polymers that will be introduced later in the module. You will also learn about a chemical reaction that is used to produce amides, a very important functional group. The naming conventions for amides are not covered.

The names, general structures, condensed formulas and suffix names of the functional groups introduced here are summarised in Table 12.28. The molecular structures, naming conventions, physical properties and chemical reactions of organic compounds that contain these functional groups will be explored in the following sections.

TABLE 12.28 Summary of Names, General Structures, Condensed Formulas and Suffix Names of Functional Groups Introduced in this Module

Functional group name	General structure	Condensed formula	Suffix in naming
Alcohol	R–O–H	ROH	–ol
Ether	R–O–R	ROR	Naming not covered
Aldehyde	R–C(=O)–H	RCHO	–al
Ketone	R–C(=O)–R	RCOR	–one
Carboxylic acid	R–C(=O)–OH	RCOOH	–oic acid
Ester	R–C(=O)–O–R	RCOOR	–ate
Primary amine	R–N(H)–H	RNH_2	–amine
Amide	R–C(=O)–N(R)–R	$RCONR_2$	Naming not covered

12.9 Alcohols

LO: Classify alcohols as primary, secondary or tertiary.

LO: Examine structural representations of alcohols and derive their IUPAC names.

LO: Interpret the IUPAC names of alcohols and draw representations for their molecular structures.

LO: Explain the physical properties of alcohols.

LO: Describe and write equations for reactions of alcohols with alkali metals.

Alcohols are organic compounds that contain a hydroxyl (–OH) group. Hydroxyl groups are also referred to as alcohol groups, or even just 'OH' (oh aitch) groups. The simplest alcohols can be thought of as alkanes that have had one hydrogen atom replaced by an –OH group. The various formulas and names of the simplest members of the alcohol family are shown in Table 12.29. See how the alcohol name is derived by dropping the final 'e' from the parent alkane name and adding the suffix 'ol'. Examine all the structural representations given for these simple alcohols and see how they relate to each other.

TABLE 12.29 The Simplest Members of the Alcohol Family

Number of C atoms (n)	1	2	3	
Parent alkane structure	H │ H—C—H │ H	H H │ │ H—C—C—H │ │ H H	H H H │ │ │ H—C—C—C—H │ │ │ H H H	
Alkane name	**methane**	**ethane**	**propane**	
Alcohol name	**methanol**	**ethanol**	**propan-1-ol***	**propan-2-ol***
Alcohol structural formula	H │ H—C—OH │ H	H H │ │ H—C—C—OH │ │ H H	H H H │ │ │ H—C—C—C—OH │ │ │ H H H	H OH H │ │ │ H—C—C—C—H │ │ │ H H H
Line bond formula	—OH	OH	OH	OH
Condensed formulas	$HO—CH_3$ $HOCH_3$ CH_3OH	$HO—CH_2—CH_3$ $HO—CH_2CH_3$ $HOCH_2CH_3$ CH_3CH_2OH	$HO—CH_2—CH_2—CH_3$ $HO—CH_2CH_2CH_3$ $HOCH_2CH_2CH_3$ $CH_3CH_2CH_2OH$	OH │ $H_3C—CH—CH_3$ OH │ H_3CCHCH_3 $CH_3CH(OH)CH_3$

*The numbers in these names describe the position of the alcohol group in the parent chain.

You may have seen in Table 12.29 that there are two alcohol structural isomers derived from propane. Structural isomers are possible for straight chain alcohols with three or more carbon atoms because there is more than one possible location for the alcohol group along the parent chain. Examine the structure of propan-1-ol and see how it has the alcohol group attached to a terminal carbon atom. The names of organic compounds that are alcohols have to describe where in the parent chain the alcohol group is attached. When numbering the parent chain of an alcohol, the lowest possible number is given to the carbon atom bonded to the alcohol group. See how the other propanol isomer has the alcohol group branching from the second carbon atom in the chain, hence the name propan-2-ol. The number describing the position of the alcohol group is placed before the suffix with hyphens separating the number from adjacent letters. In Table 12.30, the two possible structural isomers of the alcohol based on butane are shown using different structural representations. Examine the structures to see how their names describe the number of carbon atoms in the parent chain and the location of the alcohol group.

TABLE 12.30 The Two Possible Structural Isomers of the Alcohol Derived from Butane

<table>
<tr><th></th><th>Parent alkane</th><th>Alcohol isomer 1</th><th>Alcohol isomer 2</th></tr>
<tr><td>Name</td><td>butane</td><td>butan-1-ol</td><td>butan-2-ol</td></tr>
<tr><td>Structural formula</td><td>H H H H
H—C—C—C—C—H
H H H H</td><td>H H H H
H—C—C—C—C—OH
H H H H</td><td>H H OH H
H—C—C—C—C—H
H H H H</td></tr>
<tr><td>Line bond formula</td><td></td><td>OH</td><td>OH</td></tr>
</table>

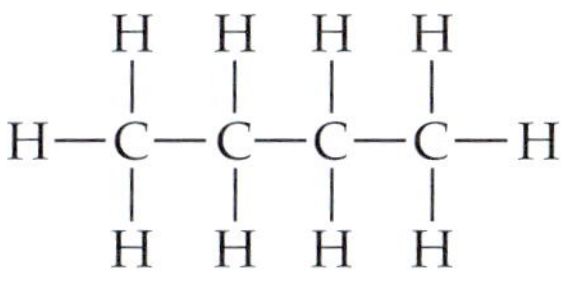

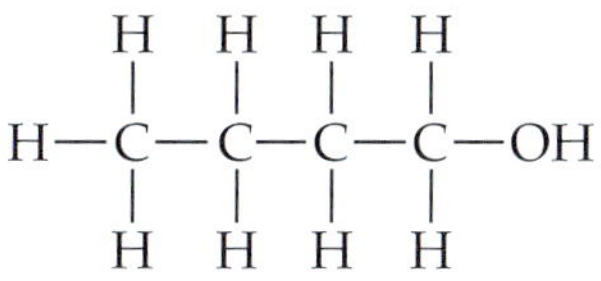

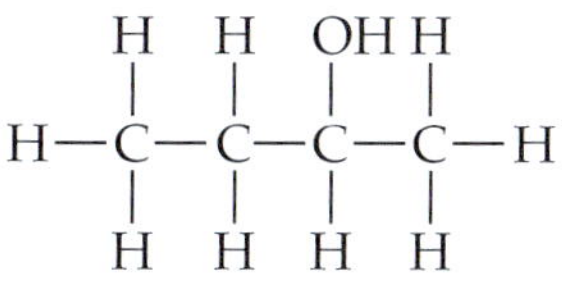

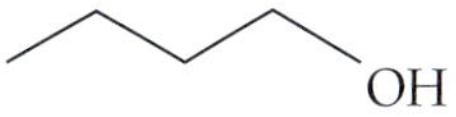

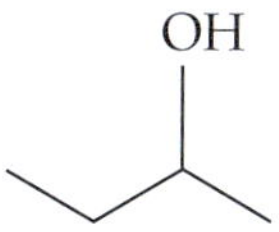

EXERCISE 12.20

There are three structural isomers of the straight chain alcohol with molecular formula $C_5H_{11}OH$ due to the different positions of the alcohol group. Draw their structural and line bond formulas, and derive their names to complete the table below.

<table>
<tr><th>Isomer</th><th>Structural formula</th><th>Line bond formula</th><th>Name</th></tr>
<tr><td>1</td><td>H H H H H
H—C—C—C—C—C—OH
H H H H H</td><td></td><td></td></tr>
<tr><td>2</td><td></td><td></td><td>pentan-2-ol</td></tr>
<tr><td>3</td><td></td><td>OH</td><td></td></tr>
</table>

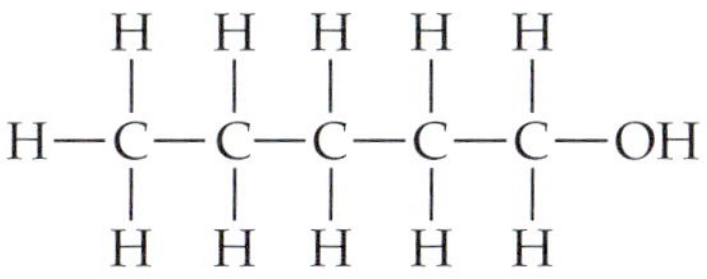

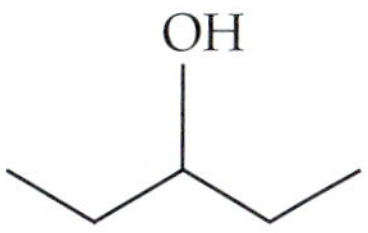

Classification of Alcohols

The molecular structure of an alcohol may be classified as primary (1°), secondary (2°) or tertiary (3°). The classification is important because the different types of alcohols have different physical properties and chemical reactivity. These influences will be described later in this section with respect to the physical properties of alcohols, and later in the module with respect to chemical reactivity. In a primary alcohol, the carbon atom bonded to the alcohol group has one R group and two hydrogen atoms attached. Secondary alcohols have two R groups and one hydrogen atom bonded to the carbon atom attached to the alcohol group. In tertiary alcohols the carbon atom bonded to the alcohol group has three R groups and no hydrogen atoms attached. Examine the general structures of each type of alcohol shown in Table 12.31 to see the difference.

TABLE 12.31 The General Structures of Primary (1°), Secondary (2°) or Tertiary (3°)Alcohols

Primary (1°) alcohol	Secondary (2°) alcohol	Tertiary (3°) alcohol
H | R—C—OH | H	H | R—C—OH | R	R | R—C—OH | R
Carbon atom bonded to the alcohol group has one R group and two hydrogen atoms attached.	Carbon atom bonded to the alcohol group has two R groups and one hydrogen atom attached.	Carbon atom bonded to the alcohol group has three R groups and no hydrogen atoms attached.

To classify an alcohol you examine the structural representation and determine the attachments of the carbon atom bonded to the alcohol group. Table 12.32 shows how to classify some of the alcohols introduced so far.

TABLE 12.32 Classifying Some Alcohols

Alcohol name	Ethanol	Butan-2-ol	Propan-2-ol
Alcohol structure	H H | | H—C—C—OH | | H H	OH (skeletal structure)	OH | H_3C—CH—CH_3
Carbon atom bonded to the alcohol group	H H | | H—C—(C)—OH | | H H	OH (skeletal structure, carbon circled)	OH | H_3C—(CH)—CH_3
Description of the carbon atom bonded to the alcohol group	One R group and two hydrogen atoms attached	Two R groups and one hydrogen atom attached	Two R groups and one hydrogen atom attached
Alcohol classification	Primary (1°)	Secondary (2°)	Secondary (2°)

EXERCISE 12.21

Classify the following alcohols.

CH_3 | HO—C—CH_3 | CH_3	CH_2OH (on benzene ring)	OH (on cyclopentane ring)	OH (skeletal structure)

Naming Alcohols

Deriving the name of an alcohol by examination of its structural representation involves application of the same sequence of steps already described for naming other organic compounds. The steps in naming organic compounds are repeated in Table 12.33 with the factors to consider when applying them to naming alcohols. Naming alcohols will be demonstrated using an example.

TABLE 12.33 The Steps to Naming Organic Compounds Applied to Alcohols

Step	Description of the step	Applying the step to naming alcohols
0	Examine the structural representation to identify the type or classification of the organic compound.	• If a hydrocarbon framework with an alcohol (–OH) group is represented, then the compound is an alcohol and the name will have the suffix **ol.** • The alcohol name is based on an alkane name, dropping the 'e' and adding 'ol'.
1	Identify the parent chain, which is the longest continuous chain of carbon atoms.	• The parent chain of an alcohol is the longest chain that contains the carbon atom bonded to the –OH group.
2	Number the carbon atoms in the parent chain.	• For alcohols the carbon atoms of the parent chain are numbered so the one bonded to the –OH group has the lowest possible number. • The name of the alcohol must describe the position of the –OH group, this is done by placing the number of the carbon atom bonded to the –OH group before the suffix.
3	Identify the substituents and the number of the parent chain carbon atoms that they branch from.	• This step is applied in the same way for alcohols.
4	Write the whole name of the compound as one word.	• This step is applied in the same way for alcohols.

ALCOHOL 1

H_3C-CH_2 (bonded to the second carbon)
$H_3C-CH-CH-CH_3$
OH (bonded to the third carbon)

Step 0 The presence of the alcohol (–OH) group means the compound is an alcohol and the suffix part of its name is **ol.**

Step 1 Identify the parent chain. For alcohols, the parent chain is the longest chain that contains the carbon atom bonded to the –OH group. Identification of the parent chain for alcohol 1 is shown below.

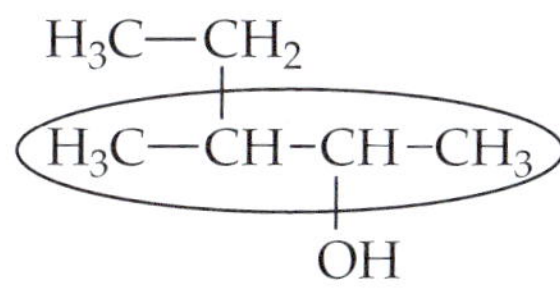

For alcohol 1 there are two carbon chains that contain the carbon atom bonded to the –OH group and one of them is circled here. This is not the parent chain because there is a longer one that contains the carbon atom bonded to the –OH group.

H_3C-CH_2
$H_3C-CH-CH-CH_3$
OH

This is the parent chain for alcohol 1 because it is the longest one that contains the carbon atom bonded to the –OH group. Since the chain is 5 carbon atoms long, the parent alkane name is pentane and in the name of the alcohol the 'e' will dropped from the alkane name to add the suffix 'ol'.

Step 2 Number the carbon atoms in the parent chain. Two options for alcohol 1 are shown below. The correct option has the lowest possible number for the carbon atom bonded to the alcohol group.

```
  5   4                          1   2
H3C—CH2                        H3C—CH2
    3|   2   1                     3|   4   5
H3C—CH-CH-CH3                  H3C—CH-CH-CH3
         |                              |
         OH                             OH
```

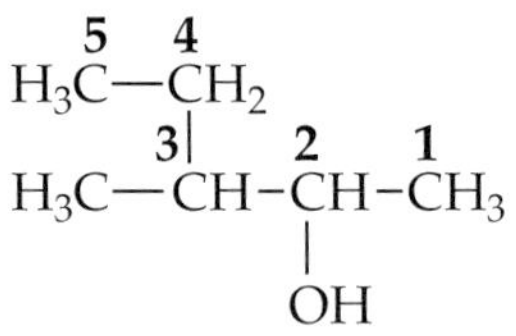

This is the correct numbering of the parent chain for alcohol 1. It gives the lowest possible number (2) to the carbon atom bonded to the –OH group.

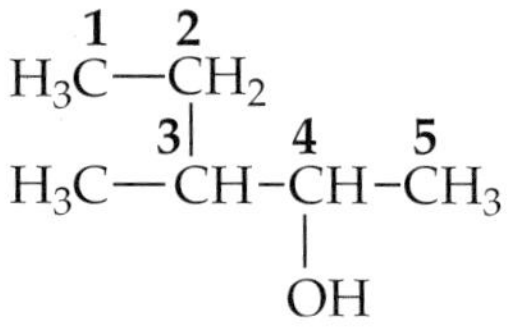

In this option the carbon atom bonded to the –OH group has a higher number (4) than the alternative, so it is not the correct parent chain numbering.

The name of an alcohol must also specify the position of the alcohol group along the parent chain. This is accomplished by placing the number of the carbon atom bonded to the alcohol group before the suffix in the name. For alcohol 1, this is second carbon atom of the chain and the number 2 is placed before the suffix and appears in the name as **2-ol**.

Step 3 Identify the substituents and the number of the parent chain carbon atom that they branch from. Alcohol 1 has a $-CH_3$ (methyl) group branching from carbon atom 3 of the parent chain. The name of the compound will have the prefix 3-methyl.

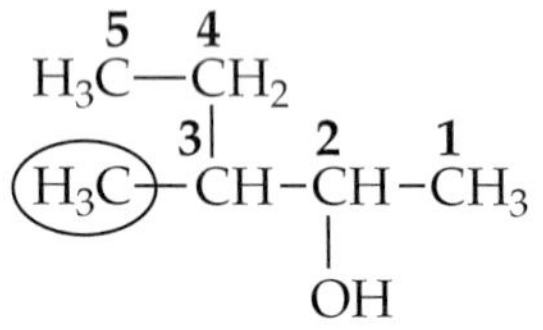

```
   5   4
 H3C—CH2
     3|   2   1
(H3C)—CH-CH-CH3
          |
          OH
```

The branching $-CH_3$ group is called methyl and it is connected to carbon atom 3 of the parent chain.
In the compound name this is written as **3-methyl**.

Step 4 Write the whole name of the compound as one word. All parts of the name for alcohol 1 have been identified and the full name can be constructed as shown below. Note the use of hyphens to separate the numbers and letters in the name, hyphens are placed before and after numbers that appear within compound names.

Alcohol 1	Components of the name identified	Compound name
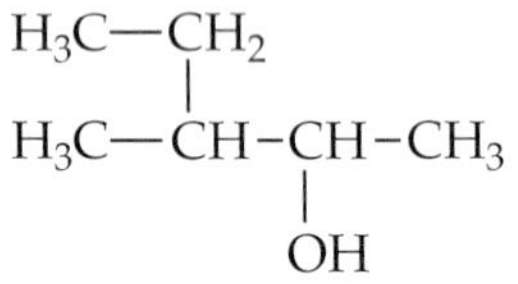 H_3C-CH_2 / $H_3C-CH-CH-CH_3$ with –OH on C2 (branch from C3)	Prefix: 3-methyl Parent alkane name: pentane (drop the 'e' to add the suffix) Suffix: 2-ol	**3-methylpentan-2-ol**

EXERCISE 12.22

Apply the steps to name the following alcohols.

$HO{-}CH_2CH(CH_2CH_2CH_2CH_2CH_2CH_3){-}CH_2CH_2CH_3$	$H_3C{-}CH(OH){-}CH(CH_3){-}CH_3$
HO	OH

EXERCISE 12.23

Draw line bond formulas for the following alcohols. You may find it helpful to draw a structural formula first.

4-ethylhexan-2-ol	4,4-dimethylpentan-1-ol
2,3-dimethylbutan-1-ol	4,5-dimethylheptan-3-ol

Properties of Alcohols

As you might have come to expect by now, the types of IMFs that are possible for alcohols determine their physical properties. One can think about alcohols as having a 'dual nature' (▼ Figure 12.15). There is a non-polar hydrocarbon portion where dispersion forces are possible, and there is the highly polar –OH group that can take part in hydrogen bonding. Both dispersion forces and hydrogen bonding are responsible for the physical properties of substances that are alcohols.

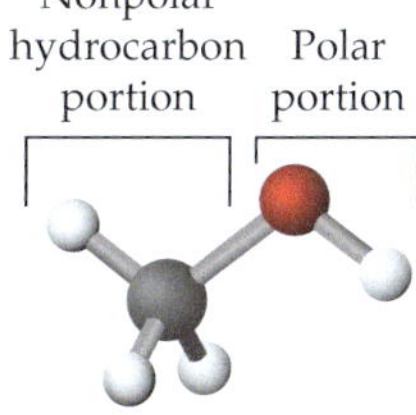

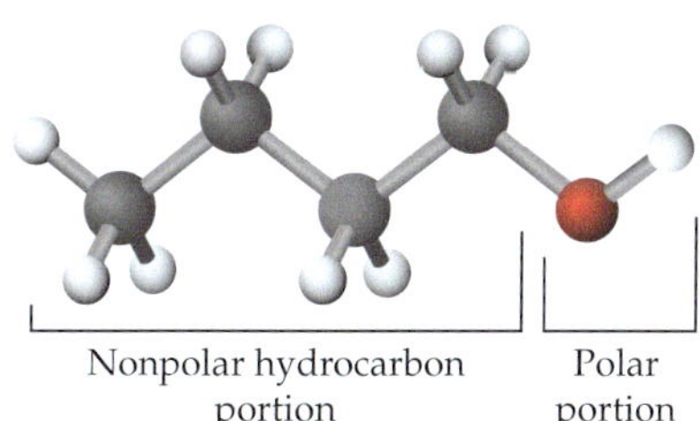

▶ **FIGURE 12.15 Alcohols** These have a 'dual nature' due to the presence of a non-polar hydrocarbon portion and the polar alcohol group.

A fact about alcohols is that they have higher boiling points than hydrocarbons with similar molar mass. When comparing two compounds with similar molar mass, we predict the extent of dispersion forces to be the same because molecules with similar molar mass have a similar number of electrons. Remember that the strength of dispersion forces increases with increasing number of electrons in molecules. When comparing the boiling points of the compounds hexane and pentan-1-ol (Table 12.34) with similar molar mass, they might be expected to have similar boiling points because the molecules experience dispersion forces to the same extent. However, the alcohol has a much higher boiling point because the molecules attract each other through hydrogen bonding. The only IMF for the non-polar hydrocarbon molecules is dispersion, which is relatively weaker than hydrogen bonding. More energy is required to overcome the stronger attractive forces between the alcohol molecules and bring about the phase change from liquid to gas that occurs during boiling. The greater energy input needed to disrupt the stronger attraction of hydrogen bonding results in a higher boiling point for alcohols compared to hydrocarbons with a similar molar mass. Hydrogen bonding between alcohol molecules is shown in ▶ Figure 12.17 (a).

TABLE 12.34 Comparison of Alkane and Alcohol Boiling Points

	Alkane	Alcohol
Name	hexane	pentan-1-ol
Structural formula	H H H H H H |				| H—C—C—C—C—C—C—H |				| H H H H H H	H H H H H |			| H—C—C—C—C—C—O—H |			| H H H H H
Line bond formula	(zigzag line structure)	(zigzag line structure) OH
Molar mass	86 g/mol	88 g/mol
Boiling point	68 °C	138 °C

▲ **FIGURE 12.16 Isocol** This household product is a mixture of water and propan-2-ol. Editorial Image, © LLC/Alamy Stock Photo

The ability to participate in hydrogen bonding is the reason behind the water solubility of low molar mass alcohols such as methanol, ethanol, propan-1-ol and propan-2-ol. It's hydrogen bonding between alcohol and water molecules that makes the property of water solubility possible. Hydrogen bonding between ethanol and water molecules is shown in ▶ Figure 12.17 (b). The household product 'Isocol' (◀ Figure 12.16) is a mixture of water and propan-2-ol (commonly called isopropanol). The alcohol can mix with water because of hydrogen bonding between the molecules. The water solubility of alcohols decreases with increasing size of the hydrocarbon portion, so higher molar mass alcohols are less able to mix with water. Alcohol molecules may be described as less polar with increasing size of their hydrocarbon component, and dispersion forces become the predominant IMF.

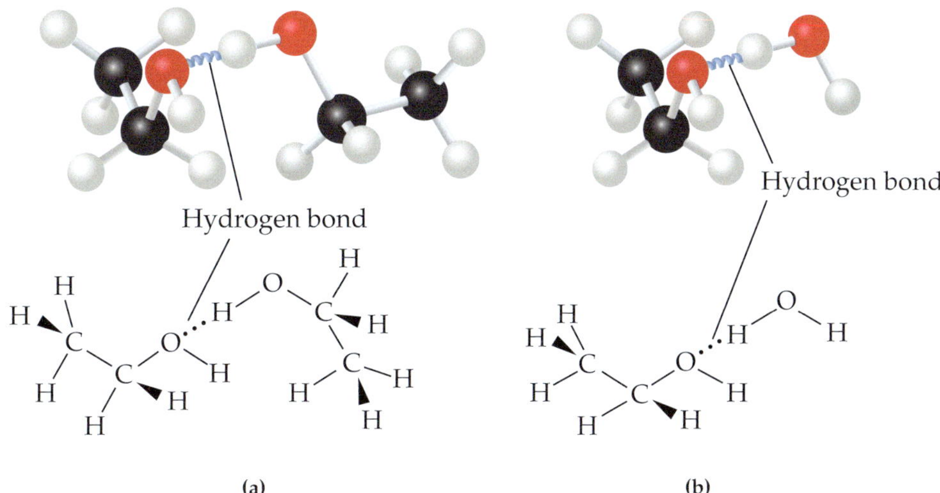

▶ **FIGURE 12.17 Hydrogen bonding (a)** Two ethanol molecules being attracted through hydrogen bonding. **(b)** The polar –OH groups of alcohols can attract through hydrogen bonding to water molecules.

Check out the trend in boiling points for the four alcohols below. Boiling point increases with increasing length of the hydrocarbon chain. This pattern has nothing to do with hydrogen bonding because each alcohol has one –OH group and no alcohol has a greater propensity for hydrogen bonding than another does. In this comparison, dispersion forces are wholly responsible for the trend. The boiling points of straight chain alcohols simply increase with increasing molar mass. Increasing molar mass means higher number of electrons, and greater electron density means stronger dispersion forces. The greater energy input required to overcome the stronger dispersion forces for the increasingly higher molar mass alcohols results in increasingly higher boiling points.

Alcohol	methanol	ethanol	propan-1-ol	butan-1-ol
Boiling point	65 °C	79 °C	97 °C	118 °C

Remember how alkane structural isomers had different boiling points and this was related to the different shapes of the molecules (Section 12.4)? Well, the same applies to alcohol structural isomers. When comparing the boiling points of two alcohol structural isomers (Table 12.35), the one with the alcohol group branching from within the carbon chain has a lower boiling point than the other with the terminal alcohol group. Both alcohols have one –OH group, so the extent of hydrogen bonding interactions is the same for both compounds. The compounds have the same molar mass, number of electrons and electron density, so the extent of dispersion forces is also expected to be the same. The difference in boiling point is related to the difference in molecular shape. Propan-2-ol has a branched structure and more spherical shape, whereas propan-1-ol has a more cylindrical shape. The more spherical isomer has a lower boiling point due to decreased surface area for attraction by dispersion forces to occur. The difference in molecular shape between the isomers is highlighted using space filled models in Table 12.35.

TABLE 12.35 Comparison of Alcohol Structural Isomer Boiling Points and Molecular Shapes

Alcohol isomer	Propan-1-ol	Propan-2-ol
Structural formula	H H H | | | H—C—C—C—OH | | | H H H	H OH H | | | H—C—C—C—H | | | H H H
Line bond formula	OH	OH
Molar mass	60 g/mol	60 g/mol
Boiling point	97 °C	82 °C
Space filled model		

In summary, the physical properties of alcohols are related to the IMFs that are possible. The presence of the –OH group means that hydrogen bonding between alcohol molecules can occur, and this is why alcohols have higher boiling points than hydrocarbons with similar molar mass. Low molar mass alcohols are soluble in water because of hydrogen bonding interaction between alcohol and water molecules. The polarity of alcohol molecules decreases with increasing size of the hydrocarbon portion and dispersion forces become the predominant IMF. Water solubility of alcohols thus decreases with increasing size of the hydrocarbon component. Increasing strength of dispersion forces with increasing length of the hydrocarbon chain is the reason for increasing boiling points for a series of straight chain alcohols. Alcohol structural isomers have different boiling points due to differences in dispersion forces related to molecular shape.

Reactions of Alcohols with Alkali Metals

The addition of a Group I (1) metal (alkali metal) to water results in a vigorous reaction that produces a fire or explosion. ◀ Figure 12.18 shows the consequence of adding sodium metal to a tub of water, the resulting fire is clear to see. This is the equation for the chemical reaction that occurs.

$$2H_2O(l) + 2Na(s) \rightarrow H_2(g) + 2NaOH(aq)$$

Addition of the reactants water (H_2O) and sodium metal (Na) produces hydrogen gas (H_2) and the salt sodium hydroxide (NaOH), which is dissolved in the excess water (aq) at the end of the reaction. Another product of the reaction is energy in the form of heat, and this ignites the hydrogen gas to cause the fire.

In terms of their structure, alcohol molecules are similar to water molecules. Alcohol molecules can be considered as water molecules where one hydrogen atom has been replaced by a hydrocarbon framework or R group. Table 12.36 shows a comparison of water and alcohol molecular structures.

▲ **FIGURE 12.18 The reaction of sodium metal with water causes a fire.**

TABLE 12.36 Comparing Water and Alcohol Molecular Structures

Water	Alcohol
H–O–H	R–O–H
Water molecules consist of an oxygen atom covalently bonded to two hydrogen atoms.	Alcohol molecules have an oxygen atom covalently bonded to one hydrogen atom and one R group.

The similarity in molecular structure means that water and alcohols can undergo the analogous reaction with Group I (1) metals (alkali metals). In the following comparison of water and alcohol reactions with the sodium (a Group I (1) metal), the formula for water is written as HOH to highlight its structural similarity to alcohols. The production of hydrogen gas (H_2) and a sodium salt in both reactions can be seen clearly by examining the equations in Table 12.37.

TABLE 12.37 Comparing Water and Alcohol Reactions with Sodium

$2HOH + 2Na \rightarrow H_2 + 2NaOH$	$2HOR + 2Na \rightarrow H_2 + 2NaOR$
The reaction of water with sodium metal produces hydrogen gas and the salt sodium hydroxide (NaOH). Sodium hydroxide is composed of sodium ions (Na^+) and hydroxide ions (OH^-).	The reaction of an alcohol with sodium metal also produces hydrogen gas, along with the salt sodium alkoxide (NaOR). Sodium alkoxide is composed of sodium ions (Na^+) and alkoxide ions (OR^-).

The equation for the reaction of the specific alcohol ethanol with sodium metal is shown below. The salt that forms in this case is called sodium ethoxide ($NaOCH_2CH_3$). If the reaction had been done using potassium (K) instead, the salt formed would be potassium ethoxide ($KOCH_2CH_3$).

$$2HOCH_2CH_3 + 2Na \rightarrow H_2 + 2NaOCH_2CH_3$$

The observations for this reaction are shown in ◀ Figure 12.19. Note that the production of hydrogen gas is observed as bubbling or effervescence, and the hydrogen gas does not ignite. The reaction of Group I (1) metals (alkali metals) with alcohols is less vigorous than their reaction with water. Note that all Group I (1) metals (alkali metals) react with alcohols in the same manner. Remember that elements in the same group of the periodic table have the same number of valence electrons and thus similar chemical reactivity.

▶ **FIGURE 12.19 The main observation for the reaction of sodium metal with an alcohol is effervescence, there is no fire**

EXERCISE 12.24

Write equations for the reactions of lithium (Li) and caesium (Cs) with ethanol.

12.10 Ethers

LO: Identify the ether functional group in representations of organic compounds.

LO: Explain the physical properties of ethers.

There are two functional groups that can be thought of as organic derivatives of water. The structure of an alcohol can be imagined as a water molecule where one hydrogen atom has been replaced by a hydrocarbon framework (R group). An organic compound that is an ether can be considered as a water molecule with both hydrogen atoms replaced by hydrocarbon frameworks (R groups). A comparison of the molecular structure of water with an alcohol and an ether is shown in Table 12.38.

TABLE 12.38 Comparison of the Molecular Structure of Water with an Alcohol and an Ether

Water	Alcohol	Ether
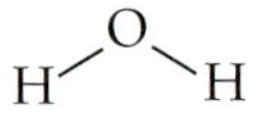 H–O–H	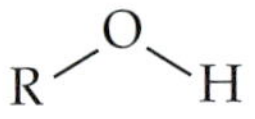 R–O–H	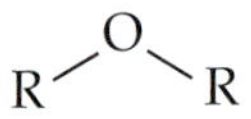 R–O–R
Water molecules consist of an oxygen atom covalently bonded to two hydrogen atoms.	Alcohol molecules have an oxygen atom bonded to one hydrogen atom and one R group.	The molecular structure of an ether consists of an oxygen atom bonded to two R groups.

Ethers are generally inert compounds, similar to alkanes in that they only undergo a limited range of chemical reactions. In this module the focus is on the molecular structure and physical properties of ethers, chemical reactions and naming conventions are not covered. After completing this section you will be able to recognise the ether functional group in representations of organic compounds, and predict or explain the physical properties of ethers compared to other organic compounds. Recognising an ether in the structural representation of an organic compound is relatively straightforward, you simply look for an oxygen atom that has two R groups attached. The R groups may be the same or different, and they may even be connected to each other to give a cyclic ether. Consider the simple ethers in Table 12.39 and note the nature of the R groups in each case.

TABLE 12.39 Examples of Simple Ethers

Ether 1		Ether 2	
Structural formula	**Line bond formula**	**Structural formula**	**Line bond formula**
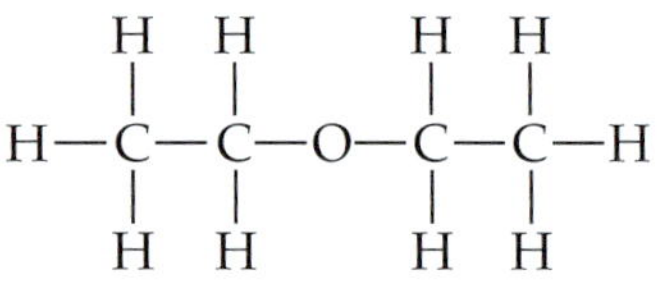 H–C(H)(H)–C(H)(H)–O–C(H)(H)–C(H)(H)–H	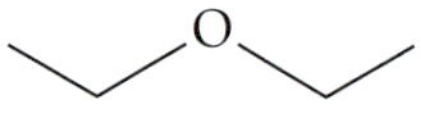 [line bond structure, O]	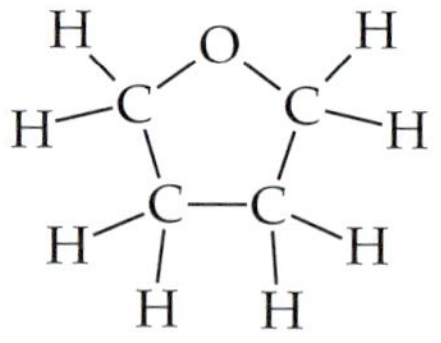 [cyclic structural formula: C, O, H]	 [five-membered ring with O]
In ether 1 the R groups are the same, they are both ethyl ($-CH_2CH_3$).		The R groups are the same in ether 2, they are also connected to each other to give a cyclic ether.	
Ether 3		**Ether 4**	
[benzene ring–O–ethyl]		[six-membered ring with two O]	
The structure of ether 3 demonstrates that the R groups can be different to each other, and they can be aliphatic and aromatic.		Here is an example of another cyclic ether. This one has two ether groups in its structure.	

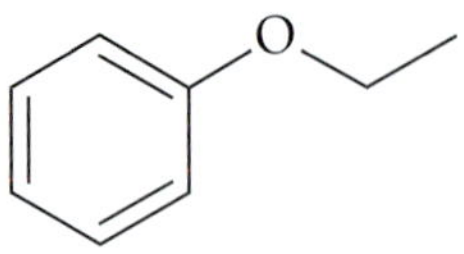

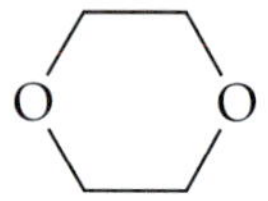

The ether functional group is a critical feature in the molecular structure of carbohydrates, which are an important class of organic compounds in biological systems. Carbohydrate molecules are made up of monosaccharides which can exist in a cyclic ether form. Cyclic forms of the commonly known monosaccharides glucose and fructose are shown in Table 12.40. Examine the structures to find the ether group in each molecule.

TABLE 12.40 Cyclic Forms of Glucose and Fructose

A cyclic form of glucose	A cyclic form of fructose

Cyclic forms of glucose and fructose can react with each other to form the disaccharide sucrose, which is commonly known as table sugar ▼ Figure 12.20 (a). The molecular structure of sucrose still has the ether groups present in the component glucose and fructose molecules, and the linkage between the two monosaccharides formed in the chemical reaction between them is also an ether group. The ether groups in the molecular structure of sucrose are highlighted in ▼ Figure 12.20 (b).

(a)

(b)

▶ **FIGURE 12.20 Sucrose**
(a) The disaccharide sucrose is commonly known as table sugar. © 917564. Shutterstock
(b) The molecular structure of sucrose contains three ether groups.

EXERCISE 12.25

The following organic compounds contain an ether functional group within their structure. Even though the compounds look reasonably complex, you can examine them to find the part of each molecule that is an ether. When you find the ether, circle the oxygen atom associated with it.

This is the structure of 'metoprolol', a drug used to treat high blood pressure. Locate the ether group present in this compound and circle its oxygen atom.	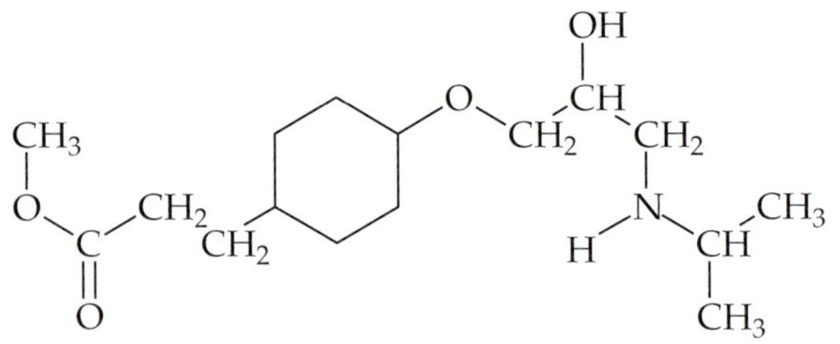 CH_3 O C O CH_2 CH_2 O CH_2 CH OH CH_2 N H CH CH_3 CH_3
This is the structure of the organic compound known as THC (tetrahydrocannabinol), which is the active constituent of cannabis. It contains an ether group, can you find it and circle its oxygen atom?	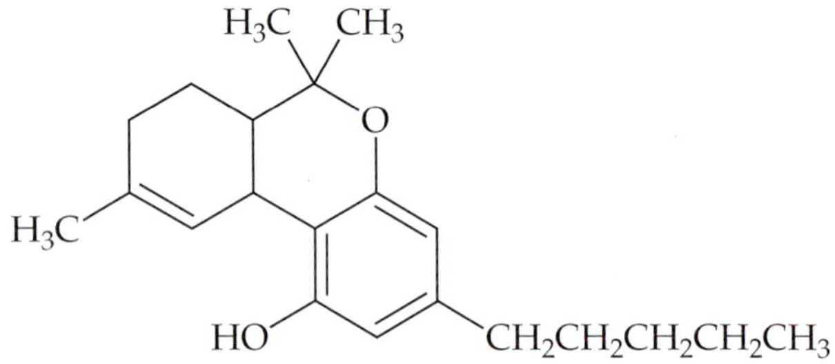 H_3C CH_3 O H_3C HO $CH_2CH_2CH_2CH_2CH_3$
Locate the ether group in the structure of morphine, an organic compound used for pain relief.	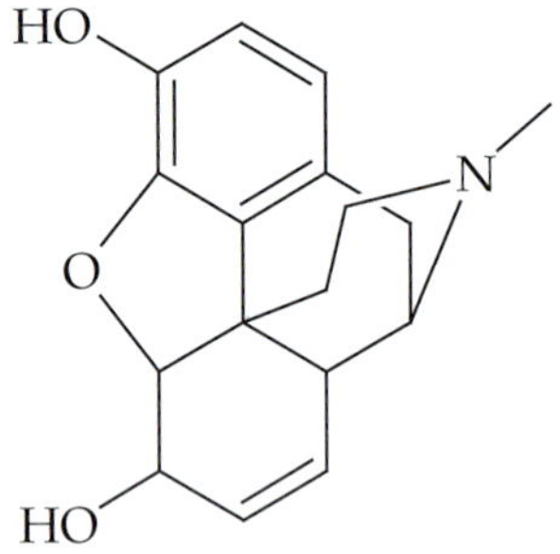 HO O N HO

The electronegativity of oxygen atoms and bent geometry about the oxygen atom within an ether molecule confers polarity, and the IMF dipole-dipole attraction is thus possible for ethers. The contrast in strength of IMFs for ethers compared to alkanes with similar molar mass can be seen in the difference in boiling point between cyclopentane (alkane) and tetrahydrofuran (ether) shown in Table 12.41. The compounds have similar molar mass, so the extent of dispersion forces is the same in each case. The higher boiling point for the ether (65 °C) compared to the alkane (49 °C) means that the attraction between the ether molecules is stronger than what is possible for the alkane molecules. More energy is required to overcome the attractive forces and bring about the phase change from liquid to gas that occurs during boiling. Since alkanes are non-polar, dispersion is the only IMF possible for cyclopentane. Dipole-dipole attraction is possible for the ether tetrahydrofuran because the molecules are polar. Dipole-dipole is a relatively stronger IMF than dispersion and the ether molecules are more strongly attracted than the alkane molecules. The difference in IMFs means that ethers generally have higher boiling points than alkanes with similar molar mass.

TABLE 12.41 Comparison of Alkane and Ether Boiling Points

Alkane		Ether	
cyclopentane, 70 g/mol, BP 49 °C		tetrahydrofuran, 72 g/mol, BP 65 °C	
Structural formula	Line bond formula	Structural formula	Line bond formula

By contrast, ethers have lower boiling points than alcohols with similar molar mass. Consider the information in Table 12.42 and see the dramatic difference in boiling point for an alcohol (butan-1-ol) and an ether (diethyl ether) with the same molar mass. The difference is again due to the strength of IMFs in each case. Hydrogen bonding is possible for the alcohol, while the ether has dipole-dipole attraction. In relative terms, hydrogen bonding is the strongest IMF, so more energy is required to overcome the attractive forces between the alcohol molecules and this results in a higher boiling point than the ether with the same molar mass.

TABLE 12.42 Comparison of Alcohol and Ether Boiling Points

Alcohol		Ether	
butan-1-ol, 74 g/mol, BP 118 °C		diethyl ether, 74 g/mol, BP 35 °C	
Structural formula	Line bond formula	Structural formula	Line bond formula
H—C—C—C—C—OH (each C bearing two H)	OH	H—C—C—O—C—C—H (each C bearing two H)	O

12.11 Aldehydes and Ketones

LO: Examine structural representations of aldehydes and ketones and derive their IUPAC names.

LO: Interpret the IUPAC names of aldehydes and ketones and draw representations for their molecular structures.

LO: Explain the physical properties of aldehydes and ketones.

There are several functional groups that contain a structural feature known as a **carbonyl group**, which consists of a carbon atom and an oxygen atom connected to each other through a double bond (C=O). Carbon atoms always form four covalent bonds in stable compounds, so the carbon atom in a carbonyl group must be bonded to another two atoms or groups. The identity of the functional group depends on what atoms or groups are attached to the carbonyl group carbon atom. In this section the nature of two functional groups that contain a carbonyl group, aldehydes and ketones, will be described. The chemical reactions of aldehydes and ketones is a rich and fascinating area of study, but it is beyond the scope of this introduction to organic compounds. The focus here will be on the molecular structure of aldehydes and ketones and how this determines their physical properties.

Naming Aldehydes

The molecules of organic compounds that are aldehydes always contain a carbonyl group, and the carbon atom of the carbonyl group is bonded to one hydrogen atom and one R group. In the simplest aldehyde, the R group is another hydrogen atom and this compound is called methanal. The name is derived from the parent alkane name (methane) by dropping the 'e' and adding the suffix 'al'. Methanal is more commonly known as formaldehyde, which exists as a gas at room temperature. However, formaldehyde readily dissolves in water and aqueous solutions of this simplest aldehyde are used to preserve biological samples (◀ Figure 12.21).

General structure of aldehydes.
The R group may be any straight, cyclic or branched alkyl group or a benzene ring.

O
‖
C
R ╱ ╲ H

In the simplest aldehyde the R group is actually another hydrogen atom.
This compound is called methanal.

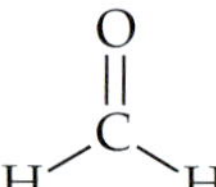

▲ **FIGURE 12.21 Formaldehyde**
Aqueous solutions of methanal (formaldehyde) are used to preserve biological samples.

The structures and derivation of names for some simple straight chain aldehydes are detailed in Table 12.43. Note that the aldehyde functional group can only occur at the end of a carbon chain, so there is no need to designate the position of the functional group in the name of the compound. The carbon atom of the carbonyl group in an aldehyde is always carbon atom number 1 and the number is not included in the name because it is always the case. Note that the aldehyde structures are drawn with the functional group to the right or left, both are correct and the carbonyl group carbon atom is always carbon 1 of the parent chain no matter how the structure is drawn.

TABLE 12.43 The Structures and Derivation of Names for the Simplest Aldehydes

Number of C atoms (n)	1	2	3	4
Parent alkane structure	H | H—C—H | H	H H | | H—C—C—H | | H H	H H H | | | H—C—C—C—H | | | H H H	H H H H | | | | H—C—C—C—C—H | | | | H H H H
Alkane name	methane	ethane	propane	butane
Aldehyde name	methanal	ethanal	propanal	butanal
Aldehyde structural formula	O ‖ C H ╱ ╲ H	H O | ‖ H—C—C—H | H	O H H ‖ | | H—C—C—C—H | | H H	H H H O | | | ‖ H—C—C—C—C—H | | | H H H
Line bond formula	O ‖ H ╱ ╲ H	O ‖ ╱ ╲ H	O ‖ ╲╱ ╲ H	O ‖ H ╱ ╲╱╲
Condensed formula	HCHO	CH_3CHO	CH_3CH_2CHO	$CH_3CH_2CH_2CHO$

Aldehydes can have branching structures and naming these follows the same conventions explained already for other classes of organic compound. The explanation below for naming a particular aldehyde, 3-chloro-4-methylpentanal, demonstrates this.

3-chloro-4-methylpentanal

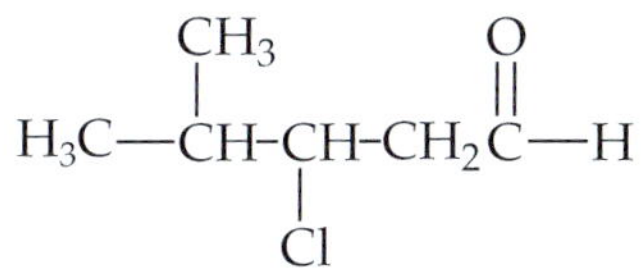

Naming explanation

- The parent name is pentanal because the longest carbon chain that contains the carbonyl group carbon atom has five carbon atoms. The 'e' was dropped from the alkane name pentane and the suffix 'al' added because the compound is an aldehyde.
- The carbonyl group carbon atom of an aldehyde is always carbon atom 1.
- The '3-chloro' prefix describes the chlorine atom branching from carbon atom number 3 of the parent chain.
- The '4-methyl' prefix describes the $-CH_3$ group branching from carbon atom number 4 of the parent chain.
- The prefixes chloro and methyl are in alphabetical order.

EXERCISE 12.26

Name the following aldehydes.

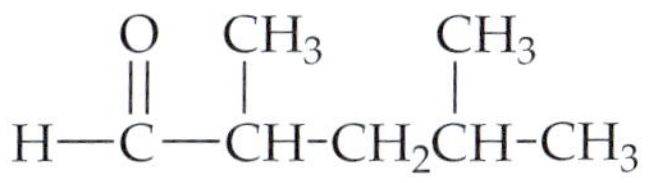	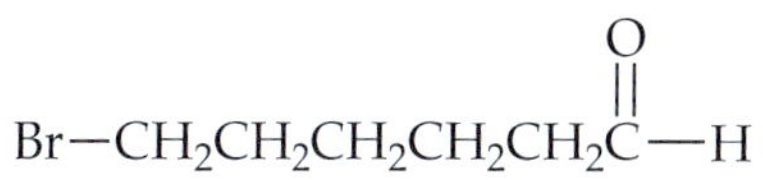
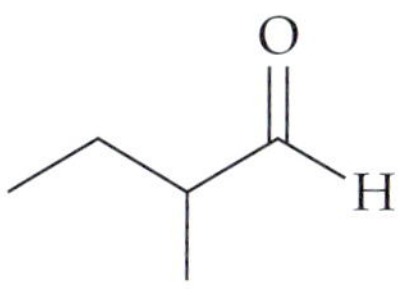	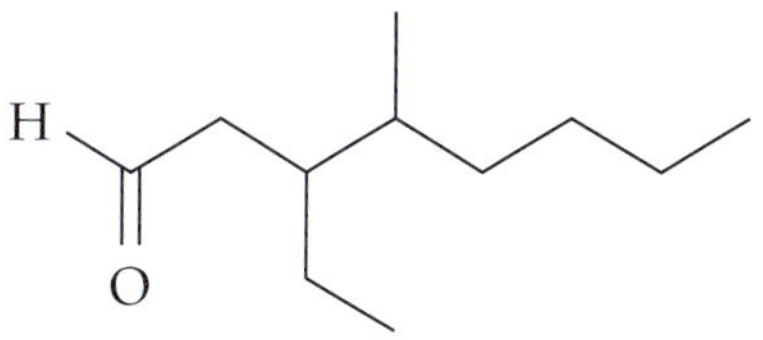

EXERCISE 12.27

Draw line bond formulas for the following aldehydes. You may find it helpful to draw a structural formula first.

2-methylpropanal	2,3-diethylheptanal
5-chloro-3 methylpentanal	3,5-dimethylhexanal

Naming Ketones

The molecular structure of ketones also involves a carbonyl group, and the carbonyl group carbon atom is bonded to two R groups. The R groups cannot be hydrogen atoms. If they were, the compound would be an aldehyde and not a ketone. In the simplest ketone, the R groups each have only one carbon atom and the name of that compound is propanone. Look at the structure of the simplest ketone shown below and see that it consists of a three carbon atom chain. The parent alkane name for three carbon atoms in a chain is propane. The ketone name is derived by dropping the 'e' from the parent alkane name and adding the suffix 'one'. You might not have heard of propanone before, it's more likely you have heard its common name, acetone. Acetone is an excellent solvent that is used for a range of purposes, including nail polish removal (◀ Figure 12.22).

General structure of ketones.
The R group may be any straight, cyclic or branched alkyl group or a benzene ring.

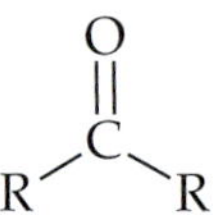

In the simplest ketone both R groups are methyl ($-CH_3$).
This compound is called propanone.

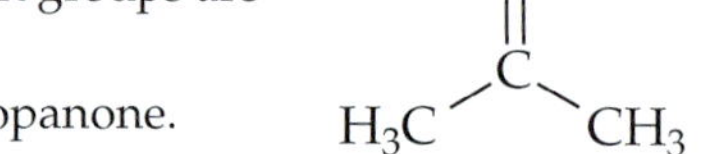

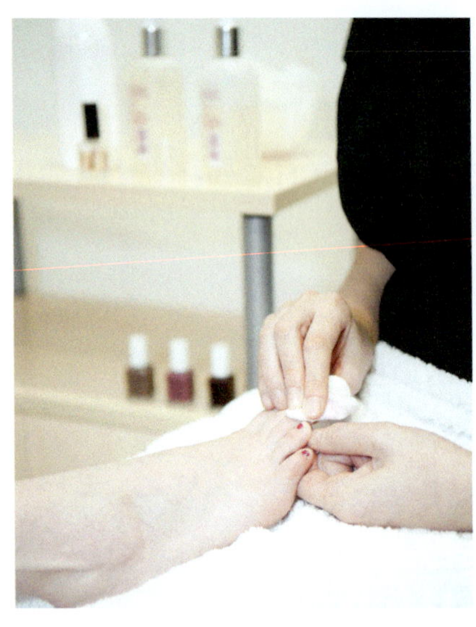

▲ **FIGURES 12.22 Acetone** The simplest ketone, propanone (acetone) is commonly used as nail polish remover.

Structural representations and derivation of names for ketones with four and five carbon atom chains are shown in Table 12.44. Note how the name for the four carbon atom ketone, butanone, does not contain a number before the suffix. This is because there is only one possible position for the carbonyl group carbon atom within the carbon chain, the second carbon atom from the end. There are two ketones that have five carbon atoms in a straight chain because there are two possible positions for the carbonyl group carbon atom along the carbon chain. The number inserted before the suffix is decided using the same principles for numbering parent chains that have been explained previously. The parent chain is numbered to give the lowest possible number to the carbonyl group carbon atom of the ketone. For ketones with five carbon atoms in a straight chain, this can be the second carbon atom in the chain (pentan-2-one) or the third (pentan-3-one). This convention is followed to name any ketone.

TABLE 12.44 Structures and Derivation of Names for the Simplest Ketones

Number of C atoms (n)	4	5	
Parent alkane structure			
Alkane name	butane	pentane	
Ketone name	butanone	pentan-2-one*	pentan-3-one*
Ketone structural formula			
Line bond formula			
Condensed formulas	$CH_3CH_2COCH_3$ $CH_3COCH_2CH_3$	$CH_3COCH_2CH_2CH_3$ $CH_3CH_2CH_2COCH_3$ $CH_3(CH_2)_2COCH_3$	$CH_3CH_2COCH_2CH_3$

*The numbers in these names describe the position of the carbonyl group carbon atom in the parent chain.

Ketones can have branching structures and naming these follows the same conventions explained already for other classes of organic compound. The explanation below for naming a particular ketone, 2-bromo-4-ethylhexan-3-one, demonstrates this.

2-bromo-4-ethylhexan-3-one

Naming explanation

- The parent name is hexanone because the longest carbon chain that contains the carbonyl group carbon atom has six carbon atoms. The 'e' was dropped from the parent alkane name hexane and the suffix 'one' added because the compound is a ketone.
- The number 3 was placed before the suffix to indicate the position of the carbonyl group within the chain. See how numbering the carbon atoms of the parent chain has been done to give the carbonyl group carbon atom the lowest possible number.
- The '2-bromo' prefix describes the bromine atom branching from carbon atom number 2 of the parent chain.
- The '4-ethyl' prefix describes the $-CH_2CH_3$ group branching from carbon atom number 4 of the parent chain.
- The prefixes bromo and ethyl are in alphabetical order.

EXERCISE 12.28

Name the following ketones.

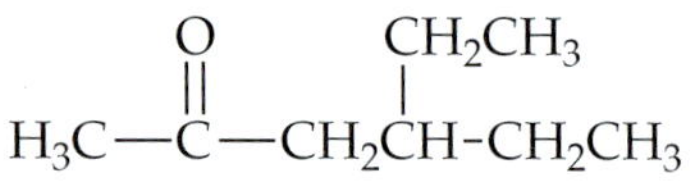 $H_3C-C(=O)-CH_2CH(CH_2CH_3)-CH_2CH_3$	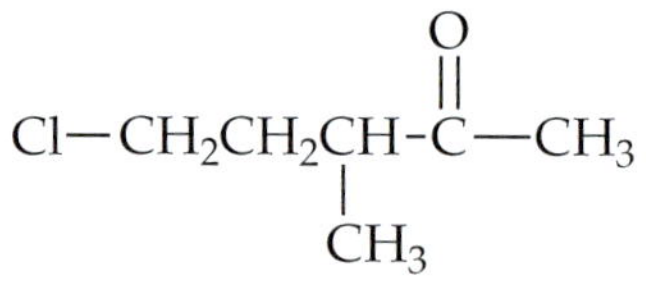 $Cl-CH_2CH_2CH(CH_3)-C(=O)-CH_3$
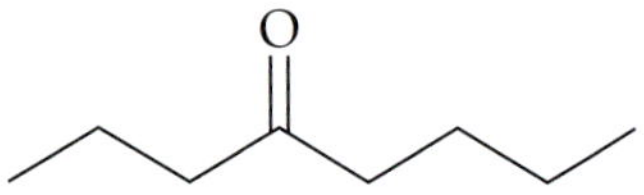 O	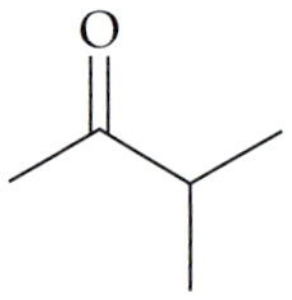 O

EXERCISE 12.29

Draw line bond formulas for the following ketones. You may find it helpful to draw a structural formula first.

1-bromopropanone	2,4-dimethylpentan-3-one
heptan-3-one	2,2-dimethylhexan-3-one

Properties of Aldehydes and Ketones

The physical properties of aldehydes and ketones are a consequence of the presence of the carbonyl group and its polarity. Oxygen is one of the most electronegative elements and when it is involved in covalent bonds with atoms of elements that have lower electronegativity, it draws electron density towards itself and away from the atom with lesser electronegativity. The higher electronegativity of oxygen relative to carbon in the carbonyl group makes it polar with partial negative charge (δ^-) near the oxygen atom and partial positive charge (δ^+) near the carbon atom as shown below. The presence of the carbonyl group and trigonal planar geometry makes aldehyde and ketone molecules polar. The IMF relevant to polar molecules is dipole-dipole attraction and this is responsible for the physical properties of aldehydes and ketones.

O δ^-

C δ^+

R R or H

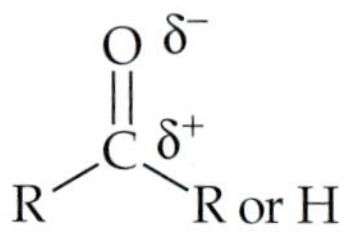

The comparison in Table 12.45 demonstrates the difference between aldehyde, alkane and alcohol boiling points. The specific compounds in the comparison have similar molar mass, so the difference between their boiling points has nothing to do with that since the number of electrons and the extent of dispersion forces would be the same for each compound. The difference relates to the type of IMF possible for each compound. The boiling point is lowest for the alkane (propane) because only the weakest IMF (dispersion) is possible. The boiling point for ethanal is higher than propane because aldehydes are polar and dipole-dipole attraction, which is stronger than dispersion, is possible. Ethanol has the highest boiling point because the presence of the highly polar –OH group allows for hydrogen bonding, which is the strongest IMF. The stronger the IMFs, the more energy input required to overcome the attraction between molecules and bring about the phase change from liquid to gas that occurs during boiling. The higher the energy required to overcome the IMFs, the higher the boiling point.

TABLE 12.45 Comparison of Aldehyde, Alkane and Alcohol Boiling Points

Type of compound	Alkane	Aldehyde	Alcohol
Compound name	**propane**	**ethanal**	**ethanol**
Formula	$CH_3CH_2CH_3$	CH_3CHO	CH_3CH_2OH
Molar mass	44 g/mol	44 g/mol	46 g/mol
Boiling point	–42 °C	20 °C	78 °C
Intermolecular forces possible	Dispersion forces	Dipole-dipole attraction	Hydrogen bonding

EXERCISE 12.30

Consider the data given for three organic compounds below and determine the type of IMF possible in each case. Then predict the order of boiling points for the compounds from lowest to highest.

Compound name	butane	propanone	propanol
Formula	$CH_3CH_2CH_2CH_3$	CH_3COCH_3	$CH_3CH_2CH_2OH$
Molar mass	58 g/mol	58 g/mol	60 g/mol
Intermolecular forces possible			
Expected order of boiling point from lowest to highest			

In summary, aldehydes and ketones have higher boiling points than alkanes with similar molar mass. This is due to stronger IMFs (dipole – dipole attraction) for polar aldehydes and ketones compared to dispersion forces for non-polar alkanes. Aldehydes and ketones have lower boiling points than alcohols with similar molar mass. This is due to stronger IMFs (hydrogen bonding) for alcohols with a highly polar –OH group compared to dipole-dipole attraction for less polar aldehydes and ketones.

12.12 Carboxylic Acids

LO: Write equations to show how carboxylic acids in aqueous solution are proton donors.

LO: Write equations for reactions of carboxylic acids with strong bases.

LO: Examine structural representations of carboxylic acids and derive their IUPAC names.

LO: Interpret the IUPAC names of carboxylic acids and draw representations for their molecular structures.

LO: Explain the physical properties of carboxylic acids.

The carbonyl group (C=O) also appears in the functional group known as carboxylic acid. The general structure below shows the two other groups attached to the carbonyl group carbon atom in a carboxylic acid, one R group and one –OH group. The –OH group will look familiar because it appears in alcohols, but its attachment to a carbonyl group carbon atom makes a completely different functional group. Take care not to confuse carboxylic acids with alcohols. If you see a structural representation with an –OH group bonded to a carbonyl group (C=O), then the compound is a carboxylic acid. If the –OH group is not bonded to a carbonyl group carbon atom, the compound is an alcohol. In the simplest carboxylic acid the R group is a hydrogen atom and its systematic name is methanoic acid. This carboxylic acid has just one carbon atom and its name is derived from the name of the alkane that has one carbon atom, methane. All carboxylic acid names are constructed by dropping 'e' from a parent alkane name, adding the suffix 'oic', and placing the word 'acid' after that. Hence the parent name of a carboxylic acid with one carbon atom is methanoic acid. Like many organic compounds, methanoic acid has a commonly used name that you may have heard, formic acid. Formic acid is an industrial chemical with various important uses, but it also occurs in nature as a component of some insect venoms (◀ Figure 12.23). As the name of the functional group suggests, compounds that are carboxylic acids have the corrosive and irritating properties of acids. The formic acid component of insect venom contributes to the uncomfortable effect of a sting.

General structure of carboxylic acids. The R group may be any straight, cyclic or branched alkyl group or a benzene ring.

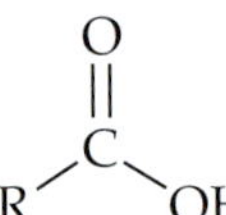

In the simplest carboxylic acid the R group is a hydrogen atom. This compound is called methanoic acid.

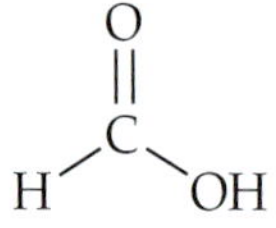

▲ **FIGURE 12.23 Formic acid** The simplest carboxylic acid, methanoic acid (formic acid), occurs in nature as a component of some insect venoms.

Recall that an acid is a substance that can donate hydrogen ions (H^+). Hydrogen ions are also called protons, so an acid may also be described as a proton donor. Another definition for an acid is a substance that produces hydrogen ions in aqueous solution. These descriptions of acids were introduced in Section 8.8. Carboxylic acids are acids because the –OH group can dissociate when the molecules are dissolved in water. They are weak acids, which means they do not dissociate completely when added to water (Section 8.8). Incomplete dissociation of a weak carboxylic acid in water is shown in chemical equations using reversible reaction arrows. Examine the following equation to see how the carboxylic acid has donated a hydrogen ion (or proton, H^+) from the –OH group to form the conjugate base (carboxylate ion, $RCOO^-$). In this reaction, the water molecule acts as a base to accept the donated hydrogen ion and form its conjugate acid (H_3O^+).

$$\underset{\text{carboxylic acid}}{R{-}C(=O){-}OH} + \underset{\text{base}}{H_2O} \rightleftharpoons \underset{\text{conjugate base}}{R{-}C(=O){-}O^-} + \underset{\text{conjugate acid}}{H_3O^+}$$

A common chemical reaction that you will have seen before is the reaction between strong acid and strong base to produce a salt and water (Section 8.8). It does not matter what strong acid and strong base are reacted, the products are always a salt and water. See how this is the case for the following examples.

$$\underset{\text{acid}}{HCl(aq)} + \underset{\text{base}}{NaOH(aq)} \rightarrow \underset{\text{salt}}{NaCl(aq)} + \underset{\text{water}}{H_2O(l)}$$

$$\underset{\text{acid}}{HNO_3(aq)} + \underset{\text{base}}{KOH(aq)} \rightarrow \underset{\text{salt}}{KNO_3(aq)} + \underset{\text{water}}{H_2O(l)}$$

When a weak carboxylic acid reacts with a strong base, the outcome is the same. The products formed in the reaction between a carboxylic acid and a strong base are a salt and water. The salts that form in the reaction between a carboxylic acid and a strong base are called carboxylate salts, and in the example given here it is a sodium carboxylate (RCOONa) that forms. Like many salts, carboxylate salts can be soluble in water. This reaction thus provides a way to convert a carboxylic acid that does not dissolve in water to its water-soluble salt. Being able to manipulate the physical properties of organic compounds in this way is very handy.

$$\underset{\text{carboxylic acid}}{RCOOH} + \underset{\text{strong base}}{NaOH} \longrightarrow \underset{\text{salt}}{RCOO^- Na^+} + \underset{\text{water}}{H_2O}$$

EXERCISE 12.31

Write an equation to show the following carboxylic acid in aqueous solution donating a proton to form its conjugate base. Show also the conjugate acid of the water molecule that accepts the proton.

$$HO{-}\overset{\overset{\large O}{||}}{C}{-}CH_2CH_3$$

EXERCISE 12.32

Write an equation to show the reaction of the following carboxylic acid with the strong base KOH.

$$C_6H_5{-}\overset{\overset{\large O}{||}}{C}{-}OH$$

Naming Carboxylic Acids

The structures and derivation of names for the first members of the straight chain carboxylic acid family are shown in Table 12.46. Note that the carboxylic acid functional group can only occur at the end of a carbon chain, so the position of the functional group does not have to be specified in the name of the compound. The carbon atom of the carbonyl group in a carboxylic acid is always carbon atom number 1 and the number is not included in the name. Note that the carbonyl group carbon atom is always carbon 1 of the parent chain no matter how the structure is drawn.

TABLE 12.46 Structures and Derivation of Names for the Simplest Carboxylic Acids

Number of C atoms (n)	1	2	3	4
Parent alkane structure	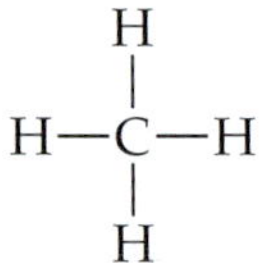 H | H—C—H | H	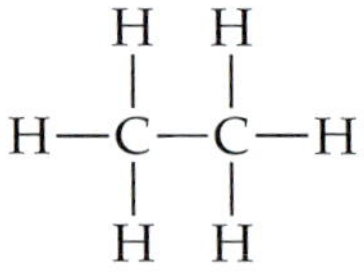 H H | | H—C—C—H | | H H	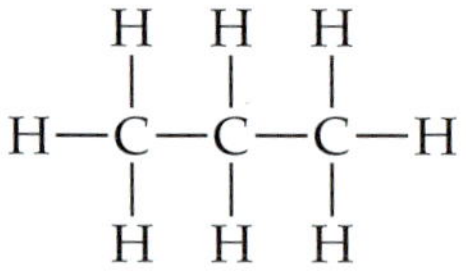 H H H | | | H—C—C—C—H | | | H H H	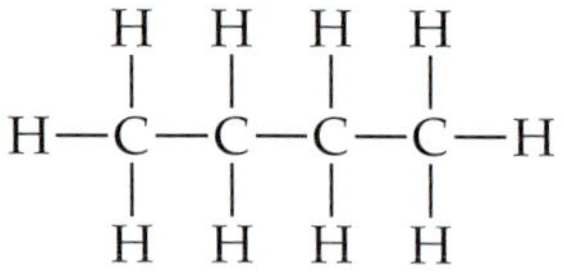 H H H H | | | | H—C—C—C—C—H | | | | H H H H
Alkane name	methane	ethane	propane	butane
Carboxylic acid name	methanoic acid	ethanoic acid	propanoic acid	butanoic acid
Carboxylic acid structural formula	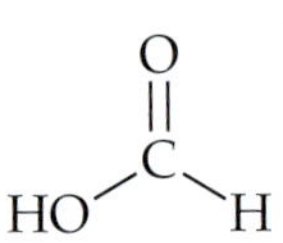 O || C HO H	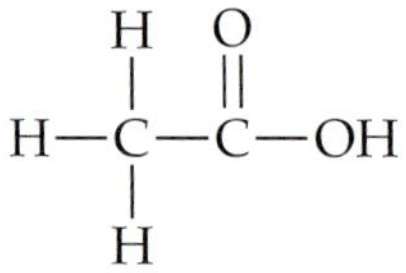 H O | || H—C—C—OH | H	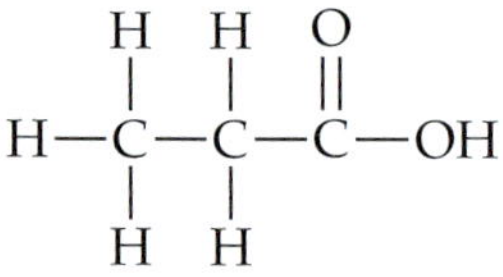 H H O | | || H—C—C—C—OH | | H H	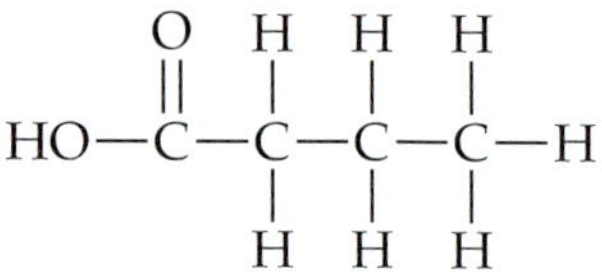 O H H H || | | | HO—C—C—C—C—H | | | H H H
Line bond formula	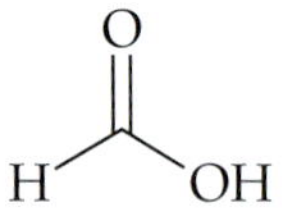 O H OH	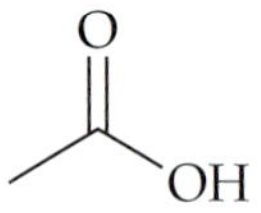 O OH	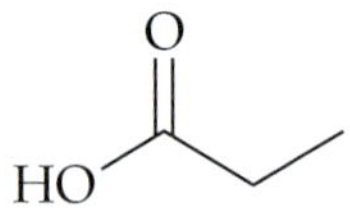 O HO	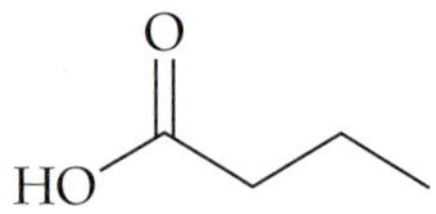 O HO
Condensed formula	HCOOH	CH_3COOH	CH_3CH_2COOH	$CH_3CH_2CH_2COOH$

Carboxylic acids can have branching structures and their naming follows the same conventions explained previously for other classes of organic compound as shown below. An explanation for naming a particular carboxylic acid, 5-chloro-3-methylpentanoic acid, demonstrates this.

5-chloro-3-methylpentanoic acid

CH_3
|
Cl—CH_2CH_2CH-CH_2C—OH
||
O

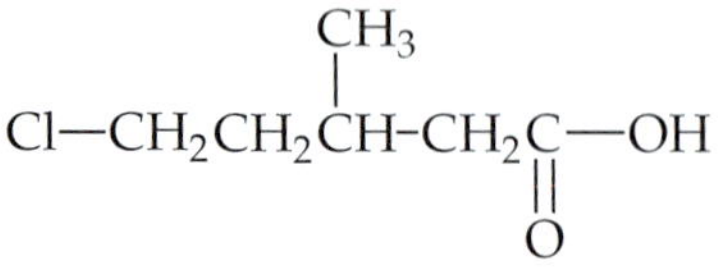

Naming explanation

- The parent name is pentanoic acid because the longest carbon chain that contains the carbonyl group carbon atom has 5 carbon atoms. The 'e' was dropped from the alkane name pentane, the suffix 'oic' and the word 'acid' added because the compound is a carboxylic acid.
- The carbonyl group carbon atom of a carboxylic acid is always carbon atom 1.
- The '5-chloro' prefix describes the chlorine atom branching from carbon atom number 5 of the parent chain.
- The '3-methyl' prefix describes the –CH_3 group branching from carbon atom number 3 of the parent chain.
- The prefixes chloro and methyl are in alphabetical order.

EXERCISE 12.33

Name the following carboxylic acids.

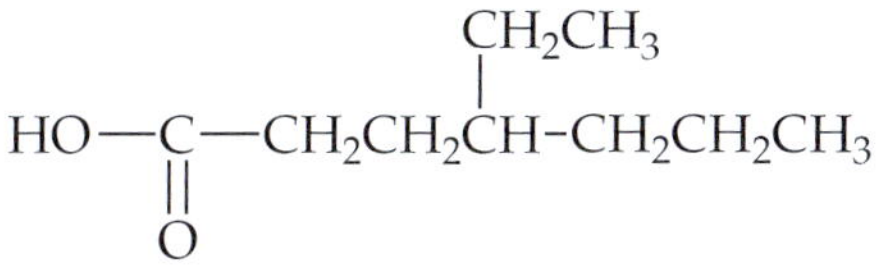	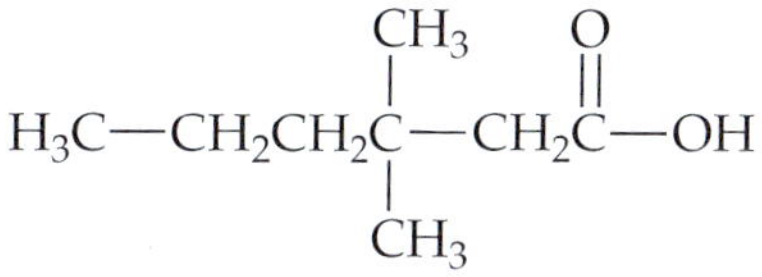
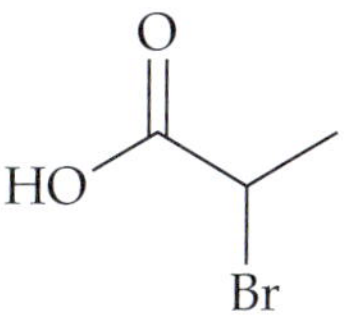	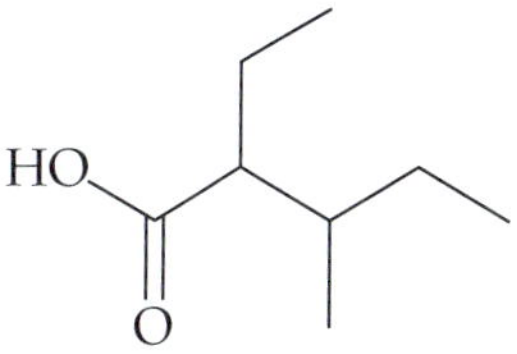

EXERCISE 12.34

Draw line bond formulas for the following carboxylic acids. You may find it helpful to draw a structural formula first.

6-ethyl-3-propyloctanoic acid	4-chlorohexanoic acid
2,3-dimethylbutanoic acid	2-propylpentanoic acid

Properties of Carboxylic Acids

The physical properties of substances that are carboxylic acids are a direct consequence of the molecular structure of the functional group. There are two polar bonds, indicated on the general structure of the carboxylic acid functional group shown below. The carbonyl group is polar because oxygen is more electronegative than carbon. The oxygen atom draws electron density towards itself, away from the carbon atom. More electron density around the oxygen atom gives a partial negative charge (δ^-) at that end of the bond, while the other end at the carbon atom has partial positive charge (δ^+). The –OH group is also polar because of the higher electronegativity of oxygen compared to hydrogen, and the acidic proton of the carboxylic acid group thus bears a partial positive charge as shown on the structure below.

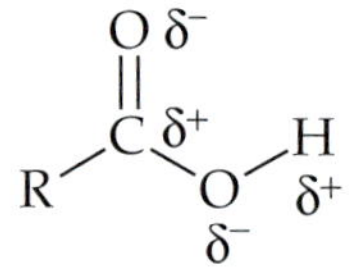

The presence of the two polar bonds means that strong attraction between carboxylic acid molecules is possible. The attraction between carboxylic acid molecules is referred to as hydrogen bonding, but it is stronger than the hydrogen bonding of alcohols because there are polar carbonyl groups and –OH groups involved. Two carboxylic acid molecules can aggregate through hydrogen bonding as shown below, and this aggregation is called a hydrogen-bonded dimer. This IMF is so strong that hydrogen bonded dimers of carboxylic acids can even be detected experimentally in the gas phase. The consequence of this strong IMF is that carboxylic acids have higher boiling points than alcohols with similar molar mass. Compare the data for propan-1-ol and ethanoic acid (Table 12.47) to see that this is the case. The compounds have the same molar mass and thus the same electron density and extent of dispersion forces. The higher boiling point for ethanoic acid compared to propan-1-ol indicates stronger IMFs. More energy is needed to overcome the attractive forces between the carboxylic acid molecules. Both compounds have hydrogen bonding as an IMF, but these interactions are stronger between carboxylic acid molecules due to the presence of the polar carbonyl group in addition to the highly polar –OH group. The carboxylic acid may be described as being more polar than the alcohol.

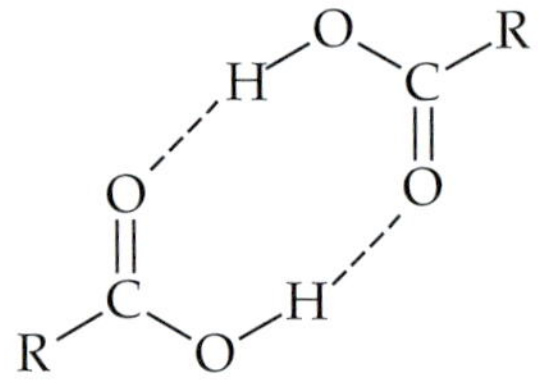

TABLE 12.47 Comparison of Alcohol and Carboxylic Acid Boiling Points

Compound name	propan-1-ol	ethanoic acid
Compound structure	$CH_3CH_2CH_2OH$	CH_3COOH
Molar mass	60 g/mol	60 g/mol
Boiling point	97 °C	118 °C

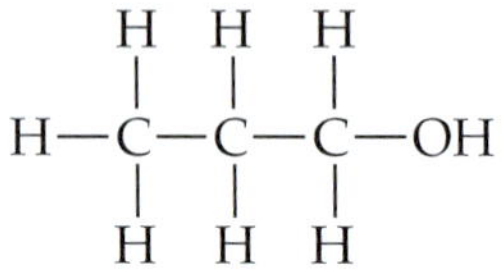

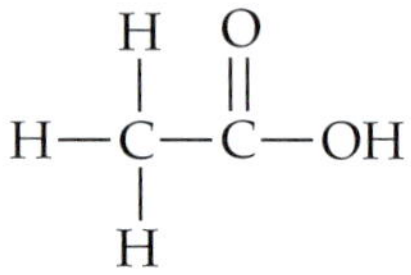

EXERCISE 12.35

The names, structures, molar masses and boiling points for three organic compounds are shown below. The compounds have similar molar mass, but their boiling points are different. Explain the reason for this difference between these compounds. Note that the small difference in molar mass is not the reason for the difference.

Compound name	propanoic acid	pentane	butanone
Compound structure	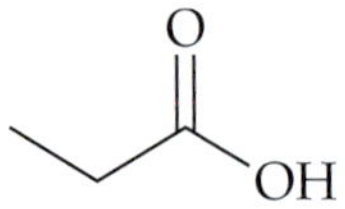		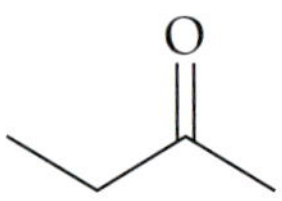
Molar mass	74 g/mol	72 g/mol	72 g/mol
Boiling point	141 °C	36 °C	80 °C
Explanation			

EXERCISE 12.36

Arrange the following compounds that have similar structures and molar mass in order of expected boiling point from lowest to highest.

	1	2	3	4
Compound structure		OH	H, O	O, OH
Molar mass	112 g/mol[1]	114 g/mol	112 g/mol	114 g/mol

12.13 Oxidation of Alcohols

LO: Describe and write equations for oxidation reactions of alcohols.

Oxidation and reduction (redox) reactions and how to balance their equations were introduced in Section 8.9. In this section we examine oxidation of alcohols, but we focus on the organic compounds that form in those reactions and don't apply the method of half reactions to balance the equations. Alcohols can be oxidised in the presence of a suitable reagent and the products of these reactions are carbonyl compounds, which were introduced in the preceding two sections. There are many oxidising agents that can be used in these reactions, but we will focus on the reaction of alcohols with aqueous potassium permanganate ($KMnO_4$, ◀ Figure 12.24) in the presence of an acid catalyst such as sulfuric acid (H_2SO_4). In equations for oxidation reactions of alcohols in this section, the presence of the oxidising agent and acid catalyst will be written as '$KMnO_4$, H^+' above the reaction arrow. The outcome of these oxidation reactions depends on the structure of the reactant alcohol.

▲ **FIGURE 12.24 Potassium permanganate** When solid potassium permanganate is dissolved in water, the resulting aqueous solution has as distinct purple colour.
© Art Directors & TRIP/Alamy Stock Photo

Primary alcohols

Primary alcohols are initially oxidised to aldehydes, but the reaction continues and the final product is a carboxylic acid. This is shown for the oxidation of a generic primary alcohol with acidified aqueous potassium permanganate in the following equation. General-purpose oxidising agents like potassium permanganate always produce a carboxylic acid as the final product in their reactions with primary alcohols. To obtain an aldehyde as the final product from oxidation of a primary alcohol, a specialised oxidising agent must be used. Another way to obtain the intermediate aldehyde is to set up the reaction vessel so the aldehyde can be removed as it forms, and before it reacts further to the carboxylic acid.

$$\underset{\text{primary alcohol}}{R{-}CH_2{-}OH} \xrightarrow{KMnO_4,\ H^+} \underset{\text{aldehyde}}{R{-}\overset{\displaystyle O}{\overset{\|}{C}}{-}H} \xrightarrow{KMnO_4,\ H^+} \underset{\text{carboxylic acid}}{R{-}\overset{\displaystyle O}{\overset{\|}{C}}{-}OH}$$

Table 12.48 shows equations for a specific primary alcohol, ethanol, undergoing this reaction to show that the R group does not change in the oxidation process. Only the attachments of the primary alcohol carbon atom bonded to the –OH group change during the reaction. Two equations are shown, one using structural formulas and the other with line bond formulas. Examine the formulas in the equations carefully and see how the framework of carbon atoms does not change in the reaction, each organic compound represented has two carbon atoms and only the functional group transforms in the reaction.

TABLE 12.48 Equations Showing Oxidation of Ethanol, a Primary Alcohol

Equation with structural formulas	$H-CH_2-CH_2-OH$	$\xrightarrow{KMnO_4,\ H^+}$	$H-CH_2-C(=O)-H$	$\xrightarrow{KMnO_4,\ H^+}$	$H-CH_2-C(=O)-OH$
Equation with line bond formulas	OH	$\xrightarrow{KMnO_4,\ H^+}$	O, H	$\xrightarrow{KMnO_4,\ H^+}$	O, OH
	primary alcohol (ethanol)		aldehyde (ethanal)		carboxylic acid (ethanoic acid)

EXERCISE 12.37

Draw the structure of the intermediate aldehyde that forms in the reaction of propan-1-ol with acidified aqueous potassium permanganate as shown in the following incomplete equation.

$$H-CH_2-CH_2-CH_2-OH \xrightarrow{KMnO_4,\ H^+} \qquad \xrightarrow{KMnO_4,\ H^+} H-CH_2-CH_2-C(=O)-OH$$

EXERCISE 12.38

Draw the structure of the carboxylic acid that forms as the final product in the oxidation of the primary alcohol shown in the following incomplete equation.

$$C_6H_5-CH_2-OH \xrightarrow{KMnO_4,\ H^+} C_6H_5-C(=O)-H \xrightarrow{KMnO_4,\ H^+}$$

EXERCISE 12.39

Draw the structures of the intermediate aldehyde and final carboxylic acid that form in the oxidation of 2-methylpropan-1-ol to complete the equation. It may be helpful to redraw the reactant using a structural formula.

$$OH \qquad \xrightarrow{KMnO_4,\ H^+} \qquad \xrightarrow{KMnO_4,\ H^+}$$

Secondary alcohols

In the presence of acidified aqueous potassium permanganate, secondary alcohols are oxidised to ketones. Unlike primary alcohols, no further reaction occurs and the final product of secondary alcohol oxidation is a ketone. Examine the general equation for this reaction shown below and see how it's only the attachments of the carbon atom bonded to the –OH group of the secondary alcohol that change in the reaction, the framework of carbon atoms does not change and the R groups of the secondary alcohol are still present in the ketone. An equation for a specific secondary alcohol, propan-2-ol, is also shown (Table 12.49) to demonstrate this.

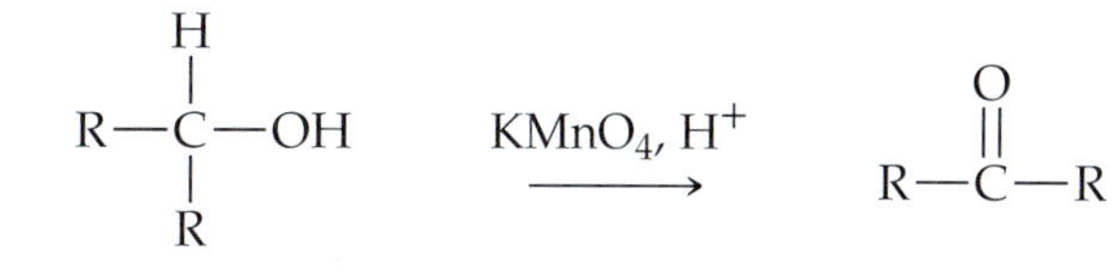

$$R-CH(OH)-R \xrightarrow{KMnO_4,\ H^+} R-C(=O)-R$$

secondary alcohol → ketone

TABLE 12.49 Equations Showing Oxidation of Propan-2-ol, a Secondary Alcohol

Equation with structural formulas	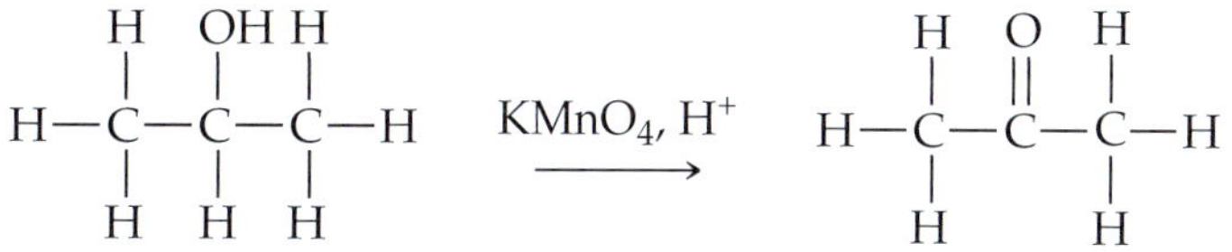 $CH_3-CH(OH)-CH_3 \xrightarrow{KMnO_4,\ H^+} CH_3-C(=O)-CH_3$
Equation with line bond formulas	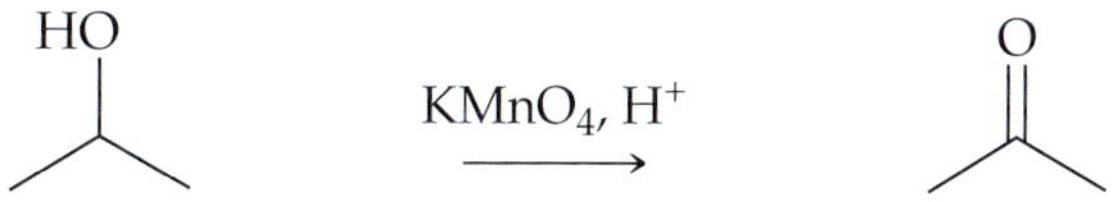 HO (line structure) $\xrightarrow{KMnO_4,\ H^+}$ O (line structure) secondary alcohol (propan-2-ol) → ketone (propanone)

EXERCISE 12.40

Complete the equation for the oxidation of the secondary alcohol shown by drawing the structure of the ketone that would form as the product.

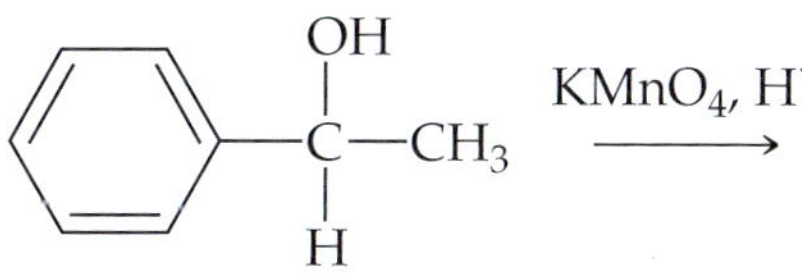

$$C_6H_5-CH(OH)-CH_3 \xrightarrow{KMnO_4,\ H^+}$$

EXERCISE 12.41

Complete the following equation by drawing the structure of the secondary alcohol required to produce the ketone product shown.

$$\xrightarrow{KMnO_4,\ H^+} \text{cyclopentanone (ring with =O)}$$

Tertiary alcohols

Tertiary alcohols are resistant to oxidation. No reaction occurs when a tertiary alcohol is in the presence of a general-purpose oxidising agent such as acidified aqueous potassium permanganate.

$$\begin{array}{c} R \\ | \\ R-C-OH \\ | \\ R \end{array} \quad \xrightarrow{KMnO_4,\ H^+} \quad \text{no reaction}$$

tertiary alcohol

The comparison of primary, secondary and tertiary alcohol general structures in Table 12.50 shows the key difference that affects their reactivity. The carbon atom bonded to the –OH group in a primary or secondary alcohol is also bonded to at least one hydrogen atom. Tertiary alcohols only have R groups and no hydrogen atoms are bonded to the carbon atom with the –OH group. In the conversion of an alcohol group to a carbonyl group, one of the attachments to the carbon atom bonded to the –OH group must be removed for the carbon-oxygen double bond to form. In primary and secondary alcohols the hydrogen atoms bonded to the carbon atom with the –OH group can leave so a carbonyl group can form, but the R groups of a tertiary alcohol cannot be removed and a carbonyl group does not form.

TABLE 12.50 Comparing Primary, Secondary and Tertiary Alcohol General Structures

Primary (1°) alcohol	Secondary (2°) alcohol	Tertiary (3°) alcohol
$\begin{array}{c} H \\ \mid \\ R-C-OH \\ \mid \\ H \end{array}$	$\begin{array}{c} H \\ \mid \\ R-C-OH \\ \mid \\ R \end{array}$	$\begin{array}{c} R \\ \mid \\ R-C-OH \\ \mid \\ R \end{array}$
• Primary and secondary alcohols have at least one hydrogen atom bonded to the carbon atom with the –OH group. • The hydrogen atoms can leave to allow formation of the carbonyl group.		• A tertiary alcohol only has R groups and no hydrogen atoms bonded to the carbon atom with the –OH group. • The R groups cannot leave to allow formation of a carbonyl group.

Now you have the knowledge to predict the outcome of any reaction of an alcohol with a general-purpose oxidising agent like acidified aqueous potassium permanganate. You can also work out what alcohol would be required to produce a particular carbonyl compound. Primary alcohols initially produce an aldehyde, but oxidation continues and the final product is a carboxylic acid. Secondary alcohols are oxidised to ketones, and tertiary alcohols are resistant to oxidation. Predicting the outcome of any alcohol oxidation relies on being able to classify the reactant alcohol (1°, 2° or 3°).

EXERCISE 12.42

The following equations show oxidation reactions for some alcohols, but they are incomplete. Draw the structure of the carbonyl compound that would form in each reaction, or the reactant alcohol as appropriate. For primary alcohols, include both the intermediate and final products. If no reaction occurs, write 'no reaction'.

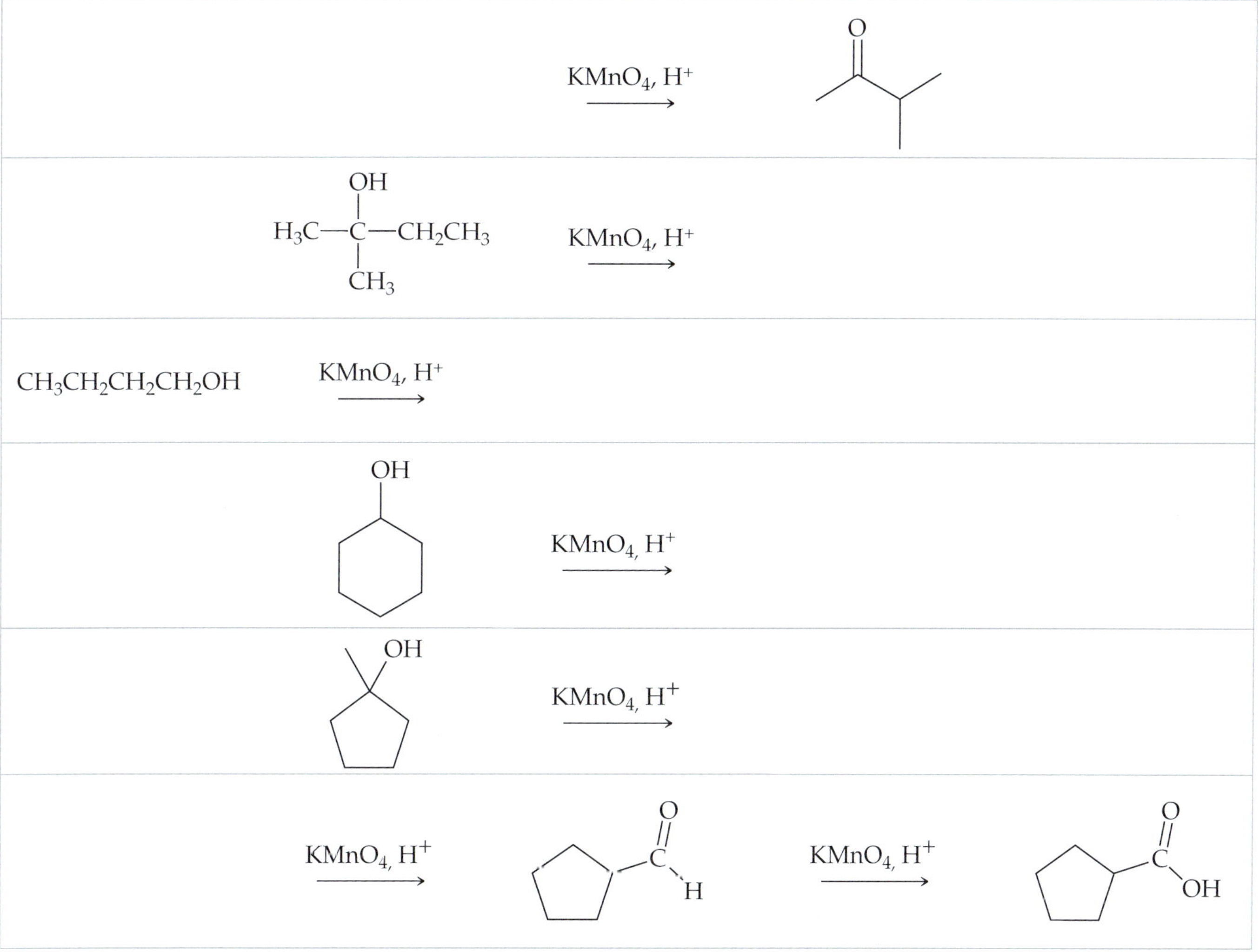

Reactant	Reagent	Product	Reagent	Product
	$\xrightarrow{KMnO_4,\ H^+}$	O (ketone)		
$H_3C{-}C(OH)(CH_3){-}CH_2CH_3$	$\xrightarrow{KMnO_4,\ H^+}$			
$CH_3CH_2CH_2CH_2OH$	$\xrightarrow{KMnO_4,\ H^+}$			
OH (cyclohexanol)	$\xrightarrow{KMnO_4,\ H^+}$			
OH (1-methylcyclopentanol)	$\xrightarrow{KMnO_4,\ H^+}$			
	$\xrightarrow{KMnO_4,\ H^+}$	O=C–H (cyclopentyl aldehyde)	$\xrightarrow{KMnO_4,\ H^+}$	O=C–OH (cyclopentyl carboxylic acid)

This section has shown that alcohols are converted by the chemical reaction, oxidation, into other functional groups. Depending on the classification of the starting alcohol, a variety of compounds with carbonyl functional groups can be prepared. The ability to convert one functional group to another using chemical reactions is a crucial tool for organic chemists. Functional group transformation is one of the methods chemists use to synthesise novel compounds, and oxidation of alcohols is just one example. The preparation of novel compounds is the continuous endeavour of thousands of chemists around the world. New organic compounds are always in demand for such applications as improved materials and more effective drug treatments for diseases.

12.14 Esters

LO: Describe and write equations for the preparation of esters.

LO: Examine structural representations of esters and derive their IUPAC names.

LO: Interpret the IUPAC names of esters and draw representations for their molecular structures.

LO: Explain the physical properties of esters.

Yet another functional group that contains a carbonyl group (C=O) is the ester. The general structure of an ester is related to that of a carboxylic acid, as shown in Table 12.51. An ester can be considered as a carboxylic acid where the hydrogen atom of the –OH group has been replaced with an R group. Esters can actually be made from carboxylic acids and are often referred to as carboxylic acid derivatives.

TABLE 12.51 Comparing the General Structures of Carboxylic Acids and Esters

General structure of carboxylic acids	General structure of esters
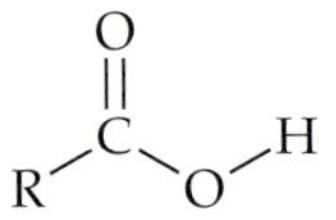 R–C(=O)–O–H	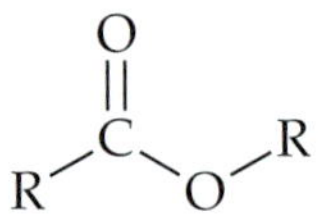 R–C(=O)–O–R
The structure of an ester is based on a carboxylic acid. Esters are often referred to as carboxylic acid derivatives. Carboxylic acids can be converted to esters.	Compare the general structure of an ester to a carboxylic acid. The hydrogen atom of the carboxylic acid –OH group is replaced with an R group, which may be the same or different to the R group bonded to the carbonyl group carbon atom.

The term 'alkoxy group' is used to describe the –OR group bonded to the carbonyl group in an ester, and this is indicated in the general structure of an ester shown below. The names of alkoxy groups are derived from alkyl group names by dropping the 'yl' and adding the 'oxy' to the end of the name. A $-CH_3$ group is called methyl, whereas an $-OCH_3$ group is called methoxy.

The –OR group bonded to the carbonyl group carbon atom in an ester is called an alkoxy group, circled in the general structure on the right	R–C(=O)–O–R	–OR is alkoxy $-OCH_3$ is methoxy $-OCH_2CH_3$ is ethoxy $-OCH_2CH_2CH_3$ is propoxy

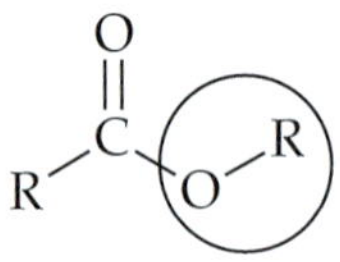

EXERCISE 12.43

Identify and name the alkoxy group in each of these esters.

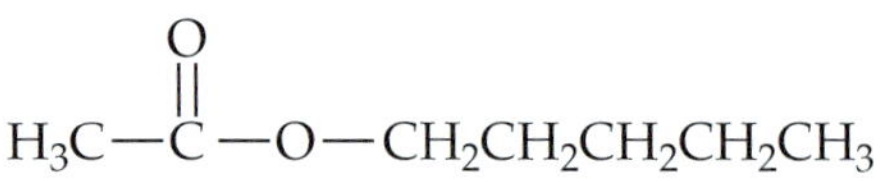 $H_3C-\overset{O}{\overset{\|\|}{C}}-O-CH_2CH_2CH_2CH_2CH_3$	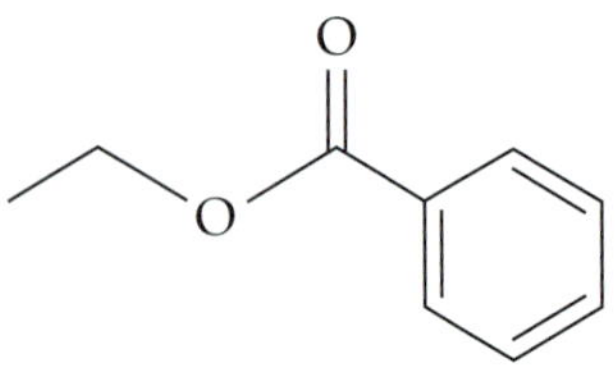 O (ethyl benzoate skeletal structure)

Preparation of Esters

There are various methods used for synthesis of esters, but their preparation from the reaction between a carboxylic acid and an alcohol is very common. This reaction requires the presence of a strong acid catalyst such as sulfuric acid (H_2SO_4) or hydrochloric acid (HCl). In this section, the presence of a strong acid catalyst in the preparation of esters will be indicated by writing H^+ above the reaction arrow. Any strong acid can be used to supply the H^+. Here is a general equation for the formation of an ester from a carboxylic acid and an alcohol. The R group of the alcohol is written at R′ so you can see clearly where it ends up in the ester product.

$$R-\overset{\overset{\displaystyle O}{\|}}{C}-O-H + H-O-R' \xrightarrow{H^+} R-\overset{\overset{\displaystyle O}{\|}}{C}-O-R' + H_2O$$

The equation is repeated to highlight the location of reactant atoms in the product. The –OR′ group of the ester came from the alcohol.

$$R-\overset{\overset{\displaystyle O}{\|}}{C}-O-H + H-\boxed{O-R'} \xrightarrow{H^+} R-\overset{\overset{\displaystyle O}{\|}}{C}-\boxed{O-R'} + H_2O$$

The rest of the ester atoms came from the carboxylic acid.

$$\boxed{R-\overset{\overset{\displaystyle O}{\|}}{C}}-O-H + H-O-R' \xrightarrow{H^+} \boxed{R-\overset{\overset{\displaystyle O}{\|}}{C}}-O-R' + H_2O$$

The formation of an ester from a carboxylic acid and an alcohol is an example of a **condensation reaction**. These reactions involve the combining of two molecules and expulsion of a small stable molecule. The small molecule expelled in this case is a water molecule. The following equation highlights where the atoms of the product water molecule come from. The –OH group from the carboxylic acid and the hydrogen atom of the alcohol combine to form a water molecule.

$$R-\overset{\overset{\displaystyle O}{\|}}{C}-\boxed{O-H} + \boxed{H}-O-R' \xrightarrow{H^+} R-\overset{\overset{\displaystyle O}{\|}}{C}-O-R' + \boxed{H_2O}$$

Here are another two equations to show the preparation of specific esters. Each equation is shown using both structural and line bond formulas. Note how the equations are written with the formula for the alcohol above the reaction arrow with the strong acid catalyst. This is another correct way to write these equations. Also, note that the product water molecule has been omitted. It is common practice to focus on the organic compound produced and exclude by products when writing equations for reactions of organic compounds. With practice, you can examine the structural representation of any ester and determine the carboxylic acid and alcohol used to prepare it.

Equation with structural formulas	$H_3C-C(=O)-OH$	$\xrightarrow{CH_3CH_2CH_2OH,\ H^+}$	$H_3C-C(=O)-OCH_2CH_2CH_3$
Equation with line bond formulas	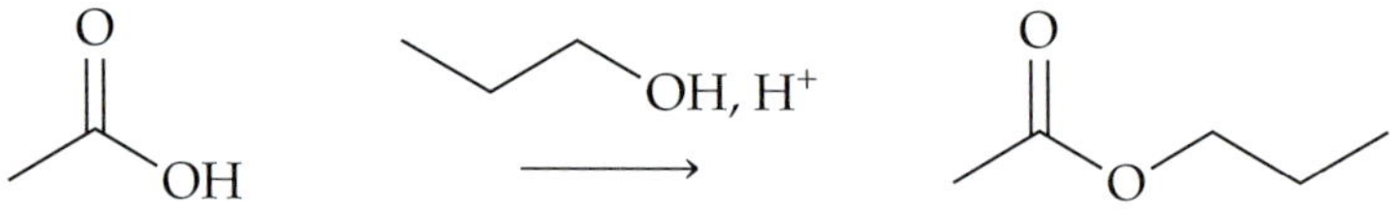 OH	$\xrightarrow{\text{OH},\ H^+}$	O
Equation with structural formulas	C, OH	$\xrightarrow{CH_3CH_2OH,\ H^+}$	C, O, CH_2CH_3
Equation with line bond formulas	OH	$\xrightarrow{\text{OH},\ H^+}$	O

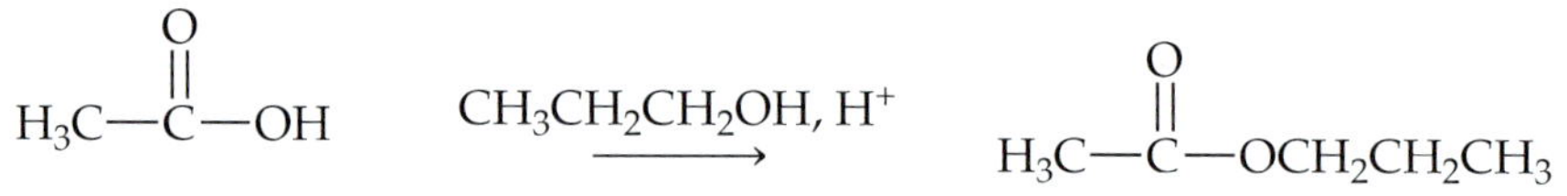

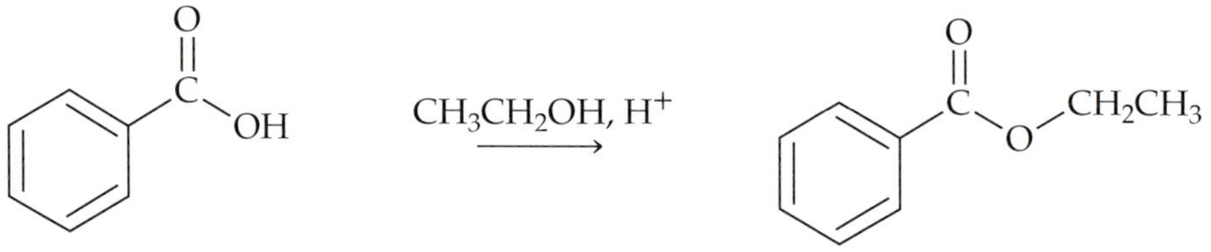

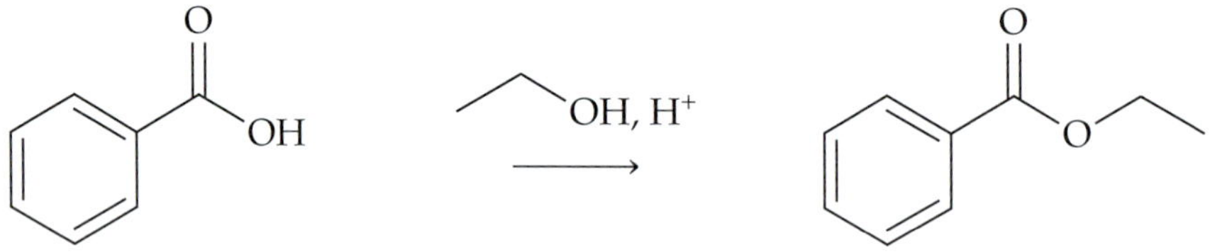

EXERCISE 12.44

Draw the structures of the carboxylic acid and alcohol used to prepare the following esters.

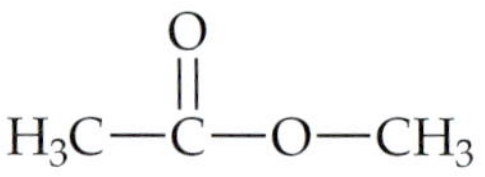

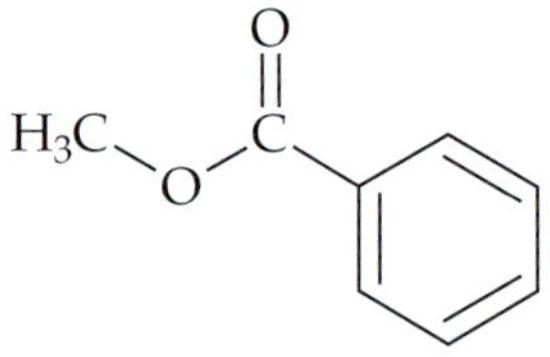

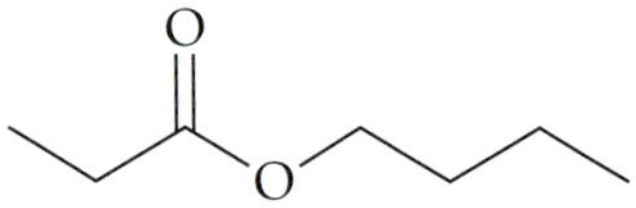

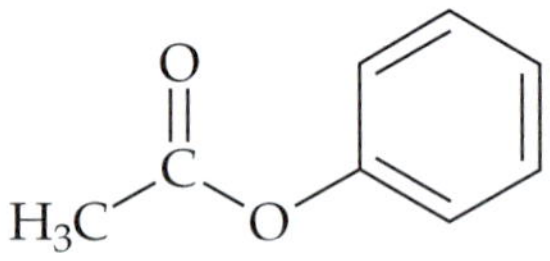

$H_3C-C(=O)-O-CH_3$	$H_3C-O-C(=O)-C_6H_5$
O, O	$H_3C-C(=O)-O-C_6H_5$

Naming Esters

The derivation of ester names relates to the common method of preparing them outlined in the previous section. The name of an ester is constructed from the names of the alcohol and carboxylic acid that would have been used to make it. The first part of an ester name is the alkyl group from the alcohol. The second part of an ester name comes from the name of the carboxylic acid. The 'ic acid' is dropped from the carboxylic acid name and the suffix 'ate' added. These two parts of an ester name are written as two words. This process is explained in Table 12.52 for a simple ester.

TABLE 12.52 Naming a Simple Ester

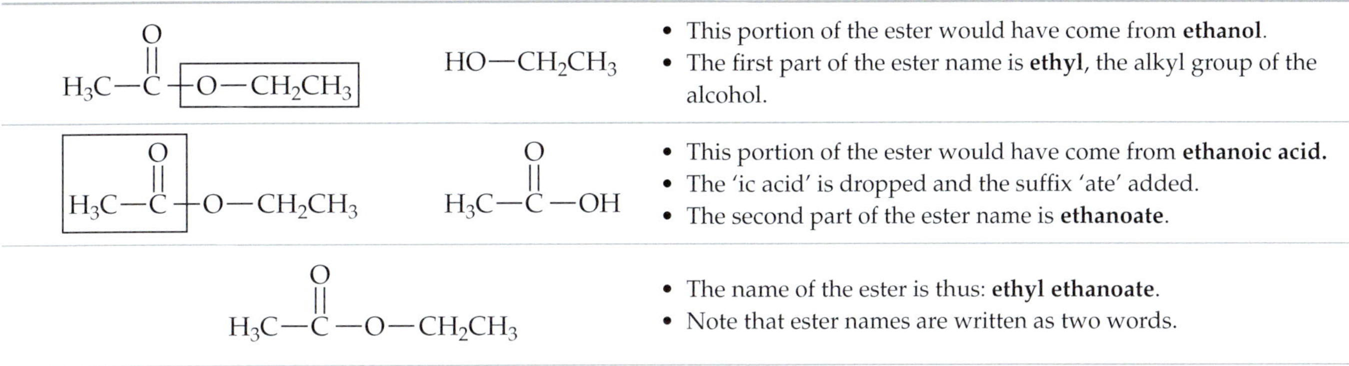

Ester	Origin	Explanation
$H_3C-C(=O)-\boxed{O-CH_2CH_3}$	$HO-CH_2CH_3$	• This portion of the ester would have come from **ethanol**. • The first part of the ester name is **ethyl**, the alkyl group of the alcohol.
$\boxed{H_3C-C(=O)}-O-CH_2CH_3$	$H_3C-C(=O)-OH$	• This portion of the ester would have come from **ethanoic acid.** • The 'ic acid' is dropped and the suffix 'ate' added. • The second part of the ester name is **ethanoate**.
$H_3C-C(=O)-O-CH_2CH_3$		• The name of the ester is thus: **ethyl ethanoate**. • Note that ester names are written as two words.

EXERCISE 12.45

Name the following esters.

$H_3C-C(=O)-O-CH_2CH_2CH_2CH_2CH_2CH_3$	$H-C(=O)-O-CH_2CH_2CH_3$
(line bond structure)	(line bond structure)

EXERCISE 12.46

Draw line bond formulas for the following esters. You may find it helpful to draw a structural formula first.

methyl propanoate	pentyl butanoate
ethyl heptanoate	butyl methanoate

Properties of Esters

Esters are generally pleasant smelling compounds that are responsible for many fruity and floral fragrances (▶ Figure 12.25), and numerous esters are used in perfumery and flavouring applications. The physical properties of esters are due to the presence of polar carbon-oxygen bonds and their molecular geometry. Esters are polar molecules and dipole-dipole attraction is the influencing IMF. When comparing the physical properties of esters to other organic compounds with similar molar mass, the type of IMFs possible in each case is the reason for the difference.

EXERCISE 12.47

By now you will have enough experience with comparing IMFs to be able to predict the order of boiling point, from lowest to highest, for the following set of organic compounds. Write a paragraph in your notes to explain the predicted order of boiling point.

Compound name	ethyl ethanoate	pentan-1-ol	hexane
Formula	$H_3C-C(=O)-O-CH_2CH_3$	$CH_3CH_2CH_2CH_2CH_2OH$	$CH_3CH_2CH_2CH_2CH_2CH_3$
Molar mass	88 g/mol	88 g/mol	86 g/mol
Expected order of boiling point from lowest to highest			

Consider the comparison of an ester and an ether with the same molar mass in Table 12.53. The boiling point of the ester (methyl ethanoate) is much higher than the ether (diethyl ether). This means that there are stronger IMFs attracting the ester molecules towards each other. More energy is required to overcome the stronger attraction and this results in a higher boiling point. With the same molar mass, the extent of dispersion forces is the same for each compound. Both compounds may be classified as polar, so dipole-dipole attraction is also relevant in both cases. Since the ester has a higher boiling point, the dipole-dipole IMF must be stronger than that of the ether. Dipole-dipole attraction is stronger for the ester because the molecules are more polar than the ether molecules. The presence of the carbonyl group in addition to the other polar carbon-oxygen bonds make the ester more polar than the ether.

TABLE 12.53 Comparing Ester and Ether Boiling points

Ester		Ether	
methyl ethanoate, 74 g/mol, BP 57 °C		diethyl ether, 74 g/mol, BP 35 °C	
Structural formula	Line bond formula	Structural formula	Line bond formula
$H_3C-C(=O)-O-CH_3$	[line bond structure: O above carbonyl, O in chain]	$H-CH_2-CH_2-O-CH_2-CH_2-H$ (all H shown)	[line bond structure with O]

© Joe Gough. Shutterstock

© aleksanderdn.123rf.com

© Zbynek Jirousek. Shutterstock

▲ **FIGURE 12.25 Esters** The pleasant fragrances of many fruits, perfumes and flowers are often due to esters.

In summary, because esters are polar and have dipole-dipole attraction as the affecting IMF, they have lower boiling points than organic compounds of similar molar mass that are capable of hydrogen bonding (alcohols and carboxylic acids). Esters have higher boiling points than organic compounds of similar molar mass that only have dispersion (hydrocarbons) as their IMF. Both esters and ethers are polar molecules, but esters are more polar and have stronger dipole-dipole attraction than ethers. Consequently, esters have higher boiling points than ethers with similar molar mass.

Hydrolysis of Esters

Hydrolysis is the specific name given to describe a chemical reaction of a compound with water. Esters can undergo hydrolysis, but the reaction of an ester with pure water is generally very slow and the most effective ester hydrolysis reactions occur if there is acid or base present. In this section, we will focus on the reaction that occurs in the presence of base. The outcomes are slightly different in the presence of acid, but we won't cover those here. The hydrolysis of an ester in the presence of a base produces a carboxylate ion ($RCOO^-$) and an alcohol, as shown in the following equation. The classification of this reaction as hydrolysis in the presence of base is indicated by writing 'H_2O, OH^-' above the reaction arrow. The hydroxide ion (OH^-) may be supplied using any strong base like sodium hydroxide (NaOH) or potassium hydroxide (KOH).

$$R{-}\overset{\overset{\displaystyle O}{\|}}{C}{-}O{-}R' \xrightarrow{H_2O,\ OH^-} R{-}\overset{\overset{\displaystyle O}{\|}}{C}{-}O^- + H{-}O{-}R'$$

The carboxylate ion ($RCOO^-$) produced in this reaction is actually the anion of a salt. The cation for the carboxylate salt that forms will be the cation from the base used in the reaction and their formulas may be written as follows.

Formula for a carboxylate salt that forms from ester hydrolysis in the presence of NaOH.

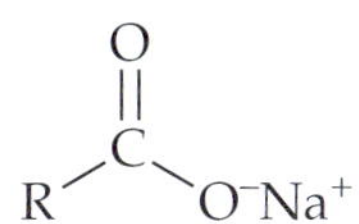

Formula for a carboxylate salt that forms from ester hydrolysis in the presence of KOH.

$$R{-}\overset{\overset{\displaystyle O}{\|}}{C}{-}O^-K^+$$

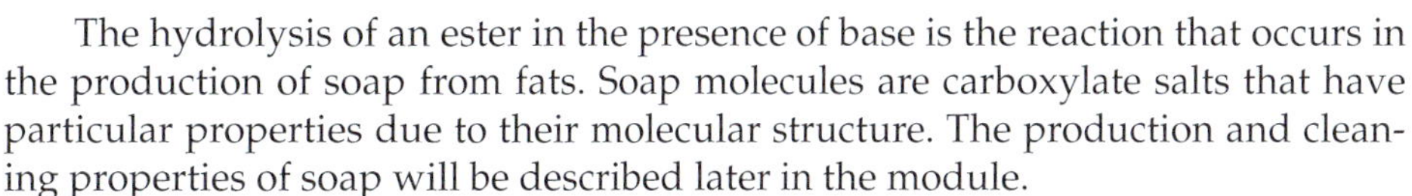

The hydrolysis of an ester in the presence of base is the reaction that occurs in the production of soap from fats. Soap molecules are carboxylate salts that have particular properties due to their molecular structure. The production and cleaning properties of soap will be described later in the module.

12.15 Amines

LO: Classify amines as primary, secondary or tertiary.

LO: Examine structural representations of primary amines and derive their IUPAC names.

LO: Interpret the IUPAC names of primary amines and draw representations for their molecular structures.

LO: Explain how amines are bases and write equations that show amines as proton acceptors.

LO: Explain the physical properties of amines.

It's now time to introduce some organic compounds that contain nitrogen. Amines are organic compounds that are based on the molecular structure of ammonia (NH_3), the simplest compound of nitrogen. The difference between ammonia and amines is the replacement of one or more hydrogen atoms by R groups. Amines can have one, two or three R groups.

Ammonia (NH_3)

H—N(—H)—H

Amines (RNH_2, R_2NH, R_3N)

R—N(—H)—H   R—N(—R)—H   R—N(—R)—R

Amines are extremely useful compounds and this functional group appears in the structures of many drug molecules. Amines are renowned for having 'fishy' and generally unpleasant odours, with the compounds 'putrescine' and 'cadaverine' being two extreme examples. As their names suggest, these related compounds have disgusting smells, and they are responsible for the odour of decomposing flesh. The structures of these compounds with their systematic and common names are shown in Table 12.54, but you do not have to memorise them. They are merely provided as examples of smelly amines, and to show that amines can have more than one of the functional group present in their molecular structure.

TABLE 12.54 Putrescine and Cadaverine are Interesting Amines

Common name	putrescine	cadaverine
Systematic name	butane-1,4-diamine	pentane-1,5-diamine
Condensed formula	$NH_2CH_2CH_2CH_2CH_2NH_2$	$NH_2CH_2CH_2CH_2CH_2CH_2NH_2$
Line bond formula	H_2N /\/\/ NH_2	H_2N /\/\/\ NH_2

The simplest amines can be considered as alkane derivatives where one hydrogen atom has been replaced with an amine ($–NH_2$) group. Table 12.55 shows the first three alkanes and the amines based on their hydrocarbon frameworks. See how the amine names are derived from the alkane name by dropping 'e' and adding the suffix 'amine'. Compare the structural representations for each amine and see how they relate to each other. The amine functional group can be bonded to a carbon atom within a chain or on the end of a chain. That is why there are two amines based on the parent structure of propane. The names propan-1-amine and propan-2-amine should make sense based on what you have learned previously for differentiating structural isomers in their naming. However, naming primary amines will be addressed in detail later in this section.

TABLE 12.55 Structures and Derivation of Names for the Simplest Amines

Number of C atoms (n)	1	2	3	
Parent alkane structure	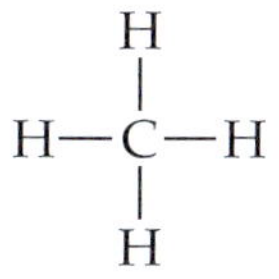 H │ H—C—H │ H	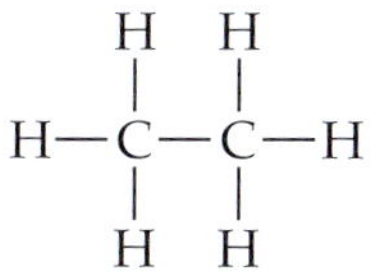 H H │ │ H—C—C—H │ │ H H	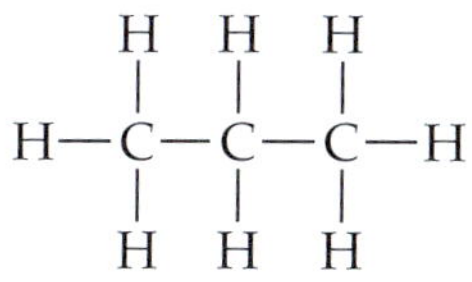 H H H │ │ │ H—C—C—C—H │ │ │ H H H	
Alkane name	methane	ethane	propane	
Amine name	methanamine	ethanamine	propan-1-amine*	propan-2-amine*
Amine structural formula	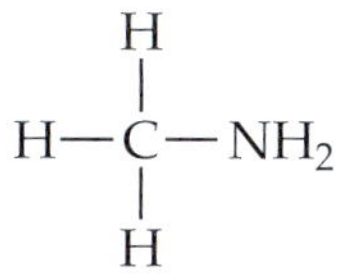 H │ H—C—NH_2 │ H	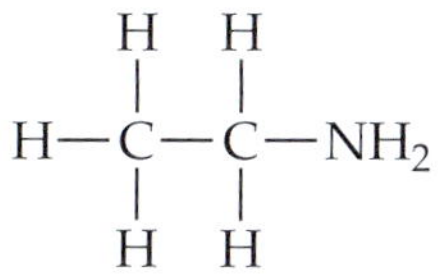 H H │ │ H—C—C—NH_2 │ │ H H	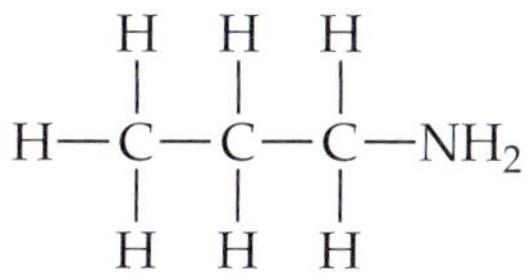 H H H │ │ │ H—C—C—C—NH_2 │ │ │ H H H	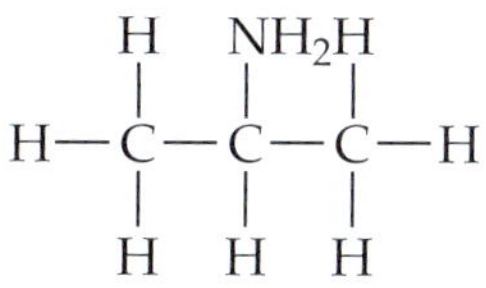 H NH_2H │ │ │ H—C—C—C—H │ │ │ H H H
Line bond formula	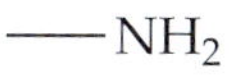 —NH_2	 NH_2	 NH_2	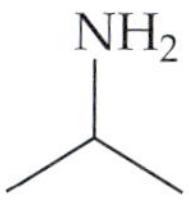 NH_2
Condensed formulas	$H_3C—NH_2$ $H_2N—CH_3$ NH_2CH_3 CH_3NH_2	$H_2N—CH_2—CH_3$ $H_2N—CH_2CH_3$ $NH_2CH_2CH_3$ $CH_3CH_2NH_2$	$H_2N—CH_2—CH_2—CH_3$ $H_2N—CH_2CH_2CH_3$ $NH_2CH_2CH_2CH_3$ $CH_3CH_2CH_2NH_2$	NH_2 │ $H_3C—CH—CH_3$ NH_2 │ H_3CCHCH_3 $CH_3CH(NH_2)CH_3$

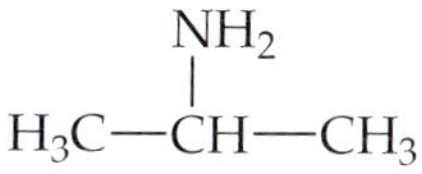

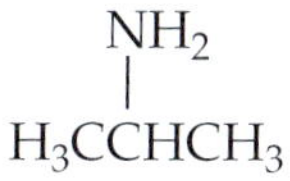

*The numbers in these names describe the position of the amine group in the parent chain.

Classification of Amines

The molecular structure of an amine can be classified as primary (1°), secondary (2°) or tertiary (3°). The classification is based on the number of R groups bonded to the amine nitrogen atom. A primary amine has one R group bonded to the amine nitrogen atom, a secondary amine has two R groups and a tertiary amine has three R groups and the differences can be seen in Table 12.56.

TABLE 12.56 The General Structures of Primary (1°), Secondary (2°) or Tertiary (3°) Amines

Primary (1°) amine	Secondary (2°) amine	Tertiary (3°) amine
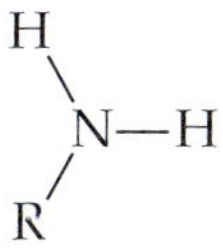 H \\ N—H / R	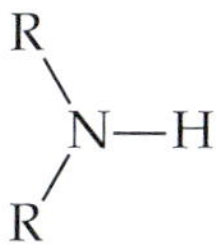 R \\ N—H / R	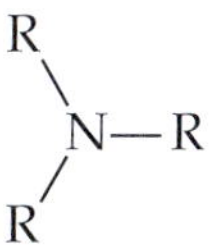 R \\ N—R / R
Amine nitrogen has one R group and two hydrogen atoms attached	Amine nitrogen has two R groups and one hydrogen atom attached	Amine nitrogen has three R groups attached

With this knowledge you can examine a structural representation of any amine and determine the classification. The classification is important because the different types of amines have different chemical reactivity. The naming of primary amines will be covered in this module. You will be able to recognise secondary and tertiary amines, but you don't have to learn how to name them in this introduction to organic compounds. Classification of amines is straightforward, it's simply a matter of examining the structural representation to determine the number of R groups bonded to the amine nitrogen atom. If the amine nitrogen atom is involved in a double bond, this counts as two R groups.

EXERCISE 12.48

Classify the following amines. Even though nicotine and morphine are reasonably complex molecules, you can still classify the amine that is present in their structures.

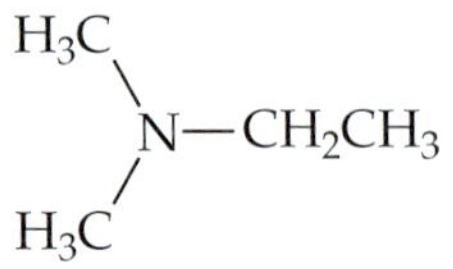

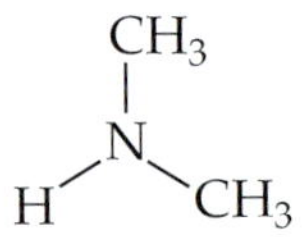

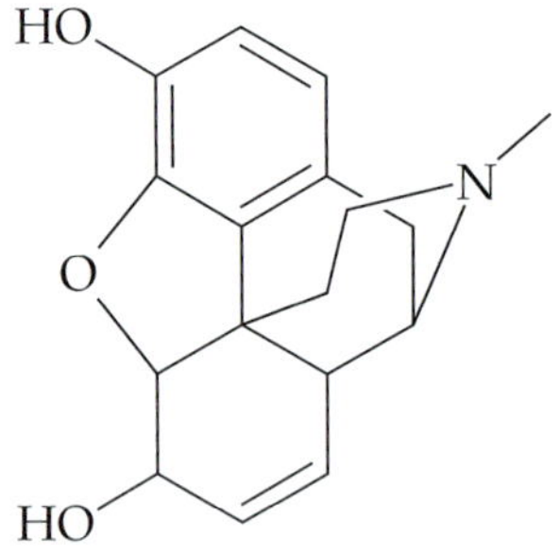

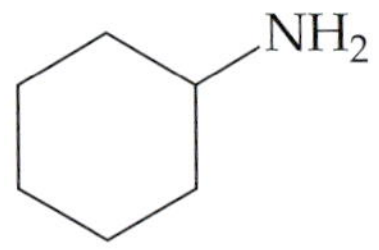

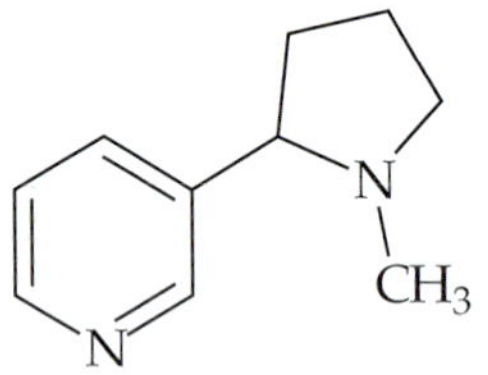

nicotine (classify both amines)

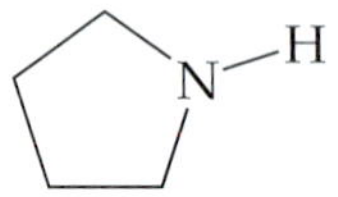

morphine

Naming Primary Amines

The process of naming any organic compound will be familiar for you now. Naming a primary amine follows the same steps that you have seen previously. Table 12.57 gives the steps for naming any organic compound with the factors to consider in naming a primary amine. An explanation for naming a particular amine will be given as an example, and then you can apply the process yourself.

TABLE 12.57 The Steps to Naming Organic Compounds Applied to Primary Amines

Step	Description of the step	Applying the step to naming primary amines
0	Examine the structural representation to identify the type or classification of the organic compound.	• If a hydrocarbon framework with a primary amine ($-NH_2$) group is represented, then the compound is a primary amine and the name will have the suffix **amine.** • The primary amine name is based on an alkane name, dropping the 'e' and adding 'amine'.
1	Identify the parent chain, which is the longest continuous chain of carbon atoms.	• The parent chain of a primary amine is the longest chain that contains the carbon atom bonded to the $-NH_2$ group.
2	Number the carbon atoms in the parent chain.	• For primary amines the carbon atoms of the parent chain are numbered so the one bonded to the $-NH_2$ group has the lowest possible number. • The name of the primary amine must describe the position of the $-NH_2$ group, this is done by placing the number of the carbon atom bonded to the $-NH_2$ group before the suffix.
3	Identify the substituents and the number of the parent chain carbon atoms that they branch from.	• This step is applied in the same way for primary amines.
4	Write the whole name of the compound as one word.	• This step is applied in the same way for primary amines.

An explanation for naming a particular primary amine, 4-ethyl-2-methylhexan-3-amine, demonstrates application of these steps.

4-ethyl-2-methylhexan-3-amine

```
      CH3
      |
H3C—CH-CH-CH-CH2CH3
         |  |
       NH2 CH2CH3
```

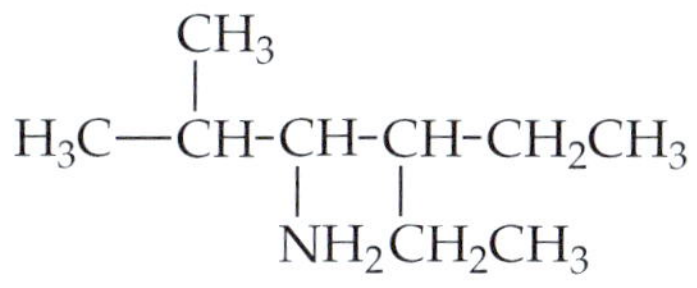

Naming explanation

- The parent name is **hexanamine** because the longest carbon chain that contains the carbon atom bonded to the $-NH_2$ group has 6 carbon atoms. The 'e' was dropped from the alkane name hexane and the suffix 'amine' added because the compound is a primary amine.
- The parent chain carbon atoms have been numbered so the one attached to the $-NH_2$ group has the lowest possible number. The number 3 has been inserted into the parent name to give **hexan-3-amine**.
- The '4-ethyl' prefix describes the $-CH_2CH_3$ group branching from carbon atom number 4 of the parent chain.
- The '2-methyl' prefix describes the $-CH_3$ group branching from carbon atom number 2 of the parent chain.
- The prefixes ethyl and methyl are in alphabetical order.

EXERCISE 12.49

Name the following primary amines.

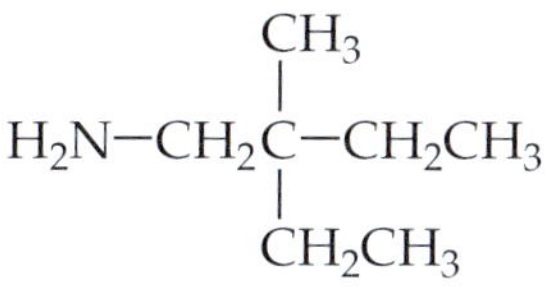 $H_2N-CH_2C(CH_3)(CH_2CH_3)-CH_2CH_3$	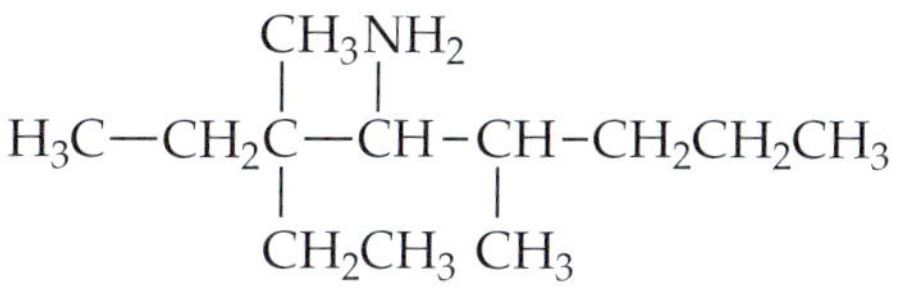 $H_3C-CH_2C(CH_3)(CH_2CH_3)-CH(NH_2)-CH(CH_3)-CH_2CH_2CH_3$
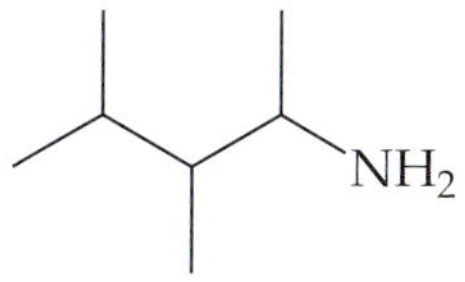 NH_2	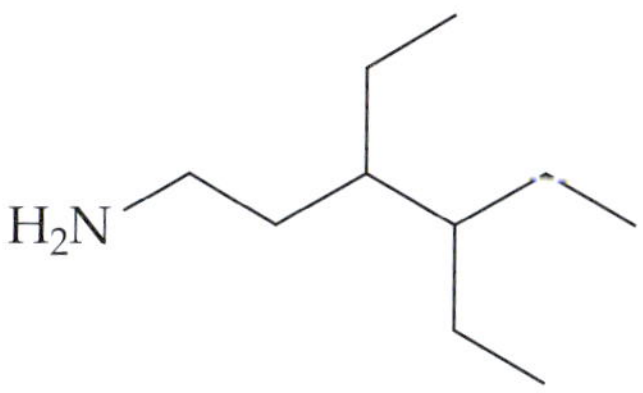 H_2N

EXERCISE 12.50

Draw line bond formulas for the following primary amines. You may find it helpful to draw a structural formula first.

butan-2-amine	3,4-diethylhexan-2-amine
2,3-dimethylpentan-1-amine	**3,4-diethylheptan-3-amine**

Amines are Bases

The nitrogen atom of an amine has a lone pair of electrons and both of these electrons can be used to form a covalent bond. General structures of primary, secondary and tertiary amines are shown below with their lone pair. All three types of amine can use the nitrogen atom lone pair to form a covalent bond.

Primary (1°) amine

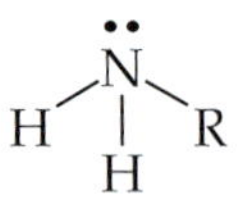

Secondary (2°) amine

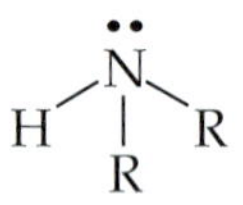

Tertiary (3°) amine

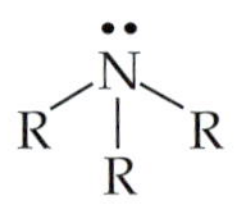

The ability of amines to use the nitrogen atom lone pair electrons to form a covalent bond with a hydrogen ion (proton, H^+) means that amines are bases. This process may be referred to as accepting a proton and this might sound familiar because one of the definitions for a base is a substance that is a proton acceptor (Section 8.8). An amine accepts a proton by using both of its nitrogen atom lone pair electrons to form a covalent bond with the proton. This process is shown in an equation (Table 12.58), which is written using both structural and condensed formulas. Note that 'H^+' shown in the equation can be from any acid (proton donor). The cation that forms after the amine base accepts a proton is the conjugate acid, which is called an ammonium ion. The equation also shows that the conjugate acid of the amine can be converted back to its free base form by reaction with a strong base such as hydroxide ion (OH^-). The equation shows a primary amine, but all amines (1°, 2° and 3°) are bases. One of the factors influencing the strength of an amine as a base is the number of R groups. This is one reason why the classification of amines as primary, secondary or tertiary is important.

TABLE 12.58 Equation Showing an Amine Accepting a Proton

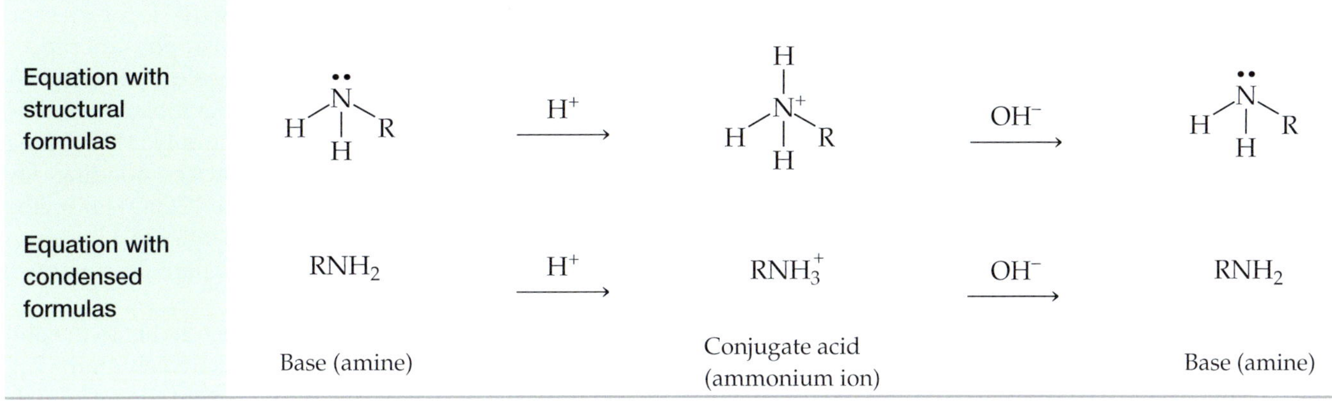

Equation with structural formulas	$\mathrm{H{-}\ddot{N}(H){-}R}$	$\xrightarrow{H^+}$	$\mathrm{H{-}N^+(H)(H){-}R}$	$\xrightarrow{OH^-}$	$\mathrm{H{-}\ddot{N}(H){-}R}$
Equation with condensed formulas	RNH_2	$\xrightarrow{H^+}$	RNH_3^+	$\xrightarrow{OH^-}$	RNH_2
	Base (amine)		Conjugate acid (ammonium ion)		Base (amine)

Recall that another definition for a base refers to substances that produce hydroxide ions (OH^-) in aqueous solution (Section 8.8). When amines are placed in water, the nitrogen atom lone pair accepts a proton from a water molecule to produce OH^-, as shown in Table 12.59. Again, this process is shown for a primary amine, but all amines can react with water to produce hydroxide ions in this way. In this reaction, water is an acid (proton donor) and the hydroxide ion produced is its conjugate base. Note that amines are weak bases. In aqueous solutions, not all of the amine molecules will accept a proton from water. The incomplete conversion of amines to their conjugate acid form in aqueous solution is indicated by using reversible reaction arrows in the equations.

TABLE 12.59 Equation Showing an Amine Producing Hydroxide Ions in Water

Equation with structural formulas	$\mathrm{H{-}\ddot{N}(H){-}R}$	+	H_2O	$\rightleftharpoons$	$\mathrm{H{-}N^+(H)(H){-}R}$	+	OH^-
Equation with condensed formulas	RNH_2	+	H_2O	$\rightleftharpoons$	RNH_3^+	+	OH^-
	Base (amine)		Acid (water)		Conjugate acid (ammonium ion)		Conjugate base (hydroxide ion)

EXERCISE 12.51

Complete the following equation to show the amine accepting a proton (H^+) to form its conjugate acid.

$$\mathrm{H{-}N(CH_3){-}CH_2CH_3} \xrightarrow{H^+}$$

EXERCISE 12.52

Write an equation to show the following amine in aqueous solution accepting a proton from water to form its conjugate acid and hydroxide ion.

NH_2

Properties of Amines

Primary and secondary amines are capable of hydrogen bonding due to the presence of the highly polar N-H bonds. Consequently, amines have similar physical properties to alcohols that also have hydrogen boding between molecules. Like alcohols, amines have higher boiling points then hydrocarbons with similar molar mass. This is because hydrocarbons are non-polar and have dispersion as the only IMF. Whereas amine molecules are polar and have the stronger IMF of hydrogen bonding. Hydrogen bonding between amine molecules is shown in ▼ Figure 12.26(a) for methanamine. In a sample of pure amine, the partially positively charged (δ^+) hydrogen atom from one amine molecule is electrostatically attracted to the partially negatively charged (δ^-) nitrogen atom of another.

The water solubility of low molar mass amines is also attributable to the ability to participate in hydrogen bonding. Hydrogen bonding between amine and water molecules allow the two substances to mix. ▼ Figure 12.26(b) shows how the δ^+ hydrogen atoms of the water molecules are attracted to the δ^- nitrogen atoms of the amine. The δ^+ hydrogen atoms of the amine molecules are also attracted to the δ^- oxygen atoms of the water molecules. Higher molar mass amines are less able to dissolve in water. Water solubility decreases with increasing size of the hydrocarbon portion because amine polarity decreases as the size of the hydrocarbon component increases and dispersion forces become the predominant IMF.

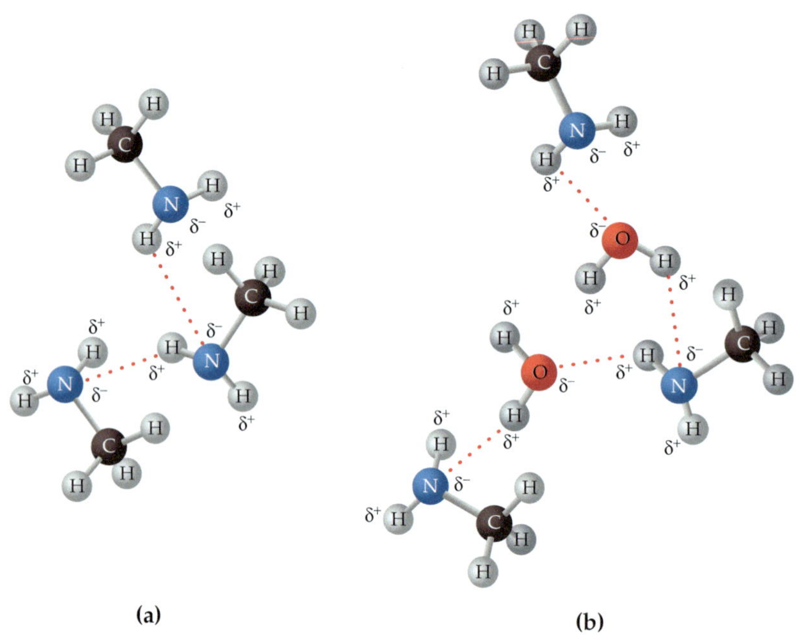

▶ **FIGURE 12.26 Amine hydrogen bonding** **(a)** Hydrogen bonding between methanamine molecules. **(b)** Hydrogen bonding between methanamine and water molecules.

The boiling points of straight chain amines increase with increasing length of the hydrocarbon chain in the same way and for the same reasons as alcohols. The boiling points of amine structural isomers decrease with increasing spherical shape of the molecules, as is the case for alcohol structural isomers.

EXERCISE 12.53

Arrange these amine structural isomers in order of expected boiling point from lowest to highest.

Amine isomer	1	2	3
Structural formula	$H-N(CH_3)-CH_2CH_3$	$CH_3CH_2CH_2NH_2$	$H_3C-N(CH_3)-CH_3$

Since alcohols and amines are both capable of hydrogen bonding, we might expect an alcohol and an amine with similar molar mass to have similar boiling points. However, this is not the case. Compare the boiling points for butan-1-amine and butan-1-ol shown in Table 12.60. Both compounds have the same straight hydrocarbon chain and a functional group that allows for hydrogen bonding, yet the amine has a much lower boiling point than the alcohol. The extent of dispersion forces is the same for each compound, so the difference in boiling point must be due to different strengths of hydrogen bonding for each compound. The higher boiling point for the butan-1-ol indicates that the hydrogen bonding is stronger for the alcohol. This comes down to the greater electronegativity of oxygen compared to nitrogen and consequent greater polarity of the alcohol O-H bonds compared to the amine N-H bonds. When comparing an alcohol and an amine with similar structure and molar mass, the alcohol is more polar than the amine. For polar molecules, the strength of IMFs increases with increasing polarity.

TABLE 12.60 Comparison of Amine and Alcohol Boiling Points

butan-1-amine	butan-1-ol
$H-CH_2-CH_2-CH_2-CH_2-NH_2$ (H–C–C–C–C–NH_2, each C bearing two H)	$H-CH_2-CH_2-CH_2-CH_2-OH$ (H–C–C–C–C–OH, each C bearing two H)
Molar mass 73 g/mol	Molar mass 74 g/mol
Boiling point 77 °C	Boiling point 118 °C

In summary, the boiling points of amines are lower than the boiling points of alcohols with similar structure and molar mass. This is because hydrogen bonding between amine molecules is not as strong as hydrogen bonding between alcohol molecules and this is due to N-H bonds being less polar than O-H bonds. The difference in bond polarity is due to the greater electronegativity of oxygen compared to nitrogen.

EXERCISE 12.54

These three organic compounds have similar structures and molar mass. All may be classified as polar, but they have different boiling points. Arrange the compounds in order of expected boiling point from lowest to highest.

	1	2	3
Compound structure	C_6H_5–C(=O)–OH	C_6H_5–$CH_2CH_2NH_2$	C_6H_5–CH_2CH_2OH
Molar mass	122 g/mol	121 g/mol	122 g/mol

12.16 Amides

LO: Identify the amide functional group in representations of organic compounds.

LO: Describe and write equations for the preparation of amides

The amide functional group is very important as it appears in the molecular structures of proteins and some polymers that will be examined later in the module. It is a very interesting functional group because it contains both oxygen in a carbonyl (C=O) group and nitrogen. All you need to know about amides is their general structure and how they can be prepared from the reaction between a carboxylic acid and an amine. You will be able to recognise an amide functional group in structural representations of organic compounds, but you will not have to learn how to name them at this stage. The general structure of an amide is shown below.

The carbon atom of the carbonyl group is bonded to one R group and a nitrogen atom.
The R groups may be hydrogen atoms or any hydrocarbon framework.
The R groups may be all the same or different.

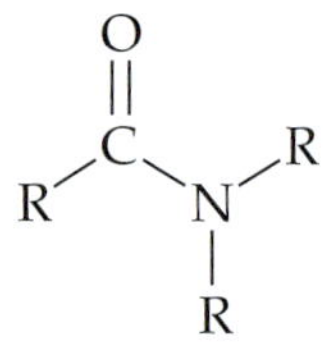

EXERCISE 12.55

Identify the amide functional group in the following structural representations of organic compounds by circling the group of atoms (O, C and N) that comprise it.

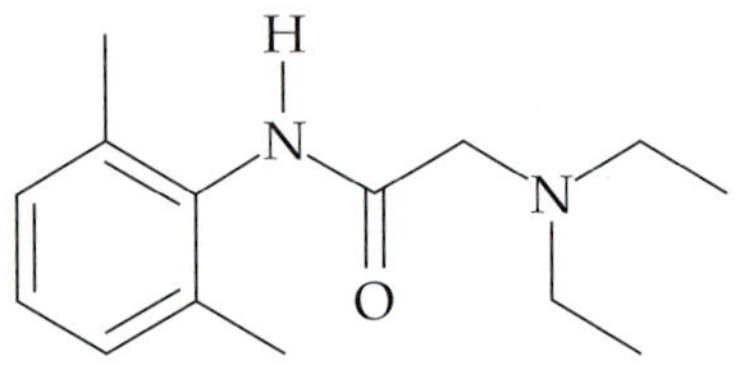 lignocaine – local anaesthetic	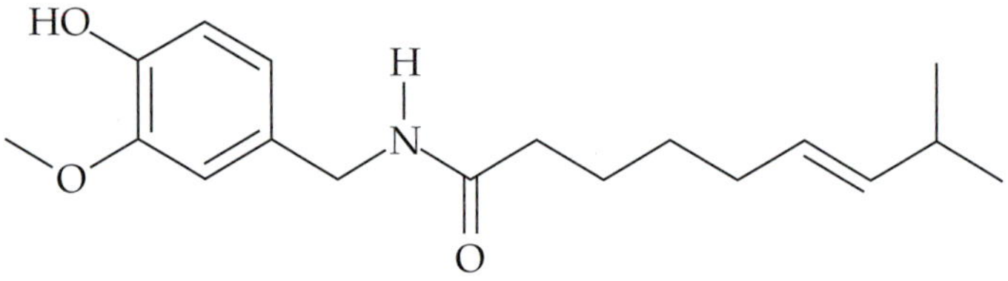 capsaicin from chillies and found in capsicum spray
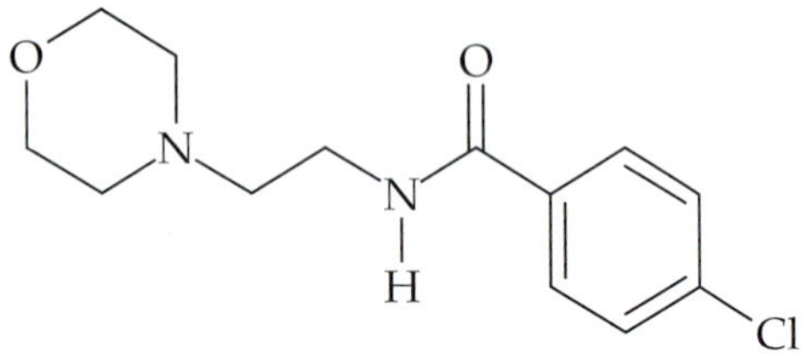 moclobemide – antidepressant drug	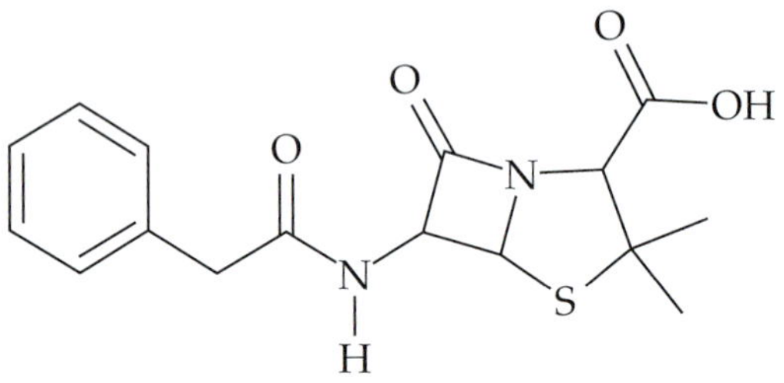 penicillin g – an antibiotic with two amide groups

Preparation of Amides

An amide may form in a condensation reaction between a carboxylic acid and an amine. Recall that **condensation reactions** involve the combining of two molecules and expulsion of a small stable molecule. Examine the general equation for the formation of an amide in a condensation reaction and see that the small molecule expelled is water. This general equation is shown for a primary amine, but secondary amines react in the same way. Tertiary amines cannot react with carboxylic acids to form amides.

$$R{-}\overset{\overset{\displaystyle O}{\|}}{C}{-}O{-}H \;+\; H{-}\underset{\underset{\displaystyle H}{|}}{N}{-}R' \;\longrightarrow\; R{-}\overset{\overset{\displaystyle O}{\|}}{C}{-}\underset{\underset{\displaystyle H}{|}}{N}{-}R' \;+\; H_2O$$

Have a close look at the equation again to see what atoms of the amine end up in the product amide.

$$R{-}\overset{\overset{\displaystyle O}{\|}}{C}{-}O{-}H \;+\; H{-}\boxed{\underset{\underset{\displaystyle H}{|}}{N}{-}R'} \;\longrightarrow\; R{-}\overset{\overset{\displaystyle O}{\|}}{C}{-}\boxed{\underset{\underset{\displaystyle H}{|}}{N}{-}R'} \;+\; H_2O$$

The R group and carbonyl group of the reactant carboxylic acid are present in the amide.

$$\boxed{R{-}\overset{\overset{\displaystyle O}{\|}}{C}}{-}O{-}H \;+\; H{-}\underset{\underset{\displaystyle H}{|}}{N}{-}R' \;\longrightarrow\; \boxed{R{-}\overset{\overset{\displaystyle O}{\|}}{C}}{-}\underset{\underset{\displaystyle H}{|}}{N}{-}R' \;+\; H_2O$$

The reactant atoms not present in the product amide, the –OH group from the carboxylic acid and one of the hydrogen atoms from the amine, combine to form a water molecule.

$$R{-}\overset{\overset{\displaystyle O}{\|}}{C}{-}\boxed{O{-}H} \;+\; \boxed{H}{-}\underset{\underset{\displaystyle H}{|}}{N}{-}R' \;\longrightarrow\; R{-}\overset{\overset{\displaystyle O}{\|}}{C}{-}\underset{\underset{\displaystyle H}{|}}{N}{-}R' \;+\; \boxed{H_2O}$$

EXERCISE 12.56

Draw structures for the amides that could form from the following pairs of carboxylic acids and amines.

Pair 1		
Carboxylic acid	Amine	
$H_3C{-}\overset{\overset{\displaystyle O}{\|}}{C}{-}O{-}H$	$H{-}\underset{\underset{\displaystyle H}{|}}{N}{-}CH_2CH_3$	
Pair 2		
Carboxylic acid	Amine	
$C_6H_5{-}\overset{\overset{\displaystyle O}{\|}}{C}{-}OH$	$H{-}\underset{\underset{\displaystyle CH_3}{|}}{N}{-}CH_3$	

Pair 3		
Carboxylic acid	Amine	
HO–C(=O)–cyclopentyl	CH_3NH_2	
Pair 4		
Carboxylic acid	Amine	
$CH_3CH_2C(=O)OH$	$CH_3CH_2CH_2NH_2$	

EXERCISE 12.57

Draw structures of the carboxylic acid and amine that would have been used as reactants to form the following amides.

$H_3C-C(=O)-N(CH_3)-CH_3$ *N*,*N*-dimethylacetamide	
HO–C_6H_4–N(H)–C(=O)–CH_3 paracetamol	

12.17 Polymers

LO: Identify and describe condensation and addition polymers.

LO: Examine structural representations of polymers and identify the monomers that were used to prepare them.

LO: Draw structural representations for polymers of particular monomers.

LO: Account for the properties of various polymeric materials.

Polymers are large molecules made up of repeating units. They are composed of small molecules called **monomers**, which react to form the large molecules. **Polymerisation** is the name of the process in which monomers react to form polymers. Two types of polymers, their constituent monomers, and the polymerisation processes used to prepare them will be introduced in this section. Some examples of polymers, which are ubiquitous in our everyday lives, are shown in ▶ Figure 12.27. The aim of this section is to introduce the molecular structure of polymers and see how this is responsible for the various properties of polymers that make them such useful materials.

▼ **FIGURE 12.27 Polymers** Some examples of polymers, which can be natural or synthetic.

(a) © 137002. Shutterstock; **(b)** © 123rf.com; **(c)** © mykeyruna. Shutterstock; **(d)** © Coleman Yuen. Pearson Education Asia Ltd; **(e)** © discpicture. Shutterstock; **(f)** © Dmitry Naumov. Shutterstock; **(g)** © 123rf.com; **(h)** © Fotokostic. Shutterstock

Natural polymers

(a) Rubber (b) Wool (c) Cellulose (d) Starch

Synthetic polymers

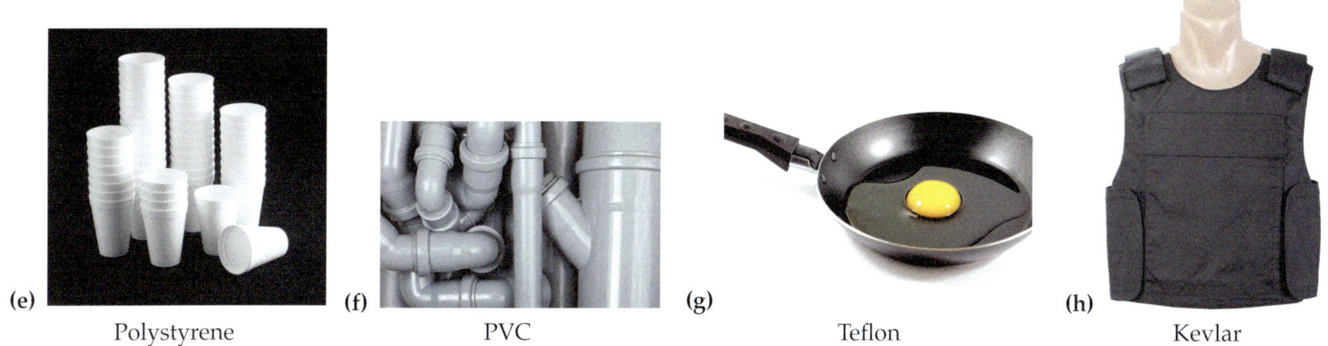

(e) Polystyrene (f) PVC (g) Teflon (h) Kevlar

There are two subsets of polymers called homopolymer and copolymer, and two major polymerisation processes called addition and condensation polymerisation. The type of polymer formed relates to the process used. A **homopolymer** is composed of just one type of monomer and is generally produced in a process called **addition polymerisation**. The monomers used in addition polymerisation are alkenes that react with each other to form a polymer. The polymerisation process involves repeating addition reactions, so the polymers that form are called **addition polymers**. A **copolymer** is made from more than one type of monomer and the process in which they are formed is known as **condensation polymerisation**. The monomers used in condensation polymerisation have two reactive functional groups and the polymerisation process involves repeated condensation reactions. The polymers that form in this way are logically called **condensation polymers**. There are examples of condensation polymers that are homopolymers, but these will not be covered in this introduction to polymers. Condensation polymers and addition polymers will be discussed in turn.

Condensation polymerisation	Addition polymerisation
• Monomers have two reactive functional groups. • Polymerisation involves condensation reactions between monomers. • Produces copolymers called condensation polymers.	• Alkene monomers react with each other to form a polymer. • Polymerisation involves repeating addition reactions. • Produces homopolymers called addition polymers.

Condensation Polymers

The two condensation reactions you have seen previously in the context of formation of esters (12.14) and amides (12.16) are relevant to the preparation of condensation polymers. While an ester forms in the reaction between a carboxylic acid and an alcohol, a **polyester** forms in the reaction between a dicarboxylic acid and a diol. A dicarboxylic acid is a carboxylic acid with two carboxylic acid groups, and a diol is an alcohol that has two –OH groups. The following equations and images show how polymerisation is possible when a dicarboxylic acid and a diol can react with each other. The polymer that forms is a polyester because the product is a long molecule with many ester groups along the structure.

This equation is a reminder of the condensation reaction between a carboxylic acid and an alcohol that produces an ester.

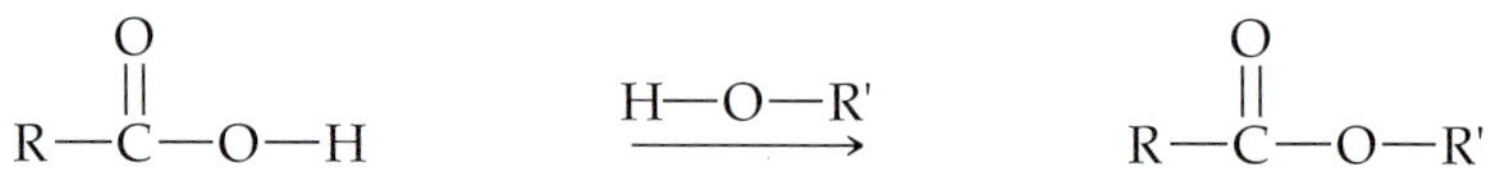

$$R{-}\overset{\overset{\displaystyle O}{\|}}{C}{-}O{-}H \xrightarrow{H{-}O{-}R'} R{-}\overset{\overset{\displaystyle O}{\|}}{C}{-}O{-}R'$$

The next equation shows a specific dicarboxylic acid and diol reacting in the same way to produce an ester. See how just one of the carboxylic acid groups reacted with just one of the alcohol groups to form the product shown.

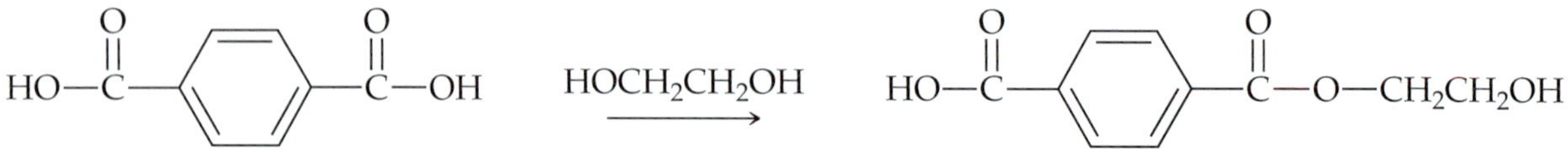

$$HO{-}\overset{\overset{\displaystyle O}{\|}}{C}{-}C_6H_4{-}\overset{\overset{\displaystyle O}{\|}}{C}{-}OH \xrightarrow{HOCH_2CH_2OH} HO{-}\overset{\overset{\displaystyle O}{\|}}{C}{-}C_6H_4{-}\overset{\overset{\displaystyle O}{\|}}{C}{-}O{-}CH_2CH_2OH$$

The thing about the product of the condensation reaction in this case is that the ends of the ester that formed still have carboxylic acid and alcohol functional groups, and further reaction can occur.

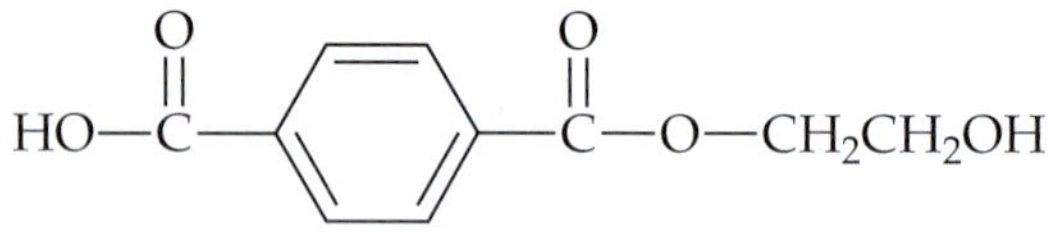

This carboxylic acid group can react with another alcohol.

$$HO{-}\overset{\overset{\displaystyle O}{\|}}{C}{-}C_6H_4{-}\overset{\overset{\displaystyle O}{\|}}{C}{-}O{-}CH_2CH_2OH$$

This alcohol group can react with another carboxylic acid.

This is the structure resulting from another reaction occurring at each end of the initial product. Note how this product still has reactive functional groups at either end.

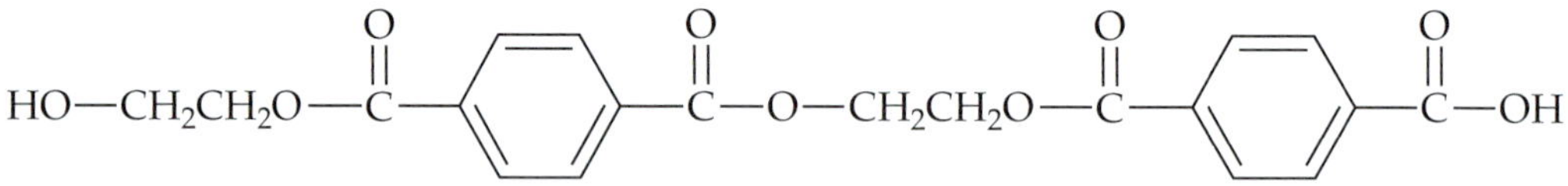

$$HO{-}CH_2CH_2O{-}\overset{\overset{\displaystyle O}{\|}}{C}{-}C_6H_4{-}\overset{\overset{\displaystyle O}{\|}}{C}{-}O{-}CH_2CH_2O{-}\overset{\overset{\displaystyle O}{\|}}{C}{-}C_6H_4{-}\overset{\overset{\displaystyle O}{\|}}{C}{-}OH$$

The reaction can continue because the ends of the growing molecule are reactive.

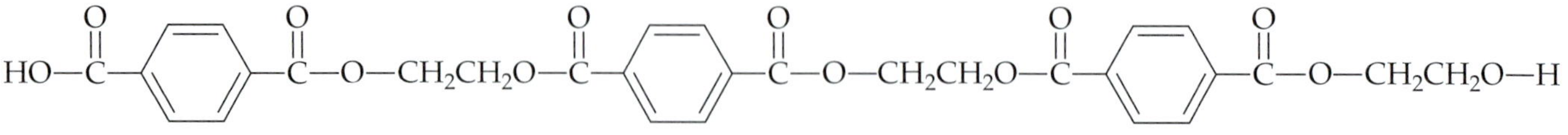

$$HO{-}\overset{\overset{\displaystyle O}{\|}}{C}{-}C_6H_4{-}\overset{\overset{\displaystyle O}{\|}}{C}{-}O{-}CH_2CH_2O{-}\overset{\overset{\displaystyle O}{\|}}{C}{-}C_6H_4{-}\overset{\overset{\displaystyle O}{\|}}{C}{-}O{-}CH_2CH_2O{-}\overset{\overset{\displaystyle O}{\|}}{C}{-}C_6H_4{-}\overset{\overset{\displaystyle O}{\|}}{C}{-}O{-}CH_2CH_2O{-}H$$

The actual product obtained at the end of such a condensation polymerisation is a number of very long molecules each with hundreds to thousands of ester groups along the structure. To represent the molecular structure of a polymer, just one repeating unit is drawn inside a bracket. The italic letter n outside the bracket means that an indeterminate number of repeating units are present. With practice, you will be able to examine structures for dicarboxylic acid and diol monomers and draw a repeating unit of the polymer they form. The repeating unit structure of a polyester can also be analysed to determine the monomers used to make it.

Dicarboxylic acid monomer — Diol monomer — Polymer (polyester) repeating unit

$$HO-\overset{O}{\overset{\|}{C}}-C_6H_4-\overset{O}{\overset{\|}{C}}-OH \qquad HOCH_2CH_2OH \qquad \left[-\overset{O}{\overset{\|}{C}}-C_6H_4-\overset{O}{\overset{\|}{C}}-O-CH_2CH_2-O-\right]_n$$

(a)

(b)

▲ **FIGURE 12.28 Polyethylene terephthalate (a)** The polyester polyethylene terephthalate can be moulded to make plastic containers. © Scisetti Alfio. Shutterstock **(b)** Polyethylene terephthalate can also be made into long fibres and woven to produce fabrics. © optimarc. Shutterstock

The example used to demonstrate condensation polymerisation above is a specific polyester called polyethylene terephthalate, which is abbreviated to PET or PETE. You may have seen these abbreviations written on plastic bottles next to the recycling symbol. This is because polyethylene terephthalate can be formed into films and moulded to form plastic containers ◀ Figure 12.28(a). The generic term 'polyester' is also used to describe polyethylene terephthalate, especially when the long fibres that can be made from it are woven to produce fabrics ◀ Figure 12.28(b).

The condensation reaction that produces an amide from a carboxylic acid and an amine can produce a **polyamide** from a dicarboxylic acid and diamine. A polyamide is a polymer with repeating amide functional groups occurring along the structure. Here is reminder of the condensation reaction that produces an amide.

$$R-\overset{O}{\overset{\|}{C}}-O-H \xrightarrow{RNH_2} R-\overset{O}{\overset{\|}{C}}-\underset{H}{\underset{|}{N}}-R$$

Compare the amide product above to that formed in the analogous reaction between a specific dicarboxylic acid and diamine. The product has reactive carboxylic acid and amine functional groups that keep reacting to produce a polymer.

$$HO-\overset{O}{\overset{\|}{C}}-(CH_2)_4-\overset{O}{\overset{\|}{C}}-OH \xrightarrow{H_2N-(CH_2)_6-NH_2} HO-\overset{O}{\overset{\|}{C}}-(CH_2)_4-\overset{O}{\overset{\|}{C}}-\underset{H}{\underset{|}{N}}-(CH_2)_6-NH_2$$

The polyamide formation can be represented as a chemical equation that shows the product as the repeating unit of the polymer. Note how the coefficient 'n' is written for each reactant to show that a large and indeterminate number of monomers are required to form the polymer.

$$n\ HO-\overset{O}{\overset{\|}{C}}-(CH_2)_4-\overset{O}{\overset{\|}{C}}-OH \xrightarrow{n\ H_2N-(CH_2)_6-NH_2} \left[-\overset{O}{\overset{\|}{C}}-(CH_2)_4-\overset{O}{\overset{\|}{C}}-\underset{H}{\underset{|}{N}}-(CH_2)_6-NH-\right]_n$$

▲ **FIGURE 12.29 Nylon** The reactants in the production of nylon are both liquids that do not mix. The solid product is drawn from the interface between the two phases where the reaction occurs.

The polyamide shown as the product of the above reaction is actually nylon, which forms at the interface of the two liquid reactants (◀ Figure 12.29). The reactants are both liquids that do not mix, but the condensation reaction occurs where the two phases contact each other. The solid product is drawn away from the interface and further reaction occurs. The nylon tread that forms has to be spooled as it forms. This is a very exciting reaction to witness.

The properties of nylon are due to its molecular structure and intermolecular forces between the polymer chains that make up the material (▼ Figure 12.30). The amide functional groups that occur along the polymer chains can participate in hydrogen bonding with those in an adjacent molecule. The hydrogen atom of the amide group is δ^+ and is attracted to the δ^- oxygen atom of an amide group in another polymer chain. This hydrogen bonding interaction is shown in Figure 12.30 as dashed lines between the atoms involved in the attraction.

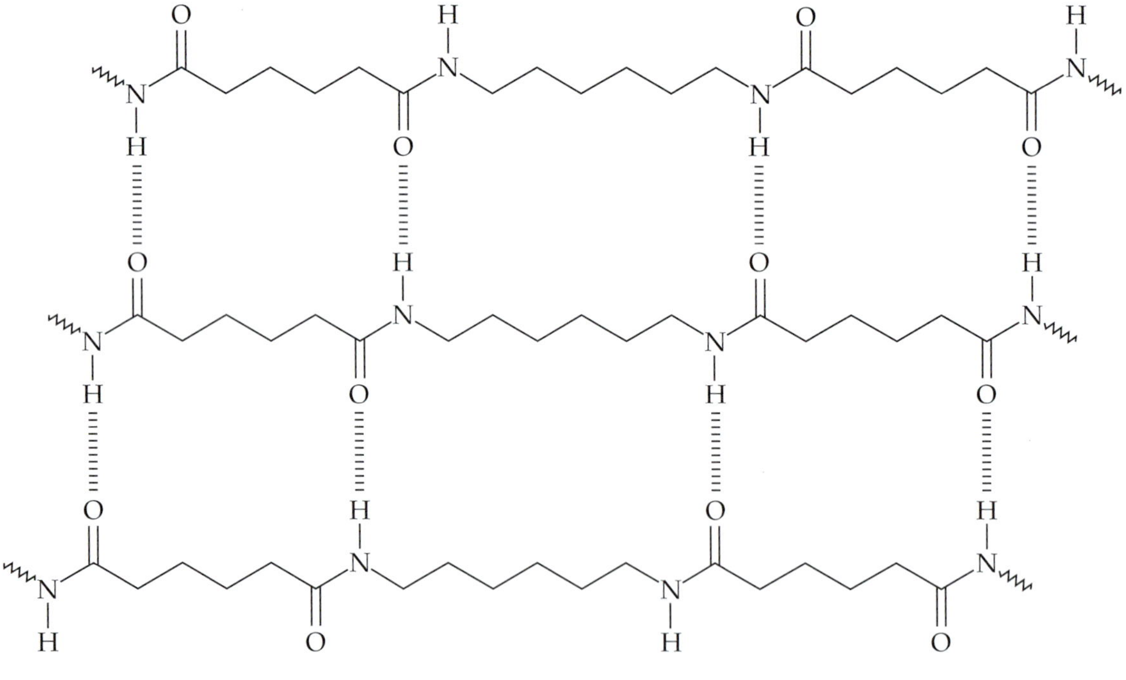

▲ **FIGURE 12.30 Molecular structure of nylon** Hydrogen bonding between nylon polymer chains is shown as dashed lines.

The material Kevlar, most commonly known for its use in bulletproof vests, is also a polyamide. Its extreme strength is wholly due to its molecular structure and intermolecular forces. ▶ Figure 12.31 shows the molecular structure of Kevlar and the hydrogen bonding IMF that occurs between particular atoms of the amide groups within adjacent polymer chains. The hydrogen bonding that occurs in Kevlar is analogous to that of nylon as they are both polyamides.

▼ **FIGURE 12.31 Molecular structure of Kevlar** The molecular structure of Kevlar and hydrogen bonding between polyamide chains, shown as dashed lines, are responsible for the properties of this material.

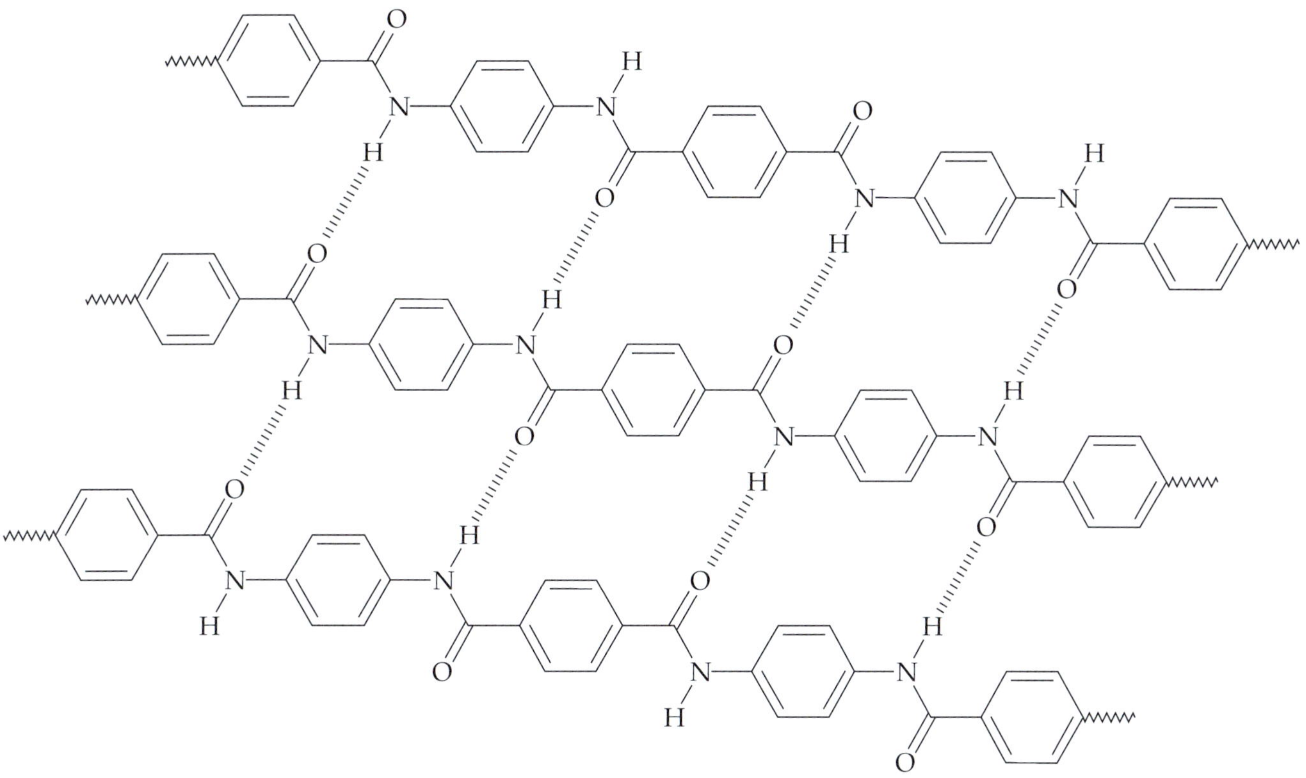

EXERCISE 12.58

Determine the monomers used to prepare the following condensation polymers, represented by their repeating units.

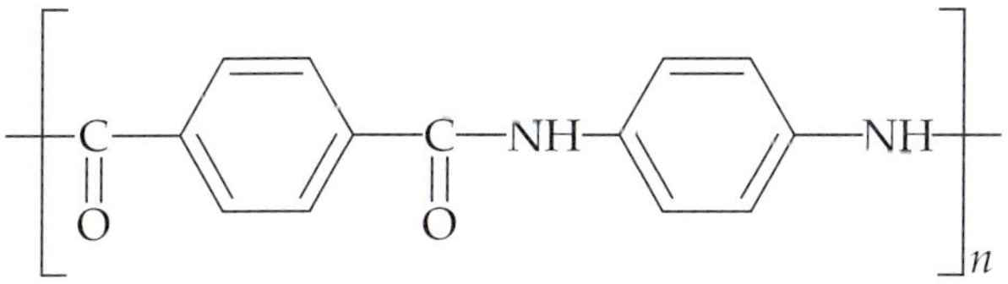 Kevlar (polyamide)	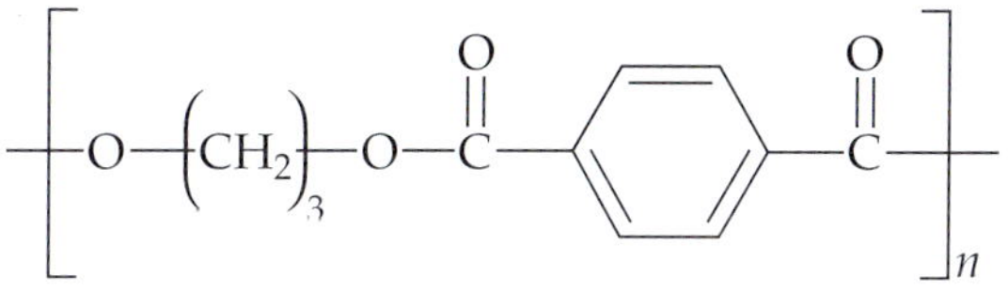 polytrimethylene terephthalate (polyester)
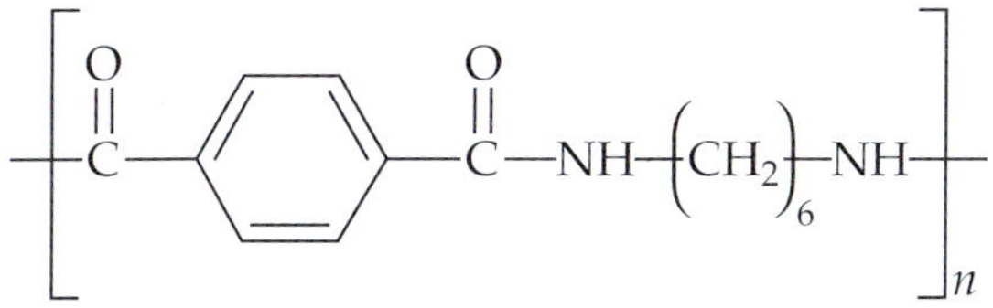 A 'polyphthalamide' (polyamide)	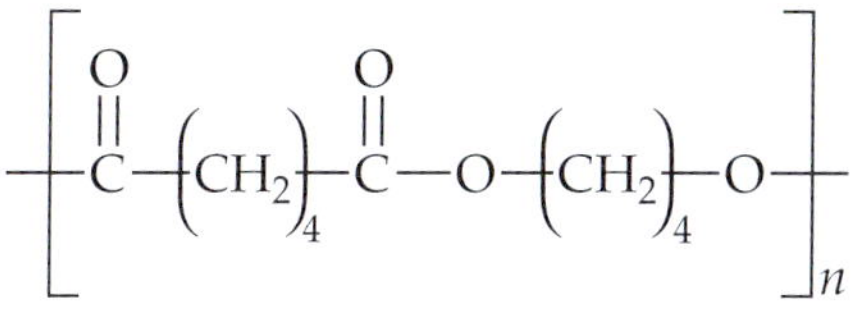 'polybutyrate adipate' (polyester)

EXERCISE 12.59

Draw the repeating unit of the condensation polymer that could form from the following pairs of monomers.

Pair 1	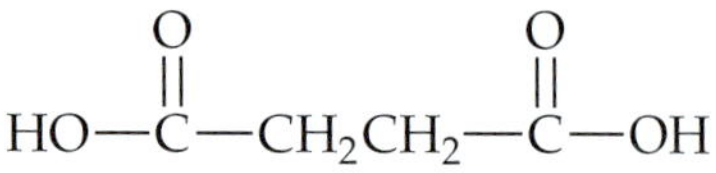 $HO-C(=O)-CH_2CH_2-C(=O)-OH$	$HO-CH_2CH_2-OH$	
Pair 2	$H_2N-CH_2CH_2-NH_2$	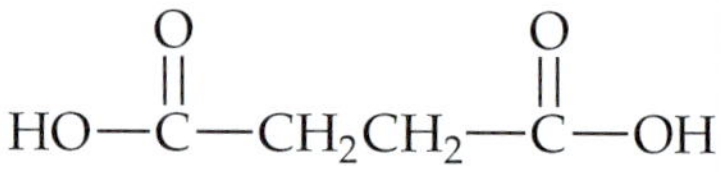 $HO-C(=O)-CH_2CH_2-C(=O)-OH$	
Pair 3	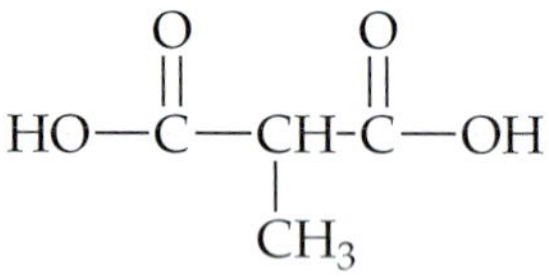 $HO-C(=O)-CH(CH_3)-C(=O)-OH$	$HO-CH_2CH_2-OH$	
Pair 4	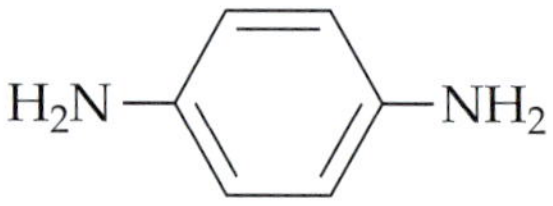 $H_2N-C_6H_4-NH_2$	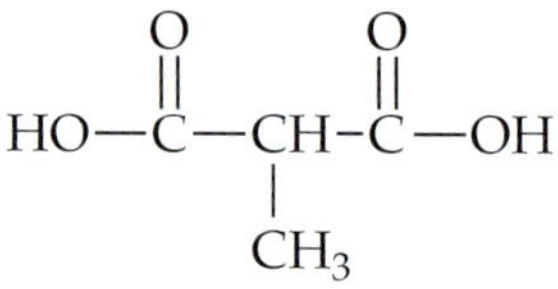 $HO-C(=O)-CH(CH_3)-C(=O)-OH$	

Addition Polymers

The monomers that make up addition polymers are always alkenes. ▼ Figure 12.32 shows molecules of the simplest alkene, ethene, reacting with each other to form a polymer. Note that ethene is often referred to by its commonly known name, ethylene, and the polymer that forms is called polyethylene. Look carefully at the representation of polyethylene formation and see that all of the atoms present in the identical reactants are present on the product side of the equation, the three molecules shown have added together. The double bonds in each reactant are not present in the product because it's the double bonds that allow the reaction to form the polymer. The dots you can see at either ends of the reactants and product indicate that there are many more reactant molecules involved, and only a portion of the very long product molecule is shown. The chemical reaction that occurs when alkenes react with each other in this way is classified as addition, hence the polymers that form in this process are called addition polymers.

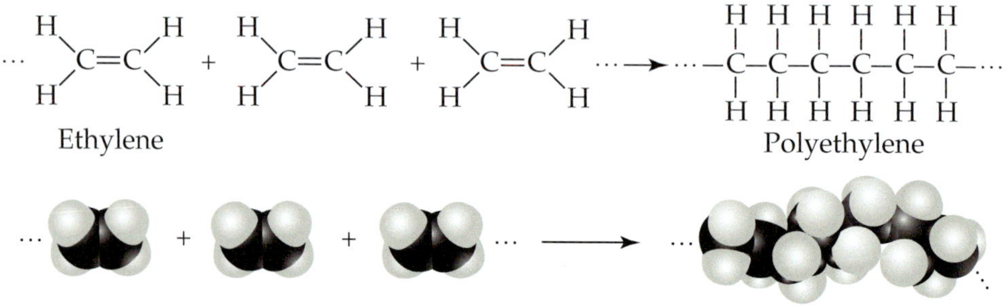

▶ **FIGURE 12.32 Polyethylene** Molecules of ethene, commonly known as ethylene, react to form the polymer.

$$\cdots H_2C{=}CH_2 + H_2C{=}CH_2 + H_2C{=}CH_2 \cdots \longrightarrow \cdots -CH_2-CH_2-CH_2-CH_2-CH_2-CH_2- \cdots$$

Ethylene   Polyethylene

The chemical reaction for the addition polymerisation of ethylene can be more efficiently represented by the following equation. Note that the detail of how this reaction occurs and the conditions required are beyond the scope of this introduction to polymers. The equation simply shows the monomer reactant and the polymer product. Note that addition polymerisation requires only one type of monomer, unlike the particular condensation polymerisations discussed earlier. The product is shown using the convention for representing polymers with a single repeating unit in brackets. The reactant coefficient n, and the n outside the brackets for the product, represent a large and indeterminate number. The actual products obtained are a number of polymer chains of varying length, containing hundreds to thousands of repeating units.

$$n\ \mathrm{H_2C{=}CH_2} \longrightarrow \left[\mathrm{-CH_2-CH_2-} \right]_n$$

Here are a couple of well-known examples of addition polymer repeating units and their monomer structures. With practice you will be able to determine the structure of the monomer used to form any addition polymer by interpreting a representation of its repeating unit, and the reverse, draw a repeating unit for an addition polymer given the monomer structure.

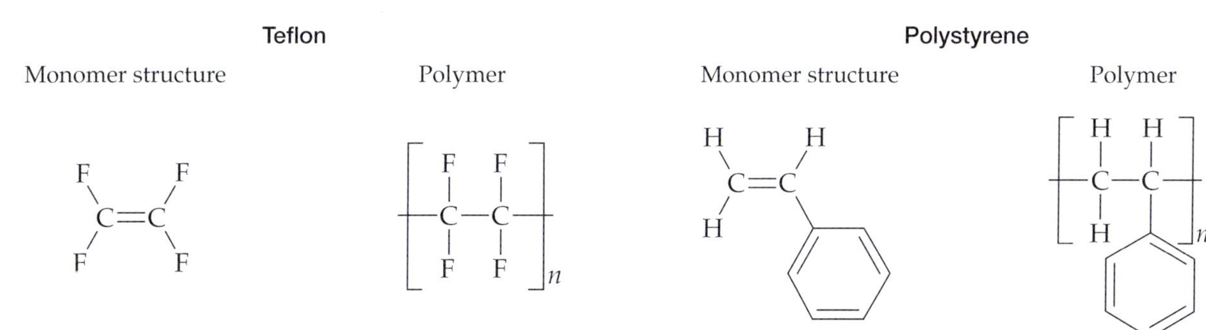

EXERCISE 12.60

Draw the structure of the monomer used to form each of the following addition polymers.

$\left[\mathrm{-CH_2-CH(CH_3)-} \right]_n$ polypropylene		$\left[\mathrm{-CH_2-CH(OCOCH_3)-} \right]_n$ polyvinyl acetate (PVA)	

EXERCISE 12.61

Draw the repeating unit for each of the addition polymers that could form from the following monomers.

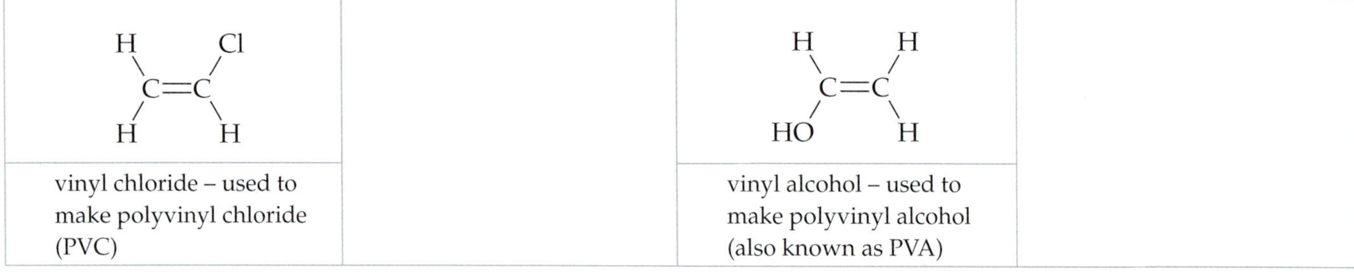

vinyl chloride – used to make polyvinyl chloride (PVC)		vinyl alcohol – used to make polyvinyl alcohol (also known as PVA)	

▲ **FIGURE 12.33 Addition polymer chains** These can be a random tangle, but there are areas of order where the non-polar chains attract.

The materials we know as polymers are made up of many polymer chains of varying lengths that form in the polymerisation process. Addition polymer chains are non-polar and interact with each other through dispersion forces. ◀ Figure 12.33 shows how the polymer chains are somewhat random and they can tangle. However, there are areas of order where sections of polymer chains attract through dispersion forces. The degree of order possible for a given polymer can be controlled to produce materials with different properties. There are two materials made from polyethylene, low-density polyethylene (LDPE) and high-density polyethylene (HDPE). In LDPE, the reaction conditions for the polymerisation produce highly branching polymer chains ▼ Figure 12.34(a). The branching structure of the polymer chains allows for fewer ordered regions and the materials made from LDPE are flexible. By contrast, the process used to produce HDPE gives linear polymer chains with very little branching ▼ Figure 12.34(b). More ordered regions are possible and this gives rise to more rigid materials.

▶ **FIGURE 12.34 Types of polyethylene** **(a)** In low-density polyethylene (LDPE) the polymer chains have extensive branching. LDPE is used to make flexible materials like plastic bags.
© Joey Chan. Pearson Education Asia Ltd
(b) In high-density polyethylene (HDPE) there is very little branching in the polymer chains. HDPE is used to produce materials that are more rigid.
© p 165 23rf.com

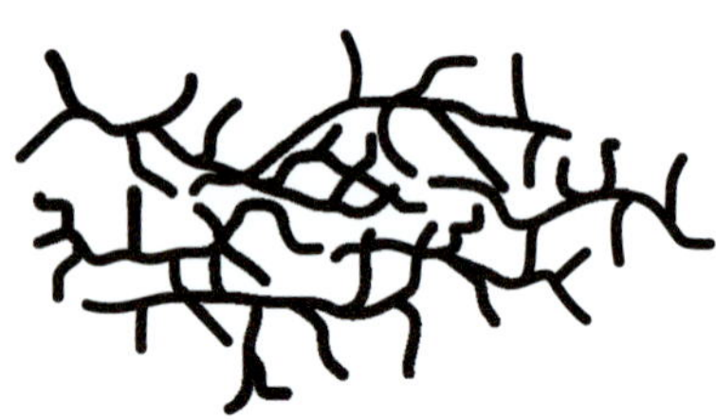

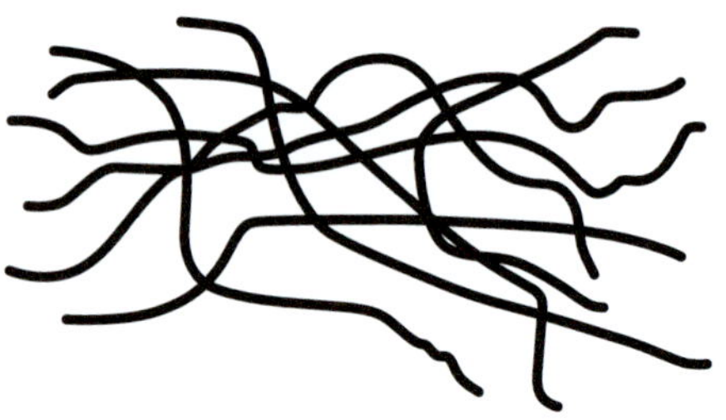

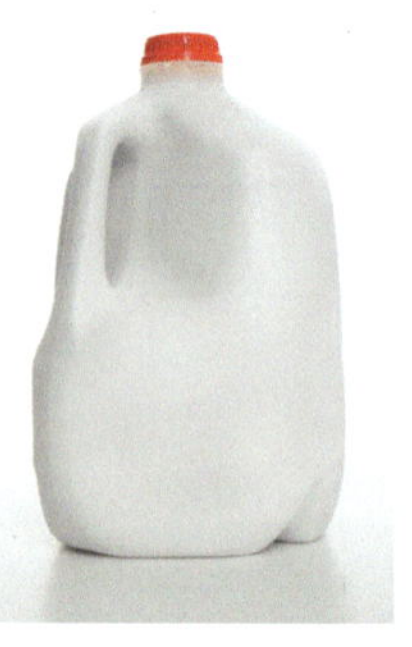

(a) **(b)**

This section has introduced condensation polymers that form from two monomers. Dicarboxylic acid and diol monomers produce polyesters while dicarboxylic acid and diamine monomers produce polyamides. The properties of any polymeric material are due to their molecular structure and the types of IMFs that are possible. Polyamide chains can attract to each other through hydrogen bonding. Non-polar polymer chains can have ordered areas due to dispersion forces facilitating their aggregation. The monomers of addition polymers are always alkenes, and addition polymers that form from one monomer were introduced. Addition polymerisation reactions can be controlled to produce highly branched or mainly linear polymer chains, which give rise to flexible or more rigid materials respectively.

12.18 Amino Acids

LO: Identify and describe the molecular structure of amino acids.

LO: Determine which ionic form of an amino acid exists at a particular pH.

LO: Explain the conditions where amino acids exist as zwitterions.

LO: Account for the physical properties of amino acids in the solid state.

Any organic compound that contains both amine and carboxylic acid functional groups is an amino acid. This section will focus on α-amino acids (alpha amino acids), which are those that have the amine and carboxylic acid groups bonded to the same carbon atom as shown in the general structure below. The carbon atom bonded to both amine and carboxylic acid functional groups is known as the α carbon (alpha carbon).

General structure of α-amino acids.
The amine and carboxylic acid groups are bonded to the same carbon atom (the α carbon).
The naturally occurring α-amino acids vary by their R group.

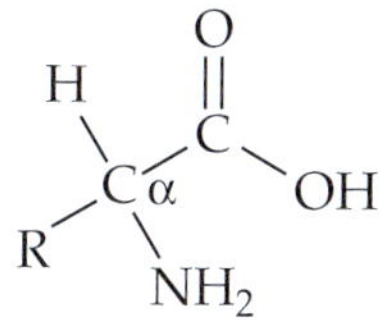

The α-amino acids are very important compounds as they combine to form proteins, as discussed in the following section. The twenty naturally occurring α-amino acids that make up proteins are given in ▶ Figure 12.35. See how their R groups are not necessarily simple hydrocarbon frameworks, and each one has a unique name and three letter code. Note that these α-amino acids are merely shown for interest. You might have to memorise their structures, names and three letter codes at some stage, but not yet.

▼ FIGURE 12.35 Structures of the twenty naturally occurring α-amino acids that make up proteins Note the diversity of the R group structures.

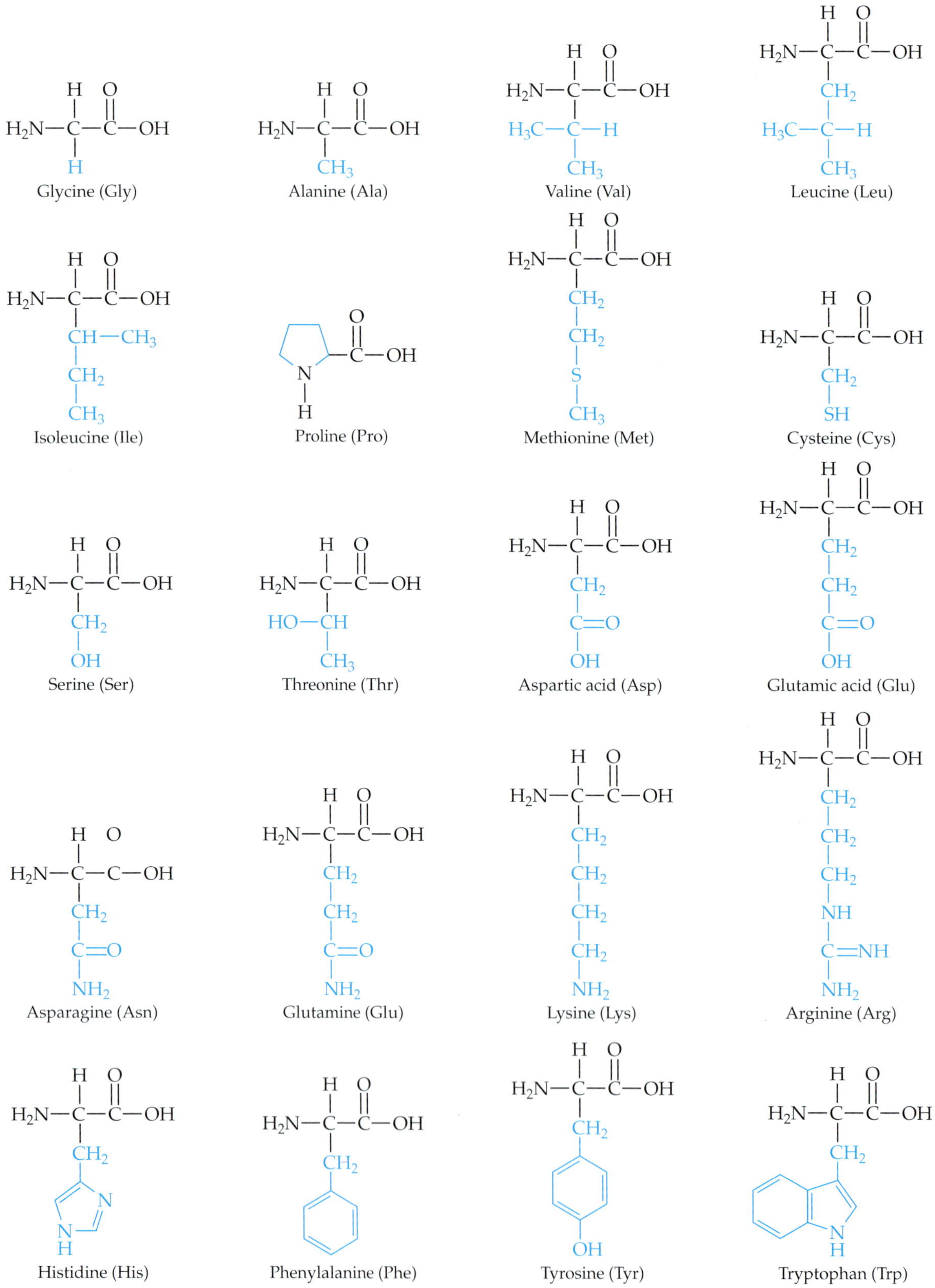

Glycine (Gly)

Alanine (Ala)

Valine (Val)

Leucine (Leu)

Isoleucine (Ile)

Proline (Pro)

Methionine (Met)

Cysteine (Cys)

Serine (Ser)

Threonine (Thr)

Aspartic acid (Asp)

Glutamic acid (Glu)

Asparagine (Asn)

Glutamine (Glu)

Lysine (Lys)

Arginine (Arg)

Histidine (His)

Phenylalanine (Phe)

Tyrosine (Tyr)

Tryptophan (Trp)

Recall from learning about amine and carboxylic acid functional groups earlier in the module that amines are weak bases (proton acceptors) and carboxylic acids are weak acids (proton donors). These functional groups can exist in ionic forms in aqueous solution depending on the pH, and the same is true when both functional groups are present in the same molecule as is the case for α-amino acids.

- When a carboxylic acid donates its proton it exists as a carboxylate ion.
- A general structure for a carboxylate ion is shown on the right.
- In neutral and basic aqueous solutions, carboxylic acids exist in this form.

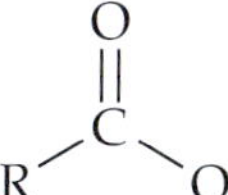

- When an amine accepts a proton it exists as an ammonium ion.
- A general structure for an ammonium ion formed from a primary amine is shown on the right.
- In neutral and acidic aqueous solutions, amines exist in this form.

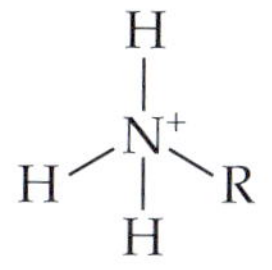

A substance that can act as an acid or a base is **amphoteric**. Since amino acids contain both acidic and basic functional groups, they are amphoteric and exist in different forms in aqueous solution, depending on the pH as follows.

Acidic solution (pH < 7) | Neutral solution (pH 7) | Basic solution (pH > 7)

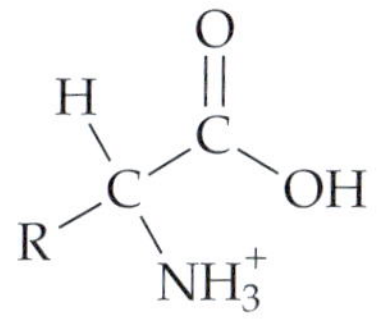

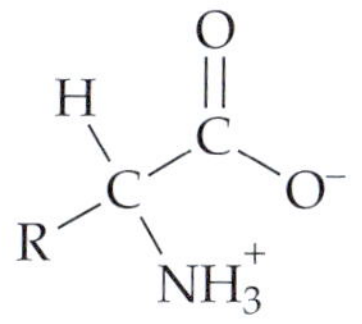

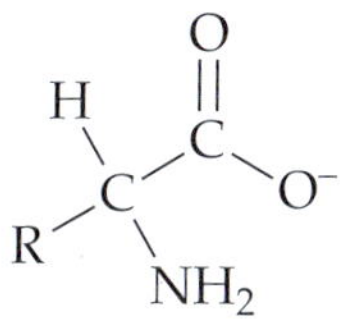

In acidic solutions (pH < 7) the amine and carboxylic acid groups are both protonated. This is because there is relatively more H^+ in acidic aqueous solution and both functional groups tend to be protonated. When both functional groups are protonated, the amine group is charged (+ve) and the carboxylic acid group is neutral. Around pH 7 (neutral), the amine group is protonated and charged (+ve) while the carboxylic acid group is deprotonated and charged (–ve). Understanding of why this is so develops with further study of chemistry. For now, it is a matter of remembering that α-amino acid amine and carboxylic acid functional groups have opposite charges in neutral solutions. In basic solutions (pH > 7), there is relatively more OH^- and this tends to keep both amine and carboxylic acid groups in their deprotonated forms. When both functional groups are deprotonated, the amine group is neutral and the carboxylic acid group is charged (–ve).

EXERCISE 12.62

In the α-amino acid alanine, the R group is methyl ($-CH_3$), the structure is shown below. Complete the table by showing the form of the alanine amine and carboxylic acid groups at the pH values given.

Alanine structure	Alanine at pH 3	Alanine at pH 9	Alanine at pH 7
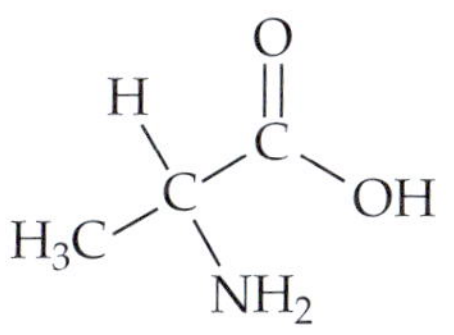			

Zwitterions

A **zwitterion** is a molecule that is neutral overall, but one part of the molecule is positively charged and another part is negatively charged. The form of an α-amino acid that exists at pH 7 (neutral) is a zwitterion, and they also exist in this form as solids. Note that, while α-amino acids exist as zwitterions in the solid state, their structures are often drawn with the amine and carboxylic acid functional groups in their neutral forms and their existence as zwitterions is understood.

The form of an α-amino acid which has the carboxylate ion (–ve charge) and ammonium ion (+ve charge) present at the same time is a zwitterion.
In the solid state, α-amino acids exist in this form.
Solid α-amino acids have similar properties to ionic compounds.

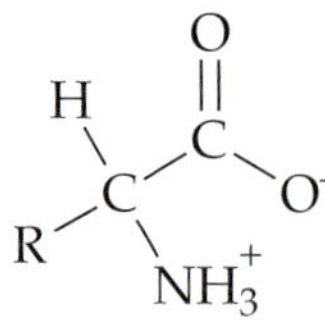

O
||
H C
\ / \
C O⁻
R / \
NH_3^+

A prominent property of ionic compounds (salts) is their high melting points, which is due to the strong electrostatic attraction between the anions (–ve charge) and cations (+ve charge) that make up the substance. The high melting points are a consequence of the amount of energy required to overcome the strong attraction between the oppositely charged ions. The zwitterion form of an α-amino acid is akin to an ionic compound with the anion and cation present in the same molecule. Like ionic compounds, zwitterion α-amino acids have high melting points because of the electrostatic attraction between them. The negative charge of one zwitterion is attracted to the positive charge of another. It takes significant energy to overcome this strong electrostatic attraction, just as it does for an ionic compound. When comparing solid α-amino acids to compounds with two polar functional groups and similar molar mass, the melting points of solid α-amino acids are higher. Based on everything you have learned about the reasons underpinning the physical properties of organic compounds, you will be able to explain why this is the case.

EXERCISE 12.63

Explain the trend in melting points for the following organic compounds that have similar molar mass and two polar functional groups present in their molecular structure.

Compound name	putrescine	oxalic acid	alanine
Molecular structure	$NH_2CH_2CH_2CH_2CH_2NH_2$	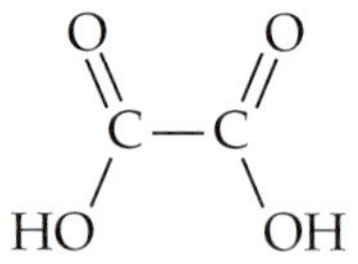 O O \\ // C—C / \ HO OH	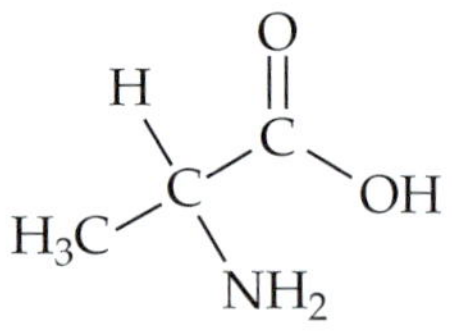 O || H C \ / \ C OH H_3C / \ NH_2
Molar mass	88 g/mol	90 g/mol	89 g/mol
Melting point	28 °C	190 °C	297 °C

In summary, solid α-amino acids have higher melting points than other organic compounds with two polar functional groups. This is because solid α-amino acids are zwitterions that have strong electrostatic attraction between them. More energy is required to overcome the electrostatic attraction between zwitterions compared to other organic compounds with two polar functional groups and similar molar mass.

12.19 Proteins

LO: Identify proteins and describe how they are formed from amino acids.

LO: Explain the four levels of protein structure.

Proteins are polymers of α-amino acids, introduced in the previous section. The importance of proteins to life cannot be overstated. People generally understand the requirement for protein in our diets and probably the most common references we make to proteins in everyday life are about dietary sources (▼ Figure 12.36). Structural tissues of organisms such as muscle, skin and hair are comprised of proteins. The natural fibres wool and silk are also proteins. Chemical reactions that sustain life on a cellular level are mediated by proteins called enzymes. Proteins are at the forefront of our metabolic processes and immunity. Without proteins there would be no life. This section describes how α-amino acids combine to form proteins and the four levels of protein structure, primary, secondary, tertiary and quaternary.

▶ **FIGURE 12.36** **Protein** Some dietary sources rich in protein

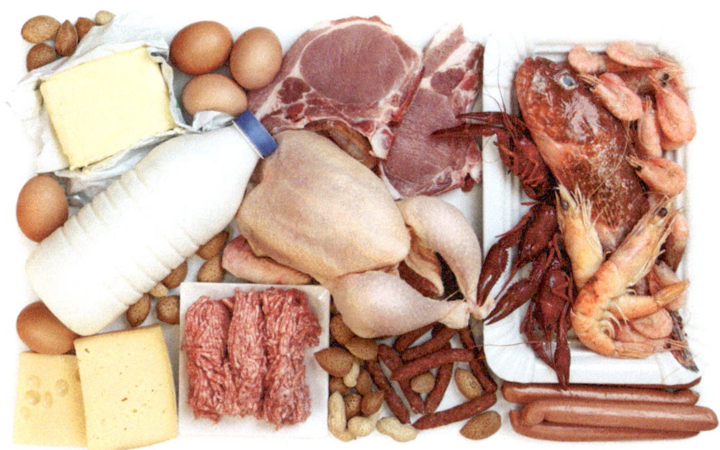

Peptides

Recall from section 12.16 that an amine and a carboxylic acid can react to form an amide. Since α-amino acids contain both amine and carboxylic acid functional groups, they can react with each other in condensation reactions to form amides. Note that the by-product water molecule is not shown in the following equation, which shows how two α-amino acids react to form an amide. The carboxylic acid group of one α-amino acid reacts with the amine group of another.

```
H    H  O           H    H  O                H    H  O     H  O
 \   |  ||           \   |  ||                \   |  ||    |  ||
  N--C--C--OH   +     N--C--C--OH   ------>    N--C--C--N--C--C--OH
 /   |               /   |                    /   |     |  |
H    R              H    R                   H    R     H  R

 α-amino acid        α-amino acid                  dipeptide
```

The amide functional group that forms in the condensation reaction of two α-amino acids

```
H    H |O    |  H  O
 \   | |||   |  |  ||
  N--C-+C--N-+--C--C--OH
 /   | |   | |  |
H    R |   H |  R
```

Polymerisation is possible because the product of the reaction between two α-amino acids has reactive functional groups at both ends. Shown below is the structure of the molecule that would form if another condensation reaction occurred at the terminal amine and carboxylic acid functional groups of a dipeptide. Note that further reaction can occur at the ends of this molecule and the condensation polymer can continue to grow in length. Proteins are polyamide condensation polymers, but they are a bit different to the ones seen in section 12.17. The polyamides introduced earlier were made from dicarboxylic acids and diamines. The monomers of proteins (α-amino acids) have the necessary amine and carboxylic acid groups in the same molecule.

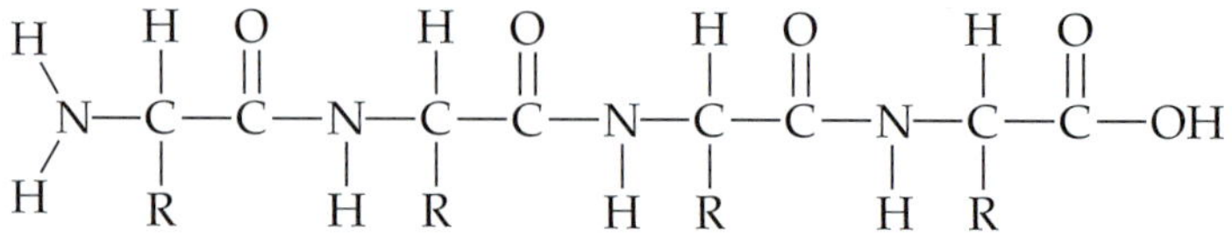

```
H     H  O        H  O        H  O        H  O
 \    |  ||       |  ||       |  ||       |  ||
  N—C—C—N—C—C—N—C—C—N—C—C—OH
 /    |       |   |       |   |       |   |
H     R       H   R       H   R       H   R
```

A molecule that forms when any group of two or more α-amino acids link together is called a **peptide**. The specific descriptor given to the peptide formed in the above equation is **dipeptide** because it consists of two α-amino acids. There are various types of peptides depending on the number of α-amino acids that have linked (Table 12.61). The amide functional groups that occur in peptides are often referred to as **peptide bonds** or peptide linkages.

Table 12.61 Types of Peptides

Peptide	Any group of two or more α-amino acids linked together.
Dipeptide	Comprised of two α-amino acids.
Oligopeptide	Contains 3 – 10 α-amino acids.
Polypeptide	More than 10 α-amino acids. Polypeptides can have hundreds of α-amino acids.

The synthesis of peptides in nature is controlled and the constituent α-amino acids are in particular sequences. Preparation of peptides in the laboratory is not so straightforward because of the presence of two functional groups in each α-amino acid reactant. Just say we wanted to prepare a dipeptide of glycine (Gly) and alanine (Ala) and we wanted them to be in the sequence Gly-Ala. If we combine the two α-amino acids under suitable conditions, there are actually four possible dipeptide products (Table 12.62). There would be some Gly-Ala, but the alternative sequence Ala-Gly would also form. The individual α-amino acids can also react with themselves to produce Gly-Gly and Ala-Ala. Fortunately, modern methods allow for the synthesis of dipeptides, oligopeptides and polypeptides with specific sequences.

Table 12.62 Dipeptides that Could Form from the α-amino Acids Gly and Ala

Reactant α-amino acids	
glycine (Gly) 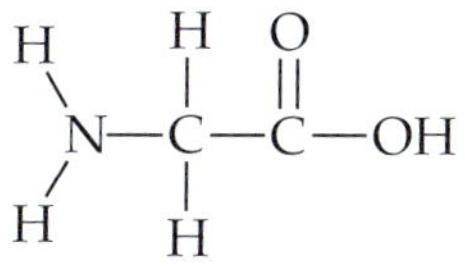	alanine (Ala) 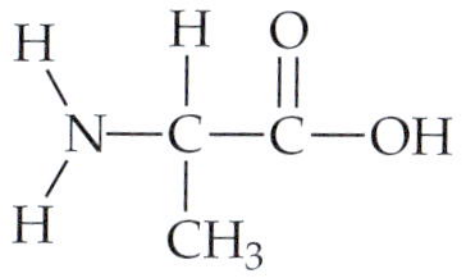
Possible dipeptide products	
Gly-Ala 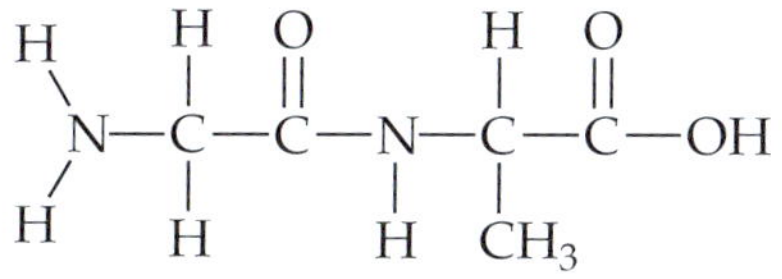	Ala-Gly 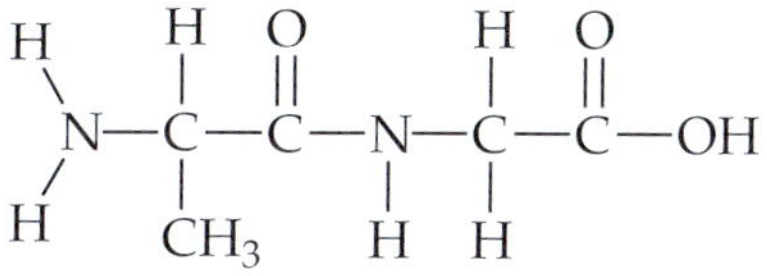
Gly-Gly 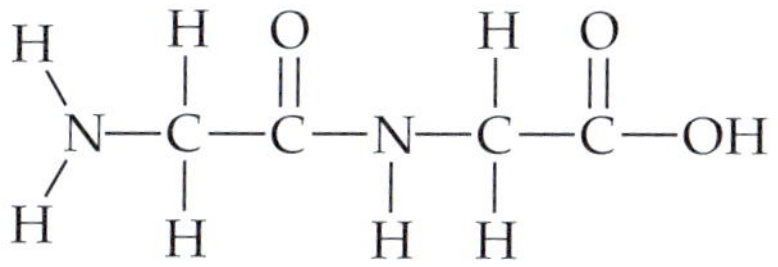	Ala-Ala 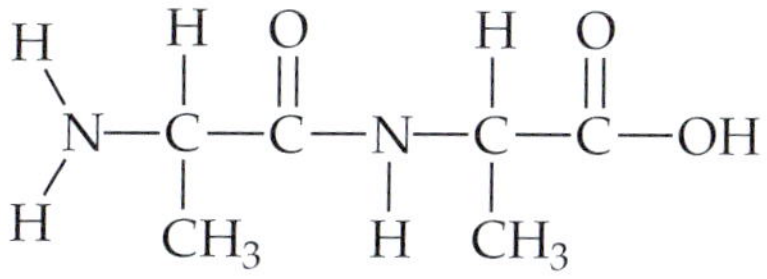

amino acids

two different proteins made from the same amino acids

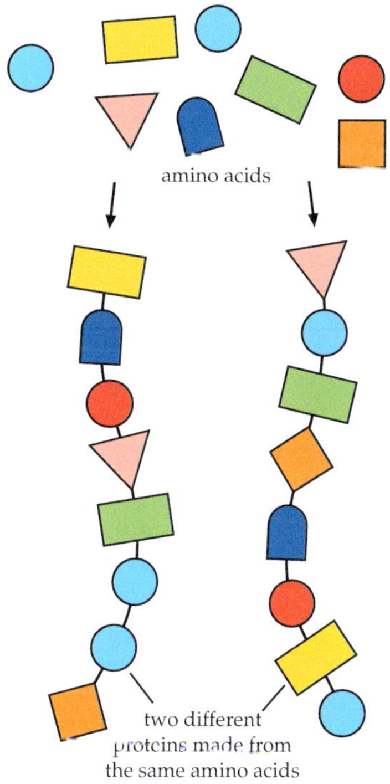

▶ **FIGURE 12.37 Two proteins made from the same α-amino acids** They are different proteins because the α-amino acids are in a different sequence in each protein.

EXERCISE 12.64

Draw structures of all the dipeptide products that could form from the α-amino acids phenylalanine (Phe) and glycine (Gly).

phenylalanine (Phe)	glycine (Gly)
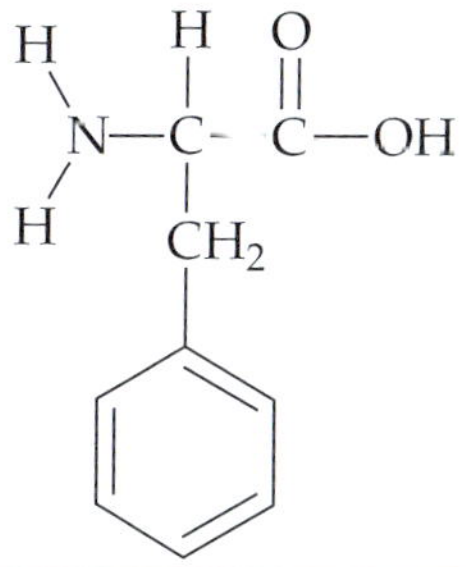	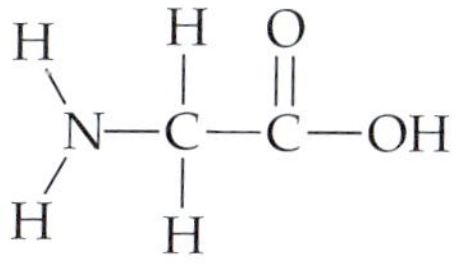

Protein Primary Structure

The order in which the α-amino acids occur in a polypeptide chain is the fundamental or primary structure of a protein. Protein **primary structure** is the sequence of α-amino acids, maintained by the peptide bonds. The sequence of α-amino acids in a protein determines the subsequent levels of structure and is therefore crucial to the protein function. Two proteins may contain the same α-amino acids connected together in different orders (◀ Figure 12.37). Even though these proteins contain the same α-amino acids, they will have completely different functions. Two proteins can have primary structures that vary only slightly, but the functions are completely different despite the similarity. See the comparison of the proteins oxytocin and vasopressin in Table 12.63. Their primary structures only vary by two α-amino acids, but they have completely different functions. Note how protein sequences are written using the α-amino acid three letter codes in the order they occur in the polypeptide chain, connected by dashes.

Table 12.63 Comparison of the Proteins Oxytocin and Vasopressin

Protein name	Primary structure (α-amino acid sequence)	Protein function
oxytocin	Cys-Tyr-Ile-Gln-Asn-Cys-Pro-Leu-Gly	A hormone with various functions related to involuntary muscles, including contraction of the uterus during childbirth.
vasopressin	Cys-Tyr-Phe-Gln-Asn-Cys-Pro-Arg-Gly	A hormone that plays an important role in water retention in the body. Only two α-amino acids different from oxytocin.

Protein Secondary Structure

Repeating structural patterns along the length of a polypeptide chain form protein **secondary structure**. The most common examples of secondary structure are the α-helix (alpha helix) and the β-pleated sheet (beta-pleated sheet). The α-helix (▼ Figure 12.38) involves winding of a polypeptide chain and the R groups of the α-amino acids point to the outside of the resulting spiral. Hydrogen bonding between atoms of amide groups along the polypeptide chain maintains the helical structure. The δ^+ hydrogen atom of one amide group attracts to the δ^- oxygen atom of another amide group a few α-amino acids away in the sequence. The hydrogen bonding attractions that occur in an α-helix are the dashed lines in Figure 12.38.

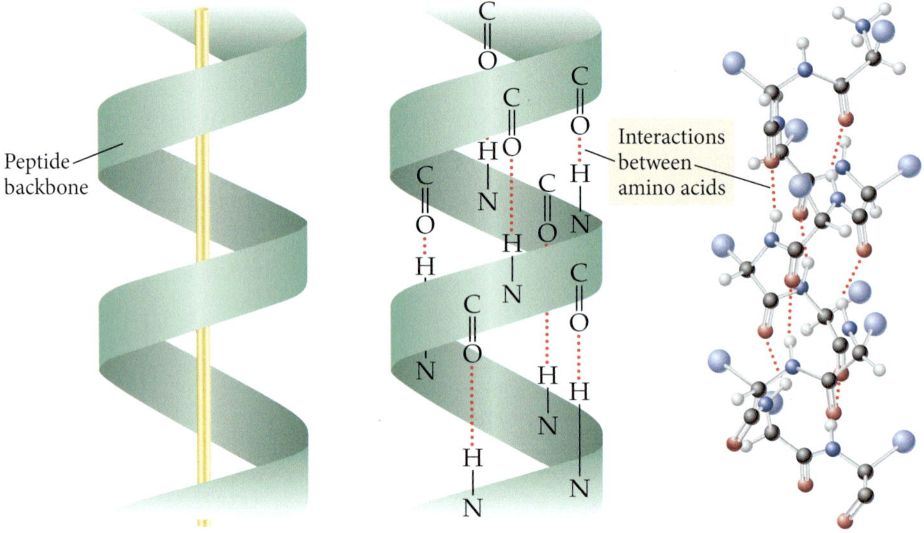

▶ **FIGURE 12.38 The α-helix** The coiling of a polypeptide chain that occurs in an α-helix is maintained by hydrogen bonding, shown as dashed lines.

In the secondary protein structure known as a β-pleated sheet, adjacent polypeptide chains interact with each other to form sheets with a regular bending pattern (▶ Figure 12.39). Hydrogen bonding between amide groups along sections of polypeptide chain that lay next to each other maintain this secondary structure. Attraction between the hydrogen atom of one amide group with the oxygen atom of another in a β-pleated sheet are the dashed lines in Figure 12.39. Note that the image in the figure looks like separate polypeptide chains forming the sheet, but it is actually sections of the same polypeptide chain that form this secondary structure.

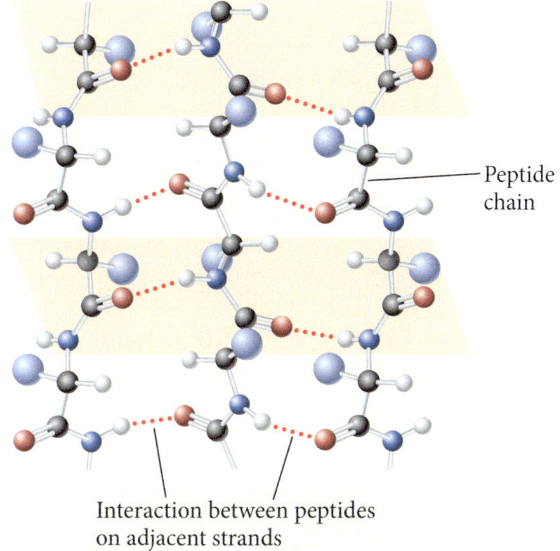

▶ **FIGURE 12.39 β-pleated sheet** The repeating pattern that forms a β-pleated sheet occurs through hydrogen bonding, shown as dashed lines, between polypeptide chain amide groups that lay next to each other.

Protein Tertiary Structure

Protein **tertiary structure** is folding of polypeptide chains that occur due to interactions between R groups far apart in the α-amino acid sequence. ▼ Figure 12.40 shows four common interactions that maintain folds in protein structures. There are α-amino acid R groups with functional groups that can participate in hydrogen bonding (alcohol, carboxylic acid, and amine). Figure 12.40 shows how serine (Ser) and threonine (Thr), which both have polar –OH groups, can attract to each other through hydrogen bonding even though they are far apart in the polypeptide sequence. Their interaction helps to maintain a fold in the protein structure. Some α-amino acids have non-polar hydrocarbon R groups and these can be attracted through dispersion forces, labelled as 'hydrophobic interaction' in Figure 12.40. Remember that proteins function in organisms, which are essentially aqueous environments. Non-polar R groups can aggregate together to exclude water and form hydrophobic or water repelling pockets in a protein structure.

In aqueous environments where proteins function, some α-amino acid R groups are charged. Figure 12.40 shows the carboxylic acid of a glutamic acid

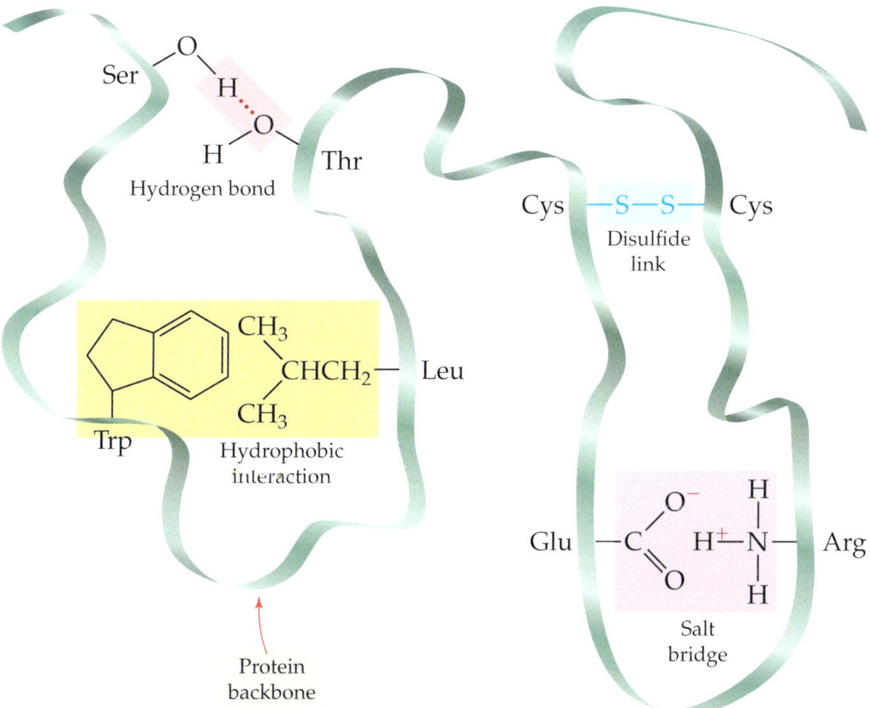

▶ **FIGURE 12.40 Protein tertiary structure** The tertiary structure of a protein is folding of the polypeptide chain that results from interactions between R groups far apart in the α-amino acid sequence.

(Glu) R group existing as carboxylate ion (–COO^-), and the amine group of arginine (Arg) in its ammonium ion (–NH_3^+) form. The oppositely charged R groups can form an ionic bond, described in Figure 12.40 as a 'salt bridge'. The term salt bridge describes an ionic bond that forms protein tertiary structure. The fourth kind of tertiary structure shown in Figure 12.40 is the disulfide bridge, which is a covalent bond between two sulfur atoms. The α-amino acid cysteine (Cys) has –SH at the end of its R group. Adjacent –SH groups can react to form an S–S covalent bond. A disulfide bridge can only form between two cysteine (Cys) α-amino acids, but they may be far apart in the polypeptide sequence.

Protein Quaternary Structure

Many proteins contain subunits, which are individual polypeptide chains that aggregate to form the functioning protein. **Quaternary structure** is the arrangement of the subunits in such proteins. The same types of interactions that maintain tertiary structure are responsible for quaternary structure. The protein collagen, a component of connective tissue, may be described as having quaternary structure. ▼ Figure 12.41 shows how collagen consists of three α-helices coiled together. Haemoglobin, the protein in red blood cells that carries oxygen to the tissues, has four subunits (▼ Figure 12.42).

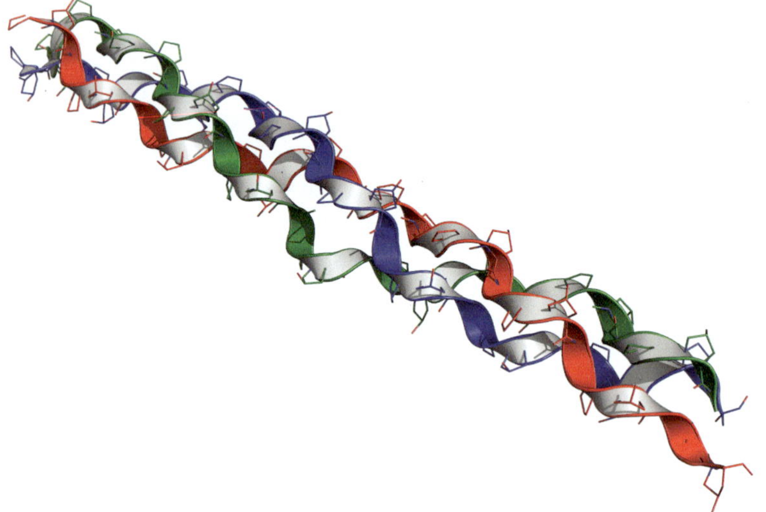

▶ **FIGURE 12.41 Collagen** The protein collagen has three α-helices, shown as different coloured ribbons, coiled together.

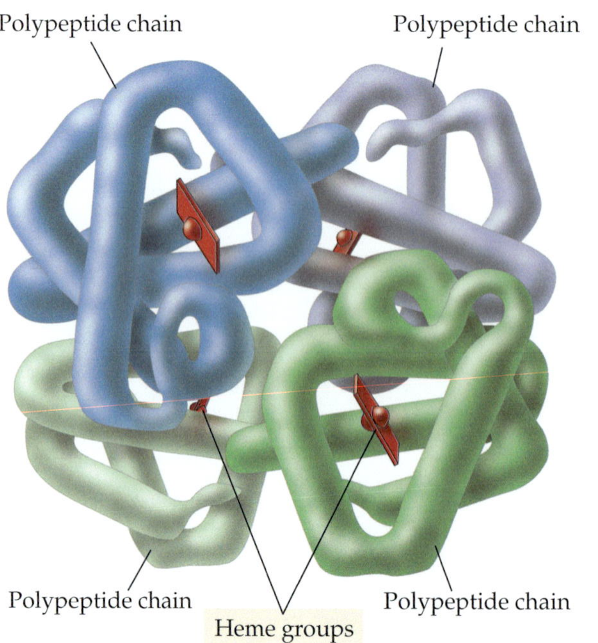

▶ **FIGURE 12.42 Haemoglobin** Haemoglobin is a protein comprised of four subunits, shown using different colours.

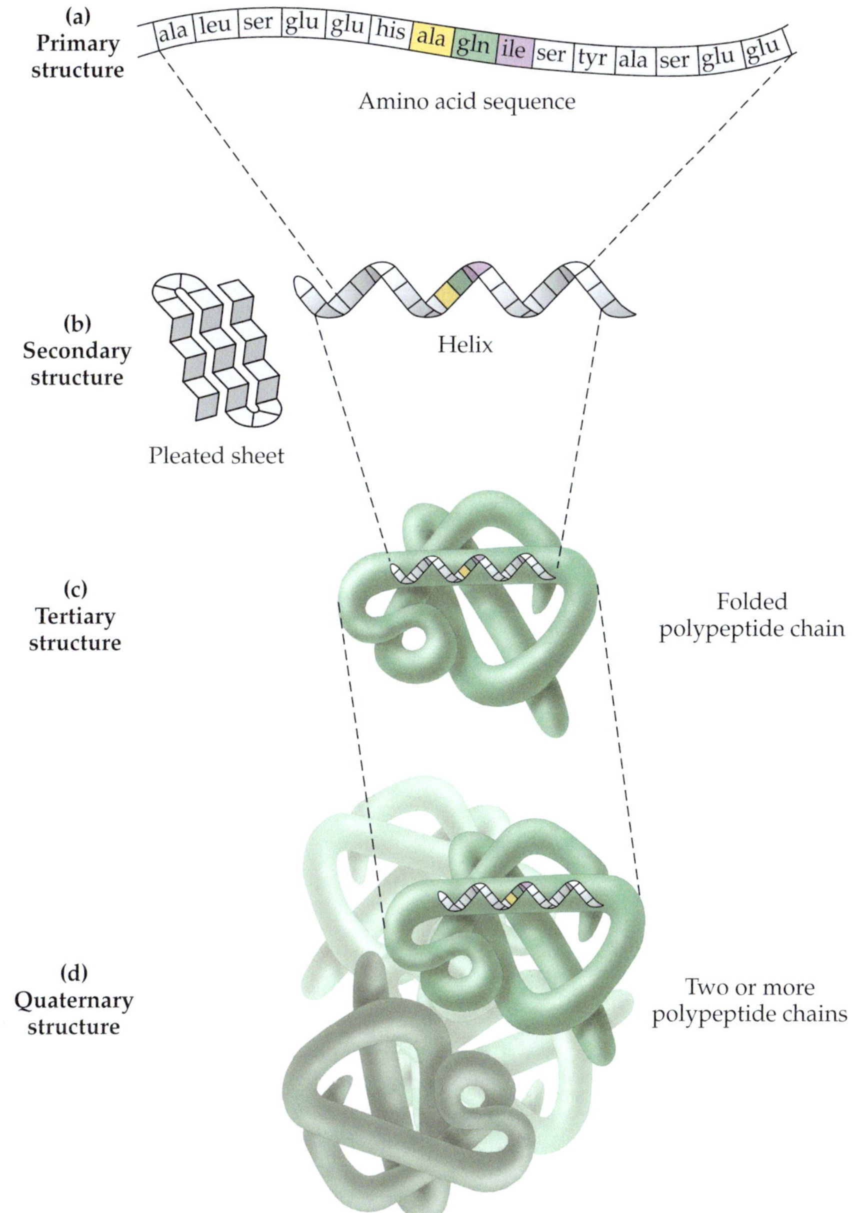

▶ **FIGURE 12.43 The four levels of protein structure** **(a)** Primary structure is the sequence of α-amino acids. **(b)** Secondary structure is repeating structural patterns. **(c)** Tertiary structure is folding of a polypeptide chain through interactions of α-amino acids far apart in the sequence. **(d)** Quaternary structure is applicable to proteins that consist of aggregated polypeptide chains.

The four levels of protein structure are summarised in ▲ Figure 12.43. Primary structure is the sequence of α-amino acids. The secondary structure of a protein are repeating patterns that occur along the polypeptide chain. Tertiary structure is folding of a polypeptide chain due to interactions between distant R groups. Proteins made up of subunits have quaternary structure, which is the arrangement of the subunits. The same type of interactions maintain both tertiary and quaternary protein structure.

EXERCISE 12.65

Classify each of the following descriptions of protein structural features as primary, secondary, tertiary or quaternary.

An ionic bond between oppositely charged α-amino acid R groups that maintains a fold in a polypeptide chain.	
Coiling of a peptide chain with the α-amino acid R groups projecting outwards.	
A particular protein that serves as an ion channel in cell membranes is an aggregate of four subunits.	
One section of a particular protein has the α-amino acid sequence: Met-Glu-Asp-Ala-Phe-His-Ser-Ile	
Hydrogen bonding that occurs between α-amino acid R groups that are far apart in the polypeptide sequence.	

12.20 Fats

LO: Identify and describe the molecular structure of fats.

LO: Account for the physical properties of fats.

LO: Describe and write equations for chemical reactions of fats.

LO: Explain the molecular structures and properties of soaps and detergents.

Fats belong to a broad class of organic compounds known as lipids, which are naturally occurring substances grouped together because of their poor ability to mix with water. Some other members of the lipid family are waxes, steroids and fat-soluble vitamins. This section will focus on just one part of the lipid family, the fats. The term fats is actually being used to describe both fats and oils because the difference between them is merely in their physical state, fats are solids and oils are liquid at room temperature (▼ Figure 12.44). Use of the term fats to describe both fats and oils is acceptable because both have the same general molecular structure. They are both esters, specifically triesters.

General structure of fats and oils, which are triesters.
See the three ester groups in the structure.
The R groups can be saturated or unsaturated hydrocarbon chains with 10 – 20 carbon atoms.
The R groups can be the same or different.

$$\begin{array}{l} R-C(=O)-O-CH_2 \\ R-C(=O)-O-CH \\ R-C(=O)-O-CH_2 \end{array}$$

▶ **FIGURE 12.44 Some samples of fats (solids) and oils (liquids)**
© JPC-PROD. Shutterstock

Like any ester, fats are considered derived from an alcohol and a carboxylic acid. The alcohol is a triol (an alcohol with three –OH groups) that has the common name glycerol. Fats are also known as **triglycerides** because of the name of their parent alcohol. The carboxylic acid components of fat triesters are called fatty acids.

glycerol – the parent alcohol of fats

$$HO-CH_2\underset{\displaystyle OH}{\underset{|}{C}}HCH_2-OH$$

fatty acids – the R groups are long hydrocarbon chains

$$R-\overset{\displaystyle O}{\overset{\|}{C}}-OH$$

Fatty acids are either saturated or unsaturated, and these terms relate to their molecular structure. The hydrocarbon chains of **saturated** fatty acids only contain carbon-carbon single bonds, while those of **unsaturated** fatty acids have at least one double bond (C=C). Naturally occurring fatty acids have even numbers of carbon atoms because they are biosynthesised from two-carbon units. Examples of saturated and unsaturated fatty acids with 18 and 20 carbon atoms are shown in Table 12.64. The saturated fatty acid with 18 carbon atoms is known as stearic acid, which occurs in both animals and plants. Oleic acid, the most abundant fatty acid in nature, is an 18 carbon unsaturated fatty acid. Note how its double bond is cis and this gives the hydrocarbon chain a distinctive bend. With only one double bond, oleic acid is a **monounsaturated** fatty acid. Compare the structure of arachidic acid, the saturated fatty acid with 20 carbon atoms, to the unsaturated counterpart arachidonic acid. Both fatty acids have 20 carbon atoms, but the structures are very different. Arachidonic acid is a **polyunsaturated** fatty acid due to the presence of more than one double bond. Note that all four of the double bonds of arachidonic acid are cis, which is generally the case for the double bonds of natural fatty acids. Arachidonic acid is an essential dietary fatty acid, converted in the body to chemicals that mediate vital biological processes such as blood clotting and immune responses.

Table 12.64 Saturated and Unsaturated Fatty Acids with 18 and 20 Carbon Atoms

Saturated fatty acid $C_{18}H_{36}O_2$ stearic acid	O HO
Unsaturated fatty acid $C_{18}H_{34}O_2$ oleic acid	O HO

Saturated fatty acid
$C_{20}H_{40}O_2$
arachidic acid

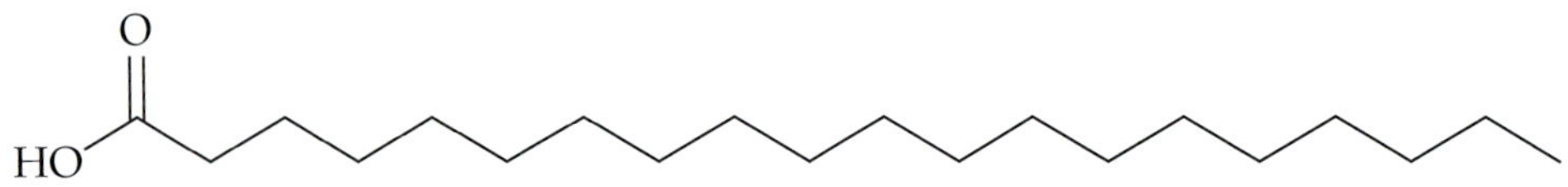

Unsaturated fatty acid
$C_{20}H_{32}O_2$
arachidonic acid

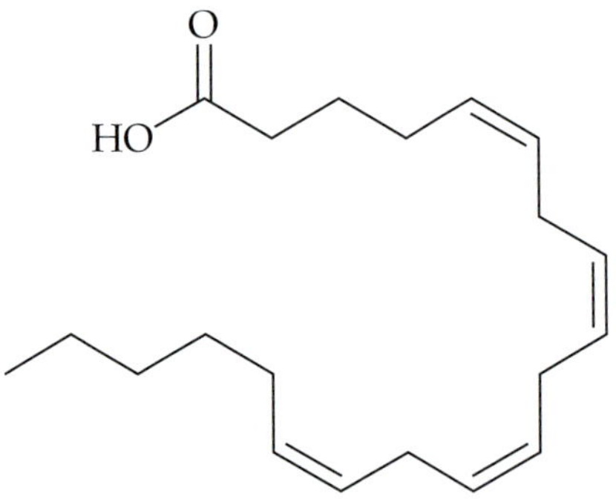

EXERCISE 12.66

Use the general structure of fats and the fatty acid structures above to draw the molecular structure of a triglyceride derived from glycerol and stearic acid.

The terms saturated and unsaturated have just been described with reference to the molecular structure of fatty acids, but they also describe fats. In **saturated fats**, which tend to be solid at room temperature, all three R groups are saturated. A fat with one or more unsaturated R groups is an **unsaturated fat**. Unsaturated fats are usually oils, which are liquid at room temperature. The physical states of saturated and unsaturated fats at room temperature, like all physical properties, are due to IMFs. The relevant IMF is attraction between the non-polar hydrocarbon chains of fats through dispersion. ▶ Figure 12.43 shows how the hydrocarbon chains of saturated fats can aggregate together due to their overall cylindrical shape. Dispersion forces between the hydrocarbon chains of unsaturated fats are not at strong due to their cis double bonds and resultant bending affecting the extent of interaction that can occur.

▶ **FIGURE 12.45 Saturated and unsaturated fats** The difference between these fats at the molecular level affects the extent of dispersion forces between the non-polar hydrocarbon chains in each case.

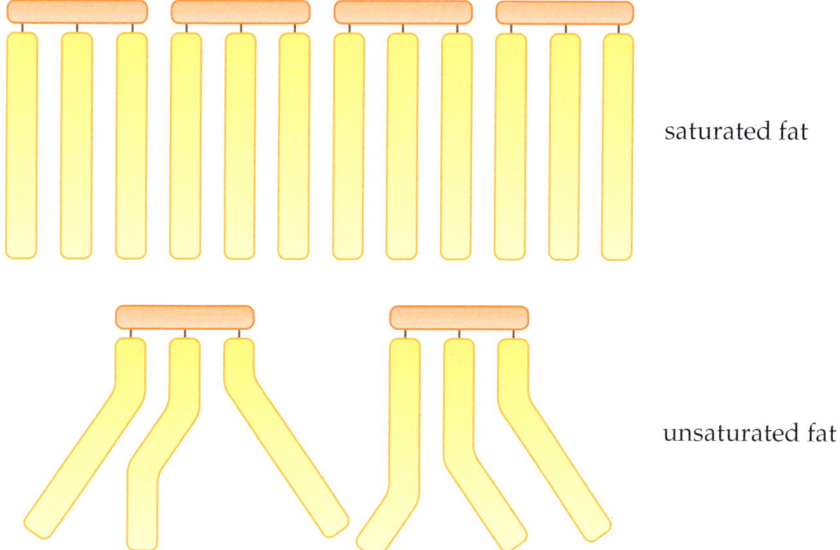

Unsaturated fats can undergo a chemical reaction to convert them to saturated fats. The process can be done on an industrial scale to change liquid oils to a solid form. This is done commercially to make the butter alternative, margarine. This chemical reaction is known as hydrogenation and an equation for this is shown in ▼ Figure 12.46. Note that a catalyst is required for this reaction. The reaction of an unsaturated fat with hydrogen gas (H_2) alone would be far too slow. The hydrogenation equation shows all the unsaturated fat double bonds converting to single bonds, but complete hydrogenation of unsaturated fats does not occur in this way. Only partial hydrogenation occurs in this process and this is sufficient to impart the desired change of state and give a more solid product. Some of the unsaturated fat double bonds, which are always cis, can be converted to trans double bonds during hydrogenation. This is the origin of the term 'trans fat', which you may have heard with reference to foods manufactured using hydrogenation. ▶ Figure 12.47 shows molecular models to highlight the structural difference between cis and trans isomers of a fatty acid. The hydrocarbon chains of trans fats have a more cylindrical overall shape like saturated fat R groups.

▼ **FIGURE 12.46 The chemical reaction that occurs in hydrogenation** The double bonds of an unsaturated fat are converted to single bonds by the addition of hydrogen. A catalyst is required for this reaction.

$$CH_3{-}CH_2{-}CH_2{-}CH_2{-}CH_2{-}CH{=}CH{-}CH_2{-}CH{=}CH{-}CH_2{-}CH_2{-}CH_2{-}CH_2{-}CH_2{-}CH_2{-}CH_2{-}\overset{\overset{O}{\|}}{C}{-}O{-}CH_2$$

$$CH_3{-}CH_2{-}CH{=}CH{-}CH_2{-}CH{=}CH{-}CH_2{-}CH{=}CH{-}CH_2{-}CH_2{-}CH_2{-}CH_2{-}CH_2{-}CH_2{-}CH_2{-}\overset{\overset{O}{\|}}{C}{-}O{-}CH \quad \xrightarrow{\text{hydrogen}}$$

$$CH_3{-}CH_2{-}CH_2{-}CH_2{-}CH_2{-}CH{=}CH{-}CH_2{-}CH{=}CH{-}CH_2{-}CH_2{-}CH_2{-}CH_2{-}CH_2{-}CH_2{-}CH_2{-}\overset{\overset{O}{\|}}{C}{-}O{-}CH_2$$

polyunsaturated (liquid at room temperature)

$$CH_3{-}CH_2{-}CH_2{-}CH_2{-}CH_2{-}CH_2{-}CH_2{-}CH_2{-}CH_2{-}CH_2{-}CH_2{-}CH_2{-}CH_2{-}CH_2{-}CH_2{-}CH_2{-}CH_2{-}\overset{\overset{O}{\|}}{C}{-}O{-}CH_2$$

$$CH_3{-}CH_2{-}CH_2{-}CH_2{-}CH_2{-}CH_2{-}CH_2{-}CH_2{-}CH_2{-}CH_2{-}CH_2{-}CH_2{-}CH_2{-}CH_2{-}CH_2{-}CH_2{-}CH_2{-}\overset{\overset{O}{\|}}{C}{-}O{-}CH$$

$$CH_3{-}CH_2{-}CH_2{-}CH_2{-}CH_2{-}CH_2{-}CH_2{-}CH_2{-}CH_2{-}CH_2{-}CH_2{-}CH_2{-}CH_2{-}CH_2{-}CH_2{-}CH_2{-}CH_2{-}\overset{\overset{O}{\|}}{C}{-}O{-}CH_2$$

saturated (solid at room temperature)

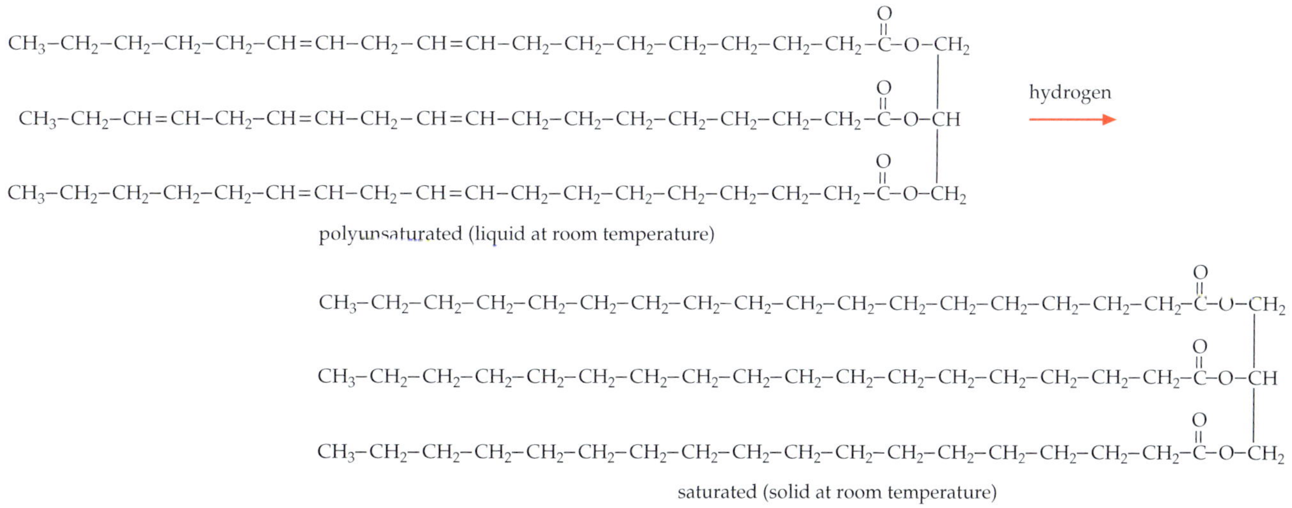

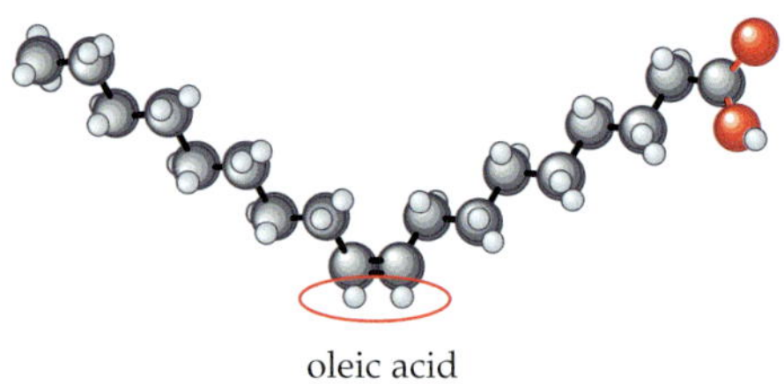

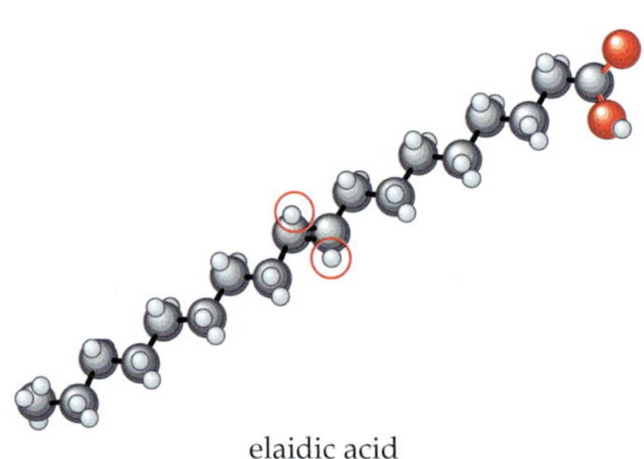

FIGURE 12.47 Molecular models of oleic acid and its trans isomer, elaidic acid The cis double bonds of unsaturated fat hydrocarbon chains can be converted to trans double bonds during hydrogenation.

Soaps

Another important chemical reaction of fats is a standard reaction of esters that produces soap, a ubiquitous substance that is familiar to everyone. Recall that hydrolysis of an ester in the presence of a base (Section 12.14) produces a carboxylate ion ($RCOO^-$) and an alcohol, as shown in the following equation. The hydroxide ion (OH^-) may be supplied using any strong base like sodium hydroxide (NaOH) or potassium hydroxide (KOH).

$$R{-}\overset{\overset{\large O}{\|}}{C}{-}O{-}R' \xrightarrow{H_2O,\ OH^-} R{-}\overset{\overset{\large O}{\|}}{C}{-}O^- + H{-}O{-}R'$$

The carboxylate ion ($RCOO^-$) produced in this reaction is actually the anion of a salt. The cation for the carboxylate salt that forms will be the cation from the base used in the reaction and their formulas may be written as follows.

Formula for a carboxylate salt that forms from ester hydrolysis in the presence of NaOH

$$R{-}\overset{\overset{\large O}{\|}}{C}{-}O^-Na^+$$

Formula for a carboxylate salt that forms from ester hydrolysis in the presence of KOH

$$R{-}\overset{\overset{\large O}{\|}}{C}{-}O^-K^+$$

Since fats are esters, they can undergo the same reaction to produce three carboxylate salt molecules and glycerol. The general equation for this reaction using sodium hydroxide (NaOH) as the base is given below. The carboxylate salts that form in this reaction are actually soap molecules. A term given to the hydrolysis of an ester in the presence of base is **saponification**, but this description is generally used for the hydrolysis reaction of fats that produces soap.

$$\begin{matrix} R{-}\overset{\overset{\large O}{\|}}{C}{-}O{-}CH_2 \\ R{-}\overset{\overset{\large O}{\|}}{C}{-}O{-}CH \\ R{-}\overset{\overset{\large O}{\|}}{C}{-}O{-}CH_2 \end{matrix} \xrightarrow{H_2O,\ NaOH} \begin{matrix} R{-}\overset{\overset{\large O}{\|}}{C}{-}O^-\,Na^+ \\ R{-}\overset{\overset{\large O}{\|}}{C}{-}O^-\,Na^+ \\ R{-}\overset{\overset{\large O}{\|}}{C}{-}O^-\,Na^+ \end{matrix} + \begin{matrix} HO{-}CH_2 \\ | \\ HO{-}CH \\ | \\ HO{-}CH_2 \end{matrix}$$

Soap molecules have physical properties due to their molecular structure (▼ Figure 12.48). One section of the molecule is charged and has properties of ionic compounds. This part of the molecule is polar and hydrophilic. The other portion of the molecule is non-polar and hydrophobic, having properties of hydrocarbons. A molecule that has both hydrophilic and hydrophobic parts is **amphipathic**.

▼ **FIGURE 12.48 Soap** A typical amphipathic soap molecule with the polar hydrophilic and non-polar hydrophobic portions highlighted.

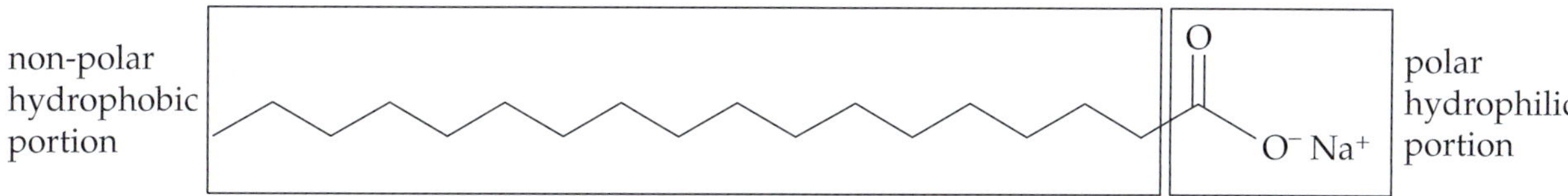

The two distinctive parts of these amphipathic soap molecules have different water solubility. When placed in water, the charged polar section dissolves in water in the same way ionic compounds do. The non-polar hydrocarbon chains of soap molecules are hydrophobic, so they aggregate together to repel as much water as possible. This results in the formation of spherical assemblies called **micelles** (▼ Figure 12.49), which have a hydrophobic region in their centre.

▶ **FIGURE 12.49 Micelles** The formation of micelles when soap molecules are placed into water.

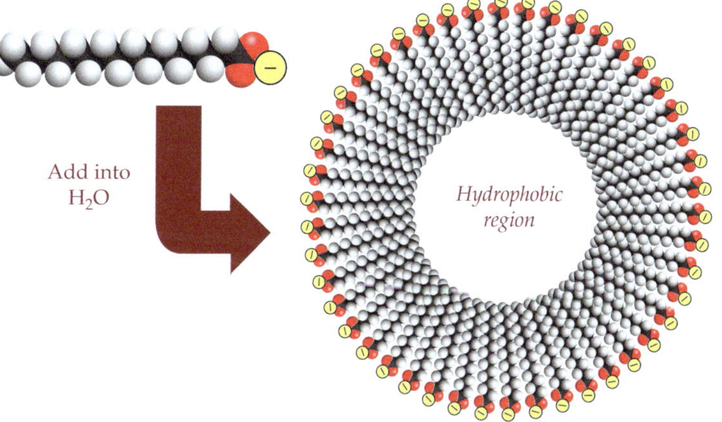

The formation of micelles is the reason why soap has cleaning properties. The substances that we like to clean away from our skin, dishes and clothing are generally non-polar and hydrophobic. Using water alone is ineffective in removing them. In the presence of soap, non-polar substances are captured in the hydrophobic region of micelles. The action of soap effectively makes non-polar hydrophobic substances water soluble, allowing them to be rinsed away. The use of warm water in conjunction with soap aids the cleaning action by softening the greasy substances to be cleaned away. Agitation is also helpful as this disrupts micelles and enhances encapsulation of non-polar substances as they reassemble. The formation of micelles and the cleaning action of soap are largely attributable to IMFs. An opportunity for you to think about and articulate the specific IMFs is provided shortly in an exercise.

It is useful to represent soap molecules and micelles with the following symbols, readily drawn by hand. The large spot symbolises the polar hydrophilic section, with the non-polar hydrophobic section represented by the wavy line projecting from the spot. Using this symbol for a soap molecule makes it straightforward to draw a micelle.

A large spot represents the polar hydrophilic portion of a soap molecule

A wavy line projecting from the spot represents the non-polar hydrophobic portion of a soap molecule

A micelle represented using several symbols for soap molecules. The polar portions are to the outside of a circle to show them dissolving in water. Draw the non-polar portions aggregating together inside the circle to show them repelling water.

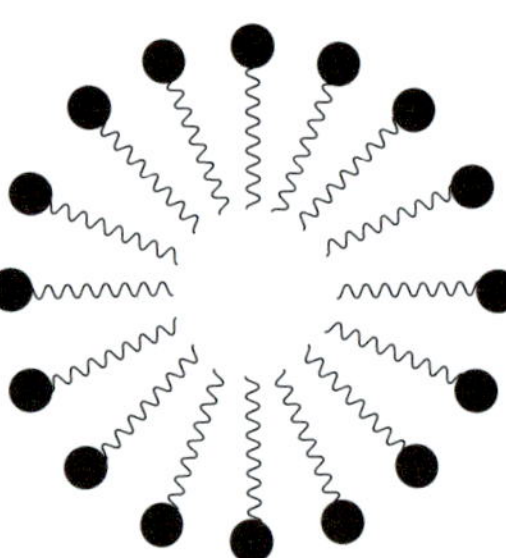

EXERCISE 12.67

Write an explanation of the cleaning action of soap with reference to the specific IMFs that (a) allow the polar sections of soap molecules to dissolve in water, (b) facilitate the aggregation of the non-polar sections of soap molecules in micelles, and (c) encapsulate non-polar substances in a micelle. Use the representations of soap molecules and micelles above in your answer.

Detergents

A limitation of soap is the formation of 'soap scum' that diminishes the cleaning action when used with hard water. Hard water contains divalent cations such as Mg^{2+} and Ca^{2+}. When such 2+ cations form salts with 1– anions, the cation must associate with two 1– anions to form a neutral compound. When soap is used in hard water the divalent cations associate with two soap molecules and they form insoluble salts that precipitate. The molecular structure of such a salt is shown below. These salts are the soap scum that is difficult to remove from surfaces. If soap molecules form insoluble salts and precipitate, then they are not available to form micelles and clean away greasy substances. Thus soaps are less effective when used in hard water.

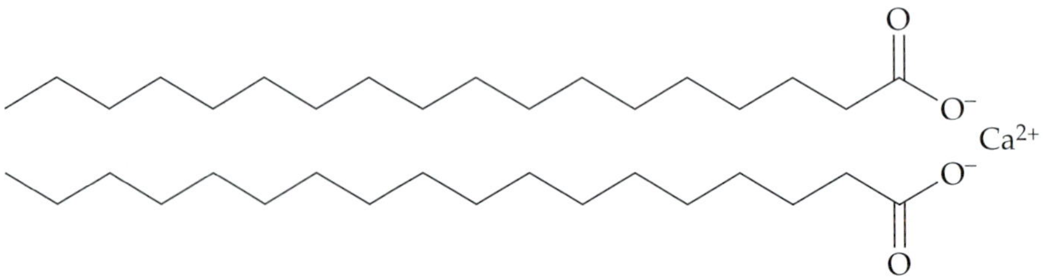

O
O⁻
Ca^{2+}
O⁻
O

Detergents overcome the problem of soap scum in hard water. Detergents are synthetic compounds that are generally salts of alkylbenzenesulfonic acids, an example is shown below. Note the molecular structure is similar to soap with a charged polar part and a non-polar hydrophobic part, which means that they form micelles in water and have the same cleaning properties as soap. The advantage of detergents is that their Mg^{2+} and Ca^{2+} salts are water soluble and they do not form soap scum in hard water.

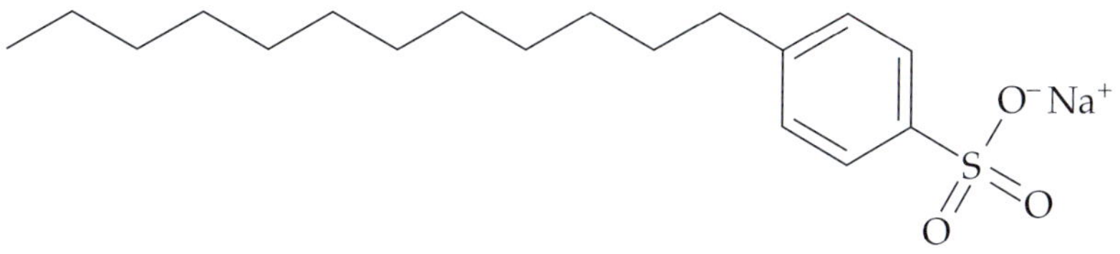

The comparison in ▼ Figure 12.50 summarises some key features of soaps and detergents. Both soap and detergent have amphipathic molecular structure that allows for the formation of micelles in water and the same cleaning action.

▼ **FIGURE 12.50 Comparison of soap and detergent**

Soap	Detergent
Solid	Liquid
Carboxylate salts	Alkylbenzenesulfonic acid salts
Made from fats by saponification	Synthesised from hydrocarbons and sulfuric acid
Formation of insoluble salts and diminished cleaning action when used with hard water	Effective in hard water

EXERCISES

QUESTIONS

68. Write a definition for organic compounds.

69. How many covalent bonds do carbon atoms form in stable organic compounds?

70. What are the possible bonding combinations and related electron group geometries of carbon atoms?

71. What causes the zigzag arrangement of carbon atoms forming chains and rings?

72. Write the structural formula (Lewis structure), condensed formulas and line bond formula for an organic compound with molecular formula C_5H_{12} in which all the carbon atoms are connected in a straight chain.

73. An organic compound with a branching structure may be represented by the condensed formula $CH_3C(CH_3)_2CH_2CH(CH_3)CH_3$. Write the structural and line bond formulas for this compound.

74. Explain the difference between saturated and unsaturated hydrocarbons.

75. Classify the following hydrocarbons as saturated or unsaturated.

(a) $H_3C-C{\equiv}C-CH_2CH_3$

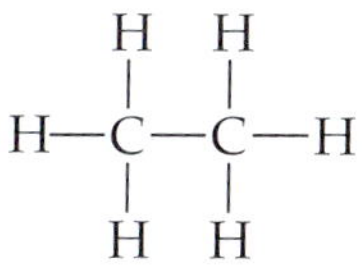

(b)

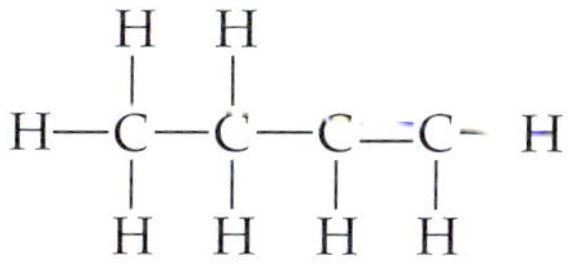

(c)

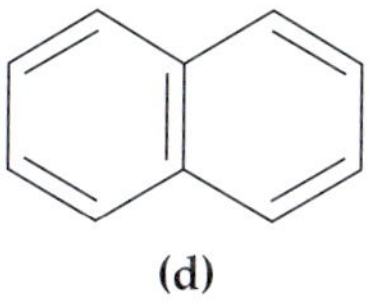

(d)

76. State why alkanes are saturated hydrocarbons.

77. Use the general formula for alkanes to derive the molecular formula for an alkane that has 11 carbon atoms.

78. What is the molecular formula of the cycloalkane with 11 carbon atoms?

79. What are structural isomers?

80. Draw structural and line bond formulas for all five structural isomers of C_6H_{14}.

81. Combustion is one of the few reactions that alkanes can undergo. Write the balanced equation for the combustion of hexane (C_6H_{14}).

82. State why alkenes are unsaturated hydrocarbons.

83. Use the general formula for alkenes to derive the molecular formula for an alkene with 7 carbons atoms. This alkene has only one double bond.

84. Explain the difference between the alkenes pent-1-ene and pent-2-ene.

85. The following alkenes are both hex-3-ene, they are geometric isomers. What should be added to each name to distinguish the isomers?

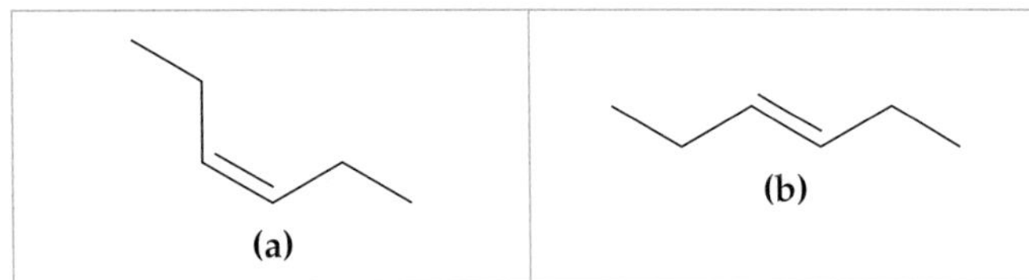

86. Which of the following alkenes could have a geometric isomer?

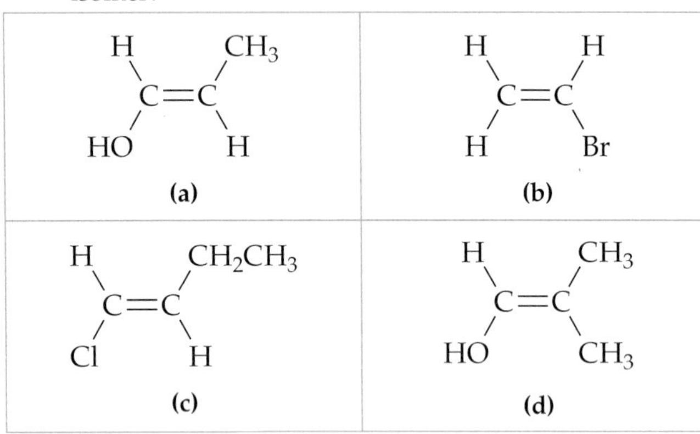

87. Classify the following hydrocarbons as aliphatic or aromatic.

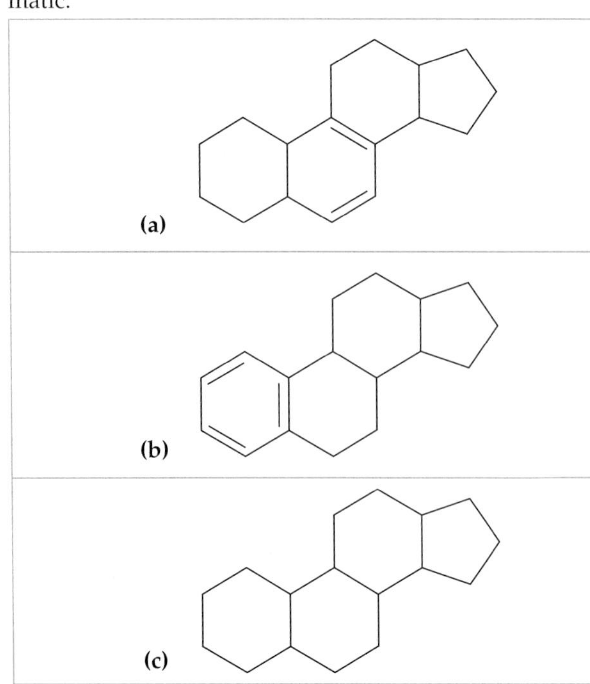

88. Classify the following alcohols as primary (1°), secondary (2°) or tertiary (3°).

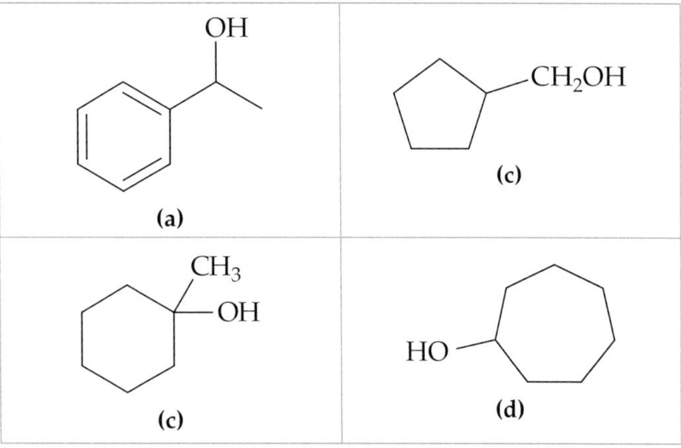

89. What is a carbonyl group?

90. Describe the difference between the molecular structures of aldehydes and ketones.

91. Write an equation to show ethanoic acid (CH_3COOH) in aqueous solution donating a proton to form its conjugate base. Also, show a water molecule accepting the donated proton in the equation.

92. Write an equation to show the reaction of ethanoic acid (CH_3COOH) with the strong base NaOH.

93. Explain the course of the reaction when a primary alcohol is oxidised in the presence of a general oxidising agent such as acidified aqueous potassium permanganate.

94. What type of organic compound is obtained when a secondary alcohol is oxidised using a general oxidising agent such as acidified potassium permanganate?

95. Explain why tertiary alcohols are resistant to being oxidised to carbonyl compounds.

96. Describe the difference between the molecular structure of carboxylic acids and esters.

97. What does the term 'alkoxy group' describe in the structure of an ester?

98. Which two functional groups can combine in a condensation reaction to form an ester?

99. Write a definition for a hydrolysis reaction.

100. What products form in the hydrolysis of an ester in the presence of a base?

101. Classify the following amines as primary (1°), secondary (2°) or tertiary (3°).

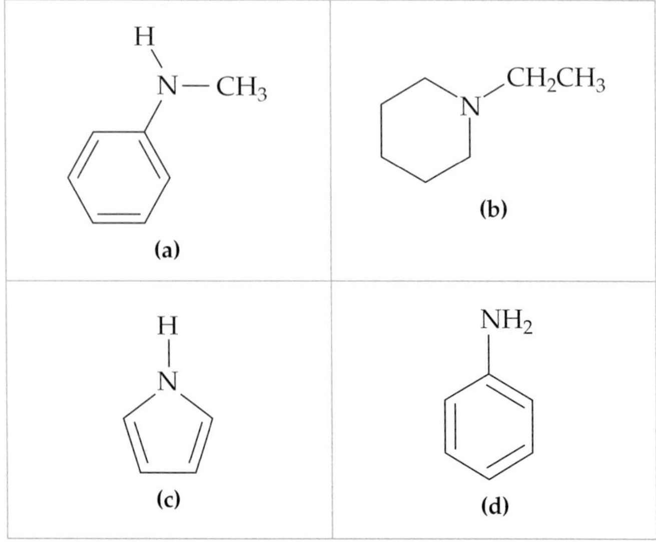

102. Write an equation to show propanamine ($CH_3CH_2CH_2NH_2$) accepting a proton (H^+) to form its conjugate acid.

103. Write an equation to show how propanamine ($CH_3CH_2CH_2NH_2$) in aqueous solution accepts a proton from water to form its conjugate acid and hydroxide ion.

104. Which two functional groups can combine in a condensation reaction to form an amide?

105. What is a polymer? What are the small molecules that make up polymers?

106. Write definitions for the following terms related to polymers: homopolymer, copolymer, addition polymer, condensation polymer.

107. List the differences between addition and condensation polymerisation.

108. What types of monomers combine to produce a polyester?

109. What types of monomers combine to produce a polyamide?

110. Describe how hydrogen bonding contributes to the physical properties of polyamides.

111. Explain how non-polar polyethylene chains can interact to form 'ordered regions'.

112. Explain the differences between low density polyethylene (LDPE) and high density polyethylene (HDPE) in terms of molecular structure, IMFs and physical properties.

114. What is an amino acid? What is an α-amino acid (alpha amino acid)?

115. Why are amino acids amphoteric?

116. When does an α-amino acid exist as a zwitterion?

117. In the α-amino acid glycine, the R group is a hydrogen atom, the structure is shown below. Redraw glycine to show how it exists in (a) acidic solution, (b) neutral solution, and (c) basic solution

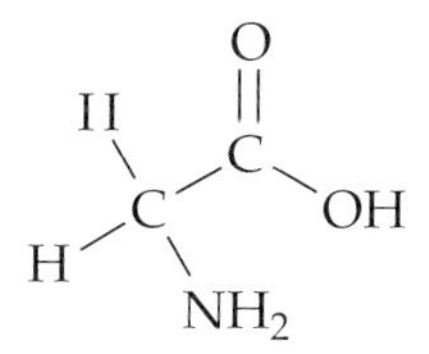

H
O
C
C
OH
H
NH$_2$

118. What are proteins?

119. Differentiate the following terms: peptide, dipeptide, oligopeptide, polypeptide.

120. Explain how two α-amino acids can react to form an amide functional group. Why are polymers of α-amino acids possible?

121. A chemist wanted to make the dipeptide Phe-Val. If the α-amino acids phenylalanine (Phe) and valine (Val) were combined under suitable conditions, what other products could form?

122. Classify each of the following descriptions of protein structural features as primary, secondary, tertiary or quaternary.

(a) Aggregation of hydrophobic α-amino acid R groups to form a water repelling pocket in a protein structure.

(b) A repeating pattern where chains of α-amino acids interact to form sheets.

(c) The way subunits fit together in proteins composed of more than one peptide chain.

(d) The –SH groups of two cysteine (Cys) α-amino acids maintain a fold in a protein by forming a S-S covalent bond, known as a disulfide bridge.

(e) The sequence of α-amino acids in a protein.

123. What is the difference between fat and oil? What are the features of molecular structure that fats and oils have in common?

124. Why are fats also called triglycerides?

125. The structure of palmitic acid is shown below. Is palmitic acid a saturated or unsaturated fatty acid?

O
OH

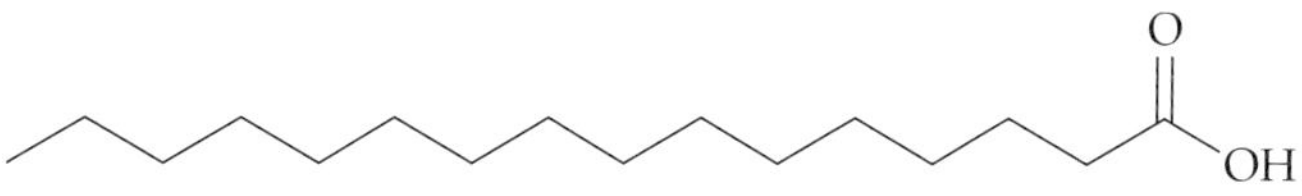

126. Is palmitoleic acid, pictured below, a monounsaturated or polyunsaturated fatty acid?

O
HO

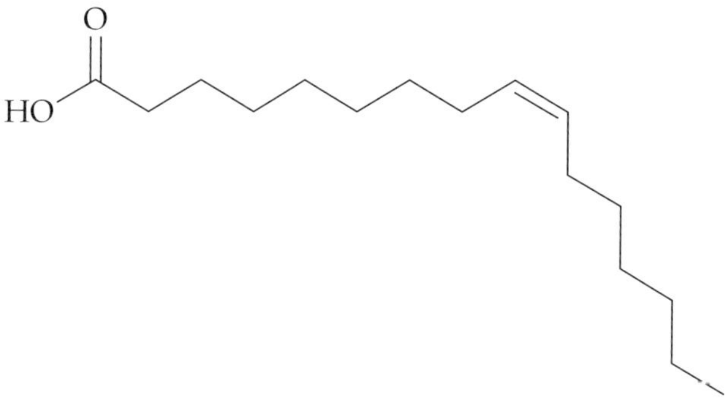

127. Describe the chemical reaction that produces soap?

128. Explain how soap molecules are amphipathic.

129. What happens when soap molecules are added to water? Explain with reference to specific IMFs that occur in this process.

130. Use diagrams to explain the cleaning action of soap.

131. What is 'soap scum', draw the structure of a soap scum molecule in your answer.

132. What is detergent? How is detergent different from soap? How is detergent similar to soap?

133. Explain why soap loses efficiency in hard water, while detergent can be used effectively in hard water?

PROBLEMS

RECOGNISING FUNCTIONAL GROUPS

134. Match each of the following general structures to its functional group name: alcohol, ether, aldehyde, ketone, carboxylic acid, ester, amine, amide.

R–C(=O)–OH **(a)**	R–C(=O)–R **(b)**	R–O–R **(b)**	R–N(H)–H **(d)**
R–C(=O)–O–R **(e)**	R–C(=O)–N(R)–R **(f)**	R–O–H **(g)**	R–C(=O)–H **(h)**

135. Identify the following functional groups in the organic compound structural formulas shown below. Circle the group of atoms that make up the functional group and clearly label it with the appropriate name. Each functional group may appear in more than one compound. The functional groups are: alcohol, ether, aldehyde, ketone, carboxylic acid, ester, amine, amide

(a) testosterone

(b) aspirin

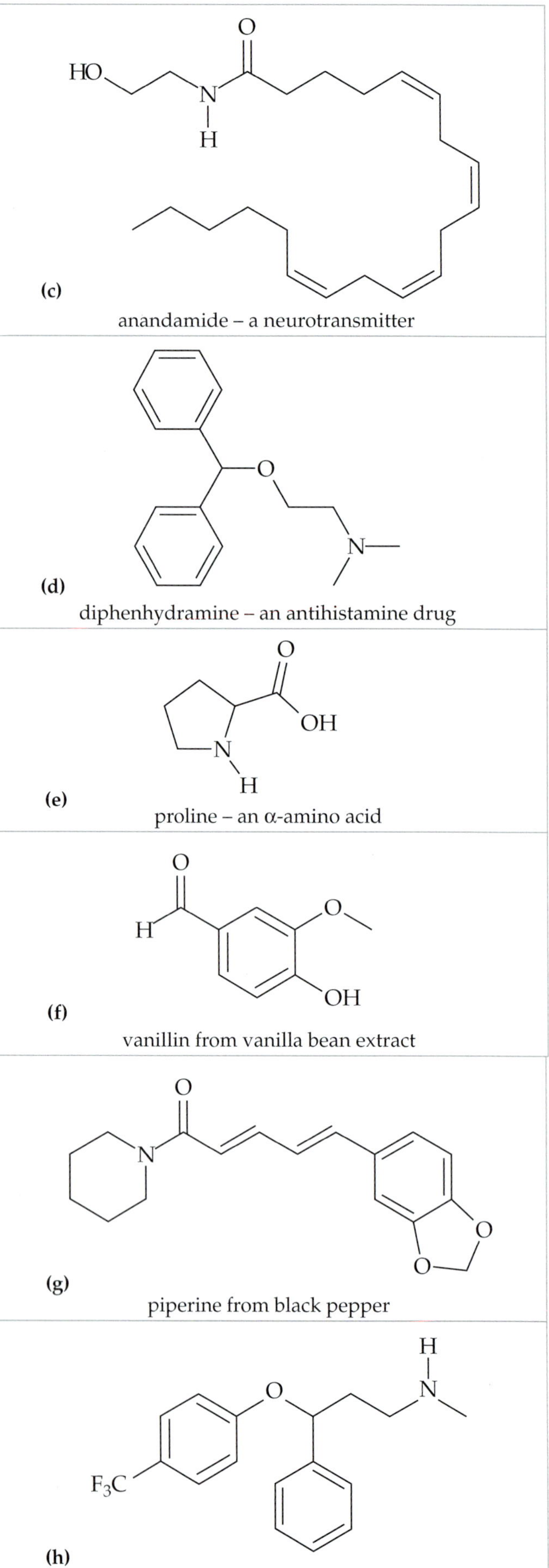

(c) anandamide – a neurotransmitter

(d) diphenhydramine – an antihistamine drug

(e) proline – an α-amino acid

(f) vanillin from vanilla bean extract

(g) piperine from black pepper

(h) prozac – an antidepressant drug

NAMING AND DRAWING ORGANIC COMPOUNDS.

136. Write the IUPAC name for each of the following organic compounds.

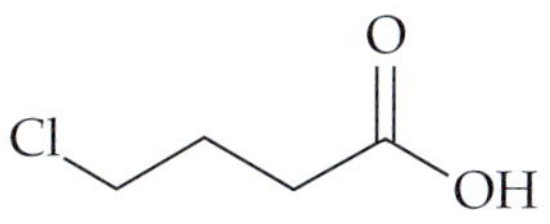 (a)	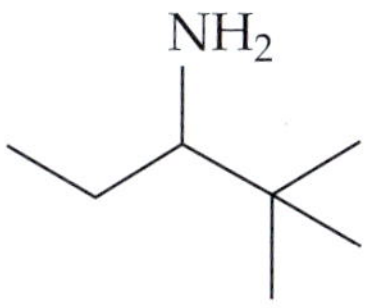 (b)
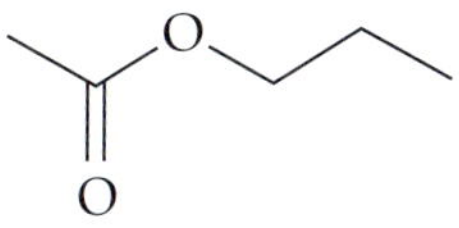 (c)	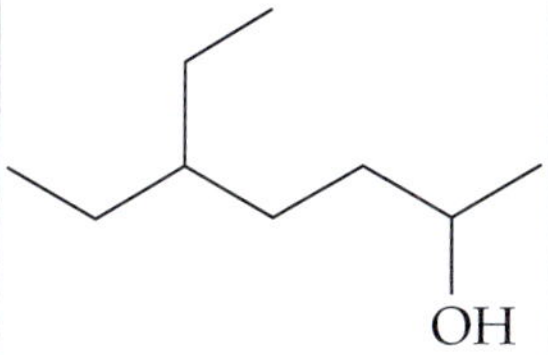 (d)
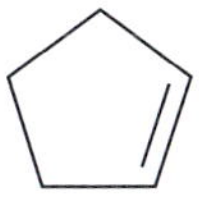 (d)	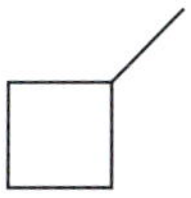 (e)
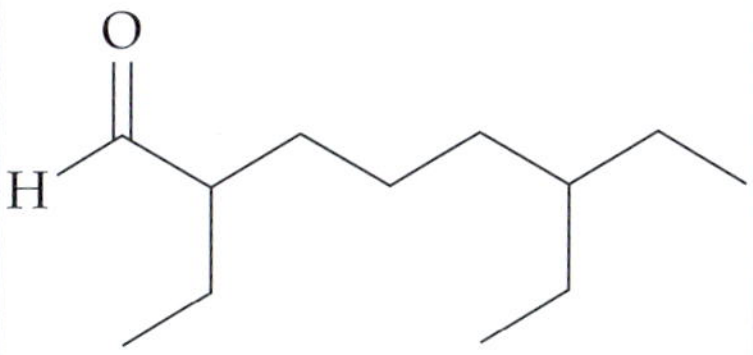 (f)	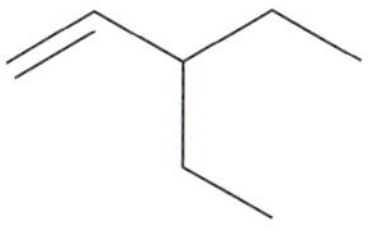 (g)
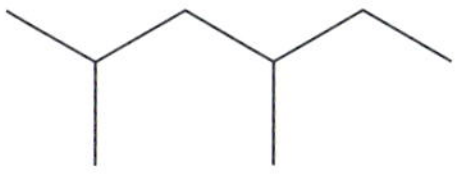 (h)	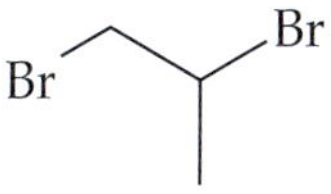 (i)
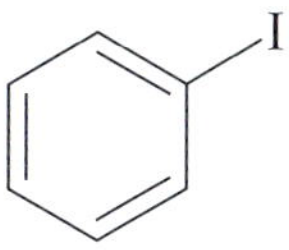 (j)	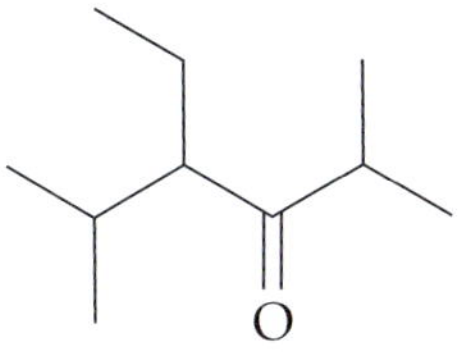 (k)

137. Draw line bond formulas to represent each of the organic compounds with the following IUPAC names.

butylbenzene	
methylcyclopentane	
5-propyloctan-4-ol	
2,2-dimethylpropanoic acid	
ethyl butanoate	
2,3-diethyl-4-methylpentanal	
iodocyclohexane	
2,3-dimethylhex-3-ene	
3,5-diethyl-4,6-dimethylhep-tan-2-one	
4-ethyl-3-methylheptane	
6-ethyl-3,5-dimethyl-2-pro-pyloctan-1-amine	
cyclohexene	

PHYSICAL PROPERTIES OF ORGANIC COMPOUNDS

138. The alkanes hexane and heptane are both non-polar and dispersion is the only relevant IMF. Explain the difference in boiling point between the two compounds.

Alkane name	hexane	heptane
Structure		
Molecular formula	C_6H_{14}	C_7H_{16}
Boiling point	69 °C	99 °C

139. The alkane structural isomers shown below are both non-polar compounds with dispersion as the only IMF possible. Explain why the straight chain isomer has a higher boiling point.

Alkane name	2,2,4-trimethylpentane 'iso-octane'	octane
Structure		
Molecular formula	C_8H_{18}	C_8H_{18}
Boiling point	99 °C	126 °C

140 The difference between the alkane and haloalkane shown below is just one atom, but there is a significant difference in their boiling points. Explain the two major reasons for the higher boiling point of the halogenated compound.

Compound name	cyclohexane	chlorocyclohexane
Structure		Cl
Molecular formula	C_6H_{12}	$C_6H_{11}Cl$
Boiling point	81 °C	142 °C

141. Describe the IMFs that allow ethanol and other low molar mass alcohols to mix with water. Representations of ethanol molecular structure are shown below for reference. Draw a diagram to show how ethanol molecules and water molecules can interact through IMFs.

H H H—C—C—OH H H	OH	CH_3CH_2OH

142. Explain the trend in boiling points for the following straight chain alcohols.

Alcohol	methanol	ethanol	propan-1-ol	butan-1-ol
Boiling point	65 °C	79 °C	97 °C	118 °C

143. Account for the difference in boiling point for the alcohol structural isomers shown below.

Alcohol name	propan-1-ol		propan-2-ol	
Structural and line bond formulas	H H H H—C—C—C—OH H H H	OH	H OH H H—C—C—C—H H H H	OH
Molar mass	60 g/mol		60 g/mol	
Boiling point	97 °C		82 °C	

144. Both ether and ester molecules are polar and dipole-dipole attraction is a relevant IMF in each case. Consider the information given below for an ether and ester that have the same molar mass and explain the difference in boiling point between the two compounds.

Compound name	diethyl ether	ethyl methanoate
Structural and line bond formulas 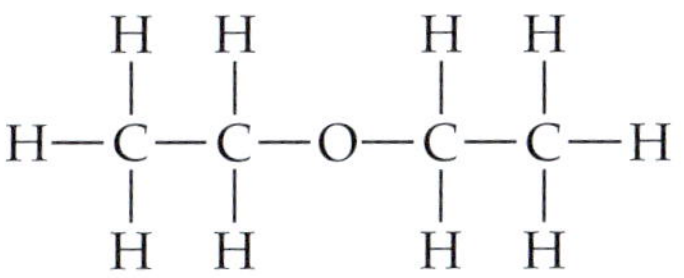 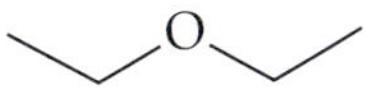 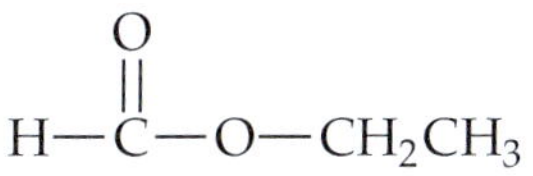 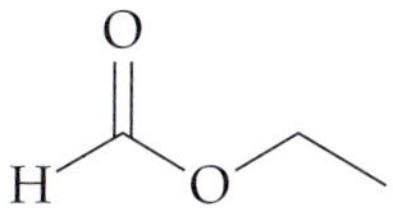	$H_3C-CH_2-O-CH_2-CH_3$ (each C drawn with its H atoms)	$H-C(=O)-O-CH_2CH_3$
Molar mass	74 g/mol	74 g/mol
Boiling point	35 °C	54 °C

145. The following organic compounds have similar molar mass but different structures. Rank the compounds in order of expected boiling point from lowest to highest.

Compound name	butanal	butan-1-ol	pentane
Structure	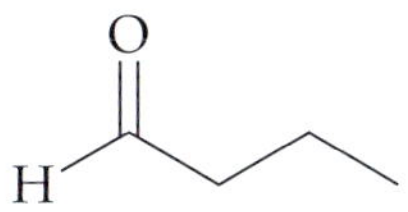 H–C(=O) line structure	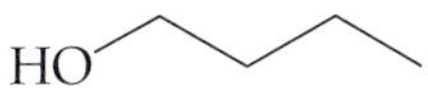 HO line structure	 line structure
Molar mass	72 g/mol	74 g/mol	72 g/mol

146. The following organic compounds have similar molar mass but different structures. Rank the compounds in order of expected boiling point from lowest to highest.

Compound name	propanoic acid	pentane	butanone
Structure	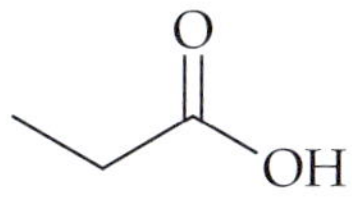 line structure with O and OH	 line structure	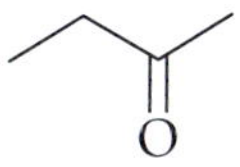 line structure with O
Molar mass	74 g/mol	72 g/mol	72 g/mol

147. The water solubility of both amines and alcohols is related to the ability of both functional groups to participate in hydrogen bonding IMF with water molecules. The water solubility at 20 °C of an amine and alcohol with similar structure and molar mass are shown below. Explain the difference in water solubility between the two compounds.

Compound name	'aniline'	'phenol'
Structure	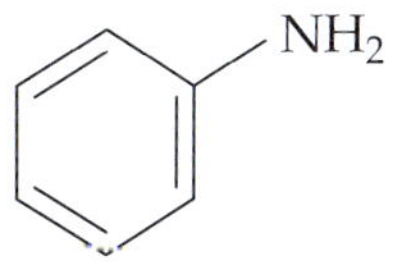 benzene ring with NH_2	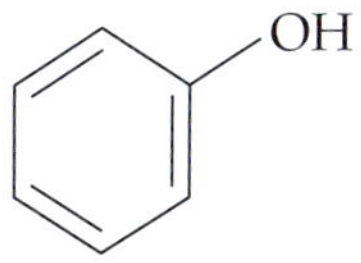 benzene ring with OH
Molar mass	93 g/mol	94 g/mol
Solubility in water at 20 °C	3.6 g/100 mL	8.3 g/100 mL

148. The α-amino acid valine and the dicarboxylic acid 2-methylpropanedioic acid have similar molar mass, and both have two polar functional groups capable of hydrogen bonding. However, the melting point of valine is much higher than that of 2-methylpropanedioic acid. Explain why this is the case.

Compound name	valine	2-methylpropanedioic acid
Structure	H_3C–CH(CH_3)–C(H)(NH_2)–C(=O)–OH	HO–C(=O)–CH(CH_3)–C(=O)–OH
Molar mass	117 g/mol	118 g/mol
Melting point	298 °C	134 °C

149. Explain the difference between saturated and unsaturated fats at the molecular level that is responsible for the existence of saturated fats as solids and unsaturated fats as liquids at room temperature.

EQUATIONS FOR REACTIONS OF ORGANIC COMPOUNDS

150. This equation shows the reaction of an alkane in which one hydrogen atom is replaced by a bromine atom. Complete the equation by writing the missing reactant and conditions required above the reaction arrow.

```
    H  H  H  H  H  H               H  H  H  H  H  H
    |  |  |  |  |  |               |  |  |  |  |  |
 H—C—C—C—C—C—C—H   ——————→   H—C—C—C—C—C—C—Br   +   HBr
    |  |  |  |  |  |               |  |  |  |  |  |
    H  H  H  H  H  H               H  H  H  H  H  H
```

151. Alkenes react with halogens (X_2) in addition reactions. Complete the equation by drawing the structure of the organic compound that would form in the reaction shown.

```
    H  H        H  H
    |  |        |  |          Cl2
 H—C—C—C=C—C—C—H      ——————→
    |  |  |  |  |  |
    H  H  H  H  H  H
```

152. Alcohols react with alkali metals to produce hydrogen gas and an alkoxide salt. Complete the equation below by writing the formula for the missing product.

$$2HOCH_2CH_2CH_3 + 2K \rightarrow H_2 +$$

153. This incomplete equation depicts a carboxylic acid in aqueous solution donating a proton to a water molecule. Draw the structure of the carboxylic acid conjugate base to complete the equation.

(cyclopentyl)–C(=O)–OH + H_2O ——→ + H_3O^+

154. The reaction of a carboxylic acid with a strong base produces a salt and a water molecule. Draw the structure of the salt that forms in the reaction of the carboxylic acid shown below with the strong base NaOH.

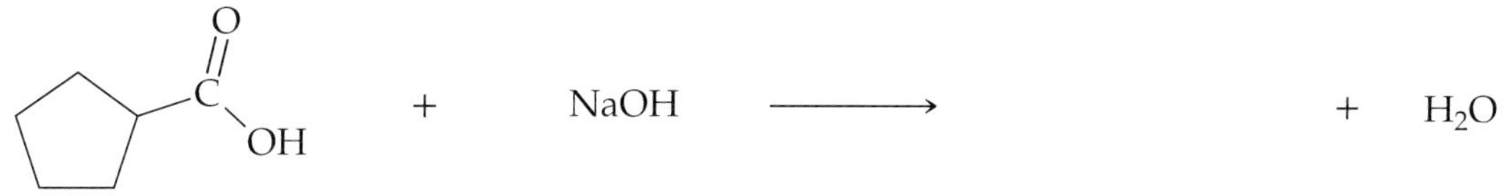

(cyclopentyl–C(=O)–OH) + NaOH ⟶ + H_2O

155. Draw the structures of the intermediate and final products that would form in the oxidation of the alcohol shown using acidified aqueous potassium permanganate.

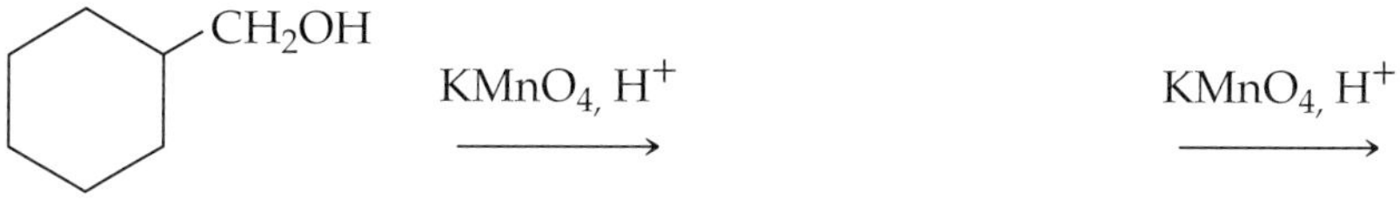

(cyclohexyl–CH_2OH) $\xrightarrow{KMnO_4,\ H^+}$ $\xrightarrow{KMnO_4,\ H^+}$

156. Draw the structure of the reactant that would be required to produce the organic compound shown in this incomplete equation.

$\xrightarrow{KMnO_4,\ H^+}$ $H_3C-C(=O)-CH_2CH_2CH_3$

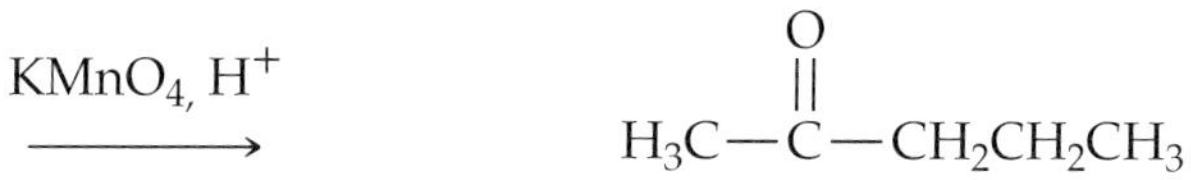

157. Compete the equation by drawing the structures of the two reactant organic compounds that would be needed to produce the ester product shown.

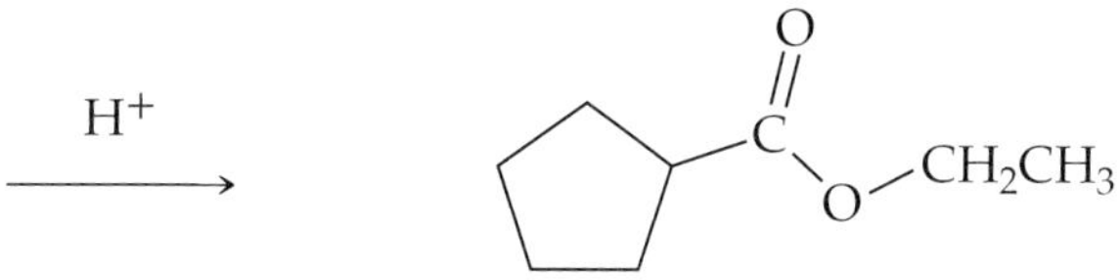

$\xrightarrow{H^+}$ (cyclopentyl–C(=O)–O–CH_2CH_3)

158. Predict the products for hydrolysis of this ester in the presence of the base shown.

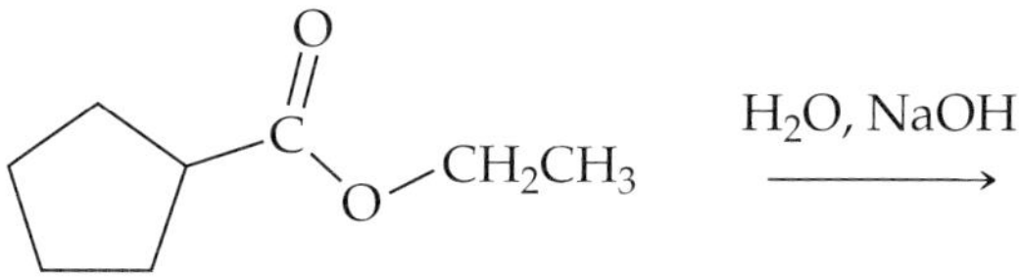

(cyclopentyl–C(=O)–O–CH_2CH_3) $\xrightarrow{H_2O,\ NaOH}$

159. Complete the equation to show the structure of the conjugate acid that forms when the amine shown accepts a proton.

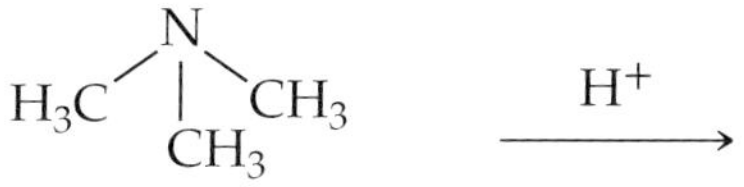

$N(CH_3)_3$ (H_3C–N(–CH_3)–CH_3) $\xrightarrow{H^+}$

160. This incomplete equation shows an amine in aqueous solution accepting a proton from water to form its conjugate acid and hydroxide ion. Draw the structure of the missing amine.

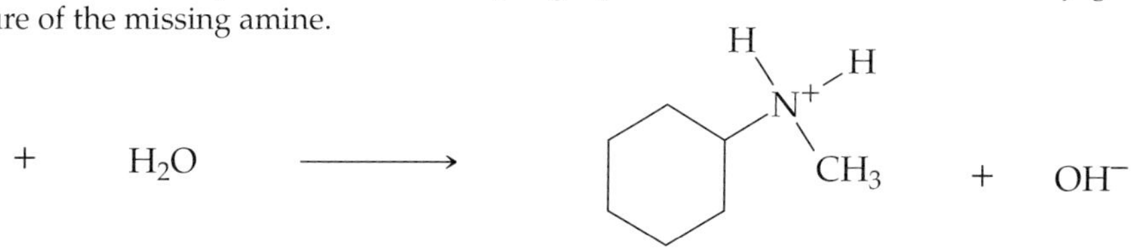

+ H_2O ⟶ (cyclohexyl–N^+(H)(H)–CH_3) + OH^-

161. Draw the structure of the organic compound product that forms in this incomplete equation for a particular condensation reaction.

$$H_3C\text{—}C(=O)\text{—}OH + C_6H_5NH_2 \longrightarrow$$

162. Draw a repeating unit of the polymer that would form from the reactant monomers shown.

$$n\,HO\text{—}C(=O)\text{—}CH_2CH_2\text{—}C(=O)\text{—}OH \xrightarrow{n\,HO\text{—}CH_2CH_2\text{—}OH}$$

163. Draw the structures of the reactant monomers that would produce the polymer depicted.

$$\longrightarrow \left[\text{—}C(=O)\text{—}C_6H_4\text{—}C(=O)\text{—}NH\text{—}C_6H_4\text{—}NH\text{—}\right]_n$$

164. Draw a repeating unit of the polymer that would form from the reactant monomer shown.

$$n\ F_2C\text{=}CF_2 \longrightarrow$$

165. Draw the structure of the reactant monomer that would produce the polymer represented.

$$\longrightarrow \left[\text{—}CH_2\text{—}CH(C_6H_5)\text{—}\right]_n$$

166. The incomplete equation below depicts hydrogenation of an unsaturated fatty acid. Draw the structure of the saturated fatty acid that would form if the reactant was completely hydrogenated.

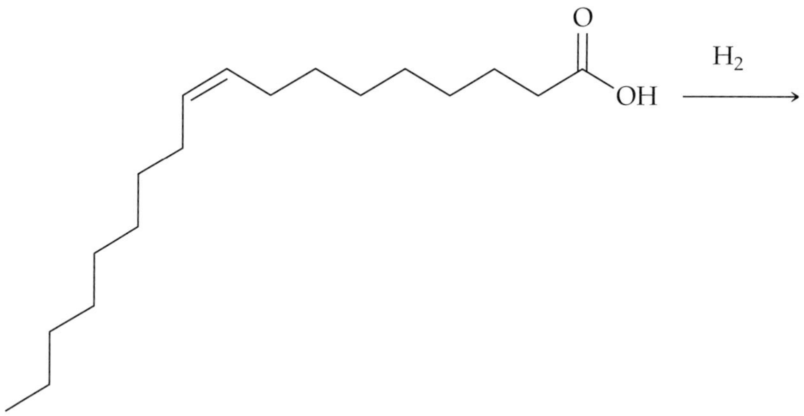

$$\xrightarrow{H_2}$$

167. This is an incomplete equation showing the production of soap from fat. Draw the structures of the missing products.

$$\xrightarrow{H_2O,\ KOH} \quad + \quad$$

HO—CH$_2$

HO—CH

HO—CH$_2$

Appendix: Mathematics Review

Basic Algebra

In chemistry, you often have to solve an equation for a particular variable. For example, suppose you want to solve the following equation for V:

$$PV = nRT$$

To solve an equation for a particular variable, you must isolate that variable on one side of the equation. The rest of the variables or numbers will then be on the other side of the equation. To solve the above equation for V, divide both sides by P.

$$\frac{PV}{P} = \frac{nRT}{P}$$

$$V = \frac{nRT}{P}$$

The Ps cancel, and you are left with an expression for V. For another example, consider solving the following equation for °F:

$$°C = \frac{(°F - 32)}{1.8}$$

First, eliminate the 1.8 in the denominator of the right side by multiplying both sides by 1.8.

$$(1.8)\ °C = \frac{(°F - 32)}{1.8}(1.8)$$

$$(1.8)°C = (°F - 32)$$

Then eliminate the –32 on the right by adding 32 to both sides.

$$(1.8)°C + 32 = (°F - 32) + 32$$

$$(1.8)°C + 32 = °F$$

You are now left with an expression for °F.

In general, solve equations by following these guidelines:

- Cancel numbers or symbols in the denominator (bottom part of a fraction) by multiplying by the number or symbol to be canceled.
- Cancel numbers or symbols in the numerator (upper part of a fraction) by dividing by the number or symbol to be canceled.
- Eliminate numbers or symbols that are added by subtracting the same number or symbol.
- Eliminate numbers or symbols that are subtracted by adding the same number or symbol.
- Whether you add, subtract, multiply, or divide, **always perform the same operation for both sides of a mathematical equation**. (Otherwise, the two sides will no longer be equal.)

For a final example, solve the following equation for x:

$$\frac{67x - y + 3}{6} = 2z$$

Cancel the 6 in the denominator by multiplying both sides by 6.

$$(\cancel{6})\frac{67x - y + 3}{\cancel{6}} = (6)2z$$

$$67x - y + 3 = 12z$$

Eliminate the +3 by subtracting 3 from both sides.

$$67x - y + 3 - 3 = 12z - 3$$
$$67x - y = 12z - 3$$

Eliminate the $-y$ by adding y to both sides.

$$67x - y + y = 12z - 3 + y$$
$$67x = 12z - 3 + y$$

Cancel the 67 by dividing both sides by 67.

$$\frac{\cancel{67}x}{\cancel{67}} = \frac{12z - 3 + y}{67}$$

$$x = \frac{12z - 3 + y}{67}$$

FOR PRACTICE **Using Algebra to Solve Equations**

Solve each of the following for the indicated variable:

(a) $P_1V_1 = P_2V_2$; solve for V_2

(b) $\frac{V_1}{T_1} = \frac{V_2}{T_2}$; solve for T_1

(c) $PV = nRT$; solve for n

(d) K = °C + 273; solve for °C

(e) $\frac{3x + 7}{2} = y$; solve for x

(f) $\frac{32}{y + 3} = 8$; solve for y

ANSWERS

(a) $V_2 = \frac{P_1V_1}{P_2}$

(b) $T_1 = \frac{V_1T_2}{V_2}$

(c) $n = \frac{PV}{RT}$

(d) °C = K 273

(e) $x = \frac{2y - 7}{3}$

(f) $y = 1$

Mathematical Operations with Scientific Notation

Writing numbers in scientific notation is covered in detail in Section 4.1. Briefly, a number written in scientific notation consists of a **decimal part**, a number that is usually between 1 and 10, and an **exponential part**, 10 raised to an **exponent**, n.

Each of the following numbers is written in both scientific and decimal notation:

$1.0 \times 10^5 = 100{,}000$ $1.0 \times 10^{-6} = 0.000001$

$6.7 \times 10^3 = 6700$ $6.7 \times 10^{-3} = 0.0067$

Multiplication and Division

To multiply numbers expressed in scientific notation, multiply the decimal parts and add the exponents.

$$(A \times 10^{m})(B \times 10^{n}) = (A \times B) \times 10^{m+n}$$

To divide numbers expressed in scientific notation, divide the decimal parts and subtract the exponent in the denominator from the exponent in the numerator.

$$\frac{(A \times 10^{m})}{(B \times 10^{n})} = \left(\frac{A}{B}\right) \times 10^{m-n}$$

Consider the following example involving multiplication:

$$\begin{aligned}(3.5 \times 10^{4})(1.8 \times 10^{6}) &= (3.5 \times 1.8) \times 10^{4+6} \\ &= 6.3 \times 10^{10}\end{aligned}$$

Consider the following example involving division:

$$\begin{aligned}\frac{(5.6 \times 10^{7})}{(1.4 \times 10^{3})} &= \left(\frac{5.6}{1.4}\right) \times 10^{7-3} \\ &= 4.0 \times 10^{4}\end{aligned}$$

Addition and Subtraction

To add or subtract numbers expressed in scientific notation, rewrite all the numbers so that they have the same exponent, and then add or subtract the decimal parts of the numbers. The exponents remain unchanged.

$$\begin{array}{r} A \times 10^{n} \\ \pm\, B \times 10^{n} \\ \hline (A \pm B) \times 10^{n} \end{array}$$

Notice that the numbers *must have* the same exponent.

Consider the following example involving addition:

$$\begin{array}{r} 4.82 \times 10^{7} \\ +3.4 \times 10^{6} \\ \hline \end{array}$$

First, express both numbers with the same exponent. In this case, rewrite the lower number and perform the addition as follows:

$$\begin{array}{r} 4.82 \times 10^{7} \\ +0.34 \times 10^{7} \\ \hline 5.16 \times 10^{7} \end{array}$$

Consider the following example involving subtraction:

$$\begin{array}{r} 7.33 \times 10^{5} \\ -1.9 \times 10^{4} \\ \hline \end{array}$$

First, express both numbers with the same exponent. In this case, rewrite the lower number and perform the subtraction as follows:

$$\begin{array}{r} 7.33 \times 10^5 \\ -0.19 \times 10^5 \\ \hline 7.14 \times 10^5 \end{array}$$

FOR PRACTICE **Mathematical Operations with Scientific Notation**

Perform each of the following operations:

(a) $(2.1 \times 10^7)(9.3 \times 10^5)$

(b) $(5.58 \times 10^{12})(7.84 \times 10^{-8})$

(c) $\dfrac{(1.5 \times 10^{14})}{(5.9 \times 10^8)}$

(d) $\dfrac{(2.69 \times 10^7)}{(8.44 \times 10^{11})}$

(e) $\begin{array}{r} 1.823 \times 10^9 \\ +1.11 \times 10^7 \\ \hline \end{array}$

(f) $\begin{array}{r} 3.32 \times 10^{-5} \\ +3.400 \times 10^{-7} \\ \hline \end{array}$

(g) $\begin{array}{r} 6.893 \times 10^9 \\ -2.44 \times 10^8 \\ \hline \end{array}$

(h) $\begin{array}{r} 1.74 \times 10^4 \\ -2.9 \times 10^3 \\ \hline \end{array}$

ANSWERS

(a) 2.0×10^{13}

(b) 4.37×10^5

(c) 2.5×10^5

(d) 3.19×10^{-5}

(e) 1.834×10^9

(f) 3.35×10^{-5}

(g) 6.649×10^9

(h) 1.45×10^4

Logarithms

The logarithm (or log) of a number is the exponent to which 10 must be raised to obtain that number. For example, the log of 100 is 2 because 10 must be raised to the second power to get 100. Similarly, the log of 1000 is 3 because 10 must be raised to the third power to get 1000. The logs of several multiples of 10 are shown as follows:

$$\begin{aligned} \log 10 &= 1 \\ \log 100 &= 2 \\ \log 1000 &= 3 \\ \log 10{,}000 &= 4 \end{aligned}$$

Because $10^0 = 1$ by definition, $\log 1 = 0$.

The log of a number smaller than 1 is negative because 10 must be raised to a negative exponent to get a number smaller than 1. For example, the log of 0.01 is –2 because 10 must be raised to the power of –2 to get 0.01. Similarly, the log of 0.001 is –3 because 10 must be raised to the power of –3 to get 0.001. The logs of several fractional numbers are as follows:

$$\begin{aligned} \log 0.1 &= -1 \\ \log 0.01 &= -2 \\ \log 0.001 &= -3 \\ \log 0.0001 &= -4 \end{aligned}$$

The logs of numbers that are not multiples of 10 can be computed on your calculator. See your calculator manual for specific instructions.

Inverse Logarithms

The inverse logarithm or invlog function (sometimes called antilog) is exactly the opposite of the log function. For example, the log of 100 is 2 and the inverse log of 2 is 100. The log function and the invlog function undo one another.

$$\log 100 = 3$$
$$\text{invlog } 3 = 1000$$
$$\text{invlog } (\log 1000) = 1000$$

The inverse log of a number is simply 10 raised to that number.

$$\text{invlog } x = 10^x$$
$$\text{invlog } 3 = 10^3 = 1000$$

The inverse logs of numbers can be computed on your calculator. See your calculator manual for specific instructions.

FOR PRACTICE **Logarithms and Inverse Logarithms**

Perform each of the following operations:

(a) $\log 1.0 \times 10^5$

(b) $\log 59$

(c) $\log 1.0 \times 10^{-5}$

(d) $\log 0.068$

(e) invlog 7.0

(f) invlog 1.44

(g) invlog −6.0

(h) invlog −0.250

(i) invlog (log 88)

ANSWERS

(a) 5.00

(b) 1.77

(c) −5.00

(d) −1.17

(e) 1×10^7

(f) 28

(g) 1×10^{-6}

(h) 0.56

(i) 88

Glossary

absolute zero The coldest temperature possible. Absolute zero (0 K or –273 °C or –459 °F) is the temperature at which molecular motion stops. Lower temperatures do not exist.

acid A molecular compound that dissolves in solution to form H^+ ions. Acids have the ability to dissolve some metals and will turn litmus paper red.

acid rain Acidic precipitation in the form of rain; created when fossil fuels are burned, which releases SO_2 and NO_2, which then react with water in the atmosphere to form sulfuric acid and nitric acid.

acid–base reaction A reaction that forms water and typically a salt.

acidic solution A solution containing a concentration of H_3O^+ ions greater than 1.0×10^{-7} M (pH < 7).

activation energy The amount of energy that must be absorbed by reactants before a reaction can occur; an energy hump that normally exists between the reactants and products.

activity series of metals A listing of metals (and hydrogen) in order of decreasing activity, decreasing ability to oxidize, and decreasing tendency to lose electrons.

actual yield The amount of product actually produced by a chemical reaction.

addition polymer A polymer formed by addition of monomers to one another without elimination of any atoms.

alcohol An organic compound containing an —OH functional group bonded to a carbon atom and having the general formula *R*OH.

aldehyde An organic compound with the general formula *R*CHO.

alkali metals The Group 1A elements, which are highly reactive metals.

alkaline battery A dry cell employing half-reactions that use a base.

alkaline earth metals The Group 2A elements, which are fairly reactive metals.

alkaloids Organic compounds that are typically found in plants and act as bases.

alkanes Hydrocarbons in which all carbon atoms are connected by single bonds. Noncyclic alkanes have the general formula C_nH_{2n+2}.

alkene A hydrocarbon that contains at least one double bond between carbon atoms. Noncyclic alkenes have the general formula C_nH_{2n}.

alkyl group In an organic molecule, any group containing only singly bonded carbon atoms and hydrogen atoms.

alkyne A hydrocarbon that contains at least one triple bond between carbon atoms. Noncyclic alkynes have the general formula C_nH_{2n-2}.

alpha particle A particle consisting of two protons and two neutrons (a helium nucleus), represented by the symbol ^4_2He.

alpha (α) radiation Radiation emitted by an unstable nucleus, consisting of alpha particles.

alpha (α)-helix The most common secondary protein structure. The amino acid chain is wrapped into a tight coil from which the side chains extend outward. The structure is maintained by hydrogen bonding interactions between NH and CO groups along the peptide backbone of the protein.

amine An organic compound that contains nitrogen and has the general formula NR_3, where *R* may be an alkyl group or a hydrogen atom.

amino acid A molecule containing an amine group, a carboxylic acid group, and an *R* group (also called a side chain). Amino acids are the building blocks of proteins.

amorphous A type of solid matter in which atoms or molecules do not have long-range order (e.g., glass and plastic).

amphoteric In Brønsted–Lowry terminology, able to act as either an acid or a base.

anion A negatively charged ion.

anode The electrode where oxidation occurs in an electrochemical cell.

aqueous solution A homogeneous mixture of a substance with water.

aromatic ring A ring of carbon atoms containing alternating single and double bonds.

Arrhenius acid A substance that produces H^+ ions in aqueous solution.

Arrhenius base A substance that produces OH^- ions in aqueous solution.

atmosphere (atm) The average pressure at sea level, 101,325 Pa (760 mmHg).

atom The smallest identifiable unit of an element.

atomic element An element that exists in nature with single atoms as the basic unit.

atomic mass A weighted average of the masses of each naturally occurring isotope of an element; atomic mass is the average mass of the atoms of an element.

atomic mass unit (amu) The unit commonly used to express the masses of protons, neutrons, and nuclei. 1 amu = 1.66×10^{-24} g.

atomic number (Z) The number of protons in the nucleus of an atom.

atomic size The size of an atom, which is determined by how far the outermost electrons are from the nucleus.

atomic solid A solid whose component units are individual atoms (e.g., diamond, C; iron, Fe).

atomic theory A theory stating that all matter is composed of tiny particles called atoms.

Avogadro's law A law stating that the volume (*V*) of a gas and the amount of the gas in moles (*n*) are directly proportional.

Avogadro's number The number of entities in a mole, 6.022×10^{23}.

balanced equation A chemical equation in which the numbers of each type of atom on both sides of the equation are equal.

base A molecular compound that dissolves in solution to form OH^- ions. Bases have a slippery feel and turn litmus paper blue.

base chain The longest continuous chain of carbon atoms in an organic compound.

basic solution A solution containing a concentration of OH^- ions greater than 1.0×10^{-7} M (pH > 7).

bent The molecular geometry in which 3 atoms are not in a straight line. This geometry occurs when the central atoms contain 4 electron groups (2 bonding and 2 nonbonding) or 3 electron groups (2 bonding and 1 nonbonding).

benzene (C_6H_6) A particularly stable organic compound consisting of six carbon atoms joined by alternating single and double bonds in a ring structure.

beta particle A form of radiation consisting of an energetic electron and represented by the symbol ${}^{0}_{-1}e$.

beta (β) radiation Energetic electrons emitted by an unstable nucleus.

beta (β)-pleated sheet A common pattern in the secondary structure of proteins. The protein chain is extended in a zigzag pattern, and the peptide backbones of adjacent strands interact with one another through hydrogen bonding to form sheets.

binary acid An acid containing only hydrogen and a nonmetal.

binary compound A compound containing only two different kinds of elements.

biochemistry The study of the chemical substances and processes that occur in living organisms.

Bohr model A model for the atom in which electrons travel around the nucleus in circular orbits at specific, fixed distances from the nucleus.

boiling point The temperature at which the vapor pressure of a liquid is equal to the pressure above it.

boiling point elevation The increase in the boiling point of a solution caused by the presence of the solute.

bonding pair Electrons that are shared between two atoms in a chemical bond.

bonding theory A model that predicts how atoms bond together to form molecules.

Boyle's law A law maintaining that the volume (V) of a gas and its pressure (P) are inversely proportional.

branched alkane An alkane composed of carbon atoms bonded in chains containing branches.

Brønsted–Lowry acid A proton (H^+ ion) donor.

Brønsted–Lowry base A proton (H^+ ion) acceptor.

buffer A solution that resists pH change by neutralizing added acid or added base.

Calorie (Cal) An energy unit equivalent to 1000 little-*c* calories.

calorie (cal) The amount of energy required to raise the temperature of 1 g of water by 1 °C.

carbohydrates Polyhydroxyl aldehydes or ketones or their derivatives, containing multiple —OH groups and often having the general formula $(CH_2O)_n$.

carbonyl group A carbon atom double bonded to an oxygen atom.

carboxylic acid An organic compound with the general formula *R*COOH.

catalyst A substance that increases the rate of a chemical reaction but is not consumed by the reaction.

cathode The electrode where reduction occurs in an electrochemical cell.

cation A positively charged ion.

cell The smallest structural unit of living organisms that has the properties associated with life.

cell membrane The structure that bounds the cell and holds the contents of the cell together.

cellulose A common polysaccharide composed of repeating glucose units linked together.

Celsius (°C) scale A temperature scale often used by scientists. On this scale, water freezes at 0 °C and boils at 100 °C at 1 atm pressure. Room temperature is approximately 22 °C.

chain reaction A self-sustaining chemical or nuclear reaction yielding energy or products that cause further reactions of the same kind.

charge A fundamental property of protons and electrons. Charged particles experience forces such that like charges repel and unlike charges attract.

Charles's law A law stating that the volume (V) of a gas and its temperature (T) expressed in kelvins are directly proportional.

chemical bond The sharing or transfer of electrons to attain stable electron configurations among the bonding atoms.

chemical change A change in which matter changes its composition.

chemical energy The energy associated with chemical changes.

chemical equation An equation that represents a chemical reaction; the reactants are on the left side of the equation, and the products are on the right side.

chemical formula A way to represent a compound. At a minimum, the chemical formula indicates the elements present in the compound and the relative number of atoms of each element.

chemical properties Properties that a substance can display only through changing its composition.

chemical reaction The process by which one or more substances transform into different substances via a chemical change. Chemical reactions often emit or absorb energy.

chemical symbol A one- or two-letter abbreviation for an element. Chemical symbols are listed directly below the atomic number in the periodic table.

chemistry The science that seeks to understand the behavior of matter by studying what atoms and molecules do.

chromosome A biological structure containing genes, located within the nucleus of a cell.

codon A sequence of three bases in a nucleic acid that codes for one amino acid.

colligative properties Physical properties of solutions that depend on the number of solute particles present but not the type of solute particles.

collision theory A theory of reaction rates stating that effective collisions between reactant molecules must take place in order for the reaction to occur.

color change One type of evidence of a chemical reaction, involving the change in color of a substance after a reaction.

combined gas law A law that combines Boyle's law and Charles's law; it is used to calculate how a property of a gas (P, V, or T) changes when two other properties are changed at the same time.

combustion reaction A reaction in which a substance reacts with oxygen, emitting heat and forming one or more oxygen-containing compounds.

complementary base In DNA, a base capable of precise pairing with a specific other DNA base.

complete ionic equation A chemical equation showing all the species as they are actually present in solution.

complex carbohydrate A carbohydrate composed of many repeating saccharide units.

compound A substance composed of two or more elements in fixed, definite proportions.

compressible Able to occupy a smaller volume when subjected to increased pressure. Gases are compressible because, in the gas phase, atoms or molecules are widely separated.

concentrated solution A solution containing large amounts of solute.

condensation A physical change in which a substance is converted from its gaseous form to its liquid form.

condensation polymer A class of polymers that expel atoms, usually water, during their formation or polymerization.

condensed structural formula A shorthand way of writing a structural formula.

conjugate acid–base pair In Brønsted–Lowry terminology, two substances related to each other by the transfer of a proton.

conservation of energy, law of A law stating that energy can be neither created nor destroyed. The total amount of energy is constant and cannot change; it can only be transferred from one object to another or converted from one form to another.

conservation of mass, law of A law stating that in a chemical reaction, matter is neither created nor destroyed.

constant composition, law of A law stating that all samples of a given compound have the same proportions of their constituent elements.

conversion factor A factor used to convert between two separate units; a conversion factor is constructed from any two quantities known to be equivalent.

copolymers Polymers that are composed of two different kinds of monomers and result in chains composed of alternating units rather than a single repeating unit.

core electrons The electrons that are not in the outermost principal shell of an atom.

corrosion The oxidation of metals (e.g., rusting of iron).

covalent atomic solid An atomic solid, such as diamond, that is held together by covalent bonds.

covalent bond The bond that results when two nonmetals combine in a chemical reaction. In a covalent bond, the atoms share their electrons.

critical mass The mass of uranium or plutonium required for a nuclear reaction to be self-sustaining.

crystalline A type of solid matter with atoms or molecules arranged in a well-ordered, three-dimensional array with long-range, repeating order (e.g., salt and diamond).

cytoplasm In a cell, the region between the nucleus and the cell membrane.

Dalton's law of partial pressure A law stating that the sum of the partial pressures of each component in a gas mixture equals the total pressure.

daughter nuclide The nuclide product of a nuclear decay.

decimal part One part of a number expressed in scientific notation.

decomposition A reaction in which a complex substance decomposes to form simpler substances; $AB \rightarrow A + B$.

density (*d*) A fundamental property of materials that relates mass and volume and differs from one substance to another. The units of density are those of mass divided by volume, most commonly expressed in g/cm^3, g/mL, or g/L.

derived unit A unit formed from the combination of other units.

dilute solution A solution containing small amounts of solute.

dimer A molecule formed by the joining together of two smaller molecules.

dipeptide Two amino acids linked together via a peptide bond.

dipole moment A measure of the separation of charge in a bond or in a molecule.

diprotic acid An acid containing two ionizable protons.

disaccharide A carbohydrate that can be decomposed into two simpler carbohydrates.

dispersion force The intermolecular force present in all molecules and atoms. Dispersion forces are caused by fluctuations in the electron distribution within molecules or atoms.

displacement A reaction in which one element displaces another in a compound; $A + BC \rightarrow AC + B$.

dissociation In aqueous solution, the process by which a solid ionic compound separates into its ions.

disubstituted benzene A benzene in which two hydrogen atoms have been replaced by an atom or group of atoms.

DNA (deoxyribonucleic acid) Long chainlike molecules that occur in the nucleus of cells and act as blueprints for the construction of proteins.

dot structure A drawing that represents the valence electrons in atoms as dots; it shows a chemical bond as the sharing or transfer of electron dots.

double bond The bond that exists when two electron pairs are shared between two atoms. In general, double bonds are shorter and stronger than single bonds.

double displacement A reaction in which two elements or groups of elements in two different compounds exchange places to form two new compounds; $AB + CD \rightarrow AD + CB$.

dry cell An ordinary battery (voltaic cell); it does not contain large amounts of liquid water.

duet The name for the two electrons corresponding to a stable Lewis structure in hydrogen and helium.

dynamic equilibrium In a chemical reaction, the condition in which the rate of the forward reaction equals the rate of the reverse reaction.

electrical current The flow of electric charge—for example, electrons flowing through a wire or ions through a solution.

electrical energy Energy associated with the flow of electric charge.

electrochemical cell A device that creates electrical current from a redox reaction.

electrolysis A process in which electrical current is used to drive an otherwise nonspontaneous redox reaction.

electrolytic cell An electrochemical cell used for electrolysis.

electromagnetic radiation A type of energy that travels through space at a constant speed of 3.0×10^8 m/s (186,000 miles/s) and exhibits both wavelike and particlelike behavior. Light is a form of electromagnetic radiation.

electromagnetic spectrum A spectrum that includes all wavelengths of electromagnetic radiation.

electron A negatively charged particle that occupies most of the atom's volume but contributes almost none of its mass.

electron configuration A representation that shows the occupation of orbitals by electrons for a particular element.

electron geometry The geometrical arrangement of the electron groups in a molecule.

electron group A general term for a lone pair, single bond, or multiple bond in a molecule.

electron spin A fundamental property of all electrons that causes them to have magnetic fields associated with them. The spin of an electron can either be oriented up $\left(+\frac{1}{2}\right)$ or down $\left(-\frac{1}{2}\right)$.

electronegativity The ability of an element to attract electrons within a covalent bond.

element A substance that cannot be broken down into simpler substances.

emission spectrum A spectrum associated with the emission of electromagnetic radiation by elements or compounds.

empirical formula A formula for a compound that gives the smallest whole-number ratio of each type of atom.

empirical formula molar mass The sum of the molar masses of all the atoms in an empirical formula.

endothermic Describes a process that absorbs heat energy.

endothermic reaction A chemical reaction that absorbs energy from the surroundings.

energy The capacity to do work.

English system A unit system commonly used in the United States.

enzymes Biological catalysts that increase the rates of biochemical reactions; enzymes are abundant in living organisms.

equilibrium constant (K_{eq}) The ratio, at equilibrium, of the concentrations of the products raised to their stoichiometric coefficients divided by the concentrations of the reactants raised to their stoichiometric coefficients.

equivalence point The point in a reaction at which the reactants are in exact stoichiometric proportions.

equivalent The stoichiometric proportions of elements and compounds in a chemical equation.

ester An organic compound with the general formula *RCOOR*.

ester linkage A type of bond with the general structure —COO—. Ester linkages join glycerol to fatty acids.

ether An organic compound with the general formula *ROR*.

evaporation A process in which molecules of a liquid, undergoing constant random motion, acquire enough energy to overcome attractions to neighbors and enter the gas phase.

excited state An unstable state for an atom or a molecule in which energy has been absorbed but not reemitted, raising an electron from the ground state into a higher energy orbital.

exothermic Describes a process that releases heat energy.

exothermic reaction A chemical reaction that releases energy to the surroundings.

experiment A procedure that attempts to measure observable predictions to test a theory or law.

exponent A number that represents the number of times a term is multiplied by itself. For example, in 2^4 the exponent is 4 and represents $2 \times 2 \times 2 \times 2$.

exponential part One part of a number expressed in scientific notation; it represents the number of places the decimal point has moved.

Fahrenheit (°F) scale The temperature scale that is most familiar in the United States; water freezes at 32 °F and boils at 212 °F at 1 atm prssure.

family (of elements) A group of elements that have similar outer electron configurations and therefore similar properties. Families occur in vertical columns in the periodic table.

family (of organic compounds) A group of organic compounds with the same functional group.

fatty acid A type of lipid consisting of a carboxylic acid with a long hydrocarbon tail.

film badge dosimeter Badges used to measure radiation exposure, consisting of photographic film held in a small case that is pinned to clothing.

fission, nuclear The process by which a heavy nucleus is split into nuclei of smaller masses and energy is emitted.

formula mass The average mass of the molecules (or formula units) that compose a compound.

formula unit The basic unit of ionic compounds; the smallest electrically neutral collection of cations and anions that compose the compound.

freezing point depression The decrease in the freezing point of a solvent caused by the presence of a solute.

frequency The number of wave cycles or crests that pass through a stationary point in one second.

fuel cell A voltaic cell in which the reactants are constantly replenished.

functional group A set of atoms that characterizes a family of organic compounds.

fusion, nuclear The combination of light atomic nuclei to form heavier ones with emission of large amounts of energy.

galvanic (voltaic) cell An electrochemical cell that spontaneously produces electrical current.

gamma radiation High-energy, short-wavelength electromagnetic radiation emitted by an atomic nucleus.

gamma rays The shortest-wavelength, most energetic form of electromagnetic radiation. Gamma ray photons are represented by the symbol ${}^{0}_{0}\gamma$.

gas A state of matter in which atoms or molecules are widely separated and free to move relative to one another.

gas-evolution reaction A reaction that occurs in solution and forms a gas as one of the products.

gas formation One type of evidence of a chemical reaction, a gas forms when two substances are mixed together.

Geiger-Müller counter A radioactivity detector consisting of a chamber filled with argon gas that discharges electrical signals when high-energy particles pass through it.

gene A sequence of codons within a DNA molecule that codes for a single protein. Genes vary in length from hundreds to thousands of codons.

genetic material The inheritable blueprint for making organisms.

glycogen A type of polysaccharide; it has a structure similar to that of starch, but the chain is highly branched.

glycolipid A biological molecule composed of a nonpolar fatty acid and hydrocarbon chain and a polar section composed of a sugar molecule such as glucose.

glycoside linkage The link between monosaccharides in a polysaccharide.

ground state The state of an atom or molecule in which the electrons occupy the lowest possible energy orbitals available.

group (of elements) Elements that have similar outer electron configurations and therefore similar properties. Groups occur in vertical columns in the periodic table.

half-cell A compartment in which the oxidation or reduction half-reaction occurs in a galvanic or voltaic cell.

half-life The time it takes for one-half of the parent nuclides in a radioactive sample to decay to the daughter nuclides.

half-reaction Either the oxidation part or the reduction part of a redox reaction.

halogens The Group 7A elements, which are very reactive nonmetals.

heat absorption One type of evidence of a chemical reaction, involving the intake of energy.

heat capacity The quantity of heat energy required to change the temperature of a given amount of a substance by 1 °C.

heat emission One type of evidence of a chemical reaction, involving the evolution of thermal energy.

heat of fusion The amount of heat required to melt one mole of a solid at its melting point with no change in temperature.

heat of vaporization The amount of heat required to vaporize one mole of a liquid at its boiling point with no change in temperature.

heterogeneous mixture A mixture, such as oil and water, that has two or more regions with different compositions.

homogeneous mixture A mixture, such as salt water, that has the same composition throughout.

human genome All of the genetic material of a human being; the total DNA of a human cell.

Hund's rule A rule stating that when filling orbitals of equal energy, electrons will occupy empty orbitals singly before pairing with other electrons.

hydrocarbon A compound that contains only carbon and hydrogen atoms.

hydrogen bond A strong dipole–dipole interaction between molecules containing hydrogen directly bonded to a small, highly electronegative atom, such as N, O, or F.

hydrogenation The chemical addition of hydrogen to a compound.

hydronium ion The H_3O^+ ion. Chemists often use $H^+(aq)$ and $H_3O^+(aq)$ interchangeably to mean the same thing—a hydronium ion.

hypothesis A theory or law before it has become well established; a tentative explanation for an observation or a scientific problem that can be tested by further investigation.

hypoxia A shortage of oxygen in the tissues of the body.

ideal gas law A law that combines the four properties of a gas—pressure (P), volume (V), temperature (T), and number of moles (n) in a single equation showing their interrelatedness: $PV = nRT$ (R = ideal gas constant).

indicator A substance that changes color with acidity level, often used to detect the end point of a titration.

infrared (IR) light The fraction of the electromagnetic spectrum between visible light and microwaves. Infrared light is invisible to the human eye.

insoluble Not soluble in water.

instantaneous dipole A type of intermolecular force resulting from transient shifts in electron density within an atom or molecule.

intermolecular forces Attractive forces that exist between molecules.

International System (SI) The standard set of units for science measurements, based on the metric system.

ion An atom (or group of atoms) that has gained or lost one or more electrons, so that it has an electric charge.

ion product constant (K_W) The product of the H_3O^+ ion concentration and the OH^- ion concentration in an aqueous solution. At room temperature, $K_w = 1.0 \times 10^{-14}$.

ionic bond The bond that results when a metal and a nonmetal combine in a chemical reaction. In an ionic bond, the metal transfers one or more electrons to the nonmetal.

ionic compound A compound formed between a metal and one or more nonmetals.

ionic solid A solid compound composed of metals and nonmetals joined by ionic bonds.

ionization The forming of ions.

ionization energy The energy required to remove an electron from an atom in the gaseous state.

ionizing power The ability of radiation to ionize other molecules and atoms.

isomers Molecules with the same molecular formula but different structures.

isoosmotic Describes solutions having equal osmotic pressure.

isotope scanning The use of radioactive isotopes to identify disease in the body.

isotopes Atoms with the same number of protons but different numbers of neutrons.

Kelvin (K) scale The temperature scale that assigns 0 K to the coldest temperature possible, absolute zero (–273 °C or –459 °F), the temperature at which molecular motion stops. The size of the kelvin is identical to that of the Celsius degree.

ketone An organic compound with the general formula *RCOR*.

kilogram (kg) The SI standard unit of mass.

kilowatt-hour (kWh) A unit of energy equal to 3.6 million joules.

kinetic energy Energy associated with the motion of an object.

kinetic molecular theory A simple model for gases that predicts the behavior of most gases under many conditions.

Le Châtelier's principle A principle stating that when a chemical system at equilibrium is disturbed, the system shifts in a direction that minimizes the disturbance.

lead-acid storage battery An automobile battery consisting of six electrochemical cells wired in series. Each cell produces 2 volts for a total of 12 volts.

Lewis structure A drawing that represents chemical bonds between atoms as shared or transferred electrons; the valence electrons of atoms are represented as dots.

Lewis theory A simple theory for chemical bonding involving diagrams showing bonds between atoms as lines or dots. In this theory, atoms bond together to obtain stable octets (8 valence electrons).

light emission One type of evidence of a chemical reaction, involving the giving off of electromagnetic radiation.

limiting reactant The reactant that determines the amount of product formed in a chemical reaction.

linear Describes the molecular geometry of a molecule containing two electron groups (two bonding groups and no lone pairs).

linearly related A relationship between two variables such that, when they are plotted one against the other, the graph produced is a straight line.

lipid A cellular component that is insoluble in water but soluble in nonpolar solvents.

lipid bilayer A structure formed by lipids in the cell membrane.

liquid A state of matter in which atoms or molecules are packed close to each other (about as closely as in a solid) but are free to move around and by each other.

logarithmic scale A scale involving logarithms. A logarithm entails an exponent that indicates the power to which a number is raised to produce a given number (e.g., the logarithm of 100 to the base 10 is 2).

lone pair Electrons that are only on one atom in a Lewis structure.

main-group elements Groups 1A–8A on the periodic table. These groups have properties that tend to be predictable based on their position in the periodic table.

mass A measure of the quantity of matter within an object.

mass number (A) The sum of the number of neutrons and protons in an atom.

mass percent composition (or mass percent) The percentage, by mass, of each element in a compound.

matter Anything that occupies space and has mass. Matter exists in three different states: solid, liquid, and gas.

melting point The temperature at which a solid turns into a liquid.

messenger RNA (mRNA) Long chainlike molecules that act as blueprints for the construction of proteins.

metallic atomic solid An atomic solid, such as iron, which is held together by metallic bonds that, in the simplest model, consist of positively charged ions in a sea of electrons.

metallic character The properties typical of a metal, especially the tendency to lose electrons in chemical reactions. Elements become more metallic as you move from right to left across the periodic table.

metalloids Those elements that fall along the boundary between the metals and the nonmetals in the periodic table; their properties are intermediate between those of metals and those of nonmetals.

metals Elements that tend to lose electrons in chemical reactions. They are found at the left side and in the center of the periodic table.

metre (m) The SI standard unit of length.

metric system The unit system commonly used throughout most of the world.

microwaves The part of the electromagnetic spectrum between the infrared region and the radio wave region. Microwaves are efficiently absorbed by water molecules and can therefore be used to heat water-containing substances.

millimetre of mercury (mmHg) A unit of pressure that originates from the method used to measure pressure with a barometer. Also called a *torr*.

miscibility The ability of two liquids to mix without separating into two phases, or the ability of one liquid to mix with (dissolve in) another liquid.

mixture A substance composed of two or more different types of atoms or molecules combined in variable proportions.

molality (m) A common unit of solution concentration, defined as the number of moles of solute per kilogram of solvent.

molar mass The mass of one mole of atoms of an element or one mole of molecules (or formula units) for a compound. An element's molar mass in grams per mole is numerically equivalent to the element's atomic mass in amu.

molar solubility The solubility of a substance in units of moles per litre (mol/L).

molar volume The volume occupied by one mole of gas. Under standard temperature and pressure conditions the molar volume of ideal gas is 22.5 L.

molarity (M) A common unit of solution concentration, defined as the number of moles of solute per litre of solution.

mole Avogadro's number (6.022×10^{23}) of particles—especially, of atoms, ions, or molecules. A mole of any element has a mass in grams that is numerically equivalent to its atomic mass in amu.

molecular compound A compound formed from two or more nonmetals. Molecular compounds have distinct molecules as their simplest identifiable units.

molecular element An element that does not normally exist in nature with single atoms as the basic unit. These elements usually exist as diatomic molecules—2 atoms of that element bonded together—as their basic units.

molecular equation A chemical equation showing the complete, neutral formulas for every compound in a reaction.

molecular formula A formula for a compound that gives the specific number of each type of atom in a molecule.

molecular geometry The geometrical arrangement of the atoms in a molecule.

molecular solid A solid whose composite units are molecules.

molecule Two or more atoms joined in a specific arrangement by chemical bonds. A molecule is the smallest identifiable unit of a molecular compound.

monomer An individual repeating unit that makes up a polymer.

monoprotic acid An acid containing only one ionizable proton.

monosaccharide A carbohydrate that cannot be decomposed into simpler carbohydrates.

monosubstituted benzene A benzene in which one of the hydrogen atoms has been replaced by another atom or group of atoms.

net ionic equation An equation that shows only the species that actually participate in a reaction.

neutral solution A solution in which the concentrations of H_3O^+ and OH^- are equal (pH = 7).

neutralization A reaction that takes place when an acid and a base are mixed; the $H^+(aq)$ from the acid combines with the $OH^-(aq)$ from the base to form $H_2O(l)$.

neutron A nuclear particle with no electrical charge and nearly the same mass as a proton.

nitrogen narcosis An increase in nitrogen concentration in bodily tissues and fluids that results in feelings of drunkenness.

noble gases The Group 8A elements, which are chemically unreactive.

nonbonding atomic solid An atomic solid that is held together by relatively weak dispersion forces.

nonelectrolyte solution A solution containing a solute that dissolves as molecules; therefore, the solution does not conduct electricity.

nonmetals Elements that tend to gain electrons in chemical reactions. They are found at the upper right side of the periodic table.

nonpolar molecule A molecule that does not have a net dipole moment.

nonvolatile Describes a compound that does not vaporize easily.

normal alkane (or *n*-alkane) An alkane composed of carbon atoms bonded in a straight chain with no branches.

normal boiling point The boiling point of a liquid at a pressure of 1 atmosphere.

nuclear equation An equation that represents the changes that occur during radioactivity and other nuclear processes.

nuclear radiation The energetic particles emitted from the nucleus of an atom when it is undergoing a nuclear process.

nuclear theory of the atom A theory stating that most of the atom's mass and all of its positive charge are contained in a small, dense nucleus. Most of the volume of the atom is empty space occupied by negatively charged electrons.

nucleic acids Biological molecules, such as deoxyribonucleic acid (DNA) and ribonucleic acid (RNA), that store and transmit genetic information.

nucleotide An individual unit of a nucleic acid. Nucleic acids are polymers of nucleotides.

nucleus (of a cell) The part of the cell that contains the genetic material.

nucleus (of an atom) The small core containing most of the atom's mass and all of its positive charge. The nucleus is made up of protons and neutrons.

observation Often the first step in the scientific method. An observation must measure or describe something about the physical world.

octet The number of electrons, 8, around atoms with stable Lewis structures.

octet rule A rule that states that an atom will give up, accept, or share electrons in order to achieve a filled outer electron shell, which usually consists of 8 electrons.

orbital The region around the nucleus of an atom where an electron is most likely to be found.

orbital diagram An electron configuration in which electrons are represented as arrows in boxes corresponding to orbitals of a particular atom.

organic chemistry The study of carbon-containing compounds and their reactions.

organic molecule A molecule whose main structural component is carbon.

osmosis The flow of solvent from a lower-concentration solution through a semipermeable membrane to a higher-concentration solution.

osmotic pressure The pressure produced on the surface of a semipermeable membrane by osmosis or the pressure required to stop osmotic flow.

oxidation The gain of oxygen, the loss of hydrogen, or the loss of electrons (the most fundamental definition).

oxidation state (or oxidation number) A number that can be used as an aid in writing formulas and balancing equations. It is computed for each element based on the number of electrons assigned to it in a scheme where the most electronegative element is assigned all of the bonding electrons.

oxidation–reduction (redox) reaction A reaction in which electrons are transferred from one substance to another.

oxidizing agent In a redox reaction, the substance being reduced. Oxidizing agents tend to gain electrons easily.

oxyacid An acid containing hydrogen, a nonmetal, and oxygen.

oxyanion An anion containing oxygen. Most polyatomic ions are oxyanions.

oxygen toxicity The result of increased oxygen concentration in bodily tissues.

parent nuclide The original nuclide in a nuclear decay.

partial pressure The pressure due to any individual component in a gas mixture.

pascal (Pa) The SI unit of pressure, defined as 1 newton per square metre.

Pauli exclusion principle A principle stating that no more than 2 electrons can occupy an orbital and that the 2 electrons must have opposite spins.

penetrating power The ability of a radioactive particle to penetrate matter.

peptide bond The bond between the amine end of one amino acid and the carboxylic acid end of another. Amino acids link together via peptide bonds to form proteins.

percent natural abundance The percentage amount of each isotope of an element in a naturally occurring sample of the element.

percent yield In a chemical reaction, the percentage of the theoretical yield that was actually attained.

period A horizontal row of the periodic table.

periodic law A law that states that when the elements are arranged in order of increasing relative mass, certain sets of properties recur periodically.

periodic table An arrangement of the elements in which atomic number increases from left to right and elements with similar properties fall in columns called families or groups.

permanent dipole A separation of charge resulting from the unequal sharing of electrons between atoms.

pH scale A scale used to quantify acidity or basicity. A pH of 7 is neutral; a pH lower than 7 is acidic, and a pH greater than 7 is basic. The pH is defined as follows: $pH = -\log[H_3O^+]$.

phenyl group The term for a benzene ring when other substituents are attached to it.

phospholipid A lipid with the same basic structure as a triglyceride, except that one of the fatty acid groups is replaced with a phosphate group.

phosphorescence The slow, long-lived emission of light that sometimes follows the absorption of light by some atoms and molecules.

photon A particle of light or a packet of light energy.

physical change A change in which matter does not change its composition, even though its appearance might change.

physical properties Those properties that a substance displays without changing its composition.

polar covalent bond A covalent bond between atoms of different electronegativities. Polar covalent bonds have a dipole moment.

polar molecule A molecule with polar bonds that add together to create a net dipole moment.

polyatomic ion An ion composed of a group of atoms with an overall charge.

polymer A molecule with many similar units, called monomers, bonded together in a long chain.

polypeptide A short chain of amino acids joined by peptide bonds.

polysaccharide A long, chainlike molecule composed of many linked monosaccharide units. Polysaccharides are polymers of monosaccharides.

positron A nuclear particle that has the mass of an electron but carries a 1+ charge.

positron emission Expulsion of a positron from an unstable atomic nucleus. In positron emission, a proton is transformed into a neutron.

potential energy The energy of a body that is associated with its position or the arrangement of its parts.

precipitate An insoluble product formed through the reaction of two solutions containing soluble compounds.

precipitation reaction A reaction that forms a solid or precipitate when two aqueous solutions are mixed.

prefix multipliers Prefixes used by the SI system with the standard units. These multipliers change the value of the unit by powers of 10.

pressure The force exerted per unit area by gaseous molecules as they collide with the surfaces around them.

primary protein structure The sequence of amino acids in a protein's chain. Primary protein structure is maintained by the covalent peptide bonds between individual amino acids.

principal quantum number A number that indicates the shell that an electron occupies.

principal shell The shell indicated by the principal quantum number.

products The final substances produced in a chemical reaction; represented on the right side of a chemical equation.

properties The characteristics we use to distinguish one substance from another.

protein A biological molecule composed of a long chain of amino acids joined by peptide bonds. In living organisms, proteins serve many varied and important functions.

proton A positively charged nuclear particle. A proton's mass is approximately 1 amu.

pure substance A substance composed of only one type of atom or molecule.

quantification The assigning of a number to an observation so as to specify a quantity or property precisely.

quantum (*pl.* quanta) The precise amount of energy possessed by a photon; the difference in energy between two atomic orbitals.

quantum number (n) An integer that specifies the energy of an orbital. The higher the quantum number n, the greater the distance between the electron and the nucleus and the higher its energy.

quantum-mechanical model The foundation of modern chemistry; explains how electrons exist in atoms and how they affect the chemical and physical properties of elements.

quaternary structure In a protein, the way that individual chains fit together to compose the protein. Quaternary structure is maintained by interactions between the *R* groups of amino acids on the different chains.

R group (side chain) An organic group attached to the central carbon atom of an amino acid.

radio waves The longest wavelength and least energetic form of electromagnetic radiation.

radioactive Describes a substance that emits tiny, invisible, energetic particles from the nuclei of its component atoms.

radioactivity The emission of tiny, invisible, energetic particles from the unstable nuclei of atoms. Many of these particles can penetrate matter.

radiocarbon dating A technique used to estimate the age of fossils and artifacts through the measurement of natural radioactivity of carbon atoms in the environment.

radiotherapy Treatment of disease with radiation, such as the use of gamma rays to kill rapidly dividing cancer cells.

random coil The name given to an irregular pattern of a secondary protein structure.

rate of a chemical reaction (reaction rate) The amount of reactant that changes to product in a given period of time. Also defined as the amount of a product that forms in a given period of time.

reactants The initial substances in a chemical reaction, represented on the left side of a chemical equation.

recrystallization A technique used to purify a solid; involves dissolving the solid in a solvent at high temperature, creating a saturated solution, then cooling the solution to cause the crystallization of the solid.

reducing agent In a redox reaction, the substance being oxidized. Reducing agents tend to lose electrons easily.

reduction The loss of oxygen, the gain of hydrogen, or the gain of electrons (the most fundamental definition).

rem Stands for *roentgen equivalent man;* a weighted measure of radiation exposure that accounts for the ionizing power of the different types of radiation.

resonance structures Two or more Lewis structures that are necessary to describe the bonding in a molecule or ion.

reversible reaction A reaction that is able to proceed in both the forward and reverse directions.

RNA (ribonucleic acid) Long chainlike molecules that occur throughout cells and take part in the construction of proteins.

salt An ionic compound that usually remains dissolved in a solution after an acid–base reaction has occurred.

salt bridge An inverted, U-shaped tube containing a strong electrolyte; completes the circuit in an electrochemical cell by allowing the flow of ions between the two half-cells.

saturated fat A triglyceride composed of saturated fatty acids. Saturated fat tends to be solid at room temperature.

saturated hydrocarbon A hydrocarbon that contains no double or triple bonds between the carbon atoms.

saturated solution A solution that holds the maximum amount of solute under the solution conditions. If additional solute is added to a saturated solution, it will not dissolve.

scientific law A statement that summarizes past observations and predicts future ones. Scientific laws are usually formulated from a series of related observations.

scientific method The way that scientists learn about the natural world. The scientific method involves observations, laws, hypotheses, theories, and experimentation.

scientific notation A system used to write very big or very small numbers, often containing many zeros, more compactly and precisely. A number written in scientific notation consists of a decimal part and an exponential part (10 raised to a particular exponent).

scintillation counter A device used to detect radioactivity in which energetic particles traverse a material that emits ultraviolet or visible light when excited by their passage. The light is detected and turned into an electrical signal.

second (s) The SI standard unit of time.

secondary protein structure Short-range periodic or repeating patterns often found in proteins. Secondary protein structure is maintained by interactions between amino acids that are fairly close together in the linear sequence of the protein chain or adjacent to each other on neighboring chains.

semiconductor A compound or element exhibiting intermediate electrical conductivity that can be changed and controlled.

semipermeable membrane A membrane that selectively allows some substances to pass through but not others.

SI units The most convenient system of units for science measurements, based on the metric system. The set of standard units agreed on by scientists throughout the world.

significant digits (figures) The non–place-holding digits in a reported measurement; they represent the precision of a measured quantity.

simple carbohydrate (simple sugar) A monosaccharide or disaccharide.

single bond A chemical bond in which one electron pair is shared between two atoms.

solid A state of matter in which atoms or molecules are packed close to each other in fixed locations.

solid formation One type of evidence of a chemical reaction involving the formation of a solid.

solubility The amount of a compound, usually in grams, that will dissolve in a certain amount of solvent.

solubility rules A set of empirical rules used to determine whether an ionic compound is soluble.

solubility-product constant (K_{sp}) The equilibrium expression for a chemical equation that represents the dissolving of an ionic compound in solution.

soluble Dissolves in solution

solute The minority component of a solution.

solution A homogeneous mixture of two or more substances.

solvent The majority component of a solution.

specific heat capacity (or specific heat) The heat capacity of a substance in joules per gram degree Celsius (J/g °C).

spectator ions Ions that do not participate in a reaction; they appear unchanged on both sides of a chemical equation.

standard temperature and pressure (STP) Conditions often assumed in calculations involving gases: $T = 0$ °C (273 K) and $P = 1$ atm.

starch A common polysaccharide composed of repeating glucose units.

states of matter The three forms in which matter can exist: solid, liquid, and gas.

steroid A biological compound containing a 17-carbon 4-ring system.

stock solution A concentrated form in which solutions are often stored.

stoichiometry The numerical relationships among chemical quantities in a balanced chemical equation. Stoichiometry allows us to predict the amounts of products that form in a chemical reaction based on the amounts of reactants.

strong acid An acid that completely ionizes in solution.

strong base A base that completely dissociates in solution.

strong electrolyte A substance whose aqueous solutions are good conductors of electricity.

strong electrolyte solution A solution containing a solute that dissociates into ions; therefore, a solution that conducts electricity well.

structural formula A two-dimensional representation of molecules that not only shows the number and type of atoms, but also how the atoms are bonded together.

sublimation A physical change in which a substance is converted from its solid form directly into its gaseous form.

subshell In quantum mechanics, specifies the shape of the orbital and is represented by different letters (*s*, *p*, *d*, *f*).

substituent An atom or a group of atoms that has been substituted for a hydrogen atom in an organic compound.

substitution reaction A reaction in which one or more atoms are replaced by one or more different atoms.

supersaturated solution A solution holding more than the normal maximum amount of solute.

surface tension The tendency of liquids to minimize their surface area, resulting in a "skin" on the surface of the liquid.

synthesis A reaction in which simpler substances combine to form more complex substances; $A + B \rightarrow AB$.

temporary dipole A type of intermolecular force resulting from transient shifts in electron density within an atom or molecule.

terminal atom An atom that is located at the end of a molecule or chain.

tertiary structure A protein's structure that consists of the large-scale bends and folds due to interactions between the *R* groups of amino acids that are separated by large distances in the linear sequence of the protein chain.

tetrahedral The molecular geometry of a molecule containing four electron groups (four bonding groups and no lone pairs).

theoretical yield The maximum amount of product that can be made in a chemical reaction based on the amount of limiting reactant.

theory A proposed explanation for observations and laws. A theory presents a model of the way nature works and predicts behavior that extends well beyond the observations and laws from which it was formed.

titration A laboratory procedure used to determine the amount of a substance in solution. In a titration, a reactant in a solution of known concentration is reacted with another reactant in a solution of unknown concentration until the reaction reaches the end point.

torr A unit of pressure named after the Italian physicist Evangelista Torricelli; also called a millimetre of mercury.

transition metals The elements in the middle of the periodic table whose properties tend to be less predictable based simply on their position in the periodic table. Transition metals lose electrons in their chemical reactions, but do not necessarily acquire noble gas configurations.

triglyceride A fat or oil; a tryglyceride is a triester composed of glycerol with three fatty acids attached.

trigonal planar The molecular geometry of a molecule containing three electron groups, three bonding groups, and no lone pairs.

trigonal pyramidal The molecular geometry of a molecule containing four electron groups, three bonding groups, and one lone pair.

triple bond A chemical bond consisting of three electron pairs shared between two atoms. In general, triple bonds are shorter and stronger than double bonds.

Type I compounds Compounds containing metals that always form cations with the same charge.

Type II compounds Compounds containing metals that can form cations with different charges.

ultraviolet (UV) light The fraction of the electromagnetic spectrum between the visible region and the X-ray region. UV light is invisible to the human eye.

units Previously agreed-on quantities used to report experimental measurements. Units are vital in chemistry.

unsaturated fat (or oil) A triglyceride composed of unsaturated fatty acids. Unsaturated fats tend to be liquids at room temperature.

unsaturated hydrocarbon A hydrocarbon that contains one or more double or triple bonds between its carbon atoms.

unsaturated solution A solution holding less than the maximum possible amount of solute under the solution conditions.

valence electrons The electrons in the outermost principal shell of an atom; they are involved in chemical bonding.

valence shell electron pair repulsion (VSEPR) A theory that allows prediction of the shapes of molecules based on the idea that electrons—either as lone pairs or as bonding pairs—repel one another.

vapor pressure The partial pressure of a vapor in dynamic equilibrium with its liquid.

vaporization The phase transition between a liquid and a gas.

viscosity The resistance of a liquid to flow; manifestation of intermolecular forces.

visible light The fraction of the electromagnetic spectrum that is visible to the human eye, bounded by wavelengths of 400 nm (violet) and 780 nm (red).

vital force A mystical or supernatural power that, it was once believed, was possessed only by living organisms and allowed them to produce organic compounds.

vitalism The belief that living things contain a nonphysical "force" that allows them to synthesize organic compounds.

volatile Tending to vaporize easily.

voltage The potential difference between two electrodes; the driving force that causes electrons to flow.

volume A measure of space. Any unit of length, when cubed, becomes a unit of volume.

wavelength The distance between adjacent wave crests in a wave.

weak acid An acid that does not completely ionize in solution.

weak base A base that does not completely dissociate in solution.

weak electrolyte A substance whose aqueous solutions are poor conductors of electricity.

X-rays The portion of the electromagnetic spectrum between the ultraviolet (UV) region and the gamma-ray region.

Index

D

E

Q

R

S

T

U

V

W

Fundamental Physical Constants

Atomic mass unit	$1\ \text{amu} = 1.660539 \times 10^{-27}\ \text{kg}$ $1\ \text{g} = 6.022142 \times 10^{23}\ \text{amu}$
Avogadro's number	$N_A = 6.022142 \times 10^{23}/\text{mol}$
Electron charge	$e = 1.602176 \times 10^{-19}\ \text{C}$
Gas constant	$R = 8.314472\ \text{J}/(\text{mol} \cdot \text{K})$ $= 0.0820582\ (\text{L} \cdot \text{atm})/(\text{mol} \cdot \text{K})$
Mass of electron	$m_e = 5.485799 \times 10^{-4}\ \text{amu}$ $= 9.109382 \times 10^{-31}\ \text{kg}$
Mass of neutron	$m_n = 1.008665\ \text{amu}$ $= 1.674927 \times 10^{-27}\ \text{kg}$
Mass of proton	$m_p = 1.007276\ \text{amu}$ $= 1.672622 \times 10^{-27}\ \text{kg}$
Pi	$\pi = 3.1415926536$
Planck's constant	$h = 6.626069 \times 10^{-34}\ \text{J} \cdot \text{s}$
Speed of light in vacuum	$c = 2.99792458 \times 10^{8}\ \text{m/s}$

Useful Geometric Formulas

Perimeter of a rectangle $= 2l + 2w$

Circumference of a circle $= 2\pi r$

Area of a triangle $= (1/2)(\text{base} \times \text{height})$

Area of a circle $= \pi r^2$

Surface area of a sphere $= 4\pi r^2$

Volume of a sphere $= (4/3)\pi r^3$

Volume of a cylinder or prism $=$ area of base $\times$ height

Important Conversion Factors

Length: SI unit = metre (m)

- $1\ \text{m} = 39.37\ \text{in.}$
- $1\ \text{in.} = 2.54\ \text{cm}$ (exactly)
- $1\ \text{mile} = 5280\ \text{ft} = 1.609\ \text{km}$
- $1\ \text{angstrom}\ (\text{Å}) = 10^{-10}\ \text{m}$

Volume: SI unit = cubic metre (m^3)

- $1\ \text{L} = 1000\ \text{cm}^3 = 1.057\ \text{qt (U.S.)}$
- $1\ \text{gal (U.S.)} = 4\ \text{qt} = 8\ \text{pt}$
$= 128$ fluid ounces
$= 3.785\ \text{L}$

Mass: SI unit = kilogram (kg)

- $1\ \text{kg} = 2.205\ \text{lb}$
- $1\ \text{lb} = 16\ \text{oz} = 453.6\ \text{g}$
- $1\ \text{ton} = 2000\ \text{lb}$
- $1\ \text{metric ton} = 1000\ \text{kg} = 1.103\ \text{tons}$
- $1\ \text{g} = 6.022 \times 10^{23}$ atomic mass units (amu)

Pressure: SI unit = pascal (Pa)

- $1\ \text{Pa} = 1\ \text{N/m}^2$
- $1\ \text{bar} = 10^5\ \text{Pa}$
- $1\ \text{atm} = 1.01325 \times 10^5\ \text{Pa}$ (exactly)
$= 1.01325\ \text{bar}$
$= 760\ \text{mmHg}$
$= 760\ \text{torr}$ (exactly)

Energy: SI unit = joule (J)

- $1\ \text{J} = 1\ \text{N} \cdot \text{m}$
- $1\ \text{cal} = 4.184\ \text{J}$ (exactly)
- $1\ \text{L} \cdot \text{atm} = 101.33\ \text{J}$

Temperature: SI unit = kelvin (K)

- $\text{K} = {}^\circ\text{C} + 273.15$
- ${}^\circ\text{C} = (5/9)({}^\circ\text{F} - 32^\circ)$
- ${}^\circ\text{F} = (9/5)({}^\circ\text{C}) + 32^\circ$